W9-AFT-665

Elements of X-RAY CRYSTALLOGRAPHY

Elements of X-RAY CRYSTALLOGRAPHY

LEONID V. AZÁROFF
Professor of Physics and Director
Institute of Materials Science
University of Connecticut

McGRAW-HILL BOOK COMPANY
New York St. Louis
San Francisco Toronto
London Sydney

Elements of X-RAY CRYSTALLOGRAPHY

Library of Congress Catalog Card Number 68-11600

02667

1234567890 HDMM 7543210698

This book is dedicated to
MARTIN J. BUERGER,
who
has found crystallography most enjoyable
and,
believing that man should enjoy his work,
has spent most of his life
making crystallography
enjoyable
for others

PREFACE

Within one year following the discovery of x-ray diffraction by crystals, the first of three papers developing the entire theory of this phenomenon was published by C. G. Darwin. A few years later, a somewhat different and equally profound analysis was presented by P. P. Ewald. Out of this was born the realization that the diffraction spectra of single crystals bear a geometrical relation to the periodic arrays in crystals; namely, they constitute a reciprocal-lattice array. Despite this, relatively little use has been made of this concept by writers of introductory textbooks on x-ray crystallography. The present book is an attempt to provide the necessary basis for a pedagogically sound development of the subject. Approaching the diffraction of x rays by first learning about the reciprocal-lattice concept can be likened to entering a foreign country after first mastering its native tongue—it may not be essential, but it is the most sensible way to proceed.

In the belief that the objective of a college course is to provide the student with a fundamental understanding of the basic principles of a subject, their validity and utility in various applications, and a stimulation of his interest for further study based on the knowledge already at his command, this book has been written to serve as a text in college courses on x-ray crystallography. This does not exclude its usefulness for self-study; in fact, some of the problems at the end of each chapter are designed to test the reader's mastery of the subject matter covered. It does limit the book's utility as a reference volume. This has been done deliberately. Original papers are referred to only in those few instances where their content is being cited in a book for the first time. Otherwise, the reader is referred to authoritative reference books listed at the end of each chapter. To present a fairly complete coverage of x-ray crystallography, it was necessary to discuss the various topics at two levels of sophistication: One is that at which most college sophomores can study, although it is usual that curriculum logistics often postpone x-ray courses until the later years. The other requires a familiarity with mathematics, physics, and related topics that is seldom achieved much earlier than the senior year. Thus the user can choose his own combinations of topics and levels of presentation. In the Table of Contents, each chapter requiring more advanced knowledge is designated by an asterisk. Similarly, an instructor should find ample material for an undergraduate course or a graduate course in x-ray crystallography. Since only the student's background varies from one academic department to another, it is hoped that the rather extensive topical coverage presented will fulfill most present and future requirements of its reader. More expanded coverage of any aspect of x-ray crystallography discussed (or omitted) can be found in the book lists accompanying each chapter and those listed in Appendix 3.

This book is intended to provide a complete introduction to all the important elements comprising the subject of x-ray crystallography. They can be classified into four categories, and this has been done in ordering the chapters. In all the discussions an attempt has been made to develop the principles necessary for a full appreciation of the subject, but not to deal with it exhaustively. This is particularly true of the description of the experimental procedures in the last part. A companion volume is currently in preparation that will supply some of the missing details, as well as fully self-contained experiments illustrating many of the principles discussed in the text.

The author is deeply indebted to three categories of contributors: his predecessors, whose contributions made this synthesis possible; his teachers, whose inspiration and instruction prepared its foundation; and his colleagues, who generously lent their time and labors to help make the final product a better book. It is possible to cite only a few of those in the first category, and an attempt has been made to identify the innovators alongside their contributions throughout the text. In the second category, the author owes a debt, first, to the late Professor I. Fankuchen, for introducing him to this fascinating subject during two memorable weeks in Brooklyn; to Professor M. J. Buerger, for putting up with his often blundering pursuit of crystallographic concepts; and to Professor B. E. Warren, for making the physics of x rays and x-ray diffraction so extremely clear. It should not surprise these gentlemen or their students that many an echo of their lectures can be found in the following pages. The third category includes many friends who kindly supplied photographs and other illustrative material specifically acknowledged in the text. It also includes those who were willing to take time out from their other labors to read and criticize parts of this book: They are Drs. W. W. Beeman, L. Birks, H. Cole, D. J. Fisher, M. E. Milberg, J. T. Norton, B. Post, and N. Stemple. The value of their contributions far exceeds what these words of appreciation can express.

In conclusion, I wish to express my appreciation to Professor R. J. Donahue and Messrs. R. R. Biederman and J. A. Reffner for their help in preparing some of the illustrations and in proofreading the typescript and to Mrs. L. Metelsky and Miss D. Stygar for their help in preparing it.

Leonid V. Azároff

NOTE TO THE READER

The aim of this book is to provide in one volume a complete introduction to the essential elements of x-ray crystallography. It is possible to group these into four categories: the geometry of crystals, the physics of x rays, how x rays interact with crystals, and finally, how such interactions can be studied experimentally. A rudimentary knowledge of vector algebra, such as the one provided in the first section in Appendix 1, suffices for the understanding of geometrical crystallography and of the reciprocal-lattice concept which lies at the core of the interaction of x rays with crystals. With its aid, it is possible to master all the experimental procedures described in the last part of this book. To analyze the physics of x rays more fully and to develop the theory of x-ray diffraction requires, additionally, the use of differential equations and Fourier theory, some of whose pertinent features are also reviewed in Appendix 1. It is possible, therefore, to distinguish discussions in this book that are mathematically more complex, and they have been denoted by affixing an asterisk to the chapter titles in the Table of Contents. The reader seeking a complete introduction to x-ray crystallography can obtain it by reading all the unmarked chapters. If a more advanced analysis of x-ray diffraction theory is sought, it is contained in the chapters marked with asterisks. Including discussions at both levels in the same book affords the reader an opportunity to pursue x-ray crystallography at his own pace.

CONTENTS

* Denotes chapters requiring a more advanced knowledge of modern physics and mathematics.

* Denotes chapters requiring a more advanced knowledge of modern physics and mathematics.

Elements of X-RAY CRYSTALLOGRAPHY

PART ONE
Elements of Crystals

ONE
Symmetry of crystals

MODES OF REPETITION

It is well known that matter is composed of atoms which may or may not be joined into distinguishable groups called molecules. The atomic aggregates are normally classified into three states of matter, the gaseous, liquid, and solid states. Of these, the present volume concerns itself primarily with certain aspects of the solid, in particular, the *crystalline state* of matter. The one feature of the atomic arrays that distinguishes the crystalline state from all others is that the atoms (or molecules) occur in a repetitive, or periodic, regular manner within the array. Thus a study of the crystalline state begins with a recognition of the repetitive patterns present in particular arrays. The manner of repetition is normally called the *symmetry* of the array, and the purpose of this first chapter is to discover what some of the possible kinds of symmetry are. Atomic arrays that are periodic in all three dimensions are called *crystals*, so that this chapter is concerned only with the symmetry that is found in crystals.

Representation of a pattern. The study of the different possibilities and laws that govern periodic arrays is called *pattern theory*, or if the periodic array is a crystal, *geometrical crystallography*. Inherently, these laws are not concerned with what is being repeated periodically, but only with the different ways of repetition. It is possible, therefore, to select any motif to represent the basic unit whose regular repetition constitutes the final pattern. For the sake of generality, however, a motif should be chosen that has no symmetry itself. As an example of an unsymmetrical object, consider the boot shown in Fig. 1-1. Note that a boot can exist in two different forms, namely, one that fits the left foot and one that fits the right foot. A pair of such boots is shown in Fig. 1-2. To distinguish the two motifs, one is usually said to be *left-handed* and the other *right-handed*. It should be realized that any asymmetric object can exist in both a left-handed and a right-handed form. A pair of objects thus related are said to be *enantiomorphous*.

Fig. 1-1 Fig. 1-2

Repetition of an object. A symbol that is almost as unsymmetrical as a boot and much easier to draw is the number 7. Figure 1-3 shows an enantiomorphous pair of 7's, one on each side of a vertical line. Note that, whereas a single 7 is asymmetric, the pattern in Fig. 1-3 is symmetric in that each side of the figure looks like a mirror image of the other. In fact, one can say that the left-handed 7 is repeated from the right-handed 7 by a *reflection* across the vertical line, and, in turn, the right-handed 7 is a repetition of the left-handed 7 by the reverse reflection, and so forth. This is a simple example of one mode of repetition, namely, reflection. It can be generalized to three dimensions by observing that the two enantiomorphous boots in Fig. 1-2 are also related by reflection across a plane passing halfway between them.

Another important aspect of the patterns discussed above is that the final pattern is formed by a continued repetition until the initial object is itself reproduced. In the illustrated case of reflection, this is accomplished by the initial reflection from left to right and then back again from right to left. Consider, however, what happens when a pattern is formed by the *rotation* of a 7 about a central point through an angle φ, as shown in Fig. 1-4. In order to assure that the pattern is completed after a finite number of rotations n, it is necessary that φ be an integral

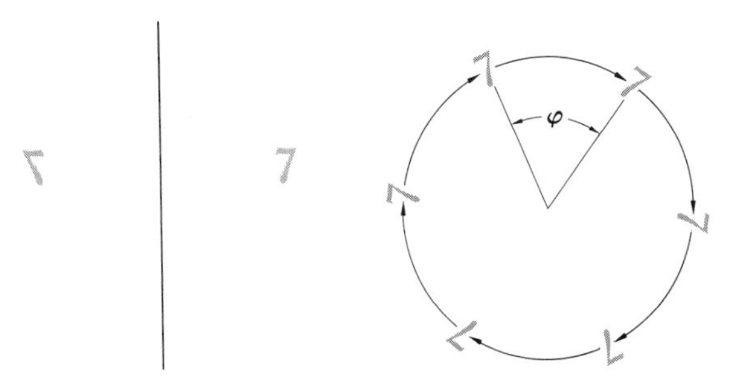

Fig. 1-3 Fig. 1-4

submultiple of a complete revolution, that is,

$$\varphi = \frac{360°}{n} \qquad n = 1, 2, 3, \ldots \tag{1-1}$$

Note that the 7's forming the pattern in Fig. 1-4 are all right-handed 7's, which are said to be *congruent* with each other.

The geometrical locus about which the operations of repetition take place is called a *symmetry element*. From the discussion so far, it is clear that one can distinguish operations relating congruent objects from those relating enantiomorphous objects. When a rotation repeats a right, a right, a right, etc., it is called a *proper* rotation, whereas if it repeats a right, a left, a right, a left, etc., it is called an *improper* rotation. The possible symmetry elements formed by such operations of repetition are considered next.

SYMMETRY ELEMENTS

Proper rotation axes. The simplest way to picture the operation of rotation depicted in Fig. 1-4 is to think of a line or axis passing through the center of and normal to the figure, so that the 7's are repeated by a rotation through an angle φ about this *rotation axis.* This is obviously a proper rotation axis, and it is common practice to call the rotation angle φ the *throw* of an axis, and the number of successive rotations n required to cause superposition, the *fold* of the axis.

A rotation is said to exist in a pattern when *all* the pattern exhibits the property of rotation symmetry about this axis. For example, a tulip is said to have 3-fold symmetry because the petals and all other parts of the flower grow in sets of three about the floral axis or stem, and a columbine has a 5-fold rotation axis passing through its stem for similar reasons. In general, a rotation axis can be 1-fold, 2-fold, . . . , n-fold, but, for reasons to be discussed in a subsequent chapter, the different kinds of rotation axis that can exist in crystals are limited to five. The five allowed proper rotation axes are shown in Fig. 1-5. The symbol used to represent a proper rotation axis is an integer corresponding to the number of repetitions in a complete revolution, that is, the fold of the axis as determined by

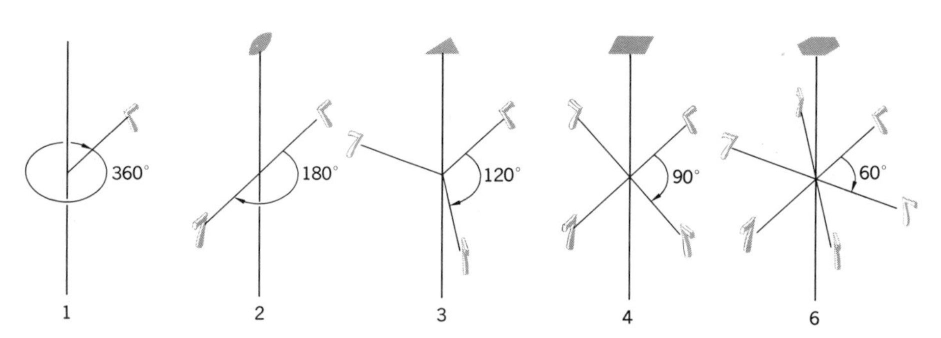

Fig. 1-5

Eq. (1-1). Thus a 1-fold axis has a throw of 360° and is called a 1; a 4-fold axis having a throw of 90° is called a 4; etc.

Improper rotation axes. As stated earlier, an improper rotation axis repeats a left-handed object from a right-handed one, and vice versa. One symmetry operation already encountered that produces such enantiomorphous sets is the operation of reflection. If a rotation is combined with a reflection into a single hybrid operation, the resulting operation is called *rotoreflection*. The corresponding symmetry element is called a *rotoreflection axis*. There exists a rotoreflection axis corresponding to each of the five proper axes. To distinguish the two kinds of axes, a *tilde* is placed over the numerical symbol of the corresponding rotation axis: $\tilde{1}$, $\tilde{2}$, $\tilde{3}$, etc.

As an illustration of a rotoreflection axis, consider $\tilde{1}$, pronounced "one-tilde." The proper rotation 1 rotates the motif representing all of space through an angle of 0 or 360°, leaving it unchanged. The rotoreflection axis $\tilde{1}$ combines this operation with a reflection to give the configuration shown in Fig. 1-6. The enantiomorphous pair of 7's is not unlike the pair already encountered in Fig. 1-3. Accordingly, it can be said that the operation of $\tilde{1}$ is equivalent to reflection through a plane, specifically, a *reflection*, or *mirror*, *plane*, symbolically represented by the letter m.

To illustrate further the operation of a rotoreflection axis, Fig. 1-7 shows a perspective view of $\tilde{2}$. Remember that the operation of rotoreflection consists of two distinct parts. First, the 7 in Fig. 1-7 is rotated by an "imaginary" 2 through an angle of 180°, but it is not left there. Instead, it is next reflected through an "imaginary" plane placed at right angles to the "imaginary" rotation axis. It should be emphasized that a 2-fold rotoreflection axis is neither a 2-fold axis nor a reflection plane, but rather a hybrid combining the operations of both into a single operation.

A very useful method of displaying a symmetry element is to show a projection of the symmetry element along some convenient direction. In the case of rotation axes, it is convenient to project along the direction of the axis. Figure 1-8 shows such projections of the proper and improper rotation axes. In illustrating the operation of an improper axis it is necessary to distinguish between an object lying above the imaginary reflection plane and one below it. If the axis is normal to the plane of the paper, the imaginary reflection plane lies in the plane of the paper. A cross is used to indicate an object lying above this plane, and a circle,

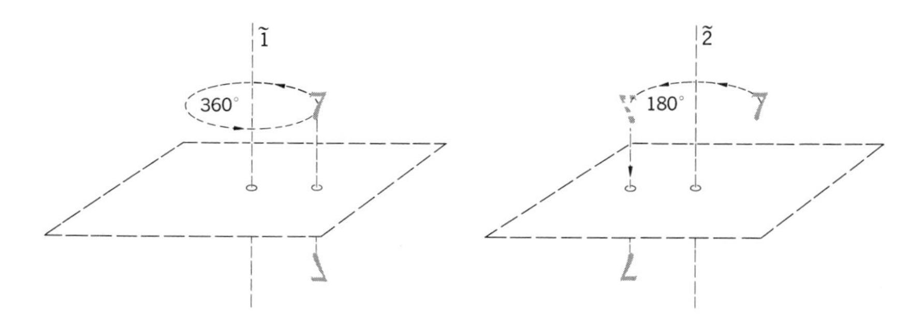

Fig. 1-6 **Fig. 1-7**

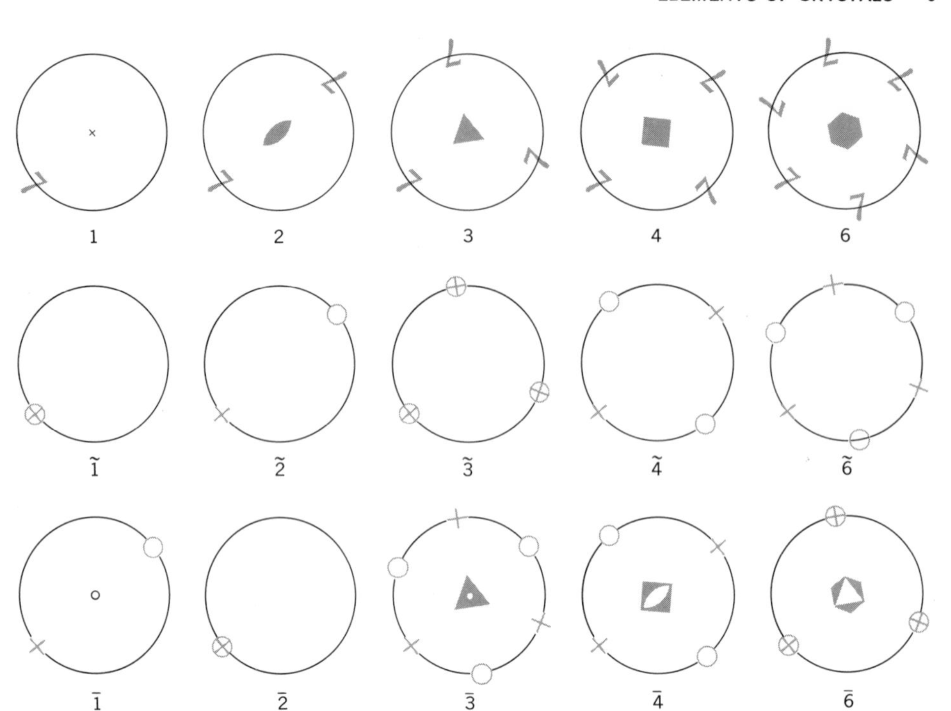

Fig. 1-8

an object below this plane. It should be realized that the cross and circle bear another important relationship to each other; namely, if the cross represents a right-handed object, the circle represents a left-handed object.

Returning to Fig. 1-7 for a moment, it can be shown that the enantiomorphous pair of 7's in that figure is related by an *inversion center* lying halfway between the two 7's. The inversion or *symmetry center* has the property of inverting all of space through a point, as illustrated in Fig. 1-9. Representing all of space by a 7, its inverted equivalent can be obtained by passing a construction line from each part of the 7 through the inversion center and continuing each construction line until its length on both sides of the center is equal. An equivalent part of the 7 is then placed at each end of every line until the enantiomorphous pair in Fig. 1-9 results.

Fig. 1-9

Table 1-1 Conventional designation of improper axes

Rotoinversion axis	*Rotoreflection axis*	*Conventional designation*	
$\bar{1}$	$\tilde{2}$	Center of symmetry	$\bar{1}$
$\bar{2}$	$\tilde{1}$	Mirror plane	m
$\bar{3}$	$\tilde{6}$	3-fold rotoinversion	$\bar{3}$
$\bar{4}$	$\tilde{4}$	4-fold rotoinversion	$\bar{4}$
$\bar{6}$	$\tilde{3}$	6-fold rotoinversion	$\bar{6}$

It is possible to combine an inversion center with a rotation axis to produce a *rotoinversion axis* in a manner analogous to the formation of a rotoreflection axis. Again, it should be realized that a rotoinversion axis is a hybrid of the two operations and is neither a proper rotation nor an inversion center. The five rotoinversion axes of crystallographic interest are shown in the bottom row of Fig. 1-8. To indicate a rotoinversion axis, a bar is placed over the corresponding proper axis symbol. For example, a 3-fold rotoinversion axis is designated $\bar{3}$, pronounced "three-bar."

If the five rotoreflection axes are compared with the five rotoinversion axes in Fig. 1-8, it becomes apparent that they are equivalent in pairs. In fact, one equivalence has already been illustrated in comparing Fig. 1-7 with Fig. 1-9, which showed $\tilde{2}$ and $\bar{1}$, respectively. Since the two kinds of improper axes are thus equivalent in pairs, it is sufficient to adopt only one kind to represent these symmetry operations. Accordingly, Table 1-1 lists the conventional designations for the equivalent pairs of improper axes.

Symmetry of a cube. It is of obvious interest to see how the symmetry elements described above can be discerned in actual crystals. As an example, consider the highly symmetric and familiar cube. Even a cursory examination makes it immediately apparent that a cube has 4-fold symmetry about an axis passing through the centers of two opposite cube faces, as shown in Fig. 1-10. What may not be as obvious is that a 3-fold axis passes through the body diagonal of the cube, as shown in Fig. 1-11. In fact, a more careful examination shows that it is the improper rotoinversion axis $\bar{3}$. To see this more clearly, consider its equivalent, the rotoreflection axis $\tilde{6}$. Figure 1-12 shows one of the cube faces rotated through an

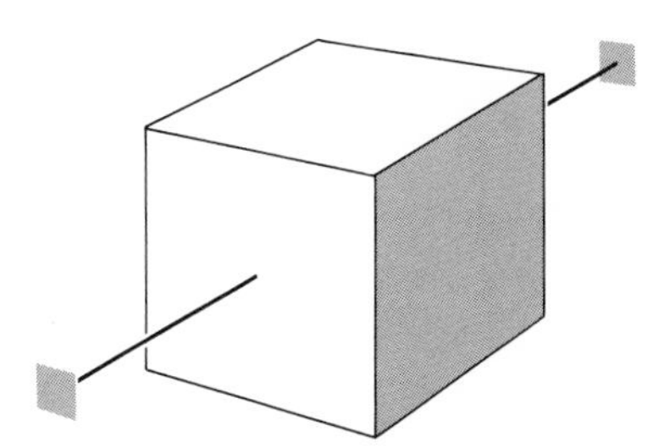

Fig. 1-10

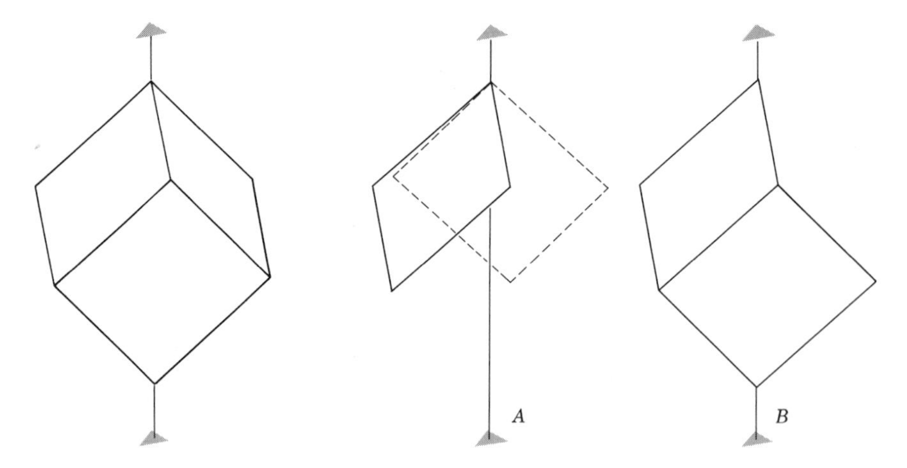

Fig. 1-11

Fig. 1-12

angle of $60°$ by the $\bar{6}$ (dashed plane in A) and then reflected across a plane perpendicular to the rotation axis (solid plane in B).

By comparison, it is a relatively simple matter to find the 2-fold axes (parallel to the face diagonals and through the cube center) and the mirror planes (through the cube center and bisecting the faces of the cube parallel to and diagonally to its edges). Finally, it is left to the reader to convince himself that a center of symmetry reposes at the center of the cube. As can be seen in Fig. 1-13, the total list of symmetry elements present in a cube includes

Three	4's
Four	$\bar{3}$'s
Six	2's
Nine	m's
One	$\bar{1}$

(1-2)

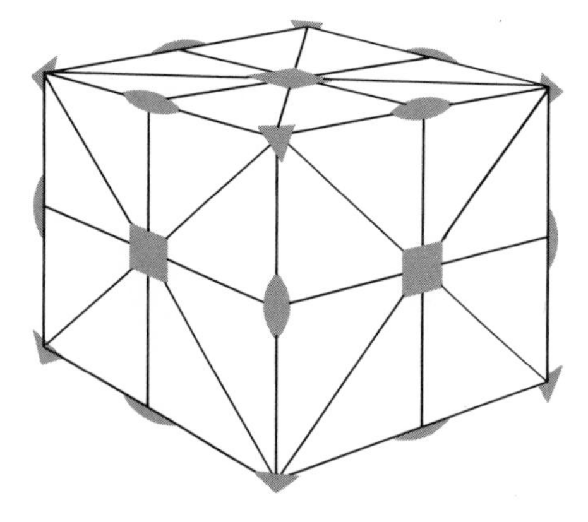

Fig. 1-13

These symmetry elements are not independent of each other since their operations repeat the entire crystal, including the other symmetry elements present. Thus the combination of a much smaller number of symmetry elements can be shown to produce the entire listing in (1-2). How this happens is considered next.

CLASSIFICATION OF CRYSTALS

As illustrated by the discussion of the cube, the external form of a crystal, called its *habit*, displays its symmetry. Early crystallographers were able to collect crystals occurring in nature, notably minerals, and to determine the symmetry elements present by studying their morphology. As more and more different kinds of crystals were examined, it became apparent that some of them had identical symmetry even though their habits may have been different. Such crystals were said to belong to the same *crystal class*. As the different possible crystal classes were established, it became evident that some of them had certain symmetry elements in common. These crystal classes were then gathered into larger groupings, called *crystal systems*.

Crystal classes. The symmetry of space about a point can be described by a collection of symmetry elements called a *point group*. In the case of crystals, the possible point groups or crystal classes must be composed of combinations of the five proper and five improper rotation axes described above. In a systematic attempt to derive all the possible combinations, one can begin by realizing that each of the five rotation axes (Fig. 1-14) and each of the five rotoinversion axes (Fig. 1-15) constitute a possible point group. All other point groups, however, must be formed from combinations of two or more rotation axes.

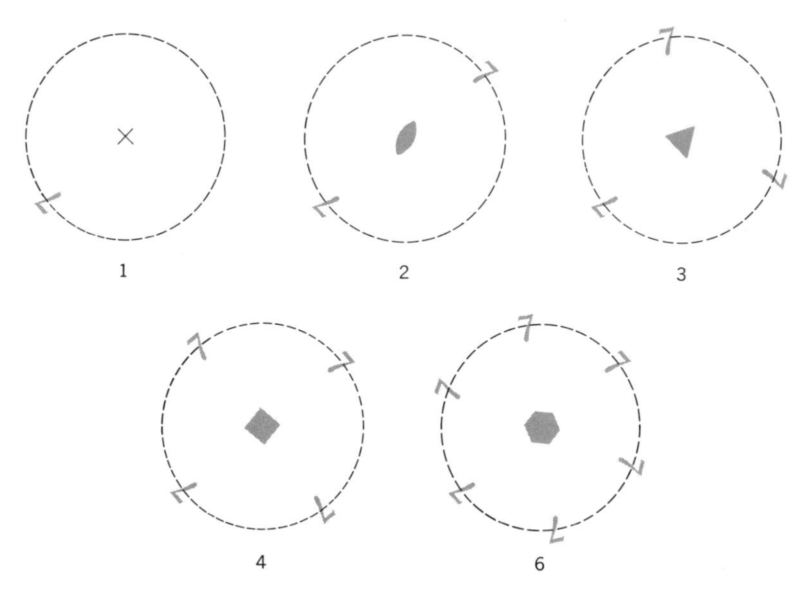

Fig. 1-14

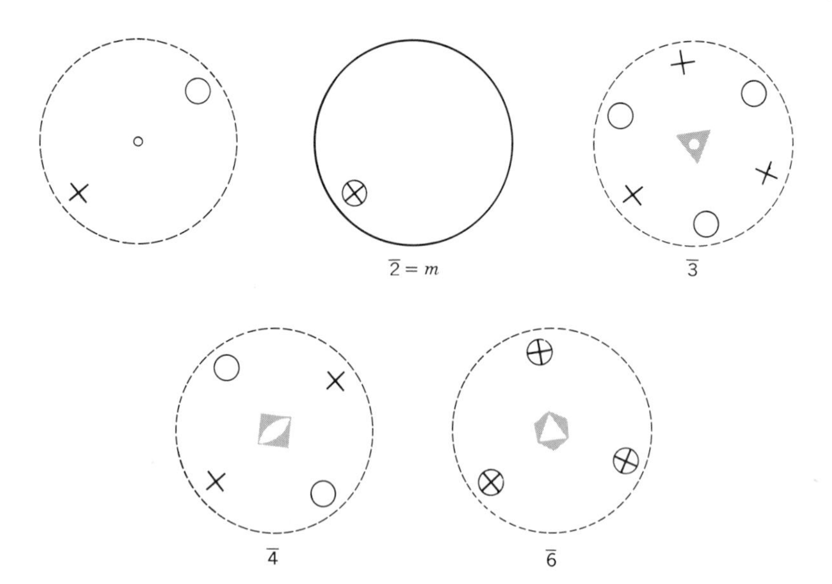

Fig. 1-15. The five improper axes and the symbols used to denote them. Note that $\overline{1}$, $\overline{2} = m$, $\overline{3}$, and $\overline{4}$ are really unique symmetries, while $\overline{6} = \frac{3}{m}$.

The simplest combination is to place a mirror plane at right angles to a proper rotation axis. Such point groups are symbolically represented by writing $\frac{1}{m}$, $\frac{2}{m}$, $\frac{3}{m}$, $\frac{4}{m}$, and $\frac{6}{m}$. When this is done, only the three new point groups shown in Fig. 1-16 result since $\frac{1}{m} = m = \overline{2}$ and $\frac{3}{m} = \overline{6}$. It is an intrinsic property of such groups *that the combination of any two symmetry elements produces a third.* In the case of an even-fold axis perpendicular to a mirror plane, the new symmetry element produced is a center of symmetry at their intersection. The presence of the center is indicated by placing a white dot at the center of the symbols, shown in Fig. 1-16, and it is left to Exercise 1-3 to prove its existence in these point groups.

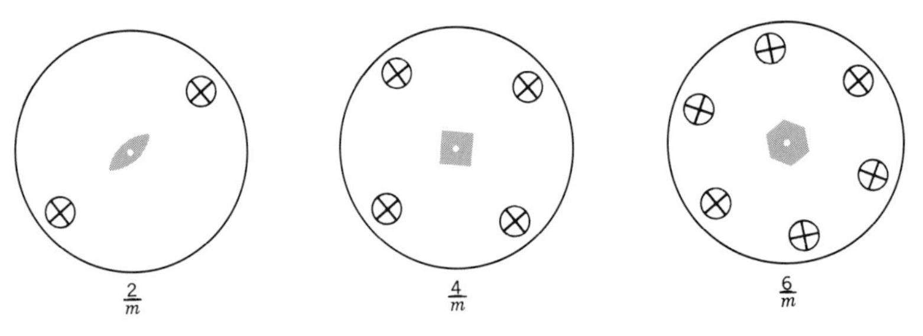

Fig. 1-16

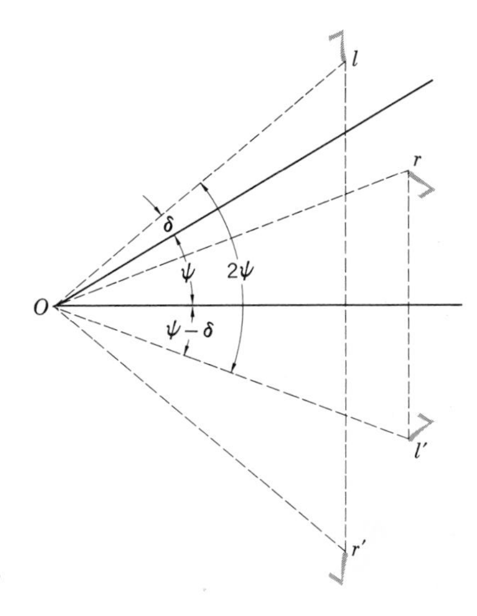

Fig. 1-17

A mirror can also be placed parallel to a proper rotation axis. In this case the third symmetry element produced is another mirror plane, also parallel to the rotation axis, and the angle between the two mirrors is equal to one-half the throw of the rotation axis. This relation is best stated by changing the order of the above operations; namely, if two mirror planes intersect at an angle ψ, a rotation axis having a throw of 2ψ arises at their intersection. A simple proof of this is shown in Fig. 1-17. Consider the two mirror planes intersecting at an angle ψ. The first one reflects a left-handed object l into a right-handed object r. The second one reflects r into l', and l into r'. That these two successive reflections are equal to a rotation about O by an amount 2ψ can be seen by adding the angles between the lines Ol and Ol' or between the lines Or and Or':

$$\delta + \psi + (\psi - \delta) = 2\psi \tag{1-3}$$

Excluding the trivial combination $1m = m$ already discussed, four additional point groups are obtained by combining a rotation axis with a parallel reflection plane: $2mm$, $3m$,† $4mm$, $6mm$ (Fig. 1-18).

† When a 3 is combined with a parallel m the angle between the mirror planes must be $120°/2 = 60°$. But this angle between the mirror planes also results by simply repeating a mirror plane every $120°$ as required by 3-fold rotation; hence an additional mirror plane is not produced, as the symbol $3mm$ would indicate.

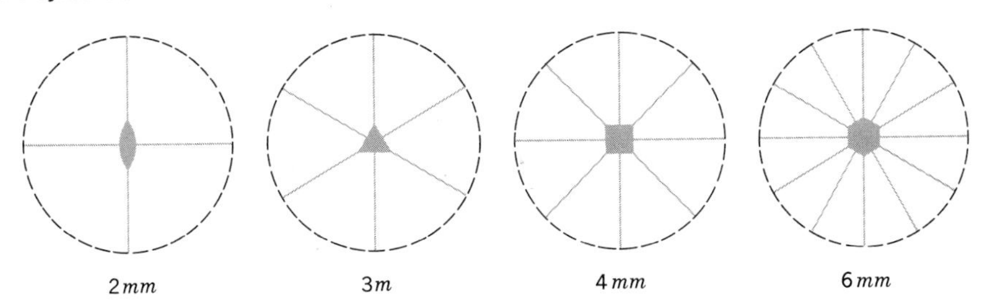

Fig. 1-18

The preceding is another example of an important concept of group theory, which can be written symbolically

$$A \cdot B = C \tag{1-4}$$

and states that when one symmetry element A is properly combined with another B, then a third element C is automatically generated. Note that it is a symbolic algorism, and should not be confused with ordinary algebra.

The symmetry that results when rotation axes are combined can be determined by a construction suggested by Euler. Because of its relative complexity, Euler's construction is not given here; however, use is made of a relation that results from the analysis of one of its spherical triangles. This relation states that

$$\cos A = \frac{\cos(\beta/2)\cos(\gamma/2) + \cos(\alpha/2)}{\sin(\beta/2)\sin(\gamma/2)} \tag{1-5}$$

Here A is the angle between two rotation axes whose respective throws are γ and β, and α is the throw of the third rotation axis. Thus Eq. (1-5) can be used to determine what combinations of rotation axes are possible and at what angles the axes are inclined to each other. The actual determination can be carried out by systematically considering all possible combinations. For example, consider the combination of a 2-fold rotation with another 2-fold rotation to give a third 2-fold rotation, symbolically expressed by $2 \cdot 2 = 2$. According to (1-5),

$$\begin{aligned} \cos A &= \frac{\cos(180°/2)\cos(180°/2) + \cos(180°/2)}{\sin(180°/2)\sin(180°/2)} \\ &= \frac{0+0}{1} \\ &= 0 \qquad (A = 90°) \end{aligned} \tag{1-6}$$

Since the three rotation axes considered above are alike, (1-6) states that three 2-fold axes can be combined at a point, provided that they are mutually perpendicular.

As a further example, consider $2 \cdot 2 = 3$. Here two interaxial angles must be determined, the angle between a 2-fold and a 3-fold axis and also the angle between the two 2-fold axes. According to (1-5),

$$\begin{aligned} \cos A &= \frac{\cos(180°/2)\cos(120°/2) + \cos(180°/2)}{\sin(180°/2)\sin(120°/2)} \\ &= \frac{0+0}{0.866} \\ &= 0 \qquad (A = 90°) \end{aligned} \tag{1-7a}$$

and

$$\begin{aligned} \cos C &= \frac{\cos(180°/2)\cos(180°/2) + \cos(120°/2)}{\sin(180°/2)\sin(180°/2)} \\ &= \frac{0+\cos 60°}{1} \\ &= \cos 60° \qquad (C = 60°) \end{aligned} \tag{1-7b}$$

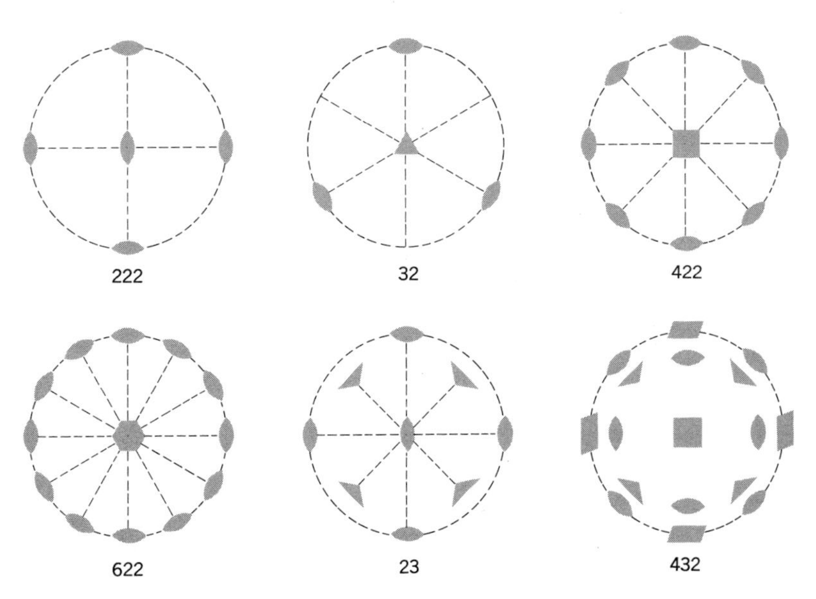

Fig. 1-19

Similarly, the other possible axial combinations can be tested. It is left to the reader to prove (see the exercises at the end of this chapter) that only six combinations of proper rotation axes intersecting at a common point are possible. These possible combinations of three proper rotation axes at a point, 222, $32(2)$,† 422, 622, $23(3)$, 432, are shown in Fig. 1-19. All other conceivable combinations, such as $6 \cdot 4 = 2$, cannot exist, as can be shown by direct substitution of appropriate values in Eq. (1-5).

Another possibility is to combine two $\bar{2}$'s with a rotation axis. For example, place one mirror plane parallel to, and the other one perpendicular to, each of the five proper rotation axes. Since the two mirror planes intersect each other at right angles, a 2-fold axis lies at their intersection, according to the theorem demonstrated in Fig. 1-17. Consequently, the combination of two such planes with a rotation axis is equivalent to the addition of these planes to the point groups shown in Fig. 1-19. The resulting six point groups, $\frac{2}{m}\frac{2}{m}\frac{2}{m}$, $\frac{3}{m}m2$,‡ $\frac{4}{m}\frac{2}{m}\frac{2}{m}$, $\frac{6}{m}\frac{2}{m}\frac{2}{m}$, $\frac{2}{m}\bar{3}$, $\frac{4}{m}\bar{3}\frac{2}{m}$, are shown in Fig. 1-20. Note that all these point groups except $\frac{3}{m}m2$ have

† This point group is conventionally written 32 instead of 322. The reason for this is that the two 2-fold axes $60°$ apart are indistinguishable from the symmetry-equivalent 2-fold axes that result from the repetition of a single axis every $120°$ by rotation about the 3-fold axis. This case is different from the point group 422, where the 4 repeats each 2 at $90°$ intervals but the two different 2's are $45°$ apart.

‡ The symbol is $\frac{3}{m}m2$, not $\frac{3}{m}\frac{2}{m}$, because the mirror plane parallel to the 3-fold axis is not perpendicular to any 2-fold axis.

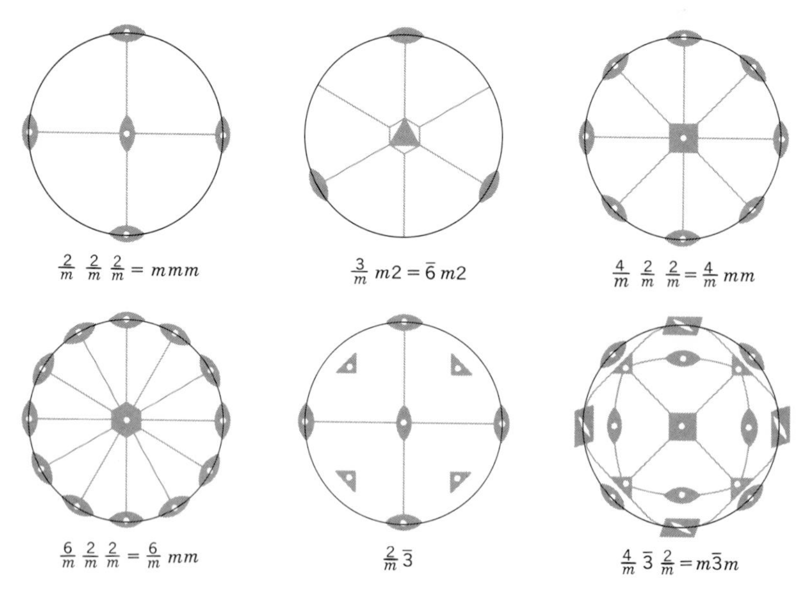

Fig. 1-20

an even-fold axis normal to a mirror plane, and hence contain an inversion center. This is the reason why, in the last two point groups listed, the 3-fold axes are rotoinversion axes.

It remains to consider the possible combinations of the improper axes $\bar{3}$ and $\bar{4}$ with proper rotation axes. Three more point groups are obtained from such combinations, namely, $\bar{4}2m$, $\bar{3}\frac{2}{m}$, and $\bar{4}3m$, shown in Fig. 1-21. This brings the total number of point groups to 32 and completes the list of possibilities. The angles between all the symmetry elements are clearly seen in the illustrations, except for the point groups containing four 3-fold axes. These interaxial angles are shown in Fig. 1-22 for the point group 432.

Crystal systems. An examination of the 32 point groups just described suggests that they can be collected into sets according to the highest-ranking rotation axis

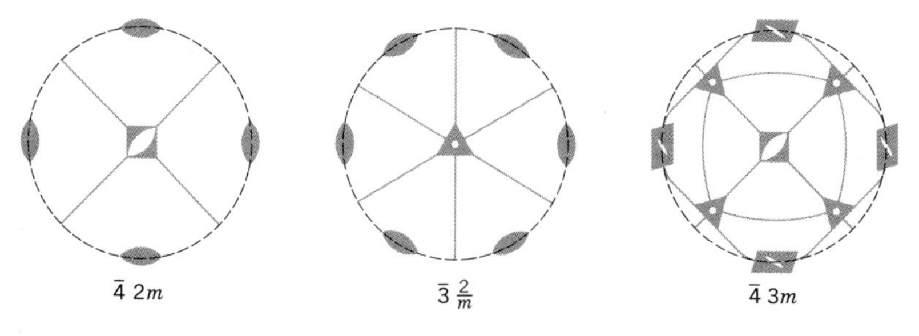

Fig. 1-21

Table 1-2 Crystal systems and crystal classes

Crystal system	*Minimal symmetry*†	*Crystal classes*
Triclinic	1 (or $\bar{1}$)	$1, \bar{1}$
Monoclinic	One 2 (or $\bar{2}$)	$2, m, \frac{2}{m}$
Orthorhombic	Three 2's (or $\bar{2}$)	$222, 2mm, mmm$
Tetragonal	One 4 (or $\bar{4}$)	$4, \bar{4}, \frac{4}{m}, 422, 4mm, \bar{4}2m, \frac{4}{m}\frac{2}{m}\frac{2}{m}$
Cubic (isometric)	Four 3's (or $\bar{3}$)	$23, 432, \frac{2}{m}\bar{3}, \bar{4}3m, \frac{4}{m}\bar{3}\frac{2}{m}$
Hexagonal	One 3 or 6 (or $\bar{3}$ or $\bar{6}$)	$3, \bar{3}, 32, 3m, \bar{3}\frac{2}{m}, 6, \bar{6}, 622, 6mm, \bar{6}m2, \frac{6}{m}, \frac{6}{m}\frac{2}{m}\frac{2}{m}$

† The symmetry elements shown in this column are the least symmetry that a crystal can have and still belong to the corresponding crystal system. In this sense, this column shows the diagnostic symmetry elements of each system.

they contain. Such an axis, say, a 6-fold rotation axis, is also the most prominent feature of the morphology of a crystal containing this axis. It is natural, therefore, that the early classifications of crystals were based on the more prominent features that different crystals shared. The scheme finally adopted was to distribute the 32 point groups among the six crystal systems listed in Table 1-2. The basis for assigning a crystal class to a particular crystal system is indicated in the column headed "minimal symmetry." For example, there are only three point groups that contain only three 2-fold axes, namely, mmm, $2mm$, 222. Accordingly, these three point groups are placed in the same system, called the *orthorhombic* system.

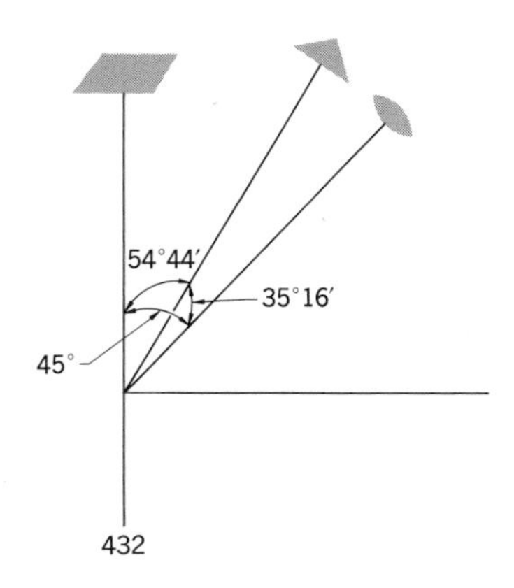

Fig. 1-22

NOTATION OF CRYSTAL FACES

Choice of crystal axes. For reasons that become self-evident as the reader becomes better acquainted with crystals and their symmetries, it is most convenient to choose the rotation axes present in a crystal as its reference axes. The three mutually noncoplanar *crystallographic axes* are usually designated a, b, and c, unless the presence of symmetry causes any of them to be equivalent to each other. In that case, it is usual to distinguish two equivalent axes by labeling them a_1 and a_2, or three equivalent axes by a_1, a_2, and a_3. In the more general case, it is necessary to indicate their disposition in space (Fig. 1-23) by specifying the interaxial angles α, β, and γ, where α is the angle between b and c (opposite the a axis), β is the angle between c and a (opposite the b axis), and γ is the angle between a and b (opposite the c axis).

It is conventional to choose the unique symmetry axis, when the crystal has one, as the c axis. Thus, in the hexagonal system, c is parallel to 6, $\bar{6}$, 3, or $\bar{3}$, while the two a axes (equivalent by symmetry) are chosen parallel to the 2-fold axes or perpendicular to the mirror planes, if present. This imposes the condition that c is perpendicular to a_1 and a_2, which, in turn, should be $120°$ apart. Similarly, in the tetragonal system, c is parallel to the 4-fold axis and perpendicular to the two equivalent a axes, which, in turn, are parallel to the symmetry-equivalent 2-fold axes, $90°$ apart. The actual choice is dictated by the highest allowed symmetry in each crystal system, as indicated in Table 1-3. It is clear from this table that the presence of certain symmetry elements severely restricts the freedom in the choice of crystallographic axes if they are to be consistent with the point-group symmetry.

As an example of such a choice, consider the crystal shown in Fig. 1-24. In addition to the rotation axes shown, it contains five mirror planes and a center of symmetry, which have been omitted from the drawing for the sake of clarity. The point group of the crystal is, obviously, $\frac{4}{m}\frac{2}{m}\frac{2}{m}$, and it belongs to the tetragonal system. Thus the c axis is chosen parallel to the 4. The two a axes can be chosen parallel to either of the two sets of 2-fold axes. Although such a choice is arbitrary,

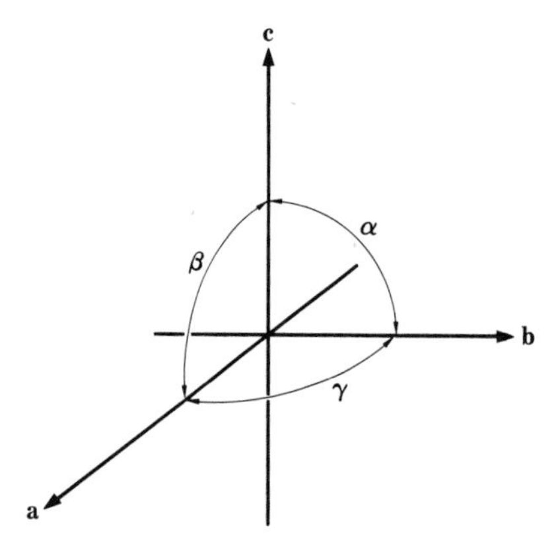

Fig. 1-23. Crystallographic axes in their conventional orientation in space: c vertical, b to the right, and a forward, thus forming a right-handed coordinate system.

Table 1-3 Orientation and relation of crystal axes

Crystal system α,β,γ	*Crystal axes and angles*†	*Orientation of axes*
Triclinic	$a \neq b \neq c$ $\alpha \neq \beta \neq \gamma$	Unspecified
Monoclinic	$a \neq b \neq c$ $\alpha = \beta = 90° \neq \gamma$	c parallel to 2‡
Orthorhombic	$a \neq b \neq c$ $\alpha = \beta = \gamma = 90°$	a, b, and c must parallel 2's
Tetragonal	$a_1 = a_2 \neq c$ $\alpha = \beta = \gamma = 90°$	c parallel to 4
Cubic (isometric)	$a_1 = a_2 = a_3$ $\alpha = \beta = \gamma = 90°$	a_1, a_2, and a_3 must parallel 4's
Hexagonal	$a_1 = a_2 \neq c$ $\alpha = \beta = 90°$; $\gamma = 120°$	c parallel to 6 (or 3)

† The unequal sign used in this column means "unequal by symmetry"; that is, the minimal symmetry of the system does not impose an equivalence. Fortuitous equalities may occur in actual crystals, e.g., $\gamma = 90°$ in a monoclinic crystal. This does not require reclassification, however, since the symmetry present, and not axial relations, determines the classification.

‡ In earlier times it was conventional to choose b unique and $\beta \neq 90°$. This custom is gradually disappearing.

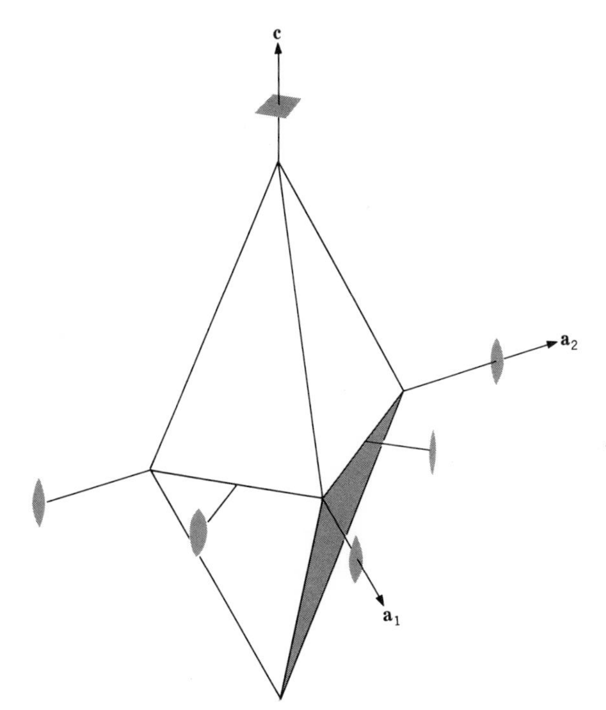

Fig. 1-24. Choice of axes for tetragonal crystal. Note that a_1 and a_2, alternately, can be chosen to be parallel to the other pair of 2-fold axes.

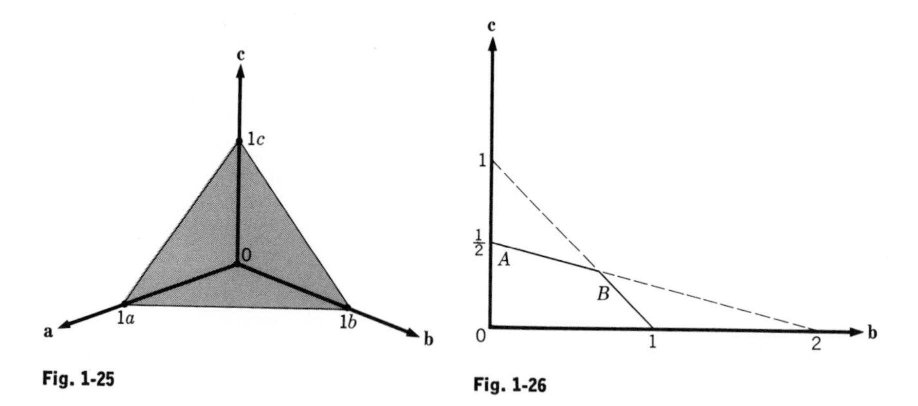

Fig. 1-25

Fig. 1-26

once made, it is necessary to refer consistently to the same axes. The importance of this will become clear in the next section.

Except for the requirements set forth in Table 1-3, no means for determining the relative magnitudes of the crystallographic axes and angles has been given as yet. Methods for determining their absolute values using x-ray diffraction are available; however, it is premature to consider them now. Instead, consider a plane that intersects all three crystallographic axes, as shown in Fig. 1-25. Using this plane, sometimes called a *parametral plane*, it is possible to define the relative lengths of the three axes by, arbitrarily, specifying that this plane intersects each axis a unit length away from an origin placed at their common intersection (usually chosen at the center of the crystal). If only one plane cuts all three axes, this is of little practical significance, but consider the case when several planes intersect the axes. Specifically, consider the cross section of two mutually intersecting planes shown in Fig. 1-26. If line B is used to define the c/b ratio by assuming that it intercepts each axis a unit length from O, then the intercepts of line A are as indicated. Since it is possible to describe each intersecting plane by noting its intercepts, the proper choice of the relative lengths of the axes and the interaxial angles has a practical significance.

Miller indices. By the middle of the eighteenth century, the study of crystals had progressed to such a high level that the Abbé Haüy concluded that crystals must be composed of small uniform blocks that are periodically repeated. This led him to develop the now famous *law of rational indices*, which states that the intercepts made on the crystallographic axes by any rational plane can be expressed by the ratios a/h, b/k, and c/l, where h, k, and l are integers, including zero. This notation was popularized in the next century by W. H. Miller, and the three integers used to index the plane bear his name. Once the relative lengths of the three crystallographic axes have been determined, for example, by selecting a parametral plane (Fig. 1-25), the three Miller indices are determined in three steps:

Step 1. Determine the intercepts of the plane along each of the three crystallographic axes.

Step 2. Note their reciprocals.
Step 3. If fractions result, multiply each by the smallest common divisor.

To illustrate the application of this procedure, assume that the planes whose cross sections are shown in Fig. 1-26 are both parallel to the a axis. Next, select the plane marked B as the parametral plane, so that it intersects both axes a unit axial length from the origin at O. Since it is parallel to a, it can be said to cut it at infinity, and it cannot define its length relative to b and c. To determine its indices, then:

		a	b	c
Step 1.	Intercepts	∞	1	1
Step 2.	Reciprocals	$1/\infty$	$1/1$	$1/1$
Step 3.	Indices	0	1	1

The three integers are placed in ordinary parentheses to designate the plane (011), pronounced "zero-one-one." By comparison, the same procedure applied to the plane labeled A yields

	a	b	c
Step 1	∞	2	$1/2$
Step 2	$1/\infty$	$1/2$	2
Step 3	0	1	4

so that the indices of this plane are (014). Incidentally, it is conventional to select the multiplier used to clear fractions in the last step in such a way that three indices (hkl) are mutually prime.

Suppose, however, that the two crystallographic axes in Fig. 1-26 are equivalent by symmetry, as will be the case if the crystal is isometric (cubic). In this case (Fig. 1-27) the indices of the two planes are determined as follows:

	Plane A			*Plane B*		
	a_1	a_2	a_3	a_1	a_2	a_3
Step 1	∞	1	$1/2$	∞	$1/2$	1
Step 2	0	1	2	0	2	1
Step 3	Not needed			Not needed		

and their respective Miller indices are (012) and (021). This illustrates the real meaning of these indices, namely, that the plane (hkl) cuts the a axis into h parts,

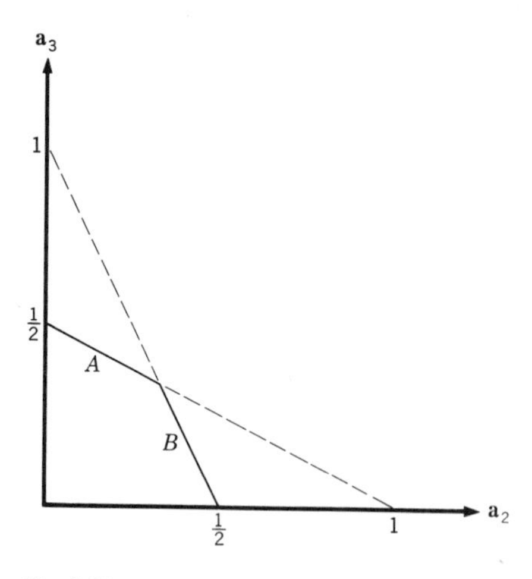

Fig. 1-27

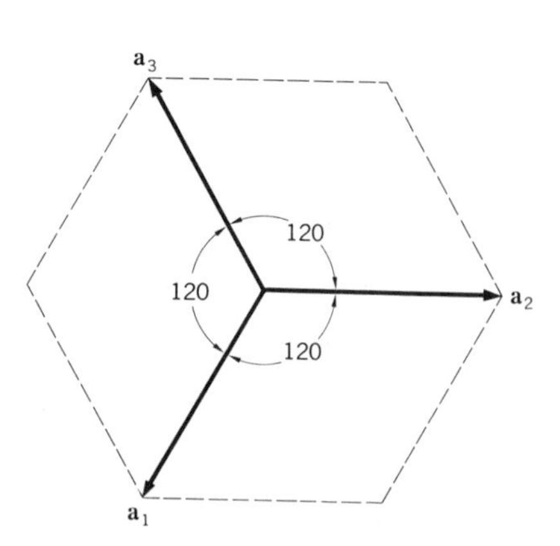

Fig. 1-28

the b axis into k parts, and the c axis into l parts. Compare this with the law of rational indices as stated above.

Hexagonal indices. A special case arises in the hexagonal system, in which there are three symmetry-equivalent a axes $120°$ apart. Looking down along the c axis, they appear as shown in Fig. 1-28. Consider a plane that is parallel to c and at least one of the a axes. Because of the symmetry, there are six such planes, as also indicated in Fig. 1-28. Note that each of the planes cuts one of the three a axes in a positive sense and one in a negative sense and is parallel to the third. If all four hexagonal axes a_1, a_2, a_3, and c are used in the indexing procedure described above, the result for, say, the uppermost plane drawn in Fig. 1-28 is as follows:

	a_1	a_2	a_3	c
Step 1	-1	∞	1	∞
Step 2	-1	0	1	0
Step 3		Not needed		

and the hexagonal indices are $(\bar{1}010)$, where $-h$ is written $\bar{h}$, by convention. This plane is called the "minus one-zero-one-zero" plane.

The above indices are also called the Bravais-Miller indices $(hkil)$ and must obey the rule that

$$h + k = -i \tag{1-8}$$

since it is easily shown (Exercise 1-11) that the Bravais-Miller axes must obey the vector relation

$$\mathbf{a}_1 + \mathbf{a}_2 = -\mathbf{a}_3 \tag{1-9}$$

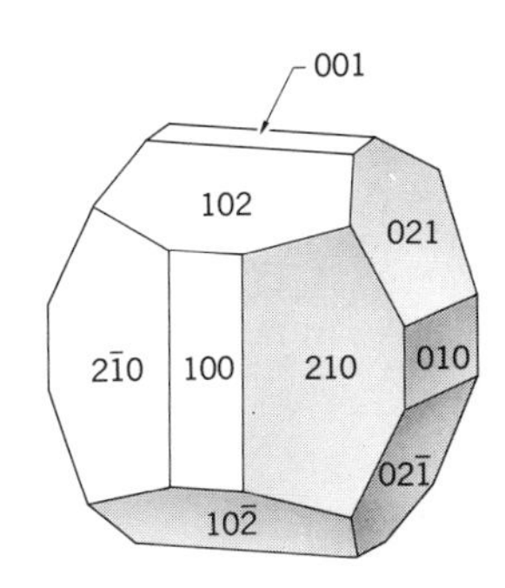

Fig. 1-29

Zones and forms. The planes shown in Figs. 1-27 and 1-28 are related to each other by symmetry elements present in the crystal. Such sets of symmetry-equivalent planes are said to constitute a *form*. The indices of the various planes belonging to a form are readily determinable, and it is sufficient to use the indices of any one plane to represent the symmetry-related set. When this is done, the indices are placed inside double parentheses; that is, a symmetrical set of planes is designated $((hkl))$.† Figure 1-29 shows the appearance of a pyrite, FeS_2, crystal and the presence of two forms, $((100))$ and $((102))$.

Another collection of planes of frequent interest is a set of planes that have one direction in common. This common direction is the direction along which the planes intersect, and it is called a *zone axis*. The planes that share this zone axis are said to belong to the same *zone*. A zone axis is designated by three integers placed in brackets $[uvw]$, defining a vector **t** (measured from the crystal origin) according to the vector equation

$$u\mathbf{a} + v\mathbf{b} + w\mathbf{c} = \mathbf{t} \tag{1-10}$$

The indices of a zone axis $[uvw]$ and a plane (hkl) in the zone must obey the algebraic relation

$$uh + vk + wl = 0 \tag{1-11}$$

Finally, the zone axis $[uvw]$ of two intersecting planes $(h_1k_1l_1)$ and $(h_2k_2l_2)$ can be determined as follows:

$$\begin{aligned} u &= k_1l_2 - k_2l_1 \\ v &= l_1h_2 - l_2h_1 \\ w &= h_1k_2 - h_2k_1 \end{aligned} \tag{1-12}$$

As an example, note that, according to (1-10), the direction, or zone axis, parallel to axis a must be specified by the indices $[100]$, parallel to b by $[010]$, and so forth, while a simple example of planes belonging to a zone is afforded by the $[001]$ zone axis and all the planes that have indices $(hk0)$. Further examples are given in the exercises.

† In older usage a form was designated by braces, $\{hkl\}$.

SUGGESTIONS FOR SUPPLEMENTARY READING

M. J. Buerger, *Elementary crystallography* (John Wiley & Sons, Inc., New York, 1956).
Charles Bunn, *Crystals: their role in nature and science* (Academic Press Inc., New York, 1964).
F. C. Phillips, *An introduction to crystallography*, 3d ed. (Longmans, Green & Co., Ltd., London, 1963).
A. F. Wells, *The third dimension in chemistry* (Oxford University Press, Fair Lawn, N.J., 1956).

EXERCISES

1-1. By means of a sketch showing a view along the axis, indicate the symmetry of space about a 5-fold and 8-fold proper rotation axis.

1-2. In the same manner as in Exercise 1-1, display the symmetry of space about a 5-fold and 8-fold rotoinversion axis.

1-3. By means of a perspective drawing, such as Fig. 1-5, show that $2 \cdot \bar{2} = \bar{1}$ if the angle between 2 and $\bar{2}$ is $0°$, that is, when they are coincident. (Note that the direction of the 2-fold rotoinversion axis is perpendicular to the plane of the resulting mirror plane.)

1-4. What point group results when a 2 is placed at $90°$ to a $\bar{2}$? (Derive the answer by making a sketch.)

1-5. Using Eq. (1-5), derive all the point groups containing a 2-fold axis. Do this by systematically considering the combinations of a 2 with a 2 to give a 1, 2, 3, etc., followed by combinations of a 2 with a 3 to give a 1, 2, 3, etc.

1-6. Derive all the point groups consistent with a 4-fold axis, using the procedure outlined in Exercise 1-5.

1-7. Examine the crystal in Fig. 1-29 carefully, and list all the symmetry elements that you find present. Next, determine the correct crystal class and system.

1-8. Consider the tetragonal crystal shown in Fig. 1-24. Using the axes labeled a_1, a_2, and c in that figure, what are the indices of the four faces visible? What is the form?

1-9. Repeat Exercise 1-8, selecting the alternative pair of 2-fold axes as a_1 and a_2. For the crystal shown, does it really matter which pair of 2-fold axes is chosen?

1-10. To what two forms do the mirror planes present in a cube (Fig. 1-13) belong? What are the zone-axis designations† of the 4-fold and 2-fold axes?

1-11. Prove relation (1-9) for hexagonal axes and relation (1-8) for hexagonal indices.

1-12. List at least six different planes belonging to the zone whose zone axis is $[1\bar{1}1]$.

† In stating the answer, note that the notation $[[uvw]]$ has the same meaning relative to $[uvw]$ as the double parentheses used to designate a form. The older usage was to place the integers between carets, $\langle uvw \rangle$, which led to confusion with similar designations of average values.

TWO
Projections of crystals

PERSPECTIVE PROJECTIONS

Spherical projection. By definition, a perspective projection consists of the placement, on a surface, of points in space as they would appear on that surface when viewed from a fixed point in space. The familiar perspective drawing of a three-dimensional object on a plane sheet is a common example. The preparation of such drawings is governed by the rules of descriptive geometry, which are designed to portray correctly the external appearance of the object, but at the sacrifice of distorting the true relative areas, angles, and other features of the projected object. In the case of crystals, this limits their usefulness to illustrations of crystal habits, but renders such drawings valueless for the detailed examination of crystallographic properties.

A most convenient way of projecting crystals is to place the point of perspective at the center of the crystal (point of intersection of the crystal axes) and draw radial lines outward, each one directed at right angles to a rational plane of the crystal. Since the direction of the normal to a plane correctly represents the orientation of the plane, a collection of such normals correctly represents the orientation of the various planes being projected. In order to intercept all the normals of projection, it is, of course, necessary to surround the point of projection by a spherical surface, preferably centered at the same point. The resulting collection of points on the sphere, called *poles*, which mark the intersections of the normals with the sphere, then constitutes a *spherical projection*. The spherical projection of the octahedral $((111))$ planes of a cubic crystal is shown in Fig. 2-1.

The spherical projection has several obvious advantages. A collection of three-dimensional planes, mutually intersecting each other in the crystal, has been reduced to a set of points (poles) clearly displayed on a single surface. The angles between the planes are correctly displayed along great circles joining the poles, as illustrated in Fig. 2-2. (Note that the true angle between two normals can be measured *only* along a great circle, not along a small circle on the sphere.) The zone axis of any two planes is easily found by noting that its pole must lie $90°$ (measured along a great circle) away from both poles (Fig. 2-2). Finally, it is

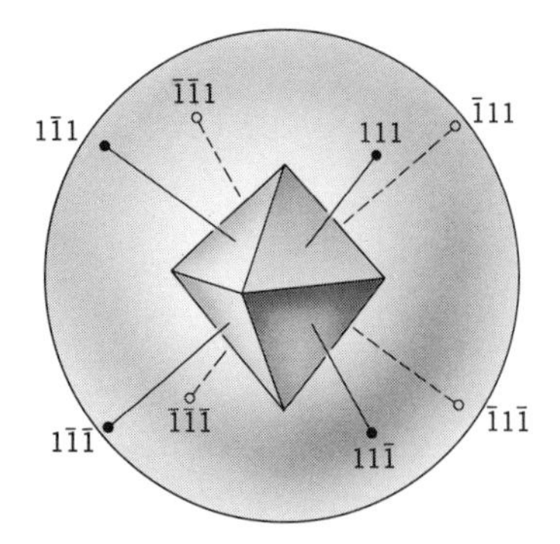

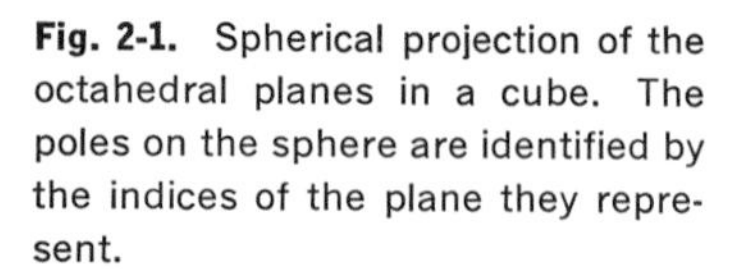
Fig. 2-1. Spherical projection of the octahedral planes in a cube. The poles on the sphere are identified by the indices of the plane they represent.

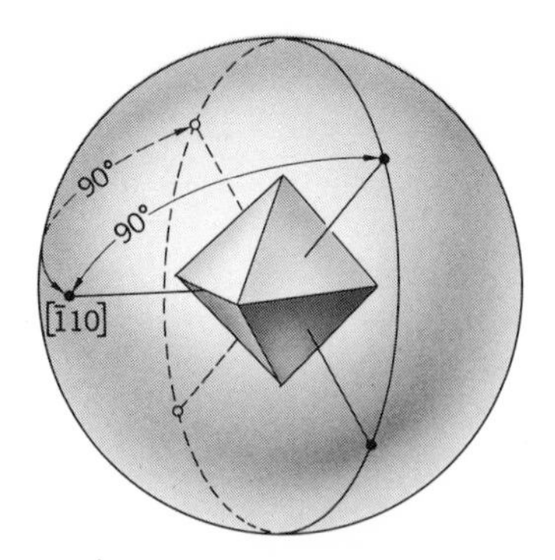

Fig. 2-2. Spherical projection of four octahedral planes belonging to a common zone. Note that the intersection of the zone axis also can be easily located on the sphere $90°$ away from the poles of the planes.

possible to mark the sphere with equally spaced great circles, called *meridians*, and small circles, called *parallels* (latitude circles), as, for example, on a globe, and to refer the various poles to such coordinates. The principal disadvantage of the spherical projection is the same as that of a globe; namely, it is not as portable as a plane sheet. Thus a more useful means of projection would be to depict the spherical projection of the crystal on a plane. Here use can be made of the many different projections developed by map makers for projecting the globe onto a plane sheet. Since, in the case of crystals, this means a two-step procedure, the spherical projection of a crystal is often called its *reference sphere*.

Types of projections. In projecting a sphere onto a plane, the first decision that must be made is how to surround the sphere by a suitable surface. The three common ways for doing this are illustrated in Fig. 2-3:

1. The *cylindrical projection*, in which the surface of projection surrounds the globe in the form of a cylinder which is tangent to the sphere at the equator. The sole advantage of this projection is that the parallels and meridians project as mutually perpendicular lines.
2. The *conical projection*, in which the surface of projection forms a cone circumscribing the globe and is tangential to a selected latitude circle. The meridians appear as straight lines radiating from the cone's vertex, while the parallels appear as concentric circles about the vertex. Note that only the selected parallel at the point of tangency appears in its true length.
3. The *zenithal projection*, in which the surface of projection is a plane made tangent to a point on the globe. The appearance of the meridians and parallels is determined by the placement of the point of tangency and the point of perspective, or *view point*, along the normal to the plane of projection. Some of the possibilities are considered next.

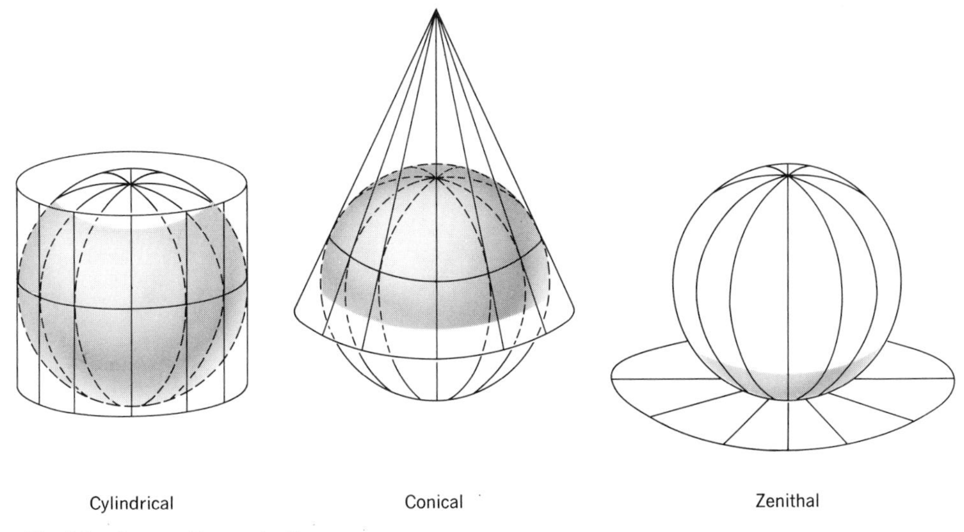

Fig. 2-3. Perspective projections.

Zenithal projections. Essentially, three different choices are available for the selection of the point of tangency for the plane of projection with the sphere:

1. One of the two geographic poles, resulting in a *polar projection*
2. A point lying on the equator, resulting in a *meridian projection*
3. Any point not lying on the geographic poles or on the equator, resulting in an *oblique projection*

The general appearance of the meridians and parallels in these three kinds of projections is shown in Fig. 2-4.

Since the plane of projection is tangent to the reference sphere in a zenithal projection, it must be perpendicular to a diameter of the sphere. The view point,

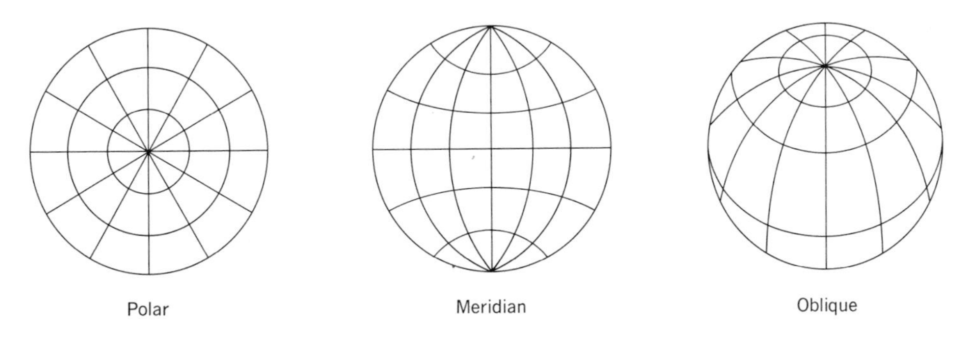

Fig. 2-4

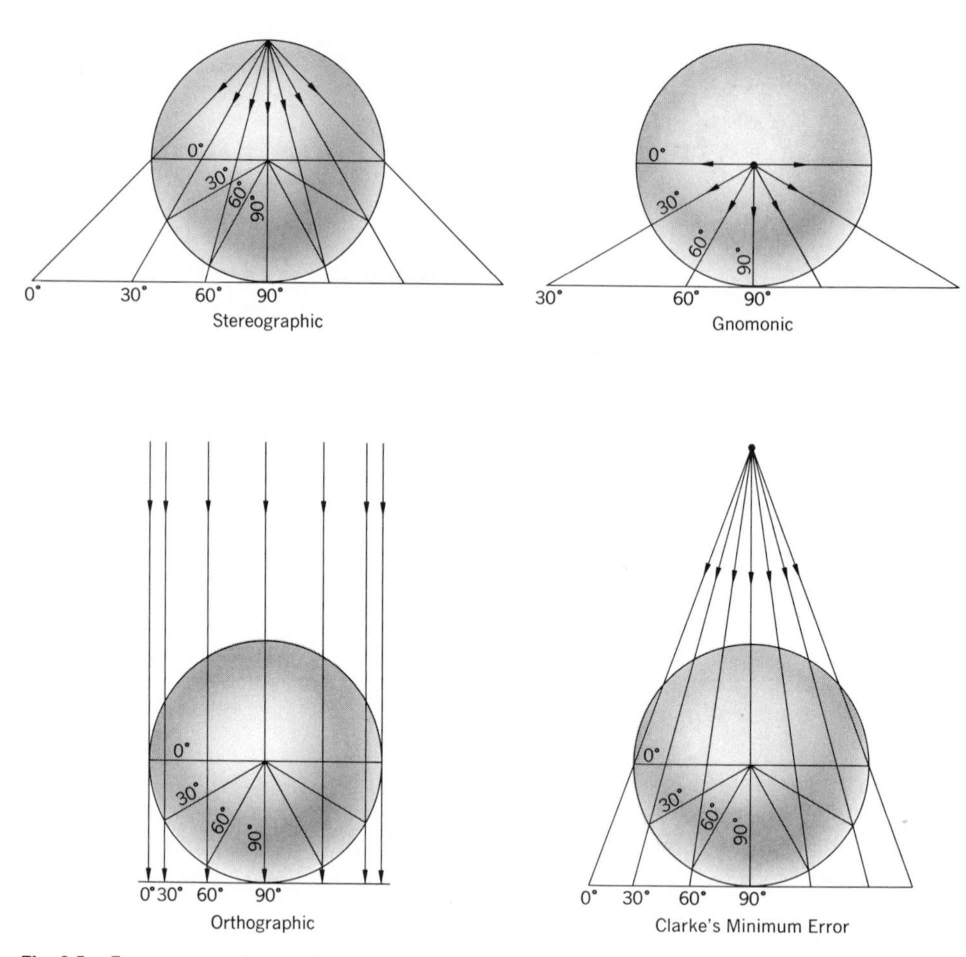

Fig. 2-5. Four common types of zenithal projections.

therefore, should be placed somewhere along this line, and four distinguishable possibilities arise:

1. The view point is placed at infinity. The lines of view, therefore, are perpendicular to the plane of projection and produce an *orthographic projection*.
2. The view point is placed on the surface of the sphere. The lines of view fan out from this point, and little more than half of the sphere can be viewed conveniently, producing the *stereographic projection*.
3. The view point is placed at the center of the sphere. This is the only projection in which no direct use need be made of the reference sphere, but only a portion of it appears on the plane in this *central* or *gnomonic projection*.
4. The view is placed outside the sphere, resulting in an *external projection*. It can be shown that for minimum distortion of selected portions of the reference sphere, the view point should be removed from the plane about 1.4 to 1.7 diameters, in which case it is called Clarke's minimum-error projection.

The four kinds of zenithal projections are illustrated in Fig. 2-5. The distortions present in the external projection render it useless in crystallography. Similarly, the crowding of the outer portions of the orthographic projection limits its utility. It is a very simple projection to visualize, however, and was used for that reason in displaying the point groups in the preceding chapter. Although the gnomonic projection has to be limited to some incomplete portion of the reference sphere by the finite size of the plane of projection, it has the virtue that all great circles appear as straight lines. Finally, as shown in Fig. 2-2, the angles between planes and their normals are correctly displayed on the reference sphere. It follows that a perspective projection that retains these angular relations without distortion is generally most useful in crystallography. This is the case in the stereographic projection, which is discussed in more detail below, following a brief description of the gnomonic projection.

GNOMONIC PROJECTION

Polar projection. As is clear from the foregoing discussions, the appearances of all polar perspective projections (Fig. 2-4) share one feature in common; namely, the meridians appear as equally spaced radial lines passing through the so-called geographic pole, and the parallels form concentric circles about the pole. The placement of the view point therefore merely determines the relative spacing of the projected parallels. It is easy to see (Fig. 2-6) that the radius of the circle r in a gnomonic polar projection of a unit reference sphere is determined by the colatitude χ according to

$$r = \tan \chi \tag{2-1}$$

It is also clear from this figure that the maximum portion of the sphere that can be projected is limited by the relative size of the plane of projection.

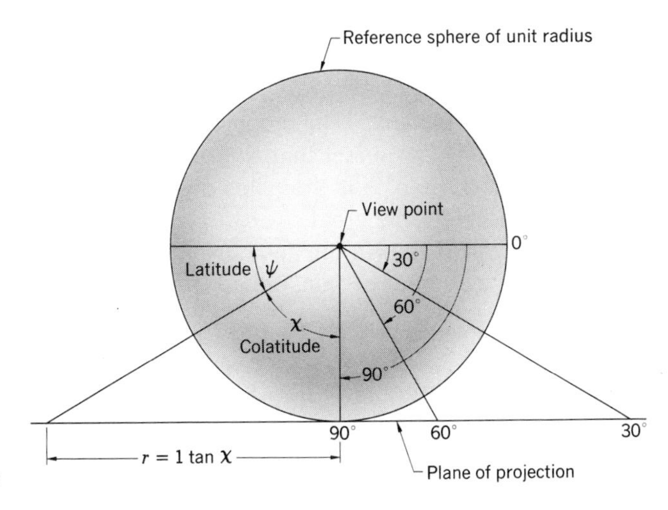

Fig. 2-6

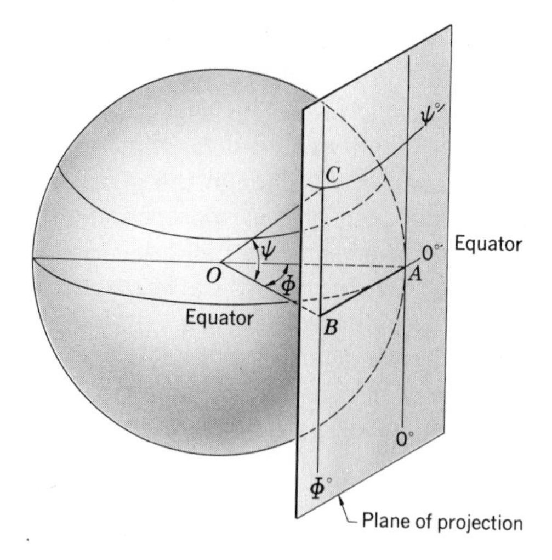

Fig. 2-7. Construction of gnomonic meridian projection.

Meridian projection. As pointed out earlier, it is an inherent property of the gnomonic projection that all great circles project as straight lines. In the meridian projection, therefore, the meridians are straight lines perpendicular to the projection of the equatorial line. The spacing of the meridians is given by a relation similar to (2-1), except that the angle to longitude Φ is used. In order to determine the intersection of a parallel at latitude Ψ, consider the two triangles OAB and OCB in Fig. 2-7. First note that BA is tangent to OA, so that, in a unit sphere,

$$OB = \sec \Phi \tag{2-2}$$

Next note that $\angle OBC$ is a right angle, so that

$$\begin{aligned} BC &= OB \tan \Psi \\ &= \sec \Phi \tan \Psi \end{aligned} \tag{2-3}$$

The intersection of a parallel with the vertical meridians, therefore, is plotted by multiplying the tangent of the latitude Ψ by the secant of the longitude Φ of the respective meridian.

Properties of projection. The property of the gnomonic projection most important in crystallography is the fact that *all* great circles project as straight lines. Going back to Fig. 2-2, note that the poles of the planes belonging to a common zone must lie along the same great circle. This means that the poles of planes in each zone form straight lines in the gnomonic projection. Use can be made of this fact in the interpretation of certain x-ray diffraction diagrams.

STEREOGRAPHIC PROJECTION

Polar projection. Because in a stereographic projection the view point is placed directly on the reference sphere, the stereographic polar projection appears like a perspective view of a transparent globe viewed along the north-south axis. As before, the meridians appear as radial straight lines (Fig. 2-4), while the parallels form circles whose radii can be determined with the aid of Fig. 2-8. The angle AOB' measures the colatitude, so that the angle subtended by the arc AB' at the view point is, from elementary geometry, equal to one-half the colatitude. The radius of the projected circle, if the radius of the reference sphere $OA = 1$, then is given by

$$r = AB = 2 \tan \frac{\chi}{2} \tag{2-4}$$

Note that the maximum value the colatitude can have is $90°$, so that the maximum value of the radius of the projection of the equatorial circle is twice the radius of the reference sphere. Also note that the equatorial circle bounds the stereographic polar projection. (Although the radius of the reference sphere is here chosen to be equal to unity, it is convenient to prepare stereographic projections having radii of 10 to 20 cm, in practice.)

Meridian projection. It is possible to construct a stereographic projection by placing the view point on the equator and proceeding as in Fig. 2-8. A much more direct procedure, however, is to make use of two properties of all stereographic projections that will be amplified in the next section, namely:

1. All angles of the reference sphere appear as true angles in the projection.
2. All circles on the sphere appear as true circles in the projection.

Use can be made of these properties in the following way: First mark off the angular intervals around the *boundary meridian*, forming the circumference of the projection, and along the *equator* and the *principal meridian*, using relation (2-4).

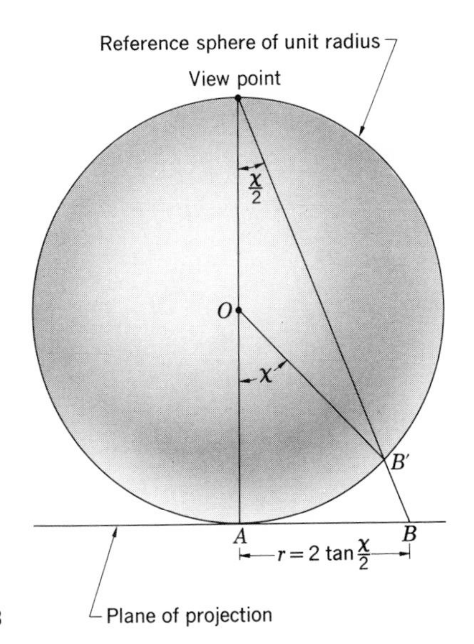

Fig. 2-8

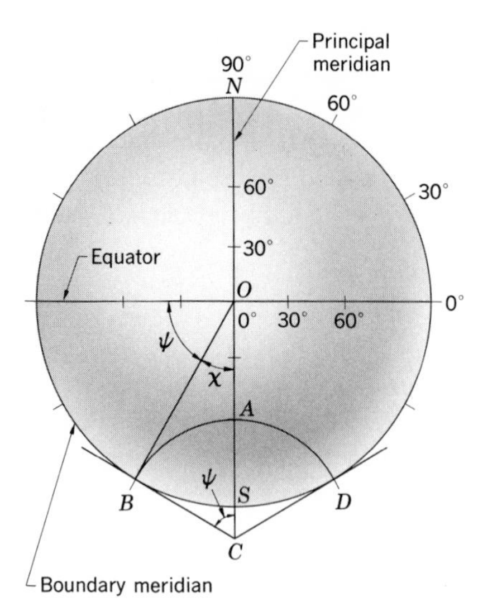

Fig. 2-9

One thus obtains three points through which each parallel and each meridian must pass. Since they will appear as circles in the projection according to property 2 above, it only remains to locate their respective centers along either the principal meridian or the equatorial line. Now, a meridian and a parallel intersect at right angles on the reference sphere, so that, according to property 1 above, the projected parallels must intersect the boundary meridian at right angles also. Consider the parallel at 60° latitude (south) shown in Fig. 2-9. Since the parallel circle is perpendicular to the boundary meridian at B and at D, the tangents to the meridian at these points, BC and DC, must intersect the principal meridian at the radius of the projected circle C. The proof of this is left to Exercise 2-5.

The centers of the projected meridians similarly must lie along the equator. The location of their loci can be determined with the aid of Fig. 2-10. Since all the meridians are great circles passing through the two geographic poles of the globe, their projections must pass through the two poles labeled N and S in the projection (Fig. 2-10). Because the projected meridians are circles also, the normal

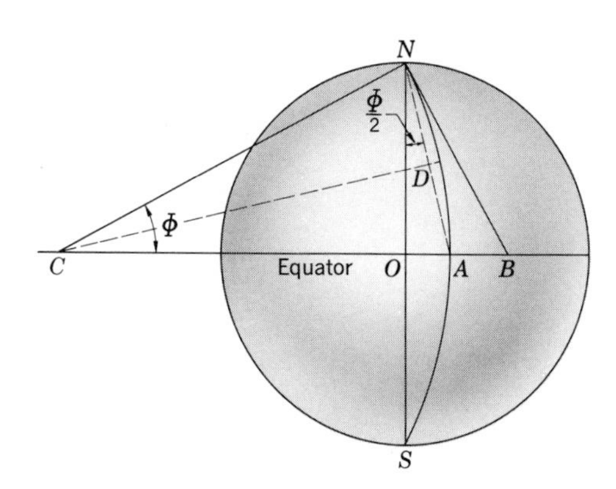

Fig. 2-10

CN to the tangent NB must cut the equator (which is normal to the tangent at A) at the center of the circle. Thus $CN = CA$ are radii of the circle whose length can be readily determined (Exercise 2-6). It is possible to determine this graphically quite easily also, as shown in Fig. 2-10. The distance $OA = \tan(\Phi/2)$, according to (2-4), so that $\angle ONA = \Phi/2$. Making use of construction line NA and its normal CD, it can be shown by considering pairs of right triangles having one angle (or side) in common that $\angle ONA = \angle DCO = \angle NCD = \Phi/2$, so that $\angle NCO = \Phi$ and its complement $\angle CNO = 90° - \Phi°$. Since this is the angle formed by the radius CN with the principal meridian NS, the location of the centers for any longitude Φ is straightforward. A stereographic meridian projection of meridians and parallels (Fig. 2-11) is called a *stereographic net*, or a *Wulff net*, after the Russian crystallographer who helped popularize its use at the beginning of this century.

Properties of projections. The proof that *all angles on the reference sphere appear as true angles in the projection* has actually been given in Fig. 2-8. It was shown there that an angle χ on the sphere projects as the angle $\chi/2$ (measured at the view point) on the stereographic projection. Since such scaling is easily allowed for when labeling the rulings in the stereographic net (Fig. 2-11), it follows that angles between poles on the sphere are conserved in the projection. It should be recalled

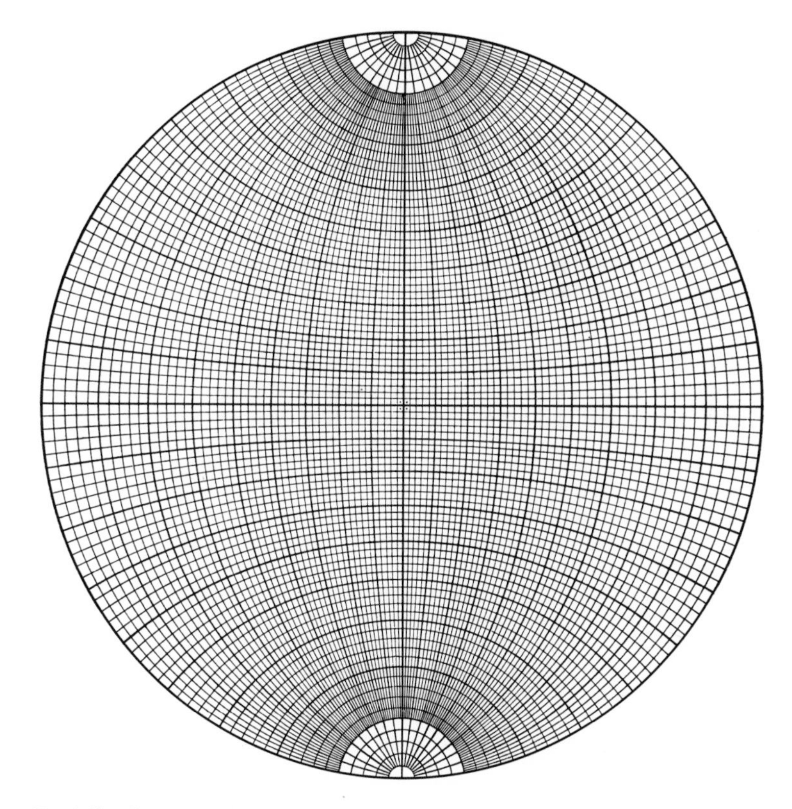

Fig. 2-11. Stereographic net.

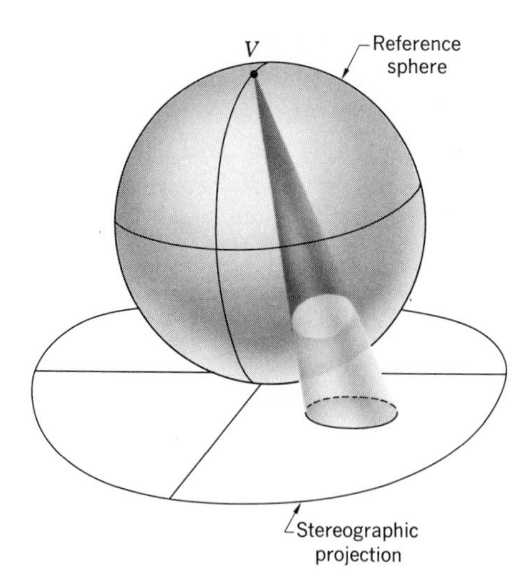

Fig. 2-12

from the introductory discussions that the true angles between two poles on a sphere can be measured only along a great circle passing through both poles (Fig. 2-2). It follows, therefore, that true angles in the stereographic projection also can be measured only along great circles. In the stereographic net shown in Fig. 2-11, the great circles, obviously, are all the meridians plus the equator.

The second important property of stereographic projection, that *all circles on the sphere appear as true circles in the projection*, is most easily proved by considering a small circle on the sphere (Fig. 2-12). All circles on a sphere are formed by planes passing through the sphere. If the intersecting plane passes through the center of the sphere, it cuts the sphere in a great circle. All planes not passing through the sphere's center intersect it in small circles; for example, the parallels of latitude are small circles. Let the drawing in Fig. 2-13 represent a vertical plane passing through the small circle in Fig. 2-12 and containing the view point V. The trace of the plane defining the small circle in Fig. 2-12 is represented by AB. Since the base of the cone represented by its trace AD is not a chord of the circle representing the sphere, the cone passing through the small circle (Fig. 2-12) is an elliptic cone, and the small circle at AB is one of its circular sections. It is left to

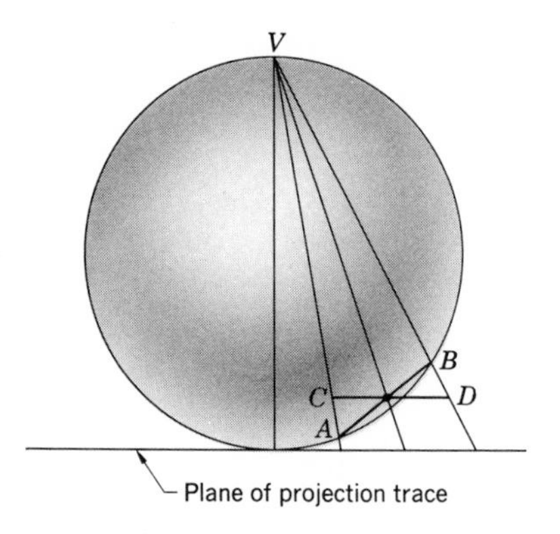

Fig. 2-13

Exercise 2-8 to prove that its conjugate circular section CD is parallel to the plane of projection. The projection of this small circle, therefore, must be circular. A more elegant proof of the two properties described above can be given by considering the stereographic projection as a form of inversion, and the interested reader is referred to the references at the end of this chapter for further details.

Standard projections. As shown in Chap. 1, the angles between all possible planes in the cubic system are fixed by symmetry. It is possible, therefore, to prepare stereographic projections for cubic crystals having known orientation relative to the view point, and such *standard projections* then are the same for all cubic crystals similarly oriented. To construct such a standard projection, it is necessary to know the angles between various planes $(h_1k_1l_1)$ and $(h_2k_2l_2)$, and these are listed in Table 2-1. It is standard practice to project only the poles lying in one hemisphere. The actual preparation of the standard projection can be greatly simplified by making use of the fact that the poles of planes belonging to the same zone must lie on the same great circle (Fig. 2-2). The standard projection along [001] in a cubic crystal is shown in Fig. 2-14. Only the poles of planes normal to rotation

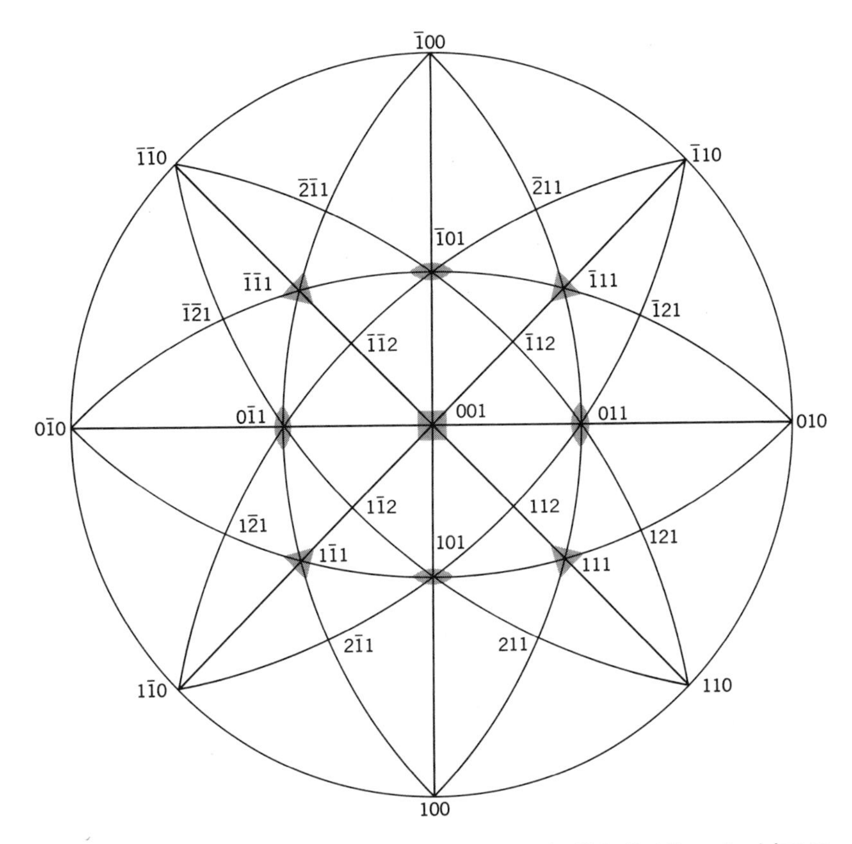

Fig. 2-14. Standard (001) projection of a cubic crystal. Note that the poles lying on the great circles indicated belong to the same zones.

Table 2-1 Angles between planes in cubic crystals†

$\{h_1k_1l_1\}$	$\{h_2k_2l_2\}$	*Possible angles (in degrees) between pairs of planes* $(h_1k_1l_1)$ *and* $(h_2k_2l_2)$				
100	100	0.00	90.00			
	110	45.00	90.00			
	111	54.74				
	210	26.56	63.43	90.00		
	211	35.26	65.90			
	221	48.19	70.53			
	310	18.43	71.56	90.00		
	311	25.24	72.45			
	320	33.69	56.31	90.00		
	321	36.70	57.69	74.50		
	322	43.31	60.98			
	331	46.51	76.74			
	332	50.24	64.76			
	410	14.04	75.96	90.00		
110	110	0.00	60.00	90.00		
	111	35.26	90.00			
	210	18.43	50.77	71.56		
	211	30.00	54.74	73.22	90.00	
	221	19.47	45.00	76.37	90.00	
	310	26.56	47.87	63.43	77.08	
	311	31.48	64.76	90.00		
	320	11.31	53.96	66.91	78.69	
	321	19.11	40.89	55.46	67.79	79.11
	322	30.96	46.69	80.12	90.00	
	331	13.26	49.54	71.07	90.00	
	332	25.24	41.08	81.33	90.00	
	410	30.96	46.69	59.04	80.12	
111	111	0.00	70.53			
	210	39.23	75.04			
	211	19.47	61.87	90.00		
	221	15.79	54.74	78.90		
	310	43.09	68.58			
	311	29.50	58.52	79.98		
	320	36.81	80.78			
	321	22.21	51.89	72.02	90.00	
	322	11.42	65.16	81.95		
	331	22.00	48.53	82.39		
	332	10.02	60.50	75.75		
	410	45.56	65.16			

Table 2-1 Angles between planes in cubic crystals† (continued)

$\{h_1k_1l_1\}$	$\{h_2k_2l_2\}$	*Possible angles (in degrees) between pairs of planes* $(h_1k_1l_1)$ *and* $(h_2k_2l_2)$
210	210	0.00 36.87 53.13 66.42 78.46 90.00
	211	24.09 43.09 56.79 79.48 90.00
	221	26.56 41.81 53.40 63.43 72.65 90.00
	310	8.13 31.95 45.00 64.90 73.57 81.87
	311	19.29 47.61 66.14 82.25
	320	7.12 29.74 41.91 60.25 68.15 75.64 82.87
	321	17.02 33.21 53.30 61.44 68.99 83.14 90.00
	322	29.80 40.60 49.40 64.29 77.47 83.77
	331	22.57 44.10 59.14 72.07 84.11
	332	30.89 40.29 48.13 67.58 73.38 84.53
	410	12.53 29.80 40.60 49.40 64.29 77.47 83.77
211	211	0.00 33.56 48.19 60.00 70.53 80.40
	221	17.72 35.26 47.12 65.90 74.21 82.18
	310	25.35 40.21 58.91 75.04 82.58
	311	10.02 42.39 60.50 75.75 90.00
	320	25.06 37.57 55.52 63.07 83.50
	321	10.89 29.20 40.20 49.11 56.94 70.89 77.40 83.74 90.00
	322	8.05 26.98 53.55 60.32 72.72 78.58 84.32
	331	20.51 41.47 68.00 79.20
	332	16.78 29.50 52.46 64.20 69.62 79.98 85.01
	410	26.98 46.12 53.55 60.32 72.72 78.58
221	221	0.00 27.27 38.94 63.61 83.62 90.00
	310	32.51 42.45 58.19 65.06 83.95
	311	25.24 45.29 59.83 72.45 84.23
	320	22.41 42.30 49.67 68.30 79.34 84.70
	321	11.49 27.02 36.70 57.69 63.55 74.50 79.74 84.89
	322	14.04 27.21 49.70 66.16 71.13 75.96 90.00
	331	6.21 32.73 57.64 67.52 85.61
	332	5.77 22.50 44.71 60.17 69.19 81.83 85.92
	410	36.06 43.31 55.53 60.98 80.69
310	310	0.00 25.84 36.87 53.13 72.54 84.26
	311	17.55 40.29 55.10 67.58 79.01 90.00
	320	15.26 37.87 52.12 58.25 74.74 79.90
	321	21.62 32.31 40.48 47.46 53.73 59.53 65.00 75.31 85.15 90.00

Table 2-1 Angles between planes in cubic crystals† (continued)

$((h_1k_1l_1))$	$((h_2k_2l_2))$	*Possible angles (in degrees) between pairs of planes* $(h_1k_1l_1)$ *and* $(h_2k_2l_2)$						
310	322	32.47	46.35	52.15	57.53	72.13	76.70	
	331	29.47	43.49	54.52	64.20	90.00		
	332	36.00	42.13	52.64	61.84	66.14	78.33	
	410	4.40	23.02	32.47	57.53	72.13	76.70	85.60
311	311	0.00	35.10	50.48	62.96	84.78		
	320	23.09	41.18	54.17	65.28	75.47	85.20	
	321	14.76	36.31	49.86	61.09	71.20	80.72	
	322	18.07	36.45	48.84	59.21	68.55	85.81	
	331	25.94	40.46	51.50	61.04	69.76	78.02	
	332	25.85	39.52	50.00	59.05	67.31	75.10	90.00
	410	18.07	36.45	59.21	68.55	77.33	85.81	
320	320	0.00	22.62	46.19	62.51	67.38	72.08	
	321	15.50	27.19	35.38	48.15	53.63	58.74	68.24
		72.75	77.15	85.75	90.00			
	322	29.02	36.18	47.73	70.35	82.27	90.00	
	331	17.36	45.58	55.06	63.55	79.00		
	332	27.50	39.76	44.80	72.80	79.78	90.00	
	410	19.65	36.18	42.27	47.73	57.44	70.35	78.36
		82.27						
321	321	0.00	21.79	31.00	38.21	44.41	49.99	64.62
		69.07	73.40	85.90				
	322	13.51	24.84	32.57	44.52	49.59	63.01	71.09
		78.79	82.55	86.28				
	331	11.19	30.85	42.63	52.18	60.63	68.42	75.80
		82.96	90.00					
	332	14.38	24.26	31.27	42.20	55.26	59.15	62.88
		73.45	80.16	83.46	86.73			
	410	24.84	32.57	44.52	49.59	54.31	63.01	67.11
		71.09	82.55	86.28				
322	322	0.00	19.75	58.03	61.93	76.39	86.63	
	331	18.93	33.42	43.67	59.95	73.85	80.97	86.81
	332	10.74	21.45	55.33	68.78	71.92	87.04	
	410	34.57	49.68	53.97	69.33	72.90		

Table 2-1 Angles between planes in cubic crystals† (continued)

$((h_1k_1l_1))$	$((h_2k_2l_2))$	Possible angles (in degrees) between pairs of planes $(h_1k_1l_1)$ and $(h_2k_2l_2)$						
331	331	0.00	26.52	37.86	61.73	80.91	86.98	
	332	11.98	28.31	38.50	54.06	72.93	84.39	90.00
	410	33.42	43.67	52.26	59.95	67.08	86.81	
332	332	0.00	17.34	50.48	65.85	79.52	82.16	
	410	39.14	43.62	55.33	58.86	62.26	75.02	
410	410	0.00	19.75	28.07	61.93	76.39	86.63	90.00

† Values given are from a more extensive table by R. J. Pearler and J. L. Lenusky, *Angles between planes in cubic crystals,* IMD Special Series Report 8 (AIME, New York).

axes have been indicated for simplicity. Similar standard projections can be prepared for other projection directions (Exercise 2-10) in the cubic system. For crystals belonging to other systems, standard projections can be prepared only when the relative lengths of the units along the crystallographic axes are known, so that the various interplanar angles can be calculated.

SUGGESTIONS FOR SUPPLEMENTARY READING

E. Boeke, *Die Anwendung der stereographische Projektion bei Kristallographischen Untersuchungen* (Borntrager, Berlin, 1911).

C. H. Deetz and O. S. Adams, *Elements of map projection,* 4th ed. (U.S. Coast and Geodetic Survey, Spec. Publ. 68, 1934).

James Mainwaring, *An introduction to the study of map projection* (Macmillan & Co., Ltd., London, 1942).

H. Tertsch, *Die stereographische Projection in der Kristallkunde* (Verlag angewande Wissenschaften, Wiesbaden, Germany, 1954).

EXERCISES

2-1. In the conical projection, is there any advantage in having the cone in Fig. 2-3 intersect the globe at two parallels? What would such a projection accomplish?

2-2. Prepare an orthographic projection of a cubic crystal along its $[001]$ direction, and show the poles of the $((100))$, $((110))$, and $((111))$ planes only. **Hint:** Compare your result with Fig. 1-19.

2-3. Making use of the construction in Fig. 2-1 and relations (2-2) and (2-3), construct a gnomonic meridian projection of a globe showing parallels and meridians spaced $10°$ apart out to a longitude of $40°$ and a latitude of $40°$.

2-4. Making use of the reference sphere in Fig. 2-1, construct the gnomonic projection of the poles of the $((111))$ form in a cubic crystal. **Hint:** Make use of the zonal relations displayed in Fig. 2-2.

2-5. Making use of theorems from plane geometry, prove that BC and DC are radii of the projected parallel in Fig. 2-9. Derive an expression, giving the length of the radius as a function of the latitude angle ψ for a unit reference sphere.

2-6. Making use of Fig. 2-10, determine the radius of the circle representing the projection of a meridian. Similarly, determine the distance from the center of the stereographic net to the center of the meridinal circle.

2-7. Why does the principal meridian, and equator, in a stereographic net appear as straight lines? Prove this with the aid of Fig. 2-10 or Exercise 2-6.

2-8. Making use of the vertical section in Fig. 2-13, show that AB and CD are traces of conjugate sections in the elliptic cone whose trace is AVD. Next, draw a construction line through B parallel to the plane of projection trace, and making use of equal arcs delimited by the construction line, prove that CD is parallel to the plane of projection.

2-9. Complete the stereographic projection of poles of the $((100))$, $((110))$, and $((111))$ planes in Fig. 2-14 by showing the projections of the poles lying in the other hemisphere of the reference sphere. Reference to Fig. 2-5 shows that they will project outside the boundary meridian in Fig. 2-14. (**Hint:** The poles of planes belonging to the same zones must lie on the same great circles, so that their projections are easily located by extending the great circles in Fig. 2-14 and noting their mutual intersections.) Label all poles projected.

2-10. Prepare the standard projections for a cubic crystal projected along $[110]$ and along $[111]$. For simplicity, show only the poles of the $((100))$, $((110))$, and $((111))$ planes.

2-11. On the standard (001) projection shown in Fig. 2-14, indicate the poles of some of the other planes belonging to the zones whose great-circle projections are indicated. For example, plot the poles (131), (021), $(\bar{1}32)$, $(\bar{3}12)$, $(\bar{3}01)$, and $(\bar{3}\bar{1}1)$ along the $[1\bar{1}2]$ zone. [**Hint:** Make use of relations (1-11) and (1-12) to determine the proper zones.] Then, using the angles given in Table 2-1, locate the poles along the proper great circle; in this case, note that it also contains the poles of (110), $(\bar{1}11)$, and $(\bar{1}\bar{1}0)$. Finally, keep in mind that angles must be measured along great circles. This is easily accomplished by superimposing the projection on a stereographic net in such a way that the great circle in the projection lies over a meridian of the net.

2-12. Using the procedure described in Exercise 2-11, prepare a standard (001) projection showing the poles of all the planes belonging to the forms $((100))$, $((110))$, $((111))$, $((210))$, $((211))$, $((221))$, $((310))$, $((311))$, $((320))$, and $((321))$. [**Hint:** Make use of relations (1-11) and (1-12) to determine two zones to which each plane belongs. Then draw the appropriate two great circles, and label their intersection at which the pole must lie.]

THREE
Crystal lattices

PERIODICITY IN CRYSTALS

Periodic repetition. The property that distinguishes crystals from other atomic aggregates is that the atomic arrays in crystals are triply periodic. It was shown in Chap. 1 how an object can be repeated in space by elements of symmetry. In the present chapter, different ways of translating an object a finite distance and then repeating it will be considered. Suppose that the distance through which an object, symbolized by the number 7, is repeatedly moved is t; then a collection of 7's results, as shown in Fig. 3-1. Such a set of 7's is congruent, and is said to be periodic with period t. One pictures the *translation* t operating indefinitely, so that the one-dimensional periodic array of 7's in Fig. 3-1 merely represents a small portion of an infinitely long row of 7's. This is not an idle observation, since, when atomic arrays in crystals are considered, the translations relating the periodically repeated units are so small that some 100,000,000 may be needed to span a crystal about 1 cm wide.

The single translation in Fig. 3-1 produces an infinite linear array of the repeated object. If such a translation, t_1, is combined with another, noncollinear translation, t_2, then a two-dimensional array (Fig. 3-2) obtains as follows: The entire linear array due to translation t_1 (Fig. 3-1) is repeated an infinite number of times by the second translation, t_2. Another way of looking at this is to say that the linear array due to t_2 is repeated by t_1.

Since the nature of the repeated objects in Fig. 3-2*A* does not affect the translation periodicity, it is conventional to represent this periodicity by replacing each object in the array with a point. The resulting collection of points shown in Fig. 3-2*B* is called a *lattice*, in this case, a two-dimensional or *plane lattice*. It should be remembered that a point is an imaginary infinitesimal spot in space, and consequently, *a lattice of points is imaginary* also. On the other hand, the array of

7 7 7 7 7 7 7

t

Fig. 3-1

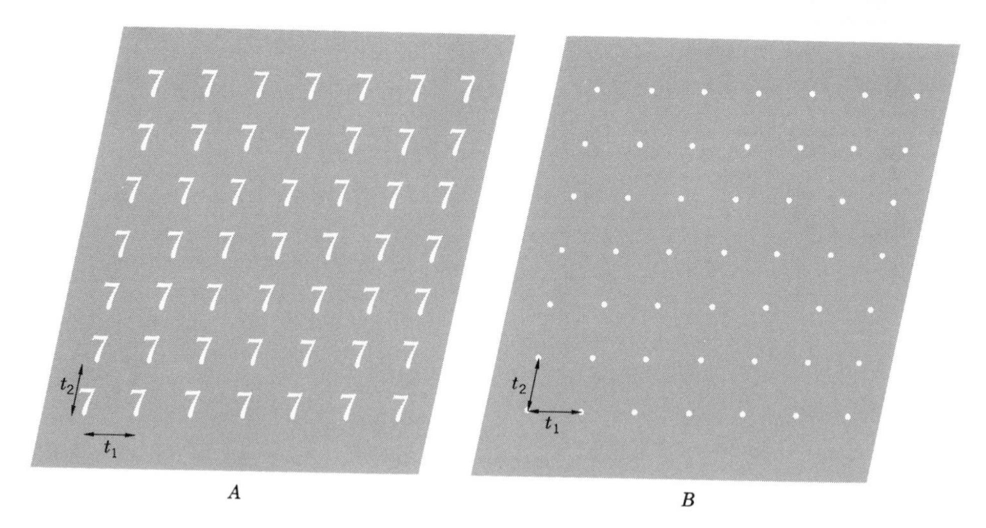

Fig. 3-2

7's in Fig. 3-2 is real. It is *not* a lattice of 7's, because a lattice is an imaginary concept; instead, it is correctly called a *lattice array* of 7's.

It is possible to add a third translation to the plane lattice in Fig. 3-2*B* or to the lattice array in Fig. 3-2*A*. In each case, the third translation repeats the entire plane at equal intervals, t_3. This third translation thus produces a *space lattice* (Fig. 3-3*B*) or a *three-dimensional lattice array* (Fig. 3-3*A*). This is an important difference, and it behooves the reader to use the words lattice and array correctly. In the case of crystals, the lattice array of atoms is properly called the *crystal structure*, and may consist of one kind of atom (elements), groups of atoms (compounds), or complex molecules, each containing a large number of atoms, repeated periodically. Yet the lattice, consisting of points, may be exactly the same in several

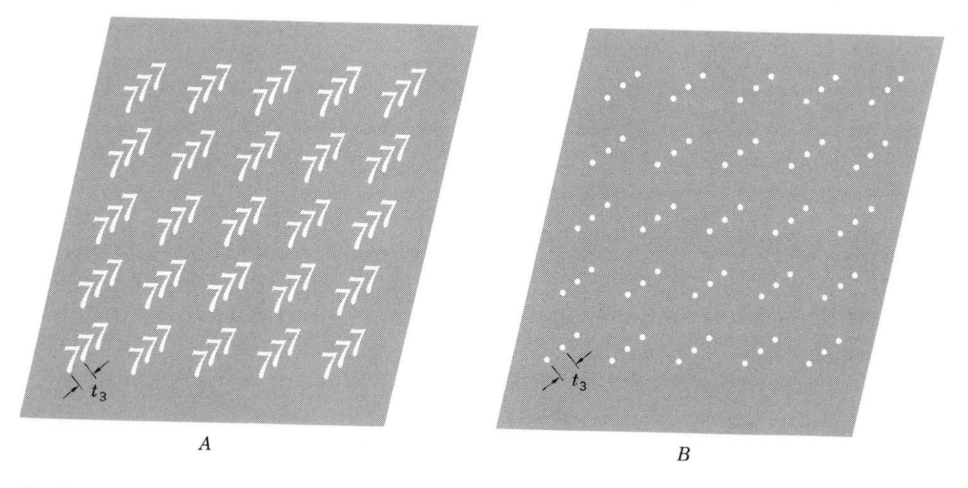

Fig. 3-3

such crystals, except for the lengths of the translations joining adjacent lattice points.

Choice of a unit cell. It has been shown above that two noncollinear translations define a plane lattice and three noncoplanar translations define a space lattice. It is natural to ask, therefore, given a particular lattice, which pair (or which triplet) of translations does one choose to describe it? Actually, there is an infinity of choices for each translation, because a line joining any two lattice points is a translation of the lattice. Figure 3-4 shows a plane lattice and some of the choices that can be made. If pairs of translations such as t_1, t_2 or t_3, t_4 are chosen, they are said to define a *primitive cell*, so called because *one* lattice point is associated with each cell. This can be seen in two ways:

1. Consider each lattice point as belonging to four adjacent cells in Fig. 3-4; thus, only a part of each point belongs to any one cell. Since each cell has four corners, it contains four such parts, whose total "area" adds up to one whole point.
2. Alternatively, displace slightly any one of the shaded primitive cells in Fig. 3-4. It becomes immediately obvious that only one lattice point can lie within the area of any one primitive cell.

On the other hand, the cell defined by t_5 and t_6 contains a lattice point within the cell in addition to the one shared at the corners, so that it is called a *multiple cell*. Several kinds of multiple cells are possible, namely, *double cells, triple cells*, etc., depending on whether they contain two, three, or more lattice points, counting all the corner points together as one point and adding one for each point contained in the interior of the cell. The translations that define a primitive cell are called *conjugate translations*.

Either a primitive or a multiple cell can be selected as a *unit cell* of the lattice. The unit cell is so named because the entire lattice can be derived by repeating this cell as a unit by means of the translations that serve as the unit cell's edges. The choice of a unit cell is dictated by convenience and convention to be that cell which best represents the symmetry of the lattice, as discussed in a later section. The three translations selected as the edges of the unit cell are the crystallographic axes **a**, **b**, **c**, already encountered in Chap. 1.

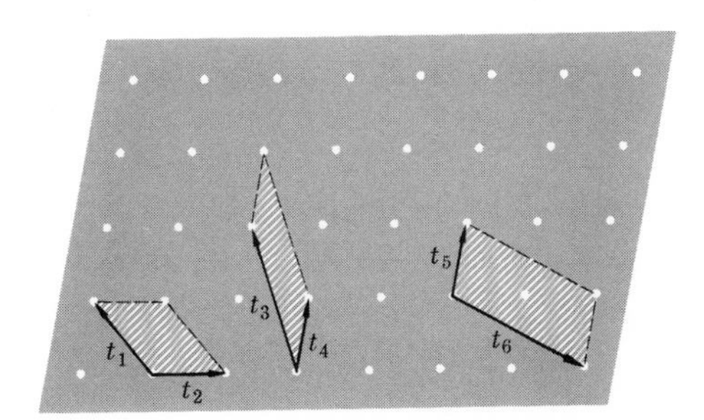

Fig. 3-4

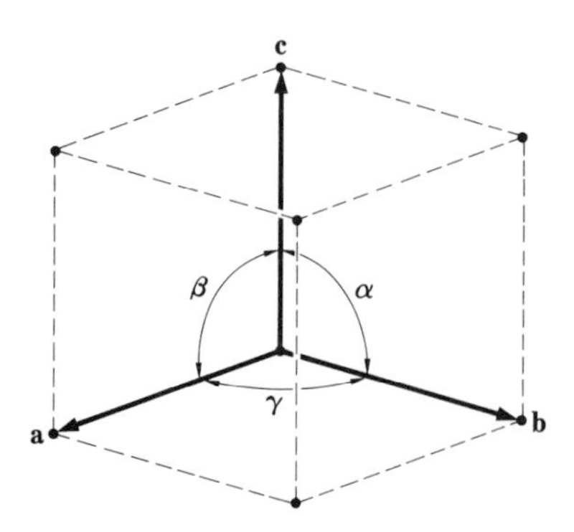

Fig. 3-5

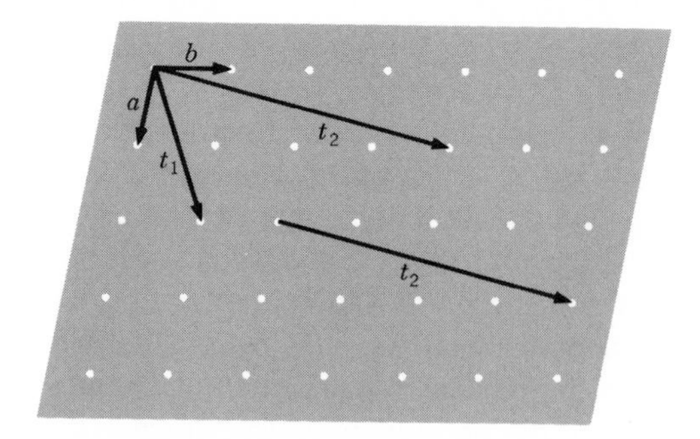

Fig. 3-6

Notation for directions and planes. Everything that was first learned in the description of crystal morphology can be applied directly to the description of directions and planes existing within the crystal. The three crystallographic axes can be thought of as three noncoplanar vectors defining a unit cell, and hence the lattice (Fig. 3-5). Any other translation in the lattice, therefore, can be represented by the appropriate vector sum of these three axes.

$$\mathbf{t} = u\mathbf{a} + v\mathbf{b} + w\mathbf{c} \tag{3-1}$$

where the three indices $[uvw]$ define the translation, and hence its direction in the lattice. This is illustrated for a plane lattice in Fig. 3-6, where

$$\begin{aligned} &\mathbf{t}_1 = 2\mathbf{a} + 1\mathbf{b} \\ \text{and} \quad &\mathbf{t}_2 = 1\mathbf{a} + 4\mathbf{b} \end{aligned} \tag{3-2}$$

Note that any translation that is parallel to $\mathbf{t}_2$ is also the translation $\mathbf{t}_2$, regardless of its origin. Examples of three-dimensional translations are taken up in the exercises at the end of this chapter.

Similarly, the designation of planes in a lattice makes use of the indexing procedure popularized by Miller. To see how it works in a lattice, consider the projection of a lattice along its c axis which is perpendicular to the ab plane ($\alpha = \beta = 90°$), shown in Fig. 3-7. Consider a plane that is parallel to c and passes through any two lattice points, for example, the plane represented in an edge view by the heavy line in Fig. 3-7. A plane passing through two or more lattice points is normally

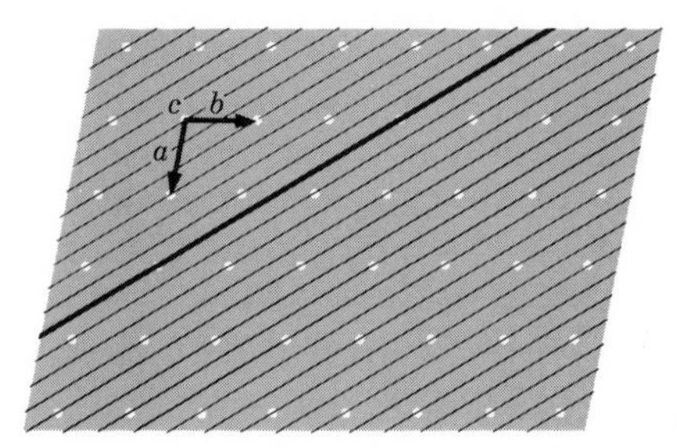

Fig. 3-7

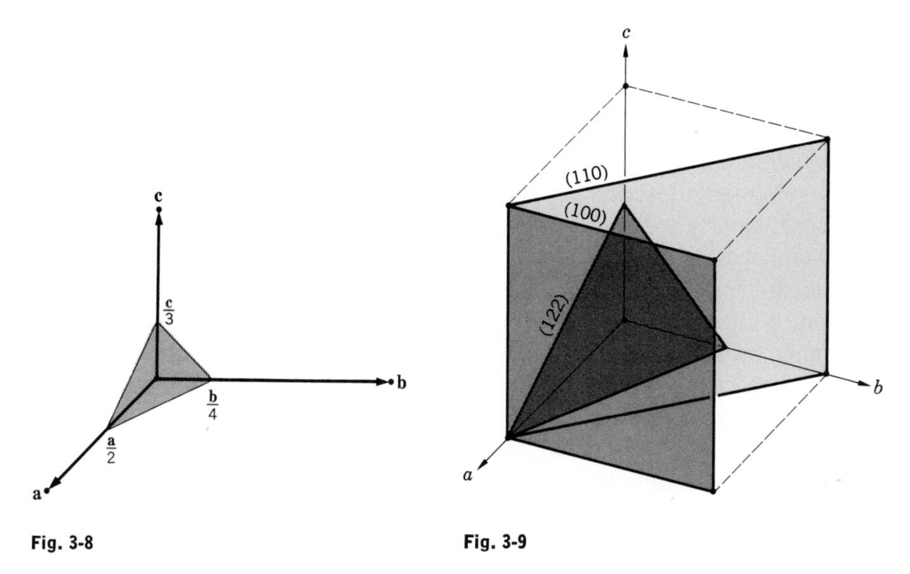

Fig. 3-8

Fig. 3-9

called a *rational* plane, because it cuts the crystallographic axes into parts expressible by rational numbers. Note that the periodicity of the lattice requires that an identical rational plane must pass through each and every lattice point (light lines in Fig. 3-7). Now, the indices of a plane are determined in the three steps described in Chap. 1, and it obviously should not matter which of the infinite stack of parallel planes in Fig. 3-7 is selected. To test the universality of the indexing scheme, consider two of the planes, the first plane mentioned (heavy line in Fig. 3-7) and the plane nearest to the origin. Clearly, both are (320) planes in the crystal. In

		"Heavy" plane			Nearest plane		
		a	b	c	a	b	c
Step 1.	Intercepts	2	3	∞	1/3	1/2	∞
Step 2.	Reciprocals	1/2	1/3	$1/\infty = 0$	3	2	0
Step 3.	Indices	3	2	0		Not needed	

fact, the plane nearest to the origin serves to illustrate the true meaning of the indices (hkl), namely, that such a plane cuts the a axis into h parts, the b axis into k parts, and the c axis into l parts. To illustrate this point further, Fig. 3-8 shows the (243) plane nearest to the origin, while Fig. 3-9 shows the three planes (100), (110), and (122) in a unit cell.

LATTICE TYPES

Plane lattices. The fundamental requirement that all crystals must have periodic structures means that the symmetry elements present in crystals must conform

A_n — t — A_n — t — A_n — t — A_n — t — A_n — t — A_n

Fig. 3-10

to their translational periodicities. This is the reason why the number of symmetry elements found in crystals is limited. As an example, consider the rotation axes already described. The reason why only 1, 2, 3, 4, and 6 are possible in crystals can be readily appreciated when the combination of an n-fold axis A_n with a translation t is considered (Fig. 3-10). As shown in Fig. 3-11, a rotation axis repeats the translation $\varphi°$ $(= 360°/n)$ away, and since n rotations cause superposition, it does not matter whether one "rotates" clockwise or counterclockwise. Beginning with the linear array in Fig. 3-10, just two such rotations produce new lattice points, labeled p and q in Fig. 3-11. Because they are equidistant from the original lattice row by construction, the two lattice points p and q must be joined by the same translation t or some integral multiple of it mt, depending on the magnitude of φ. This means that the allowed values that φ can have can be determined directly from the construction in Fig. 3-11.

$$mt = t + 2t \cos \varphi \qquad m = 0, \pm 1, \pm 2, \pm 3, \ldots \tag{3-3}$$

$\pm m$ is used, depending on whether the rotation is clockwise or counterclockwise. Dividing both sides of (3-3) by t, rearranging terms, and noting that, since m is an integer, $m - 1 = N$ is also an integer (positive or negative), one can solve for

$$\cos \varphi = \tfrac{1}{2}N \qquad N = 0, \pm 1, \pm 2, \pm 3, \ldots \tag{3-4}$$

Since the magnitude of the cosine in (3-4) must be less than unity, the only possible values that N can have are listed in Table 3-1, along with the five allowed rotation axes listed in the last column.

Table 3-1 Determination of rotation axes allowed in a lattice

N	$\cos \varphi$	φ, deg	n
-2	-1	180	2
-1	$-\frac{1}{2}$	120	3
0	0	90	4
$+1$	$+\frac{1}{2}$	60	6
$+2$	$+1$	360 or 0	1

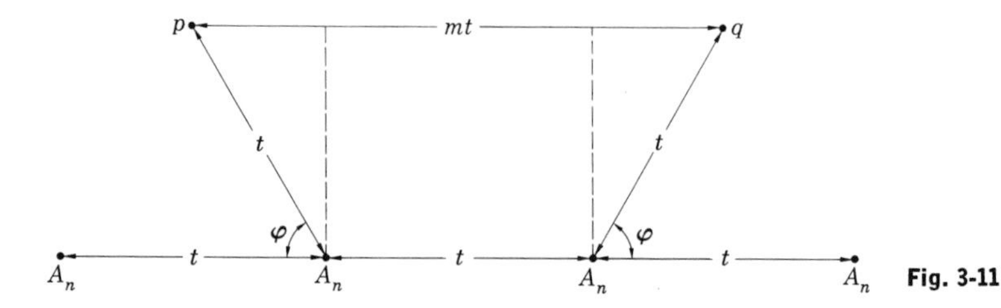

Fig. 3-11

Effect of symmetry. Just as the presence of a lattice limits the kinds of symmetry elements crystals can have, the presence of a particular symmetry element in a crystal controls the kind of lattice it can have. This is demonstrated below by considering the five distinguishable plane lattices shown in Fig. 3-12. The *parallelogram lattice* is the most general kind of lattice possible in two dimensions and can accommodate either a 1 or a 2. A $\bar{2} = m$ requires that the lattice rows must be normal to the mirror reflection, so that a *rectangular lattice* or a *diamond lattice* is required. Finally, a 3 or 6 requires that the angle between the equal axes be $120°$, producing a *triangular lattice,* while a 4 similarly requires a *square lattice.*

Plane groups. When a rotation axis is combined with a translation at right angles to it, a new rotation axis results. This can be expressed symbolically:

$$A \cdot t = B \tag{3-5}$$

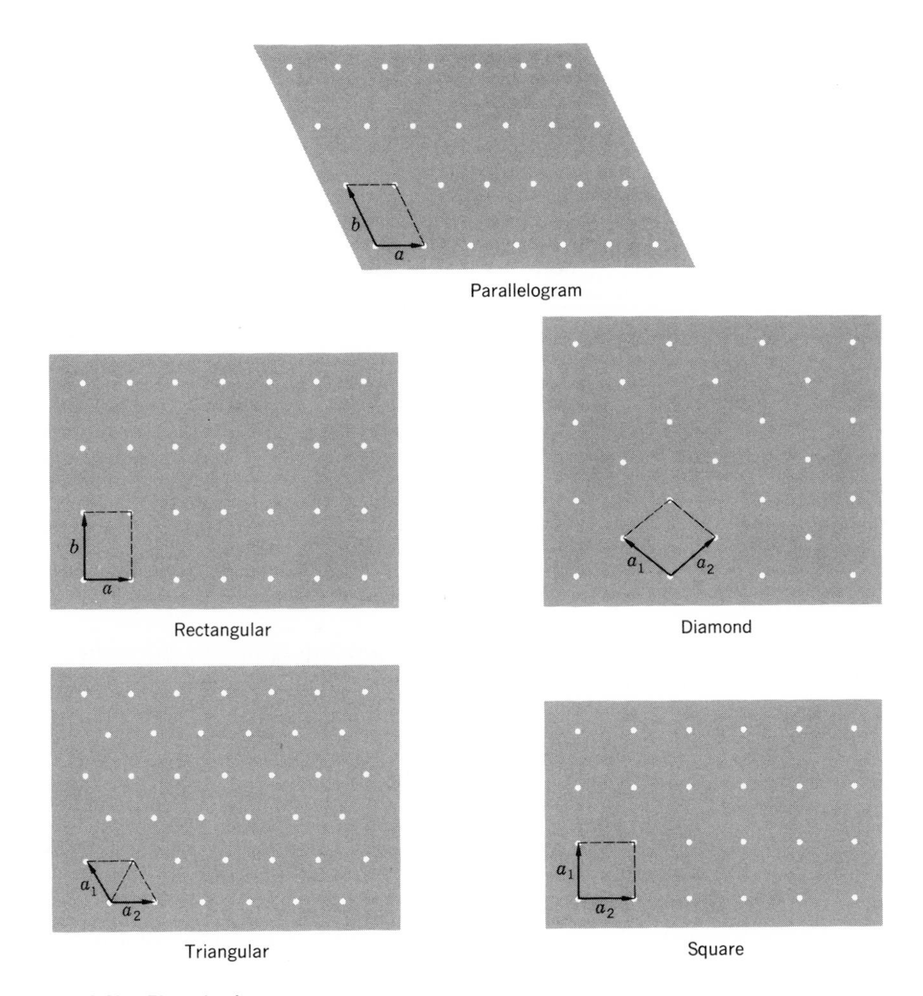

Fig. 3-12. Plane lattices.

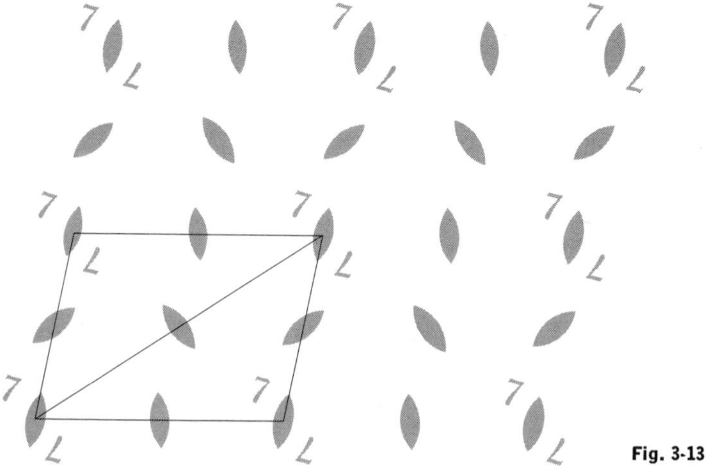

Fig. 3-13

To see what happens when a 2 is combined with an orthogonal translation, place a 2 at the lattice points of the parallelogram lattice of the two-dimensional lattice array shown in Fig. 3-13. Next, surround each translation equivalent 2, that is, each 2-fold axis connected by a lattice translation, with a pair of symmetry-related motifs (7's in Fig. 3-13), thereby constructing a lattice array. It is now easy to see that new 2-fold axes "spring up" halfway between translation-equivalent 2-fold axes.

Similarly, the combination of a 3 with an orthogonal translation produces a new 3, and the combination of a 4 with an orthogonal translation produces a new 4. In the latter case note that a 4 also includes the operation of a 2; therefore new 2-fold axes are also produced. The combinations of symmetry elements with plane lattices are called *plane groups.* The pure axial plane groups are illustrated in Fig. 3-14, where the symbol p is used to denote a primitive plane lattice. Note in that figure that 2-fold axes arise halfway between translation equivalent 2's; 3-fold axes are 60° away from the translation joining two 3's; and so forth. It can be shown that two translation-equivalent mirror reflections produce another mirror halfway between them (Exercise 3-3), but a complete derivation of the 17 possible plane groups is considered to lie outside the scope of this book.

Space lattices. The restrictions placed on a plane lattice by different symmetry elements have been shown above. Since a three-dimensional lattice can be thought of as a periodic stack of plane lattices, the possible space lattices can be deduced by considering the different ways of stacking each of the four primitive plane lattices consistent with the symmetries of the 32 point groups. An alternative procedure is, first, to deduce the uniquely different lattice types, and then to observe what restrictions are placed on these lattices by the point groups. If the second procedure is followed, it can be shown that there are only five unique space-lattice types, listed in Table 3-2.

To see how these lattices are distributed among the six crystal systems, it is sufficient to consider the restrictions placed on each lattice in turn by the minimal symmetry of the crystal systems listed in Table 1-2. In the triclinic system, 1 or

Table 3-2 Space-lattice types

Name	*Location of nonorigin points*	*Symbol*
Primitive	. . . .	P
Side-centered	Center of A face or (100) if A-centered	A
	Center of B face or (010) if B-centered	B
	Center of C face or (001) if C-centered	C
Face-centered	Centers of A, B, and C faces	F
Body-centered	Center of each cell	I
Rhombohedral	If primitive rhombohedron is referred to hexagonal cell, at $\frac{1}{3}\,\frac{2}{3}\,\frac{1}{3}$ and $\frac{2}{3}\,\frac{1}{3}\,\frac{2}{3}$ of that cell (see Fig. 3-16)	R

$\bar{1}$ places no restrictions on the space lattice. Hence there is no advantage in selecting a nonprimitive lattice. The primitive triclinic or *general lattice* can therefore be described by a unit cell with $a \neq b \neq c$ and $\alpha \neq \beta \neq \gamma$.

In the monoclinic system, a 2-fold axis passing through each lattice point of a parallelogram lattice requires that successive plane lattices have their lattice points fall on 2-fold axes. Since $p2$ has four independent sets of 2-fold axes, the stacking sequence of two successive nets can go as shown in Fig. 3-15. (Open

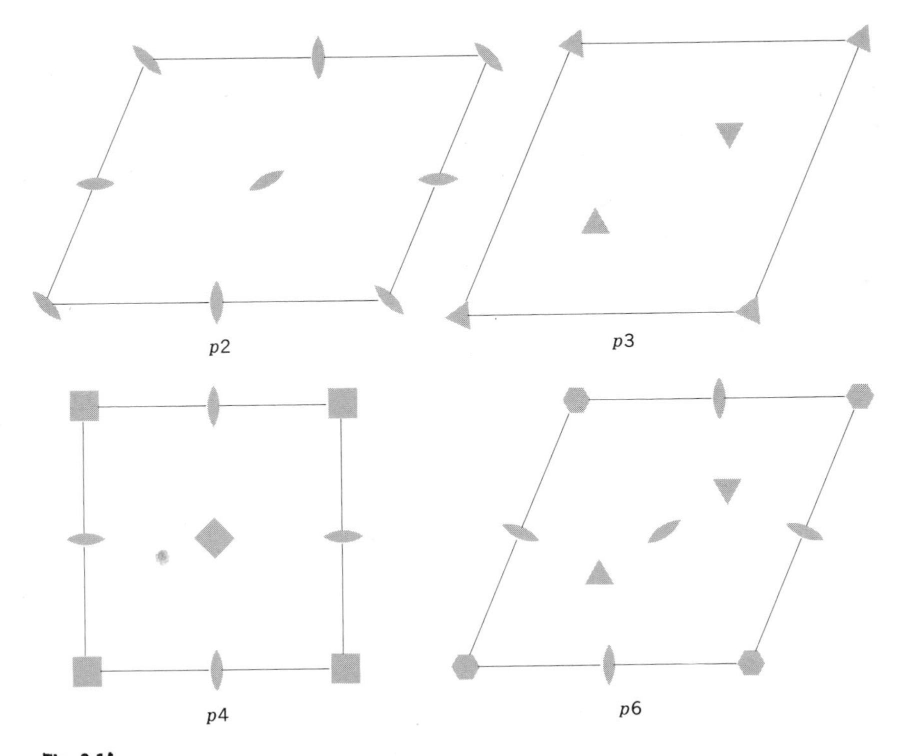

Fig. 3-14

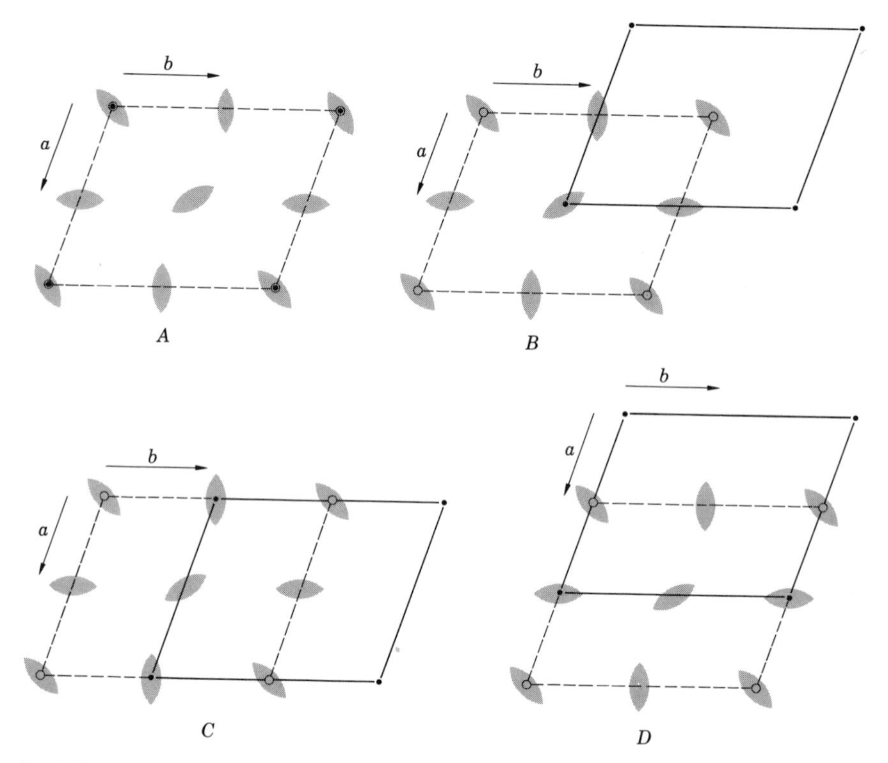

Fig. 3-15

circlets and dots are used to distinguish the two nets.) It is clear from this figure that the stacking in A gives a primitive lattice, the stacking in B a body-centered lattice, and C and D give side-centered lattices. Moreover, it is easy to show that the side-centered lattices in C and D become body-centered if the diagonal of the parallelogram is chosen as one cell edge. Thus there are only two uniquely different lattices in the monoclinic system, namely, P and I. In each of these lattices the presence of a 2-fold axis requires that the cell edge parallel to this axis be normal to the other two; therefore the unit cell has $a \neq b \neq c$, $\alpha = \beta = 90° \neq \gamma$.

Similar reasoning leads to the conclusion that 222 in the orthorhombic system requires an orthogonal lattice whose unit cell can have unequal sides and that all the four lattice types are possible. It can also be shown that only two lattices can occur in the tetragonal system, and three in the isometric or cubic system. The hexagonal system differs slightly from the others in that two unique rotation axes, 3 or 6, are possible. In the case of a 6, it is easy to see from Fig. 3-14 that there are 6-fold axes only at the lattice points of the triangular lattice, and successive plane lattices therefore must superimpose to form a primitive space lattice. In the case of $p3$, however, there are three sets of 3-fold axes, and hence three stacking sequences are possible, as shown in Fig. 3-16. Of these, A gives a primitive hexagonal lattice, and B and C give three-layer-high stacks which are identical, as can be seen by rotating either by 180° about a normal to the plane of the drawing. Figure

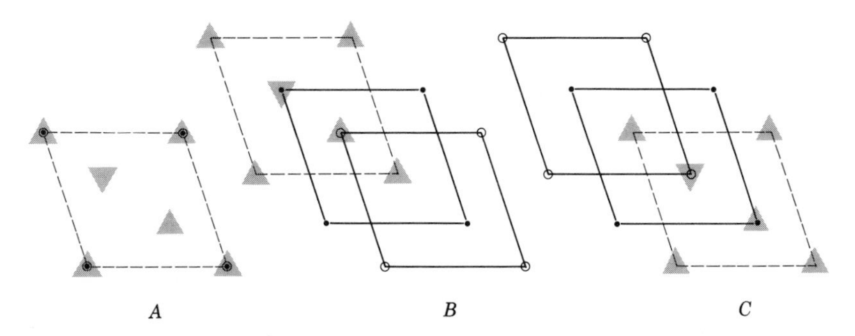

Fig. 3-16

3-17 shows a perspective drawing of this stacking sequence. As can be seen in this figure, the primitive cell in the resulting lattice is a rhombohedron having three equal edges and three equal interaxial angles ($a_1 = a_2 = a_3$, $\alpha_1 = \alpha_2 = \alpha_3$). A more convenient hexagonal cell can be chosen, however, as shown by the heavy lines in Fig. 3-17. Such a cell contains two additional, equally spaced lattice points along the long body diagonal of the cell. It is this uniqueness in the lattice that has led to a suggestion for the adoption of a *rhombohedral* subsystem of the hexagonal system, chiefly in the United States. The objection to such a designation is that a lattice type, rather than the minimal symmetry, would be the basis for classification. An alternative subdivision, called the *trigonal* subsystem, includes all point groups containing a 3 (or $\bar{3}$), and is most popular in the United Kingdom. It differs from the above proposition in that a 3 or $\bar{3}$ is equally compatible with a P or R lattice. Because of the difficulties in allocating certain space groups, it is recommended practice nowadays to refer both cases to a hexagonal lattice, and hence the subsystem designations should not be used.

The 14 space lattices thus derived and their distribution among the six crystal

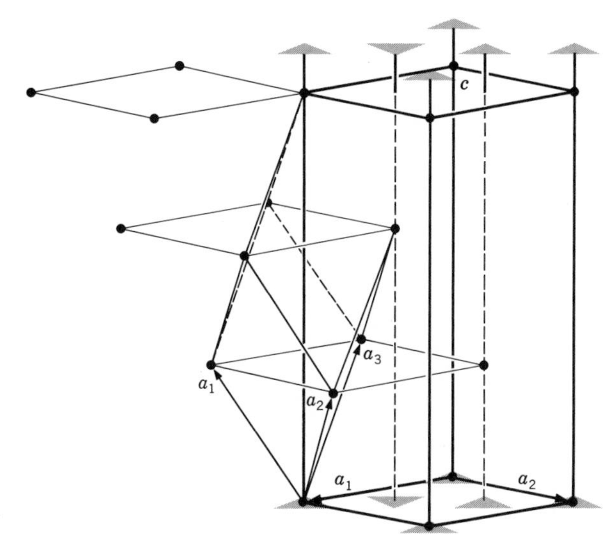

Fig. 3-17

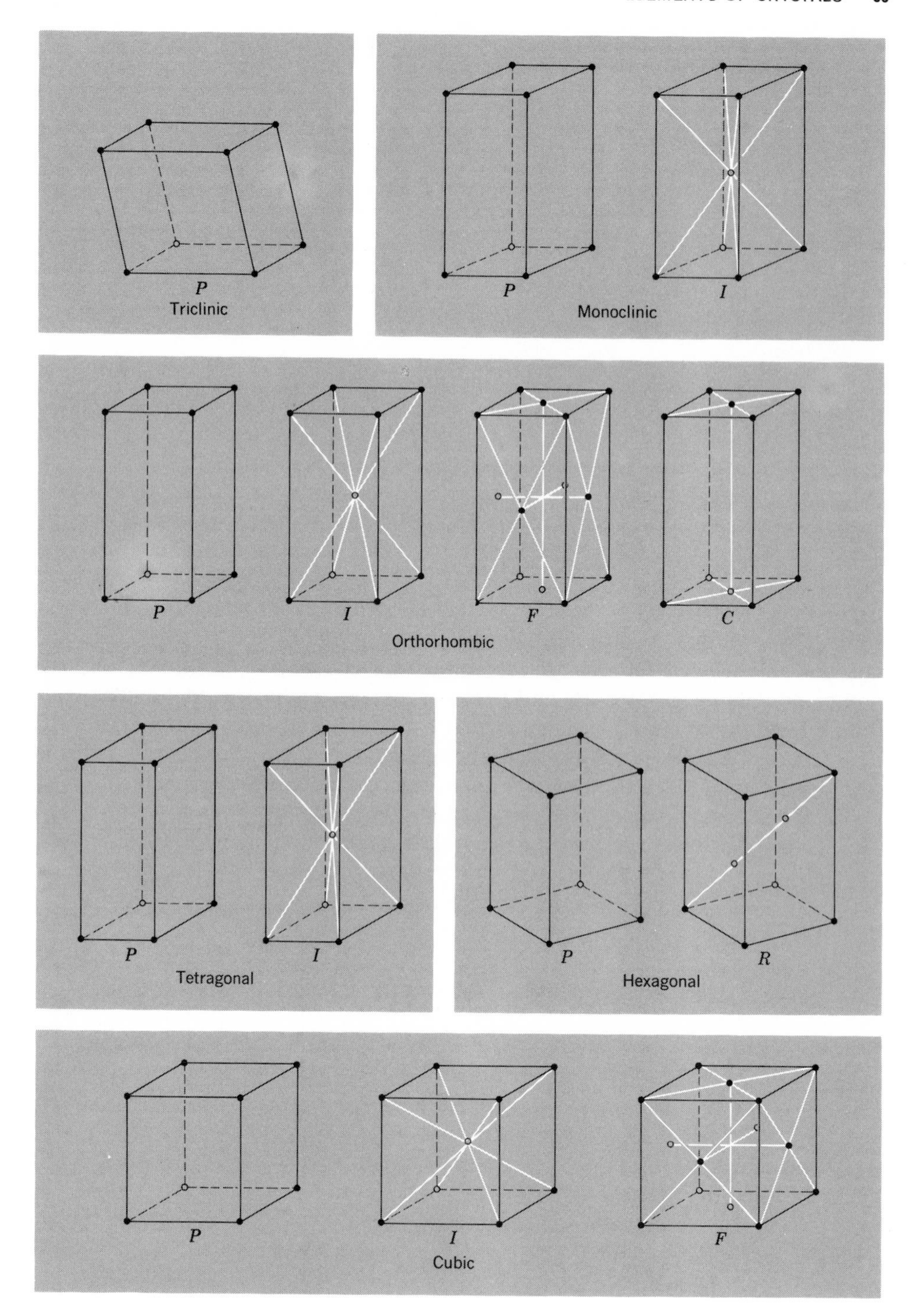
P
Triclinic
P
I
Monoclinic
P
I
F
C
Orthorhombic
P
I
Tetragonal
P
R
Hexagonal
P
I
F
Cubic

Fig. 3-18

Table 3-3 Bravais lattices

Crystal system	Lattice types: P	$A(B,C)$	F	I	R†	Unit-cell dimensions
Triclinic	X	...	...	...	...	$a \neq b \neq c$ $\alpha \neq \beta \neq \gamma$
Monoclinic	X	...	...	X	...	$a \neq b \neq c$ $\alpha = \beta = 90° \neq \gamma$
Orthorhombic	X	X	X	X	...	$a \neq b \neq c$ $\alpha = \beta = \gamma = 90°$
Tetragonal	X	...	...	X	...	$a = b \neq c$ $\alpha = \beta = \gamma = 90°$
Isometric	X	...	X	X	...	$a = b = c$ $\alpha = \beta = \gamma = 90°$
Hexagonal	X	...	...	...	X	$a = b \neq c$ $\alpha = \beta = 90°; \gamma = 120°$

† The symbol R denotes the rarely used primitive rhombohedral lattice shown in Fig. 3-17.

systems are given in Table 3-3. They are sometimes called the 14 *Bravais lattices,* and are depicted in Fig. 3-18. The relative unit-cell dimensions characteristic of each lattice type are also given in Table 3-3. Note that they are the same as those derived from symmetry considerations in Table 1-3. This is not surprising since the minimal symmetry of a crystal determines both the crystal system in which it is classified and its lattice type.

TRANSFORMATION THEORY

Transformation of crystallographic axes. Although there are well-defined rules for selecting the crystallographic axes in a lattice, it nevertheless has happened that alternative sets of axes have been chosen by different investigators. Such alternative selections may have been made because a different convention for designating axes was prevalent at the time the choice was made, or it may have been advantageous to compare a primitive unit cell with a larger-centered cell to illustrate some aspect of the atomic array. Whatever the reason for considering two or more alternative sets of crystallographic axes in a lattice, the actual transformation from one set to the other is carried out quite simply by making use of the vector properties of the axes. Thus, according to (3-1), three new axes, $\mathbf{a}_n$, $\mathbf{b}_n$, $\mathbf{c}_n$, can be defined in terms of the three original axes, $\mathbf{a}_o$, $\mathbf{b}_o$, $\mathbf{c}_o$, by three simultaneous equations:

$$\begin{aligned} \mathbf{a}_n &= u_1\mathbf{a}_o + v_1\mathbf{b}_o + w_1\mathbf{c}_o \\ \mathbf{b}_n &= u_2\mathbf{a}_o + v_2\mathbf{b}_o + w_2\mathbf{c}_o \\ \mathbf{c}_n &= u_3\mathbf{a}_o + v_3\mathbf{b}_o + w_3\mathbf{c}_o \end{aligned} \tag{3-6}$$

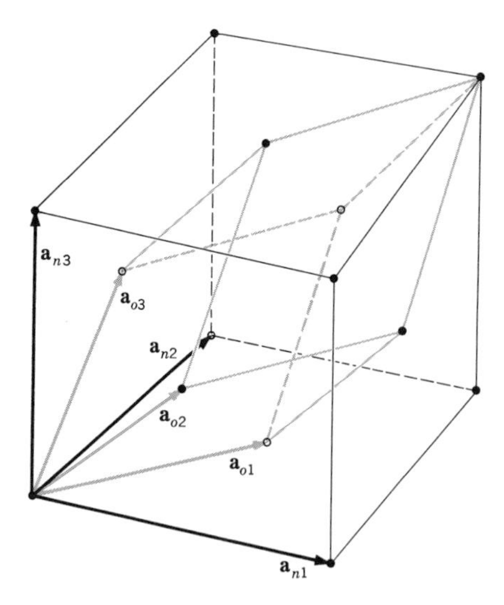

Fig. 3-19

It is self-evident that the form of (3-6) is the same for any choice of the new axes, so that the specific definition of the new axes is given by the coefficients

$$\begin{Vmatrix} u_1 & v_1 & w_1 \\ u_2 & v_2 & w_2 \\ u_3 & v_3 & w_3 \end{Vmatrix} \tag{3-7}$$

whose array in (3-7) displays the geometrical form of the transformation, and when placed between double vertical bars, is called the *matrix of the transformation.*

As a simple example, consider the transformation matrix of the new axes $\mathbf{t}_1$ and $\mathbf{t}_2$ in (3-2), which is

$$\begin{Vmatrix} 2 & 1 \\ 1 & 4 \end{Vmatrix}$$

A practically more important example is the transformation from the unit-cell axes of a primitive cell to that of a centered cell displaying the symmetry of the crystal better. Consider the primitive rhombohedral cell in a face-centered cubic lattice denoted by $\mathbf{a}_{o1}$, $\mathbf{a}_{o2}$, $\mathbf{a}_{o3}$ in Fig. 3-19. The three cubic axes $\mathbf{a}_{n1}$, $\mathbf{a}_{n2}$, $\mathbf{a}_{n3}$ are obviously given by

$$\begin{aligned} \mathbf{a}_{n1} &= 1\mathbf{a}_{o1} + 1\mathbf{a}_{o2} - 1\mathbf{a}_{o3} \\ \mathbf{a}_{n2} &= 1\mathbf{a}_{o1} - 1\mathbf{a}_{o2} + 1\mathbf{a}_{o3} \\ \mathbf{a}_{n3} &= -1\mathbf{a}_{o1} + 1\mathbf{a}_{o2} + 1\mathbf{a}_{o3} \end{aligned}$$

whose matrix of transformation is

$$\begin{Vmatrix} +1 & +1 & -1 \\ +1 & -1 & +1 \\ -1 & +1 & +1 \end{Vmatrix} \tag{3-8}$$

Quite similarly, the three hexagonal axes $\mathbf{a}_1$, $\mathbf{a}_2$, $\mathbf{c}$ in Fig. 3-17 can be derived from the rhombohedral ones with the aid of the matrix (Exercise 3-10)

$$\begin{Vmatrix} +1 & 0 & -1 \\ -1 & +1 & 0 \\ +1 & +1 & +1 \end{Vmatrix} \tag{3-9}$$

Transformation of indices of a plane. The selection of three new axes $\mathbf{a}_n$, $\mathbf{b}_n$, $\mathbf{c}_n$ obviously changes the intercepts which a set of parallel planes makes with the axes defining the lattice. In order to examine the nature of this change, consider the two-dimensional example in Fig. 3-20. The parallel lines divide $\mathbf{a}_o$ into three parts and $\mathbf{b}_o$ into two parts, so that $h_o = 3$ and $k_o = 2$. Suppose a new axis,

$$\mathbf{a}_n = 2\mathbf{a}_o + 1\mathbf{b}_o$$

is chosen, as shown in Fig. 3-20. An examination of the figure shows that lines dividing the vectors $2\mathbf{a}_o$ and $1\mathbf{b}_o$ project onto the new vector, so that it is cut into $2 \cdot 3 + 1 \cdot 2 = 8$ parts. Another way of stating this is to observe that the number of parts into which the new axis is divided, h_n, is the sum of the number of parts into which its projection on the old axis is divided. This can be expressed by the relation

$$h_n = u_1 h_o + v_1 k_o$$

By considering the lines in Fig. 3-20 to be the edge views of planes that are perpendicular to the plane of the drawing (as in Fig. 3-7), it is possible to generalize the above result to three dimensions. The new axis $\mathbf{a}_n$ then becomes

$$\mathbf{a}_n = 2\mathbf{a}_o + 1\mathbf{b}_o + 0\mathbf{c}_o$$

and the new index,

$$h_n = 2h_o + 1k_o + 0l_o$$

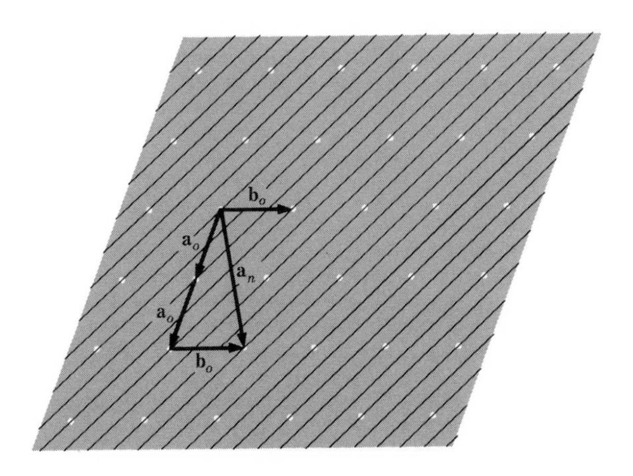

Fig. 3-20

It should be apparent that similar relations also can be derived for the other two indices, referred to new axes, so that the general form of the index transformation can be written

$$\begin{aligned} h_n &= u_1 h_o + v_1 k_o + w_1 l_o \\ k_n &= u_2 h_o + v_2 k_o + w_2 l_o \\ l_n &= u_3 h_o + v_3 k_o + w_3 l_o \end{aligned} \tag{3-10}$$

and *the matrix of the transformation for the indices is identical with the matrix of the transformation for the axes* (3-7).

It follows from the above that, for example, the indices of a plane referred to the hexagonal axes h_H, k_H, l_H (Fig. 3-17) can be obtained directly from the rhombohedral indices h_R, k_R, l_R with the aid of the matrix of transformation (3-9):

$$\begin{aligned} h_H &= +1h_R + 0 - 1l_R \\ k_H &= -1h_R + 1k_R + 0 \\ l_H &= +1h_R + 1k_R + 1l_R \end{aligned} \tag{3-11}$$

and the fourth hexagonal index $i = -(h_H + k_H)$. Note one interesting consequence of the above transformation: When the three hexagonal indices in (3-10) are combined algebraically according to

$$h_H - k_H + l_H = 3h_R \tag{3-12}$$

their algebraic sum must be some integer divisible by 3. Thus, when the planes in a rhombohedral lattice are referred to hexagonal axes, only certain combinations of the hexagonal indices are allowed, according to (3-12). Since such restrictions do not exist in the primitive hexagonal lattice, it becomes possible to distinguish the two cases by examining a list of indices of the planes. This has a real importance in the study of crystals by x-ray diffraction methods, and some further examples are considered in the exercises at the end of this chapter.

Unit-cell volume changes. One of the obvious consequences of selecting new axes in a lattice is the possibility that the unit-cell volume is changed thereby. As a simple illustration of how such volume changes take place, consider the area of a plane lattice defined by two axes, $\mathbf{a}_o \times \mathbf{b}_o$. Suppose two new axes are chosen according to the matrix

$$\begin{Vmatrix} u_1 & v_1 \\ u_2 & v_2 \end{Vmatrix}$$

so that the area of the new cell is

$$\begin{aligned} \mathbf{a}_n \times \mathbf{b}_n &= (u_1\mathbf{a}_o + v_1\mathbf{b}_o) \times (u_2\mathbf{a}_o + v_2\mathbf{b}_o) \\ &= u_1u_2(\mathbf{a}_o \times \mathbf{a}_o) + u_1v_2(\mathbf{a}_o \times \mathbf{b}_o) + v_1u_2(\mathbf{b}_o \times \mathbf{a}_o) + v_1v_2(\mathbf{b}_o \times \mathbf{b}_o) \\ &= (u_1v_2 - u_2v_1)\mathbf{a}_o \times \mathbf{b}_o \end{aligned} \tag{3-13}$$

since $\mathbf{a} \times \mathbf{a} = \mathbf{b} \times \mathbf{b} = 0$ and $\mathbf{a} \times \mathbf{b} = -\mathbf{b} \times \mathbf{a}$, according to the rules of vector multiplication. (See Appendix 1.)

The vector product on the right of (3-13) is the area of the old cell, and the quantity in parentheses is the *determinant of the transformation*, so that (3-13) can be written

$$\mathbf{a}_n \times \mathbf{b}_n = \begin{vmatrix} u_1 & v_1 \\ u_2 & v_2 \end{vmatrix} \mathbf{a}_o \times \mathbf{b}_o \tag{3-14}$$

As proved by M. J. Buerger, a parallel relation obtains in three dimensions where the unit-cell volume is determined by the triple product.

$$\mathbf{a}_n \times \mathbf{b}_n \cdot \mathbf{c}_n = \begin{vmatrix} u_1 & v_1 & w_1 \\ u_2 & v_2 & w_2 \\ u_3 & v_3 & w_3 \end{vmatrix} \mathbf{a}_o \times \mathbf{b}_o \cdot \mathbf{c}_o \tag{3-15}$$

The determinant of the transformation is an integer whenever the transformation is to a larger cell, or a simple fraction if the transformation is to a smaller cell. It is called the *modulus of the transformation*, and some examples of its magnitude are given in the exercises.

SUGGESTIONS FOR SUPPLEMENTARY READING

M. J. Buerger, *Elementary crystallography* (John Wiley & Sons, Inc., New York, 1956).
X-ray crystallography (John Wiley & Sons, Inc., New York, 1942), especially chaps. 2 and 4.

EXERCISES

3-1. Periodicity is not limited to crystals. Two-dimensional periodicity can be found in patterns on neckties, dresses, wallpapers, etc. Reproduce two such designs, show the lattice type and unit cell, and attempt to determine the correct plane group.

3-2. Select a centered cell in the diamond lattice (Fig. 3-12). Is there any advantage to such a cell? What kinds of symmetry elements can it accommodate? Show how the plane group $p2$ would look with this lattice.

3-3. Consider the rectangular lattice in Fig. 3-12. Pass mirror reflections through the lattice rows parallel to a and b. Are any new mirror reflections produced? What is the plane group in this case? **Hint:** Draw in a symbolic 7 and repeat it by the mirror reflections; then find any additional symmetry present.

3-4. Draw a unit cell like the one in Fig. 3-9. Then indicate in it the directions $[100]$, $[011]$, $[111]$, and $[112]$.

3-5. What are the indices $[uvw]$ for the directions of the three kinds of rotation axes in a cube?

3-6. Extending the crystallographic axes into negative directions also, indicate in the same way as in Exercise 3-4 the directions $[\bar{1}20]$, $[\bar{1}0\bar{1}]$, and $[\bar{1}\bar{1}1]$. **Hint:** Instead of redrawing the unit cells, move the origin of the cell to one of the other lattice points.

3-7. In the cubic system, the plane (hkl) is normal to the direction $[hkl]$. Illustrate this for (100) and $[100]$; (101) and $[101]$; and (111) and $[111]$. Then show that this is not the case for (110) and $[110]$ in the orthorhombic system.

3-8. Show, by trying to fit pentagons together, that 5-fold symmetry is incompatible with periodicity in two dimensions.

3-9. Show by means of drawings similar to Fig. 3-15 that in the tetragonal system $F = I$ and $C = P$.

3-10. Show by means of drawings similar to Fig. 3-15 that in the orthorhombic system it is possible to have a one-face-centered and an all-face-centered lattice, but not a two-face-centered lattice. What is the correct lattice type in that case?

3-11. With the aid of Fig. 3-17 derive the transformation matrix (3-8). **Hint:** Place the origin of the rhombohedral and hexagonal axes at the same lattice point. Next, determine the modulus of the transformation.

3-12. Prove that the only allowed indices in a face-centered cubic lattice are those for which h, k, and l are all even or all odd. What is the modulus of the transformation matrix (3-7)?

FOUR
Group theory applications

SPACE GROUPS

Space-group symmetry. The discussions in the preceding chapters have illustrated how the combination of two symmetry elements gives rise to a third element (point groups) and how the combination of a symmetry element with two translations gives rise to new symmetry elements in a plane group. It remains only to combine the 32 point groups with the 14 space lattices to obtain all possible *space groups.*† It turns out that symmetry elements existing in space lattices can include translation components in their operations of repetition. This can be illustrated quite simply by considering the plane group cm shown in Fig. 4-1. When vertical mirror reflections pass through the lattice points of a diamond lattice,

† These arguments can be extended to more than three dimensions, but such considerations clearly lie outside the scope of this book. Also, recently, it has been shown that there are operators which have the property of repeating a white motif into a black motif, etc., giving rise to derivatives of the regular space groups called *color space groups.* Since they do not find regular application in x-ray crystallography, color space groups are not further discussed here, despite their proven usefulness in neutron diffraction, magnetic studies, and elsewhere.

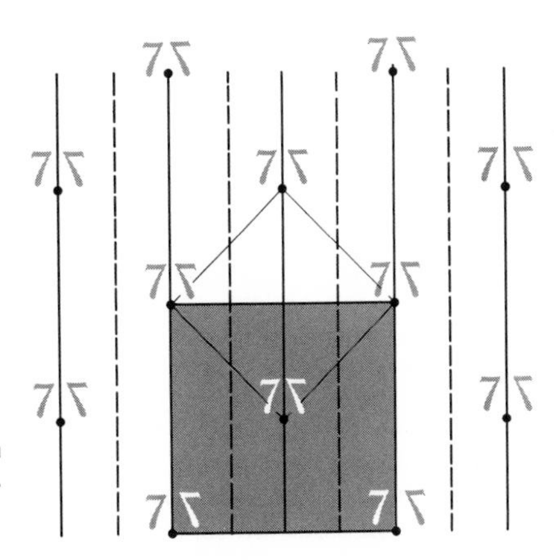

Fig. 4-1. Combination of a centered-plane lattice c with a mirror reflection m produces glide reflections g halfway between, so that $cm = cg$.

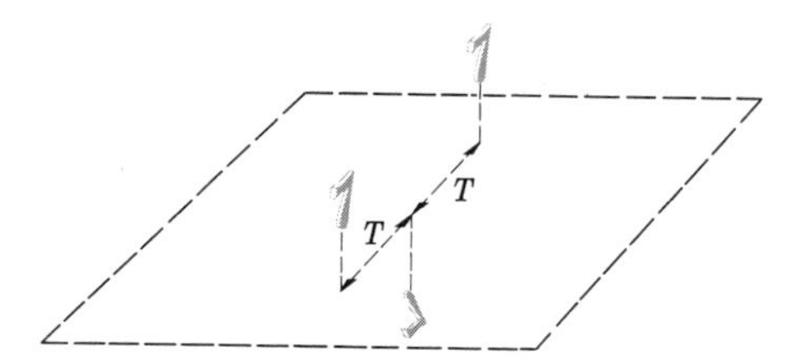

Fig. 4-2

the motif is repeated as illustrated by the 7's in Fig. 4-1. Halfway between the mirrors a new symmetry element is generated, called a *glide reflection*, which combines the operation of a reflection with a translation into a single hybrid operation of symmetry. The glides are shown by dashed lines, and it is easy to see that they first reflect the 7's and next translate them by one-half the regular translation distance and then "set them down" to complete the operation. This is also illustrated in three dimensions in Fig. 4-2, and the symmetry element depicted is called a *glide plane*. The translation component of T of a glide plane is equal to one-half of the normal translation of the lattice in the direction of the glide. Thus a glide along the **a** axis has $\mathbf{T} = \frac{1}{2}\mathbf{a}$, and is called an a glide. Similarly, a diagonal glide can have $\mathbf{T} = \frac{1}{2}\mathbf{a} + \frac{1}{2}\mathbf{b}$ or $\frac{1}{2}\mathbf{b} + \frac{1}{2}\mathbf{c}$, etc., noting that these are vector additions. The different possible glide planes and their translation components are listed in Table 4-1.

Returning to Fig. 4-1, it is clear that a glide results when two mirrors are repeated by a translation that is inclined to them. Note that the orthogonal centered cell, shown stippled in Fig. 4-1, has a mirror halfway between mirrors joined by a perpendicular translation. Also note, however, that if the edges of the shaded cell are chosen as the basic translations (heavy lines), the centered cell

Table 4-1 Possible glide planes

Type of glide	*Symbol*	*Translation component* T
Axial glide	a	$\frac{a}{2}$
Axial glide	b	$\frac{b}{2}$
Axial glide	c	$\frac{c}{2}$
Diagonal glide	n	$\frac{a}{2}+\frac{b}{2}$, $\frac{b}{2}+\frac{c}{2}$, or $\frac{c}{2}+\frac{a}{2}$
Diamond glide†	d	$\frac{a}{4}+\frac{b}{4}$, $\frac{b}{4}+\frac{c}{4}$, or $\frac{c}{4}+\frac{a}{4}$

† The translation component T in the diamond glide is actually one-half of the true translation along the face diagonal of a centered plane lattice.

contains subtranslations (light lines) joining nearest-neighbor lattice points. It should not be surprising, therefore, that centered cells normally contain symmetry elements that also incorporate subtranslations.

Just as it is possible to combine a reflection with a subtranslation to produce a glide, it is possible to combine a proper rotation with a subtranslation to produce a *screw axis.* Such an operation is shown in Fig. 4-3, where a rotation from e to f by an amount φ, combined with a translation from f to g by an amount T, is equivalent to a screw motion from e to g. Recalling that proper rotation axes must be parallel to some translation of a lattice, it follows that, after n rotations through an angle φ and n translations T along the screw axis, the cumulative translation distance must equal some integral multiple m of the lattice translation t. In other words,

$$nT = mt \tag{4-1}$$

where both n and m are integers. Equation (4-1) can be rearranged to determine the values that T can have:

$$T = \frac{mt}{n} \tag{4-2}$$

According to (4-2), the different values that the translation component, or *pitch,* of a screw axis can have depend on the fold n of the axis. Since $m = 0, 1, 2, 3, \ldots$, the values that T can have, for the various axes, are those given in Table 4-2. Because $\frac{3}{2}t = t + \frac{1}{2}t$ and $\frac{5}{4}t = t + \frac{1}{4}t$, etc., the only unique screw axes according to Table 4-2 are 2_1, 3_1, 3_2, 4_1, 4_2, 4_3, 6_1, 6_2, 6_3, 6_4, 6_5, where the subscript is the value of m in Eq. (4-2). Note that $m = 0$ and $m = n$ correspond to pure rotations, which can be thought of as special cases of screw axes. The 11 screw axes are shown in perspective in Fig. 4-4. It can be seen in this figure that pairs of axes such as 3_1 and 3_2 or 6_1 and 6_5 are identical, except for the sense of the screw motion.

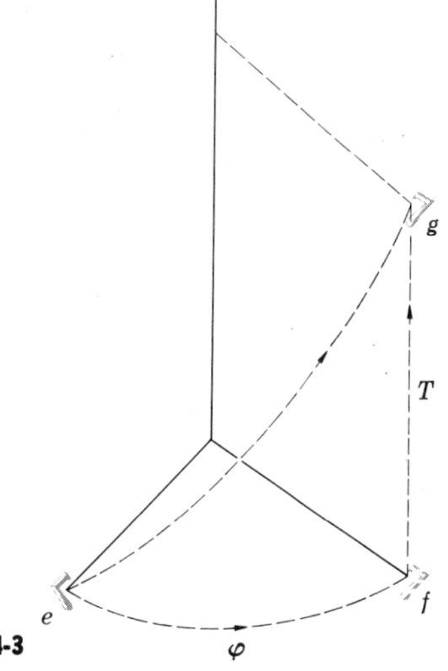

Fig. 4-3

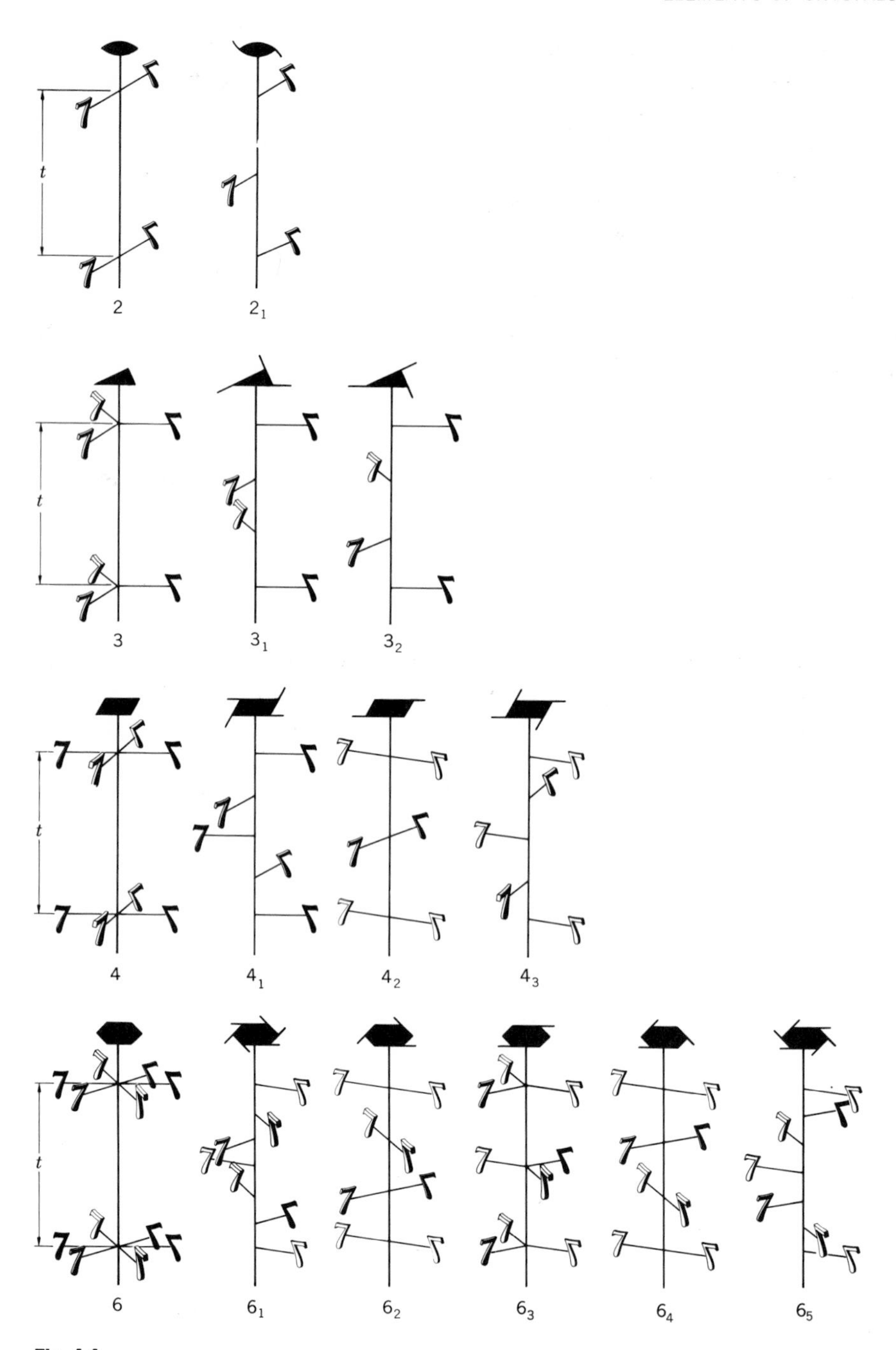

Fig. 4-4

Table 4-2 Possible values of the pitch T of a screw axis

Fold of axis	*Possible values of the pitch T*
1	$0t$, $1t$, $2t$, etc.
2	$0t$, $\frac{1}{2}t$, $\frac{2}{2}t$, $\frac{3}{2}t$, etc.
3	$0t$, $\frac{1}{3}t$, $\frac{2}{3}t$, $\frac{3}{3}t$, $\frac{4}{3}t$, etc.
4	$0t$, $\frac{1}{4}t$, $\frac{2}{4}t$, $\frac{3}{4}t$, $\frac{4}{4}t$, $\frac{5}{4}t$, etc.
6	$0t$, $\frac{1}{6}t$, $\frac{2}{6}t$, $\frac{3}{6}t$, $\frac{4}{6}t$, $\frac{5}{6}t$, $\frac{6}{6}t$, $\frac{7}{6}t$, etc.

Such pairs are therefore enantiomorphous; the individual axes are commonly distinguished as being left- or right-handed.†

In trying to determine the possible combinations of point groups with space lattices, it is necessary to consider not only the 32 point groups already derived, but also the symmetry groups having identical angular relationships but containing screw axes in place of pure rotation axes and glide planes in place of pure reflection planes. If these *isogonal* symmetry groups are included, it turns out that there are only 230 uniquely different ways to combine such symmetry with lattices. Some illustrations of how such combinations are formed are considered next.

Monoclinic space groups. Since the derivation of all the possible space groups is a lengthy undertaking, a few simple examples are given to illustrate the full symmetry that develops from the combination of symmetry elements with three noncoplanar translations. In the triclinic system, there are only two combinations possible, namely, $P1$ and $P\bar{1}$. In the monoclinic system, on the other hand, there are three point groups to consider and two lattice types. As an example, consider the possible combinations of the point group 2, the isogonal symmetry 2_1, and the two lattices P and I. Figure 4-5 shows a view along the c axis of the four possible combinations $P2$, $P2_1$, $I2$, $I2_1$. Similarly to the procedure used in finding all the symmetry elements present in a plane group, motifs (circlets in Fig. 4-5) are used to represent the operations of the symmetry elements and the translations. The actual height of the object above the plane of the drawing is indicated by the fraction placed within the circlets in Fig. 4-5. It is easy to see that $I2$ and $I2_1$ are the same space group; only the origins are different. It is therefore concluded that three different space groups belong to the crystal class 2. Some additional examples are considered in the exercises at the end of this chapter.

Space-group nomenclature. The labels in Fig. 4-5 are the international symbols used to denote the appropriate space group.‡ The international space-group

† Conventionally, a right-handed axis is one whose screw motion is clockwise when looking up along the axis.

‡ A more terse notation due to Schoenflies may be encountered in earlier writings on crystallography. A complete listing of equivalences can be found in the *International tables*. (See the literature list at the end of the chapter.)

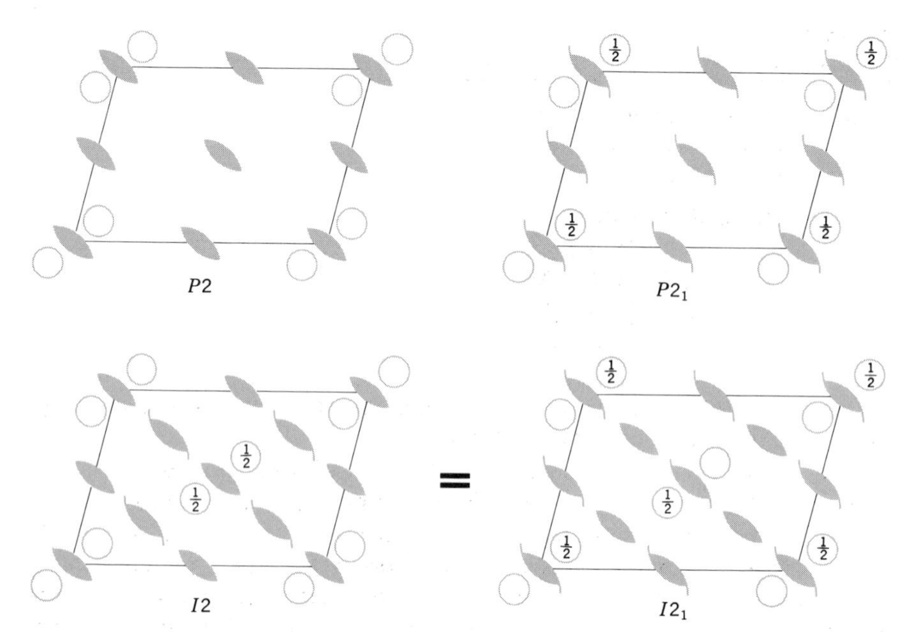

Fig. 4-5

symbol is a shorthand notation describing the symmetry elements present, preceded by the appropriate lattice-type symbol. The description of the symmetry elements is not unlike that used in point-group notation. There is one important difference; namely, in the space-group symbol the particular sequence of the symmetry elements listed describes their orientation in space relative to the three crystallographic axes. In the *triclinic system* this point is trivial; however, in all other systems it is not. Unfortunately, because of precedents arbitrarily arrived at, a consistent notational system for all crystal systems does not exist.

In the *monoclinic system* there is a choice between calling the unique axis c or b. In giving the complete space-group symbol, if the symmetry elements are listed in the sequence abc, the two symbols for $P2$ are, respectively, $P112$ and $P121$. These two possibilities are known as the *first setting* and *second setting*, respectively.

In the *orthorhombic system*, it is conventional to list the symmetry elements in the order abc. This order has a very real importance, as can be seen in the space groups belonging to the crystal class $2mm$, which are properly presented with c as the unique axis, namely, $Pmm2$. The nontrivial nature of the need for consistency in this representation becomes apparent in the two different space groups $Pmna$ and $Pnma$, whose full symbols are $P\frac{2}{m}\frac{2}{n}\frac{2_1}{a}$ and $P\frac{2_1}{n}\frac{2_1}{m}\frac{2_1}{a}$, respectively.

In the *tetragonal system*, the c axis is the 4-fold axis. The sequence for listing the symmetry elements is c, a, $[110]$, since the two crystallographic axes orthogonal to c are equivalent. For example, $P\bar{4}m2$ states that the unique c axis in a primitive tetragonal lattice has the symmetry $\overline{4}$, the two a axes each are parallel to m, and the $[110]$ direction has the symmetry 2.

The above rule for listing the symmetry elements also applies to the *hexagonal system*, since here again c is the unique axis and $a_1 = a_2$. The lattice symbol P denotes the primitive hexagonal lattice, whereas R denotes the "centered" hexagonal lattice in which the primitive rhombohedral cell has been chosen as the unit cell.

In the *cubic system* all three crystallographic axes are equivalent, and the order of listing the symmetry elements is a, $[111]$, $[110]$. Since $[[111]]$ must have the symmetry 3 or $\overline{3}$, the appearance of a 3 in the second position serves to distinguish the cubic system from the hexagonal system.

DERIVATIVE SYMMETRY

Supergroups and subgroups. As will be demonstrated in subsequent parts of this book, the interpretation of x-ray diffraction results is often aided by recognizing the relations between similar symmetry groups. For example, consider the addition of a symmetry element to a group or the suppression of a symmetry element present in a group. If the addition of the symmetry element to an existing group produces a new group, this new group is said to be a *supergroup* of the former group. Similarly, if the suppression of symmetry in a particular group produces a new group, the resulting group is called a *subgroup* of the old group. As an example of these relationships, consider the point group 1, which has the lowest possible symmetry and is therefore a subgroup of each of the other 31 point groups. Conversely, the point groups $\frac{6}{m}mm$ and $m3m$ can have no supergroups because it is not possible to obtain a new crystallographic point group by the addition of symmetry elements to either group. As a final example, the point groups 1, 2, m, $2mm$, and 4 are the subgroups of $4mm$, and $\frac{4}{m}mm$ and $m3m$ are its two supergroups. It is clear from the above that the subgroups of a point group are the group-theoretical analogs of the algebraic factors of a number.

Derivative structures. The terms sub- and supergroup are usually limited to point groups. It is often of interest, however, to investigate similar relationships between space groups also. This is so because two crystal structures may be related to each other in a simple way. For example, consider the body-centered cubic unit cell in Fig. 3-18, and suppose it has one atom of cesium occupying each lattice point. This results in the body-centered cubic structure of cesium shown in Fig. 4-6, whose space group is $Im3m$. Next, examine the structure of $CsCl$ shown in Fig. 4-7 very carefully. The cesium chloride structure obviously consists of eight cesium atoms at the corners of a cube and the larger chlorine atom at its center. (The eight corner atoms are shared by eight unit cells meeting at the corners, so that, on the average, there is one $CsCl$ unit per cell.) Thus the body-

Fig. 4-6

centering translation is suppressed in going from the body-centered cubic lattice of Cs to the primitive cubic lattice of CsCl. It is an easy matter to confirm that none of the symmetry elements has been suppressed, however, so that the space group of CsCl is $Pm3m$. The Cs structure is an example of an atomic array having one atom per lattice point, and CsCl illustrates a structure having two atoms per lattice point. This is also a simple example of the importance of distinguishing the structure, or atomic array, from the lattice of crystals.

The relationships between one crystal structure and another derived from it have been discussed by Buerger, who proposed the name *derivative structure* for the latter structure and *basic structure* for the former. A derivative structure can be obtained by suppressing one or more symmetry operations in the basic structure or by suppressing a translation in the basic structure. Combinations of these two cases are possible also, as illustrated by the relations between the structures of KN_3 and KO_3 shown in Fig. 4-8.

Although it is clearly outside the scope of this book to consider the various physical reasons why crystal structures should resemble each other, one particular relationship is often of great importance in x-ray diffraction analyses. This is the case when one or more different kinds of atoms are substituted for a specific kind

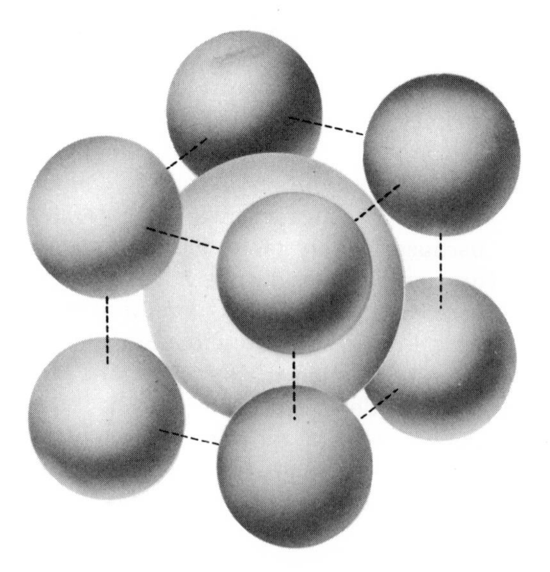

Fig. 4-7

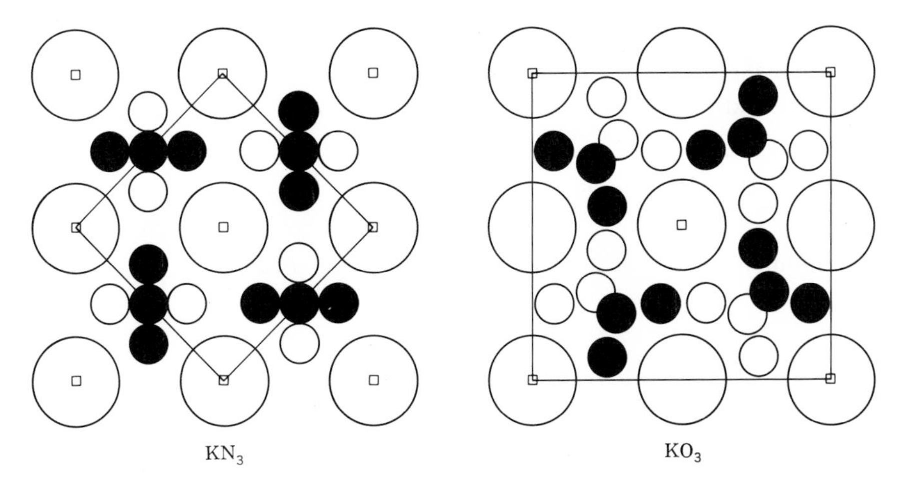

Fig. 4-8. The derivative structure KO_3 shown on the right is derived from the basic structure of KN_3 on the left by suppressing one set of 4-fold axes and one of the translations. Physically, this is caused by a bending of the ozonide ions, O_3^-, as compared with the linear azide ions, N_3^-. The dark and light complex ions lie, respectively, above and below the planes containing the larger K^+ ions. [From L. V. Azároff and I. Corvin, *Proceedings of the National Academy of Science,* vol. 49 (1963), pp. 1–5.]

of atom in the basic structure, resulting in a *solid solution* or a *substitution structure.* To help see the relationship between some of the geometric relations discussed above and actual crystal structures, consider the basic structure of sphalerite, ZnS, and two of its derivative structures shown in Fig. 4-9. When all the zinc atoms are replaced by equal numbers of copper and iron atoms, the crystal

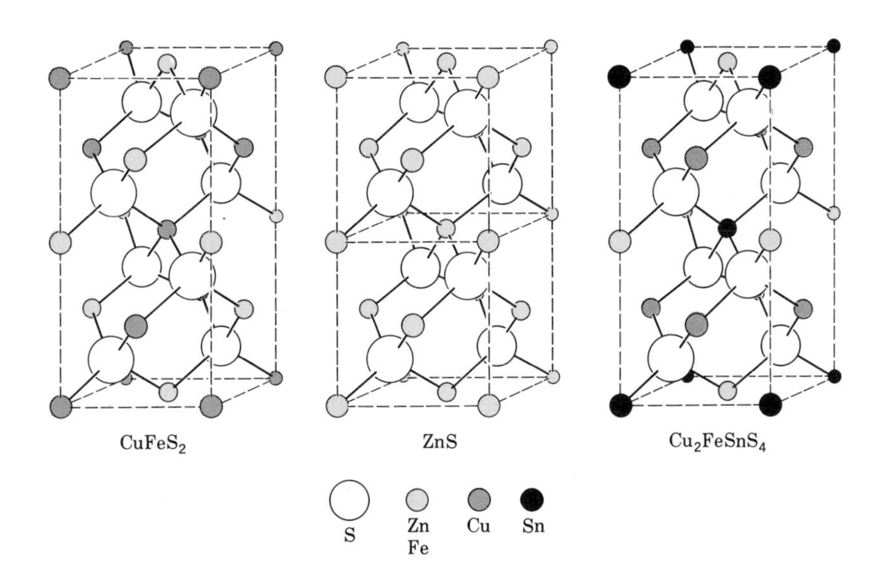

Fig. 4-9

structure of $CuFeS_2$ results, shown to the left. Similarly, substitution of $2Cu + Fe + Sn$ for four Zn atoms leads to the stannite structure on the right. These relations and their effect on the space groups are summarized below.

Basic structure, sphalerite	Zn	S	$(Fm3m)$
Substitution structure, chalcopyrite	$Cu_{\frac{1}{2}}$ $Fe_{\frac{1}{2}}$	S	$(I\bar{4}2d)$
Substitution structure, stannite	$Cu_{\frac{1}{2}}$ $Fe_{\frac{1}{4}}$ $Sn_{\frac{1}{4}}$	S	$(I\bar{4}2m)$

The tetragonal unit cells of the two derivative structures above are compared with two cubic unit cells of sphalerite in Fig. 4-9. Note that Cu_2FeSnS_4 is also a derivative of $CuFeS_2$.

EQUIVALENT POSITIONS IN A UNIT CELL

Plane groups. The application of some of the group-theoretical results discussed above to real crystals requires an appreciation of the constraints that symmetry elements impose on all the space surrounding them. Thus, if an atom is placed at some point xyz in a unit cell, it will be repeated a number of times, determined by the aggregate of symmetry elements present and their disposition relative to the point xyz. Another way of stating this is to realize that, in real crystals, there are like atoms that are related to each other (made equivalent) by the symmetry of the crystal. The set of positions occupied by such symmetry-equivalent atoms is called an *equipoint set* of the crystal, and the number of equivalent points in the set is called its *rank*. Finally, all the distinguishable sets that exist in one cell are called the equipoints of a space group.

To simplify the determination of the possible equipoints, begin by considering plane groups, for example, the plane group $p4$. Figure 4-10 shows a point having the general coordinates xy repeated by the symmetry of $p4$. It is usual to give the coordinates of a point in a unit cell in terms of fractions of the cell edges, thereby making them independent of actual cell dimensions. The value of the fraction x is the actual distance along the a axis, from the origin to the point, divided by the actual length of a. Similarly, the fraction y is the actual distance along the b axis divided by the length of b. Thus one complete translation along a is $a/a = 1$, etc. Proceeding in a clockwise manner, the coordinates of the next point in Fig. 4-10 are the fraction y along the a axis and $1 - x = 1 + \bar{x} = \bar{x}$ along the b axis. Since it is customary to list the coordinate along a first and along b next, the coordinates of this point are $y\bar{x}$. The coordinates of the other equivalent points are also indicated in Fig. 4-10. Because there are four such points related by the symmetry of $p4$, the rank of the equipoint xy is 4 in this plane group. Since each of the

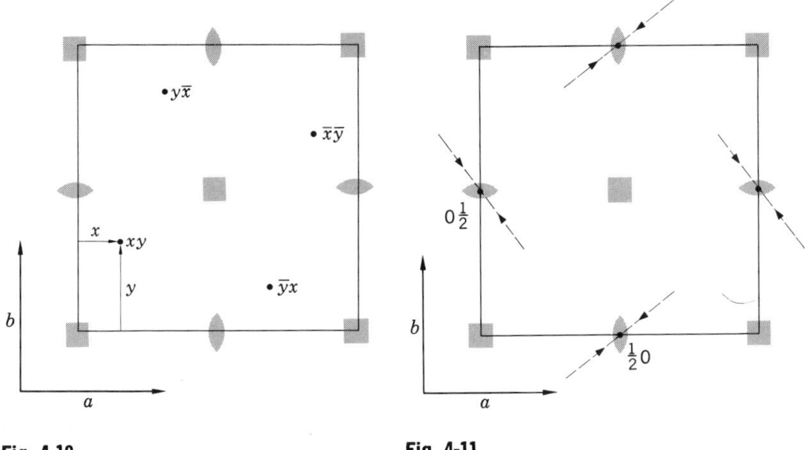

Fig. 4-10

Fig. 4-11

coordinates of this equipoint, x and y, is variable between 0 and 1, this position in the cell is called a *general position*.

Next, consider the pairs of points related by the 2-fold axes in $p4$, and let these points coalesce to single points on each 2, as indicated in Fig. 4-11 by the dashed lines. The coordinates of these equivalent points (the black dots in Fig. 4-11) are $\frac{1}{2}0$ and $0\frac{1}{2}$, and there are two such points in each cell. Finally, the points on the two different 4-fold axes, at 00 and $\frac{1}{2}\,\frac{1}{2}$, each have the rank 1. Since the x and y values of these equipoints are fixed, they are called *special positions.* Table 4-3 lists the various equipoints in $p4$ and their symmetry. Note, for example, that the equipoint 00 must have 4-fold symmetry since it occupies a 4-fold axis. Also note that the arithmetic product of the first two columns in Table 4-3 is a constant. This is so because a point lying on a symmetry axis is not repeated by that axis; hence its rank is decreased by the fold of the axis. This relationship holds true whether the axis is a proper or an improper axis.

Space groups. The equipoints of a space group can be determined in a similar manner. As a simple extension of the above example, consider the space group $P4$. The equipoints of this space group are obtained by adding the z coordinate

Table 4-3 Equipoints of plane group $p4$

Rank of equipoint	*Symmetry of location*	*Coordinates of equivalent points*
1	4	00
1	4	$\frac{1}{2}\,\frac{1}{2}$
2	2	$0\frac{1}{2}$, $\frac{1}{2}0$
4	1	xy, $y\bar{x}$, $\bar{x}\bar{y}$, $\bar{y}x$

Table 4-4 Equipoints of space group $P4$

Rank of equipoint	*Wyckoff notation*	*Symmetry of location*	*Coordinates of equivalent points*
1	a	4	$00z$
1	b	4	$\frac{1}{2}\,\frac{1}{2}z$
2	c	2	$0\frac{1}{2}z$, $\frac{1}{2}0z$
4	d	1	xyz, $y\bar{x}z$, $\bar{x}\bar{y}z$, $\bar{y}xz$

to the equipoints of $p4$, as shown in Table 4-4. Since the symmetry of $P4$ does not affect z, this coordinate is not specified for any equipoint. Note that Table 4-4 contains an additional column headed "Wyckoff notation." This notation was proposed in one of the first published tabulations of such data, and it has become common practice to refer to a position in a unit cell by the rank and letter of the corresponding equipoint.

As a further illustration, consider the space group $P\bar{4}m2$ shown in Fig. 4-12. The arrows in the plane of this figure indicate the 2-fold axes along $[[110]]$; a two-sided arrowhead denotes a 2, and a single-sided arrowhead denotes a 2_1. The general position xyz is indicated in this drawing by a circlet; a $+$ near the circlet means that it lies above the plane of the drawing, and a $-$ means that the circlet lies an equal distance below the plane. Finally, a comma enclosed within a circlet denotes an enantiomorphous relationship to a symmetry-related circlet containing no comma. This notation is used here in order to acquaint the reader with the similar notation used in *International tables for x-ray crystallography.*

The equipoints of space group $P\bar{4}m2$ are listed in Table 4-5. The general position xyz is shown in Fig. 4-12. The special positions are obtained, for example, by "moving" pairs of points related by the two kinds of mirrors intersecting at $00z$ and $\frac{1}{2}\,\frac{1}{2}z$, respectively, onto these mirror planes. This gives the special positions $4j$ and $4k$. Alternatively, pairs of points can coalesce onto the diagonal 2-fold

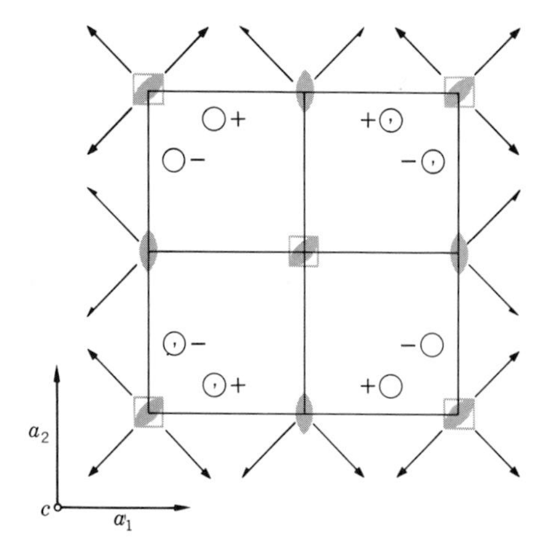

Fig. 4-12

Table 4-5 Equipoints of space group $P\bar{4}m2$

Rank of equipoint	*Wyckoff notation*	*Symmetry of location*	*Coordinates of equivalent points*
1	a	$\bar{4}2m$	000
1	b	$\bar{4}2m$	$\frac{1}{2}\frac{1}{2}0$
1	c	$\bar{4}2m$	$00\frac{1}{2}$
1	d	$\bar{4}2m$	$\frac{1}{2}\frac{1}{2}\frac{1}{2}$
2	e	mm	$00z,\ 00\bar{z}$
2	f	mm	$\frac{1}{2}\frac{1}{2}z,\ \frac{1}{2}\frac{1}{2}\bar{z}$
2	g	mm	$0\frac{1}{2}z,\ \frac{1}{2}0\bar{z}$
4	h	2	$xx0,\ \bar{x}\bar{x}0,\ x\bar{x}0,\ \bar{x}x0$
4	i	2	$xx\frac{1}{2},\ \bar{x}\bar{x}\frac{1}{2},\ x\bar{x}\frac{1}{2},\ \bar{x}x\frac{1}{2}$
4	j	m	$x0z,\ \bar{x}0z,\ 0x\bar{z},\ 0\bar{x}\bar{z}$
4	k	m	$x\frac{1}{2}z,\ \bar{x}\frac{1}{2}z,\ \frac{1}{2}x\bar{z},\ \frac{1}{2}\bar{x}\bar{z}$
8	l	1	$xyz,\ \bar{x}\bar{y}z,\ \bar{x}yz,\ x\bar{y}z,$ $\bar{y}x\bar{z},\ y\bar{x}\bar{z},\ yx\bar{z},\ \bar{y}\bar{x}\bar{z}$

axes to produce $4h$ and $4i$. Next, the points lying along the lines of intersection of the mirror planes can have the coordinates of the equipoints $2e$, $2f$, or $2g$. Note that these points occur in pairs, differing only in whether the z coordinate is positive or negative. This is because of the 2-fold axes along [[110]] which intersect the c axis at $z = 0$ and $z = \frac{1}{2}$. (Compare these with equipoints $1a$ and $1b$ in $P4$.) Finally, if the points in $2e$, $2f$, and $2g$ are made to coalesce in pairs onto the positions $z = 0$ and $z = \frac{1}{2}$, the four special positions $1a$, $1b$, $1c$, and $1d$ result.

It is possible to make use of relations between equipoints to determine what the coordination of any atom in a crystal is in terms of the equipoints occupied by the other atoms. For example, suppose that a crystal structure whose space group is $P\bar{4}m2$ has an A atom in equipoint $1d$ surrounded by two B atoms in $2f$ and four B atoms in $4i$. The resulting sixfold coordination of the A atom is said to be *octahedral* because the polyhedron formed by connecting the centers of the six B atoms surrounding it is an octahedron, as shown in Fig. 4-13. The center of the A atom is shown at the body center of the unit cell, $\frac{1}{2}\frac{1}{2}\frac{1}{2}$, and the six B atoms are grouped around it according to the equipoint listing in Table 4-5. It is easy to see in this figure that the interatomic distances between the central A atom and the two B atoms in $2f$ are equal to $(\frac{1}{2} - z)c$. Similarly, the four B atoms lying at the corners of the square in the $xy\frac{1}{2}$ plane are each equidistant from the central A atom. This interatomic distance is given by the vector sum of $(\frac{1}{2} - x)\mathbf{a} + (\frac{1}{2} - x)\mathbf{b}$, or, since $a = b$ in the tetragonal system, the interatomic distance is, simply, $\sqrt{2(\frac{1}{2} - x)^2}\, a$.

The interatomic distances can be determined directly from the equipoints without recourse to a model of the actual structure. This is particularly easy if the reference or crystal axes are mutually orthogonal. Using the above example, convert the coordinates of each equipoint to absolute dimensions by multiplying

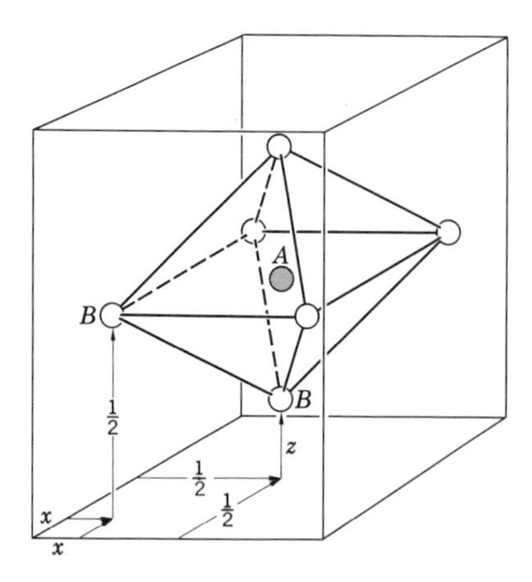

Fig. 4-13

x by **a**, y by **b**, etc. (If an equipoint set has a rank higher than 1, the use of any one of its equivalent points produces the same final result.) The absolute coordinates of the three atoms of interest are as follows:

$$\begin{aligned} &\text{Absolute coordinates of } A \text{ atom in } 1d \text{ are } \tfrac{1}{2}\mathbf{a},\ \tfrac{1}{2}\mathbf{b},\ \tfrac{1}{2}\mathbf{c} \\ &\text{Absolute coordinates of } B \text{ atom in } 2f \text{ are } \tfrac{1}{2}\mathbf{a},\ \tfrac{1}{2}\mathbf{b},\ z\mathbf{c} \\ &\text{Absolute coordinates of } B \text{ atom in } 4i \text{ are } x\mathbf{a},\ x\mathbf{b},\ \tfrac{1}{2}\mathbf{c} \end{aligned} \tag{4-3}$$

The interatomic distances between any two atoms is simply the vector difference between their absolute coordinates. For example, the interatomic distance between the A atom and the B atom in $2f$ is

$$(\tfrac{1}{2}\mathbf{a} + \tfrac{1}{2}\mathbf{b} + \tfrac{1}{2}\mathbf{c}) - (\tfrac{1}{2}\mathbf{a} + \tfrac{1}{2}\mathbf{b} + z\mathbf{c}) = (\tfrac{1}{2} - z)c \tag{4-4}$$

Similarly, the other A-B distance is

$$(\tfrac{1}{2}\mathbf{a} + \tfrac{1}{2}\mathbf{b} + \tfrac{1}{2}\mathbf{c}) - (x\mathbf{a} + x\mathbf{b} + \tfrac{1}{2}\mathbf{c}) = (\tfrac{1}{2} - x)\mathbf{a} + (\tfrac{1}{2} - x)\mathbf{b} \tag{4-5}$$

in full agreement with the result derived earlier from Fig. 4-13. In carrying out such calculations, it should be borne in mind that the operations in Eqs. (4-4) and (4-5) are vector operations. Thus, if the crystallographic axes are not mutually orthogonal, the angular relations between the vectors must be kept in mind when the interatomic distances are computed. It is usually helpful in such cases first to make an accurate sketch or model of the structure. (See Exercise 4-10.)

Determination of atomic arrangements. The equipoints of all 230 space groups are tabulated in full detail in *International tables for x-ray crystallography.* If the space-group symmetry of the crystal is known, a knowledge of its chemical composition makes it possible to determine how the atoms must be distributed among the available equipoints, that is, the crystal structure. It will be shown later how x-ray diffraction can be used to help determine the space group of a crystal and the size of the unit cell. Before the atomic distribution among the equipoint sets can be considered, however, it is necessary to determine the number of atoms or formula weights contained in the unit cell. After the density D of the crystal has been determined, this is a simple matter because

$$D = \frac{\text{mass of cell}}{\text{volume of cell}} = \frac{nM}{V} \tag{4-6}$$

where n is the number of formula weights or molecules having the mass M that are contained in the unit-cell volume V.

The usual list of atomic weights found in tables is based on oxygen equal to 16 atomic mass units (amu), which can be converted to grams by dividing by Avogadro's number (6.023×10^{23} molecules/g mole) or multiplying by its reciprocal, 1.660×10^{24} g/amu. Putting in all units in Eq. (4-6) after rearranging it to solve for n, the number of formula weights M contained in a cell is given by

$$n = \frac{D\ (\text{g/cm}^3) \times V\ (\text{cm}^3)}{M\ (\text{amu}) \times 1.660 \times 10^{-24}\ (\text{g/amu})} \tag{4-7}$$

Ideally, n should be an integer; however, because of errors in the experimental determination of the density or defects present in the crystal, n may deviate slightly from an integer. Usually, it is somewhat smaller.

After rounding off the experimentally determined value of n to an integer, it is possible to consult the appropriate space-group table in order to determine the atomic distribution. For example, assume that the space group of a crystal is $P4$ and that it contains two molecules in a unit cell. By consulting Table 4-4 it can be seen that the two molecules can occupy the special position $2c$ or they can occupy special positions $1a$ and $1b$. It is obvious that they cannot have the general coordinates of $4d$ because the symmetry would require that there be four identical molecules in each cell. Such reasoning, supplemented by additional knowledge of chemical or physical properties, is frequently sufficient to determine simple crystal structures. For instance, in the above example, if the configuration and symmetry of the two molecules are known from stereochemical reasoning, equipoints $2c$ or $1a$ and $1b$ can be eliminated, depending on whether the known symmetry of the molecule is 2-fold or 4-fold.

As a further example, assume that the space group is $P\bar{4}m2$ and that the unit cell contains one formula weight, $A(BC_4)_2$. It can be seen from Table 4-5 that the A atom can occupy any one of the four special positions $1a$, $1b$, $1c$, or $1d$, but no others. It should be noted that this choice is arbitrary in that these four positions are simply alternative choices for the origin of coordinates in the unit cell. For simplicity, therefore, the A atom can be placed in $1a$ at 000. This choice, however, then fixes the relative positions of the B and C atoms. The B atoms can occupy any of the positions of rank 2 or any pair of unoccupied positions of rank 1. The C atoms, in turn, can occupy the positions $8l$, or any pair of positions of rank 4, or any position of rank 4 and any pair of unoccupied positions of rank 2, etc. It can be seen, even in the above relatively simple example, that a choice has to be made between a large number of permutations. Moreover, to complete the actual crystal structure, the unspecified values of the xyz coordinates must be determined. The sum total of these operations constitutes the problem of crystal-structure analysis, which is further discussed in later chapters.

SUGGESTIONS FOR SUPPLEMENTARY READING

N. V. Belov, A class-room method for the derivation of the 230 space qroups (English translation by V. Balashov), *Proceedings Leeds Philosophical Society,* vol. 8 (1950), pp. 1–46.

M. J. Buerger, Derivative crystal structures, *Journal of Chemical Physics,* vol. 15 (1947), pp. 1–16.

Elementary crystallography (John Wiley & Sons, Inc., New York, 1965).

Norman F. M. Henry and Kathleen Lonsdale, *International tables for x-ray crystallography,* vol. 1. (The Kynoch Press, Birmingham, England, 1952).

H. Hilton, *Mathematical crystallography and the theory of groups of movements* (Clarendon Press, Oxford, 1903, republished by Dover Publications, Inc., New York, 1963).

F. C. Phillips, *An introduction to crystallography,* 2d ed. (Longmans, Green & Co., Ltd., London, 1963).

EXERCISES

4-1. Combine two mutually orthogonal glides with a plane-centered lattice. Derive thereby the plane group $c2gg = c2mm$ analogously to the procedure typified by Fig. 4-1.

4-2. By means of a projection along the axis, indicate the symmetry of space about a 3_1 and 3_2. To do this, start the rotations in a clockwise manner, and mark the elevation of each congruent motif by a fraction representing successive subtranslations. Show that 3_1 and 3_2 are enantiomorphous. **Hint:** Remember that adding or subtracting unit translations t does not change the value of T (Table 4-2).

4-3. Derive the four monoclinic space groups belonging to the crystal class m. Using the first setting, do this by considering the following possible combinations: Pm, Pa, Pn, Im, Ia, In.

4-4. Derive the orthorhombic space groups that result from the combinations of three orthogonal 2-fold axes with a side-centered lattice. Do this by combining a C lattice with the isogonal groups 222, 2_122, 22_12, 222_1, 2_12_12, 2_122_1, 22_12_1, and $2_12_12_1$. Which of the space groups so derived are unique?

4-5. Identify the crystal systems to which the following space groups belong: $Ima2$, $I4_122$, $I2_12_12_1$, $I4_132$, $P3_12$, $I23$, $F43c$, and $P622$.

4-6. What are the subgroups of $6mm$ and mmm? What are the supergroups of 32 and $\frac{4}{m}$? **Hint:** Consult the point-group listings in Chap. 1.

4-7. Consider the crystal structure of nickel which has one nickel atom per lattice point of a face-centered cubic lattice. When one-third of the nickel atoms in each cell are replaced by aluminum, the crystal structure of Ni_3Al obtains. If you are told that this is a substitution structure of nickel and that it also belongs to the cubic system, how must the aluminum atoms be distributed without significantly changing the length of a? What are the space groups of Ni and of Ni_3Al?

4-8. Derive the equipoints of plane groups $p6$, cm, and $p2mm$.

4-9. Derive the equipoints of space group $P3m1$ and $P4bm$.

4-10. The crystal structure of zinc is hexagonal, $P\frac{6_3}{m}mc$, and there are two atoms per unit cell at 000 and $\frac{1}{3}\frac{2}{3}\frac{1}{2}$. Calculate the interatomic distances between zinc atoms if the unit-cell dimensions are $c = 4.945 \times 10^{-8}$ cm and $a = 2.664 \times 10^{-8}$ cm. **Hint:** First sketch a unit cell and its contents, and do not forget that the cell is not orthogonal.

4-11. Carbon crystallizes in two possible modifications, as diamond (cubic, $a = 7.91 \times 10^{-8}$ cm) and graphite (hexagonal, $a = 2.46 \times 10^{-8}$ cm, $c = 6.70 \times 10^{-8}$ cm). If the density of diamond is 3.51 g/cm^3, and of graphite, 2.25 g/cm^3, how many atoms of carbon does each of the two kinds of cells contain?

4-12. Look up the space group $Fm3m$ in *International tables for x-ray crystallography*, and, by considering the equipoints listed, determine the crystal structure of diamond based on the results of Exercise 4-11.

PART TWO
Elements of X-ray Physics

FIVE
Historical note

RÖNTGEN'S DISCOVERY

In the second half of the nineteenth century, numerous physicists were studying the then recently discovered cathode rays in order to establish their properties and to decide whether they were particles, that is, rays that travel in straight lines and cast sharp shadows, or some kind of electromagnetic field that moves with a wavelike motion. By the end of the century it had been established that cathode rays were readily absorbed by matter, that their absorption was inversely related to the accelerating voltage, and that glass, and particularly certain crystals, when struck by the cathode rays, emitted visible light, a process called *fluorescence.*† By 1896, J. J. Thomson clearly demonstrated at Cambridge University that cathode rays indeed were composed of small negatively charged particles having a mass approximately $1/1{,}800$ that of the smallest atom, hydrogen. Following an earlier proposal of J. Stoney, the name electron was adopted for this particle. In 1910, Robert Millikan, in his famous oil-drop experiment at the University of Chicago, succeeded in measuring the absolute charge of an electron, which has the currently accepted value of 4.803×10^{-10} esu, or 1.601×10^{-19} C.

In the summer of 1895, a physics professor at the University of Wurzburg, in Bavaria, Wilhelm Conrad Röntgen, constructed a cathode-ray tube and enclosed it in a light-tight cardboard box. The exact experiments that Röntgen planned to pursue during the following fall and winter never became known, because, by the end of the first week of November, he observed a peculiar phenomenon. Every time he sent a pulse of cathode rays through his tube, a screen made from barium platinocyanide crystals and placed some distance away from the tube would light up in fluorescence. It was quite clear to Röntgen that the fluorescence was not caused by cathode rays, which would have been easily absorbed by the glass envelope of the tube, the cardboard box surrounding it, and the air in the room. A rapid succession of experiments showed that the radiation responsible for this fluorescence emanated from that part of the glass envelope of the tube where the

† It is more correct nowadays to call this process luminescence, or cathodoluminescence.

cathode rays struck it, that the rays traveled in straight lines (cast sharp shadows), and that they were absorbed by matter, although much less so than the cathode rays. Accordingly, Röntgen called these mysterious new rays the *x* rays. By the time the first communication of his results was prepared, in time for distribution as New Year's greetings to some of his professional colleagues, Röntgen had also demonstrated that x rays passed through flesh more readily than through bone, by preparing the world's first radiograph of a human hand, and that the x rays could be produced more efficiently if the cathode rays were allowed to strike a metal target in place of the tube's glass walls. He also attempted several experiments to see whether his x rays could be reflected, refracted, or diffracted, but without success.

The news of Röntgen's discovery spread like wildfire, and very soon the principal application of x rays, radiography, was being utilized by hospitals, and later, by industry all over the world. Röntgen himself continued his researches, and discovered that an anode made of heavy metal like platinum emitted x rays more copiously than one made of a light element like aluminum; that x rays blackened a photographic film and ionized a gas through which they passed; that the penetrability, or "hardness," of the rays increased with increasing tube voltage; and many other characteristics of the rays. For his momentous discoveries, Röntgen was awarded the first Nobel prize in physics in 1901. Meanwhile, other physicists, as well as medical doctors, were busy examining the new rays. In 1896 A. Winkelmann and R. Straubel placed a photographic film on top of a crystal of fluorite, CaF_2, and observed that the film was much blacker than the incident x-ray dose warranted (Fig. 5-1). To examine the nature of the "backscatter" from the crystal, they placed a sheet of black paper between the film and the crystal, with the result that the film's blackening was reduced to very nearly what it would be in the absence of the fluorite crystal. Placing the same paper in the direct path of the incident x rays did not have the same effect, so that it became clear that the incident x rays were not reflected backward by the fluorite crystal, but instead produced a new kind of x ray in the crystal which was "softer," that is, more readily absorbed by paper. This discovery was further extended by C. G. Barkla, who had already discovered in 1905 that x rays, like visible light, became polarized upon scattering, a strong indication that the new rays behaved like transverse waves. By 1909,

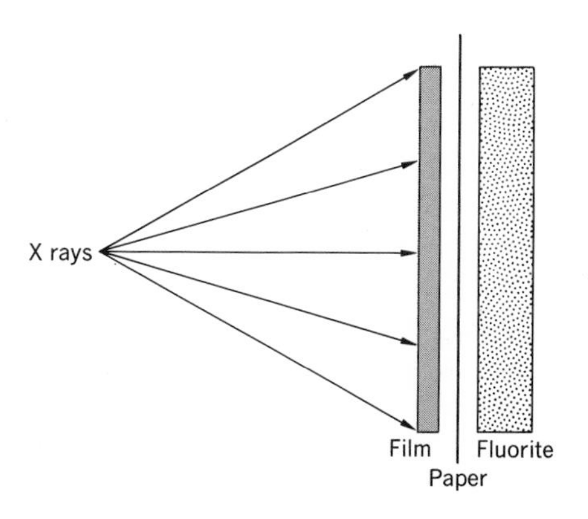

Fig. 5-1

Barkla further discovered the presence of strong homogeneous components in the x rays emitted by his tubes when they were operated under suitable conditions. He realized that these components were characteristic of the target metals employed, and suggested that the two groups of emission lines be designated as the K and L spectra. Later, these spectra were related to the K and L electrons of the target atoms, in accordance with the subsequently introduced model of an atom devised by Niels Bohr. For his contribution to the elucidation of the properties and characteristics of x rays, Barkla received the Nobel prize in 1917. A more complete understanding of x rays, however, had to await the discovery of x-ray diffraction.

LAUE'S DISCOVERY

In 1900 Röntgen was appointed to the chair of experimental physics at Munich University, where he headed a research institute housed in a separate three-story building. Munich University's staff already included the famous mineralogist and crystallographer P. Groth, who was responsible for the first classification of minerals according to their chemical compositions, as well as according to their geographic sites of occurrence, and who later pioneered the correlation of crystallography with chemistry in his monumental compilation *Chemische kristallographie.* The university also housed a theoretical physics group headed by A. Sommerfeld, whose staff included P. Debye, and was joined in 1909 by Max Laue, a graduate from the University of Berlin, where he had worked under Max Planck. In January of 1912, one of Sommerfeld's graduate students, P. P. Ewald, met with Laue to discuss the conclusions from his theoretical analysis of the propagation of light through a crystal which Ewald was in the process of writing up for his doctor's thesis. Instead of commenting on the results of this study, Laue was far more interested in the fact that Ewald used as his model of a crystal, small oscillators, periodically spaced in three dimensions, approximately 10^{-8} cm apart. Laue already knew from experiments conducted in Röntgen's laboratories and elsewhere that the wavelength of x rays was of the order of 10^{-8} cm. He made the bold guess, therefore, that a crystal could serve as an ideal grating for the diffraction of x rays. A presentation of the idea to Professor Sommerfeld, however, encountered numerous objections. Nevertheless, Laue was able to persuade one of Sommerfeld's assistants, W. Friedrich, and a recent graduate of Röntgen's, P. Knipping, to carry out the experiment anyway. After one abortive attempt, Friedrich and Knipping succeeded in recording the diffraction diagram of a copper sulfate crystal in the spring of 1912. Upon verifying the results, Laue applied his knowledge of the diffraction theory of light by one-dimensional and two-dimensional gratings to the problem of x-ray diffraction by a three-dimensional grating in a crystal, and, on his way home that very day, he formulated the necessary concepts in his mind. Laue's theory of x-ray diffraction was published in 1912, and won him the Nobel prize in 1914. It is interesting to note that Ewald learned of Laue's discovery in June of 1912, after he had left Munich for a post at Göttingen. On his way home from attending a lecture at which the visitor, Professor Sommerfeld, had described Laue's findings,

Ewald realized that the diffraction of x rays by crystals could be explained by a theory similar to the one in which he had vainly tried to interest Laue earlier that year. The results of his analysis of x-ray diffraction were published by Ewald in 1913, and include the extremely useful reciprocal-lattice concept discussed in later chapters.

W. H. AND W. L. BRAGG'S CONTRIBUTION

Bragg spectrometer. Shortly after his appointment to a professorship in physics at Leeds University in 1908, W. H. Bragg became interested in the study of the ionization of gases by x rays. These studies convinced him that x rays were particles rather than waves, because the ionization seemed to proceed in a hit-or-miss fashion, involving only a few of the atoms present in the gas. This view was shared by his son, W. L. Bragg, who was completing his studies at Cambridge University so that, during the summer of 1912, father and son collaborated in an analysis of Laue's paper on the diffraction of x rays by sphalerite (zincblende), ZnS, and attempted to explain the observed diffraction spots as being produced by x-ray corpuscles passing through tunnels formed by atomic rows in the crystal. Later on, however, W. L. Bragg became convinced of the correctness of Laue's postulate of the wave nature of x rays. In fact, by the winter of 1912, W. L. Bragg presented an alternative explanation of Laue's observations in terms of the reflection of x-ray "waves" by atomic planes in the crystal. The next step is best described by quoting Bragg's own words from a much later lecture, recorded in *Science in Britain* (Institute of Physics, London, 1943): "It remained to explain why certain of these atomic mirrors in the zincblende crystal reflected more powerfully than others, a difficulty which led Laue to postulate a group of definite wave-lengths. Pope and Barlow had a theory that the atoms in simple cubic compounds like ZnS were packed together, not unlike balls at the corners of a stack of cubes, but in what is called cubic close-packing, where the balls are also at the centre of the cube faces. I tried whether this would explain the anomaly—and it did! It was clear that the arrangement of atoms in zincblende was of the face-centered type. I was careful to call my paper on the structure of zincblende 'The Diffraction of Short Electromagnetic Waves by a Crystal' because I was still unwilling to relinquish my father's view that the X-rays were particles; I thought they might possibly be particles accompanied by waves."

The work of W. L. Bragg attracted the attention of C. T. R. Wilson of the Cavendish Laboratory at Cambridge, who suggested to young Bragg that he try reflecting x rays from the cleavage face of a crystal. The successful reflection of x rays by crystals was then utilized by W. H. Bragg to construct an x-ray spectrometer, permitting the quantitative measurement of x-ray intensities. This instrument, depicted in Fig. 5-2, measures the intensity of the x rays reflected at various angles by the atomic planes of a crystal parallel to its surface. As demonstrated below, this is really a measurement of the intensity at various x-ray wavelengths. A typical curve of the x-ray intensity obtained from a platinum target tube after reflection by a sodium chloride crystal is shown in Fig. 5-3. It is evident in this

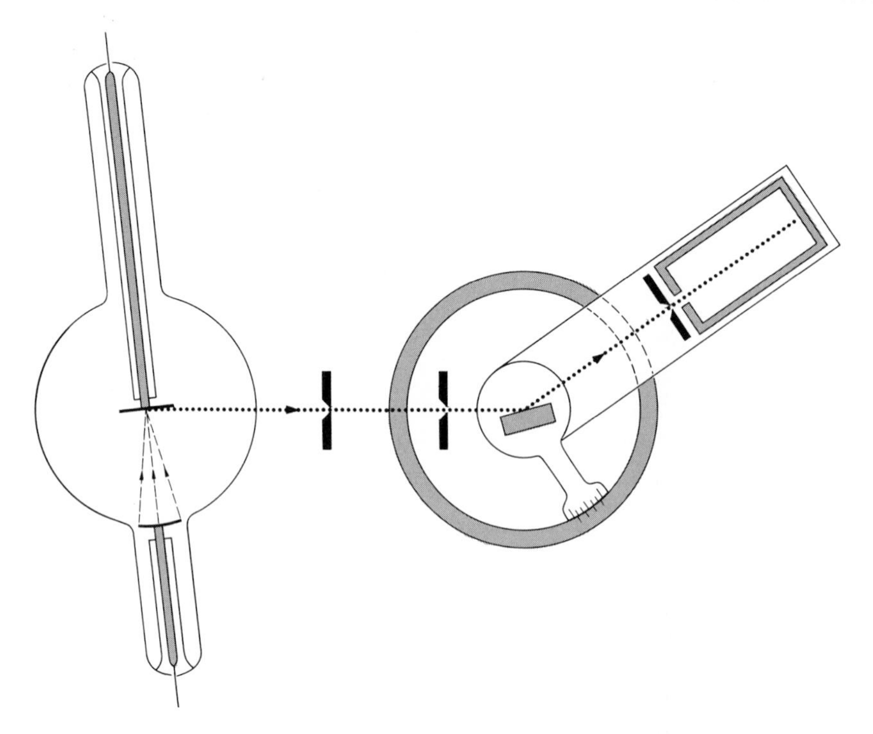

Fig. 5-2. Bragg spectrometer.

figure that the x-ray intensity does not vary uniformly with angle (wavelength), but consists of sharp lines superimposed on a gradually varying background. The explanation of such observations and their relation to the atomic arrays in crystals was carried out in close collaboration between father and son, for which W. H. Bragg and W. L. Bragg were awarded the Nobel prize in 1915.

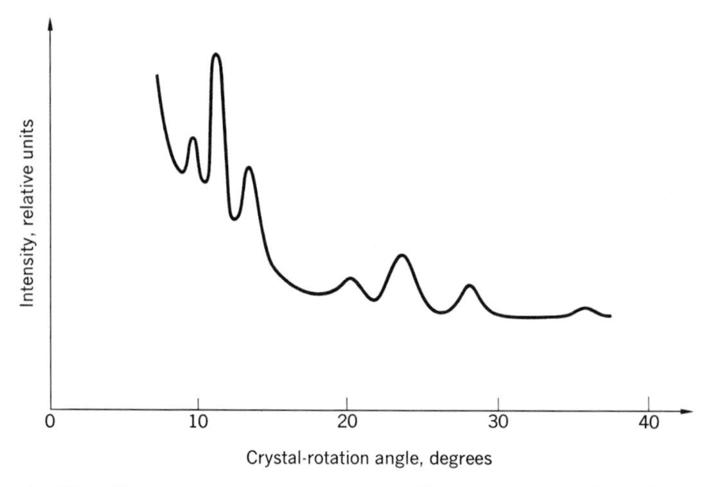

Fig. 5-3. X-ray spectrum of platinum. *(After Compton and Allison.)*

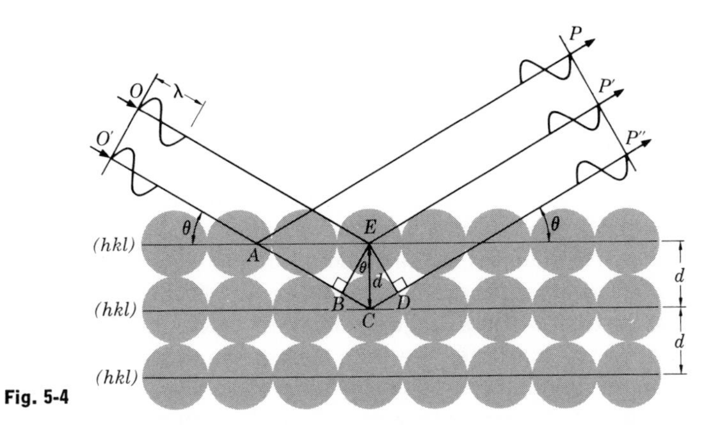

Fig. 5-4

Bragg law. One of the limitations that had prevented physicists from characterizing x rays better, following Röntgen's discovery, was their inability to correlate the measured x-ray intensities to their energies. Using Planck's postulate that $E = h\nu = hc/\lambda$, it remained for Bragg to establish a relationship between the angle at which x rays are reflected and their wavelength, λ. If one considers a crystal as made up of parallel planes of atoms periodically repeated by lattice translations (Chap. 3) and spaced by a distance d from each other, then the crystal structure viewed along the planes looks like Fig. 5-4. Next, suppose that a beam of parallel x rays impinges on the crystals so as to form the angle θ with it. If the x rays are treated as waves, the incident beam has a common wave front; that is, all the incident rays are in phase with each other at the normal to the rays $O'O$. This matter of "being in phase" is fundamental to the understanding of the interaction of waves with each other, and is illustrated in Fig. 5-5. On the left side, three

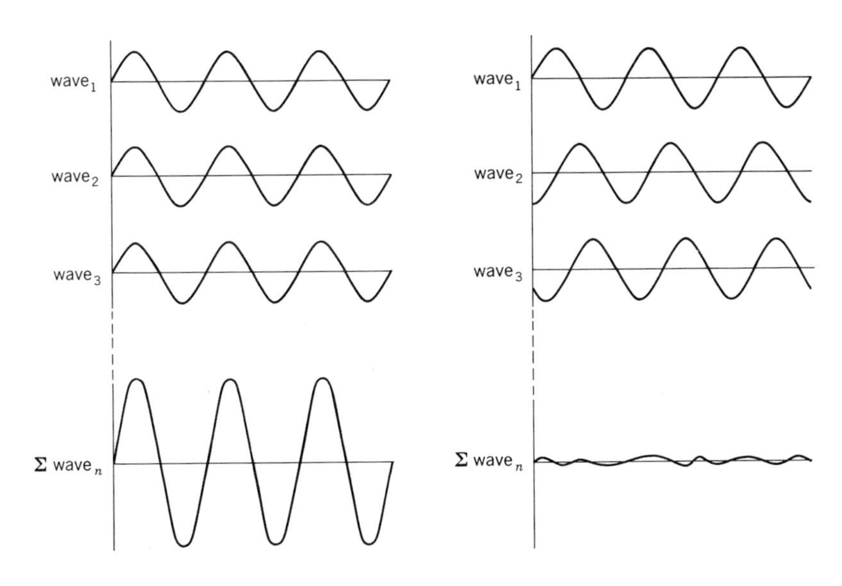

Fig. 5-5

waves are drawn starting at a common wave front. This is the case for *constructive interference*, when the waves are in phase with each other (they have zero and maximum amplitudes at the same points along the abscissa), and the addition of the individual amplitudes of n such waves produces a similar wave having a much larger amplitude, as illustrated by the wave designated Σ wave$_n$. If, on the other hand, the component waves have different starting points, as shown on the right of Fig. 5-5, the addition of the amplitudes at any point along the abscissa leads to *destructive interference*, since some of the waves have positive amplitudes at the same time that others have negative amplitudes. It can be demonstrated quite readily that when a large number of waves having completely random phases relative to each other are thus added, the net wave has virtually zero amplitude. If the incident x rays in Fig. 5-4 behave like waves, therefore, the way that they will interfere with each other at different scattering angles depends on the phase relations between them.

Consider two of the incoming x rays OE and $O'A$ (Fig. 5-4) inclined at an angle θ with the parallel stack of planes (hkl). Suppose that they are scattered in the direction of AP and EP', which also forms the angle θ with the top plane. Since the pathlengths of the rays OEP' and $O'AP$ are the same, they arrive at PP' in phase with each other and again form a common wave front. This is the condition for scattering in phase by one plane in the crystal. (Note that the above argument applies for any value of the angle θ, which is the case for true reflection by a single plane.) To see what restrictions the other planes in the stack make, consider next the incoming ray $O'C$ and the scattered ray CP''. If EB is parallel to the wave front $O'O$ and, similarly, ED is parallel to PP'', then the total pathlength $O'CP''$ is longer than that of ray OEP' (or $O'AP$) by an amount

$$\begin{aligned} \Delta &= BCD \\ &= 2BC \end{aligned} \tag{5-1}$$

by construction. Examination of the right triangle EBC next shows that $BC = d \sin \theta$, so that the path difference in (5-1) can be written

$$\begin{aligned} \Delta &= 2BC \\ &= 2d \sin \theta \end{aligned} \tag{5-2}$$

It is quite easy to see that if the ray $O'CP$ scattered by the second plane is to arrive at $PP'P''$ in phase with the rays $O'AP$ and OEP' scattered by the first plane, that is, if the two planes are to scatter in phase with each other, the path difference Δ must equal an integral number of wavelengths $n\lambda$, where $n = 0, 1, 2, 3, \ldots$. Thus the condition for in-phase scattering by a set of regularly spaced parallel planes in a crystal can be written

$$n\lambda = 2d \sin \theta \tag{5-3}$$

This is the relation which Bragg used to explain the diffraction experiments of Laue, and it has become common practice to use the terms x-ray reflection and diffraction interchangeably. It should be realized, however, that true reflection (interference in two dimensions rather than in three dimensions) is not limited to the specific angles determined by the Bragg law (5-3) for the diffraction of x rays by crystals.

Recasting (5-3) to solve for the diffraction angle,

$$\theta = \sin^{-1} \frac{n\lambda}{2d} \tag{5-4}$$

shows that, for a given wavelength of the x-ray beam, diffraction is possible only at certain angles determined by the interplanar spacings d. Recalling from Chap. 3 that the magnitude of the spacings d is determined by the relative dimensions of the crystal's lattice, it follows that *the lattice of a crystal determines the diffraction angles.*

To see what effect the atomic arrangement in the crystals has on the total x-ray intensity diffracted at these angles, it is necessary to see how the atoms are disposed among the diffracting planes. Consider the relatively simple structure composed of only two kinds of atoms relatively disposed, as shown in Fig. 5-6. The larger atoms all lie in parallel planes (solid lines) spaced d apart, and the small atoms lie in parallel planes (dashed lines), which, in crystals, must also be spaced d apart. Consequently, at some angle θ, given by (5-4), all the large atoms will scatter in phase with each other, and all the small atoms will scatter in phase with each other. Because the pathlengths in Fig. 5-6 of x rays scattered by the two arrays differ by an amount Δ that is less than one wavelength, however, the two atomic arrays do not scatter in phase with each other. This leads to partial, but not complete, destructive interference, so that the amplitude of the diffracted wave is less than what it would be for in-phase scattering by all the atoms. As shown in later chapters, the intensity of the diffracted beam is directly proportional to the square of the net amplitude of the scattered rays, so that it can be deduced from Fig. 5-6 and the foregoing discussion that *the intensity is determined by the actual atomic array, or crystal structure.*

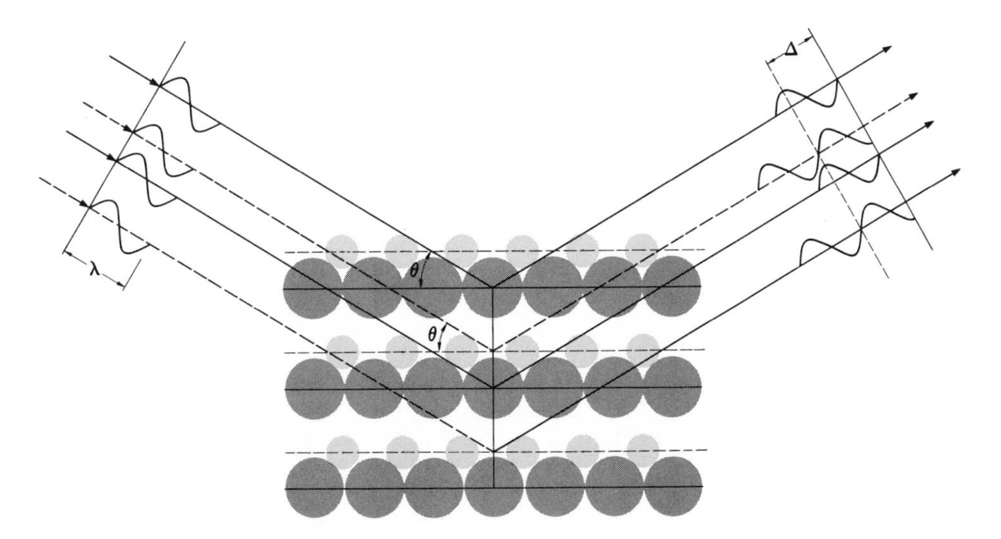

Fig. 5-6

Determination of x-ray wavelengths. An examination of Fig. 5-3 shows that if the angles for corresponding maxima in each of the two triplets of characteristic lines shown are substituted in the Bragg equation (5-3), pairs of simultaneous equations result which can be satisfied by assigning $n = 1$ and $n = 2$ to the two sets of lines (Exercise 5-1). This is an example of the familiar first-order and second-order diffraction of radiation. Unlike the field of spectroscopy, however, in x-ray diffraction studies it is conventional to incorporate the order integer n with the interplanar spacing d by dividing both sides of (5-3) by n.

$$\lambda = 2\frac{d}{n}\sin\theta \tag{5-5}$$

The ratio d/n has a very simple meaning in crystallography because it is possible to assign the indices (nh,nk,nl) to fictitious planes whose spacing is d_{hkl}/n (Exercise 5-2). These new planes are said to be fictitious because they do not pass through lattice points, but, as shown in later chapters, this representation is very useful in the interpretation of x-ray diffraction experiments. In fact, it is normal practice to omit n entirely from the equation. To distinguish this new form of the Bragg equation it is customary to write it

$$\lambda = 2d_{hkl}\sin\theta$$

or $$\lambda = 2d\sin\theta_{hkl} \tag{5-6}$$

in order to indicate that the diffraction is by the specific planes (hkl) or at the particular angle θ_{hkl}, often called the *Bragg angle* of the plane (hkl). The choice of

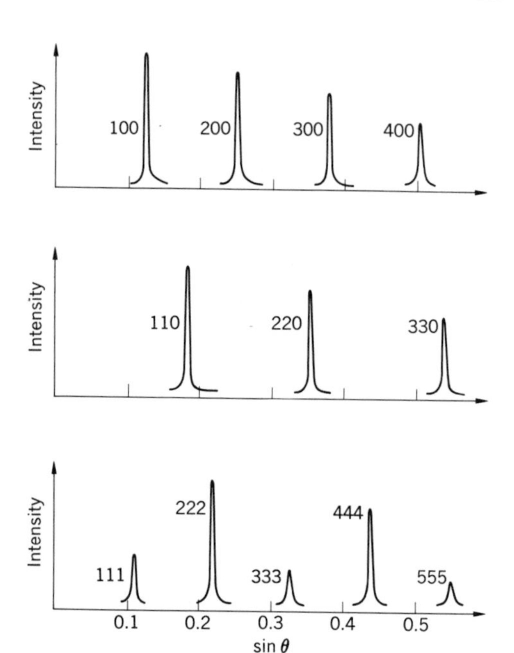

Fig. 5-7. Diffraction intensities (schematic) of the first few orders from the (100), (110), and (200) planes of NaCl. As discussed in the text, the indices assigned above to the 100 and 110 reflections are incorrect.

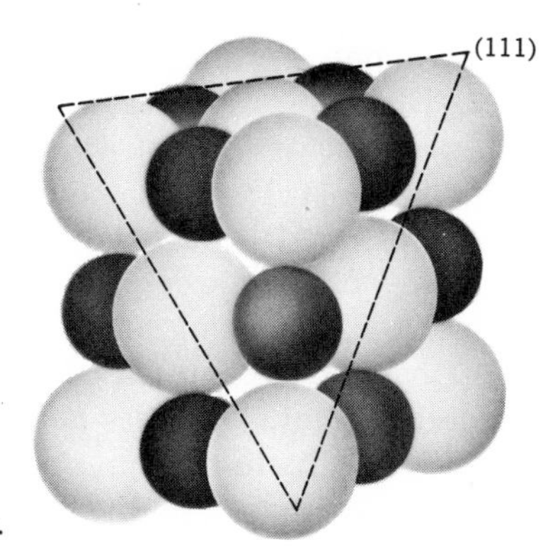

Fig. 5-8. Sodium chloride structure.

using one of the two alternative representations in (5-6) depends on whether one is concerned with relating the measured angle θ to the interplanar spacing d_{hkl}, as is the case in diffraction experiments in which λ is known, or to the unknown wavelength, as is the case in x-ray spectroscopic experiments in which the crystal spacing is known.

When the Braggs first used their spectrometer (Fig. 5-2) to study the spectra of x rays reflected from a single crystal, neither d nor λ was known. To overcome this difficulty, W. L. Bragg prepared three crystals of sodium chloride cut so that they could be aligned readily to reflect from the (100), (110), and (111) planes. The diffraction intensities measured from each crystal are schematically indicated in Fig. 5-7. Note that, unlike the other two sets, the 111 reflections are alternately weak and strong. This suggests that there are alternate atomic planes parallel to (111) in the crystal that are composed of unlike atoms. [The like atoms in each set of (111) planes scatter in phase with each other but are out of phase with the other kind of atom.] Such a structure model for NaCl already had been proposed by W. L. Barlow (Fig. 5-8) and, at the suggestion of Professor Pope of Cambridge University, Bragg tested it against his x-ray measurements. Substituting the $\sin\theta$ values for reflections from (100), (110), and (111) planes given in Fig. 5-7 into the original Bragg equation (5-3),

$$n_1\lambda = 2d_{100}(0.126) \qquad n_2\lambda = 2d_{110}(0.178) \qquad n_3\lambda = 2d_{111}(0.109) \tag{5-7}$$

From the known geometry of the cubic lattice, $d_{100} = \sqrt{2}\, d_{110}$, $d_{100} = \sqrt{3}\, d_{111}$, and $d_{110} = \sqrt{\frac{3}{2}}\, d_{111}$. Forming these ratios between the relations in (5-7) eliminates λ and gives

$$\begin{aligned} \frac{d_{100}}{d_{110}} &= \frac{(0.178)n_1}{(0.126)n_2} = \sqrt{2} \\ \frac{d_{100}}{d_{111}} &= \frac{(0.109)n_1}{(0.126)n_3} = \sqrt{3} \\ \frac{d_{110}}{d_{111}} &= \frac{(0.109)n_2}{(0.178)n_3} = \sqrt{\frac{3}{2}} \end{aligned} \tag{5-8}$$

An examination of these simultaneous equations shows that the last two ratios can be satisfied only if $n_1 = 2$, $n_2 = 2$, and $n_3 = 1$. This indicates that the observed reflections come from planes whose indices are all even, like (200) and (220), or all odd, like (111), so that the indices assigned to the maxima in the upper two graphs of Fig. 5-7 must be doubled. Recalling from Exercise 3-12 that this is a consequence of indexing on the basis of a face-centered rather than a primitive lattice, it is interesting to note that an examination of Fig. 5-7 indeed shows that NaCl has a face-centered cubic lattice with two atoms, one Na and one Cl atom per lattice point. This has been, therefore, an example of how Bragg was able to deduce the structures of crystals for the very first time, a contribution that has had an enormous impact on the development of modern science.

Once the crystal structure was established, it was a simple matter to make use of the known density of NaCl and relation (4-7) to determine the size of the unit-cell volume $a^3 = (2d_{200})^3$ and, finally, the wavelength of the x rays used (Exercise 5-3). It turned out that the x-ray wavelength, as expected, was of the order of 10^{-8} cm, so that it is convenient to define a new unit, called an *angstrom unit,* $1\ \text{Å} = 10^{-8}$ cm. In order to fix it on a standard scale, Bragg suggested that the unit-cell edge of NaCl be standardized at $a = 5.628\ \text{Å}$. This, in turn, permitted the measurement of the wavelengths of the characteristic x rays emitted by different target metals. By 1930 it became possible to measure x-ray wavelengths using ruled gratings, and it soon became apparent that a systematic difference of about 0.2% existed between the same wavelengths measured in the two different ways. In order to reconcile the two kinds of measurements, it was later decided that the cell edge of NaCl should be set equal to 5,628 x units, or 5.628 kxu (kilo-x-units). Accordingly, an angstrom unit (10^{-8} cm) is related to the x units by

$$1\ \text{Å} = 1.00202\ \text{kxu} \tag{5-9}$$

It should be noted, therefore, that publications preceding 1950 may present values labeled angstroms when they really are kxu.

SUGGESTIONS FOR SUPPLEMENTARY READING

G. E. Bacon, *X-ray and neutron diffraction* (Pergamon Press, Oxford, England, 1966).

P. P. Ewald, *Fifty years of x-ray diffraction* (N.V.A. Oosthoek's Uitgeversmaatschappij, Utrecht, The Netherlands, 1962).

EXERCISES

5-1. Read from the curve in Fig. 5-3 the diffraction angles at the centers of each of the three peaks in the two groups of characteristic lines shown. Making use of the Bragg law, show that these are the first- and second-order reflections from the same planes in the crystal. Can you determine either λ or d from this knowledge? How?

5-2. By preparing suitable sketches, show that a set of planes parallel to (110) and spaced $d_{110}/2$ apart produces a new set of fictitious planes (220). Repeat for the planes (330), (420), and (400). **Hint:** Note that these planes can be represented easily in two dimensions by an edge view along the c axis.

5-3. One formula weight of NaCl equals 58.45 amu, and the density of a NaCl crystal is 2.15 g/cm^3. What is the length of a cubic cell edge? What is the length of the x-ray wavelength responsible for the diffraction maxima in Fig. 5-7?

5-4. The wavelength of characteristic x rays from a copper target is 1.54 Å. Prepare a list of the Bragg angles for the first six reflections from a NaCl crystal, $a = 5.63$ Å. **Hint:** Remember that the lattice is face-centered cubic, and consider all allowed values of hkl.

5-5. The structure of aluminum is cubic, with one atom per lattice point of a face-centered cubic lattice. Would you expect the intensity of alternate 111 reflections to be nearly equal or similar to the face-centered cubic structure of NaCl? Why?

5-6. One of the early and still popular applications of x-ray diffraction has been to the determination of the correct molecular weights (compositions) of complex molecules. The density of $(CH_3 \cdot C_6HSO_3)Mg \cdot nH_2O$ is 1.42 g/cm^3, and the monoclinic unit cell has $a = 25.2$ Å, $b = 6.26$ Å, $c = 6.95$ Å, $\beta = 91°54'$. If the number of water molecules n lies between 4 and 8, what is the correct formula of magnesium p-toluene sulfonate hydrate? **Hint:** The number of molecules in one unit cell must be integral.

5-7. Using the results of Exercise 5-6, calculate what the density of magnesium p-toluene sulfonate hydrate should be for the correct number of water molecules. This value should turn out to be slightly larger than the measured one, and is normally called the x-ray density. Why does one expect experimentally measured densities to be smaller?

SIX
Properties of x rays

SCATTERING OF X RAYS

The visible light that surrounding objects scatter in the direction of an observer enables him to see the world around him. In a similar way the scattering of x rays by matter discloses the atomic arrays within that matter. Because the scattering process is governed by the energy of the electromagnetic radiation being scattered, a study of the scattering process helps elucidate the properties of the radiation involved. Albert Einstein demonstrated early in the twentieth century that all electromagnetic radiation has the dual property of behaving like a wave *and* like a particle. This dual nature becomes particularly prominent in the case of x radiation. As shown below, an individual electron can interact with an x-ray beam whose electromagnetic-wave field imparts to the electron a certain acceleration. The accelerated electron, in turn, becomes a source of electromagnetic radiation that is identical with that incident on it. This kind of interaction, therefore, is called *coherent scattering.* Another kind of interaction between an electron and an x-ray photon is possible. This interaction can be likened to a collision between two billiard balls in which the energy and momentum of each ball change. Because the x-ray quantum participating in this process has its energy modified by the collision, it produces *incoherent scattering.*

Still a third kind of interaction between electromagnetic radiation and matter is possible. Suppose that an incident photon has sufficient energy to eject an inner electron from an atom. (If the ejected electron remains inside the material, this is called the internal photoelectric effect.) The energy of the atom is raised by an amount equal to the work done in ejecting its electron, so that, when it returns to its initial state by recapturing an electron, a photon of corresponding energy must be emitted. Because the time elapsed between these two events is of the order of only 10^{-16} sec, it appears as if this also is a "scattering" process. The x rays radiated in all directions by this means have energies that are strictly characteristic of the emitting atoms, so that this process is called fluorescence scattering, a name derived from the generic term *fluorescence* that is applied to all processes in which characteristic electromagnetic radiation is emitted by a substance simultaneously with the external excitation.

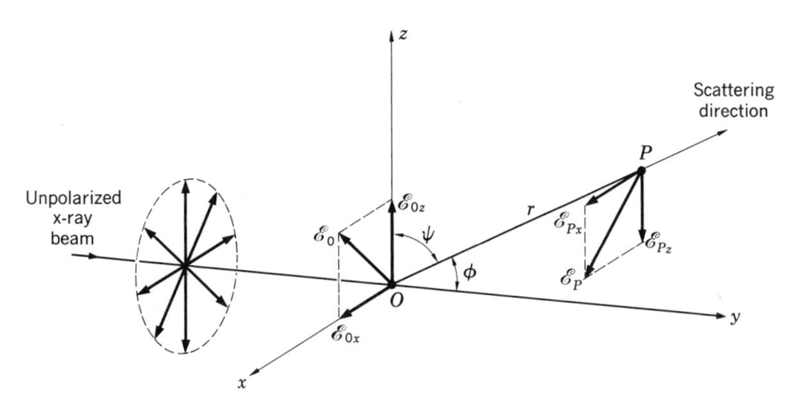

Fig. 6-1

Coherent scattering. Consider an unpolarized x-ray beam propagating along the y direction, as shown in Fig. 6-1. (By unpolarized it is meant that the electric vector $\mathcal{E}$ has the same magnitude in all directions transversely to the y axis.) Suppose that it encounters an electron of charge e and mass m at O. If the electron is moving slowly compared with the velocity of light c, it will receive an acceleration a from the electric field, which can be calculated with the aid of Newton's first law of motion.

$$\mathbf{a} = \frac{\mathbf{F}}{m} = \frac{e}{m}\mathcal{E} \tag{6-1}$$

According to J. C. Maxwell's classical electrodynamic theory, the accelerated electron must, in turn, radiate in all directions a transverse wave whose velocity

$$c = \frac{\text{electromagnetic unit of charge}}{\text{electrostatic unit of charge}} \tag{6-2}$$

so that the resultant field reaches some point P (Fig. 6-1) after a time r/c.

Following an analysis first carried out by Thomson, consider the vector $\mathcal{E}_0$ acting on the electron at O. The components of this vector along the x and z axes are indicated in Fig. 6-1. The transverse wave propagating toward point P, lying in the yz plane, can be similarly analyzed. In the scattered beam, the electric vector parallel to x,

$$\mathcal{E}_{Px} = \frac{\mathbf{a}e}{rc^2} = \frac{\mathcal{E}_{0x}e^2}{rmc^2} \tag{6-3}$$

according to Maxwell's equations. The component lying in the yz plane also must be perpendicular to OP, so that

$$\mathcal{E}_{Pz} = \frac{\mathbf{a}e}{rc^2}\sin\psi = \frac{\mathcal{E}_{0z}e^2}{rmc^2}\sin\psi \tag{6-4}$$

or, expressing it in terms of the angle ϕ formed with the incident beam's direction,

$$\mathcal{E}_{Pz} = \frac{\mathcal{E}_{0z}e^2}{rmc^2}\cos\phi \qquad (6\text{-}5)$$

The energy per unit volume flowing past the point P is proportional to

$$\mathcal{E}_P{}^2 = \mathcal{E}_{Px}{}^2 + \mathcal{E}_{Pz}{}^2 = \frac{e^4}{r^2m^2c^4}(\mathcal{E}_{0x}{}^2 + \mathcal{E}_{0z}{}^2\cos^2\phi) \qquad (6\text{-}6)$$

For an unpolarized incident beam, however, the average magnitude of the field in all directions is the same, so that

$$\bar{\mathcal{E}}_{0x}{}^2 = \bar{\mathcal{E}}_{0z}{}^2 = \tfrac{1}{2}\bar{\mathcal{E}}_0{}^2 \qquad (6\text{-}7)$$

and $$\bar{\mathcal{E}}_P{}^2 = \frac{\bar{\mathcal{E}}_0{}^2e^4}{r^2m^2c^4}\left(\frac{1+\cos^2\phi}{2}\right) \qquad (6\text{-}8)$$

It is more convenient to deal with the intensity of the beam $I = (c/4\pi)\mathcal{E}^2$, which measures the energy crossing a unit area per unit time. Multiplying both sides of (6-8) by $c/4\pi$ then gives the familiar Thomson equation for the intensity of coherent scattering by one electron,

$$I_e = I_0\frac{e^4}{r^2m^2c^4}\left(\frac{1+\cos^2\phi}{2}\right) \qquad (6\text{-}9)$$

The quantity in parentheses in (6-9) is called the *polarization factor,* since it is related to the change in magnitude of the electric vector lying in the plane containing the scattering angle ϕ. In other words, the scattered beam no longer has $\mathcal{E}$ of equal magnitude in all directions about its propagation direction, and is said to be polarized. Note that the Thomson equation (6-9) predicts an intensity distribution that has rotational symmetry about the incident-beam direction, and mirror symmetry about a plane perpendicular to this direction.

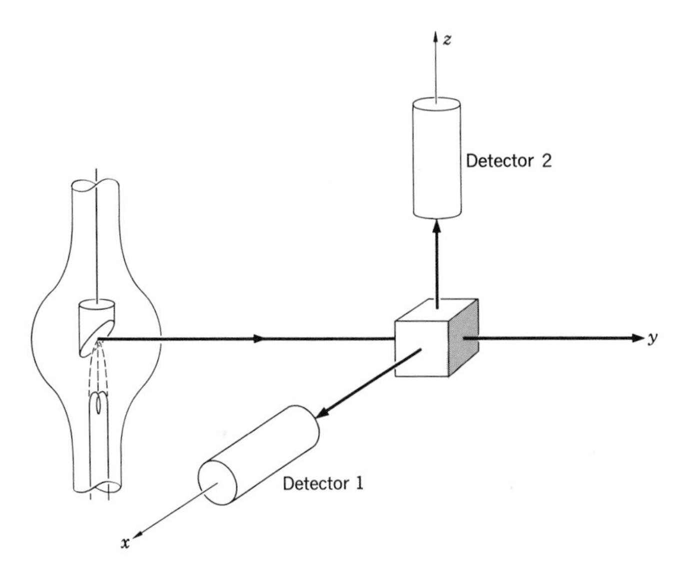

Fig. 6-2

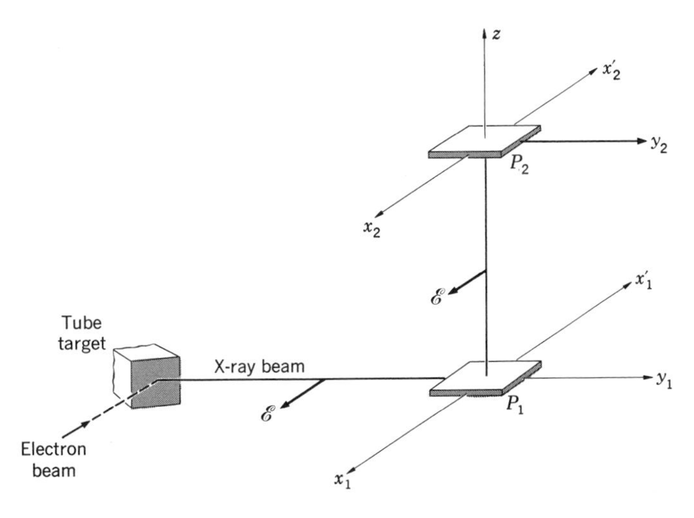

Fig. 6-3

Polarization of x rays. To test the polarizability of x rays, Barkla used an arrangement in which the scattered intensity of the x-ray beam was measured in two mutually orthogonal directions, as shown in Fig. 6-2. The x-ray tube first was positioned so that the flow of electrons in the tube was parallel to the z axis, after which the experiment was repeated with the tube axis parallel to x. Denoting the intensity of the x-ray beam scattered in a parallel direction to the tube axis by $I_{\parallel}$ (detector 2 in Fig. 6-2) and that at right angles (detector 1) by $I_{\perp}$, it is possible to define a polarization P by

$$P = \frac{I_{\perp} - I_{\parallel}}{I_{\perp} + I_{\parallel}} \tag{6-10}$$

Barkla found that the primary beam was partly polarized, with its maximum magnitude lying parallel to the tube axis. This was a most important discovery, because it demonstrated conclusively that x rays were a form of electromagnetic radiation (waves), a fact which had been suspected but had not been proved previously. A spectral analysis of (6-10) shows that polarization is complete at the shortest-wavelength x rays emitted, but declines as the wavelength increases. (The characteristic line spectrum from the target is always unpolarized.)

In a refinement of this experiment, A. H. Compton and C. F. Hagenow used two vanishingly thin metal foils, positioned as shown in Fig. 6-3, and carefully collimated beams. According to Barkla's experiment, the primary x-ray beam produced in an x-ray tube whose axis is parallel to x should be polarized so that its electric vector $\mathcal{E}$ is also parallel to x. This means that the electrons in the thin metal foil at P_1 can be accelerated only along $x_1x'_1$, so that a scattered beam should be observable in the z direction but not parallel to x. Since the primary x-ray beam is not completely polarized, the electric field may have some component along z, and one would expect to observe merely a much larger intensity along z than along x, as originally observed by Barkla. Consider next the x-ray beam scattered along

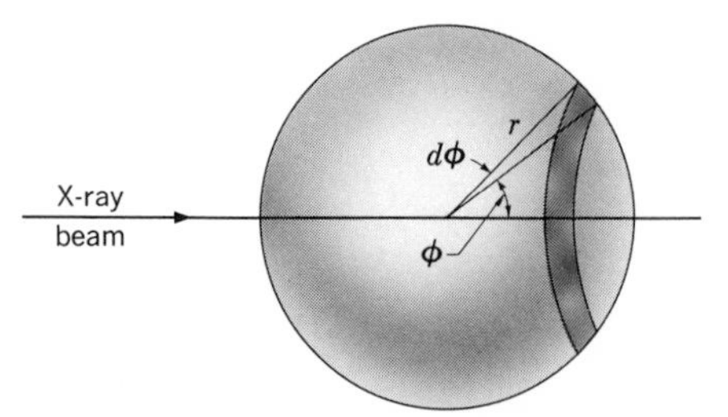

Fig. 6-4

z. The electrons in the foil placed at P_2 can be accelerated only parallel to x_2x_2' since the beam scattered in this direction can have $\mathcal{E}$ parallel to x only. Thus the intensity of the twice-scattered beam should go from a maximum along y to zero along x_2x_2'. This was actually observed in the experiments of Compton and Hagenow within an experimental error of less than 2%. This is an important result because it serves to validate the electromagnetic-wave treatment of Thomson and clearly demonstrates the polarization of x rays accompanying coherent scattering.

Scattering coefficients. The Thomson equation tells the intensity I_e of the x-ray beam scattered by an electron at an angle ϕ to the incident-beam direction. The total power scattered by one electron P_e can be determined next by measuring the intensity crossing a surface completely surrounding the electron, that is, by integration over the surface of a sphere of radius r. Consider the annular ring shown shaded in Fig. 6-4. The area of this ring is given by the product of the circumference $2\pi r \sin \phi$ with its width $r\,d\phi$, so that the total power is given by

$$\begin{aligned} P_e &= \int_0^\pi I_e \times 2\pi r^2 \sin \phi \, d\phi \\ &= I_0 \frac{\pi e^4}{m^2c^4} \int_0^\pi (1 + \cos^2 \phi) \sin \phi \, d\phi \\ &= I_0 \frac{8\pi e^4}{3m^2c^4} \end{aligned} \tag{6-11}$$

after substituting (6-9) for I_e and evaluating the definite integral.

Dividing both sides of (6-11) by the incident beam's intensity gives a constant

$$\sigma_e = \frac{P_e}{I_0} = 6.66 \times 10^{-25}\ \text{cm}^2 \tag{6-12}$$

called the *classical scattering cross section* or *coefficient* for an electron. Assuming that a material contains n electrons per unit mass that scatter x rays independently of each other and according to the Thomson relation (6-9), it is possible to define a *mass scattering coefficient.*

$$\sigma_m = n\sigma_e \tag{6-13}$$

It should be realized that electrons do not scatter x rays independently of each other, and the Thomson theory also fails to take quantum effects into account. Nevertheless, a fair agreement between the value in (6-12) and the actual scattering by light elements is observed. (See Exercise 6-12.) More pronounced deviations from the above value are observed when the total power scattered by heavy atoms is measured. The deviations from Thomson's classical theory become still more pronounced when x rays having large energies ($\lambda < 0.2$ Å) are employed. Fortunately, the energies of x rays used in diffraction experiments are sufficiently low so that it is possible to use the Thomson equation after correction for quantum effects, considered next.

Incoherent scattering. Suppose a carefully monochromatized beam of x rays is scattered by a block of carbon at some angle ϕ. An analysis of the spectral composition of the scattered beam shows that it consists of two distinct components, like those in Fig. 6-5. One component is fairly sharp and has the same energy (wavelength) as the incident beam, whereas the second component is broader and has a longer wavelength. In order to explain the origin of the second component, Compton boldly applied the then newly emerging quantum theory to the x-ray scattering problem.

Suppose an x-ray quantum of energy $E_0 = h\nu_0 = hc/\lambda_0$ is incident on a "free" electron of mass m that is assumed to be initially at rest. If both the electron and the x-ray photon can be considered to be particles, then energy and momentum should be conserved in the resulting collision. The energy of the photon scattered at some angle ϕ from its original direction (Fig. 6-6) is decreased to a new energy $E_\phi = h\nu_\phi$, while the resulting kinetic-energy increase of the electron E_e is given by the relativistic formula

$$E_e = mc^2\left(\frac{1}{\sqrt{1 - \beta^2}} - 1\right) \tag{6-14}$$

where $\beta = v/c$ expresses the ratio between the electron's velocity v and the velocity

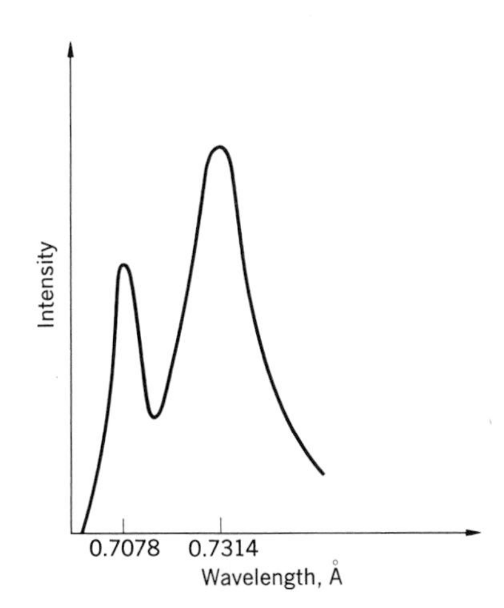

Fig. 6-5. Spectrum of an initially monochromatic x-ray beam ($\lambda = 0.7078$ Å) after scattering by carbon. (*After Compton.*)

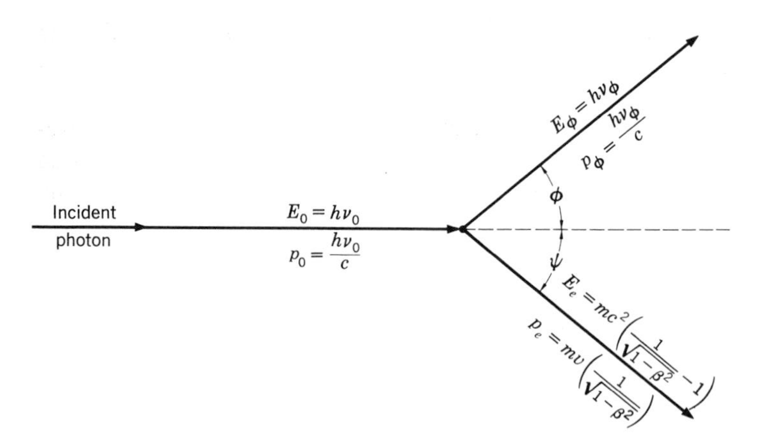

Fig. 6-6

of light c. The law of conservation of energy then requires that

$$E_0 = E_\phi + E_e \tag{6-15}$$

Similarly, the momentum of the photon before and after collision can be expressed by $p_0 = h\nu_0/c$ and $p_\phi = h\nu_\phi/c$, respectively, while that of the electron after collision is $p_e = mv(1 - \beta^2)^{-\frac{1}{2}}$. The conservation of momentum (a vector quantity) can be analyzed in terms of the component parallel to the initial direction of the photon, which, according to Fig. 6-6, is

$$\frac{h\nu_0}{c} = \frac{h\nu_\phi}{c} \cos \phi + \frac{mv}{\sqrt{1 - \beta^2}} \cos \psi \tag{6-16}$$

and perpendicular,

$$0 = \frac{h\nu_\phi}{c} \sin \phi + \frac{mv}{\sqrt{1 - \beta^2}} \sin \psi \tag{6-17}$$

It is left to Exercise 6-3 to show that the simultaneous satisfaction of relations (6-15) to (6-17) leads to the relation that the wavelength of the scattered photon λ differs from that of the original photon λ_0 by

$$\Delta\lambda = \lambda - \lambda_0 = \frac{h}{mc}(1 - \cos \phi) \tag{6-18}$$

Expressing the Compton wavelength shift in angstroms,

$$\Delta\lambda = 0.0243(1 - \cos \phi) \qquad \text{Å} \tag{6-19}$$

According to the foregoing analysis, the collision of an x-ray photon with a really free electron causes the photon to be scattered at an angle ϕ with an increase in wavelength that is a function of the scattering angle. Because the electron is normally not at rest, its initial momentum must be included in Eqs. (6-16) and (6-17), leading to $\Delta\lambda$ values slightly different from those predicted by (6-18), which can be thought of as predicting the most probable Compton shift. This is, quite obviously, the reason why the second component in the scattered radiation (Fig. 6-5), called the *Compton modified line,* is normally broader than the unmodified line.

Relation (6-18) also can be derived directly from quantum mechanics, as was first pointed out by E. Schrödinger. It predicts that the x rays scattered at right angles to the primary beam will always have wavelengths that are 0.0243 Å longer than those of the original photons, regardless of the initial wavelength or the nature of the scatterer. The general validity of this relation for x rays has been proved by several investigators, who confirmed the wavelength shift of the modified line for various scattering angles, and others who measured the energy and momenta of the recoil electrons and found them to agree with values predicted by relations (6-16) and (6-17). A more detailed analysis, based on quantum-mechanical considerations of the forces that bind electrons to the nuclei in atoms, however, shows that an energy dependence should exist. Thus, for incident photons of relatively low energy (long wavelength) like those in visible light, the electrons cannot be considered to be really free, and collisions of the type considered in this section are unlikely. Consequently, low-energy radiation is scattered coherently. As the wavelength of the incident photon decreases to a size commensurate with atomic dimensions, coherent scattering decreases, particularly at large angles, and the modified line grows in prominence. Finally, at very short wavelengths (high-energy x rays or γ rays), Compton scattering becomes the dominant scattering mechanism.

The polarization is the same for modified and unmodified scattered x rays. The main difference is in the scattering mechanisms. The modified x rays bear no relation of specific wavelength or phase to each other or to the primary beam, so that they are scattered incoherently and cannot build up constructive interferences like those required by Bragg's law. The unmodified x rays, on the other hand, can be thought of as generated by identical oscillatory motions of electrons induced by the incident electromagnetic wave field. The unmodified x rays therefore bear definite phase relations to each other and the incident beam, and all have the same wavelength, so that constructive interference or diffraction is possible only for them. Because the modified x rays do not form diffraction maxima, they contribute instead a uniformly varying background intensity, as discussed further in later chapters.

Fluorescence. As discussed in the preceding chapter, Röntgen had noticed that when a fluorspar or fluorite crystal, CaF_2, was placed behind a film being irradiated with x rays, the exposure was increased. He concluded from this that the x rays transmitted through the film were scattered backward by the fluorite crystal. Shortly thereafter, Straubel and Winkelmann showed that the radiation emanating from the crystal was more easily absorbed than the primary radiation, and called it fluorspar radiation. A more exacting experiment was performed subsequently by Barkla, who let a well-collimated beam of x rays fall on a scatterer and measured the attenuation in the scattered beam reaching his detector (Fig. 6-7) when a thin absorber was placed alternately in positions 1 and 2. Barkla found no difference when the scatterer was a light element like carbon, in agreement with the Thomson scattering equation. (The slightly increased attenuation of modified x rays in position 2 was beyond the detectability of Barkla's instrumentation.) When the

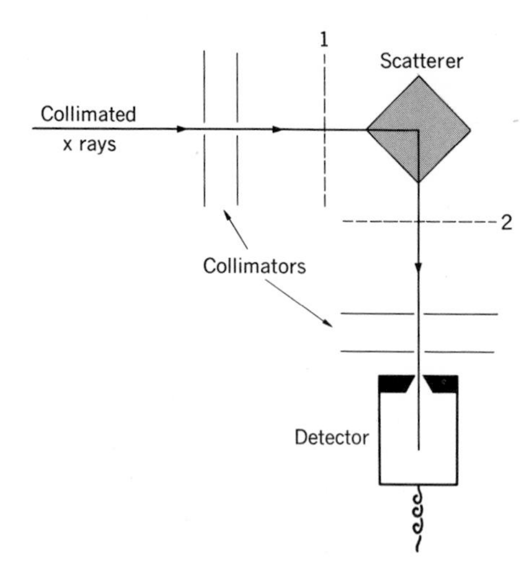

Fig. 6-7

scatterer was a heavier element like copper, however, Barkla observed a definitely increased attenuation in position 2. From this he concluded that the heavier scatterer gave rise to secondary x rays that were softer or less penetrating than the primary scattered rays. By repeating this experiment using various scatterers, he found that the penetrating power, or *hardness,* of the secondary rays was characteristic of the scatterer. As discussed in a later section, this characteristic, or fluorescence radiation, is produced inside an atom and, strictly speaking, is not scattered radiation at all. It is given off by an atom uniformly in all directions and, like Compton modified radiation, can contribute to the background intensity but does not partake in constructive interference.

To further his understanding of the phenomenon he was observing, Barkla used foils made from different metals, as well as a series of successively heavier scatterers. By comparing the attenuation in, say, a foil of copper placed in front of the detector (position 2 in Fig. 6-7) with that in an aluminum foil placed in the same position, it is possible to construct a curve similar to the one shown in Fig. 6-8. As the hardness of the fluorescence radiation increases (from left to right in Fig. 6-8), the attenuation in copper decreases slightly more rapidly than that in aluminum. The lowest point occurs when radiation from the same metal as the foil falls on the foils. When the fluorescence radiation is slightly more penetrating than that of a

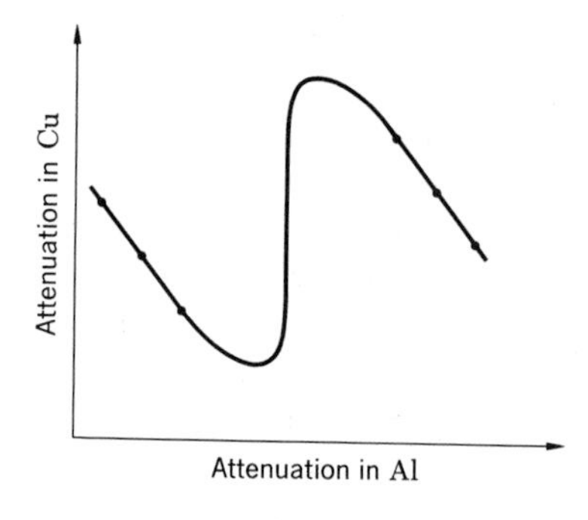

Fig. 6-8. Experimental curve showing the relative attenuation factors of a copper foil and an aluminum foil for different scatterers in an arrangement like that in Fig. 6-7. The successive points are plotted in order of increasing atomic number of the scatterers.

copper scatterer (in Fig. 6-8), the attenuation in the copper foil increases abruptly, without a corresponding increase in the aluminum foil. The same kind of behavior is observed in an iron foil when fluorescence radiation slightly more penetrating than that from an iron scatterer is employed, and so forth. This series of measurements clearly shows that fluorescence radiation is characteristic of the irradiated material producing it, and suggests several interesting possibilities regarding the absorption of x rays by matter, considered next.

ABSORPTION OF X RAYS

Classical theory. When an x-ray beam passes through a material, the transmitted beam is attenuated by the scattering of x rays in all directions by electrons present in the material. In order to estimate the size of the loss, suppose the material contains n electrons per gram of mass and, for convenience, has a unit cross-sectional area perpendicular to the beam and a thickness x parallel to it (Fig. 6-9). Consider next a slice (shaded in Fig. 6-9) of thickness dx and volume

$$dV = 1 \times 1 \times dx \qquad \text{cm}^3$$

If the density of the material is ρ, the slice contains a total of $n\rho\, dV$ electrons. The total scattered radiation going out in all directions from this slice of volume dV is

$$dP = I\sigma_e n\rho\, dV \tag{6-20}$$

The decrease in the transmitted beam's intensity, dI, due to scattering by this slice,

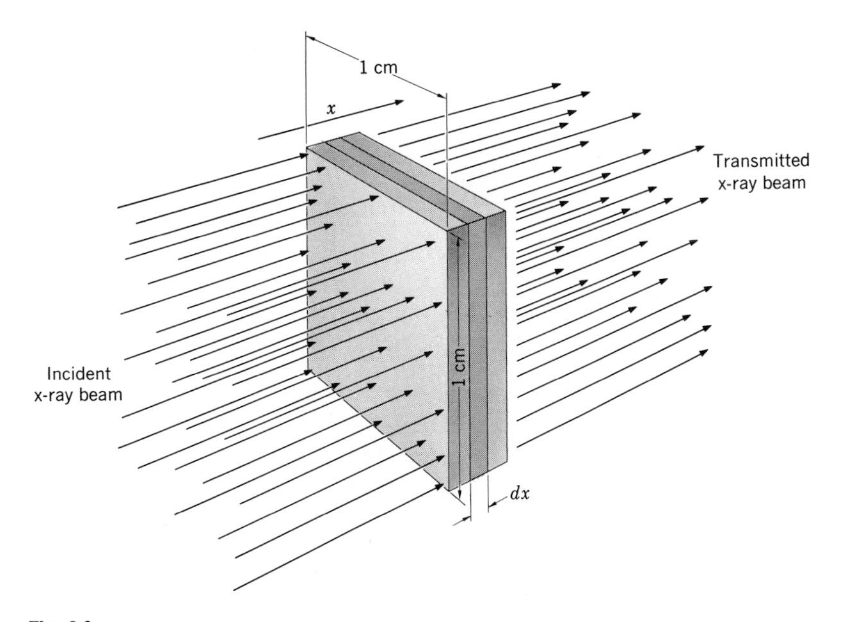

Fig. 6-9

then is

$$dI = -I\sigma_e n\rho(1)\,dx \tag{6-21}$$

where the minus sign denotes that the intensity is subtracted from that of the direct beam.

Combining like terms in (6-21) and integrating both sides give

$$\int \frac{dI}{I} = -\int \sigma_e n\rho\,dx$$

$$\ln I = -\sigma_e n\rho x + \text{const} \tag{6-22}$$

The constant of integration in (6-22) can be evaluated by noting that, at the front surface in Fig. 6-9, when $x = 0$, the incident-beam intensity is I_0, so that

$$\ln I_0 = 0 + \text{const}$$

Substituting this value in (6-22) and rearranging the terms yield the relation

$$\frac{I}{I_0} = e^{-\sigma_m \rho x} \tag{6-23}$$

after substituting the mass scattering coefficient σ_m for $n\sigma_e$ in accord with (6-13).

The relation (6-23) has the form of the classical absorption equation, which states that the fraction of the incident beam's intensity I_0 that is transmitted through a substance of thickness x is inversely proportional to an exponential raised to the power of x times a linear attenuation coefficient [$= \sigma_m \rho$ in Eq. (6-23)]. In actual fact, an x-ray beam passing through any material is attenuated by other processes in addition to the coherent scattering considered above, namely, Compton modified scattering and fluorescence.† Lumping all these factors together, it is possible to define a linear absorption coefficient μ_l in terms of a mass absorption coefficient μ_m.

$$\mu_l = \mu_m \rho \tag{6-24}$$

where ρ is the density of the material. The absorption relation (6-23) then can be written in its more familiar form,

$$\frac{I}{I_0} = e^{-\mu_l x} \tag{6-25}$$

Suppose a material is composed of several different kinds of elements whose fractional parts (by weight) are $f_1, f_2, f_3, \ldots$, so that $\sum_i f_i = 1$. Next, suppose that the total irradiated material has a mass per unit area $M = \rho x$ g/cm². The various elements composing this material can be thought of as forming successive layers that are transverse to the incident-beam direction (like the shaded layer in Fig. 6-9), each containing a mass per unit area $m_i = f_i M$. Since the order in which the different atoms absorb the incident x-ray beam should not affect the

† In addition, energy can be lost by the process of pair production. Because the formation of positron-electron pairs requires rather large energies, this process can be neglected for the x-ray energies used in diffraction studies.

total absorption, suppose that all the atoms for which $i = 1$ form a layer of thickness x_1, all atoms denoted by $i = 2$ form a layer of thickness x_2, etc. According to (6-25), the intensity transmitted through the first layer then is

$$I_1 = I_0 e^{-\mu_{m_1} f_1 M} \tag{6-26}$$

where μ_{m_1} is the mass absorption coefficient for element $i = 1$. I_1, however, is the intensity of the beam incident on the second layer, containing the mass m_2, etc., so that after transmission through all the layers constituting the total thickness x of the material,

$$\begin{aligned} I &= I_0 e^{-\mu_{m_1} f_1 M} e^{-\mu_{m_2} f_2 M} e^{-\mu_{m_3} f_3 M} \cdots \\ &= I_0 e^{-(f_1 \mu_{m_1} + f_2 \mu_{m_2} + f_3 \mu_{m_3} + \cdots) \rho x} \end{aligned} \tag{6-27}$$

The quantity in parentheses in the exponent in (6-27) is the mass absorption coefficient of the material, which is equal to the sum of the mass absorption coefficients of each constituent element multiplied by the mass fraction of that element present (Exercises 6-4 and 6-5). The mass absorption coefficient then is multiplied by the density of the material, which takes into account the state of aggregation of the elements composing the material, and yields the linear absorption coefficient (6-24) needed to calculate the final reduction in intensity in accord with (6-25). It should be noted here that the mass absorption coefficient of an element is a function of the energy of the incident x rays, as discovered by Barkla (Fig. 6-8), so that appropriate values must be used when applying relation (6-27).

Energy dependence. In order to determine the energy dependence of absorption by an element, a thin absorber of pure metal can be placed in front of the detector in an x-ray spectrometer, and the ratio I/I_0 for a highly monochromatic beam measured at a succession of wavelengths. This is carried out most easily on an x-ray spectrometer, although quantitative measurements require instrumentation that is more exacting than conventional equipment. Defining an atomic absorption coefficient,

$$\mu_a = \frac{A}{N_0} \mu_m \tag{6-28}$$

where A is the atomic weight of the element and N_0 is Avogadro's number, the variation of the absorption coefficient with wavelength is shown for lead atoms in Fig. 6-10. Two characteristics of this curve are noteworthy: the discontinuities are called *absorption edges,* and occur at wavelengths that are characteristic of the absorbing element. Barkla supposed that they are related to the energies required to knock out K, L, etc., electrons from the atom, and labeled the edges accordingly. The second feature is that the gradual rise in the absorption curve appears to be proportional to the third power of the wavelength. Replotting the atomic absorption coefficient of lead as a function of λ^3 does yield straight lines, as shown in Fig. 6-11. The dependence of the absorption coefficient on wavelength can be described, therefore, by a linear equation,

$$\mu_a = k\lambda^3 + b \tag{6-29}$$

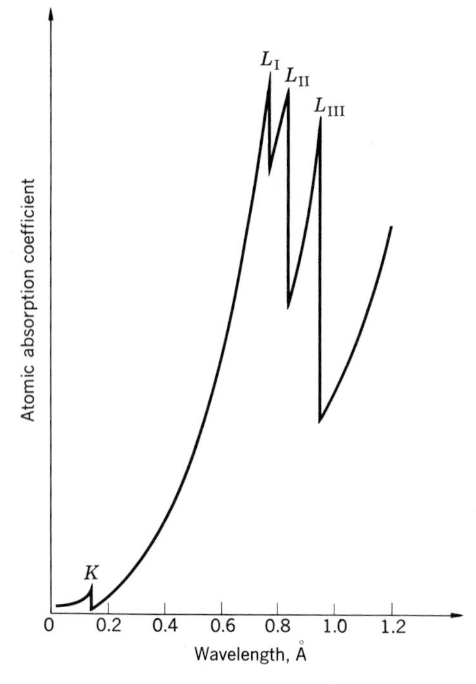

Fig. 6-10. Variation of the atomic absorption coefficient of lead as a function of incident x-ray wavelength. (*After Richtmeyer.*)

where k is the slope of the straight-line portions and takes on different values on either side of the absorption edge, and b is the constant intercept that both lines appear to make on the vertical axis.

When similar measurements are made for other kinds of atoms, similar curves are obtained, except that the wavelengths at which the absorption edges occur are characteristically different for each element. When the atomic absorption coefficients of a series of elements, measured at a constant x-ray wavelength, are plotted against their atomic numbers, a linear relationship appears between μ_a and Z^4. This suggests that $k \propto Z^4$ in (6-29), so that it is possible to express an empirical dependence by

$$\mu_a = \begin{cases} c_a\lambda^3Z^4 + b & \text{for } \lambda < \lambda_k \\ c_a'\lambda^3Z^4 + b' & \text{for } \lambda > \lambda_k \end{cases} \tag{6-30}$$

where $c_a = 2.25 \times 10^{-2}$ cm and $c_a' = 0.33 \times 10^{-2}$ cm when λ is measured in centimeters. The constants b and b' have nearly, but not exactly, the same values, and are usually equated to the atomic scattering coefficient σ_a. This is only approximately correct since the coherent scattering of x rays is independent of energy only for relatively long wavelengths (low energies). The magnitude of b, moreover, is quite small at most wavelengths of practical interest when compared with the second term, which is called the fluorescence term because it is related to the production of fluorescence radiation, as shown in a later section. It should be noted here that, although the relations (6-30) are violated very close to the absorption edge, their universality permitted E. Jönsson to construct an empirical curve which allows one to calculate the atomic absorption coefficient of any element at any wavelength. (See Exercise 6-8.) The values of mass absorption coefficients for all elements at selected wavelengths are given in Table A-5 (Appendix 2), and the wavelengths of their K and L edges in Tables A-3 and A-4.

Absorption edges. Having established the empirical dependence of the absorption coefficient on wavelength, it remains to explain the origin of the abrupt discontinuities evident in Fig. 6-10. Consider the modern concept of the structure of an atom in which a central positively charged nucleus is surrounded by Z negative electrons traveling in more or less defined orbitals. The electrons can be grouped according to the energy of attraction that binds them to the nucleus. Barkla called the electrons most tightly bound to the nucleus the K electrons, the next group the L electrons, and so forth, in agreement with the energies being calculated for atoms by Niels Bohr and other atomic physicists at that time. It is interesting to note that the center of the alphabet was chosen as a starting point because the early investigators could not be certain that more tightly bound electrons might not be discovered later. For convenience, then, denote by

W_K the energy required to remove a K electron
W_L the energy required to remove an L electron
W_M the energy required to remove an M electron, etc.

Suppose an x-ray photon incident on the atom has an energy $E = h\nu$ such that $E > W_L$. Then it is possible that this photon will knock out an L electron from the

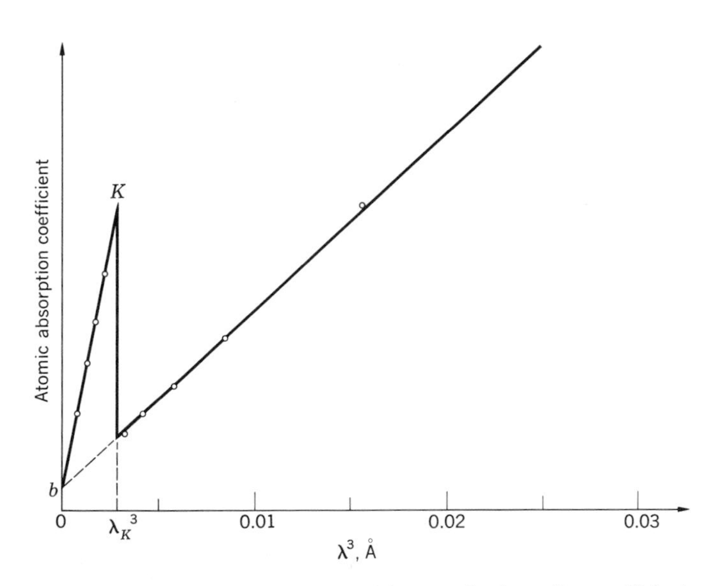

Fig. 6-11. Wavelength dependence of the atomic absorption coefficient of lead, showing the linear relation between μ_a and λ^3. (*After Richtmeyer, Kennard, and Lauritsen.*)

atom by a process similar to the Compton collision discussed above, except that the photon becomes absorbed by the atom in the process, its energy being transferred to the kinetic energy of the photoelectron. According to quantum-mechanical considerations, the probability that such a collision will take place is proportional to the x-ray wavelength cubed. With this in mind, consider the absorption curve in Fig. 6-10. Beginning on the right and proceeding to the left, as λ decreases, the energy of the x-ray photon increases. On the extreme right of Fig. 6-10, the energy of a photon $E < W_L$, so that absorption is largely due to the ejection of M electrons. As the photon energy increases, moreover, the absorption decreases, because the collision probability decreases in proportion to λ^3. When $E = W_L$, an incident photon has just enough energy to remove an L electron so that μ increases abruptly as x-ray photons are absorbed in the process of ejecting L electrons. As the photon energy increases further, the absorption resumes declining in accord with the diminishing collision probability. Note that there are three discontinuities evident in Fig. 6-10, labeled L_{I}, L_{II}, L_{III}, even though, according to quantum mechanics, there are only two distinguishable kinds of L electrons, called the $2s$ and $2p$ electrons, respectively, by spectroscopists. Clearly, there must be three different ways to eject these electrons from an atom, each having its characteristic energy, $W_{L_{\text{I}}}$, $W_{L_{\text{II}}}$, and $W_{L_{\text{III}}}$. This observation, first made with the aid of x-ray spectra, is noteworthy and is further discussed in a later chapter.

Further increase in the photon energy again causes μ to decline in proportion to λ^3, until $E \geq W_K$, when x-ray absorption by the ejection of K electrons becomes possible. The fact that there is only one K edge testifies to the fact that there is only one way to eject a K electron. It should be clear from the foregoing discussion that the determination of x-ray absorption edges can be used to determine the different energy states that electrons can have in atoms and their respective binding or ionization energies. Obviously, these energies are different in unlike atoms, so that the wavelengths at which their absorption edges occur are different also. This property can be utilized in the selective absorption of portions of the x rays emitted by an x-ray tube, considered next.

X-ray filters. As already discussed in the preceding chapter, the x-ray spectrum emitted by a conventional x-ray tube consists of a continuous spectrum on which are superimposed sharp maxima at wavelengths that are characteristic of the target metal (Fig. 6-12*A*). Since most x-ray diffraction experiments require a virtually monochromatic beam, some way has to be found for isolating the most intense component (labeled α in Fig. 6-12) from the rest of the spectrum. In view of the sawtooth shape of an absorption curve near the K edge (Fig. 6-12*B*), it proves possible to select an element whose K absorption edge lies at a wavelength slightly less than that of the desired x-ray component. As shown in Fig. 6-12*C*, the insertion of a foil fashioned from this element, called an *x-ray filter,* makes it possible to absorb quite strongly x rays having shorter wavelengths, while transmitting the desired component with relatively little attenuation. It must be kept in mind that such a filter merely decreases the intensity of certain spectral regions in accord with (6-25), but does not entirely absorb any of them. A convenient practical rule is to select a filter that reduces the intensity of the β component relative to that of the α component in the ratio of 1:600. Some of the filter elements most suitable for use with conventional x-ray tube targets are listed in Table 6-1.

It should be realized that a single filter, like one of those listed in Table 6-1, only serves to enhance the relative intensity of the α component, but does not eliminate the other components. Thus such a beam is said to be filtered but not monochromatic. It is possible to achieve virtual monochromatization, however, by employing two filters, as first pointed out by P. A. Ross. Suppose the intensity of x rays from a molybdenum target is measured after passage through a zirconium filter. As can be seen in Fig. 6-13, a zirconium filter absorbs most heavily x-rays

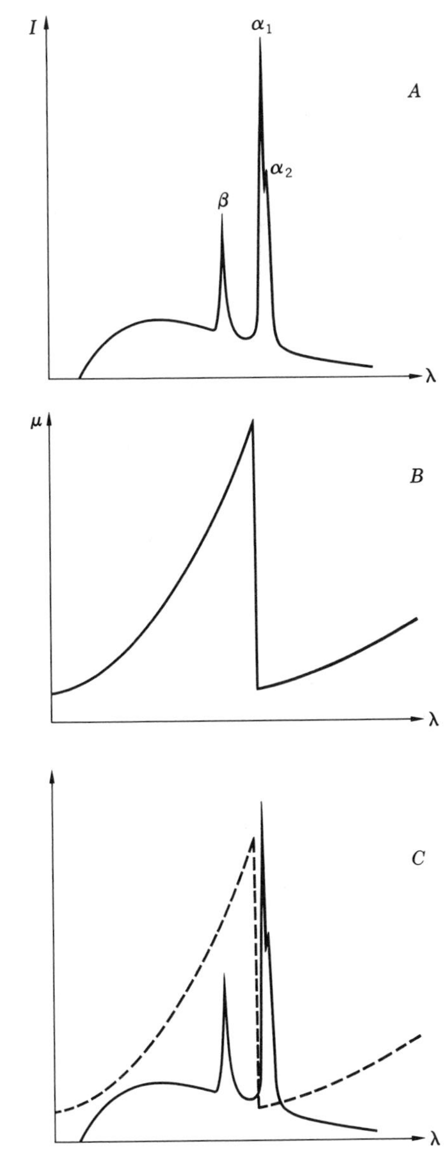

Fig. 6-12

Table 6-1 Commercial x-ray tube target wavelengths and suitable filters

Target element	$K\alpha$* λ, Å	*Filter element*	*Density of filter element,* g/cm²	*Optimum thickness,*‡ mm
Cr	2.2909	V	0.009	0.016
Fe	1.9373	Mn	0.012	0.016
Co	1.7902	Fe	0.014	0.018
Ni	1.6591	Co	0.014	0.013
Cu	1.5418	Ni	0.019	0.021
Mo	0.7107	Zr	0.069	0.108
Ag	0.5598	Pd	0.030	0.030

* This wavelength is determined by the formula $(2\lambda_{K\alpha_1} + \lambda_{K\alpha_2})/3$.
‡ This thickness reduces the intensity of $K\beta/K\alpha$ to $1/600$.

having wavelengths less than 0.6874 Å, to the left of the $K\alpha$ line of molybdenum (0.7107 Å). Next, suppose the intensity of the same x rays is measured after passage through an yttrium filter. As also shown in Fig. 6-13, this filter absorbs most strongly x rays whose wavelength is less than 0.7255 Å. Provided that the two filters have their thicknesses adjusted so that they both absorb equally in the short-wavelength region of the continuous spectrum, resulting in what are called *balanced filters,* then the difference in the intensity measured by the above two methods must be due, almost exclusively, to x rays whose wavelengths lie in the

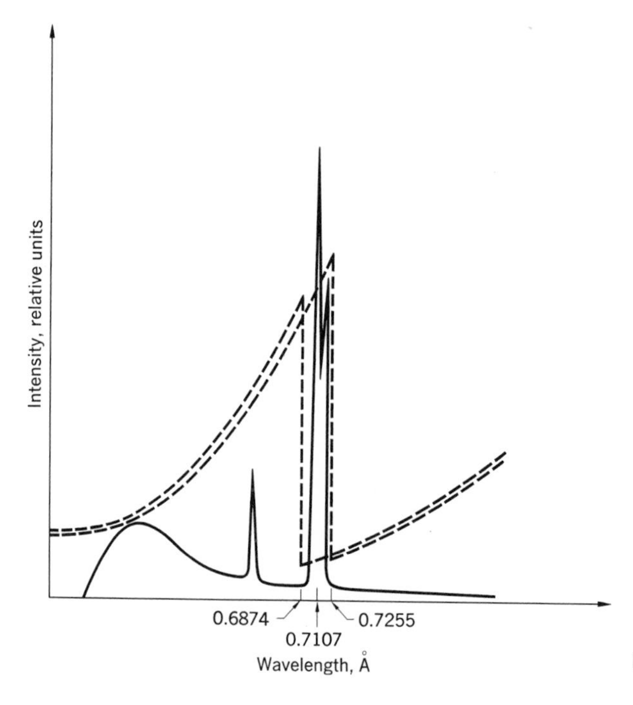

Fig. 6-13

Table 6-2 Balanced filters for commonly used x radiations†

Target element	Filter elements A	Filter elements B	Filter A Density, g/cm²	Filter A Thickness, mm	Filter B Density, g/cm²	Filter B Thickness, mm
Cr	V	Ti	0.0052	0.0087	0.0058	0.0128
Fe	Mn	Cr	0.0064	0.0086	0.0070	0.0098
Co	Fe	Mn	0.0068	0.0087	0.0078	0.0105
Ni	Co	Fe	0.0075	0.0084	0.0081	0.0103
Cu	Ni	Co	0.0082	0.0092	0.0093	0.0104
Mo	Zr	Y	0.024	0.036	0.026	0.0477
Mo	Zr	Sr	0.024	0.036	0.028	0.1074
Ag	Pd	Mo	0.018	0.0147	0.022	0.0214

† Based on values calculated by B. W. Roberts.

region between the two absorption edges. In the example cited, these are primarily the $K\alpha$ rays of molybdenum. Suitable filter elements for Ross filters are listed in Table 6-2, along with suggested thicknesses for proper balancing. Note that, in practice, a foil that is somewhat thinner than the recommended values in Table 6-2 can be placed in the path of the beam, and its effective thickness can be adjusted by changing its inclination relative to the beam direction. The use of such filters in conjunction with direct ionization detectors provides a most convenient way for obtaining an effectively monochromatic x-ray beam.

EMISSION OF X RAYS

Continuous spectrum. In order to gain an insight into the x-ray spectra emitted by metal targets, consider what happens to the spectrum from a tungsten target as the tube voltage is increased successively from 20 to 50 kV (Fig. 6-14). In addition to observing the general form of the resulting continuous spectrum, which systematically grows in intensity as the voltage increases, several specific features are noteworthy. Although the continuous spectrum tails off gradually at longer wavelengths, it is terminated quite abruptly at the *short-wavelength limit,* usually abbreviated swl. The maximum in the intensity distribution lies at approximately three-halves times the wavelength of the short-wavelength limit, and both shift to shorter wavelengths as the voltage increases. Converting the wavelength to a frequency, $\nu_{\text{swl}} = c/\lambda_{\text{swl}}$ it can be seen in Fig. 6-15 that the limit frequency varies linearly with the voltage applied to the tube.

$$\nu_{\text{swl}} = 2.43 \times 10^{14} V \tag{6-31}$$

where V is the voltage applied to the x-ray tube, in volts.

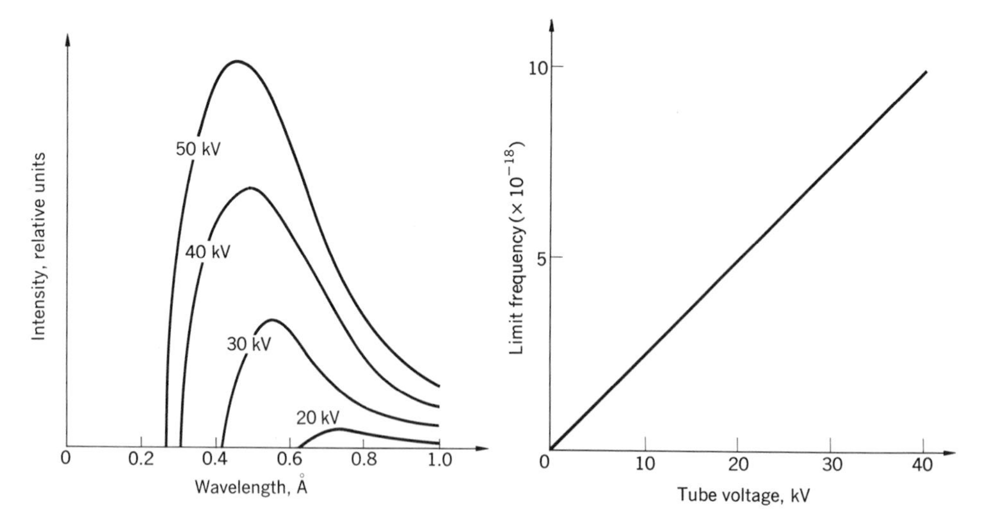

Fig. 6-14. Relative intensity of continuous spectrum from tungsten tube at several voltages. (*After Ulrey.*)

Fig. 6-15

In attempting to explain the above features, one can begin by asking, what happens when a cathode electron strikes the target metal? Basically, two different possibilities can be distinguished: the electron is deflected by the field of the target atoms without loss of energy, or it is deflected with an accompanying loss of energy. In the latter case, the energy lost may be reemitted as an x-ray photon whose energy is $\Delta E = h\nu$. It is clear from this that the maximum energy that the x-ray photon can have cannot exceed the maximum energy of the cathode rays, which, in turn, is determined by the voltage applied to the tube. Thus

$$\Delta E_{\text{max}} = eV_{\text{max}} = h\nu_{\text{swl}} = hc/\lambda_{\text{swl}} \tag{6-32}$$

Rearranging the terms in (6-32) gives, first of all,

$$\nu_{\text{swl}} = \frac{e}{h} V \tag{6-33}$$

which is plotted in Fig. 6-15. Obviously, a careful determination of the slope of this line permits one to determine the ratio e/h, a very important constant in many atomic problems.

Next, rearranging the terms in (6-32) to solve for λ_{swl} and substituting for the necessary constants give

$$\lambda_{\text{swl}} = \frac{12{,}390}{V} \quad \text{Å} \tag{6-34}$$

provided that V is expressed in volts (Exercise 6-10).

The value of λ_{swl} in (6-34) corresponds to the maximum energy that an electron can lose in a single collision. Actually, it is more probable that the cathode electrons give up less energy in individual collisions and most probable that they undergo several collisions before coming to rest in the target. This is the reason, of course, why the other x-ray photons have lower energies and explains the origin of the rest of the continuous spectrum. The intensity maximum at about $1.5\lambda_{swl}$ simply reflects the statistically most probable energy loss. By comparison, should the electron strike a metal target so thin that multiple scattering is not likely, quantum-mechanical calculations indicate that the more likely interactions involve either total energy loss or none at all. The continuous spectrum from a very thin metal foil placed transversely to the electron beam (Fig. 6-16) shows the expected intensity maximum at the short-wavelength limit and a tailing off at increasing wavelengths. It has been suggested that the intensity distribution in Fig. 6-14 can be thought of as a superposition of a succession of curves like Fig. 6-16, modified by self-absorption in the target of incident electrons and emitted x-ray photons. It should be noted also that the form of the curves in Fig. 6-14 does not appear to depend on the target metal. Similarly, it is clear from (6-34) that the short-wavelength limit does not depend on Z either and is a function of the applied voltage only.

When the continuous spectra from different target tubes operated at the same tube voltage and current are compared with each other, it is discovered that the intensity increases at all wavelengths with increasing atomic number. As can be seen from Fig. 6-14, the total radiation, or area, under the intensity curve for a fixed tube current and a particular metal target increases in proportion to the square of the tube voltage. Similar empirical comparisons permitted H. Kulenkampff to formulate an empirical formula for the curves in Fig. 6-14. The intensity, as a function of wavelength and target element Z, is given accurately by

$$I_\lambda = CZ\frac{1}{\lambda^2}\left(\frac{1}{\lambda_{swl}} - \frac{1}{\lambda}\right) + BZ^2\frac{1}{\lambda^2} \tag{6-35}$$

where C and B are constants and $C \gg B$. As further discussed in Chap. 12, only approximately 1% of the energy absorbed by the tube target is reemitted in the form of x rays.

Quantum-mechanical theories for the continuous spectrum have been developed by H. A. Kramers, modified by G. Wentzel, and independently analyzed by Sommerfeld. The theories are in general agreement with the foregoing discussion,

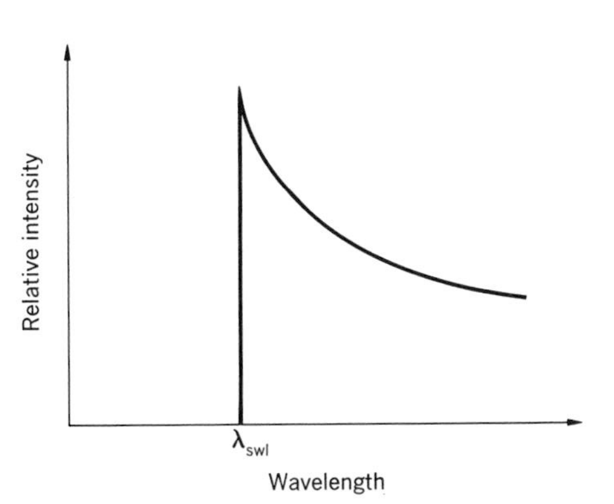

Fig. 6-16

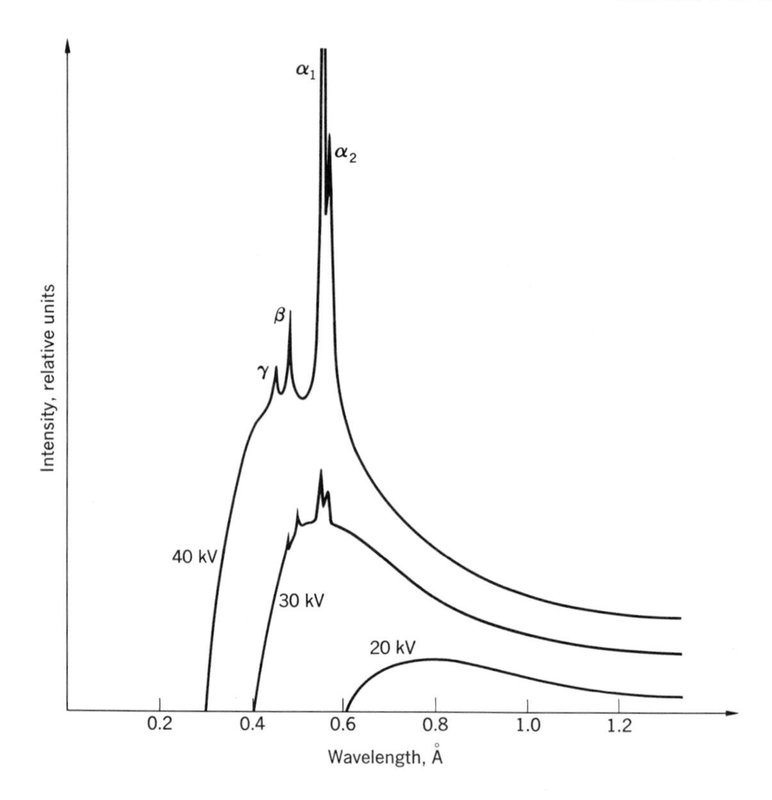

Fig. 6-17

and produce relations between I_λ and the other parameters that are similar to those above. In addition, they show that the continuous spectrum from a metal target should be polarized with a maximum intensity in a direction ranging from 30 to $60°$ to that of the incident cathode rays, depending on the tube voltage. A quantitative test of these theories is most difficult because of the need to make proper allowance for such experimental factors as self-absorption by the target and absorption by the analyzing crystal, tube windows, detector window, etc. The problem is further aggravated by the nonlinear dependence on x-ray wavelength of absorption, crystal reflecting power, and detector response, so that even a semi-quantitative evaluation is quite difficult.

Characteristic spectrum. In order to examine the conditions leading to the generation of characteristic spectra, consider the x-ray spectrum emitted by a silver-target x-ray tube as the voltage is increased systematically (Fig. 6-17). At 20 kV, the tube emits a continuous spectrum very similar to the lowermost curve in Fig. 6-14. This similarity persists until slightly more than 25 kV is placed across the tube, at which point three small pips appear, respectively, at 0.4860, 0.4962, and 0.5597 Å. As the voltage is increased further, say, to 30 kV, these pips grow in prominence with the one at 0.5597 Å, beginning to split into a resolvable doublet, until, at 40 kV, the second peak also begins to show a doublet of two lines at 0.4950

and 0.4967 Å, although it is not resolvable on the scale of Fig. 6-17. According to Barkla's surmise, cited earlier, these pips represent x rays emitted by the target atoms following the ejection of a K electron. This postulate can be easily tested by recalling that the work required to knock out a K electron is inversely proportional to the wavelength of the K edge. For silver,

$$W_K = \frac{hc}{\lambda_K} = \frac{6.62 \times 10^{-27}\,(\text{erg-sec}) \times 3.0 \times 10^{10}\,(\text{cm/sec})}{0.485 \times 10^{-8}\,(\text{cm})} = 4.0 \text{ ergs} \qquad (6\text{-}36)$$

The maximum energy that an electron accelerated by a 25-kV potential applied to the tube can have is

$$E_{\max} = \frac{hc}{\lambda_{\text{swl}}} = eV = \frac{25 \times 10^3\,(\text{V}) \times 4.8 \times 10^{-10}\,(\text{esu})}{300\,(\text{V/esu})} = 4.0 \text{ ergs} \qquad (6\text{-}37)$$

It is clear from the above discussion that cathode rays accelerated by a 25-kV tube potential have just enough energy to knock out K electrons in silver, and the above demonstration substantiates the premise that before any of the characteristic x rays, called the K spectrum by Barkla, can be emitted, a K electron first must be ejected from the atom.

The discovery of x-ray diffraction by crystals provided x-ray spectroscopists with their first tool for measuring wavelengths precisely. In the years 1913–1914, H. G. I. Moseley discovered that the frequency of the strongest characteristic line emitted by an element increased systematically as its atomic weight increased. Guided by the then developing Bohr theory of the atom, according to which the frequency of a spectral line is given by

$$\nu = \frac{2\pi^2 me^4}{h^3} Z^2 \left(\frac{1}{n_2^2} - \frac{1}{n_1^2}\right) \qquad (6\text{-}38)$$

where n_2 and n_1 are the principal quantum numbers of the final and initial states of the atom, respectively, and the other constants have their usual meanings, Moseley plotted the square root of the frequency of the most intense line against the atomic number Z of the target element. Such a plot is shown in Fig. 6-18, where it is seen that a straight line results which does not pass through the origin of coordinates

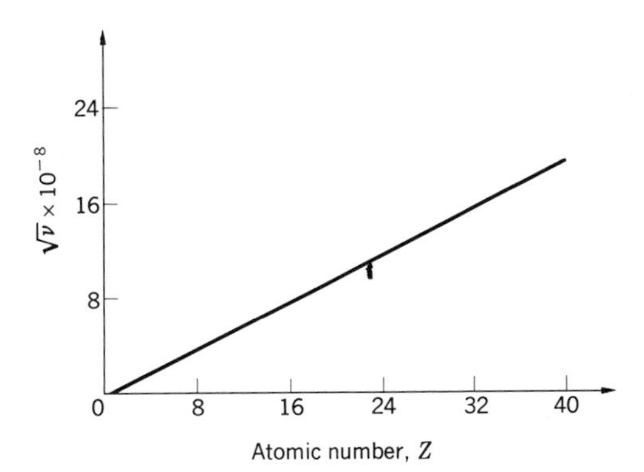

Fig. 6-18

but goes to zero at $Z = 1$. It follows that this line can be represented by

$$\nu = 0.248 \times 10^{16}(Z - 1)^2 \tag{6-39}$$

The similarity between these two relations suggests that the origin of the x-ray spectra can be related to the transitions from one excited state to another by an atom. Relation (6-39) is often called Moseley's law, and is based on his measurements employing a rather crude one-crystal spectrometer and photographic recording. Thus Moseley could not resolve the doublet in the most intense line, and called it the $K\alpha$ line, preceded by the $K\beta$ and $K\gamma$ lines (Fig. 6-17), also first named by Moseley. Nevertheless, Moseley's linear equation remains a very close approximation to the truth, even when the improved modern-day spectrometers are used to test it (Exercise 7-12).

Making use of the Bohr model of an atom, it is possible to construct the following picture of the process by which characteristic x-ray spectra are emitted. First an atom must "lose" an inner electron. When the ejected electron is a K electron, the atom is said to be in the *K quantum state of the atom;* when it is an L electron, the atom is in the L quantum state; and so forth. Obviously, more work must be done to remove the more tightly bound K electron so that the energy of the K quantum state is higher than that of the L quantum state, as diagrammatically displayed in Fig. 6-19. Note that the L quantum state of the atom actually consists of three discrete energy levels, because, as indicated by the L absorption spectra, there are three different binding energies for L electrons. Suppose that a K electron is ejected from an atom by the absorption of energy equal to W_K. This can be depicted in an energy-level diagram like Fig. 6-19 by drawing an arrow upward from the atom's ground state to its K state. The resulting vacancy in the K shell can be filled when an electron "falls" into it. The nearest electron, and therefore the most likely to do this, is an L electron. If it is an L electron of the type labeled L_{III} in Fig. 6-19, the new excited state of the atom is the L_{III} quantum state. Note that the atom now has a vacancy in the L shell that is exactly the same as if it had been produced by the ejection of the L_{III} electron by some external means. Thus

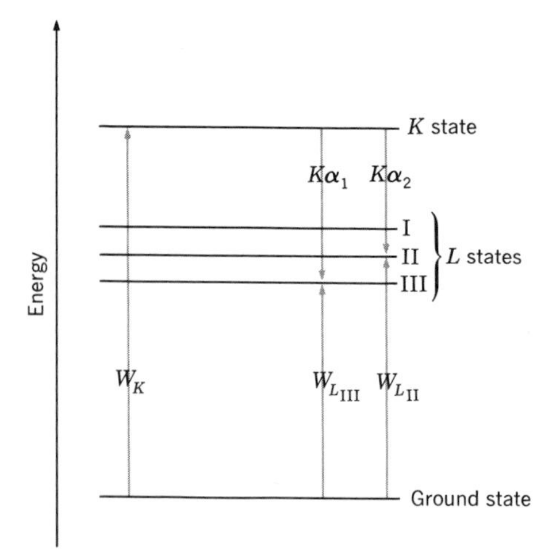

Fig. 6-19

the new state of the atom can be represented in Fig. 6-19 by an upward arrow to the L_{III} level.

In going from the K state to the L_{III} state the atom's energy has decreased, so that the conservation of energy requires that a quantum of energy be given off. Calling this quantum $K\alpha_1$, its energy

$$E_{K\alpha_1} = W_K - W_{L_{\mathrm{III}}} \tag{6-40}$$

Similarly, the other line in this doublet has an energy determined by the transition $K \rightarrow L_{\mathrm{II}}$, namely,

$$E_{K\alpha_2} = W_K - W_{L_{\mathrm{II}}} \tag{6-41}$$

It follows that the values of the energy levels can be determined by relations like (6-36), making use of measured absorption edges.

An examination of Fig. 6-19 suggests that still another alpha line, corresponding to the transition $K \rightarrow L_{\mathrm{I}}$, should be possible. Such a line never has been observed (Exercise 6-14), so that it can be assumed to be a forbidden transition. As discussed further in the next section, this is in complete agreement with quantum mechanics, which establishes specific *selection rules* for atomic-energy transitions.

Quantum theory. Although the detailed role that the study of x-ray spectra played in the evolution of modern physics has been alluded to, it is not possible to give a full account of it here, despite its great importance.† By the same token, it is outside the scope of the present volume to undertake a very detailed discussion of quantum-mechanical analyses of atomic structure, etc. For present purposes it is sufficient to recall that the electrons in an atom can be described by four quantum numbers, of which the principal quantum number n $(= 1, 2, 3, \ldots)$ specifies the shell ($n = 1$ corresponds to K shell, $n = 2$ to L shell, etc.) and determines the energy of the electron; the azimuthal quantum number l $(= 0, 1, 2, \ldots, n-1)$ determines its orbital angular momentum; the magnetic quantum number m_l $(= 0, \pm 1, \pm 2, \ldots, \pm l)$ specifies the orientation of the electron's orbital in a magnetic field; and the spin quantum number m_s $(= \pm\frac{1}{2})$ describes the orientation of the spin direction. The specific quantum numbers assigned to the electrons are determined by thermodynamic considerations that require the occupation of states having lowest energies first, and the Pauli exclusion principle, which forbids more than one electron in an atom to have the same four quantum numbers.

When discussing atomic spectra, it is necessary to take into account the influence that interactions between the magnetic moments of the spin and the orbital motion of an electron have on its energy in a many-electron atom. To best describe the consequences of these spin-orbit interactions, it is convenient to define two different quantum numbers, namely, the quantum number $j = l \pm m_s$ and m, which can have all integrally spaced values from $+j$ to $-j$. Their distribution

† A more complete discussion can be found in other texts. Especially recommended: F. K. Richtmeyer, E. H. Kennard, and T. Lauritsen, *Introduction to modern physics*, 5th ed. (McGraw-Hill Book Company, New York, 1955). In addition to a complete discussion of modern physics and a carefully interwoven historical narrative, this book contains one of the best short discussions of x rays, in chap. 8, to be found anywhere.

Table 6-3 Allowed combinations of quantum numbers

Shell	K	L			M				
n	1	2			3				
l	0	0	1		0	1		2	
j	$\frac{1}{2}$	$\frac{1}{2}$	$\frac{1}{2}$	$\frac{3}{2}$	$\frac{1}{2}$	$\frac{1}{2}$	$\frac{3}{2}$	$\frac{3}{2}$	$\frac{5}{2}$
m	$\pm\frac{1}{2}$	$\pm\frac{1}{2}$	$\pm\frac{1}{2}$	$\pm\frac{3}{2} \pm \frac{1}{2}$	$\pm\frac{1}{2}$	$\pm\frac{1}{2}$	$\pm\frac{3}{2} \pm \frac{1}{2}$	$\pm\frac{3}{2} \pm \frac{1}{2}$	$\pm\frac{5}{2} \pm \frac{3}{2} \pm \frac{1}{2}$
Number of electrons	2	2	2	4	2	2	4	4	6

among the lowest energy states is indicated in Table 6-3. Note that the usual picture of the electronic structure of an atom is not changed in that the K shell contains two s electrons, the L shell contains two s and six p electrons, and the M shell contains two s, six p, and ten d electrons. An examination of Table 6-3, however, indicates why there are three different energy levels for the L electrons, in that, not only can one distinguish the s from the p electrons by their l value, but one can also distinguish two kinds of p electrons, depending on whether j is larger or smaller than l. (It turns out that m is unimportant in determining the energy, so that it is not considered further here.)

The total angular momentum of an atom can be determined by summing vectorially the orbital and spin moments of the individual electrons in essentially two ways: Following a procedure first used by H. N. Russell and F. A. Saunders, all the spin vectors are summed first to give the resultant vector **S** and the orbital vectors to give **L**, followed by the vector addition of the two resultants to give the atomic resultant **J**. In another procedure, called jj coupling, the individual **j** vectors of each electron are formed first, and then added vectorially to give the resultant **J**. The two processes give somewhat different values for the final angular momentum, which is of consequence when transitions from one state to another are considered in spectroscopy. For present purposes, however, it is sufficient to realize that, whether Russell-Saunders or jj coupling is employed, the total angular momentum of an atom having *all* its electrons is markedly different from **J** when calculated after removing an inner electron. In fact, it is convenient to describe the quantum state of the atom by the quantum numbers of the missing electron. Following this scheme, the relative energy levels for the quantum states that a silver atom can have are diagramed in Fig. 6-20. The vertical arrows pointing downward indicate the observed transitions, that is, the observed characteristic lines whose energies agree with the transitions indicated, according to a calculation like (6-40) or (6-41). These transitions are also in agreement with quantum-mechanical calculations of the transition probabilities. It turns out that transitions

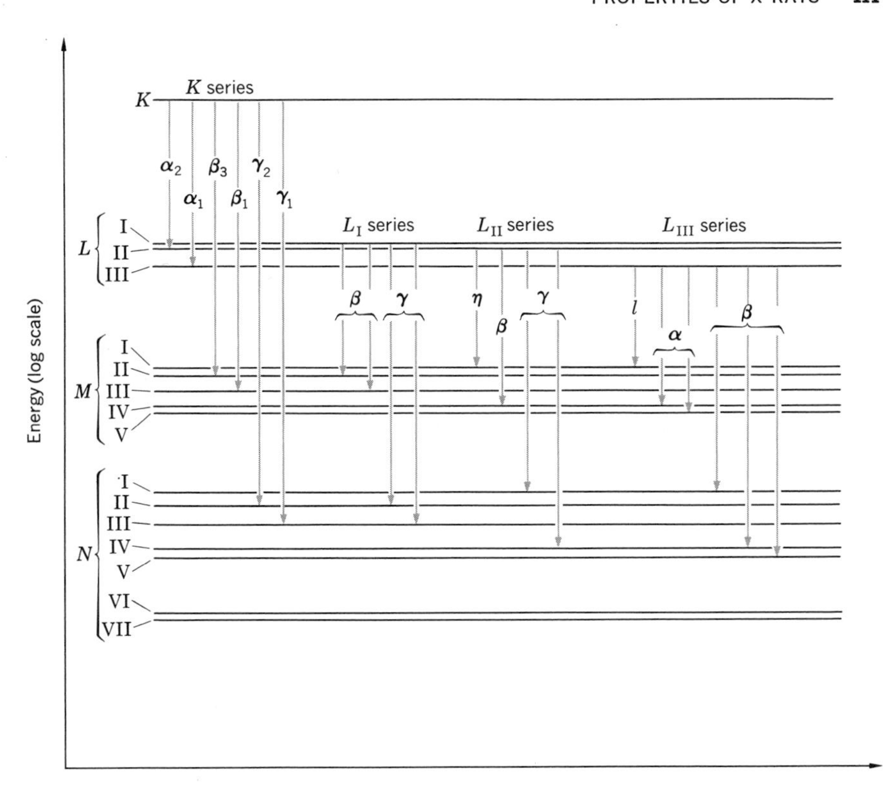

Fig. 6-20. Energy-level diagram of a silver atom with the allowed transitions indicated by arrows. (The lines are labeled after a convention proposed by Siegbahn, except that $K\gamma$ is used in place of $K\beta_2$.)

are allowed (most probable) when changes in the quantum numbers in going from one state to the other obey the conditions

$$\Delta l = \pm 1$$

$$\text{and} \qquad \Delta j = 0 \text{ or } \pm 1 \tag{6-42}$$

(If Russell-Saunders coupling is employed, the same conditions apply to ΔL and ΔJ.) The observed characteristic lines obeying these selection rules are sometimes called *diagram lines* to distinguish them from much weaker, but nevertheless distinguishable, maxima which cannot be explained by the direct application of these selection rules. Such *nondiagram lines* can be measured accurately only with x-ray spectrometers considerably more sensitive than those employed by Barkla, Moseley, and other early investigators, so that the self-consistent picture of x-ray spectra presented above was not disturbed prematurely, before the then modern quantum theory could become more firmly established and more generally accepted.

Nondiagram lines. The presence of weak but distinct maxima, usually on the short-wavelength side of diagram lines, gradually became more noticeable as the quality of x-ray spectrometers was improved in the 1920's. These maxima were called *satellites,* or nondiagram lines, because they could not be explained by simple allowed transitions on energy-level diagrams like Fig. 6-20. In seeking an explanation for their origins, a correlation was sought with the excitation potential needed to produce specific satellites. It turned out, for example, that to produce the satellites on the short-wavelength side of a $K\alpha$ line, the cathode electrons had to be imparted an energy sufficient to knock out a K and an L electron, that is, $W_K + W_L$. Although quantitatively difficult to measure because the intensity of satellites is so weak, such observations led Wentzel to postulate that the production of nondiagram lines required an initial *double* ionization of the atom, which can be denoted as the KL, KM, etc., quantum state of an atom. Denoting by $K\alpha_{3,4}$ the satellite produced by a transition from a KL state to an LL state by the transition of an L electron to the K vacancy, the frequency $\nu_{K\alpha_{3,4}}$ appears to obey the relation

$$h\nu_{K\alpha_{3,4}} = W_{KL} - W_{LL} \tag{6-43}$$

as compared with the frequency of the diagram line $K\alpha_{1,2}$, which can be represented by

$$h\nu_{K\alpha_{1,2}} = W_K - W_L \tag{6-44}$$

The first term on the right of (6-43) must represent the energy required to knock out a K electron, W_K, plus the energy required to remove an L electron bound to a nucleus that is shielded by one electron less than normally. Let this be equivalent to the work that must be done to remove an L electron from an atom whose atomic number is higher by 1, that is, $(W_L)_{Z+1}$. Similarly, let $W_{LL} = W_L + (W_L)_{Z'+1}$, where $Z' + 1$ denotes that the nuclear charge is somewhat less screened because it is an L, not a K, electron that is missing; in other words, $(W_L)_{Z+1} > (W_L)_{Z'+1}$. With this representation, it is possible to make an order-of-magnitude estimate of the energy required in (6-43).

$$\begin{aligned} h\nu_{K\alpha_{3,4}} &= [W_K + (W_L)_{Z+1}] - [W_L + (W_L)_{Z'+1}] \\ &= W_K - W_L + (W_L)_{Z+1} - (W_L)_{Z'+1} \\ &= h\nu_{K\alpha_{1,2}} + \delta \end{aligned} \tag{6-45}$$

after substituting (6-44) for the first two terms and letting δ represent the small but positive difference between the last two terms. It is clear from (6-45) that the energy of the satellite lines is larger than the accompanying diagram lines. The above calculation is similar to one carried out by M. J. Druyvesteyn in an extension of Wentzel's postulate. It should be noted that very approximate quantum-mechanical calculations appear to support the double-ionization model. A triple ionization has also been proposed, but such a model appears too difficult to verify experimentally or theoretically at present.

The most likely process by which double ionization seems possible is one in which the incident cathode ray has sufficient energy to knock out two electrons from the same atom. The short lifetime of x-ray excited states (about 10^{-16} sec) precludes the likelihood that two successive electrons striking the same atom are responsible. This conclusion agrees with the above analysis and serves as the basic model for theoretical estimates of the dependence of the intensity of satellite lines on atomic number. The theoretical considerations indicate that the nondiagram lines accompanying the K spectrum, called the $K\alpha$ satellites, should decrease in intensity relative to the diagram-line intensity as Z increases. This is actually observed to be the case, taking into consideration the difficulty in separating the satellite intensities from general background. Similarly, theory predicts that the intensity of $L\alpha$ satellites should decrease with increasing Z. Experimental observations show, however, a fairly rapid decline in intensity up to $Z = 50$, followed by a virtual absence of satellites up to $Z = 75$, at which point their intensity abruptly starts to increase with atomic number. This unexpected behavior can be explained by assuming that, in the range $Z = 50$ to $Z = 75$, the transitions involving doubly ionized atoms are radiationless.

Auger effect. Radiationless transitions in atoms were first clearly identified by P. Auger, who was studying the x-ray fluorescence of argon gas in a Wilson cloud chamber. The x-ray photons incident on the gas had more than enough energy to knock out K electrons in argon atoms, and the ejected photoelectrons left clearly visible tracks. A fairly large portion of the tracks, however, were accompanied by a second, wider track, characteristic of slower, lower-energy electrons coming seemingly from the same origin. By a systematic study of their frequency of occurrence, ejection direction, track length, etc., under varying cloud-chamber conditions, Auger was able to construct the following explanation for his observations: After absorbing a photon in the process of excitation to the K quantum state, the atom can undergo a transition to the L quantum state without giving up an energy quantum in the form of a K x ray. Instead, the energy is expended in a radiationless process in which another inner electron is ejected from the atom. Such a radiationless transition is possible, of course, only if the energy of the quantum liberated in the atomic transition is larger than the sum of the energies required to knock out the second electron plus its resultant kinetic energy.

For example, suppose the energy released in a transition from the K to the L state is used up in knocking out another L electron, thus resulting in a doubly ionized atom in the LL state. Denoting the kinetic energy of the Auger electron $\frac{1}{2}mv^2$, the conservation of energy requires that

$$h\nu_{K\alpha} = (W_L)_{Z'+1} + \tfrac{1}{2}mv^2 \tag{6-46}$$

where $(W_L)_{Z'+1}$ denotes the energy necessary to remove the second L electron.

In considering the possible transitions responsible for nondiagram lines and the parallel kind of transitions that take place in the Auger effect, it is evident that, depending on the relative magnitudes of the terms in relations like (6-47), Auger transitions may or may not occur. When Auger transitions are possible, the intensity of satellite lines can be expected to be decreased correspondingly. This turns out to be the principal reason for the absence of $L\alpha$ satellites when Z ranges from 50 to 75 (Exercise 6-16).

SOLID-STATE SPECTRA

The discussion of x-ray spectra so far has been limited to processes taking place in isolated atoms. When the actual spectra of isolated atoms in a gas are examined with the aid of modern spectrometers, certain irregularities are observed that are not explainable by the preceding theories. In the case of absorption spectra, for example, a careful analysis of the dependence of μ on λ right at the edge shows marked oscillations extending for about 5 eV from the short-wavelength (high-energy) side of the main edge in an inert gas like argon. In Cl_2 gas this *fine structure* extends about 20 eV, while in $GeCl_4$, gas or liquid, a fine structure extending for several hundred electron volts is observed. In the case of crystals, fine structures extending for more than 1,000 eV have been recorded! Similarly, changes in the fine structures occurring right at the main edge are observed, and also shifts in the wavelength values at which the edges (or emission lines) lie when the spectra of the same element in different compounds are compared. Although only tentative explanations have been offered so far for these differences, it is clear that they must be related to the atomic environments. Thus the x-ray spectroscopy of materials offers a potentially powerful tool for the study of atomic behavior in such materials. The principal obstacle to its full exploitation at the present time is the lack of rigorous theory. Nevertheless, certain deductions are possible, some of which are indicated below, following a brief description of some of the instruments developed for such studies. The reader is alerted to the fact that much of the ensuing discussion will presuppose a familiarity with solid-state physics not required elsewhere in this book, and that a number of arguments propounded will not be understandable without such prior knowledge.

X-ray spectrometers. The nature of an appropriate x-ray spectrometer is determined, first of all, by the hardness or energy of the x rays to be measured. For wavelengths shorter than $2\frac{1}{2}$ Å, absorption is not a serious problem, so that crystal analyzers operating in air and ordinary x-ray detectors can be used. For $\lambda < 20$ Å, called the soft x-ray region, it becomes necessary to minimize absorption by using less absorbent crystals, specialized detectors, and vacuum paths for the x rays. Finally, for ultrasoft x rays having wavelengths longer than 20 Å, crystal analyzers must be replaced by gratings, and special instruments must be constructed in which the tube, analyzer, and detector are housed in a single evacuated chamber. The instrumentation for ultrasoft x-ray spectroscopy is often so complicated that the interpretation of the results is strongly dependent on one's ability to make proper corrections for purely instrumental factors. Although such corrections must be made for all spectral measurements, regardless of the energy range, both the corrections and the instrumentation needed are better understood for the soft x-ray region, so that only these instruments are considered below.

The first spectrometer employing a single crystal was built by W. L. Bragg. When x rays emanating from a point source strike a flat crystal (Fig. 6-21), individual rays form different angles with the reflecting planes parallel to its surface. The rays reflected at the corresponding Bragg angles, therefore, must have different wavelengths, resulting in a dispersion of the reflected rays. Suppose that the

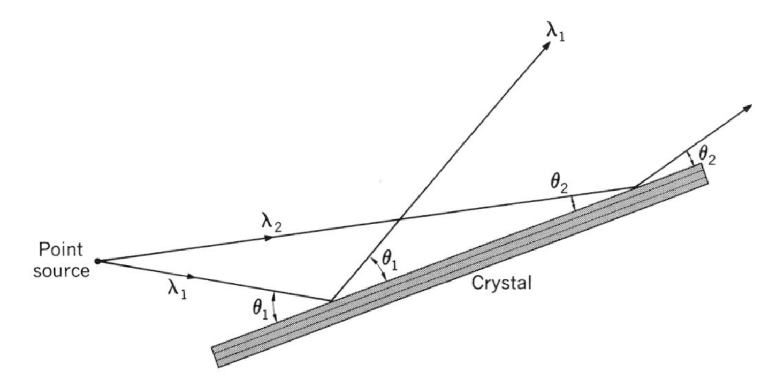

Fig. 6-21. Dispersion by a single crystal of radiation emanating from a point source.

crystal next is rotated by a small amount about a direction parallel to the reflecting surface and perpendicular to the spectrometer plane. If one ray formed the Bragg angle θ for a particular wavelength when the crystal was in the first position, then another ray forms the same angle θ when the crystal is in the second position. As shown in Fig. 6-22, both of these rays are successively reflected to the same point F on the spectrometer circle, leading to a kind of focusing of the reflected rays. It is not a true focusing, as was pointed out by J. C. M. Brentano, who proposed the term *parafocusing,* because it does not actually increase the intensity

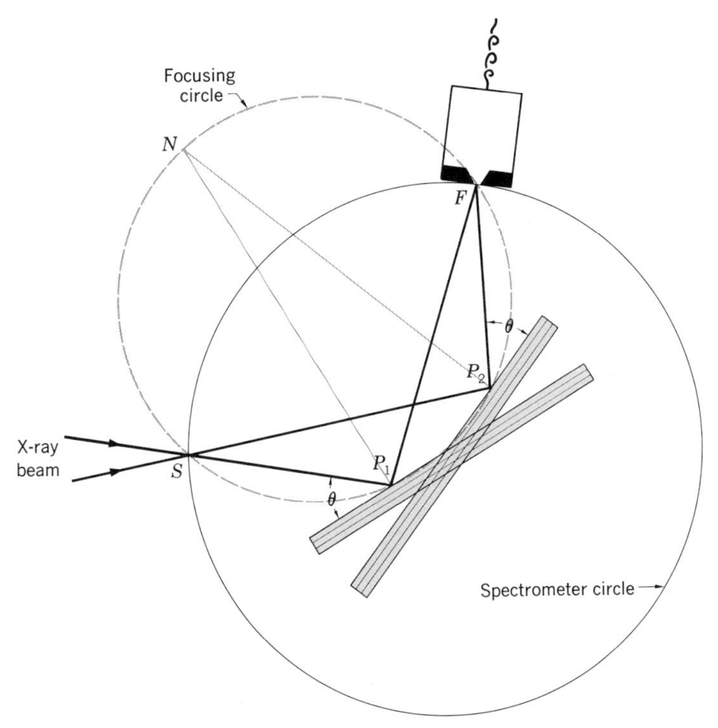

Fig. 6-22. Parafocusing action of a one-crystal spectrometer.

of the reflected beams, but merely superimposes successive reflections. The parafocusing action can be seen most easily by passing a circle (shown dashed in Fig. 6-22) through the x-ray source S, the focal point F, and the reflecting point P_1 on the crystal's surface in its first position. Next, draw a normal P_1N to the reflecting plane as shown. When the crystal is rotated to its second position, point P_2 on its surface touches the focusing circle, so that P_2N is the normal to the reflecting planes at this point. Since they both subtend the same arc on the circle, $\angle SP_1N$ and $\angle SP_2N$ are equal, so that the incident ray SP_2 forms the same Bragg angle with the reflecting planes. This means that the two reflected rays P_1F and P_2F must have the same wavelength. The consequence of this parafocusing geometry is that all the rays having the same wavelength are reflected to the same point on the spectrometer circle by the single crystal. Thus the one-crystal spectrometer combines the dispersion due to divergent radiation with a parafocusing action.

In an actual x-ray tube, the x-ray source is finite, so that the rays diverge from many adjacent points, destroying the focusing effect described above. It is necessary, therefore, to place a defining slit at S on the spectrometer circle. Because of the parafocusing action described above, a second slit in front of the detector is not necessary. Furthermore, during its rotation, the crystal "sees" all the adjacent points constituting the x-ray source. The intensity emitted by each point is successively superimposed at the appropriate focal point, so that any intensity variation in adjacent regions of the x-ray tube target are automatically equalized for each spectral component. It should be noted, however, that the reflecting planes are not perfectly parallel throughout the crystal, as discussed in later chapters. This produces a slight broadening in the reflected beam at F. The finite height of the crystal H (or finite length of the slit) also causes a widening of the reflected beam. The broadening or widening w of the beam is related to the pathlengths of the incident and reflected beams D_1 and D_2 at the Bragg angle θ by

$$w = \frac{H^2}{8(D_1 + D_2)\cos\theta} \tag{6-47}$$

Actually, the chief limitation of the one-crystal spectrometer is that the need for narrow slits severely cuts down the intensity of the x-ray beam reaching the

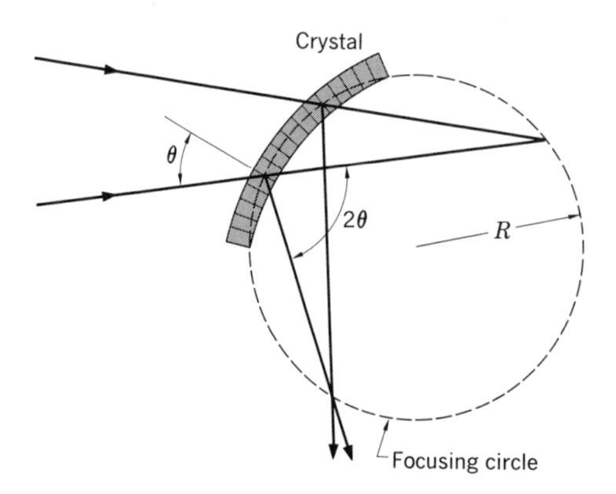

Fig. 6-23. Transmission-focusing arrangement.

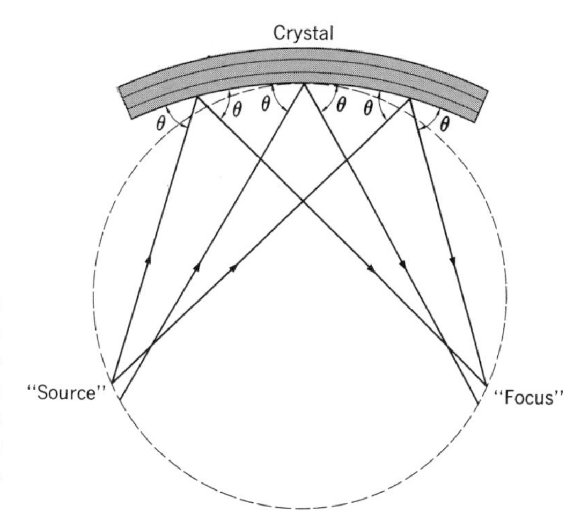

Fig. 6-24. Johann bent crystal. The rays drawn from the "source" all make the same angle θ with the reflecting planes of the crystal. Note that a parafocusing condition can be satisfied for two equidistant points along the crystal, but the "source" and "focus" must be displaced along the "focusing circle" for each such construction.

analyzing crystals. The slit width, of course, controls its resolving power, so that, in practice, one reaches a compromise between maximum resolution and adequate intensities.

In order to increase the intensity of the x-ray beam without having to broaden it at the focus, it is possible to bend the crystal so as to cause a geometric focusing of the rays. This can be done in a transmission arrangement (Fig. 6-23) in which the reflecting planes of an elastically bent crystal like mica cause the reflected rays to converge to a point on the focusing circle. The crystal's radius of curvature must be changed (Exercise 6-17) as the Bragg angle (wavelength) changes. Although a relatively large source of x rays can be used with such an arrangement, the intensity of the beam reaching the detector is decreased by passage through the bent crystal. This limitation can be avoided by using a bent crystal in reflection. The first such arrangement was devised by H. H. Johann, and is shown in Fig. 6-24. Note that the rays reflected from different points along the bent surface do not reflect to the same point on the focusing circle, drawn tangent to the central point of the crystal. This is similar to the situation in the transmission arrangement in Fig. 6-23 and leads to a certain amount of achromatic broadening at the focal point. As suggested first by J. W. M. DuMond and H. A. Kirkpatrick and, independently, by T. Johansson, it is possible to eliminate the achromatic broadening by bending the crystal to a radius $2R$ and grinding its surface to fit the focusing circle of radius R. Johansson actually built the first spectrometer using this principle, and, as can be seen in the arrangement in Fig. 6-25, it leads to perfect focusing, symmetrically, on both sides of the diverging direct beam. The focusing geometry requires that

$$D_1 = D_2 = 2R \cos\left(\frac{\pi}{2} - \theta\right) \tag{6-48}$$

Combining this relation with the Bragg law (5-6) determines the amount of bending,

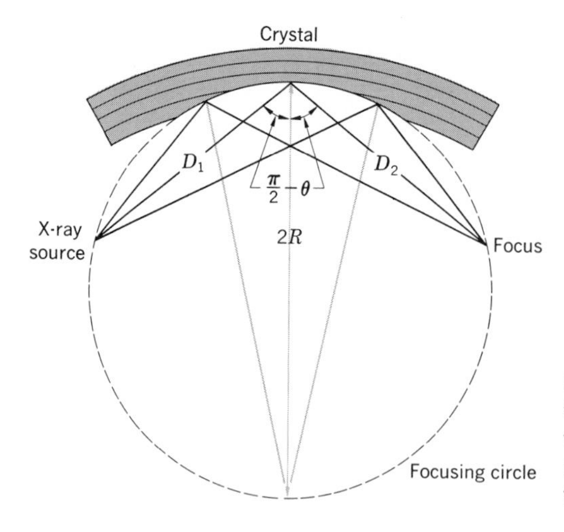

Fig. 6-25. Johansson bent and ground crystal. Note that the x-ray source can have any width and still satisfy the focusing geometry, provided it is moved inside the circle within the boundaries of the rays.

$2R$, required for a crystal having an interplanar spacing d to reflect x rays of wavelength λ.

$$R = D\frac{d}{\lambda} \tag{6-49}$$

A somewhat different approach had been adopted much earlier by Compton. In place of using a slit to define the incident-beam width, Compton reflected the x-ray beam first from another crystal. Theoretical analyses by W. Ehrenberg and H. Mark of the twice-reflected beams suggested that this two-crystal arrangement provides a most reliable means for studying the structures and widths of spectral curves. More recently, L. G. Parratt and his collaborators have shown that the two-crystal spectrometer is the only one for which reliable instrumental corrections are presently possible. It also turns out that the theoretical resolving power of this instrument is superior to that of a one-crystal instrument (Exercise 6-20). Of the various relative dispositions possible for the two crystals in this instrument, the two shown in Fig. 6-26 are most commonly used. In the first arrangement, the two crystals are arrayed to reflect from opposite faces, and this is called the *parallel setting*. If both crystals are turned so as to form the same Bragg angle with the

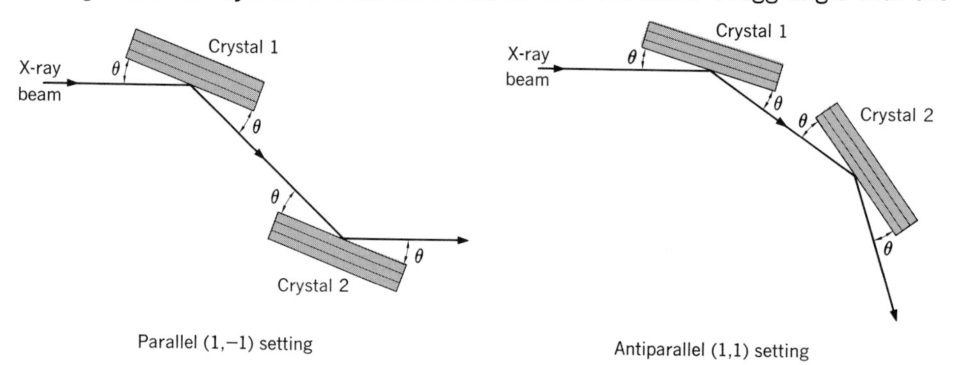

Fig. 6-26

x-ray beam, it is clear that they must be rotated in opposite directions, so that the parallel position is also called the $(1,-1)$ setting. The other, or *antiparallel setting*, in Fig. 6-26 is then denoted the $(1,1)$ setting, since both crystals have to rotate together. [The numbers in parentheses denoted the orders of the reflections for the first and second crystals. Thus $(3,-3)$ means that both crystals in the parallel setting are reflecting in the third order.] In actual practice, it is easier to construct an instrument in which only one of the two crystals is rotated at one time. In this case it is easily seen that the parallel position allows all the rays reflected by the first crystal to be reflected by the second crystal during its rotation. This is a valuable arrangement for studying the intensities of x rays diffracted by the second crystal, and the name *two-crystal diffractometer* was proposed for this arrangement by S. Weissman. On the other hand, any dispersion caused by the first crystal (Fig. 6-22) in the antiparallel setting has the result that rays other than the central rays shown in Fig. 6-26 completely miss the second crystal. This increased dispersion is mainly responsible for the larger resolving power of this *two-crystal spectrometer* arrangement.

The ultimate resolving power of an x-ray spectrometer depends on the crystals employed. The choice is controlled, first of all, by the spectral range to be investigated, since it is clear from the Bragg law (5-6) that long wavelengths require large d values. For a constant λ, however, the dispersion increases with decreasing d (Exercise 6-21). The relative perfection of a crystal determines the angular range over which it can reflect x rays, as discussed in more detail in Chap. 9. Obviously, the narrower this range, the better the resolution. As in the case of slits, however, the narrower the reflection range, the less total power is reflected by the crystals. This also suggests another criterion that must be considered in the selection of suitable crystals, namely, the relative intensity of the reflection chosen. Some typical crystals having relatively high intensities for the reflections indicated are listed in Table 6-4. It should be realized that these crystals also can be used in x-ray diffraction arrangements whenever a monochromatic beam of x rays is

Table 6-4 Crystals useful as monochromators

	Reflecting plane	
Crystal	*Indices*	d, Å
Topaz	(101)	1.353
LiF	(100)	2.013
Calcite	$(10\bar{1}1)$	3.029
Si	(111)	3.135
Ge	(111)	3.266
Quartz	$(10\bar{1}1)$	3.336
Quartz	$(10\bar{1}0)$	4.244
Gypsum	(010)	7.579
Mica	(001)	9.927

required. Thus any of the arrangements described in this section and the crystals listed in Table 6-4 are equally suitable as *crystal monochromators.*

When the profile of a twice-reflected x-ray beam is examined, it is found to depend on the kinds of crystals employed in the diffractometer or spectrometer arrangement discussed above. This is so because each crystal superimposes a characteristic "distortion," as discussed further in later chapters. The emerging beam normally exhibits, in addition to a central peak, more or less extensive wings which gradually merge with the background at some distance from the center of the peak. These wings can be very bothersome when it is desired to measure the total intensity of the emerging beam, because their full extent is often difficult to evaluate. Similarly, they are a nuisance when high resolution is desired, because the wings overlap neighboring reflections. Although corrections for these effects can be computed, their usefulness is limited by the extent to which the instrumental distortions can be properly formulated. Recently, it was demonstrated by U. Bonse and M. Hart that it is possible to produce a monochromatic x-ray beam that has virtually negligible wings on the main peak. A very perfect single crystal is so cut that a channel is formed by the same crystallographic planes serving as opposite sides of the channel. When one of these walls forms the correct Bragg angle with the incident x-ray beam, the reflected beam forms the correct angle for reflection by the opposite wall of the channel, and so on. The "sharpening" of the x-ray beam by such successive reflections is a characteristic of the reflection of x rays by highly perfect crystals (Chap. 9). Although the overall intensity of the beam is reduced by each reflection, the virtual elimination of the wings by as few as three successive reflections makes such an arrangement most useful for many experimental purposes.

Emission spectra. To gain an understanding of the interpretive difficulties in x-ray spectroscopy, consider the energy-level diagram in Fig. 6-20. This diagram was based on the assumption that, when an electron is ejected from an atom, the other electrons remain unaffected by its removal; that is, their relative energy levels do not change in the diagram. This is obviously not a reasonable expectation, even in view of the very short lifetime of an excited state ($\approx 10^{-16}$ sec). After ejection of an inner electron, the other electrons find themselves in a modified Coulomb field. In solids, this change is further complicated by the presence of surrounding atoms and the various possibilities of the final location of the ejected electron. Parratt has proposed the name *valence-electron-configuration,* or simply, VEC, excitation states to describe the possible energy levels of the outer electrons in an atom having a missing inner electron. Because the number of different configurations that can be envisaged is quite large and depends on the different ways an inner vacancy can be created, plus the different environments (coordinations) in which the atom may find itself, it should become evident that the construction of an accurate energy-level diagram becomes a well-nigh impossible undertaking, particularly since information regarding the exact possibilities is not available. Because of this, current interpretations are frequently based on one-electron energy-level diagrams like that in Fig. 6-20, but their shortcomings must be clearly recognized.

Assuming that the number of states available for a transition in an energy interval between E and $E + dE$ is known, the experimental emission curve, after suitable correction for all instrumental factors, represents a quantity proportional to the density of states dN/dE in that interval times the transition probability $T(E)$ to such states. The general form of this relation, from quantum mechanics, is

$$\frac{dN}{dE} T(E) \propto \frac{dN}{dE} \left| \int \Psi_f^* \frac{\delta}{\delta x} \Psi_i \, d\tau \right|^2 \tag{6-50}$$

where Ψ_f and Ψ_i are the wave functions of the final and initial states of the electron, and the integration is over all space. The difficulty in the application of (6-50) arises from the fact that the initial and, particularly, the final states in a system perturbed by a previously ejected electron cannot be specified with certainty. As stated above, it thus becomes necessary to utilize wave functions like those in unperturbed atoms, and this leads to transition probabilities that are expressed by the simple selection rules already discussed.

The density of states dN/dE in solids is not the same as that in isolated atoms, and, as is well known, the presence of closely packed atoms leads to overlapping electron orbitals, which leads to a spreading of the discrete energy levels (Fig. 6-20) into a multitude of closely spaced levels called energy bands. When the band theory of solids was first being developed in the early 1930's, a number of experimental studies of the x-ray spectra of solids were undertaken in order to test the new theories. These were largely successful, and established soft x-ray spectroscopy as an important tool in such studies. As an example of the kind of curves observed and the interpretations that have been proposed, consider Fig. 6-27. The top part shows the $K\alpha_{2,5}$ emission line (superposition of $K\alpha_2$ with $K\alpha_5$) of copper as measured by W. W. Beeman and H. Friedman on a two-crystal spectrometer, without corrections for instrumental effects. Assuming, nevertheless, that it is proportional to the quantity in (6-50), it is of interest to compare it with the density-of-states curve recently calculated by G. A. Burdick and shown at the bottom part of Fig. 6-27. (Beeman and Friedman originally compared their curves with a similar dN/dE curve calculated by E. Rudberg and J. C. Slater.) The Fermi energy E_0 is indicated, and the allowed states below this energy, normally occupied, are marked

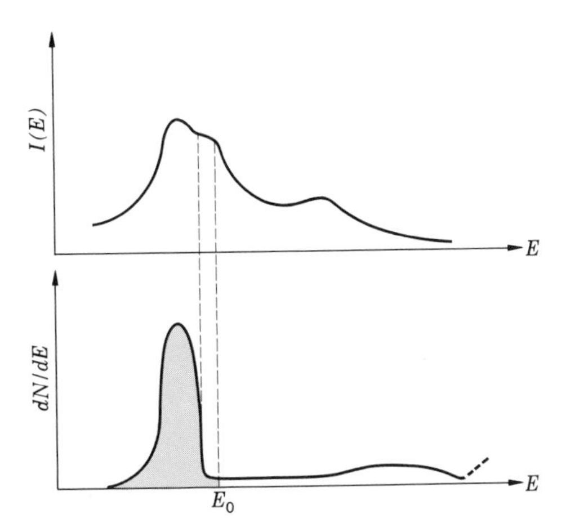

Fig. 6-27. $K\beta_{2,5}$ emission from copper [*from W. W. Beeman and H. Friedman, Physical Review, vol.* 56 (1939), *pp.* 392–412] compared with the density of states calculated for copper by Glenn A. Burdick [*Physical Review, vol.* 129 (1963), *pp.* 138–150].

by shading. According to Burdick's density-of-states curve, the shaded states have largely $3d$ symmetry, but there is an increasing admixture of $4s$ and $4p$ symmetry as E_0 is approached. About 4 eV above E_0 (to the right), the $4s$ symmetry becomes most predominant, while at larger energies it becomes purely $4p$.

To utilize (6-50) in the interpretation, it is necessary to realize that the initial state responsible for the $K\alpha_{2,5}$ line is the K or $1s$ state of an excited copper atom. If the atomic transition probabilities are assumed to be governed by the simple dipole selection rules, the final state must have p-type symmetry. Returning to Fig. 6-27, it is seen that, immediately to the left of E_0, the relatively large emission is caused by transitions to states having strongly admixed p symmetry. While the density of states increases at still lower energies, the admixture of p symmetry decreases, so that the intensity curve does not continue to rise. By comparison, as the energy increases to the right of E_0, the emitted intensity declines, but does not go to zero. This falling off is due partly to the rapidly declining number of electrons occupying energy states above E_0, and partly to some of the other factors cited at the beginning of this section, for example, VEC states, not taken into account in the present instance. It can be seen in Fig. 6-27 that the gross features of the dN/dE curve are clearly reflected in the emission curve. The width of the emission peak is related to the calculated bandwidth, although it is often difficult to locate the limiting points in the experimental curves precisely. In making quantitative correlations it is necessary also to take into account the width of the initial K state.

The emission curves of the same element in different compounds or of the same metal in different alloys may or may not show characteristic differences. For example, the manganese $K\alpha_1$ peak in MnO is shifted by $+0.06$ eV, in MnO_2 by $+0.59$ eV, and in $KMnO_4$ by $+1.18$ eV, as compared with its value in pure manganese metal. At the same time the $K\beta_1$ peak is shifted in MnO and MnO_2 by -1.8 eV, while in $KMnO_4$ it is shifted by 0.97 eV. By comparison, the $Al\ K\alpha_1$ line is shifted by as much as $+2.23$ eV in Al_2O_3 from its value in the pure metal. Despite the absence of a rigorous theory explaining the magnitudes and directions of these shifts, it is clear that they represent changes in the energy-level separations in the emitting atoms. It is possible, therefore, to make certain qualitative deductions about the electron structures or the valence states of elements in various compounds. At the very least, one can make use of such systematic shifts to identify an element's valence state in any compound, once an empirical relationship between the nature of the shift and the valence state in a known case has been established. This is, essentially, a fingerprinting method, and it has proved to be of considerable importance.

It should be noted that sometimes no change occurs, or the change is so small that it escapes detection. This is the case, for example, for nickel metal and nickel oxide, in which only the $Ni\ K\gamma$ line shows a detectable shift, in this case by $+1.66$ eV. As a final example, consider the $M_{II,III}$ emission bands from several copper-nickel alloys (Fig. 6-28). The metals Cu and Ni adjoin in the periodic table and have so closely similar sizes and electronic structures that they form continuous solid solutions in which the atoms are more or less randomly arrayed for all atomic ratios. Without attempting to explain the detailed fine structure evident in the

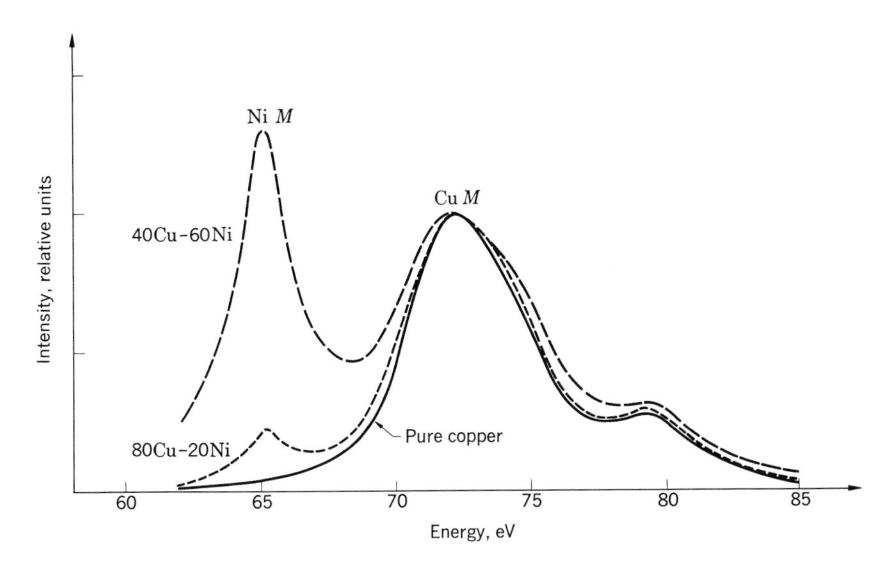

Fig. 6-28. *M* spectra from copper-nickel solid solutions. [*From J. Clift, C. Curry, and B. J. Thompson, Philosophical Magazine, vol.* 8 (1963), *pp.* 593–604.]

curves, it is noteworthy that, as the nickel content increases, the emission curve simply shows a rise in the relative intensity of the nickel line. (The curves in Fig. 6-28 have been normalized to the copper emission line, so that its magnitude is shown unchanged.) In fact, as Clift, Curry, and Thompson have demonstrated, the experimental curves for any Cu/Ni ratio can be reproduced by simple superposition of suitably weighted emission curves of pure copper and pure nickel. These results appear to indicate that the x-ray spectra of the atoms in such solid-solution alloys are primarily a function of the emitting atom's electronic structure, as modified by its immediate environment, and not necessarily of the density-of-states distribution as deduced for the alloy as a whole.

Absorption spectra. Making use of the one-electron approximation, the complete energy-band model of a typical metal may look like Fig. 6-29. In this diagram, the energy plotted is that of an electron corresponding to the quantum states listed. Suppose an L electron is knocked out with sufficient additional energy to raise it to one of the normally unoccupied quantum states, but not enough to eject it completely. The smallest amount of additional energy needed is that which will raise the electron to a state whose energy just exceeds the Fermi level E_0. Since the conventional energy-band model (Fig. 6-29) plots electron energy increasing upward, such a transition can be indicated by an upward sense on the arrow, shown to the left in the diagram. If the resulting hole in the L band is then filled by one of the valence-conduction electrons in the metal, whose energy band is near the top in Fig. 6-29, the transition can be indicated by either of the two downward arrows in the diagram. As clearly seen in Fig. 6-29, the energies of the two x-ray photons emitted in these two cases differ by nearly the width of the

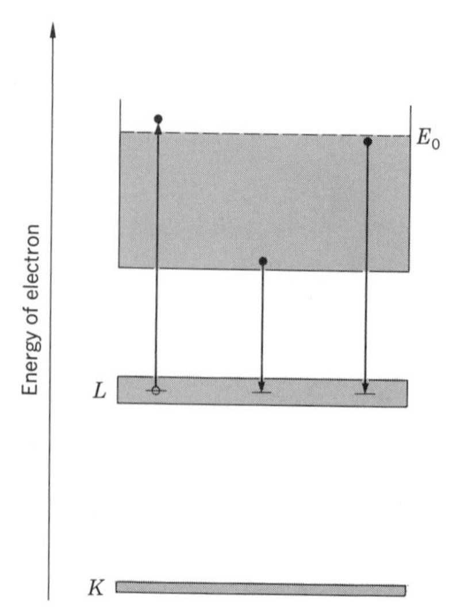

Fig. 6-29

valence-conduction band, which is the underlying cause of the relative widths of the emission lines discussed in the preceding section.

Note that the energy required to raise an electron to a state just above E_0 is quite similar to that released when an electron falls from an energy state just below E_0. The upward process corresponds to the absorption of energy, for example, an x-ray photon, while the reverse transition corresponds to the release of an x-ray photon. From this discussion, therefore, it appears that the appropriate emission and absorption spectra in metals should overlap. Such an overlap is clearly seen in the case of copper $K\alpha_{2,5}$ emission and K absorption spectra, as shown in Fig. 6-30. The fine structure evident in the main absorption edge can be explained in terms of the density-of-states curve (Fig. 6-27) by assuming that the transition probability (6-50) reduces to the familiar dipole selection rules (6-43). On this basis, the initial rise corresponds to the onset of unoccupied quantum states having energies exceeding E_0 and p-type symmetry. It will be recalled from the preceding section that about 4 eV above E_0 the symmetry of the empty states assumes a predominantly s character, so that the transition probabilities decline at this energy, causing a decline in the absorption. At slightly higher energies, however, the states assume virtually pure $4p$ character, and the absorption coefficient increases rapidly

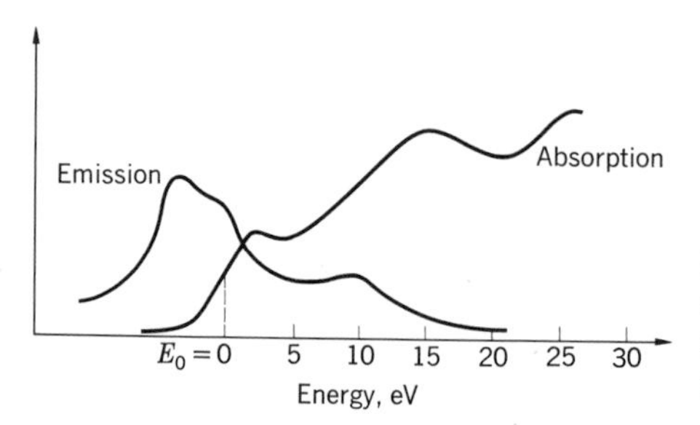

Fig. 6-30. Copper $K\beta_{2,5}$ emission band and K absorption spectrum drawn to the same energy scale. [*From W. W. Beeman and H. Friedman, Physical Review, vol.* 56 (1939), *pp.* 392–412.]

again. The subsequent maxima, similarly, are believed to be due to transitions to $5p$ and higher p states.

The above model ignores the fact that when a K electron is ejected, the potential distribution in the immediate vicinity of the absorbing atom is altered, so that the simple diagram is no longer valid. Without attempting to describe them quantitatively, Parratt has drawn attention to the possible existence of such perturbed states by suggesting the name *bound-ejected-electron,* or simply, BEE, states for such cases. Attempts to calculate the energy-band models for similar possible states in metals have been made by J. Friedel, but direct experimental correlation is still lacking. This is simply another example of the interpretive difficulties in x-ray spectroscopy. In this connection it should be noted that the conventional band model (Fig. 6-30) can be thought of in another way also. The energy bands shown are composed of many closely spaced energy levels corresponding to the actual energy values that electrons can have near some particular atom. Thus, for a given atom, the available possible BEE states to which the ejected (but remaining bound) electron can be excited are determined largely by what is available in its immediate environment.

This approach is helpful in understanding what happens to the fine structure of the copper K edge when nickel is alloyed with it. Because copper has one more electron than nickel, it is very likely that for each nickel atom added, one of the copper valence electrons is "shared" with an adjacent nickel atom, thus making it more "like" the copper atoms in the structure. If a K electron in a copper atom that gave up its $4s$ electron then is ejected, the empty states in its vicinity are increased in number by the removal of this valence electron. The result should be a slight rise in the first absorption peak of copper as more nickel is added because the number of copper atoms having such additional vacant states is increased thereby. Such an increase is actually observed, as shown in Fig. 6-31. Correspondingly, the electron transferred to the nickel atom should decrease the density of empty states near it, so that the first maximum in the nickel absorption edge should show a corresponding decrease. Although relatively small, a systematic decrease is clearly evident when the curves of several solid-solution alloys containing increasing amounts of copper are compared.

The curves shown in Fig. 6-31 were obtained on a two-crystal spectrometer and have been corrected for instrumental effects. In order to enable a direct comparison between the curves derived from alloys containing different amounts of copper, the quantity plotted is the atomic absorption coefficient of copper μ_{Cu} given by the equation

$$\mu_{\text{Cu}} = \rho_{\text{Cu}}\mu'_{\text{Cu}} = \frac{\rho_{\text{Cu}}}{p_{\text{Cu}}}\left(\frac{\mu_{\text{alloy}}}{\rho_{\text{alloy}}} - p_{\text{Ni}}\mu'_{\text{Ni}}\right) \tag{6-51}$$

where the prime denotes the mass absorption coefficient; ρ is the density; and p is the weight fraction of each constituent. After measuring the absorption-coefficient values of the other elements in the alloy at the energies (wavelengths) employed, relation (6-51) can be used to convert the linear absorption coefficient of the alloy μ_{alloy} to the absorption coefficient of the element desired. Having thus normalized the alloy curves, they can be compared directly with the absorption curve of the

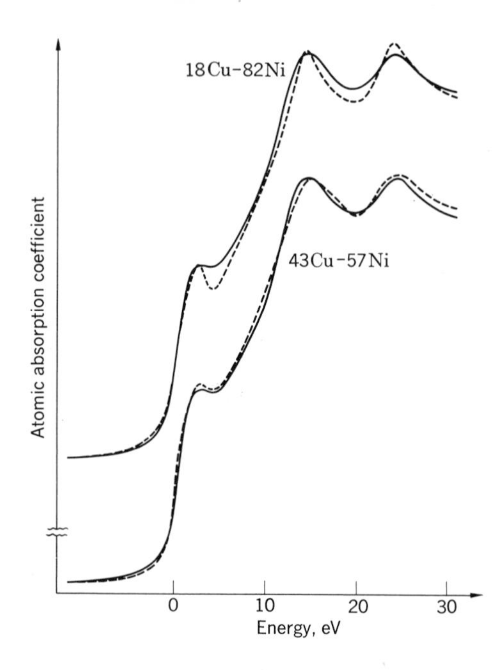

Fig. 6-31. Copper-K-edge fine structure in two nickel alloys (dashed curves) superimposed on that of pure copper (solid curves) in both cases. [*From L. V. Azároff and B. N. Das, Physical Review, vol.* 134 (1964), *pp.* A747–A751.]

pure metal. This is done in Fig. 6-31, where the absorption curve of pure copper is shown by the solid curves. It is repeated several times, with the copper absorption curves in the alloys superimposed on each solid curve to make it easier to compare the relative magnitudes of the derivations produced by alloying.

The foregoing proposed explanation of the systematic rise in the first maximum near the copper absorption edge is based on the so-called rigid-band model for solid-solution alloys. If this explanation is correct, the reverse situation should be encountered when zinc is alloyed to copper. According to the rigid-band model, the copper and zinc atoms share their $4s$ electrons statistically in solid solutions. Since copper normally has one $4s$ electron while zinc has two, such electron sharing causes a decline in the number of empty $4(s\text{-}p)$ states in the vicinity of copper atoms. This, in turn, should cause a decline in the number of $4s$ states that an ejected K electron can transfer to, with a commensurate decline in the first absorption maximum of copper. That this is indeed the case can be seen by comparing the curves shown in Fig. 6-32. The progressive decline in the integrated absorption in this energy region can be followed semiquantitatively. By dividing the absorption curves in Fig. 6-32 into appropriate energy intervals, it is possible to measure the areas of each region for each alloy and to observe the actual changes taking place. Since alloying affects only the occupation of states lying close to the Fermi energy (initial rise in absorption), it turns out that integrated absorption changes only in the region corresponding to the valence band. In the case of nickel, for example, the addition of copper (or zinc) causes a progressive filling of the normally empty $3d$ states, so that a decline in the area of this portion of the curve is observed. Assuming that the integrated absorption in this energy interval is proportional to 0.6 holes per atom in pure nickel and to zero holes per atom in a solid-solution alloy containing 80% copper, it is possible to plot the declining density of $3d$ holes

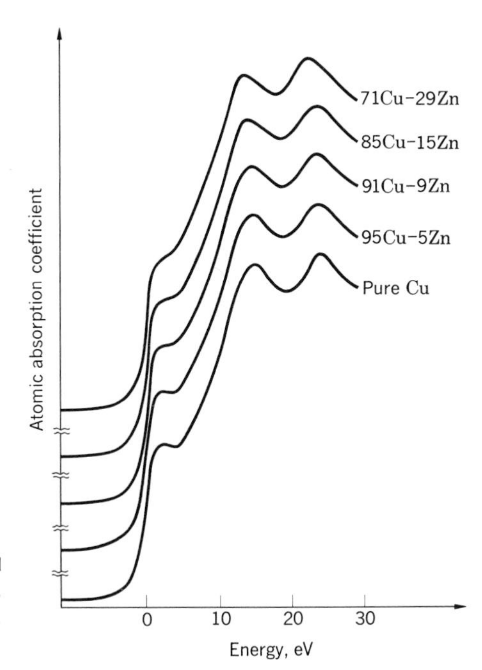

Fig. 6-32. Copper-K-edge fine structure in several copper-zinc solid solutions. [*From H. C. Yeh and L. V. Azároff, Applied Physics Letters, vol.* 6 (1965), *pp.* 207–208.]

as a function of composition (Fig. 6-33). Quite similarly, one can follow the decline in the $3d$-hole density of nickel when zinc is added (Fig. 6-34). It is significant to note that saturation sets in close to 60 at.% copper and 30 at.% zinc, in accord with similar extrapolations based on magnetic and other measurements.

The fine structure occurring within a few tens of electron volts from the main edge is frequently called the *Kossel structure,* and the fine structure extending for hundreds of electron volts is called the *Kronig structure.* The distinction is primarily due to the absorption processes involved. As already described, the near-in structure is caused by variously excited states of the atom, whereas it seems unlikely that the extended fine structure is greatly influenced by the atomic states. The reason for this is that the increase in the kinetic energy of the electron is

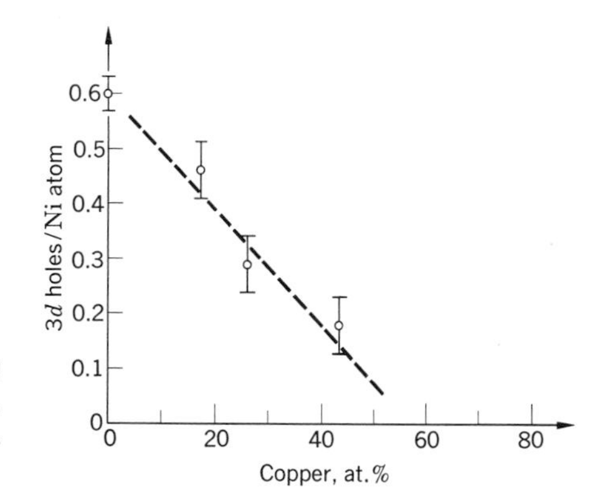

Fig. 6-33. Decline of $3d$-hole density at nickel atoms in copper solid solutions. [*From L. V. Azároff, Journal of Applied Physics, vol.* 37 (1967), *pp.* 2809–2812.]

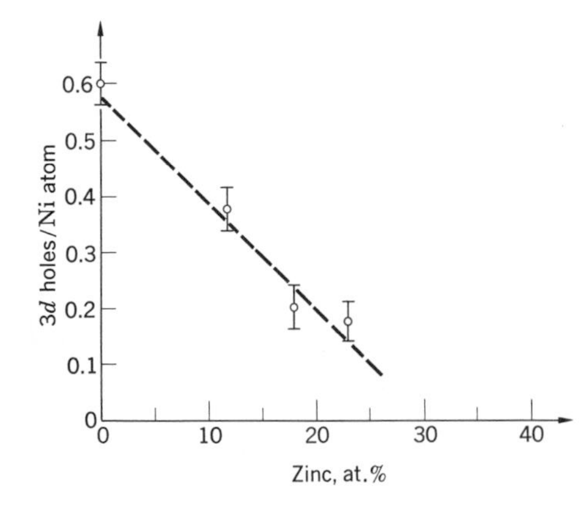

Fig. 6-34. Decline of $3d$-hole density at nickel atoms in zinc solid solutions. [*From H. C. Yeh and L. V. Azároff, Journal of Applied Physics, vol.* 37 (1967), *pp.* 4034–4038.]

sufficient to remove it from the immediate "sphere of influence" of the absorbing atom. Such a suggestion was first put forth by R. de L. Kronig in 1931, who based his argument on concepts evolved in the development of the band theory of solids. Assuming that an electron ejected from an atom can be represented by a plane wave, Kronig suggested that the electron wave undergoes Bragg reflection by planes (hkl), similarly to the treatment of conduction-electron waves in metals, with the result that zones of allowed- and forbidden-energy values are formed. The maxima in an absorption curve like that in Fig. 6-35 then correspond to electron transitions to allowed energy states, while the minima occur at forbidden-energy values. An approximately accurate agreement between the allowed-energy values in electron volts,

$$E = 37.5 \frac{h^2 + k^2 + l^2}{a^2} \qquad \text{eV} \tag{6-52}$$

and the location of the maxima and minima is in fact observed for a number of

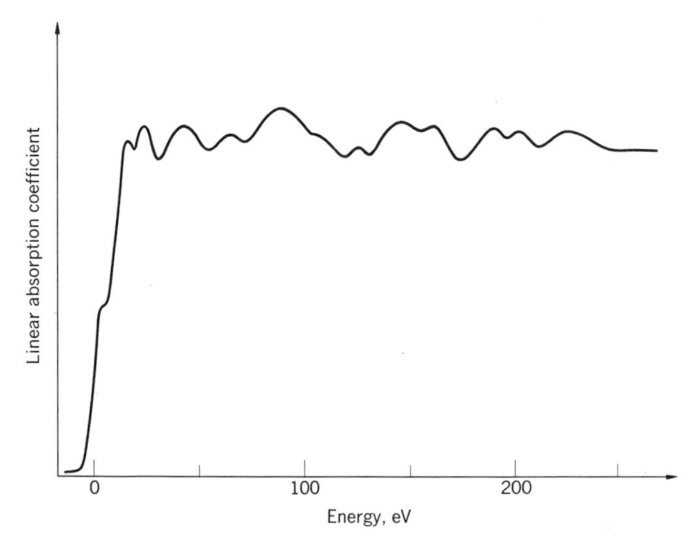

Fig. 6-35. Extended fine structure at copper K edge.

cubic metals. The predicted inverse dependence of the relative spacings of the maxima on a^2, where a is the cell edge in a cubic lattice, is also observed.

A somewhat different interpretation has been proposed by T. Hayasi, who assumes that the plane wave describing the ejected electron is reflected backward so that it overlaps the wave function of the initial state. This obviously causes an increased overlap of the two functions in (6-50), and hence corresponds to an increased transition probability. For cubic crystals, the energy values are still given by (6-52), except that now the absorption maxima correspond to energy states, called quasi-stationary states, lying in the forbidden-energy regions of the band model. Their relative spacing is similar in both theories, and because the true zero along the energy scale cannot be located in an absorption study with sufficient accuracy, it is not possible to discern the relative validity of either theory from a simple experiment.

Both the Kronig and Hayasi models are connected to the presence of long-range order, that is, regular crystal periodicity, although it may be possible to reformulate the Hayasi theory without such a restriction. Several experiments have demonstrated quite convincingly that the extended fine structure is the same in various compounds whether they are in a crystalline or a glassy (noncrystalline) state. Thus it appears that the long-range periodicity is not essential for an extended fine structure to form. A number of treatments for absorption by solids have been developed that are based on atomic models requiring only a short-range order or grouping of atoms about the absorbing atom. Chief among the investigators are A. I. Kostarev, M. Sawada, and A. I. Kozlenkov. Actually, their mathematical treatments are based on original calculations developed for molecules by Kronig and modified by H. Petersen. Because of the attendant mathematical difficulties, most of these analyses include a number of simplifying assumptions, so that they do not actually predict the shape of the absorption curve with the same reliability that they purport to locate the absorption maxima along the energy scale. To see how they compare with each other, Table 6-5 lists the energy values at which the maxima are observed to occur within the first 300 eV of the extended fine structure of copper (Fig. 6-35), along with the values predicted by (6-52), in the column headed "Hayasi," as well as the theoretical predictions of the three main short-range-order calculations. Since all the theories calculate only relative peak locations with reliability, the values have been "normalized" by fixing the maximum observed at 103 eV from the main edge (labeled E) to have the same value in all calculations. It is clear from Table 6-5 that the various calculations appear to agree equally well with the experimental values, regardless of differences in their initial formulations. A similar correspondence exists between the energies predicted and observed for the absorption minima. The inescapable conclusion from this comparison is that, until a theory is developed that correctly predicts the shape as well as the positions of the undulations in the absorption curve, it is not possible to refine present notions regarding what the actual absorption mechanisms are that produce the extended fine structure. It also should be noted that these theories do not take account of the effect that the inner vacancy has on the wave functions, nor do they consider such processes as multiple ionization or plasma interactions for the ejected electrons. Plasma oscillations are particularly impor-

Table 6-5 Experimental and theoretical values of the energy (in electron volts) of absorption maxima in the extended fine structure of copper†

Peak	*Experiment*	*Kronig*	*Hayasi*	*Kostarev*	*Sawada*	*Kozlenkov*
A	26.2	...	23	24	10	21
B	34.3	...	34	34	39	31
C	56.5	69	57	52	63	52
D	77.6	78	80	80	77	75
E	103.0	103	103	103	103	103
F	122.0	122	115	125	121	122
G	143.2	149	126	...	147	149
H	163.3	166	...	161	165	168
I	176.0	196	172	182	...	...
J	203.5	208	...	...	...	209
K	213.0	218	219	210	211	217
L	235.2	241	...	241	231	242
M	282.3	...	287	308	286	298

† After Leonid V. Azároff, Theory of extended fine structure of x-ray absorption edges, *Reviews of Modern Physics*, vol. 35 (1963), pp. 1012–1022.

tant in metals, although it is not certain whether they could affect the fine structure beyond the first 50 eV.

REFRACTION OF X RAYS

Dispersion. The term dispersion generally is used in physics to mean the separation of electromagnetic radiation into its component wavelengths. In ordinary optics, dispersion is observed by a differential bending of these components on passage through a prism. The first theoretical treatment of this phenomenon was carried out by Maxwell, who considered the influence of a medium containing oscillators having characteristic frequencies on the light transmitted through it. The first treatment devoted specifically to x rays was by Compton, who adapted the theories developed earlier for visible light by H. A. Lorentz.

For present purposes it is sufficient to realize that the electrons in atoms can be considered to act like oscillators having resonance frequencies determined by their binding forces, that is, depending on whether they are in the K, L, etc., shells of an atom. When the frequency of the incident radiation is far removed from these characteristic frequencies, the presence of such oscillators simply causes a change in the velocity of the transmitted wave which is usually related to the refractive index μ_r of the medium. When the frequency of the incident radiation is close to the characteristic frequency of the oscillators, a phase shift also takes place, and this phenomenon is called *anomalous dispersion*, and is briefly discussed further in Chap. 8. Experimentally, it is found that the index of refraction for x rays normally used in diffraction experiments is less than unity by a few parts per million in most materials.

Modification of Bragg law. As early as 1919, W. Stenström observed that the apparent wavelength of molybdenum $L\beta_1$ radiation changed when reflected by different orders of the same plane in a single crystal. Designating the Bragg angle in air θ (corresponding to λ) and within the crystal θ' (corresponding to the changed wavelength λ'), the refractive index for x rays μ_r can be defined by the ordinary law of optics, by noting, in Fig. 6-36, that the Bragg angle θ is the complement of the conventional angle of incidence.

$$\mu_r = \frac{\lambda}{\lambda'} = \frac{\cos\theta}{\cos\theta'} \tag{6-53}$$

Within the crystal the Bragg law becomes

$$n\lambda' = 2d\sin\theta' \tag{6-54}$$

which can be substituted in (6-53) to give

$$\begin{aligned}\mu_r &= \frac{\lambda}{\lambda'} = \frac{\lambda}{2(d/n)\sin\theta'} \\ &= \frac{n\lambda}{2d[1-(\cos^2\theta)/\mu_r{}^2]^{\frac{1}{2}}}\end{aligned} \tag{6-55}$$

The terms in (6-55) can be rearranged to give a modified form of the Bragg equation,

$$n\lambda = 2d\sin\theta\left[1 + \frac{(\mu_r+1)(\mu_r-1)}{\sin^2\theta}\right]^{\frac{1}{2}} \tag{6-56}$$

Since the refractive index for x rays is so close to unity, $\mu_r + 1 \cong 2$, while $1 - \mu_r \equiv \delta$ is very nearly zero. Expanding the numerator in the last term in (6-56) in powers of δ and retaining only the first term, it can be simplified to

$$n\lambda = 2d\sin\theta\left(1 - \frac{1-\mu_r}{\sin^2\theta}\right) \tag{6-57}$$

which is the standard form of the *modified Bragg equation.*

The refraction of x rays has been studied by passage through $90°$ prisms, with the incident rays virtually grazing the calcite prism used and by total reflection below a critical glancing angle, as discussed further in the next section. The refractive index also can be determined by measuring the reflecting angles for two orders of reflection from the same plane in a crystal and forming a ratio between the two resulting forms of (6-57) as outlined in Exercise 6-23. Finally, x-ray wavelengths unaffected by refraction can be determined by using ruled gratings instead

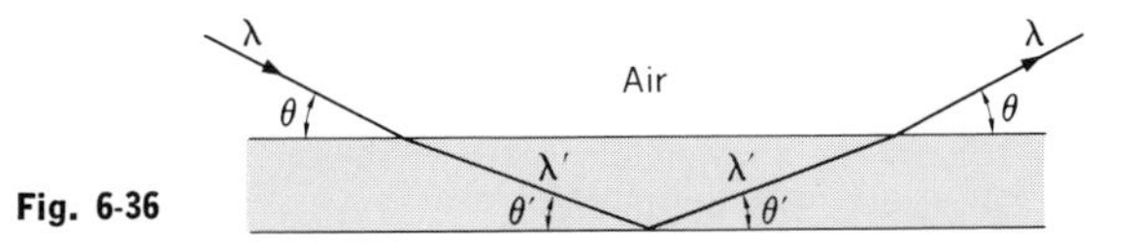

Fig. 6-36

of crystals. This method has the added advantage that any errors introduced by variations in crystal perfection are also excluded.

Total reflection. An examination of (6-53) shows that, since $\mu_r < 1$, the Bragg angle within the crystal is always less than the glancing angle θ of the incident beam. As the glancing angle decreases, there must come a point when $\theta' = 0°$ while θ remains finite. This is the familiar critical angle θ_c below which total reflection sets in. According to elementary optics, the critical angle is related to the refractive index by

$$\theta_c^2 = 2(1 - \mu_r) = 2\delta \tag{6-58}$$

For glancing angles less than this, none of the rays are transmitted; that is, all the rays incident at $\theta < \theta_c$ are reflected at the surface. Some reflection, of course, takes place at the interface between two media having different refractive indices at all glancing angles. To estimate the critical angle at which total reflection sets in, consider $\mathrm{Mo}\ K\alpha$ radiation striking a polished glass surface. For most glasses $\delta \approx 2 \times 10^{-6}$, so that, from (6-58), $\theta_c \approx 7'$. In view of the smallness of the critical angle, it is clear that extremely flat surfaces are necessary for determinations of the refractive index of x rays by the total-reflection method. On the other hand, it is possible to devise reflection experiments which serve to characterize the surface smoothness. It also has been shown that it is possible to produce interference effects in the films deposited on glass substrates by the total reflection at the boundaries of the metal film. As discussed in Exercise 6-24, the interference effects are related to film thickness, and also can be used to measure the refractive index of the film.

SUGGESTIONS FOR SUPPLEMENTARY READING

Leonid V. Azároff, Theory of extended fine structure of x-ray absorption edges, *Reviews of Modern Physics*, vol. 35 (1963), pp. 1012–1022.

Mikhail A. Blokhin, *The physics of x-rays*, 2d ed. (State Publishing House of Technical-Theoretical Literature, Moscow, 1957), translated into English by U.S. Atomic Energy Commission, document number AEC-tr-4502.

Y. Cauchois, *Les spectres de rayons x et la structure électronique de la materière* (Gauthier-Villars, Paris, 1948).

Arthur H. Compton and S. K. Allison, *X-rays in theory and experiment*, 2d ed. (D. Van Nostrand Company, Inc., Princeton, N.J., 1935).

Maurice de Broglie, *X-rays* (E. P. Dutton & Co., Inc., New York, 1930), translated into English by J. R. Clarke.

L. G. Parratt, Electronic band structure of solids by x-ray spectroscopy, *Reviews of Modern Physics*, vol. 31 (1959), pp. 616–645.

F. K. Richtmeyer, The multiple ionization of inner electron shells of atoms, *Reviews of Modern Physics*, vol. 9 (1937), pp. 391–402.

Arne Eld Sandström, Experimental methods of x-ray spectroscopy: ordinary wavelengths, in S. Flügge (ed.), *Handbuch der Physik* (Springer-Verlag OHG, Berlin, 1957), vol. 30, pp. 78–245.

Werner Schaafs, Erzeugung von Röntgenstrahlen, in S. Flügge (ed.), *Handbuch der Physik* (Springer-Verlag OHG, Berlin, 1957), vol. 30, pp. 1–77.

C. H. Shaw, The x-ray spectroscopy of solids, in *Theory of alloy phases* (American Society for Metals, Cleveland, 1956), pp. 13–62.

Seymour Town Stephenson, The continuous x-ray spectrum, in S. Flügge (ed.), *Handbuch der Physik* (Springer-Verlag OHG, Berlin, 1957), vol. 30, pp. 337–370.
Diran H. Tomboulian, The experimental methods of soft x-ray spectroscopy and the valence band spectra of the light elements, in S. Flügge (ed.), *Handbuch der Physik* (Springer-Verlag OHG, Berlin, 1957), vol. 30, pp. 246–304.

EXERCISES

6-1. Consider an initially unpolarized x-ray beam that is reflected in succession by two crystals, like the arrangement in Fig. 6-26. Show that the polarization factor for the twice-reflected beam is $(1 + \cos^2 2\theta_1 \cos^2 2\theta_2)/(1 + \cos^2 2\theta_1)$, where the subscripts distinguish the Bragg angles for the first and second crystal, respectively. Note that this relation is valid only provided that the x-ray beam is confined to remain within the same plane.

6-2. The value of the mass scattering coefficient for carbon was determined by Barkla to be approximately 0.2 cm²/g. Assuming that the atomic weight of carbon is 12 amu, prove that a carbon atom has six electrons by calculating the number of electrons per gram of carbon and dividing this by the number of carbon atoms in 1 g. This was the first experimental determination of the electronic structure of carbon.

6-3. Derive Eq. (6-19), giving the Compton shift in angstroms. **Hint:** Squaring and combining equations like (6-16) and (6-17) permits the elimination of undesired trigonometric functions.

6-4. What is the attenuation of a monochromatic beam of x radiation after passage through 1 m of air if it comes from a chromium target ($\lambda_{K\alpha} = 2.291$ Å)? Assume that air is composed of $\frac{1}{5}$ oxygen and $\frac{4}{5}$ nitrogen and that the density of air is 1.45×10^{-3} g/cm³. Through how much air would a monochromatic beam from a molybdenum target ($\lambda_{K\alpha} = 0.711$ Å) have to travel to suffer the same amount of attenuation?

6-5. Calculate the mass and linear absorption coefficients of pyrite, FeS_2, for molybdenum radiation ($\lambda_{K\alpha} = 0.711$). The density of pyrite is 4.9 g/cm³, and the weight fractions can be calculated from the known atomic weights.

6-6. Making use of the mass absorption coefficients in Table A-5, prepare a graph plotting the atomic absorption coefficient of nickel as a function of λ^3. Next, plot μ_a vs Z^4 for the elements S through Zn for, say, Mo $K\alpha$ radiation. (Look up their densities in a handbook.) How well do these curves agree with relations (6-30) in the text?

6-7. Using the wavelength values of the absorption edges listed in Tables A-3 and A-4, calculate the energy required to knock out a K electron or one of the L electrons from a tungsten atom. How high a potential must be applied to the x-ray tube to excite both the K and L spectra of tungsten?

6-8. A complete discussion of the Jönsson empirical curve is given on pp. 538–542 in Compton and Allison's book (see literature list at end of Chap. 6). After reading that section, determine the mass absorption coefficient of nickel from Jönsson's curve at the same wavelengths as those given in Table A-5, and compare the two sets of values.

6-9. Consider the nickel filter recommended for use with copper radiation in Table 6-1. By how much does it actually attenuate the $K\alpha$ component? By how much does this filter attenuate a component in the continuous spectrum having the same wavelength as Mo $K\alpha$ radiation?

6-10. Check the numerical coefficients in relations (6-31) and (6-34) by appropriate combinations of physical constants.

6-11. The variation of the intensity of x rays of constant wavelength emitted from an x-ray target with changing tube voltage is called an *isochromat*. By consulting appropriate references (see list at end of Chap. 6), discover what this variation is and prepare a short report on its role in the physics of x rays.

6-12. Making use of the wavelength values in Table A-3, prepare a Moseley diagram by plotting the wavelengths of the K edges and the $K\alpha_1$, $K\alpha_2$, and $K\beta$ lines of the first 40 elements against their atomic numbers. To simplify the construction, plot the wavelengths given for each third element on the same piece of graph paper. What is the reason for the joint convergence of all these lines at $Z \cong 1$?

6-13. Using the wavelength values of the K edges in Table A-3, calculate the wavelengths of the $K\alpha_1$, $K\alpha_2$, and $K\beta$ lines of nickel and copper. Compare these values to the tabulated wavelengths.

6-14. Supposing that the transition $W_K \rightarrow W_{L_\mathrm{I}}$ were possible, calculate the wavelength of the corresponding emission line for silver. Would it be possible to resolve this line from the rest of the K spectrum of silver using an instrument capable of resolving the $\alpha_1\alpha_2$ doublet?

6-15. Calculate the wavelength of the $K\alpha_{3,4}$ satellite for copper.

6-16. The absence of L satellites in the range $50 < Z < 75$ has been attributed to the existence of Auger transitions. Assuming that, initially, the quantum state of the atom is L_I and following the Auger transition it is $L_\mathrm{III}M_\mathrm{IV}$, show that such transitions are possible in atoms like barium $(Z = 56)$ and tungsten $(Z = 74)$, but not possible in silver $(Z = 47)$. **Hint:** Derive a relation like (6-45), and show that the transition energy is positive only in the indicated range. Would you expect the presence of these Auger transitions to affect the diagram lines of the L spectrum?

6-17. Consider the bent-crystal monochromator in the transmission arrangement shown in Fig. 6-23. If the radius of the focusing circle is R and the crystal-to-focus distance is D, show that they are related to the Bragg angle by the relation $2R = D/\cos\theta$. What provisions must be made to enable the use of a single crystal for measuring different wavelength components in the x-ray beam in such an arrangement?

6-18. Consider the Johann arrangement in Fig. 6-24. By constructing a suitable drawing, show that focusing circles having different radii are needed to obtain perfect focusing for rays reflected at the same Bragg angle by different points along the crystal's surface. How does this condition differ from that of the transmission arrangement in Fig. 6-23?

6-19. Consider a quartz crystal whose elastic properties make it most suitable for bent-crystal monochromators. Selecting the strongly reflecting planes $(10\bar{1}1)$ for the construction of a Johansson monochromator, what are the bending radii necessary for its use with $\mathrm{Mo}\ K\alpha$ or with $\mathrm{Cu}\ K\alpha$ radiation for a crystal-to-focus distance of 10 cm?

6-20. There are different ways to define the resolving power of a spectrometer. The most convenient is $\lambda/d\lambda = \lambda/Dw$, where D is the dispersion of the instrument and w the instrumental broadening of the line. The dispersion in a two-crystal spectrometer in the antiparallel arrangement is twice that of a single crystal. If the line broadening by successive reflection from two identical crystals is $\sqrt{2}$ larger than that due to a single crystal, what is the theoretical increase in the resolving power of a two-crystal spectrometer?

6-21. Defining the resolving power of a spectrometer $\lambda/d\lambda$, make use of the Bragg law to demonstrate that, for a given wavelength, the resolving power increases with increasing θ, that is, with decreasing interplanar spacing d.

6-22. Express the energy at the K edge of copper in electron volts. Next, assume that it is desired to measure the variation of μ with energy in 5-eV steps. With what precision must one measure relative wavelength values? If a silicon crystal analyzer is used $(d_{111} = 3.135\ \text{Å})$, what is the corresponding angular increment that must be measured?

6-23. One of the most direct methods for measuring the index of refraction in a crystal is to measure the apparent wavelengths satisfying the Bragg law for two different orders of reflection and then forming a ratio between the two resulting modified equations (6-57). For the case of gypsum $(d = 7.57907\ \mathrm{kxu})$ and $\mathrm{Fe}\ K\alpha_1$, E. Hjalmar found the wavelength value $1.9341\ \mathrm{kxu}$ for the first-order reflection and $1.9306\ \mathrm{kxu}$ for the sixth-order reflection. On the basis of these observations, what is the refractive index of gypsum for $\mathrm{Fe}\ K\alpha_1$ radiation?

6-24. It can be shown by an analysis similar to that used in deriving the Bragg law that monochromatic x rays reflected (true optical reflection is meant here) at two opposite surfaces of a thin film will interfere constructively to produce maxima when

$$n\lambda = 2t\sqrt{\theta^2 - \theta_c^2}$$

where n is an integer; t is the film thickness; and θ is the glancing angle of the incident beam. Supposing that the interference maxima for two orders n_1 and n_2 can be measured, derive an expression that permits the determination of the film thickness t and another that measures the film's refractive index.

PART THREE
Elements of Diffraction Theory

SEVEN
Reciprocal-lattice concept

ELEMENTARY CONSIDERATIONS

Whenever it is necessary to consider sets of planes in a crystal, there is a conceptual advantage in thinking in terms of their normals (one-dimensional lines) instead of the two-dimensional planes. This advantage already should be apparent from Chap. 2, where it was shown how each plane can be represented by a pole (point) in a projection. Recalling from Chap. 5 that Bragg analyzed the diffraction of x rays by crystals in terms of reflections by internal (hkl) planes, it can be surmised that a simplified representation of the planes in a crystal would be of considerable value in the interpretation of x-ray diffraction experiments. The relative disposition of poles in a projection serves to disclose the interplanar angles and to describe the relative dispositions of the crystal planes. When considering the different directions in which x rays are diffracted by a crystal, however, it is not enough to know the disposition of the planes relative to each other. Their interplanar spacings d also must be known, because they determine the reflection angles θ. How the relative orientations of the planes can be displayed by their normals is already familiar. Consider, next, how the interplanar spacing of each plane also can be represented by these normals after appropriately limiting their lengths. If, as before, the normals to each plane are drawn from a common origin and their lengths are made proportional to the reciprocals of the respective interplanar spacings, the points at the ends of these normals form a lattice array. Because distances in this array are reciprocal to distances in the crystal, the array is called the *reciprocal lattice* of the crystal. It has the big advantage that each parallel stack of planes in a crystal is represented by a single reciprocal-lattice point, greatly simplifying thereby the task of keeping track of the various mutually interpenetrating planes in the crystal.

Graphical construction. Before proceeding to a mathematical analysis of the reciprocal-lattice concept, it is instructive to recapitulate how a reciprocal lattice can be constructed graphically. For simplicity, consider all the planes belonging to a single zone. Since these planes all are parallel to the zone axis, their normals

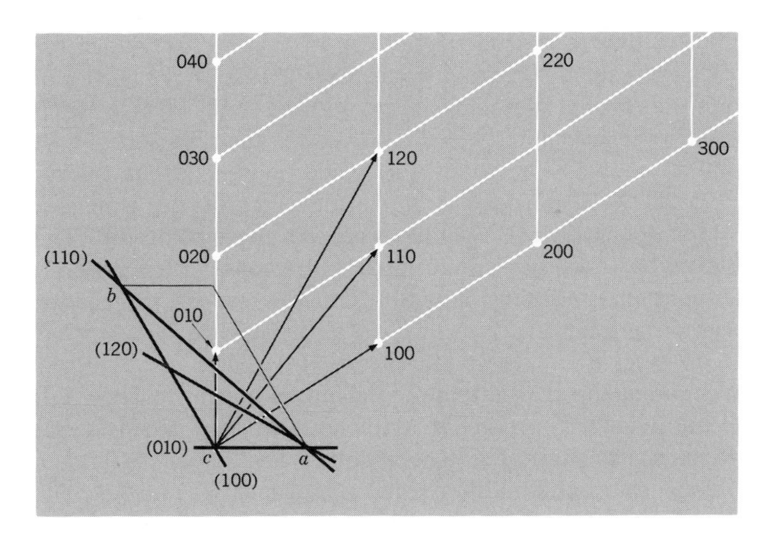

Fig. 7-1

all must lie in the same plane (normal to the zone axis), so that these planes can be represented by a two-dimensional reciprocal lattice. Such a situation is pictured in Fig. 7-1, which shows the unit cell of a monoclinic crystal looking along its unique axis, here designated c. The cell edges seen in Fig. 7-1 are, accordingly, a and b. The illustration also shows the edge views of four (hkl) planes belonging to the $[001]$ zone, namely, (100), (110), (120), and (010).

To locate the reciprocal-lattice points corresponding to these planes, one proceeds as follows:

1. From a common origin, draw a normal to each plane.
2. Place a point on the normal to each plane (hkl) at a distance from the origin equal to $1/d_{hkl}$.

Note that each point preserves all the important characteristics of the parallel stack of planes it represents. The direction of the point from the origin preserves the orientation of the plane, and the distance of the point from the origin preserves the interplanar spacing of that stack of planes.

The reader should perform (Exercise 7-1) the actual construction diagramed in Fig. 7-1. In doing this, it will become necessary to decide on some scale factor to use in the actual drawing. For convenience, therefore, define the length of the normal in step 2 above by

$$\sigma_{hkl} = K \frac{1}{d_{hkl}} \tag{7-1}$$

where K is the scale factor chosen for the reciprocal-lattice construction. Next, it should be clear that the planes (100), (200), (300), etc., all have parallel normals. The lengths of their interplanar spacings bear the relation $d_{100} = 2d_{200} = 3d_{300}$ . . . , so that $\sigma_{100} = \frac{1}{2}\sigma_{200} = \frac{1}{3}\sigma_{300}$ . . . , and similarly for all other (hkl) planes.

Thus a divider can be used to lay off quickly successive reciprocal-lattice points along the same normal. When the normals to the (hkl) planes shown in Fig. 7-1 are thus drawn, it becomes apparent that the points labeled 100, 110, 120 lie along a straight line parallel to the normal of the crystal plane (010) along which the points labeled 010, 020, etc., lie (Exercise 7-2). When the reciprocal-lattice points of all the $(hk0)$ planes are thus located, the set of points is found to lie on a two-dimensional lattice. A simple proof that these points *must* lie on a two-dimensional lattice has been given by Buerger, whereas a vector-algebraic proof that a three-dimensional reciprocal lattice is similarly formed in a real crystal was presented earlier by Ewald.

Vector-algebraic discussion. Having demonstrated how the reciprocal lattice is constructed, it now will be demonstrated that the collection of points in three dimensions conforming to conditions 1 and 2 above is indeed a lattice. First, the relationship between the normal to a plane (7-1) and the crystallographic axes a, b, and c must be developed. This is done most easily by considering the primitive unit cell shown in Fig. 7-2. The volume of this cell is equal to the area of the shaded base whose sides are b and c times the height of the cell, which is d_{100}. Accordingly,

$$V = \text{area} \cdot d_{100}$$

so that

$$\frac{1}{d_{100}} = \frac{\text{area}}{V} \tag{7-2}$$

In vector algebra (see Appendix 1), the normal to a plane can be represented by the unit vector $\mathbf{n}$, so that (7-1) can be written

$$\boldsymbol{\sigma}_{hkl} = K \frac{1}{d_{hkl}} \mathbf{n} \tag{7-3}$$

An area is represented by the vector product of its sides, so that (7-2) can be written

$$\frac{1}{d_{100}} \mathbf{n} = \frac{\mathbf{b} \times \mathbf{c}}{V} \tag{7-4}$$

For convenience in mathematical manipulations, the scale factor in (7-1) is set equal to unity. Thus, for $K = 1$, combining (7-3) with (7-4) and expressing the volume

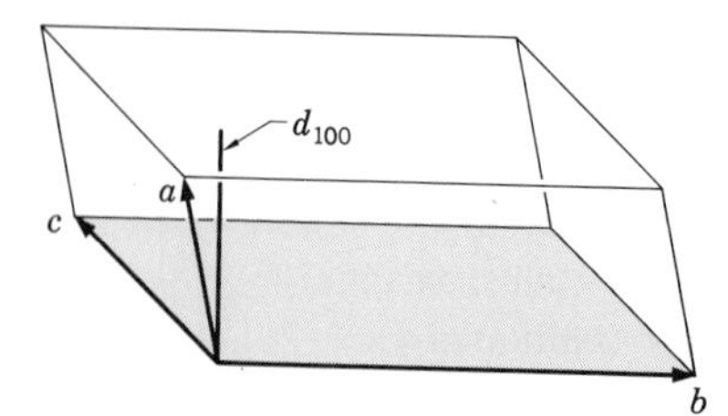

Fig. 7-2

in vector form, one obtains

$$\boldsymbol{\sigma}_{100} = \frac{1}{d_{100}} \mathbf{n} = \frac{\mathbf{b} \times \mathbf{c}}{\mathbf{a} \cdot \mathbf{b} \times \mathbf{c}} \tag{7-5}$$

By analogy, similar expressions can be written for $\boldsymbol{\sigma}_{010}$ and $\boldsymbol{\sigma}_{001}$, and these three vectors are chosen as the three reciprocal axes for defining the three-dimensional reciprocal lattice. It is customary to indicate a reciprocal-lattice axis by adding a "star" to the usual symbol (a^* is pronounced "a-star"), so that the three reciprocal axes are defined

$$\begin{aligned} \mathbf{a}^* &\equiv \boldsymbol{\sigma}_{100} = \frac{\mathbf{b} \times \mathbf{c}}{\mathbf{a} \cdot \mathbf{b} \times \mathbf{c}} \\ \mathbf{b}^* &\equiv \boldsymbol{\sigma}_{010} = \frac{\mathbf{c} \times \mathbf{a}}{\mathbf{a} \cdot \mathbf{b} \times \mathbf{c}} \\ \mathbf{c}^* &\equiv \boldsymbol{\sigma}_{001} = \frac{\mathbf{a} \times \mathbf{b}}{\mathbf{a} \cdot \mathbf{b} \times \mathbf{c}} \end{aligned} \tag{7-6}$$

The reciprocal axes bear a simple relationship to the crystal axes, which follows from the vector notation of (7-6):

$\mathbf{a}^*$ is normal to $\mathbf{b}$ and $\mathbf{c}$

$\mathbf{b}^*$ is normal to $\mathbf{c}$ and $\mathbf{a}$ (7-7)

$\mathbf{c}^*$ is normal to $\mathbf{a}$ and $\mathbf{b}$

These conditions require the following vector relations:

$$\begin{aligned} \mathbf{a}^* \cdot \mathbf{b} &= 0 \qquad & \mathbf{a}^* \cdot \mathbf{c} &= 0 \\ \mathbf{b}^* \cdot \mathbf{c} &= 0 \qquad & \mathbf{b}^* \cdot \mathbf{a} &= 0 \\ \mathbf{c}^* \cdot \mathbf{a} &= 0 \qquad & \mathbf{c}^* \cdot \mathbf{b} &= 0 \end{aligned} \tag{7-8}$$

These relations can also be derived by forming the scalar product of both sides of the first relation of (7-6) with $\mathbf{b}$ and with $\mathbf{c}$, etc. A corollary to the three conditions in (7-8) are the three relations

$$\begin{aligned} \mathbf{a}^* \cdot \mathbf{a} &= 1 \\ \mathbf{b}^* \cdot \mathbf{b} &= 1 \\ \mathbf{c}^* \cdot \mathbf{c} &= 1 \end{aligned} \tag{7-9}$$

These relations can also be derived by forming the scalar product of both sides of the first relation of (7-6) with $\mathbf{a}$, etc.

If a lattice is constructed using the reciprocal-lattice vectors of (7-6), it follows that successive points in the $\mathbf{a}^*$ direction represent successive submultiples h of the spacing of (100); in the $\mathbf{b}^*$ direction, successive submultiples k of the spacing of (010); and in the $\mathbf{c}^*$ direction, successive submultiples l of the spacing of (001).

That this is so is clearly evident from (7-3), since

$$\begin{aligned}
\mathbf{a}^* &= \boldsymbol{\sigma}_{100} = \frac{1}{d_{100}}\,\mathbf{n} \\
2\mathbf{a}^* &= 2\boldsymbol{\sigma}_{100} = \frac{2}{d_{100}}\,\mathbf{n} \\
&= \boldsymbol{\sigma}_{200} = \frac{1}{d_{200}}\,\mathbf{n} \\
3\mathbf{a}^* &= 3\boldsymbol{\sigma}_{100} = \frac{3}{d_{100}}\,\mathbf{n} \\
&= \boldsymbol{\sigma}_{300} = \frac{1}{d_{300}}\,\mathbf{n}
\end{aligned} \tag{7-10}$$

To reach any reciprocal-lattice point hkl, one goes h units along $\mathbf{a}^*$, k units along $\mathbf{b}^*$, and l units along $\mathbf{c}^*$. Accordingly, the reciprocal-lattice vector $\boldsymbol{\sigma}_{hkl}$ can be written in vector notation.

$$\boldsymbol{\sigma}_{hkl} = h\mathbf{a}^* + k\mathbf{b}^* + l\mathbf{c}^* \tag{7-11}$$

It can now be shown that the set of points at the ends of the collection of vectors $\boldsymbol{\sigma}_{hkl}$ conforms to conditions 1 and 2. The simplest way to prove that $\boldsymbol{\sigma}_{hkl}$ is normal to the crystal plane (hkl) is to show that the scalar products of $\boldsymbol{\sigma}_{hkl}$ and vectors lying in the (hkl) plane vanish. The plane (hkl), shown in Fig. 7-3, intercepts $\mathbf{a}$ at $\mathbf{a}/h$, $\mathbf{b}$ at $\mathbf{b}/k$, $\mathbf{c}$ at $\mathbf{c}/l$. Consider the vector

$$\mathbf{C} = \frac{\mathbf{a}}{h} - \frac{\mathbf{b}}{k} \tag{7-12}$$

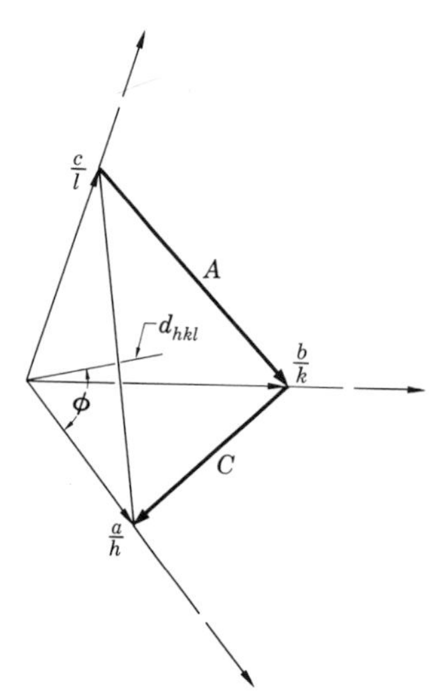

Fig. 7-3

lying in this plane. The scalar product of $\mathbf{C}$ in (7-12) with $\boldsymbol{\sigma}_{hkl}$ in (7-11) is

$$\begin{aligned}\mathbf{C} \cdot \boldsymbol{\sigma}_{hkl} &= \left(\frac{\mathbf{a}}{h} - \frac{\mathbf{b}}{k}\right) \cdot (h\mathbf{a}^* + k\mathbf{b}^* + l\mathbf{c}^*) \\ &= \frac{\mathbf{a}}{h} \cdot (h\mathbf{a}^* + k\mathbf{b}^* + l\mathbf{c}^*) - \frac{\mathbf{b}}{k} \cdot (h\mathbf{a}^* + k\mathbf{b}^* + l\mathbf{c}^*) \\ &= \frac{h}{h} + 0 + 0 - 0 + \frac{k}{k} + 0 \\ &= 1 - 1 \\ &= 0 \end{aligned} \tag{7-13}$$

Similarly, the scalar product of the vector $\mathbf{A} = \mathbf{b}/k - \mathbf{c}/l$ with $\boldsymbol{\sigma}_{hkl}$ vanishes also:

$$\begin{aligned}\mathbf{A} \cdot \boldsymbol{\sigma}_{hkl} &= \left(\frac{\mathbf{b}}{k} - \frac{\mathbf{c}}{l}\right) \cdot (h\mathbf{a}^* + k\mathbf{b}^* + l\mathbf{c}^*) \\ &= \left(0 + \frac{k}{k} + 0\right) - \left(0 + 0 + \frac{l}{l}\right) \\ &= 1 - 1 \\ &= 0 \end{aligned} \tag{7-14}$$

Since, according to (7-13) and (7-14), $\boldsymbol{\sigma}_{hkl}$ is normal to $\mathbf{C}$ and $\mathbf{A}$, it is normal to the plane containing $\mathbf{C}$ and $\mathbf{A}$, that is, to the plane (hkl). In view of this, $\mathbf{n}$, the unit vector normal to (hkl), is parallel to $\boldsymbol{\sigma}_{hkl}$, so that (7-11) can be written

$$|\boldsymbol{\sigma}_{hkl}|\mathbf{n} = h\mathbf{a}^* + k\mathbf{b}^* + l\mathbf{c}^* \tag{7-15}$$

from which it follows that

$$\mathbf{n} = \frac{h\mathbf{a}^* + k\mathbf{b}^* + l\mathbf{c}^*}{|\boldsymbol{\sigma}_{hkl}|} \tag{7-16}$$

The length of the interplanar spacing of the plane shown in Fig. 7-3 is, obviously,

$$\begin{aligned} d_{hkl} &= \frac{a}{h} \cos \varphi \\ &= \frac{\mathbf{a}}{h} \cdot \mathbf{n} \\ &= \frac{\mathbf{a}}{h} \cdot \frac{h\mathbf{a}^* + k\mathbf{b}^* + l\mathbf{c}^*}{|\boldsymbol{\sigma}_{hkl}|} \\ &= \frac{1}{|\boldsymbol{\sigma}_{hkl}|} \end{aligned} \tag{7-17}$$

verifying relation (7-1).

Table 7-1 Relations between the dimensions of the direct and reciprocal cells

Angular parameters	*Linear parameters*
$\cos \alpha^* = \dfrac{\cos \beta \cos \gamma - \cos \alpha}{\sin \beta \sin \gamma}$	$a^* = \dfrac{bc \sin \alpha}{V}$
$\cos \beta^* = \dfrac{\cos \gamma \cos \alpha - \cos \beta}{\sin \gamma \sin \alpha}$	$b^* = \dfrac{ca \sin \beta}{V}$
$\cos \gamma^* = \dfrac{\cos \alpha \cos \beta - \cos \gamma}{\sin \alpha \sin \beta}$	$c^* = \dfrac{ab \sin \gamma}{V}$
$\cos \alpha = \dfrac{\cos \beta^* \cos \gamma^* - \cos \alpha^*}{\sin \beta^* \sin \gamma^*}$	$a = \dfrac{b^*c^* \sin \alpha^*}{V^*}$
$\cos \beta = \dfrac{\cos \gamma^* \cos \alpha^* - \cos \beta^*}{\sin \gamma^* \sin \alpha^*}$	$b = \dfrac{c^*a^* \sin \beta^*}{V^*}$
$\cos \gamma = \dfrac{\cos \alpha^* \cos \beta^* - \cos \gamma^*}{\sin \alpha^* \sin \beta^*}$	$c = \dfrac{a^*b^* \sin \gamma^*}{V^*}$

Volume

$$V^* = a^*b^*c^* \sqrt{1 - \cos^2 \alpha^* - \cos^2 \beta^* - \cos^2 \gamma^* + 2 \cos \alpha^* \cos \beta^* \cos \gamma^*}$$
$$V = abc \sqrt{1 - \cos^2 \alpha - \cos^2 \beta - \cos^2 \gamma + 2 \cos \alpha \cos \beta \cos \gamma}$$

It is a relatively simple matter to show that expressions similar to (7-6) can be used to express the crystal-lattice axes in terms of the reciprocal-lattice axes. Consider the reciprocal of the reciprocal-lattice vector.

$$(\mathbf{a}^*)^* \equiv \frac{\mathbf{b}^* \times \mathbf{c}^*}{\mathbf{a}^* \cdot \mathbf{b}^* \times \mathbf{c}^*} \tag{7-18}$$

Multiplying the right member by $\mathbf{a} \cdot \mathbf{a}^*$, which, according to (7-8), is unity,

$$(\mathbf{a}^*)^* = \mathbf{a} \cdot \mathbf{a}^* \cdot \frac{\mathbf{b}^* \times \mathbf{c}^*}{\mathbf{a}^* \cdot \mathbf{b}^* \times \mathbf{c}^*}$$
$$= \mathbf{a} \cdot \frac{\mathbf{a}^* \cdot \mathbf{b}^* \times \mathbf{c}^*}{\mathbf{a}^* \cdot \mathbf{b}^* \times \mathbf{c}^*}$$
$$= \mathbf{a} \tag{7-19}$$

Combining (7-18) and (7-19), it follows that

$$\mathbf{a} = \frac{\mathbf{b}^* \times \mathbf{c}^*}{\mathbf{a}^* \cdot \mathbf{b}^* \times \mathbf{c}^*}$$

and, similarly,

$$\mathbf{b} = \frac{\mathbf{c}^* \times \mathbf{a}^*}{\mathbf{a}^* \cdot \mathbf{b}^* \times \mathbf{c}^*} \tag{7-20}$$

$$\mathbf{c} = \frac{\mathbf{a}^* \times \mathbf{b}^*}{\mathbf{a}^* \cdot \mathbf{b}^* \times \mathbf{c}^*}$$

There are other useful relationships between the elements of the crystal lattice and their counterparts in the reciprocal lattice. A detailed derivation of them is beyond the scope of this book. Since some of them may be necessary in the chapters that follow, Table 7-1 lists these relations, expressed for the most general lattice. The relations for more symmetrical lattices are easily derived from these by substitution of appropriate values for α, β, and γ.

Relation to interplanar spacings. For reasons to be discussed in later chapters, it is often desirable to know the relation between interplanar spacings and the cell edges of a crystal lattice. This relation is particularly easy to determine in the reciprocal lattice by simply combining (7-1) with (7-11). First form the dot product of the reciprocal-lattice vector with itself.

$$\begin{aligned}\boldsymbol{\sigma}_{hkl} \cdot \boldsymbol{\sigma}_{hkl} &= (h\mathbf{a}^* + k\mathbf{b}^* + l\mathbf{c}^*) \cdot (h\mathbf{a}^* + k\mathbf{b}^* + l\mathbf{c}^*) \\ &= hh\mathbf{a}^* \cdot \mathbf{a}^* + hk\mathbf{a}^* \cdot \mathbf{b}^* + hl\mathbf{a}^* \cdot \mathbf{c}^* \\ &\qquad + kh\mathbf{b}^* \cdot \mathbf{a}^* + kk\mathbf{b}^* \cdot \mathbf{b}^* + kl\mathbf{b}^* \cdot \mathbf{c}^* \\ &\qquad + lh\mathbf{c}^* \cdot \mathbf{a}^* + lk\mathbf{c}^* \cdot \mathbf{b}^* + ll\mathbf{c}^* \cdot \mathbf{c}^* \end{aligned} \tag{7-21}$$

Carrying out the dot products (Appendix 1) and collecting like terms finally yields

$$\begin{aligned}\sigma_{hkl}^2 = \frac{1}{d_{hkl}^2} &= h^2a^{*2} + k^2b^{*2} + l^2c^{*2} + 2hka^*b^* \cos\gamma^* \\ &\qquad + 2klb^*c^* \cos\alpha^* + 2lhc^*a^* \cos\beta^* \end{aligned} \tag{7-22}$$

which is the general expression applicable to the triclinic system.

When crystals having higher symmetry are considered, the form of (7-22) becomes greatly simplified. For example, in the cubic system, $a^* = b^* = c^*$ and all the angles are $90°$, so that

$$\frac{1}{d_{hkl}^2} = (h^2 + k^2 + l^2)a^{*2} \tag{7-23}$$

Similarly, in the hexagonal system, $a^* = b^* \neq c^*$ and $\alpha^* = \beta^* = 90°$ while $\gamma^* = 60°$, so that

$$\frac{1}{d_{hkl}^2} = (h^2 + hk + k^2)a^{*2} + l^2c^{*2} \tag{7-24}$$

Still further relations are explored in Exercise 7-4.

Quite often it is desirable to know the relation between d_{hkl} and the unit-cell edges of the crystal lattice. These relations can be derived directly from (7-22) by inserting the proper transformation equations from Table 7-1. For example, consider the simple transformations of relation (7-23) and (7-24). In the cubic system, $a^* = 1/d_{100} = 1/a$, so that

$$\frac{1}{d_{hkl}^2} = \frac{h^2 + k^2 + l^2}{a^2} \tag{7-25}$$

while in the hexagonal system $a^* = 1/d_{100} = 1/a \cos 30° = \sqrt{3}/2a$ and $c^* = 1/d_{001} = 1/c$

so that

$$\frac{1}{d_{hkl}^2} = \frac{4(h^2 + hk + k^2)}{3a^2} + \frac{l^2}{c^2} \tag{7-26}$$

Additional transformations are considered in Exercise 7-5.

RELATION TO X-RAY DIFFRACTION

Interpretation of Bragg's law. In addition to providing a relatively simple means for displaying the geometrical relations between planes in a crystal, the reciprocal lattice is also extremely useful in dealing with x-ray diffraction phenomena. To help see how this comes about it will be recalled from Chap. 5 that the diffraction of x rays was controlled by the Bragg law, which can be written in a form suitable for determining the diffraction angle θ_{hkl} in terms of the other variables involved.

$$\sin \theta_{hkl} = \frac{\lambda/2}{d_{hkl}} = \frac{1/d_{hkl}}{2/\lambda} \tag{7-27}$$

Since any triangle that is inscribed in a circle and has a diameter as its hypotenuse is a right triangle, draw a circle whose diameter is $2/\lambda$ and inscribe in it a right triangle having $1/d_{hkl}$ as one of its legs, as shown in Fig. 7-4. The angle opposite to this leg must be θ according to (7-27), so that Fig. 7-4 is a graphical representation of Bragg's law.

Recalling that $1/d_{hkl} = \sigma_{hkl}$, it is possible to give the construction in Fig. 7-4

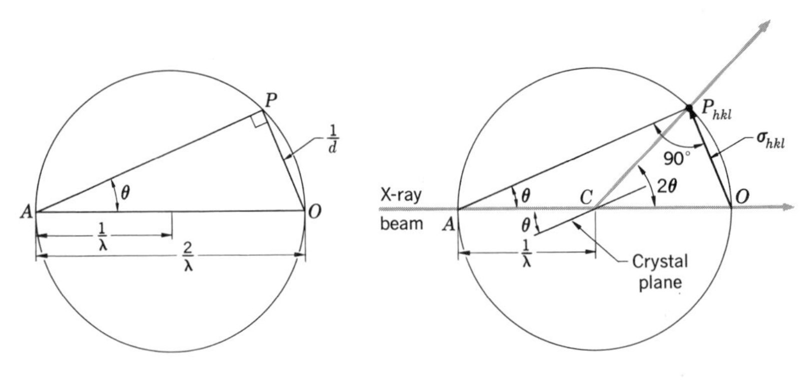

Fig. 7-4

Fig. 7-5

a physical meaning. Let the horizontal diameter AO represent the direction of the incident x-ray beam. Since the line AP forms the Bragg angle θ with the incident beam, it has the slope of a crystal plane that is in position to diffract the x-ray beam, as indicated by the plane drawn in Fig. 7-5 at the center C of the circle. Finally, OP is normal to the crystal plane (and AP) and has the length σ_{hkl}, while $\angle OCP = 2\angle OAP = 2\theta$, so that CP is the direction of the x-ray beam diffracted by the plane at the glancing angle θ (Fig. 7-5). Thus Fig. 7-5 is a graphical interpretation of the Bragg law in terms of the reciprocal-lattice vectors $\boldsymbol{\sigma}_{hkl}$. To recapitulate:

1. The crystal can be depicted as being located at the center C of a circle (sphere in three dimensions) of radius $1/\lambda$.
2. The point O where the x-ray beam leaves the circle after passing through the crystal is the origin of the reciprocal lattice of the crystal.
3. Whenever a reciprocal-lattice point [at the end of a vector $\boldsymbol{\sigma}_{hkl}$, which is normal to the (hkl) plane] lies on the circle (sphere in three dimensions), the Bragg law (7-27) is satisfied, and a diffracted x-ray beam passes through the reciprocal-lattice point.
4. X-ray diffraction can occur only when a reciprocal-lattice point lies on the circle, which, in three dimensions, becomes a sphere, and is called either the *Ewald sphere* or *sphere of reflection*. (The construction in Fig. 7-5 represents a diametral plane of the sphere.)

It is clear from the foregoing discussion that the reciprocal lattice can be used in combination with the sphere of reflection to explain any x-ray diffraction experiment. Because it is conceptually easy to imagine the reciprocal lattice centered at the point of emergence of the incident x-ray beam from the sphere of reflection, a construction like Fig. 7-5 can be used to examine the different ways in which a crystal can be made to satisfy the Bragg reflection condition for its various planes. In fact, as subsequent discussions will demonstrate even more clearly, because diffraction occurs only when a reciprocal-lattice point intersects the sphere of reflection and then the diffracted beam passes through the point of intersection, a photographic record of the diffracted x-ray beams is nothing more than a photographic record of the reciprocal lattice. It is precisely because *one normally records the reciprocal lattice of crystals in x-ray diffraction experiments* that the reciprocal-lattice concept is so essential to their straightforward interpretation.

Laue diffraction condition. Before considering some actual experimental arrangements devised to satisfy the above conditions for x-ray diffraction by various planes in a crystal, it is of interest to relate the original analysis of x-ray diffraction carried out by Laue to the reciprocal-lattice concept. First, consider a periodic row of atoms (Fig. 7-6) spaced a apart and an incident x-ray beam whose direction is given by the unit vector $\mathbf{S}_0$. Next, consider the x-ray beam scattered by the row of atoms in the direction of the unit vector $\mathbf{S}$ (Fig. 7-6). In order that successive atoms along the row scatter in phase with each other, the path difference of the two rays in Fig. 7-6 must equal an integral number of wavelengths. This requirement can be expressed

$$\mathbf{a} \cdot \mathbf{S} - \mathbf{a} \cdot \mathbf{S}_0 = \mathbf{a} \cdot (\mathbf{S} - \mathbf{S}_0) = h\lambda \tag{7-28}$$

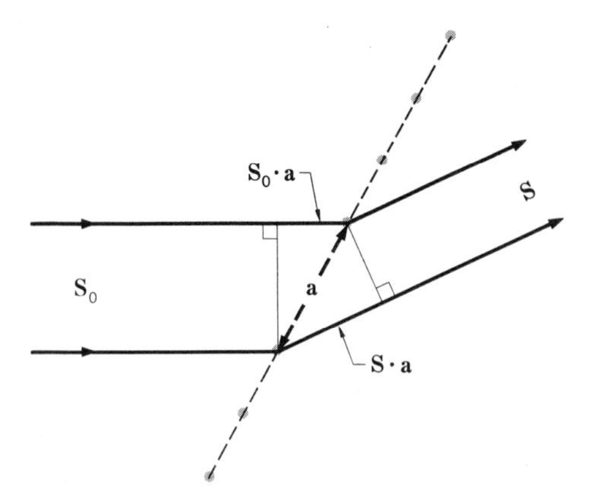

Fig. 7-6

where h is any integer and $\mathbf{a} \cdot \mathbf{S}_0$ and $\mathbf{a} \cdot \mathbf{S}$ are the extra distances each ray travels in Fig. 7-6 from the common wave fronts (normals to both rays) to the atomic row. The dot products have the magnitude of a (hypotenuse) times the sine of the angle defined by the x rays since $\mathbf{S}_0$ and $\mathbf{S}$ have unit magnitudes. It should be noted that $\mathbf{S}$, $\mathbf{S}_0$, and $\mathbf{a}$ need not be coplanar, and, in general, the diffracted beam $\mathbf{S}$ can be thought of as forming a cone about the atomic row for a fixed direction of the incident beam $\mathbf{S}_0$.

The above conditions can be extended to a two-dimensional array of atoms quite simply. If the atoms along one direction of the plane array have the period a and along another direction the period b, then the condition that the atoms in the first row scatter in phase is simply (7-28). The condition that the atoms along the second row scatter in phase is, similarly,

$$\mathbf{b} \cdot (\mathbf{S} - \mathbf{S}_0) = k\lambda \tag{7-29}$$

where k, again, is any integer. When conditions (7-22) and (7-23) are satisfied simultaneously, the entire plane array scatters in phase. This leads to the situation depicted in Fig. 7-7, where it is seen that conditions (7-28) and (7-29) each produce cones of allowed rays, but can simultaneously be satisfied only along two directions, namely, the two directions along which the cones intersect.

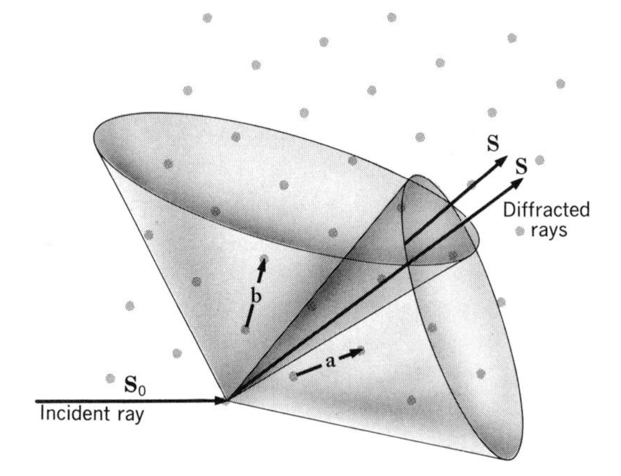

Fig. 7-7

It should be a simple matter now to see that, for a triply periodic array having three mutually noncoplanar translations a, b, and c, the condition that the entire array scatter in phase requires that the following three conditions be satisfied simultaneously:

$$\begin{aligned} \mathbf{a} \cdot (\mathbf{S} - \mathbf{S}_0) &= h\lambda \\ \mathbf{b} \cdot (\mathbf{S} - \mathbf{S}_0) &= k\lambda \\ \mathbf{c} \cdot (\mathbf{S} - \mathbf{S}_0) &= l\lambda \end{aligned} \tag{7-30}$$

The conditions in (7-30) are called the *three Laue equations.* When all three are satisfied simultaneously, the diffracted beam has only one allowed direction, since three cones (by analogy with Fig. 7-7) can mutually intersect along only one line.

The three Laue conditions (7-30) can be related to the reciprocal lattice by making use of the important property of a lattice that any vector in that lattice can be expressed in terms of the three lattice axes. For example, consider a reciprocal-lattice vector

$$\mathbf{R} = p\mathbf{a}^* + q\mathbf{b}^* + r\mathbf{c}^* \tag{7-31}$$

The three coefficients p, q, r may have any values which can be determined by making use of the vector-algebraic relations (7-8) and (7-9), namely:

$$\begin{aligned} \mathbf{a} \cdot \mathbf{R} &= \mathbf{a} \cdot (p\mathbf{a}^* + q\mathbf{b}^* + r\mathbf{c}^*) \\ &= p \\ \text{and, similarly,} \quad \mathbf{b} \cdot \mathbf{R} &= q \\ \text{and} \quad \mathbf{c} \cdot \mathbf{R} &= r \end{aligned} \tag{7-32}$$

Next, suppose that the vector $\mathbf{R}$ represents the difference between the unit vectors $\mathbf{S}$ and $\mathbf{S}_0$ multiplied by $1/\lambda$ (Fig. 7-8), namely,

$$\mathbf{R} = \frac{\mathbf{S} - \mathbf{S}_0}{\lambda} \tag{7-33}$$

Substituting the right side of (7-33) for $\mathbf{R}$ in (7-32) and making use of the three Laue

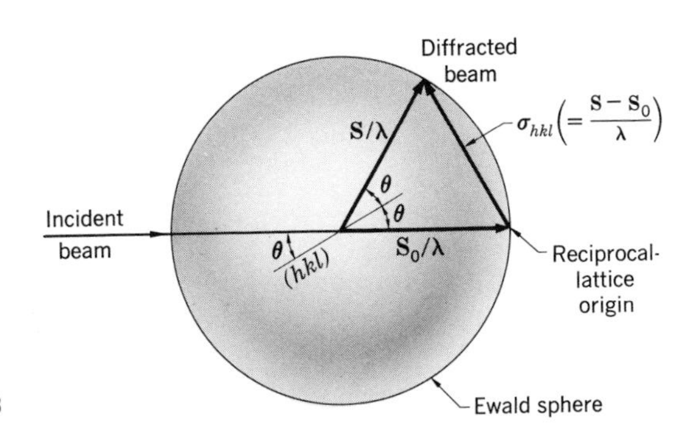

Fig. 7-8

equations (7-30) gives

$$p = \mathbf{a} \cdot \left(\frac{\mathbf{S} - \mathbf{S}_0}{\lambda}\right) = h$$

$$q = \mathbf{b} \cdot \left(\frac{\mathbf{S} - \mathbf{S}_0}{\lambda}\right) = k \qquad (7\text{-}34)$$

$$r = \mathbf{c} \cdot \left(\frac{\mathbf{S} - \mathbf{S}_0}{\lambda}\right) = l$$

Since h, k, and l in (7-30) must be integers, substituting (7-33) and (7-34) in (7-31) gives the very important result

$$\frac{\mathbf{S} - \mathbf{S}_0}{\lambda} = h\mathbf{a}^* + k\mathbf{b}^* + l\mathbf{c}^*$$

$$= \boldsymbol{\sigma}_{hkl} \qquad (7\text{-}35)$$

Equivalence of Bragg and Laue conditions. The meaning of relation (7-35) is illustrated in Fig. 7-8. Since the two vectors on both sides of (7-35) are equal to each other, it means that the two vectors must be parallel and their lengths must be equal. Their parallelism means that both vectors are perpendicular to the plane (hkl) in the crystal, so that the incident beam represented by $\mathbf{S}_0/\lambda$, and the reflected beam represented by $\mathbf{S}/\lambda$ form the same angle θ with the (hkl) plane. The equivalence of their lengths can be stated vectorially:

$$\frac{1}{\lambda}|\mathbf{S} - \mathbf{S}_0| = |\boldsymbol{\sigma}_{hkl}| \qquad (7\text{-}36)$$

It is easily seen in Fig. 7-8 that (see also Exercise 7-6)

$$\mathbf{S} - \mathbf{S}_0| = 2 \sin \theta \qquad (7\text{-}37)$$

The magnitude of $\boldsymbol{\sigma}_{hkl}$ is given by (7-17), so that the relation (7-36) becomes

$$\frac{1}{\lambda}(2 \sin \theta) = \frac{1}{d_{hkl}}$$

or $$2d_{hkl} \sin \theta = \lambda \qquad (5\text{-}3)$$

which is the Bragg law previously encountered. Thus it is obvious that the three Laue conditions are directly equivalent to the Bragg reflection condition and that both can be related to the reciprocal-lattice concept in a straightforward manner.

RELATION TO DIFFRACTION EXPERIMENTS

Original Laue experiment. An examination of (7-35) shows that if the diffraction conditions for different h, k, l values are to be satisfied by a crystal (having fixed a, b, c values), either $|\mathbf{S} - \mathbf{S}_0|$ or λ must be variable. In other words, it is necessary to introduce at least one additional degree of freedom if the three Laue equations

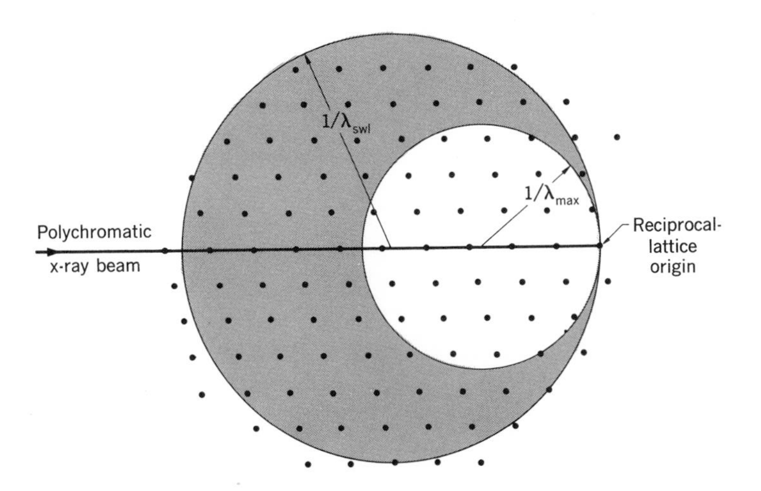

Fig. 7-9. Cross section of reciprocal-lattice construction for experimental Laue arrangement. The shaded portion of the reciprocal lattice, lying between the two spheres of reflection indicated, satisfies the diffraction condition (7-29) for different λ values.

(7-30) are to be satisfied simultaneously for different crystal planes. As briefly mentioned in Chap. 5, this extra degree of freedom was inadvertently introduced in the first x-ray diffraction experiment ever performed, because neither Friedrich, Knipping, nor Laue knew what the exact wavelengths of x rays were nor how to secure an x-ray beam having only one wavelength, that is, a monochromatic beam. Their experiment can be represented in terms of the reciprocal-lattice concept, as shown in Fig. 7-9. The shortest-wavelength x radiation present in any beam is that occurring at the short wavelength limit (6-34), so that only the reciprocal-lattice points lying inside the sphere of reflection with radius $1/\lambda_{swl}$ can satisfy the diffraction condition. For each $\lambda > \lambda_{swl}$, one can picture another sphere of reflection drawn tangent to the origin of the reciprocal lattice, which, of course, lies at the point of emergence of the incident beam. Assuming that there is some maximum value of λ beyond which the intensity in the direct or diffracted beams can no longer be detected, the successive spheres of reflection lie in the shaded region in Fig. 7-9, limited by some sphere whose radius is $1/\lambda_{max}$.

To determine the directions of the diffracted beams it is necessary to pass the appropriate sphere of reflection through the reciprocal-lattice point (Exercise 7-8) and then join that point to the sphere's origin, as in Fig. 7-5. It is clear from the construction in Fig. 7-9 that, because all the reciprocal-lattice points within the shaded region satisfy the diffraction condition concurrently, diffracted x-ray beams go off in all directions. Although in their original experiment Friedrich and Knipping, not knowing exactly what to expect, surrounded their crystal on virtually all sides with photographic plates, it is usual procedure nowadays to place a flat plate at right angles to the x-ray beam so as to intercept the x rays diffracted in either a forward or a backward direction. The practical aspects of the *Laue method* are further discussed in Chap. 14.

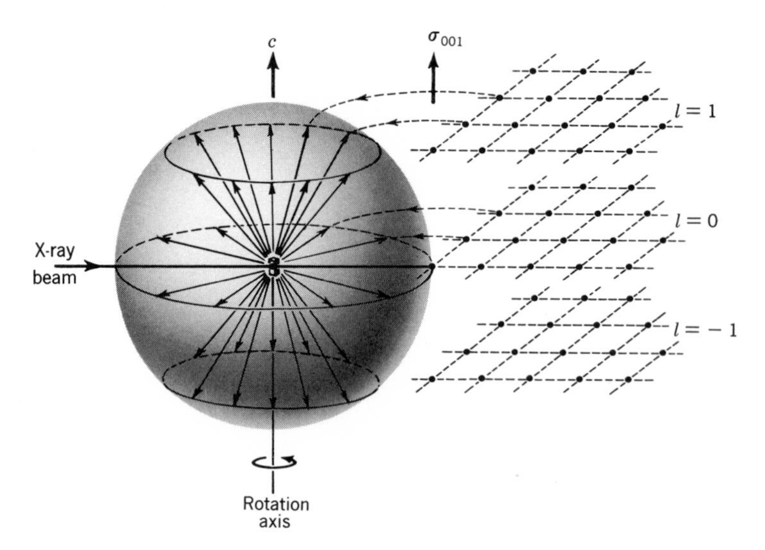

Fig. 7-10. Rotating-crystal arrangement. A tetragonal crystal is shown at the center of the sphere of reflection. A portion of the reciprocal lattice is shown, and the way it rotates about σ_{001}, while the crystal rotates about c, is indicated by a couple of points in the $l = 0$ and the $l = 1$ level.

Rotating-crystal arrangement. One of the principal limitations of the Laue arrangement is that it is not possible to determine the exact wavelength that satisfies the diffraction condition unless the unit-cell constants are known in advance. Thus, in many cases in practice, it is necessary to use a monochromatic beam whose wavelength is known quite accurately. As already discussed, it is customary to select the $K\alpha$ line of the x-ray tube target for this purpose. Returning to relation (7-35), it is clear that, because λ is now fixed, $\mathbf{S} - \mathbf{S}_0$ must be varied. Since both vectors have unit magnitude, the only way that this can be accomplished is by changing the angle between $\mathbf{S}$ and $\mathbf{S}_0$, which has the effect of changing θ, the angle of incidence. The simplest way that this can be done is by rotating the crystal about a line that is perpendicular to the incident x-ray beam direction. The reciprocal lattice, whose origin lies at the point of emergence of the x-ray beam from the Ewald sphere, can be thought of as rotating about a parallel direction, as shown in Fig. 7-10. Whenever a reciprocal-lattice point intersects the sphere of reflection, the diffraction condition is satisfied for the corresponding plane in the crystal (at the center of the sphere) and a diffracted beam passes through the point.

Figure 7-10 shows the *rotating-crystal arrangement* in which a tetragonal crystal has been set to rotate about its c axis. A portion of its reciprocal lattice is shown, and the way that some of the points in the $l = 1$ layer are caused to intersect the Ewald sphere is indicated. When the rotating crystal is so oriented relative to the incident beam that the rotation axis is perpendicular to a set of parallel reciprocal-lattice layers, these layers intersect the Ewald sphere along circles called *layer lines.* The distance between layer lines is obviously proportional to the spacing between the reciprocal-lattice layers, so that rotating-crystal photographs can be used to determine the lattice constants of single crystals. To intercept the diffracted beam, a flat film placed at right angles to the incident beam can be used, although it is more usual to form a cylinder out of the film and surround the entire sphere of reflection with just one film. Further details concerning the rotating-crystal arrangement are given in Chap. 15.

Moving-film methods. As can be seen in Fig. 7-10, the condensation of all the reciprocal-lattice points of one layer into one line complicates the subsequent unraveling of the individual points, particularly when the exact nature of the reciprocal lattice is not previously known. It is highly desirable, therefore, to devise some means for photographing the reciprocal lattice without distortion. A simple way that this can be done is illustrated in Fig. 7-11. Suppose a metal plate containing an annular opening is so placed relative to the rotating crystal that only one of the diffraction cones shown in Fig. 7-10 can pass through it. (Such a plate is called a *layer-line screen*.) A photographic film is then positioned so as to be parallel to this screen and to the corresponding reciprocal-lattice layer (not shown in Fig. 7-11). When the crystal is rotated about its axis, the film is rotated synchronously about what would be the axis of rotation of the reciprocal lattice (Fig. 7-10). Whenever successive reciprocal-lattice points pass through the Ewald sphere, the photographic film accompanies them and records the resulting diffracted beams at a point on the film that lies directly below the reciprocal-lattice point. Thus the reciprocal-lattice layer is photographed without any distortion whatever, except for a scale factor proportional to the crystal-to-film distance. The experimental scheme illus-

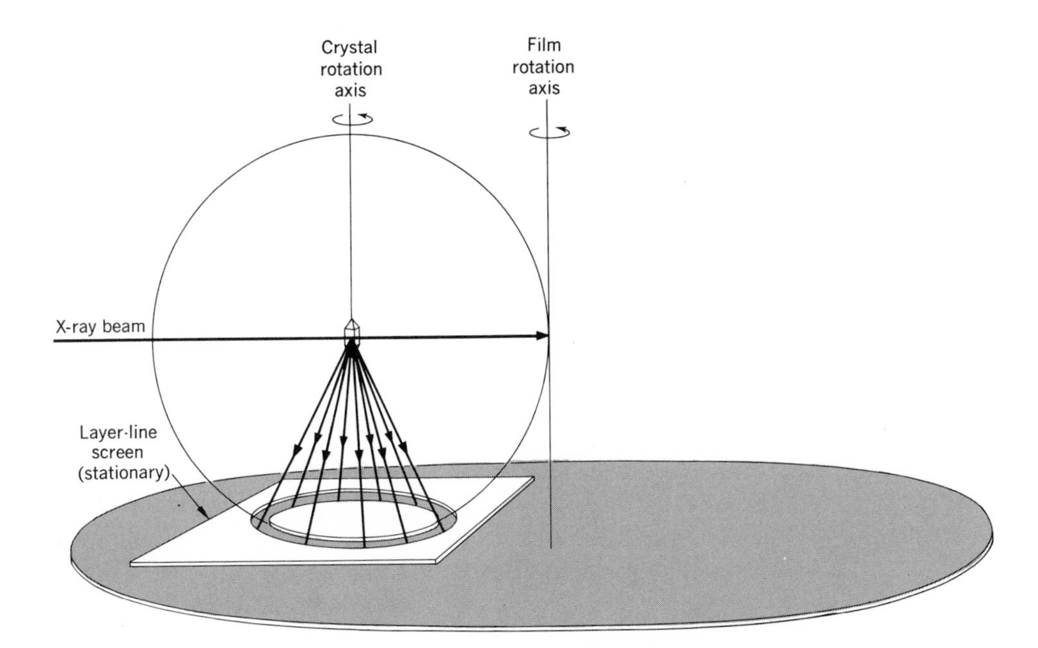

Fig. 7-11. Moving-film arrangement for photographing one layer of the reciprocal lattice without distortion. (Compare with Fig. 7-10.)

trated in Fig. 7-11 is quite similar to one actually devised for this purpose by W. F. de Jong and J. Bouman. Because the film, as well as the crystal, rotates during the exposure, this is one example of what are called *moving-film methods,* further discussed in Chap. 16. These methods have been devised to minimize the distortion of the reciprocal lattice, which, it will be recalled, is what is actually being recorded in any x-ray diffraction experiment.

Powder method. The experimental arrangements described so far were devised to study single crystals. Most crystalline materials do not normally occur in sufficiently large single crystals, so that a powder of a large number of very tiny crystals must be utilized instead. The characteristics of x-ray diffraction effects from a fine-grained crystalline aggregate were first studied by P. Debye and P. Scherrer in Germany in 1916 and, independently, by A. W. Hull in the United States. The reciprocal-lattice construction for a powder is self-evidently a superposition of the reciprocal lattices of the individual crystallites comprising it. Consider just one reciprocal-lattice vector, $\boldsymbol{\sigma}_{hkl}$. If the number of crystallites present in the powder is very large and the crystallites are randomly oriented, such reciprocal-lattice vectors in each crystallite point in all possible directions, so that the corresponding reciprocal-lattice points lie on the surface of a sphere of radius $|\boldsymbol{\sigma}_{hkl}|$. Obviously, a separate sphere exists for each conceivable value of σ_{hkl}, so that the reciprocal lattice of a powder is, simply, a set of concentric spheres. Since the origin of the reciprocal lattice of each crystallite must lie at the point of emergence of the incident x-ray beam, each reciprocal-lattice sphere cuts the sphere of reflection, provided that $\sigma_{hkl} < 2/\lambda$. The mutual intersection for one set of planes (hkl) is illustrated in Fig. 7-12. As can be seen therein, the diffraction condition is satisfied by a collection of planes whose respective reciprocal-lattice points lie on the sphere of reflection, so that the diffracted rays emanating from the powder (at the center of the Ewald sphere) appear to form a continuous cone. Since the distribution of reciprocal-lattice points along the surface of the sphere (Fig. 7-12) is uniform for a truly random polycrystalline aggregate, the distribution of the x rays along the diffraction cone is also uniform, so that it is sufficient to intercept a representative portion of each cone. A typical placement of a narrow strip of film to accomplish

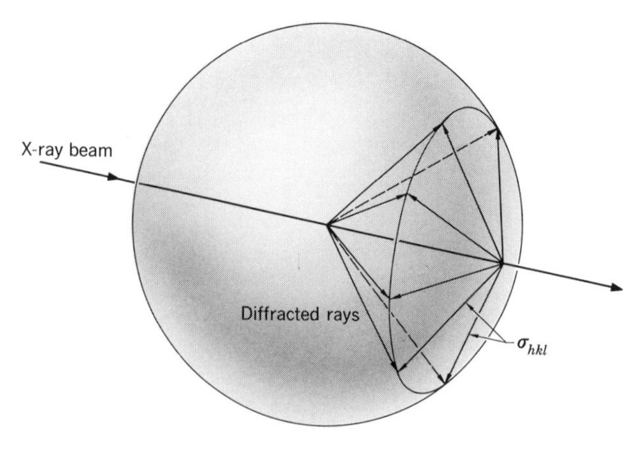

Fig. 7-12. Reciprocal-lattice construction for one set of planes in a powder sample. Some of the vectors σ_{hkl} satisfying the condition for x-ray diffraction are indicated.

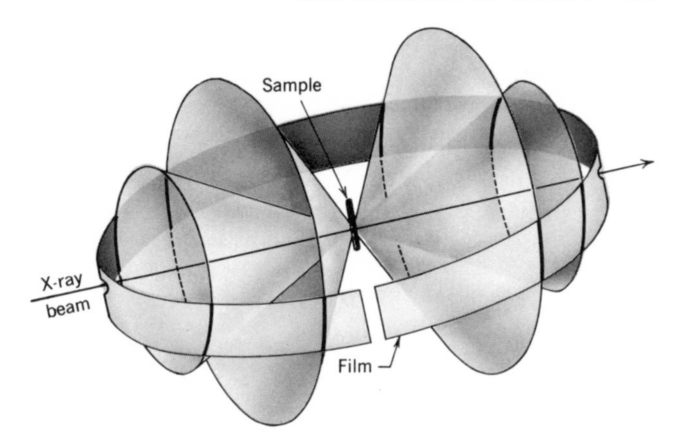

Fig. 7-13. Powder method. The diffraction cones emanating from a usually cylindrical sample are shown intersecting a film strip, also cylindrically placed so as to intercept all the cones produced.

this is shown in Fig. 7-13. Further discussion of the experimental details of the *powder method* is postponed until Chap. 17.

EXERCISES

7-1. Prepare a drawing of the $hk0$ layer in the reciprocal lattice of a monoclinic crystal for which $a = 2.42$ Å, $b = 3.63$ Å, $c = 5.12$ Å, and $\gamma = 120°$. Select a suitable scale constant to show all the reciprocal-lattice points out to $h = 6$ and $k = 9$.

7-2. Making use of similar triangles in Fig. 7-1, prove that the distance from the origin to the reciprocal-lattice points 100, 110, and 120 are indeed reciprocal to the appropriate interplanar spacings. **Hint:** Note that the distance to the row along which these points lie is proportional to $1/a$.

7-3. It has been shown in Chap. 3 that when a centered lattice is chosen to describe a crystal, certain combinations of h, k, and l cannot exist. As shown in Exercise 3-12, the rule for a face-centered lattice is that only those planes for which $h + k + l = 2n$, where n is any integer, are allowed. This means that those reciprocal-lattice points for which $h + k + l \neq 2n$ will be missing. Show by means of a suitable sketch that the reciprocal lattice in this case is body-centered.

7-4. Derive the relation between d_{hkl} and the reciprocal-lattice axes for the tetragonal, orthorhombic, and monoclinic systems.

7-5. Derive the relations between d_{hkl} and the crystal axes for the three crystal systems in Exercise 7-4.

7-6. Draw S_0/λ and S/λ ahead of an (hkl) plane so as to show the "reflection" arrangement. Next show that $(1/\lambda)(S - S_0)$ must be perpendicular to (hkl) if the two unit vectors form the same angle θ with the plane. Finally, prove relation (7-37) in the text from this construction.

7-7. Draw the $hk0$ layer of the reciprocal lattice of a cubic crystal having $a = 4.5$ Å. Next, assume that an x-ray tube operated at 20 kV is used in the Laue arrangement and that the intensity of the continuous spectrum drops to virtually zero at $\lambda_{max} = 2\lambda_{swl}$. If the incident x-ray beam direction is parallel to an a axis of the crystal, which of the following reflections can be observed with this arrangement: 110, 120, 130, 410, 420, 530? Which of these reflections fall in the front-reflection region $(\theta < 45°)$?

7-8. Prove, graphically or analytically, that the directions for the reflections from the planes (110), (220), and (330) in any single crystal must be parallel in the Laue arrangement. What does this mean in terms of recording them on a Laue photograph?

7-9. Using moving-film methods, the reciprocal lattice of a triclinic crystal was found to have the following unit-cell constants: $a^* = 0.176$ Å^{-1}, $b^* = 0.213$ Å^{-1}, $c^* = 0.268$ Å^{-1}, $\alpha^* = 89.5°$, $\beta^* = 49.0°$, and $\gamma^* = 90.0°$. What are the corresponding unit-cell constants of the crystal lattice?

7-10. When preparing rotating-crystal photographs of a tetragonal crystal whose unit-cell dimensions are not known, it is necessary to use three mountings of the crystal, namely, rotation about c and about a and $[110]$, since the latter two cannot be distinguished from the crystal's morphology. If the spacing between layer lines for the last two mountings gives reciprocal-lattice vectors of, respectively, 0.177 Å^{-1} and 0.250 Å^{-1}, which is the setting corresponding to rotation about the true a axis of the crystal?

7-11. In the arrangement shown in Fig. 7-13, the diffraction cones from the powder sample intercept the film in arcs symmetrically disposed about the incident x-ray beam. Derive a simple relation between the distance L between such pairs of arcs on the film and the Bragg angle θ (in radians) for a specimen-to-film distance R.

EIGHT
X-ray scattering by atoms

ATOMIC SCATTERING FACTOR

Classical definition. The way that an individual electron scatters x rays incident on it is described correctly by the classical Thomson equation (6-9). A collection of electrons in an atom, therefore, can be expected to scatter Z times as strongly if Z is the number of electrons present, that is, its atomic number. Because the distances between electrons in an atom are commensurate with the wavelength of x rays, the waves they scatter interfere with each other, so that the preceding statement is literally correct only for the forward direction of scattering. For scattering in some other direction, the differences in pathlengths from scatterers located at different points in an atom produce partially destructive interference, so that the total scattering amplitude decreases with increasing scattering angle. It is convenient to denote a ratio between the amplitude scattered by an atom E_a and that by an isolated electron E_e, under identical conditions, by an *atomic scattering factor*

$$f = \frac{E_a}{E_e} \tag{8-1}$$

Clearly, the maximum value that f can have is Z, the magnitude of the scattering factor when all the electrons scatter in phase with each other.

To see how the scattering factor depends on the scattering angle, it is necessary to consider the actual distribution of electrons in an atom. Suppose that this distribution is spherically symmetric and correctly represented by a charge density $\rho(r)$ (Fig. 8-1). The amount of charge contained in a volume element dV of the atom then is

$$dq = \rho\, dV \tag{8-2}$$

The ratio between the amplitudes scattered by this volume element, dE_a, and that scattered by a single electron in the same direction must be the same as that between their charges. Consequently,

$$\frac{dE_a}{E_e} = \frac{dq}{e} = \frac{\rho\, dV}{e} \tag{8-3}$$

Consider now what happens when an x-ray beam incident on the atom in a direction denoted by the unit vector $\mathbf{S}_0$ is scattered in a direction denoted by the unit vector $\mathbf{S}$, as shown in Fig. 8-1. Each volume element dV scatters an amplitude dE_a, given by (8-3). In summing the contributions from the different volume elements, however, it is necessary to take into account the pathlength differences for each point in the atom. As shown in Appendix 1, this can be done most conveniently mathematically by using a *phase factor* relating the orientation of the incident and scattered beams $(\mathbf{S} - \mathbf{S}_0)$ to the position vector $\mathbf{r}$ of the volume element. Accordingly,

$$df = \frac{\rho(r)}{e}\, e^{(2\pi i/\lambda)(\mathbf{S}-\mathbf{S}_0)\cdot\mathbf{r}}\, dV \tag{8-4}$$

where the exponential term expresses the phasal relationship between the different parts of the atom.

The total contribution of the atom is obtained by integrating (8-4) over its entire volume. In doing this, it is convenient to employ spherical coordinates since the charge density is assumed to be spherically symmetric. The volume element in this case is chosen to be an annular ring (Fig. 8-2) of radius $r \sin \varphi$, width $r\, d\varphi$, and thickness dr, so that

$$dV = 2\pi r^2 \sin \varphi\, d\varphi\, dr \tag{8-5}$$

Recalling from Eq. (7-37) that $|\mathbf{S} - \mathbf{S}_0| = 2 \sin \theta$, the dot product in the exponent in (8-4) can be written

$$(\mathbf{S} - \mathbf{S}_0) \cdot \mathbf{r} = |\mathbf{S} - \mathbf{S}_0|\, r \cos \varphi = 2r \sin \theta \cos \varphi$$

and the volume integral in (8-4) becomes

$$f = \frac{1}{e} \int_{r=0}^{r=\infty} \int_{\varphi=0}^{\varphi=\pi} \rho(r) e^{ikr \cos \varphi}\, 2\pi r^2 \sin \varphi\, d\varphi\, dr \tag{8-6}$$

where $$k = \frac{4\pi \sin \theta}{\lambda} \tag{8-7}$$

and the integration over r can be extended to infinity, since only the charge density within the atom can make a contribution to the integral.

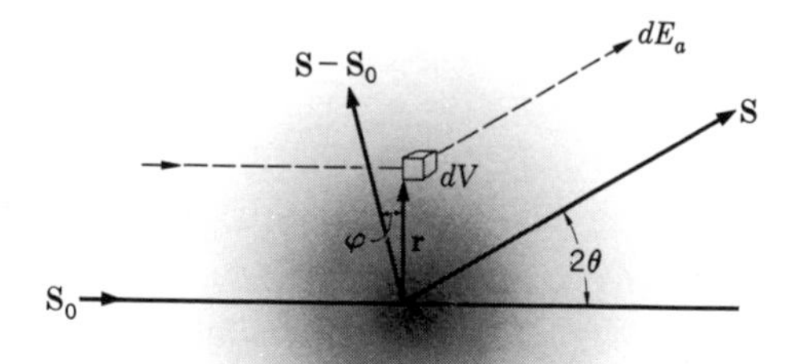

Fig. 8-1. Contributions to scattering in an isolated atom whose charge density can be depicted by a radially symmetric function that gradually goes to zero as **r** increases.

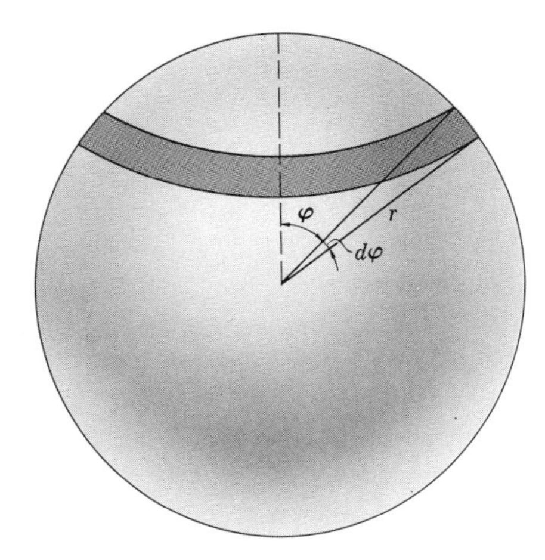

Fig. 8-2

Integrating first with respect to φ,

$$\int_0^\pi e^{ikr\cos\varphi}\sin\varphi\, d\varphi = \frac{1}{ikr}\, e^{ikr\cos\varphi}\Big]_0^\pi$$

$$= \frac{2}{kr}\,\frac{e^{ikr} - e^{-ikr}}{2i}$$

$$= \frac{2\sin kr}{kr} \tag{8-8}$$

Substituting this back into (8-6) gives the expression

$$f = \frac{4\pi}{e}\int_0^\infty r^2\rho(r)\,\frac{\sin kr}{kr}\, dr \tag{8-9}$$

The atomic scattering factor given by (8-9) can be evaluated provided that the charge density $\rho(r)$ in the atom is known. Although not previously stressed, this expression is strictly correct only when the charge distribution is spherically symmetric and provided that the energy of the x rays is not close to an absorption edge of the atom. When this is not the case, the atomic scattering factor becomes complex, as discussed in a later section in this chapter.

Quantum-mechanical approach. In seeking a suitable functional form for $\rho(r)$, early investigators turned to the Bohr model of an atom, according to which the electrons are confined to spherical shells of radius r_K, r_L, etc., about the nucleus. It is possible to describe such a distribution quite easily mathematically and to calculate the atomic scattering curve. It does not give a satisfactory result, however, because such a curve (Exercise 8-1) contains undulations, whereas the experimental curves are found to be smooth. It also has another drawback, in that the electrons are presumed to scatter independently and classically, whereas the existence of the Compton effect clearly demonstrates that quantum effects should be considered in x-ray scattering.

According to quantum mechanics, one cannot think of discrete electrons in definable orbits, but instead must think in terms of wave functions Ψ which are solutions of Schrödinger's equation for the atom. Furthermore, the uncertainty principle does not permit one to locate electrons in space very precisely, and it is only possible to calculate a probability for finding an electron at a given point in the atom. Max Born showed in 1926 that this probability of finding an electron in a volume element dV is proportional to $|\Psi|^2\, dV$, so that the magnitude of the wave functions squared represents a kind of average probability density throughout space. In fact, one can express the charge density by $e|\Psi|^2$ and use (8-9) forthwith. Such a procedure still assumes that each volume element scatters independently (classically), but this is not an unreasonable assumption for evaluating coherent scattering (unmodified). The problem reduces itself, therefore, to finding appropriate wave functions, that is, solutions of the Schrödinger equation.

The Schrödinger equation can be solved exactly only for the hydrogen atom having one electron which moves in the spherically symmetric Coulomb field of its nucleus. For the ground state,

$$\Psi_H = \frac{1}{\sqrt{\pi a_B{}^3}}\, e^{-r/a_B} \tag{8-10}$$

where $a_B = 0.53$ Å is called the Bohr radius of hydrogen. The probability density, then, is

$$|\Psi_H|^2 = \frac{e^{-2r/a_B}}{\pi a_B{}^3}$$

and since $\rho(r) = e|\Psi|^2$, the atomic scattering factor for hydrogen, according to (8-9), can be written

$$\begin{aligned} f_H &= 4\pi \int_0^\infty r^2\, \frac{e^{-2r/a_B}}{\pi a_B{}^3}\, \frac{\sin kr}{kr}\, dr \\ &= \frac{4}{a_B{}^3 k} \int_0^\infty r e^{-2r/a_B} \sin kr\, dr \end{aligned} \tag{8-11}$$

The integral in (8-11) is of the form

$$\int_0^\infty x e^{-ax} \sin bx\, dx = \frac{2ab}{(a^2 + b^2)^2}$$

as evaluated in the integral tables for definite integrals. Accordingly,

$$\begin{aligned} f_H &= \frac{1}{[1 + (a_B k/2)^2]^2} \\ &= \frac{1}{\left[1 + \left(\dfrac{2\pi a_B \sin\theta}{\lambda}\right)^2\right]^2} \end{aligned} \tag{8-12}$$

after substituting (8-7) for k. Note that the atomic scattering factor of hydrogen depends only on $(\sin\theta)/\lambda$, since the other terms are constants. In the case of crystals, $(\sin\theta)/\lambda = \frac{1}{2}d_{hkl}$, according to the Bragg law, so that the scattering factor takes on a constant value for each crystal plane and is actually independent of wavelength.

When an atom containing more than one electron is considered, the Schrödinger equation cannot be solved exactly, and certain approximations must be made. The first successful approach was that of D. R. Hartree, who developed, in 1928, the self-consistent field approximation now bearing his name. A detailed consideration of this calculation is clearly outside the scope of this book, but its essential features are quite simple. One of the electrons in the atom is considered at a time. The charge distribution due to all the other electrons is averaged to give a spherically symmetric field, which then acts on the electron whose wave function is being calculated. The individual wave functions thus determined are, in turn, used to calculate the central field, so that this is a process of successive approximations. When the wave functions calculated by the above procedure reproduce the atomic field employed in their determinations, that is, when a self-consistent set of solutions is obtained, the process is terminated. The actual calculations are generally carried out using appropriate variation methods, and the criterion testing the physical soundness of the result obtained is that the total energy of the atom should be minimized. This is important when comparing various refinements that have been proposed to improve the original Hartree self-consistent field approach. It should be noted that, although the criterion of minimum energy is invariably used in atomic calculations, it never has been proved that the resulting charge density yields the most reliable scattering-factor curves for x-ray scattering. In the absence of evidence to the contrary, however, this criterion has been deemed valid for x-ray scattering.

Numerical calculations. Using the Hartree model, R. W. James and G. W. Brindley calculated the atomic scattering factors for many frequently encountered atoms and ions in 1931. Since then several modifications and alternative schemes have been proposed, both in order to improve the calculated values and to extend the calculations to heavier atoms, for which the early calculations proved to be too difficult. A brief discussion of the procedures still employed, along with complete tables of scattering-factor values currently believed to be the most reliable, can be found in Volume 3 of *International tables for x-ray crystallography.* The interested reader will also find further references to original papers there. For present purposes it is deemed sufficient to describe the general nature of the calculations that have been carried out. Basically, two methods have been followed: The first is derived from the Hartree model, except that the atomic field calculations include electron correlation effect, exchange interactions, and other possible refinements. The second method employs a statistical approach in which the electrons in an atom are treated like a degenerate electron gas obeying Fermi-Dirac statistics. This approach has the advantage that, instead of seeking approximate solutions to the Schrödinger equation, the charge density in an atom is calculated with the aid of Poisson's equation. As in Hartree's model, the potential of the electrons is assumed to be spherically symmetric. Because of the relatively simpler mathematical expressions, however, it is far easier to carry out exact calculations. In fact, once the charge distribution for one atom has been determined, that for other atoms is easily derivable by simple scaling procedures. As one might expect, the statistical approach works best for relatively heavy atoms having many electrons. It should be noted here that L. Pauling and J. Sherman made some rather approximate calculations in 1932 for light and heavy atoms in which they assumed that each electron moved in a simple Coulomb field like the one electron in hydrogen. Thus they were able to use the known hydrogenic wave functions for all the electrons, but their scattering curves are consequently less accurate. With the current availability of high-speed computers, ease of computation is no longer as important a criterion, so that good practice nowadays is to carry out the more exact calculations based on the two methods described above.

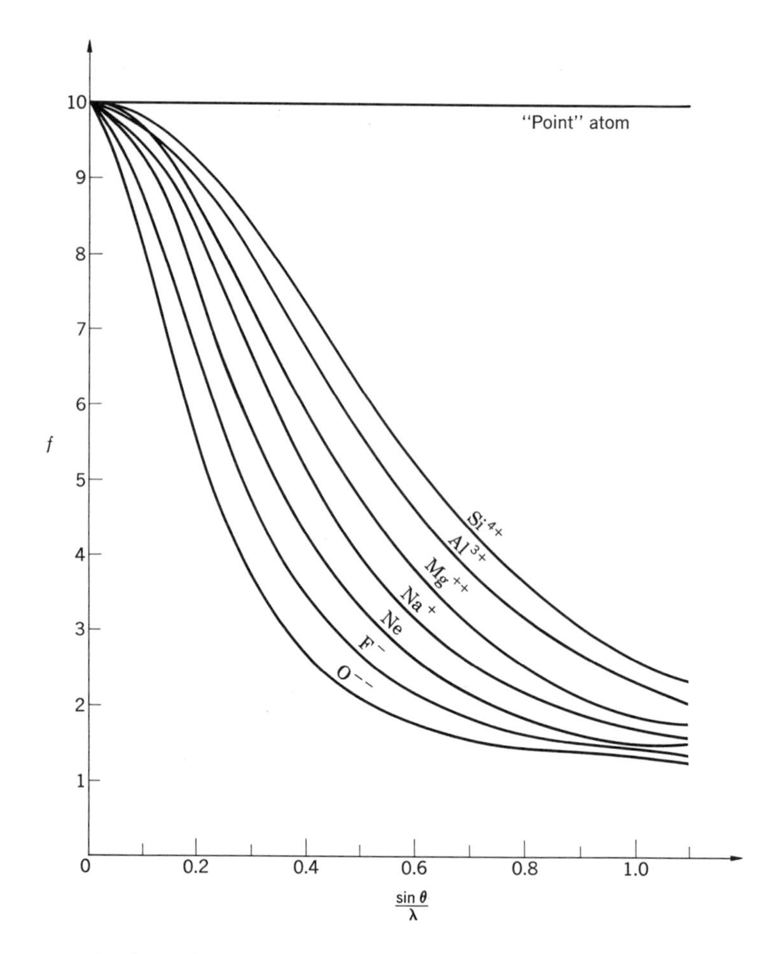

Fig. 8-3. Scattering-factor curves for six ions having a total of ten electrons surrounding the nucleus. The atomic scattering factors for neon ($Z = 10$) and a "point" atom containing ten electrons at its center are also shown for comparison purposes.

In an actual calculation, the charge density for each electron is used to calculate the amplitude of coherent scattering by that electron. Next, the amplitudes for all the electrons are summed, to give the coherent-scattering amplitude for the atom, from which f is then determined, according to (8-1). As an illustration of the dependence of the resulting atomic-scattering-factor curves on the scattering angle, Fig. 8-3 compares the curves for six ions and one neutral atom, all of them having ten electrons surrounding the nucleus. It is clearly seen that, as the nuclear charge increases, drawing the electrons closer in to the nucleus and decreasing differences in pathlength thereby, the scattering curves deviate successively less and less from the horizontal line representing the scattering factor for a point atom (zero path differences). This figure illustrates another important point also. The outermost, more weakly bound electrons make an appreciable contribution to the total scattering only at smaller angles. This means that, in any atom, the innermost, or K, electrons make a contribution at all angles, the contribution of the L electron falls off more rapidly, and so forth, while the valence electrons make a very small total contribution which is limited primarily to small scattering angles. Before concluding this discussion it should be noted that, in the case of atoms in molecules, the valence electrons are not usually distributed in a spherically symmetrical manner, but are probably localized along bonding directions between adjacent atoms. Because of the relatively small contribution that these outer electrons make to the scattering factors of heavy atoms, such deviations from spherical symmetry are normally negligible in x-ray diffraction. On the other hand, in organic molecules containing light atoms, the valence electrons constitute a major fraction of the total number present. The effect of directional bonding and the possibility of calculating scattering factors for bonded atomic aggregates (molecules) have been examined, and suitable scattering-factor tables have been prepared, primarily by R. McWeeny.

INTENSITY OF SCATTERING

Hydrogen atom. Since hydrogen has only one electron, the total power scattered by this electron can be calculated from the Thomson equation (6-9). According to the quantum theory, however, it must be comprised of both coherent and incoherent scattering. The intensity of coherent scattering follows directly from (8-1),

$$(E_a)^2 = (fE_e)^2$$

so that

$$I_a = f^2 I_e \tag{8-13}$$

where I_e is just the classical intensity scattered by an electron according to (6-9). It is often convenient to express the intensity of scattering in *electron units*, abbreviated eu. This is done by dividing the intensity by I_e, so that

$$\frac{I_a}{I_e} = f^2 = (I_{\text{eu}})_{\text{coh}} \tag{8-14}$$

Thus it is seen that the atomic scattering factor squared is equal to the intensity of coherent scattering by an atom when that intensity is expressed in electron units. For hydrogen, f is given by (8-12), so that f_H^2 must decline from a maximum value of 1.0 as $(\sin \theta)/\lambda$ increases from zero.

The total intensity scattered by hydrogen's one electron is given by (6-9), as already noted. As also stated above, it must equal the sum of the intensity of coherent or *unmodified* scattering plus the intensity of Compton modified scattering, so that

$$I_e = (I_e)_{\text{mod}} + (I_e)_{\text{unmod}} \tag{8-15}$$

Expressing the intensities in (8-15) in electron units by dividing through by I_e gives

$$1 = (I_{\text{eu}})_{\text{mod}} + (I_{\text{eu}})_{\text{unmod}} \tag{8-16}$$

For hydrogen, the intensity of unmodified or coherent scattering is $f_H^2 = f_e^2$, where f_e is the scattering factor for a single electron, so that the Compton modified-scattering intensity from (8-16) is

$$(I_{\text{eu}})_{\text{mod}} = 1 - f_e^2 \tag{8-17}$$

The total intensity of x-ray scattering and its two components are shown, plotted as a function of $(\sin \theta)/\lambda$ in Fig. 8-4. As can be seen, at zero scattering angle, all the intensity is due to coherent scattering, which then declines as the intensity of modified scattering increases. At large $(\sin \theta)/\lambda$ values, incoherent scattering becomes predominant. This has been noted already in Chap. 6, where it was pointed out that the Compton effect became dominant for x rays having very high energies (short wavelengths).

Other atoms. As an example of the intensity of x-ray scattering by an atom containing several electrons, consider a lithium atom having two K electrons $(1s)$

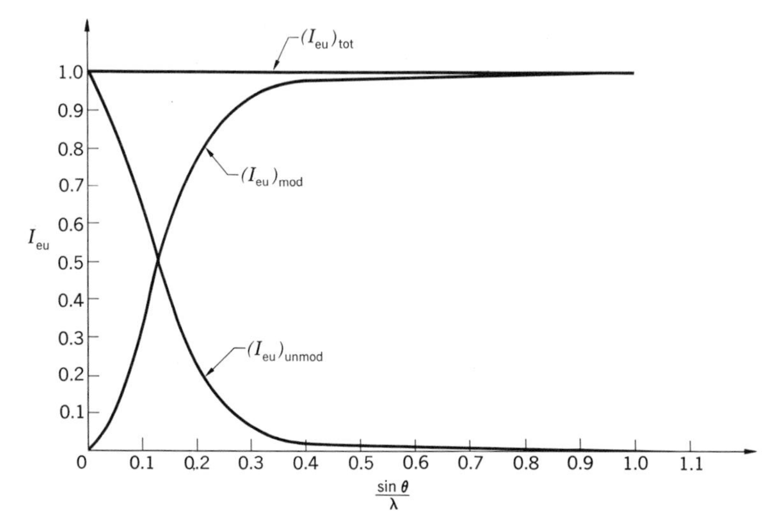

Fig. 8-4. Intensity of x-ray scattering by hydrogen atom.

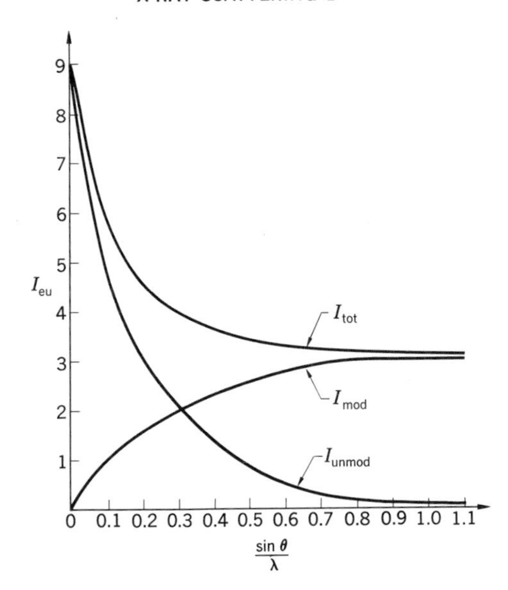

Fig. 8-5. Intensity of x-ray scattering by lithium atom.

and one L electron $(2s)$. Because the three electrons are all in s orbitals having spherical symmetry, the Hartree self-consistent field approximation yields highly accurate results for this atom. The atomic-scattering-factor curve is calculated by adding up the scattering amplitudes for the three electrons, so that the coherent scattering intensity $(= f_{\text{Li}}^2)$ begins at 9 eu, as shown in Fig. 8-5, and declines toward zero as $(\sin\theta)/\lambda$ increases. Each of the three electrons scatters modified radiation incoherently, so that the intensity of the Compton modified scattering by each electron is given by expressions like (8-17). Denoting the coherent scattering factor for one of the K electrons by $(f_e)_K$ and for the L electron by $(f_e)_L$, the total modified intensity is the sum of the three intensities. For lithium

$$(I_{\text{eu}})_{\text{mod}} = 2(1 - f_e^2)_K + (1 - f_e^2)_L \tag{8-18}$$

and is also shown plotted in Fig. 8-5. It is clear from (8-18) that the maximum value the intensity of modified scattering from lithium can have is 3 eu. Thus the total intensity of x-ray scattering by lithium does not remain constant, as was the case for hydrogen, but ranges from a maximum of 9 eu to a minimum of 3 eu.

The foregoing discussion obviously can be generalized to include atoms having any number of electrons. The intensity of coherent scattering is determined by summing the scattering amplitudes of all the electrons and then squaring the sum. The Compton modified-scattering intensity for the atom is the sum of the modified-scattering intensities of the individual electrons. Thus

$$(I_{\text{eu}})_{\text{unmod}} \leq Z^2$$

while $$(I_{\text{eu}})_{\text{mod}} \leq Z \tag{8-19}$$

Incoherent (modified) scattering. In most x-ray diffraction studies of crystals, incoherent scattering is not taken into account explicitly because it simply contributes to the background intensity and cannot be readily distinguished from other causes of background scattering experimentally. As can be seen from (8-19) and Fig. 8-5, it is not a very large factor for x-ray wavelengths normally employed in diffraction studies $[(\sin\theta)/\lambda \leq 1.0]$. Nevertheless, in a number of x-ray experiments it must be considered explicitly. In studies involving crystals, this is the case whenever the intensity distribution between reciprocal lattice points is measured. As discussed further in Chap. 10, a great deal of information can be learned about the mechanical properties and perfection of crystals by studying what is called the diffuse scattering of crystals. Since the theoretical analysis of the measurements considers coherent-scattering effects only, it is necessary to subtract the Compton modified scattering before the experimental data can be processed. Similarly, when the scattering from gases, liquids, or vitreous solids is studied, proper corrections must be made for the modified scattering before the experimental values can be utilized. All this means that the proper calculation of the incoherent scattering by atoms deserves the same careful study that is devoted to coherent scattering.

From the discussion in the preceding section, it appears that the intensity of modified scattering, in electron units, can be calculated for any atom by the following formula:

$$(I_{\mathrm{eu}})_{\mathrm{mod}} = \sum_{n=1}^{Z} (1 - f_e^2)_n = Z - \sum_n (f_e)_n^2 \tag{8-20}$$

where the index n also denotes whether the electron is a K, L, etc., electron. Using quantum mechanics, Wentzel calculated the electronic scattering factor using the expression

$$(f_e)_n = \int \Psi_n^* \, e^{(2\pi i/\lambda)(\mathbf{S}-\mathbf{S}_0)\cdot\mathbf{r}} \, \Psi_n \, dV \tag{8-21}$$

where Ψ_n is the wave function of the nth electron, and Ψ_n^* its complex conjugate.

A more rigorous analysis by I. Waller, taking into account the Pauli exclusion principle, which forbids electronic transitions to already occupied states, leads to an expression for the intensity of incoherent scattering.

$$(I_{\mathrm{eu}})_{\mathrm{mod}} = Z - \sum_n (f_e)_n^2 - \sum_{m \neq n}\sum (f_e)_{mn}^2 \tag{8-22}$$

where $$(f_e)_{mn} = \int \Psi_m^* e^{(2\pi i/\lambda)(\mathbf{S}-\mathbf{S}_0)\cdot\mathbf{r}} \, \Psi_n \, dV \tag{8-23}$$

are the nondiagonal terms in the determinant expressing the total atomic wave function. Finally, experimental measurements of the incoherent scattering suggest that the value in (8-22) is consistently too large by a small factor. This factor turns out to be Breit-Dirac recoil factor

$$R = \left(\frac{\nu}{\nu_0}\right)^3 \tag{8-24}$$

where ν and ν_0 are the frequencies of the scattered and incident radiations, and can be derived by an explicit consideration of the recoil electron's velocity (G. Breit) or from quantum-mechanical considerations (P. A. M. Dirac).

As in the case of coherent-scattering calculations, it is first necessary to describe correctly the atomic field so that the individual electron wave functions can be determined. This has been done using Hartree self-consistent field calculations and also Fermi-Dirac statistical distributions for the heavier atoms. Tables giving the incoherent-scattering intensity (8-22) as a function of $(\sin \theta)/\lambda$ appear in Volume 3 of *International tables for x-ray crystallography.* The tabulated values should be corrected by multiplying by the appropriate Breit-Dirac factor (8-24) for the experimental conditions employed. Because these values were calculated for isolated atoms, one might expect to observe deviations when experiments are carried out on crystals since the atomic field now should take into account crystal-field effects. Two kinds of deviations have been reported so far. It was observed in a direct comparison of the values calculated for free atoms and those measured using diamond, aluminum, and several halide crystals that some disagreement arises at small scattering angles while good agreement exists at larger $(\sin \theta)/\lambda$ values. Thus it appears that, within the accuracy currently attainable, the calculated values can be used safely except at very small scattering angles. In another study, K. Das Gupta reported observing fine structure, which he related to the transitions allowed in a solid for the recoil electron, analogously to the explanations proposed for the extended fine structure observed at the absorption edges of atoms in solids discussed in Chap. 6. These measurements were repeated and extended by R. J. Weiss, who failed to duplicate the results reported, however, even though he employed more sensitive detectors. The possible existence of such Raman scattering of x rays was investigated theoretically by T. Suzuki who concluded that it should, in fact, occur with an angular dependence given by $(1 + \cos^2 \theta) \sin^2 \frac{1}{2}\theta$. He tested his theory using polycrystalline lithium, beryllium, boron, and carbon and confirmed the existence of a Raman band which tended to overlap the Compton peak, consistent with the predicted dependence on scattering angle.†

ANOMALOUS DISPERSION

Classical dipole oscillator. In the discussion so far it has been assumed that the electrons in atoms are relatively free and scatter x rays according to the Thomson treatment. Actually, the electrons are bound to the nucleus by forces that vary with the atomic field strength and the quantum state of the electron. Provided that the force exerted by the external electromagnetic field is much larger than these binding forces, the above assumption is justifiable. (The force exerted by the alternating field is derivable from its energy, and hence is proportional to its frequency.) When the external force (frequency) approaches in magnitude the binding forces (natural frequencies) in the atom, however, this assumption is no longer valid. The classical differential equation describing the motion of a particle of a mass m and charge e in an alternating field intensity $\mathcal{E}_0 e^{i\omega t}$ is

$$\frac{d^2x}{dt^2} + \gamma \frac{dx}{dt} + \omega_0{}^2 x = \frac{e}{m} \mathcal{E}_0 e^{i\omega t} \tag{8-25}$$

† Tadasu Suzuki, X-ray Raman scattering. Experiment I, *Journal of the Physical Society of Japan,* vol. 22 (1967), pp. 1139–1150.

where ω_0 is the natural angular frequency of the vibrating particle, and $\gamma(dx/dt)$ expresses a damping term resulting from the emission of radiation by a vibrating particle according to classical mechanics. Relation (8-25) is, of course, the familiar harmonic-oscillator equation, whose solution has the form

$$x(t) = \frac{e\mathcal{E}_0 e^{i\omega t}/m}{\omega_0{}^2 - \omega^2 + i\gamma\omega} \tag{8-26}$$

as can be verified by substituting (8-26) in (8-25).

The displacement in (8-26), when multiplied by the charge of the electron, gives the polarizability (moment) $\alpha(t) = ex(t)$ of each dipole, so that one can calculate directly from (8-26) the electrical susceptibility of a collection of N uncoupled dipoles,

$$\chi_e^* = \frac{Nex(t)}{\mathcal{E}_0 e^{i\omega t}} = \frac{Ne^2/m}{\omega_0{}^2 - \omega^2 + i\gamma\omega} \tag{8-27}$$

where the asterisk denotes that the susceptibility is a complex quantity. In fact, it is instructive to divide the complex susceptibility into a real and an imaginary part,

$$\begin{aligned} \chi_e^* &= \chi' - i\chi'' \\ &= \frac{Ne^2}{m}\left[\frac{\omega_0{}^2 - \omega^2}{(\omega_0{}^2 - \omega^2)^2 + \gamma^2\omega^2} - \frac{i\gamma\omega}{(\omega_0{}^2 - \omega^2)^2 + \gamma^2\omega^2}\right] \end{aligned} \tag{8-28}$$

The real and imaginary parts are shown plotted in Fig. 8-6 as a function of the frequency of the incident radiation. Note that the real part is equal to zero when $\omega = \omega_0$, while the imaginary part has a maximum at this frequency. (This is called resonance absorption.) At the same time, the real part of the susceptibility is strongly frequency-dependent near this frequency and undergoes a change in sign. This is the phenomenon of *anomalous dispersion.*

Returning to (8-26), the electric field produced by the dipole oscillator, at a distance R that is large compared with its displacements, has a magnitude that is ω^2/c^2R times its dipole moment, so that

$$\mathcal{E}(t) = \frac{\omega^2 ex(t)}{c^2R} = \frac{\omega^2 e^2\mathcal{E}_0/mc^2R}{\omega_0{}^2 - \omega^2 + i\gamma\omega}\, e^{i\omega t} \tag{8-29}$$

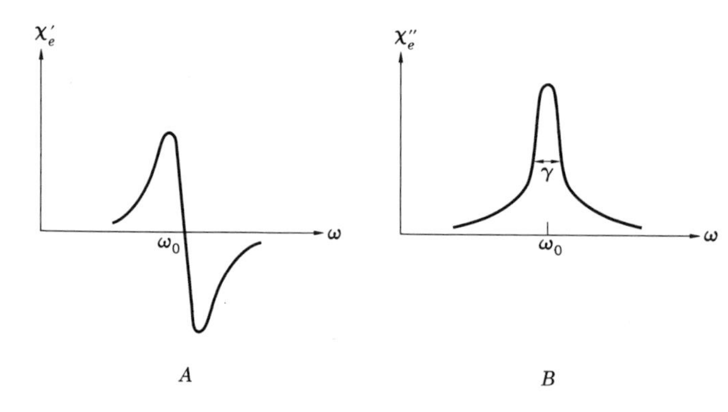

Fig. 8-6. Dependence of the complex electrical susceptibility on frequency near $\omega = \omega_0$. (A). The real part of the susceptibility. (B). The imaginary part of the susceptibility.

Strictly speaking, the expression in (8-29) should be multiplied by a polarization factor, but for reasons to be made clear shortly, it can be omitted from the present discussion.

Suppose a free electron had occupied the same point in space and had been acted upon by the same incident electric field. Since $\gamma = 0$ and $\omega_0 = 0$ for a free electron, the scattered amplitude for this electron at the same point, a distance R removed, is

$$\mathcal{E}'(t) = \frac{\omega^2 e^2 \mathcal{E}_0 / mc^2 R}{-\omega^2} \tag{8-30}$$

where the same polarization factor has been omitted as in (8-29). It is now possible to define a scattering factor for an electron.

$$f_e \equiv \frac{\mathcal{E}(t)}{\mathcal{E}'(t)} = \frac{-\omega^2}{\omega_0{}^2 - \omega^2 + i\gamma\omega} = \frac{\omega^2}{\omega^2 - \omega_0{}^2 - i\gamma\omega} \tag{8-31}$$

Note the minus sign in (8-31). This means that the wave scattered by the free electron in the forward direction has an opposite phase to that of the incident wave.

Effect on atomic scattering. One important consequence of (8-31) has already been noted; namely, from the definition of the scattering factor here employed, it follows that the x-ray wave scattered by an electron undergoes a phase shift of π rad $(= 180°)$. This phase shift occurs whenever f_e, as defined in (8-31), is positive. Separating the real and imaginary terms,

$$\begin{aligned} f_e &= f_e' + if_e'' \\ &= \frac{\omega^2(\omega^2 - \omega_0{}^2)}{(\omega^2 - \omega_0{}^2)^2 + \gamma^2\omega^2} + i\,\frac{\gamma\omega^3}{(\omega^2 - \omega_0{}^2)^2 + \gamma^2\omega^2} \end{aligned} \tag{8-32}$$

It is seen from the above that the imaginary term is always positive, whereas the real term becomes negative when $\omega < \omega_0$ and is positive when $\omega > \omega_0$. Thus, for incident energies that are larger than the natural absorption energies, the scattering is always out of phase with the incident radiation. In fact, when $\omega \gg \omega_0$, f_e' tends to unity, while f_e'' goes to zero. (It should be noted that $\gamma \ll \omega$, so that the imaginary term can be neglected, except when $\omega \simeq \omega_0$.) This means that, when the incident radiation has a relatively large energy, the electrons scatter in accordance with the Thomson model. When $\omega \ll \omega_0$, then f_e' becomes negative, with a magnitude proportional to $\omega^2/\omega_0{}^2$, and the atom scatters in phase with the incident radiation. This is the basis of the anomalous scattering at frequencies close to the natural frequencies; that is, not only does the magnitude of the scattering factor change, but the phase also changes. The imaginary term still can be neglected, except near ω_0. It turns out that, physically, the imaginary term corresponds to the ejection of an inner electron, so that it is equal to zero for energies too small to knock out a bound electron, that is, for $\omega < \omega_0$, and has a nonzero magnitude only when $\omega > \omega_0$.

Calculation of anomalous-scattering factor. As indicated above, the natural frequency figuring in the oscillator equations discussed so far is relatable to the absorption edges of the atom being considered. Quantum-mechanically speaking, the ejection of an electron from a K, L, etc., shell is accompanied by its transition to a higher energy state. In keeping with the Pauli principle, this higher state must be unoccupied. It turns out that the most likely transitions are to the continuum of states having positive energies. Each such transition corresponds to a possible value of ω_0, so that it can be thought of as a virtual oscillator. The problem of evaluating the scattering factor for an atom thus reduces to calculating the density of allowed states times the respective transition probabilities. This is the *oscillator strength,* named the K, L, etc., oscillator strength, to correspond with the initial state of the particular absorption edge being considered.

It is possible to estimate the oscillator strengths from observations of the photoelectric absorption in the vicinity of an absorption edge. This leads to approximate values whose accuracy is too low to be of much usefulness in practice. A fairly rigorous quantum-mechanical evaluation was carried out by H. Hönl using hydrogen-like atomic wave functions. Because hydrogenic wave functions are probably fairly accurate for the innermost K electrons, Hönl's treatment is quite satisfactory for the anomalous scattering by K electrons, and suitable corrections have been tabulated by James. Complete tables for the L oscillator strengths were not prepared because it was believed that Hönl's calculation was not sufficiently valid for this purpose. Considering the anomalous scattering due to K electrons only, the magnitude of the atomic scattering factor can be expressed

$$|f|^2 = (f_0 + \Delta f'_K)^2 + (\Delta f''_K)^2 \tag{8-33}$$

where f_0 is the frequency-independent part of the atomic scattering factor (the value normally used and already discussed at the beginning of this chapter); $\Delta f'_K$ is the real part of the anomalous scattering by K electrons; and $\Delta f''_K$ is the imaginary part. It follows that the scattered wave lags by an amount $\pi - \phi$ behind the incident wave. The phase angle is given by

$$\tan \phi = \frac{\Delta f''_K}{f_0 + f'_K} \tag{8-34}$$

and is equal to zero for $\omega < \omega_K$ since $\Delta f''_K$ is zero for energies less than that of the K absorption edge. This is in agreement with the roles of the real and imaginary parts discussed in the preceding section.

It is possible to employ (8-33) to estimate the difference in the magnitudes that should be observed between the experimentally determined scattering factor f and the tabulated value f_0. Thus $|f - f_0|$ represents the frequency-dependent scattering amplitude, and can be used to measure the dispersion as a function of wavelength. The form of the curve that one obtains using Hönl's correction for K virtual oscillators only is shown by the dashed curve in Fig. 8-7. It displays a large dip at the K absorption edge, but is nearly horizontal ($|f - f_0| = \text{const}$) for all other wavelengths. More recently, estimates of the effect of including L and M virtual oscillators were made by Parratt and Hempstead, and the form of the dis-

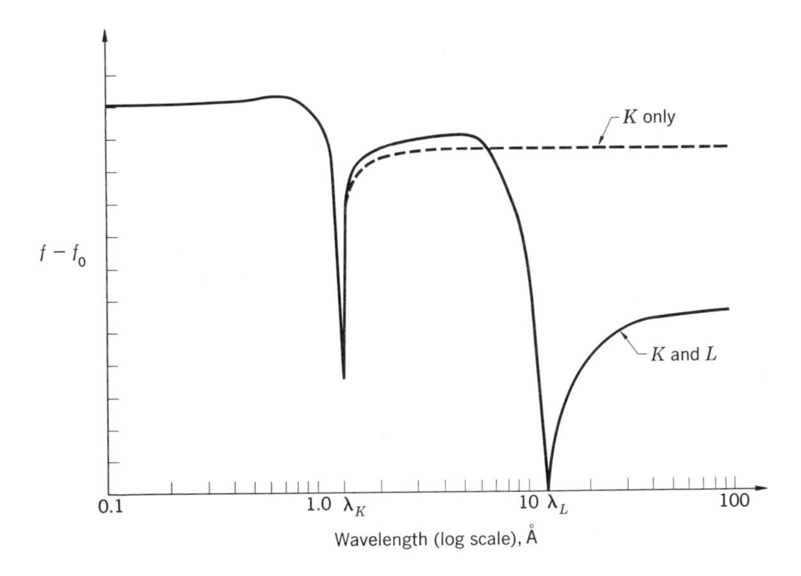

Fig. 8-7. Dispersion curve for an atom. The solid curve represents the dispersion curve, including K and L virtual oscillators, while the dashed curve shows the dispersion when K oscillators only are included. *(After Parratt and Hempstead.)*

persion that results when K and L oscillators are included is shown in Fig. 8-7 by the solid curve. The inclusion of M virtual oscillators introduces still another dip and other details in the dispersion curves, all of which serves to point out that there are no regions of what may be called normal dispersion, that is, where $|f - f_0|$ truly is zero or constant. Instead, it is clearly seen that anomalous dispersion is present at all wavelengths.

The oscillator strengths used in these calculations were derived from the wavelength dependencies of the atomic absorption coefficients. Unfortunately, this introduces some arbitrariness in the oscillator-strength calculations, so that it is difficult to estimate the actual magnitude of any errors that such calculations entail. Nevertheless, dispersion corrections for many atoms based on the above procedure have been calculated by Dauben and Templeton for several wavelengths, corresponding to the $K\alpha$ radiations commonly used in x-ray diffraction experiments. An extension of their tables to include several more atoms and the effect of N virtual oscillators, where necessary, is reproduced in Volume 3 of *International tables for x-ray crystallography*. In these tables, the real part $\Delta f'$ and the imaginary part $\Delta f''$ of the frequency-dependent correction are listed for Cr, Cu, and Mo $K\alpha$ radiations, so that relations (8-33) and (8-34) can be used to evaluate the necessary corrections to the scattering-factor amplitude and the phase angle, respectively. The tabulated corrections are given as a function of $(\sin\theta)/\lambda$, because the contributions of the outer, more loosely bound electrons are dependent on the scattering angle, as already discussed in this chapter. Until satisfactory experimental verification of the tabulated values becomes available, the corrections should be used with circumspection. On the other hand, Fig. 8-7 clearly demonstrates the need for correcting for anomalous dispersion at virtually all wavelengths whenever a direct comparison between experimental and theoretical intensities is made.

Effect of phase lag. The phase lag occasioned by anomalous scattering produces several interesting effects in x-ray diffraction experiments. For example, the atomic array, or crystal structure, of the mineral sphalerite, ZnS, is such that alternate layers parallel to (111) are comprised of all zinc or all sulfur atoms. When the (111) plane is set for reflection, the incident x-ray beam may be scattered first by a zinc-atom layer and next by a sulfur-atom layer. Conversely, when this crystal is turned around so that $(\bar{1}\bar{1}\bar{1})$ is in reflecting position, the topmost layer consists of sulfur atoms above a zinc-atom layer. Now suppose that the wavelength of the monochromatic x-ray beam used is slightly longer than that of the zinc K edge but still considerably shorter than that of any sulfur edges. The sulfur atoms will display relatively "normal" dispersion ($\lambda < \lambda_K$ in Fig. 8-7), while the zinc atoms will not. Recalling from Chap. 5 that all the atoms in a lattice array scatter in phase with each other but that two such arrays may be slightly out of phase with each other, let the phase difference between the scattering by sulfur atoms and by zinc atoms be δ. This is the phase difference, therefore, caused by the relative displacements (path differences) of the two atomic arrays. When x-ray diffraction by the $(\bar{1}\bar{1}\bar{1})$ planes of sphalerite is observed, the sulfur atoms scatter normally; that is, they scatter x rays that are $\pi/2$ rad out of phase with the incident rays. Next, the zinc-atom layers scatter anomalously; that is, they cause an additional phase shift ϕ given by (8-34), so that the total phase shift is $\delta + \pi/2 - \phi$. On the other hand, when the (111) planes are diffracting, the scattering is first by zinc-atom planes, followed by sulfur-atom planes. The total phase difference in this case is $\delta + \phi - \pi/2$, and the two are obviously different (unless $\phi = \pi/2$). The phase difference produces a difference in the x-ray intensity reflected by these two sides, so that they are not equal. By employing the appropriate wavelength, therefore, it is possible to distinguish one side from the other because of the anomalous scattering by zinc atoms. This was first demonstrated by D. Koster, K. S. Knol, and J. A. Prins in 1930, who used gold $L\alpha_1$ and $L\alpha_2$ radiations because the K edge of zinc lies midway between them. More recently, identical experiments have been used to distinguish the (111) from the $(\bar{1}\bar{1}\bar{1})$ faces in several semiconducting selenides and sulfides and to correlate them to characteristically different etch figures that the pairs of external faces display. As discussed further in Chap. 11, the anomalous-scattering effect also has been used to help establish the phase differences between the scattering of x rays by different atomic arrays present in the same crystal.

SCATTERING BY ATOMIC AGGREGATES

It has been shown in the early parts of this chapter that interference effects arise between x rays scattered by different parts of the electron cloud surrounding the nucleus in an individual atom. This interference is responsible for the decline in the atom's scattering power with increasing scattering angle as measured by the scattering factor f. It should not be particularly surprising to learn, therefore, that x rays scattered by different atoms in the same material also can interfere with each other. This was probably first clearly realized by Debye, who, in 1915, demonstrated that even the relatively limited interactions between x rays scattered by a monatomic gas produce definite interference effects. Essentially, the interactions can be described as follows: A group of atoms, say, in a molecule, usually form a more or less fixed spatial array, so that there should be a similar interference effect produced by each such group. Debye called this the *internal interference effect.* Concurrently, these groups are disposed in some kind of spatial distribution relative to each other. The most random possible array would be that in a gas, in which the freedom of atomic motion prohibits the prescription of a definite arrangement. But even in a gas, the finite size of the atoms and the repulsive forces that keep two atoms from occupying the same space at the same time impose a kind of limitation on the arbitrariness that one might expect. All this leads to interference effects between the atomic groups, and this Debye called the *external interference effect.* The relative importance of these two effects becomes a matter of degree. Beginning with the simplest possible situation, a monatomic gas, the external interference is the dominant effect since each "group" contains only one atom. Going on to diatomic and polyatomic gases, the contributions from the internal interference effect grow in importance. In a sense, the above statement says that, because an atomic array in a gas actually is forced to deviate from being truly and completely random, certain interference effects between x rays scattered by the atoms are produced. This was also independently discovered by P. Ehrenfest in 1915, although it was Debye who extended and developed the theory from its early beginnings.

When the gas is condensed to form a liquid, "fixed" average interatomic distances are produced which strongly enhance the external interference effect. Again, if the liquid is composed of polyatomic groups, internal interference effects also arise. It should be noted that increasing the viscosity of certain liquids to the point where they are normally considered to be solids, variously denoted by the terms glassy, vitreous, or amorphous, does not significantly alter these interference effects. Except in detail, therefore, the x-ray diffraction (scattering) by liquids and by vitreous solids is quite similar. It turns out, moreover, that the atomic groupings in many liquids and glasses show a very close resemblance to similar groupings encountered in crystalline arrays. This is particularly true when the arrays extending over short distances from an atom only are considered, and is usually called *short-range order.*

When the atoms form a three-dimensional periodic array, the material is called a crystal, and for many purposes it is sufficient to consider the interference effects produced by the atoms in a single unit cell. The sequential development of the interference effects brought on by increasing the order in the atomic arrays is considered in the rest of this chapter. The interference effects produced by crystals, of course, are treated further in considerable detail in the rest of this book. Only some of the simplest aspects of x-ray diffraction by noncrystals are

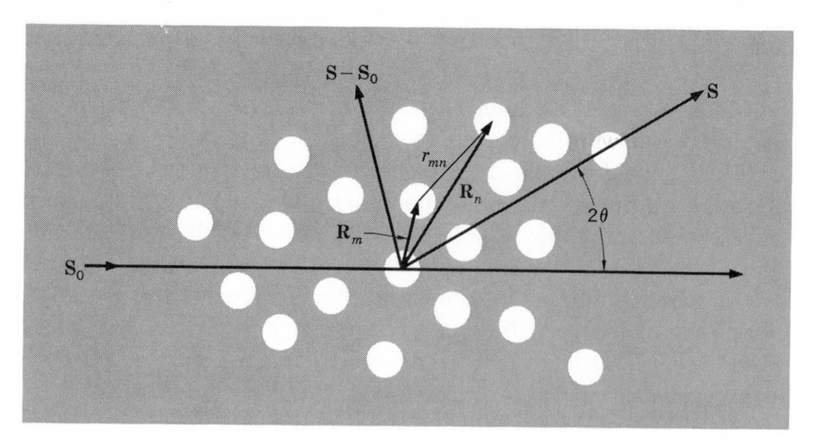

Fig. 8-8

considered in this chapter. The reader interested in learning more about such effects is referred to the books listed at the end of this chapter. It is of interest to note before proceeding that the consequences of short-range order in otherwise random atomic arrays is that the coherently scattered x rays produce diffraction cones which, when intercepted by a photographic film, form recognizable, although broad, rings, often called diffraction halos. When the short-range order is sufficiently extensive for the ordered groups to form recognizable crystallites which, however, are still randomly arrayed with respect to each other, then the rings sharpen into those characteristic of diffraction by polycrystalline aggregates. In fact, the consideration of the role that the size and the shape of molecules play in the interference effects produced was what led Debye and Scherrer to discover in 1916 what is nowadays called the powder method.

Monatomic gases. The general scattering problem of the intensity scattered in a direction denoted by S, forming an angle 2θ with the incident-beam direction denoted by S_0, can be formulated most easily by analyzing the geometry shown in Fig. 8-8. Suppose that this figure shows correctly the relative disposition of like atoms in a gas at one particular instant of time. The intensity of x rays, expressed in electron units to obviate carrying along the constants in Thomson's equation (6-9), scattered by any one of the atoms in the direction S, is established by the value of the scattering factor f for the appropriate scattering angle. (For the sake of simplicity in this treatment, it is assumed that a monochromatic x-ray beam is employed throughout and that its wavelength is such that the scattering factors are real; that is, $\Delta f'' = 0$.) At the instant of time being considered, the atoms comprising the gas lie at various distances from the atom placed at the origin. These distances can be denoted by radius vectors R_1, R_2, R_3, . . . , R_m, R_n, etc. The phase difference between the x rays scattered by the atom m a distance R_m from the origin atom can be expressed, as before, by $e^{(2\pi i/\lambda)(\mathbf{S}-\mathbf{S}_0)\cdot\mathbf{R}_m}$, so that the total intensity scattered by all the atoms relative to the origin atom is simply the sum of the intensities scattered by each of the atoms $\sum_m f_m e^{(2\pi i/\lambda)(\mathbf{S}-\mathbf{S}_0)\cdot\mathbf{R}_m}$. This sum correctly expresses the interference effects between x rays scattered by the origin atom in Fig. 8-8 and all the other atoms, but does not consider the fact that each atom in the gas could have equally well been placed at the origin. In other words, to obtain the complete formulation of the total external interference effect, it is necessary to repeat the above summation procedure for each of the atoms in the gas. This leads to the double summation

$$\begin{aligned} I_{\text{eu}} &= \sum_m f_m e^{(2\pi i/\lambda)(\mathbf{S}-\mathbf{S}_0)\cdot\mathbf{R}_m} \sum_n f_n e^{(-2\pi i/\lambda)(\mathbf{S}-\mathbf{S}_0)\cdot\mathbf{R}_n} \\ &= \sum_m \sum_n f_m f_n e^{(2\pi i/\lambda)(\mathbf{S}-\mathbf{S}_0)\cdot(\mathbf{R}_m-\mathbf{R}_n)} \end{aligned} \tag{8-35}$$

The difference between the radius vectors in the exponent of (8-35) is just the interatomic vector

$$\mathbf{r}_{mn} = R_m - R_n \tag{8-36}$$

and substituting this vector in (8-35),

$$I_{\text{eu}} = \sum_m \sum_n f_m f_n e^{(2\pi i/\lambda)(\mathbf{S}-\mathbf{S}_0)\cdot\mathbf{r}_{mn}} \tag{8-37}$$

The intensity of scattering by a monatomic gas in the above expression correctly describes what happens at one instant of time; however, experimentally, one measures the intensity over a finite time (power), during which the atomic array depicted in Fig. 8-8 is continuously changing. This changing configuration can be described most conveniently for our purposes in terms of the interatomic vectors $\mathbf{r}_{mn}$, as shown in Fig. 8-9. Let the angle formed by the interatomic vector and the diffraction vector $(\mathbf{S} - \mathbf{S}_0)$ be denoted by ϕ. It follows immediately, then, that the magnitude of the exponent in (8-37) is

$$\frac{2\pi i}{\lambda}(\mathbf{S} - \mathbf{S}_0) \cdot \mathbf{r}_{mn} = ikr_{mn} \cos \phi \tag{8-38}$$

where k has been previously defined in (8-7). This is a more convenient form for the exponent in this case because the direction of each possible interatomic vector is constantly changing in a gas, while its permissible lengths are, of course, the same. Saying this another way, since the individual atoms in a monatomic gas can be considered to be indistinguishable, one way to produce all other possible configurations would be to rotate the one in Fig. 8-8 until it assumes all possible orientations. During such a complete rotation, each interatomic vector, like the one in Fig. 8-9, sweeps out a sphere of radius r_{mn}, while the angle it forms with

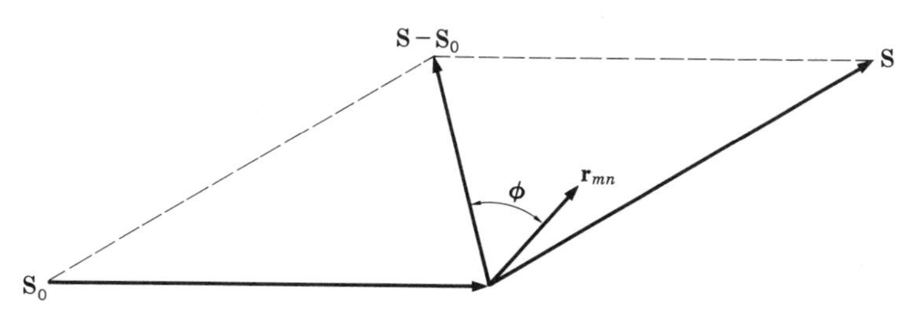

Fig. 8-9. Instantaneous dispositions of incident and scattered rays denoted by the unit vectors $\mathbf{S}_0$ and $\mathbf{S}$ relative to an interatomic vector $\mathbf{r}_{mn}$ joining the mth and nth atoms.

$(\mathbf{S} - \mathbf{S}_0)$ ranges from zero to π. Thus, in order to calculate the time-averaged intensity scattered in the direction of $\mathbf{S}$, it is sufficient to determine the averaged value of the exponential term relating the relative phases of scattered x rays.

$$\overline{e^{(2\pi i/\lambda)(\mathbf{S}-\mathbf{S}_0)\cdot\mathbf{r}_{mn}}} = \overline{e^{ikr_{mn}\cos\phi}}$$

The usual procedure for determining average values of functions is

$$\bar{y} = \int y \frac{dA}{A}$$

The required area in this case is that of the sphere of radius r_{mn}, while

$$dA = 2\pi r_{mn}{}^2 \cos\phi \, d\phi$$

(see Fig. 8-2), so that

$$\overline{e^{ikr_{mn}\cos\phi}} = \int_0^\pi \frac{e^{ikr_{mn}\cos\phi}\, 2\pi r_{mn}{}^2 \cos\phi \, d\phi}{4\pi r_{mn}{}^2}$$

$$= \frac{\sin kr_{mn}}{kr_{mn}} \tag{8-39}$$

similarly to the derivation of (8-8). Substituting (8-39) into (8-37) gives the time-averaged intensity of scattering by a monatomic gas.

$$\overline{I_{\text{eu}}} = \sum_m \sum_n f_m f_n \frac{\sin kr_{mn}}{kr_{mn}} \tag{8-40}$$

Expression (8-40) is called the Debye scattering equation, and it is plotted in Fig. 8-10, superimposed on a constant background such as might be produced by incoherent scattering and other causes. This curve has been calculated assuming a fixed value of r_{mn}, so that it does not represent the actual intensity distribution observed for x-ray scattering by gases. Instead, it is presented to indicate what the general form of such scattering curves should be like. It is clear that the strongest intensity occurs at very small $(\sin\theta)/\lambda$ values. More importantly, however, note that the functional form of (8-40) imposes an undulation on the scattering

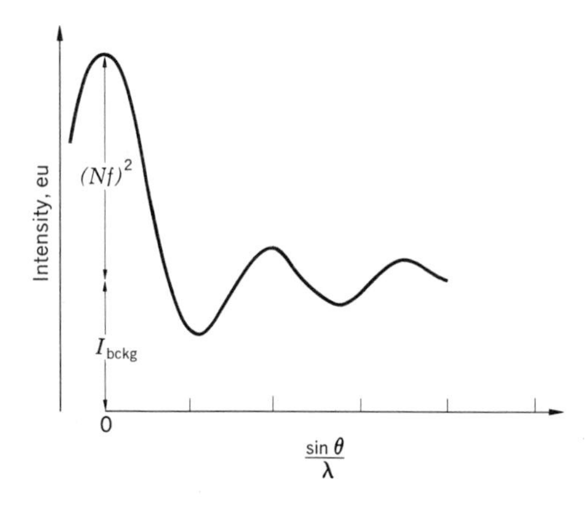

Fig. 8-10. Form of x-ray scattering intensity for a gas. The intensity of coherent scattering, plotted for a fixed value of r_{mn}, is shown superimposed on a constant background intensity.

intensity; that is, interference maxima and minima are produced even by a random assemblage of atoms in a monatomic gas.

When $\theta = 0$, $I_{eu} = N^2f^2$ since $f_m = f_n = f$ in a monatomic gas. This merely demonstrates that, in any kind of atomic assembly, be it a random gas or a triply periodic array in a crystal, all the atoms scatter in phase in the forward beam direction. As the scattering angle decreases, the intensity of coherent scattering rapidly drops down to zero. Consequently, the most pronounced deviation in the intensity distribution is a broadening of the direct beam. The undulations shown in Fig. 8-10 are difficult to detect in practice because they become washed out by an averaging effect when all possible values that r_{mn} can have are considered. The broadening of the x-ray beam, however, may be so slight that it is usually impossible to detect it. The effect can be enhanced, of course, by decreasing the wavelength of the radiation so as to expand the $(\sin \theta)/\lambda$ scale. This is not a practical procedure in the case of x rays because the scattering efficiency (intensity) declines proportionately to λ^3. By comparison, electrons accelerated by a potential of 50 kV have wavelengths of the order of 0.05 Å and give rise to more clearly resolved variations, so that electron diffraction is usually employed in the study of gases.

Polyatomic gases. It is clear that the scattering by a monatomic gas involves only the external interference effect of Debye. When a polyatomic gas is irradiated, the internal interference effect also comes into play. Consider, first, a diatomic gas such as O_2, N_2, etc. If the interatomic distance between the atoms in each molecule has an average value of r_{12}, then the intensity, in electron units, scattered by a gas containing N irradiated molecules, can be written in the following way: First, let one of the atoms in the molecule be the atom m in (8-40). The n atoms then are the atom itself ($m = n = 1$); the second atom in the same molecule ($n = 2$) that is r_{12} distant; and the remaining ($N - 2$) atoms a distance r_{mn} away. The double summation in (8-40) then can be expanded to

$$\overline{I_{eu}} = \sum_m f_m \left(f_1 + f_2 \frac{\sin kr_{12}}{kr_{12}} + \sum_{n=3}^{N} f_n \frac{\sin kr_{1n}}{kr_{1n}} \right) \tag{8-41}$$

The sum over m simply serves to repeat the expression in (8-41) m times, so that, realizing that all atoms in this gas are alike and $f_m = f_n = f$,

$$\overline{I_{eu}} = N \left(f^2 + f^2 \frac{\sin kr_{12}}{kr_{12}} + f^2 \sum_{n=3}^{N} \frac{\sin kr_{1n}}{kr_{1n}} \right) \tag{8-42}$$

From a glance at Fig. 8-10 it is clear that the last term in (8-42) makes a significant contribution only at extremely small scattering angles, so that, except for a contribution of $(N - 2)f^2$ to the intensity of the forward scattered beam, it can be neglected. The remaining two terms then express the true angular dependence of the scattered intensity,

$$I_{eu} = Nf^2 \left(1 + \frac{\sin kr_{12}}{kr_{12}} \right) \tag{8-43}$$

and, as shown in Exercise 8-6, give rise to relatively weak and broad diffraction maxima at small scattering angles.

As a further illustration of the internal interference effect between the atoms within a molecule, consider a gas composed of carbon tetrachloride molecules. From chemical information it is known that the CCl_4 molecule has a central carbon atom tetrahedrally surrounded by the four chlorine atoms. In expanding the summations in (8-40), first let m be a carbon atom, while n is every other atom, including itself, and next let n be a chlorine atom, while m is every other atom, including itself. The expanded sums then are

$$I_{\text{eu}} = Nf_{\text{C}}\left(f_{\text{C}} + 4f_{\text{Cl}}\frac{\sin kr_{\text{C-Cl}}}{kr_{\text{C-Cl}}} + \sum_{p=2}^{N} f_p \frac{\sin kr_{\text{C-}p}}{kr_{\text{C-}p}}\right) + 4Nf_{\text{Cl}}\left(f_{\text{Cl}} + f_{\text{C}}\frac{\sin kr_{\text{C-Cl}}}{kr_{\text{C-Cl}}} + 3f_{\text{Cl}}\frac{\sin kr_{\text{Cl-Cl}}}{kr_{\text{Cl-Cl}}} + \sum_{p=2}^{N} f_p \frac{\sin kr_{\text{Cl-}p}}{kr_{\text{Cl-}p}}\right) \tag{8-44}$$

where N is the total number of irradiated molecules, while the summations over p include interactions with all the other $(N - 1)$ molecules. These summations produce the external interference effect, which, for gases, has been seen to be small, so that they are dropped from subsequent equations. Combining like terms, the intensity of scattering due to the internal interference effect can be expressed

$$I_{\text{eu}} = N\left(f_{\text{C}}^2 + 4f_{\text{Cl}}^2 + 8f_{\text{C}}f_{\text{Cl}}\frac{\sin kr_{\text{C-Cl}}}{kr_{\text{C-Cl}}} + 12f_{\text{Cl}}^2\frac{\sin kr_{\text{Cl-Cl}}}{kr_{\text{Cl-Cl}}}\right) \tag{8-45}$$

From the known geometry of a tetrahedron, it is possible to relate the length of the Cl—Cl distance to the C—Cl distance, $r_{\text{C-Cl}} = \sqrt{\frac{3}{8}}\, r_{\text{Cl-Cl}}$, so that (8-45) can be made a function of one parameter only.† The experimentally measured scattering curve then can be approximated by trying out several values of this parameter. The one that gives the best fit then determines the correct interatomic distance. As already noted, the increased angular dispersion of electron diffraction renders it preferable to x-ray diffraction for such studies.

In the derivation of the Debye scattering equation (8-40) it was assumed that the averaging over all possible configurations was best done by rotating any arbitrary array into all possible orientations, so that the "average" distribution was represented by concentric spheres of atoms of radius r_{mn}. Had it been assumed instead that the atoms were dimensionless, that is, that the gas consisted of point atoms, it would have been necessary to carry out a volume-averaging rather than a surface-averaging process, and the interference function would have had the form

$$\frac{3}{(kr)^2}\left(\frac{\sin kr}{kr} - \cos kr\right) \tag{8-46}$$

This function is plotted in Fig. 8-11 against $(\sin\theta)/\lambda$, using the same vertical scale and value for r as that employed in constructing the curve in Fig. 8-10. A direct

† By placing a tetrahedron in a cube $r_{\text{Cl-Cl}}$ becomes a face diagonal while $r_{\text{C-Cl}}$ is one-half of the body diagonal.

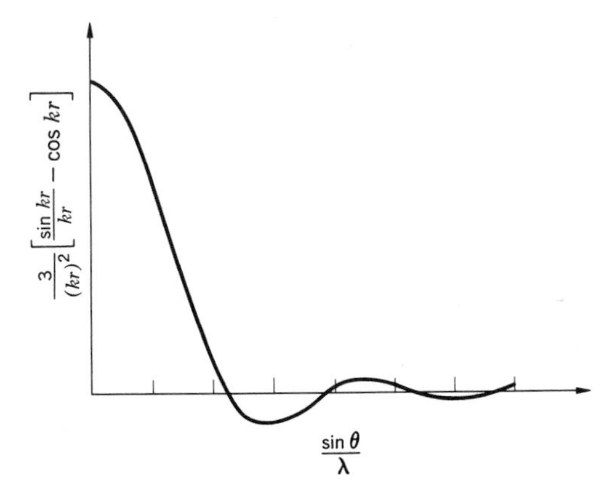

Fig. 8-11

comparison of these two curves clearly shows that they are quite similar, except that the one in Fig. 8-11 has a slightly broader maximum about $\theta = 0$, but it still declines quite rapidly, and the accompanying ripples at slightly larger angles are much less pronounced. Because the atoms in a gas do have a finite size, however, the more restricted relation (8-40) better describes the scattering of x rays by the atoms in a gas.

Monatomic liquids. The condensed state of the atoms in a liquid imposes still further restrictions on their possible arrays. Consider one atom in a simple monatomic liquid. The impenetrability of the atom (due to repulsion of like charges) sets a lower limit on the value that the interatomic distance between it and its neighbors can have, namely, twice the atomic radius. By the same token, the atoms surrounding it cannot move very far away because they are repelled by the atoms surrounding them, and so on. In fact, it really begins to make sense to talk about averaged interatomic distances in liquids. The existence of such short-range order in liquids makes it possible to describe the atomic array in terms of an average density of atoms per unit volume ρ_0 and an actual density $\rho(r)$ which measures the number of atoms per unit volume a distance r away from an atom placed at the origin. In a monatomic liquid, it is quite reasonable to assume that the radial density of atoms surrounding any atom is spherically symmetric, so that it is convenient to talk about the number of atoms per unit volume contained in a spherical shell of thickness dr surrounding the origin atom. The volume of this shell is the difference between the volumes of two spheres whose radii are r and $r + dr$, so that the density of atoms in this shell is

$$4\pi r^2 \rho(r)\, dr \tag{8-47}$$

where the quantity $4\pi r^2 \rho(r)$ is called the *radial distribution function* for the atoms.

The Debye scattering equation (8-40) can be simplified to

$$I_{\text{eu}} = Nf^2 \left(1 + \sum_m{}' \frac{\sin kr_{mn}}{kr_{mn}}\right) \tag{8-48}$$

for a monatomic liquid because the quantity in parentheses represents the terms in one of the sums, while the second sum simply serves to multiply each of these terms by Nf. It should be noted that the primed summation sign Σ' is a conventional way of indicating that the origin term has been moved outside the summation sign. Since the summation remaining to be carried out considers only the atoms surrounding the origin atom, it is possible to replace it by an integration after multiplying this term by the differential density of (8-47). This assumes, of course, that the environment of any atom in the liquid (when it is made the origin atom) is the same, on the average, as that of any other atom. Since this is a most reasonable assumption for a monatomic liquid, the intensity of x-ray diffraction by such a liquid can be expressed by

$$I_{\text{eu}} = Nf^2 \left[1 + \int 4\pi r^2 \rho(r) \frac{\sin kr}{kr} \, dr \right] \tag{8-49}$$

where the integration goes from $r = 0$ to the outer limits of the liquid being irradiated, and the continuously variable r, rather than discrete interatomic distances, is considered, since $\rho(r)$ can have appreciable magnitudes only when r takes on the values of the interatomic distances for nearest, next-nearest, next-next-nearest neighbors, and so forth. In fact, an examination of the radial distribution function shows that $\rho(r)$ very rapidly approaches the value of the average density ρ_0 as the radial distance increases.

For practical reasons, it is convenient to substitute $\rho(r) = [\rho(r) - \rho_0] + \rho_0$ in (8-49), which then becomes

$$I_{\text{eu}} = Nf^2 \left\{ 1 + \int 4\pi r^2 \left[\rho(r) - \rho_0 \right] \frac{\sin kr}{kr} \, dr + \int 4\pi r^2 \rho_0 \frac{\sin kr}{kr} \, dr \right\} \tag{8-50}$$

The upper limit for the integration is determined by the radius of the irradiated sample. Suppose that the last term in (8-50) is integrated from $r = 0$ to $r = R$, some large radius value.

$$4\pi\rho_0 \int_0^R r^2 \frac{\sin kr}{kr} \, dr = \frac{4}{3} \pi \rho_0 R^3 \left[\frac{3}{(kR)^2} \left(\frac{\sin kr}{kr} - \cos kr \right) \right] \tag{8-50a}$$

and it is clear that the quantity in parentheses is the function previously encountered in (8-46) and plotted in Fig. 8-11. Since this function has appreciable magnitudes only very close to the direct beam (for $\text{Cu}\ K\alpha$ radiation and $R = 1$ mm, this function first becomes zero when $\theta = 0.02$ second of arc), it is reasonable to neglect it in further considerations.

Rearranging the remaining terms in (8-50),

$$\frac{I_{\text{eu}}/N - f^2}{f^2} = 4\pi \int_0^\infty r^2 [\rho(r) - \rho_0] \frac{\sin kr}{kr} \, dr \tag{8-51}$$

where the upper limit of integration can be set at infinity because $\rho(r) \rightarrow \rho_0$ as r increases, so that the integral goes to zero. The left side of (8-51) is a function of

$(\sin \theta)/\lambda$ or k only, so that, let the function

$$i(k) = \frac{I_{eu}/N - f^2}{f^2} \tag{8-52}$$

and after a final cancellation of like terms,

$$k[i(k)] = 4\pi \int_0^\infty r[\rho(r) - \rho_0] \sin kr \, dr \tag{8-53}$$

It is now possible to make use of the well-known Fourier integral theorem, Appendix 1, which states that if

$$\varphi(k) = 4\pi \int_0^\infty f(r) \sin kr \, dr$$

then $$f(r) = \frac{1}{2\pi^2} \int_0^\infty \varphi(k) \sin rk \, dk \tag{8-54}$$

Making use of this theorem, it is possible to write the companion equation to (8-53).

$$r[\rho(r) - \rho_0] = \frac{1}{2\pi^2} \int_0^\infty k[i(k)] \sin rk \, dk \tag{8-55}$$

or, expressing it in terms of the radial distribution function,

$$4\pi r^2 \rho(r) = 4\pi r^2 \rho_0 + \frac{2r}{\pi} \int_0^\infty k[i(k)] \sin rk \, dk \tag{8-56}$$

The foregoing mathematical procedure was first developed by F. Zernicke and G. Prins in 1927, but it was not applied for about four years, until Debye, assisted by H. Menke, used it to determine the radial distribution function of liquid mercury. In carrying out such analyses, it should be remembered that the intensity in the above equations is the intensity of coherent scattering expressed in electron units. When one measures the intensity of x-ray scattering experimentally, one measures a composite of coherent and incoherent scattering, modified by polarization and absorption and superimposed on a background intensity due to various causes. The latter can be minimized by eliminating air scattering by evacuating the diffraction camera and by using crystal-monochromatized radiation. The polarization correction in this case becomes

$$\frac{1 + \cos^2 2\theta_M \cos^2 2\theta}{1 + \cos^2 2\theta_M} \tag{8-57}$$

where θ_M is the Bragg angle of the monochromator crystal. The absorption correction, in turn, must be made in accordance with the sample geometry employed. A further discussion of the experimental aspects of such analyses can be found in the book by J. T. Randall and the review paper by N. S. Gingrich, listed at the conclusion of this chapter.

Note that $i(k)$ is essentially a function of $(\sin kr)/kr$, as can be seen by returning to the original Debye equation (8-48). As shown in Fig. 8-10, $i(k)$ tends to zero as k increases, so that, at large $(\sin\theta)/\lambda$ value,

$$\frac{I_{\mathrm{eu}}/N - f^2}{f^2} = 0$$

or

$$\frac{I_{\mathrm{eu}}}{N} = f^2 \tag{8-58}$$

This means that the coherent scattering at large scattering angles should be equal to Nf^2, while the incoherent scattering can be calculated with the aid of (8-20) or (8-22). Their sum, at large angles, must equal the total intensity of scattering, that is, the experimentally measured intensity suitably corrected for absorption and polarization effects. The way that these different intensities are related to each other is illustrated in Fig. 8-12. It is clear from this that it is possible to establish a scale for the experimental intensities by trial-and-error curve fitting, according to the principles outlined above.

A more exact analysis of the various factors affecting experimental curves was carried out by C. Finback, who concluded that a slightly better result can be obtained by using an electronic, rather than an atomic, radial distribution function. The principal difference (apart from some multiplicative constants) is that the angular

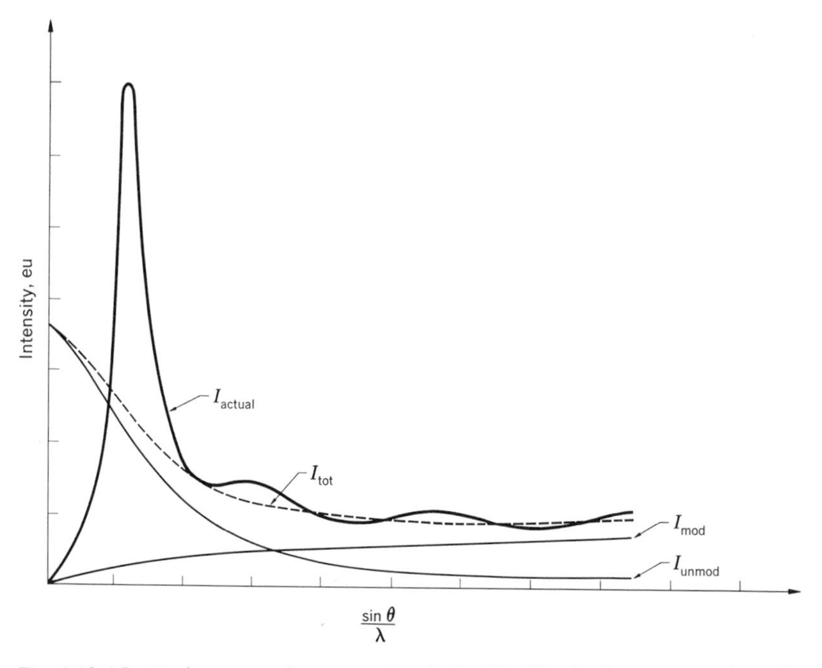

Fig. 8-12. Scattering curve for a monatomic liquid. The broken curves show the angular dependence of $I_{\mathrm{tot}} = I_{\mathrm{mod}} + I_{\mathrm{unmod}}$ for N independently scattering atoms per unit volume. Note that, at large scattering angles, the actual scattering curve is nearly equal to I_{tot}. *(After Klug and Alexander.)*

function becomes

$$i(k) = I_{eu} - Nf^2 \tag{8-59}$$

instead of the form given in (8-52). As Finback has shown, this functional relation eliminates some small spurious peaks that had been observed in the radial distribution functions calculated for a number of molten metals.

Vitreous solids. Quite similarly to the case of polyatomic gases, the introduction of several kinds of atoms, or such atomic groups as complex ions or molecules, requires that the Debye scattering equation in (8-40) be considered term by term. Whether a liquid or a vitreous solid is considered, there is no directional order present, so that it is still possible to use the concept of a radial distribution function which, statistically speaking, has spherical symmetry about a central atom. Because the origin atom may actually have different nearest-neighbor configurations, depending on which atom in a molecule is placed at the origin, the use of such a radial distribution function blurs the distinction between the various atoms, so that the peaks in a radial distribution curve serve primarily to indicate the interatomic distances in the liquid or vitreous solid. Without considering the procedure in detail, the utilization of the Fourier integral theorem (8-54) to interpret the experimental scattering curves means that the final equation for the intensity (in electron units) scattered by a polyatomic liquid or a vitreous solid is quite similar to (8-56), except that $i(k)$ has a slightly modified form, and the radial distribution $\rho(r)$ represents a kind of "averaged" distribution for the different atoms.

The actual interpretation then proceeds by using the $i(k)$ curve determined experimentally (like the suitably scaled curve in Fig. 8-12) to provide the necessary information for a numerical integration of an expression like (8-56). The integral is evaluated for each value of r (in preselected intervals), and the radial distribution function is plotted as a function of the radial distance. A typical curve is shown for vitreous silicon dioxide (silica) in Fig. 8-13 and represents one of the earliest applications of these procedures to the study of the "structure" of glasses by Warren and his students. The first peak has a maximum at about 1.62 Å, a value quite close to the $\mathrm{Si{-}O}$ distance in crystalline silicates. The area under this peak is proportional to the products of the scattering powers of the central atom and the atoms coordinating it, so that it is indicative of the number of nearest-neighbor atoms at this distance. In the case of silica, it is known that there are four oxygen atoms tetrahedrally bonded to each silicon atom in an $(SiO_4)^{4-}$ group. An analysis of the area under the first peak yields the number 4.3 instead of the expected 4.0. The next interatomic distance occurring in silica should be the oxygen-oxygen distance, which, from the known geometry of a tetrahedron, should be $\sqrt{\frac{8}{3}}\, r_{\mathrm{Si-O}} = 2.65$ Å. A peak having the requisite area for oxygen-oxygen interactions actually can be resolved in the curve in Fig. 8-13 about the point $r = 2.65$ Å. Finally, assuming that the tetrahedra are joined to each other so that the $\mathrm{Si{-}O{-}Si}$ distance separating two neighboring silicon atoms is twice $r_{\mathrm{Si-O}}$, then a peak representing the interaction between silicon atoms should appear at about 3.24 Å, and such a maximum is again apparent in Fig. 8-13. By such means it is possible to interpret the radial

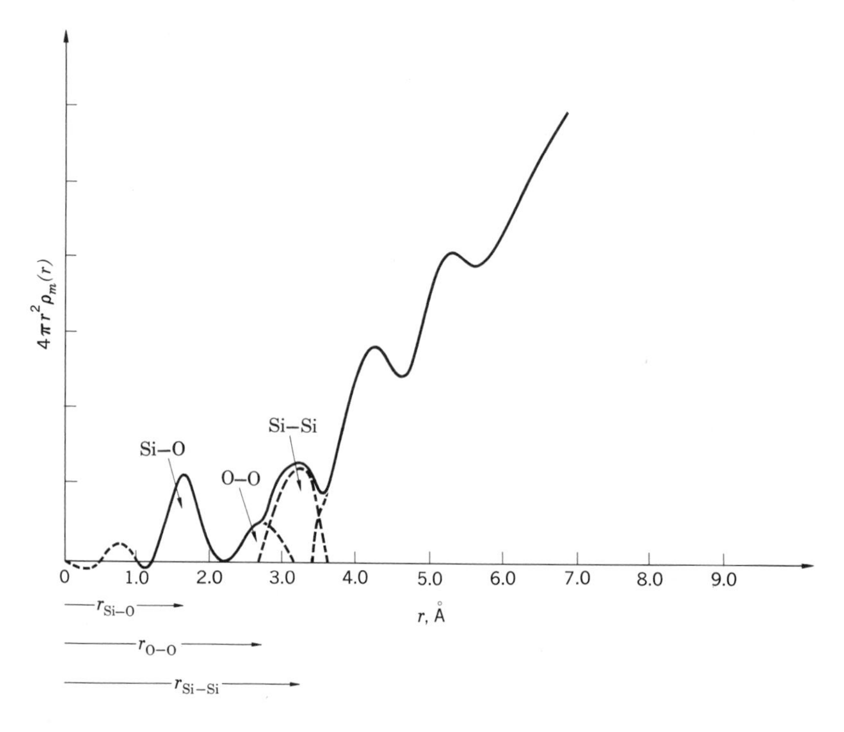

Fig. 8-13. Radial distribution function for vitreous silica. *(After Warren.)*

distribution curves and confirm (or refute) certain possible models for the atomic configurations in liquids and vitreous solids.

More recently, Warren has suggested that the ambiguities in assigning areas to the peaks in a radial distribution curve, caused by uncertainties in the width of the peaks and their tails, can be largely eliminated by using a reversed procedure. A model of the atomic array in the glass is postulated, and a calculated radial distribution function is matched to the observed one. (In a somewhat more rigorous formulation, Warren makes use of a "pair function" which expresses directly the phasal interactions between an ith and jth atom.) It should be noted that, although such analyses are approximate, a more exact description of the atomic distribution is without real meaning because such a more regular distribution does not, in fact, exist in vitreous solids. It is, of course, possible for a greater degree of order to exist, both in the liquid and the solid state. This is the case, for example, when large organic molecules are present whose asymmetric shape dictates a more or less regular arrangement. This leads to a material that is part way between being a crystal and a noncrystal. A brief further discussion of x-ray diffraction by such materials can be found in Chap. 20.

Crystals. As stated frequently already, the chief characteristic that distinguishes crystalline aggregates from noncrystalline ones is the presence of the three-dimensional periodicity in the atomic arrays of crystals. In calculating the intensity scattered by a crystal, therefore, it is sufficient to calculate the intensity scattered by the periodically repeated atomic aggregate (unit-cell contents), which is relatable to the internal interference effect of Debye. The radiation scattered by each unit cell, whose interaction expresses the external interference effect of Debye, is then in phase for scattering in certain directions and has a random phase relationship for all other directions. This leads to the results already discussed in Chap. 7 and considered more fully in the next chapter. For the present, it is interesting to see what happens when the general scattering equation (8-35) is applied to this problem.

The positions of the atoms in a crystal can be denoted by a vector $\mathbf{R}$ drawn from the origin chosen at a lattice point in the crystal. If the unit cell contains N atoms distributed at the ends of vectors $\mathbf{r}_n$ drawn from the corresponding lattice point in the unit cell, the desired vector consists of two parts, $\mathbf{r}_n$ and the vector from the crystal origin point to the desired unit cell, namely,

$$m_1\mathbf{a} + m_2\mathbf{b} + m_3\mathbf{c} \tag{8-60}$$

where $\mathbf{a}$, $\mathbf{b}$, $\mathbf{c}$, are the unit-cell vectors, and m_1, m_2, m_3 represent the necessary integer multipliers. For this reason, the vector from the common origin to the atom n in the unit cell m_1, m_2, m_3 is defined

$$\mathbf{R}^n_{m_1m_2m_3} = m_1\mathbf{a} + m_2\mathbf{b} + m_3\mathbf{c} + \mathbf{r}_n \tag{8-61}$$

This vector is illustrated in two dimensions in Fig. 8-14.

Substituting this vector in the exponents in (8-35), the summation over all the atoms in the crystal now requires four sums, respectively, over n (to include all the atoms in one unit cell) and over m_1, m_2, and m_3 (to include all the unit cells). Accordingly, the intensity expression for a crystal can be written

$$\begin{aligned} I_{\text{eu}} &= \sum_n \sum_{m_1} \sum_{m_2} \sum_{m_3} f_n e^{(2\pi i/\lambda)(\mathbf{S}-\mathbf{S}_0)\cdot\mathbf{R}^n_{m_1m_2m_3}} \sum_{n'} \sum_{m_1'} \sum_{m_2'} \sum_{m_3'} f_{n'} e^{(-2\pi i/\lambda)(\mathbf{S}-\mathbf{S}_0)\cdot\mathbf{R}^{n'}_{m_1'm_2'm_3'}} \\ &= \sum_n \sum_{n'} \sum_{m_1} \sum_{m_1'} \sum_{m_2} \sum_{m_2'} \sum_{m_3} \sum_{m_3'} f_n f_{n'} e^{(2\pi i/\lambda)(\mathbf{S}-\mathbf{S}_0)\cdot(\mathbf{R}^n_{m_1m_2m_3} - \mathbf{R}^{n'}_{m_1'm_2'm_3'})} \end{aligned} \tag{8-62}$$

In the general case, when the scattering factors f_n are complex, f_n and $f_{n'}$ are complex conjugate quantities.

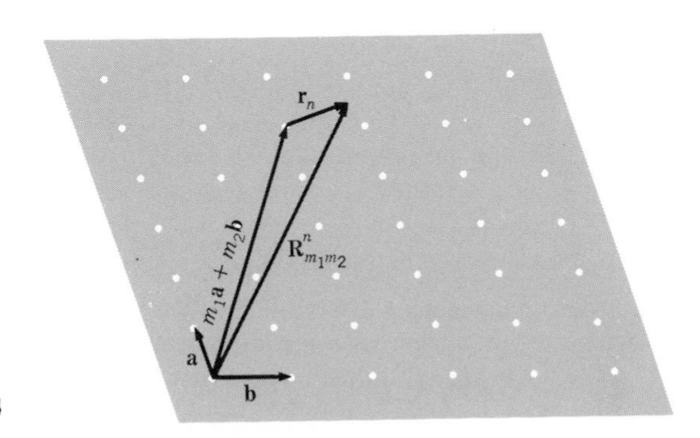

Fig. 8-14

It is possible to define a term

$$F = \sum_{n=1}^{N} f_n e^{(2\pi i/\lambda)(\mathbf{S}-\mathbf{S}_0)\cdot\mathbf{r}_n} \tag{8-63}$$

which is called the *structure factor* of the crystal and acts to represent the complex amplitude scattered by one unit cell. Substituting (8-61) in (8-62), separating like terms, and realizing that the complex conjugate of F, denoted by F^*, has a negative exponent in the expression (8-63), it is possible to write (8-62) as follows:

$$I_{\text{eu}} = F \cdot F^* \sum_{m_1} \sum_{m_1'} \sum_{m_2} \sum_{m_2'} \sum_{m_3} \sum_{m_3'} e^{(2\pi i/\lambda)(\mathbf{S}-\mathbf{S}_0)\cdot[(m_1-m_1')\mathbf{a}+(m_2-m_2')\mathbf{b}+(m_3-m_3')\mathbf{c}]} \tag{8-64}$$

The product $F \cdot F^* \equiv F^2$ then represents the internal interference effect, and the remaining summations represent the external interference effect of Debye. As shown in the next chapter, the external interference effect leads to zero scattering intensity, except when the three Laue conditions are simultaneously obeyed by the dot product in the exponent of (8-64). The structure factor (8-63) is also encountered in the next chapter, but a fuller discussion of its use is postponed until Chap. 11.

SUGGESTIONS FOR SUPPLEMENTARY READING

J. Bouman (ed.), *Selected topics in x-ray crystallography* (Interscience Publishers, Inc., New York, 1951), pp. 191–210 and 263–296.

K. Furukawa, The radial distribution curves of liquids by diffraction methods, *Reports on Progress in Physics*, vol. 25 (1962), pp. 395–440.

N. S. Gingrich, The diffraction of x-rays by liquid elements, *Reviews of Modern Physics*, vol. 15 (1943), pp. 90–110.

R. W. James, *The optical principles of the diffraction of x-rays* (G. Bell & Sons, Ltd., London, 1950).

Harold P. Klug and Leroy E. Alexander, *X-ray diffraction procedures* (John Wiley & Sons, Inc., New York, 1954), pp. 586–633.

J. T. Randall, *The diffraction of x-rays and electrons by amorphous solids, liquids, and gases* (John Wiley & Sons, Inc., New York, 1934).

B. E. Warren, The basic principles involved in the glassy state, *J. Applied Physics*, vol. 13 (1942), pp. 602–610.

EXERCISES

8-1. Assume that the radial distribution for a monovalent sodium ion is such that $4\pi r^2\rho(r)/e$ equals 2 when $r = r_K$, and 8 when $r = r_L$ for sodium. The atomic scattering factor then can be expressed by summing the contributions from the K and L electrons according to

$$f = 2\frac{\sin kr_K}{kr_K} + \frac{\sin kr_L}{kr_L}$$

Plot the scattering-factor curves for K and L electrons separately as a function of k, and then add them together to obtain the resultant f curve.

8-2. Helium has two electrons, both of which are K electrons. On the same graph paper, plot $(I_{\text{eu}})_{\text{mod}}$, $(I_{\text{eu}})_{\text{unmod}}$, and $(I_{\text{eu}})_{\text{tot}}$ as a function of $(\sin\theta)/\lambda$.

8-3. Assuming that the wave function for a K electron is correctly given by (8-10), calculate $(I_{eu})_{mod}$ for the two electrons in helium, using Waller's equation (8-22). Plot the modified intensity as a function of $(\sin \theta)/\lambda$, and compare it with the result of Exercise 8-2.

8-4. Look up the values of the dispersion terms in *International tables for x-ray crystallography,* Volume 3, for copper and zinc. If you wanted to maximize the difference between the scattering factors of Cu ($Z = 29$) and Zn ($Z = 30$) and you had your choice of x-ray tubes having Cr, Co, Cu, Mo, and Ag targets, which target x-ray tube would you select? Since one can use a crystal monochromator to select any wavelength component whatever, which characteristic line would you select for this purpose?

8-5. The phase lag accompanying dispersion, discussed in connection with the sphalerite structure in this chapter, has several practical applications in crystal-structure analysis and associated problems. The sphalerite structure, however, lacks a center of symmetry. Is it possible to perform a similar experiment to distinguish (222) from $(\bar{2}\bar{2}\bar{2})$ planes in the sodium chloride structure which consists of alternating Na and Cl planes but in centrosymmetric array? What is the effect of anomalous dispersion in a centrosymmetric structure?

8-6. The interatomic separation in O_2 molecules is 1.21 Å. Accordingly, plot the intensity (in electron units) scattered by a gas of oxygen atoms as a function of $(\sin \theta)/\lambda$. To make the plot independent of gas pressure, plot I_{eu}/Nf^2.

8-7. When the electrons in an atom are concentrated in a very small volume, they can be assumed to scatter x rays fully in phase with each other, at all angles. Consider a CCl_4 molecule comprised of such point atoms. Next, suppose that $r_{C-Cl} = 1.76$ Å. Calculate the intensity in electron units scattered by a gas of CCl_4 molecules, and plot I_{eu}/N as a function of $(\sin \theta)/\lambda$. How would this plot differ if liquid CCl_4 were examined?

8-8. The phosphorus molecule P_4 is known to have a tetrahedral structure with a P—P distance of 2.20 Å. Consider an x-ray diffraction diagram of P_4 vapor, and plot the unmodified intensity in electron units per molecule through the range of $(\sin \theta)/\lambda$ from 0.0 to 0.5.

8-9. The location of an atom can be specified by its fractional coordinates in a unit cell, that is, its equipoint. This makes it convenient to express the atomic vector $\mathbf{r} = x\mathbf{a} + y\mathbf{b} + z\mathbf{c}$, where x, y, z are the three fractional coordinates. Substituting this vector in (8-63), what is the structure-factor expression when the three Laue conditions are satisfied?

NINE
X-ray diffraction by ideal crystals

KINEMATICAL THEORY

Diffraction by a crystal. Consider a monochromatic x-ray beam incident on a crystal such that the beam's cross section is larger than that of the crystal. If the beam is composed of parallel x rays, they form a common wave front, as shown to the left in Fig. 9-1, that is normal to the propagation direction of the incident beam, denoted by the unit vector $\mathbf{S}_0$. Next consider the beam scattered in a direction denoted by the unit vector $\mathbf{S}$. In this section, it is of interest to examine under what conditions the x rays scattered in this direction interfere constructively to form a common wave front normal to $\mathbf{S}$, as shown to the right in Fig. 9-1. As in the preceding chapter, it is convenient to represent the position of any atom in the crystal by a vector from a common origin.

$$\mathbf{R}_m{}^n = m_1\mathbf{a} + m_2\mathbf{b} + m_3\mathbf{c} + \mathbf{r}_n \tag{9-1}$$

where the first three terms define the vector to the origin of the mth cell (actually, cell $m_1m_2m_3$), and $\mathbf{r}_n$ is the vector from the origin of a unit cell to the nth atom within the cell. The electrons in the nth atom scatter the incident electric field according to the appropriate atomic scattering factor f_n, as discussed in Chap. 8. To see how x rays scattered in the direction $\mathbf{S}$ by a periodic array of atoms interfere with each other, it is necessary to examine the pathlengths the various rays travel before they reach a point P lying a distance R from the crystal that is very large compared with atomic dimensions.

Suppose that the x rays in the incident monochromatic beam have the wavelength λ and form a common wave front at the origin O of the crystal. By the time this wave front reaches the atom lying at the end of the vector $\mathbf{R}_m{}^n$, it has traveled a distance x_1 (Fig. 9-1), so that the instantaneous value of the electric field at this point is

$$\mathcal{E}_0 = E_0 e^{2\pi i[(c/\lambda)t-(1/\lambda)x_1]} \tag{9-2}$$

where E_0 is the electric field amplitude of the incident beam. As discussed in the preceding chapter, the atom scatters the incident field in the direction of $\mathbf{S}$ in such

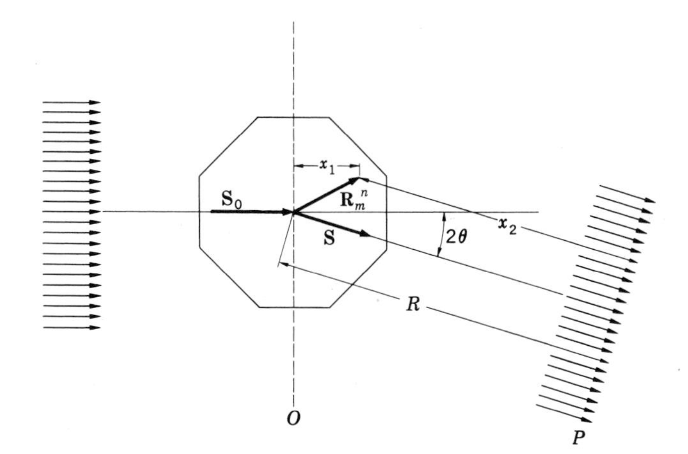

Fig. 9-1

a way that its value, when it reaches P, a distance x_2 from the atom being considered, is

$$\mathcal{E}_P = f_n \frac{e^2 E_0}{mc^2 R} e^{2\pi i[(c/\lambda)t - (1/\lambda)(x_1+x_2)]} \tag{9-3}$$

where e is the charge and m the mass of an electron; c is the velocity of light; and f_n is the atomic scattering factor.

The total pathlength $x_1 + x_2$ that the scattered ray travels from the origin plane at O to the detector plane at P can be evaluated by considering the construction in Fig. 9-1. Clearly,

$$x_1 = \mathbf{R}_m{}^n \cdot \mathbf{S}_0$$

and

$$x_2 = R - \mathbf{R}_m{}^n \cdot \mathbf{S} \tag{9-4}$$

so that

$$x_1 + x_2 = R - (\mathbf{S} - \mathbf{S}_0) \cdot \mathbf{R}_m{}^n$$

Substituting this distance in (9-3),

$$\mathcal{E}_P = \frac{e^2 E_0}{mc^2 R} f_n e^{2\pi i\left(\frac{c}{\lambda}t - \frac{1}{\lambda}[R - (\mathbf{S}-\mathbf{S}_0)\cdot\mathbf{R}_m{}^n]\right)} \tag{9-5}$$

It is seen from Fig. 9-1 that the scattering of the incident x-ray beam by an atom in the crystal has the effect of deflecting it in the scattering direction $\mathbf{S}$. As a result, its pathlength differs from that of the central ray passing through the origin, and such path differences produce phase differences between the various rays arriving at the plane at P. The resulting field at P can be evaluated by summing the fields "scattered" by each atom and expressed by (9-5). In view of the form of vector $\mathbf{R}_m{}^n$, it is most convenient to first sum over the n atoms in a unit cell and then to add up the contributions from all the unit cells in the crystal by summing over m_1, m_2, and m_3. The instantaneous value of the field at P due to scattering by the entire crystal then can be expressed

$$\mathcal{E}_P = \frac{e^2E_0}{mc^2R} \sum_n \sum_{m_1} \sum_{m_2} \sum_{m_3} f_n e^{2\pi i\left[\frac{c}{\lambda}t - \frac{R}{\lambda} + \left(\frac{\mathbf{S}-\mathbf{S}_0}{\lambda}\right)\cdot(\mathbf{r}_n + m_1\mathbf{a} + m_2\mathbf{b} + m_3\mathbf{c})\right]} \tag{9-6}$$

Separating terms depending on n, m_1, m_2, and m_3 only and recalling the definition of the structure factor expressing the scattering by one unit cell from the previous chapter,

$$F = \sum_n f_n e^{(2\pi i/\lambda)(\mathbf{S}-\mathbf{S}_0)\cdot\mathbf{r}_n} \tag{8-63}$$

it is possible to rewrite (9-6) after regrouping like terms.

$$\mathcal{E}_P = \frac{e^2E_0}{mc^2R} e^{(2\pi i/\lambda)(ct-R)} F \sum_{m_1} e^{(2\pi i/\lambda)(\mathbf{S}-\mathbf{S}_0)\cdot m_1\mathbf{a}} \sum_{m_1} e^{(2\pi i/\lambda)(\mathbf{S}-\mathbf{S}_0)\cdot m_2\mathbf{b}} \sum_{m_3} e^{(2\pi i/\lambda)(\mathbf{S}-\mathbf{S}_0)\cdot m_3\mathbf{c}} \tag{9-7}$$

In order to carry out the three summations in (9-7), it is convenient to consider the crystal as having the shape of a parallelepipedon whose sides are parallel to the cell edges. The length of each crystal edge is then simply the number of unit cells lying along that edge times the cell-edge length, respectively, M_1a, M_2b, M_3c. As will become clear in subsequent discussion, the assumption of such a simple form for the crystal facilitates the above summations without limiting the generality of the arguments that follow. According to this model of the crystal, it contains a total

$$M = M_1 \times M_2 \times M_3 \tag{9-8}$$

unit cells, while the sums over m have a common form that can be expressed

$$\sum_{m_1=0}^{m_1=M_1-1} e^{(2\pi i/\lambda)(\mathbf{S}-\mathbf{S})\cdot m_1\mathbf{a}} = e^0 + e^{(2\pi i/\lambda)(\mathbf{S}-\mathbf{S})\cdot 1\mathbf{a}} + e^{(2\pi i/\lambda)(\mathbf{S}-\mathbf{S})\cdot 2\mathbf{a}} + \cdots + e^{(2\pi i/\lambda)(\mathbf{S}-\mathbf{S})\cdot(M_1-1)\mathbf{a}} \tag{9-9}$$

The sum in (9-9) is a geometric series of the form

$$\sum_{m=0}^{M-1} ar^m = a + ar + ar^2 + \cdots + ar^{M-1} = \frac{ar^M - a}{r-1} \tag{9-10}$$

Thus it is possible to replace each of the three sums in (9-7) by their convergence limit ratios to give

$$\mathcal{E}_P = \frac{e^2E_0}{mc^2R} e^{\frac{2\pi i}{\lambda}(ct-R)} F \frac{e^{\frac{2\pi i}{\lambda}(\mathbf{S}-\mathbf{S}_0)\cdot M_1\mathbf{a}} - 1}{e^{\frac{2\pi i}{\lambda}(\mathbf{S}-\mathbf{S}_0)\cdot\mathbf{a}} - 1} \frac{e^{\frac{2\pi i}{\lambda}(\mathbf{S}-\mathbf{S}_0)\cdot M_2\mathbf{b}} - 1}{e^{\frac{2\pi i}{\lambda}(\mathbf{S}-\mathbf{S}_0)\cdot\mathbf{b}} - 1} \frac{e^{\frac{2\pi i}{\lambda}(\mathbf{S}-\mathbf{S}_0)\cdot M_3\mathbf{c}} - 1}{e^{\frac{2\pi i}{\lambda}(\mathbf{S}-\mathbf{S}_0)\cdot\mathbf{c}} - 1} \tag{9-11}$$

The ultimately desired quantity is the intensity of the x-ray beam reaching the point of observation I_P. Recalling that the intensity is proportional to the square of the amplitude of the electric field and that squaring a complex quantity requires multiplication by its complex conjugate,

$$E_P{}^2 = |\mathcal{E}_P|^2 = \mathcal{E}_P \times \mathcal{E}_P^* \tag{9-12}$$

Thus, to obtain $I_P = (c/8\pi)E_P{}^2[(1 + \cos^2 2\theta)/2]$, it is necessary to multiply (9-11) by its complex conjugate, which involves the formation of products between the quotients of the type

$$\frac{e^{iMx} - 1}{e^{ix} - 1} \times \frac{e^{-iMx} - 1}{e^{-ix} - 1} = \frac{e^0 - e^{iMx} - e^{-iMx} + 1}{e^0 - e^{ix} - e^{-ix} + 1} = \frac{4 \sin^2 \frac{1}{2}Mx}{4 \sin^2 \frac{1}{2}x} \tag{9-13}$$

so that the intensity at P can be expressed

$$I_P = I_e F^2 \frac{\sin^2 \frac{\pi}{\lambda}(\mathbf{S} - \mathbf{S}_0) \cdot M_1\mathbf{a}}{\sin^2 \frac{\pi}{\lambda}(\mathbf{S} - \mathbf{S}_0) \cdot \mathbf{a}} \frac{\sin^2 \frac{\pi}{\lambda}(\mathbf{S} - \mathbf{S}_0) \cdot M_2\mathbf{b}}{\sin^2 \frac{\pi}{\lambda}(\mathbf{S} - \mathbf{S}_0) \cdot \mathbf{b}} \frac{\sin^2 \frac{\pi}{\lambda}(\mathbf{S} - \mathbf{S}_0) \cdot M_3\mathbf{c}}{\sin^2 \frac{\pi}{\lambda}(\mathbf{S} - \mathbf{S}_0) \cdot \mathbf{c}} \tag{9-14}$$

letting $I_e = e^4 I_0(1 + \cos^2 2\theta)/2m^2c^2R^2$, and assuming that the incident beam of intensity I_0 is unpolarized.

The three quotients in (9-14) are of the form $(\sin^2 Mx)/\sin^2 x$, which is shown plotted as a function of x in Fig. 9-2. Note that this function is virtually zero everywhere, except at the points $x = q\pi$ (q is any integer including zero), where it rises to a maximum value of M^2. Thus each of the three quotients in (9-14) is virtually equal to zero unless the arguments of the sine terms in the denominators are equal to integral multiples of π. The intensity I_P equals the product of these three quotients, so that it will equal zero unless all three quotients take on their maximum values at the same time. This means that the three arguments must be simultaneously equal to integer multiples of π; that is, the three denominators must vanish at the same time. This takes place when the following three equations are

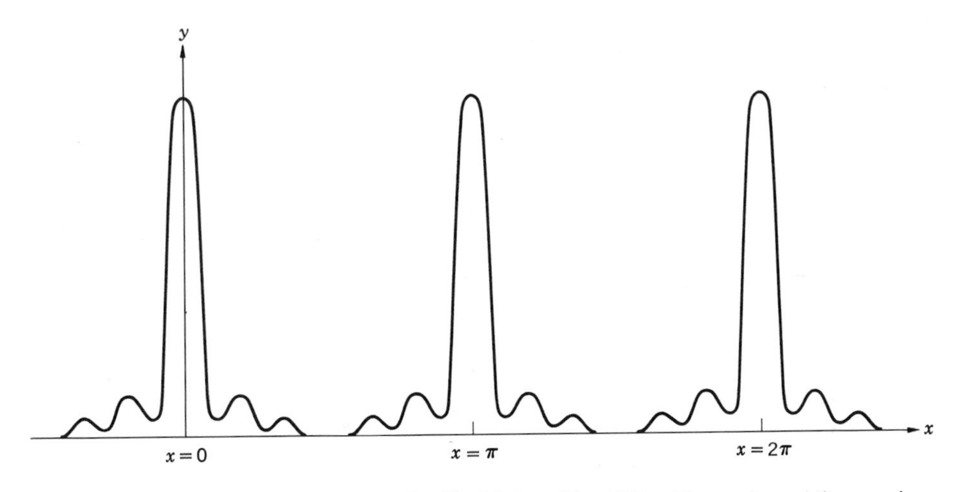

Fig. 9-2. Graphical representation of $y = \sin^2 Mx/\sin^2 x$. (The width of the peaks and the prominence of the ripples are inversely proportional to M.)

satisfied simultaneously:

$$(\mathbf{S} - \mathbf{S}_0) \cdot \mathbf{a} = h\lambda$$
$$(\mathbf{S} - \mathbf{S}_0) \cdot \mathbf{b} = k\lambda \qquad (9\text{-}15)$$
$$(\mathbf{S} - \mathbf{S}_0) \cdot \mathbf{c} = l\lambda$$

where h, k, and l can be positive or negative integers, including zero. These are, of course, the already familiar Laue equations (7-30) derived from consideration of the geometry of diffraction in Chap. 7, so that the three integers h, k, l can be identified immediately as the Miller indices of the reflecting plane. When the three Laue conditions (9-15) are obeyed, the quotients in (9-14) take on the values M_1^2, M_2^2, M_3^2, so that

$$I_P = I_e F^2 M^2 \qquad (9\text{-}16)$$

where M is the total number of unit cells in the crystal, according to (9-8).

It is seen from (9-16) that the intensity reaching the point P is proportional to the square of the total number of unit cells in the crystal and the magnitude of the structure factor squared. It also follows from the foregoing discussion that the intensity of coherently scattered x rays in any direction normally is zero because of the destructive interference between the rays scattered from different points in the crystal. Only for the finite number of directions defined by the three simultaneous equations in (9-15) is constructive interference possible. Thus the diffracted-beam directions are determined by the three crystallographic axes **a**, **b**, and **c**, and conversely, the lattice of a crystal can be determined by discovering what these diffraction directions (Bragg angles) are. The intensities of the diffracted beams can be used, in turn, to determine the magnitudes of the structure factors, which, as discussed in the next section, depend on the atomic arrangements within the unit cell.

Crystal structure factor. When the three Laue conditions are satisfied, the vector $(\mathbf{S} - \mathbf{S}_0)/\lambda$ coincides with the reciprocal-lattice vector $\boldsymbol{\sigma}_{hkl}$, as demonstrated by (7-35). In this case, the structure factor becomes

$$F = \sum_n f_n e^{(2\pi i/\lambda)(\mathbf{S} - \mathbf{S}_0)\cdot\mathbf{r}_n} = \sum_n f_n e^{(2\pi i/\lambda)\boldsymbol{\sigma}\cdot\mathbf{r}_n} \qquad (9\text{-}17)$$

The reciprocal-lattice vector $\boldsymbol{\sigma}$ is defined in terms of the Miller indices of the reflecting plane and the reciprocal-lattice axes according to (7-11), and the vector $\mathbf{r}_n$ to the nth atom within the unit cell can be defined

$$\mathbf{r}_n = x_n\mathbf{a} + y_n\mathbf{b} + z_n\mathbf{c} \qquad (9\text{-}18)$$

where x_n is the fractional distance from the cell origin measured parallel to **a** (= actual distance along **a** divided by length of **a**), and y_n and z_n, similarly, are the fractional coordinates along **b** and **c**, respectively. By making use of these fractional coordinates it is possible to obtain quite general expressions for the structure factors and to utilize the tabulated equipoint coordinates discussed in Chap. 4. How this is done is further demonstrated by actual examples in Chap. 11.

Substituting (7-11) and (9-18) into the exponent of (9-17) and recalling the orthogonality condition for multiplying reciprocal vectors (7-8) and (7-9), the structure factor can be written

$$F_{hkl} = \sum_{n=1}^{N} f_n e^{2\pi i(hx_n+ky_n+lz_n)} \tag{9-19}$$

which shows that its magnitude depends only on the relative disposition of the N atoms in the unit cell and on their respective scattering powers f_n.

The meaning of (9-19) can be understood most easily by writing out the sum term by term:

$$F_{hkl} = f_1 e^{2\pi i(hx_1+ky_1+lz_1)} + f_2 e^{2\pi i(hx_2+ky_2+lz_2)} + \cdots + f_N e^{2\pi i(hx_N+ky_N+lz_N)} \tag{9-20}$$

Each of the terms in (9-20) can be considered to represent a wavelet having an amplitude f_n and a phase $\varphi_n = 2\pi(hx_n + ky_n + lz_n)$ which expresses the path-length for each scattered wavelet. The structure factor F then is simply the resultant of the wavelets scattered by the N atoms in a unit cell. This can be seen directly by representing each wavelet by an appropriate vector in the complex plane, that is, by constructing an Argand diagram as shown in Fig. 9-3. It is clear from this construction that the structure factor normally is a complex quantity having a magnitude and a phase angle. According to (9-16), the intensity of x rays diffracted by the crystal is proportional to $F^2 = F \cdot F^*$, so that, experimentally, it is possible to determine only the amplitude F of the structure factor, but not its phase. This epitomizes what is called the *phase problem* in crystal-structure analysis, because it is the phase factors in (9-20) that are directly related to the atomic positions within the unit cell, that is, the crystal structure. The main substance of crystal-structure analysis therefore consists in deducing the missing phases. Some of the ways that this can be done are described in Chap. 11.

Width of diffraction maxima. It will be recalled from Chap. 7 that, when the three Laue equations in (9-15) are satisfied, the following equivalent relation,

$$(\mathbf{S} - \mathbf{S}_0) = \lambda \boldsymbol{\sigma}_{hkl} \tag{7-35}$$

is valid. According to the derivation of (9-16), based on the form of the quotients in (9-14), the diffracted intensity rapidly drops to zero when this condition is violated. In order to estimate the angular range over which the diffracted intensity retains an appreciable magnitude, let the diffracted beam direction deviate slightly

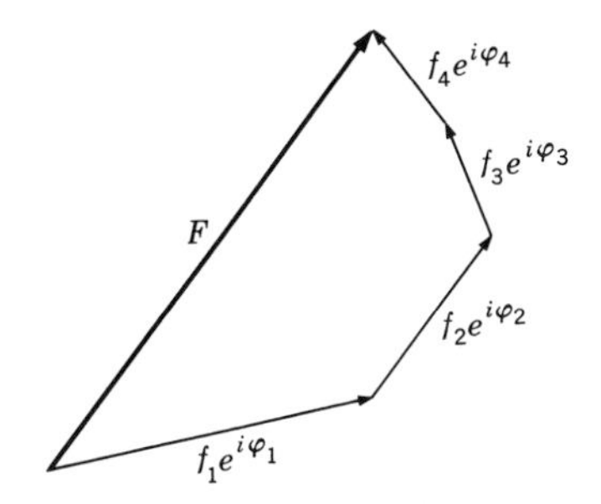

Fig. 9-3. Argand diagram showing the summation of wavelets comprising the structure factor F for $N = 4$.

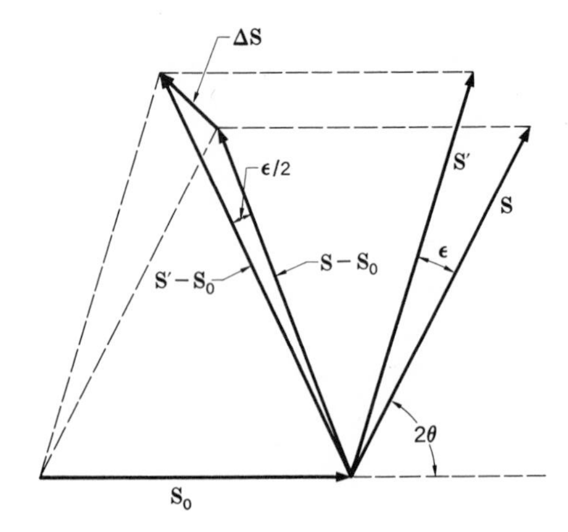

Fig. 9-4

from 2θ, say, by the small angle ϵ shown in Fig. 9-4. The new diffracted beam direction can be represented by the unit vector $\mathbf{S}'$, while the difference vector (7-35) is displaced by $\Delta\mathbf{S}$, so that

$$(\mathbf{S}' - \mathbf{S}_0) = (\mathbf{S} - \mathbf{S}_0) + \Delta\mathbf{S}$$
$$= \lambda\boldsymbol{\sigma}_{hkl} + \Delta\mathbf{S} \tag{9-21}$$

Substituting this relation for $(\mathbf{S} - \mathbf{S}_0)$ in Eq. (9-14) for the intensity I_P, the sine terms in the three quotients take on the form

$$\sin^2\frac{\pi}{\lambda}(\mathbf{S}' - \mathbf{S}_0)\cdot M\mathbf{a} = \sin^2\frac{\pi}{\lambda}(\lambda\boldsymbol{\sigma} + \Delta\mathbf{S})\cdot M\mathbf{a} = \sin^2\left(\pi Mh + \frac{\pi}{\lambda}\Delta\mathbf{S}\cdot M\mathbf{a}\right) \tag{9-22}$$

where $\pi Mh = \pi\boldsymbol{\sigma}\cdot M\mathbf{a} = \pi(h\mathbf{a}^* + k\mathbf{b}^* + l\mathbf{c}^*)\cdot M\mathbf{a}$. The term on the right in (9-22) can be expanded according to $\sin^2(A + B) = (\sin A\cos B + \sin B\cos A)^2$ to give

$$\sin^2\left(\pi Mh + \frac{\pi}{\lambda}\Delta\mathbf{S}\cdot M\mathbf{a}\right) = \left[\sin(\pi Mh)\cos\left(\frac{\pi}{\lambda}\Delta\mathbf{S}\cdot M\mathbf{a}\right) + \sin\left(\frac{\pi}{\lambda}\Delta\mathbf{S}\cdot M\mathbf{a}\right)\cos(\pi Mh)\right]^2$$
$$= \sin^2\left(\frac{\pi}{\lambda}\Delta\mathbf{S}\cdot M\mathbf{a}\right) \tag{9-23}$$

because $\sin\pi Mh = 0$ since M and h are integers, and $\cos\pi Mh = \pm 1$ for the same reason.

It is thus possible to rewrite the intensity relation (9-14)

$$I_P = I_e F^2 \frac{\sin^2[(\pi/\lambda)\,\Delta\mathbf{S}\cdot M_1\mathbf{a}]}{\sin^2[(\pi/\lambda)\,\Delta\mathbf{S}\cdot\mathbf{a}]}\frac{\sin^2[(\pi/\lambda)\,\Delta\mathbf{S}\cdot M_2\mathbf{b}]}{\sin^2[(\pi/\lambda)\,\Delta\mathbf{S}\cdot\mathbf{b}]}\frac{\sin^2[(\pi/\lambda)\,\Delta\mathbf{S}\cdot M_3\mathbf{c}]}{\sin^2[(\pi/\lambda)\,\Delta\mathbf{S}\cdot\mathbf{c}]} \tag{9-24}$$

The oscillations at the toes of the quotients in (9-24), seen in Fig. 9-3, can be smoothed out by using the approximation

$$\frac{\sin^2 \frac{1}{2}Mx}{\sin^2 \frac{1}{2}x} \cong M^2 e^{-M^2x^2/4\pi} \tag{9-25}$$

where the function on the right has the same maximum value at $x = 0$ and the same area as the quotient on the left. Before making this substitution in (9-24), it is necessary to evaluate the magnitude of the arguments in (9-24). For example,

$$\begin{aligned} |\Delta\mathbf{S} \cdot M_1\mathbf{a}| &= |\Delta\mathbf{S}|\,|M_1\mathbf{a}| \cos 0° \\ &= [|\mathbf{S}' - \mathbf{S}_0| - |\mathbf{S} - \mathbf{S}_0|]M_1a \\ &= \left[2 \sin\theta - 2\sin\left(\theta + \frac{\epsilon}{2}\right)\right] M_1a \\ &= \epsilon M_1 a \cos\theta \end{aligned} \tag{9-26}$$

as demonstrated in Exercise 9-5. In these terms, the general expression for the intensity can be written

$$\begin{aligned} I_P &\cong I_eF^2M^2e^{-\frac{1}{4\pi}\left(\frac{4\pi^2}{\lambda^2}\epsilon^2\cos^2\theta\right)(M_1{}^2a^2+M_2{}^2b^2+M_3{}^2c^2)} \\ &\cong I_eF^2M^2e^{-(\pi/\lambda^2)\epsilon^2D^2\cos^2\theta} \end{aligned} \tag{9-27}$$

where $D = (M_1{}^2a^2 + M_2{}^2b^2 + M_3{}^2c^2)^{\frac{1}{2}}$ is an average diagonal dimension in the crystal.

As already noted, when $\epsilon = 0$, the intensity has a maximum value

$$(I_P)_{\max} = (I_P)_{\epsilon=0} = I_eF^2M^2$$

It is customary to measure the width of a diffraction peak at the point where the intensity has fallen to one-half of its maximum value. This is called the *half-maximum width*, or simply, the *halfwidth*, $\epsilon_{\frac{1}{2}}$. In the present case

$$\frac{(I_P)_{\epsilon=\epsilon_{\frac{1}{2}}}}{(I_P)_{\epsilon=0}} = \tfrac{1}{2} \cong e^{-(\pi/\lambda^2)\epsilon_{\frac{1}{2}}{}^2D^2\cos^2\theta}$$

so that $$-\ln 2 \cong -\frac{\pi}{\lambda^2}\epsilon_{\frac{1}{2}}{}^2 D^2 \cos^2\theta$$

and the halfwidth

$$\epsilon_{\frac{1}{2}} \cong \left(\frac{\ln 2}{\pi}\right)^{\frac{1}{2}} \frac{\lambda}{D\cos\theta} \tag{9-28}$$

Even though this is only an estimation of the true diffraction halfwidth, it clearly shows that the peaks are very sharp for all reasonable values of the diameter D. A similar conclusion could have been reached on qualitative grounds by noting that M in (9-14) represents the number of unit cells along some direction in the crystal, so that for a typical linear dimension of 10^{-2} cm and a cell dimension of 10 Å, the value of M is $10^{-2}/10^{-7} = 10^5$. As evident in Fig. 9-3, the peak heights increase with M^2, so that the large M values encountered in real crystals guarantee that sharp diffraction maxima are obtained.

Integrated reflecting power. The finite width of the diffracted beam poses a serious problem if it is attempted to measure the structure-factor magnitudes directly from the maximum intensity in (9-16). Because the halfwidth is of the order of seconds in ideal crystals, an excessively narrow slit would be required to transmit only $(I_P)_{\max}$, that is, the intensity value given by (9-16) when the three Laue conditions are satisfied exactly. The experimental problem is further complicated in practice by the fact that the incident beam cannot be rendered perfectly parallel, so that the incident rays deviate somewhat from $\mathbf{S}_0$. Concurrently, real crystals normally contain imperfections which cause the atoms in different parts of the crystal to become slightly displaced from the ideally perfect lattice array. One can think of these imperfections as having the effect of bending, or tilting, a set of reflecting planes so that the individual planes are not perfectly parallel throughout all parts of the crystal. What this all adds up to is that it is not possible to satisfy the three Laue equations simultaneously for all portions of a single crystal and all incident-ray directions. To overcome these difficulties, the experimental arrangements normally used progressively change the orientation of the crystal while continuously recording the intensity of the beam diffracted during such a motion. Thus one actually measures the total energy diffracted by the crystal during some time interval. As shown below, this quantity also can be evaluated theoretically by integrating the diffracted intensity over a range of angular deviations from the ideal Bragg angle, including all nearby regions of the reciprocal lattice, where the intensity has nonzero values. For this reason, it is commonly called the *integrated intensity*.

The simplest experimental arrangement for measuring the integrated intensity of a crystal employs an almost, but not quite, parallel incident beam that has been rendered suitably monochromatic. The beam is intercepted by a crystal that is set so that a set of planes (hkl) form the correct Bragg angle for diffraction by these planes. The crystal is then slowly rotated with a uniform angular velocity ω about a line that is parallel to the reflecting planes and perpendicular to the plane containing the incident and diffracted beams. This rotation is normally begun, say, $1°$ before the ideal Bragg-angle value and terminated $2°$ later, so that every irradiated part of the crystal has an equal opportunity to reflect every nonparallel ray incident on it. During the rotation of the crystal the intensity of the diffracted beam is continuously recorded by a detector large enough to intercept every diffracted ray.

The equivalent-ray diagram, helpful in calculating the total energy diffracted by the crystal during its rotation, is shown in Fig. 9-5. The reflecting planes of the crystal are set parallel to the cartesian coordinate axes X and Y. The unit vector $\mathbf{S}$ represents the corresponding diffracted-beam direction. (Both $\mathbf{S}_0$ and $\mathbf{S}$ lie in the XZ plane.) The detector surface is normal to the diffracted-beam vector $\mathbf{S}$, a distance R removed from the crystal's origin O. The direction parallel to Y on the detector surface is denoted Y', while the third coordinate direction, perpendicular to $\mathbf{S}$ and Y', is denoted Z'. For convenience in analysis, instead of rotating

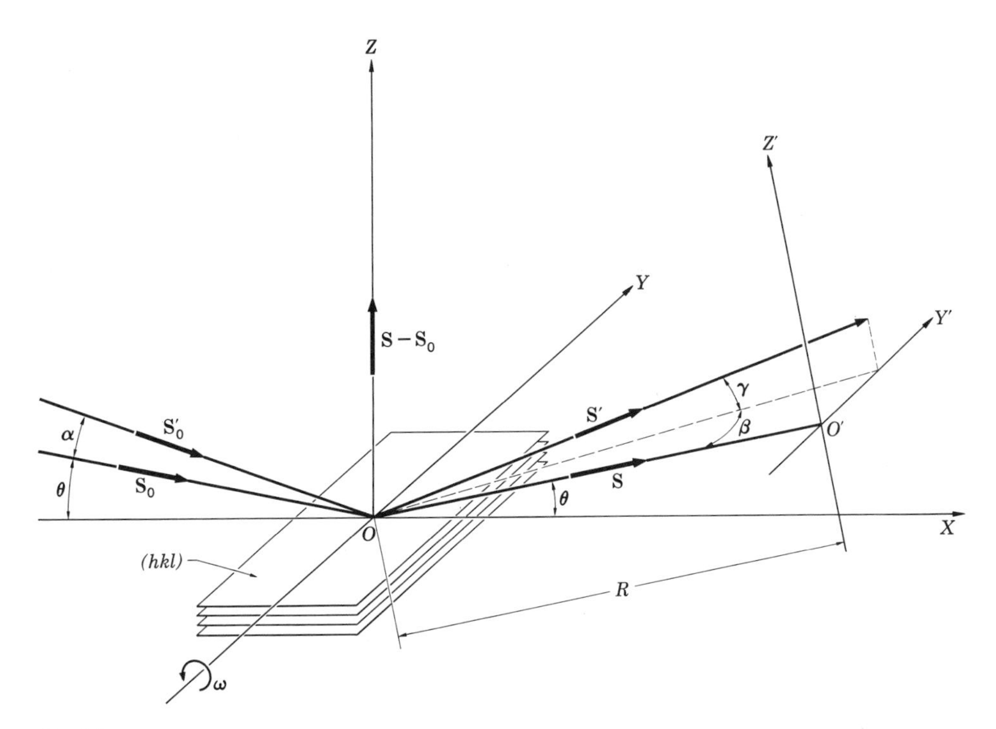

Fig. 9-5

the crystal with a uniform angular velocity about Y, the incident-beam direction is assumed to have moved (in the opposite direction but at the same speed) in the XZ plane through an angle α to a new instantaneous direction denoted by the unit vector $\mathbf{S}_0'$. Finally, to take into account nonparallelism in the incident beam ($\mathbf{S}_0'$ not lying in the XZ plane) and possible nonparallelism in the reflecting planes, the diffracted beam is displaced from the XZ plane to a direction marked by the unit vector $\mathbf{S}'$. This displacement can be traced on the detector surface in terms of two angles, β and γ, along the two mutually orthogonal directions indicated to the right in Fig. 9-5.

Recalling that intensity is a measure of the energy crossing a unit area per unit time, it is possible to evaluate the total energy crossing the detector surface $Y'Z'$ by integrating the intensity at that surface over time and over the area of the detector surface. Accordingly, the total energy, or the integrated intensity of a reflection by the set of planes (hkl),

$$\mathcal{I}_{hkl} = \int\int I_P \, dt \, dA \tag{9-29}$$

The time it takes for the incident-beam vector $\mathbf{S}_0'$ to move from α to $\alpha + d\alpha$ is

$$dt = \frac{d\alpha}{\omega} \tag{9-30}$$

since the beam (crystal) is rotating about the Y axis at a uniform angular speed ω.

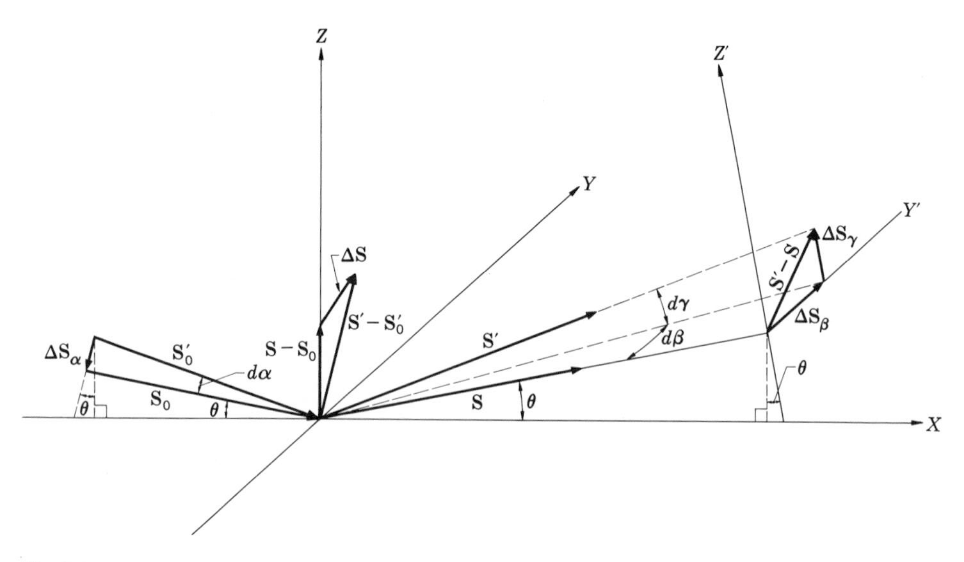

Fig. 9-6

Similarly, the area element dA on the detector surface can be expressed in terms of the displacement of S′ during this time.

$$dA = dY'\,dZ' = R^2\,d\beta\,d\gamma \tag{9-31}$$

Substituting (9-30) and (9-31) in the expression for the integrated intensity,

$$\mathcal{I}_{hkl} = \iiint I_P \frac{R^2}{\omega}\,d\alpha\,d\beta\,d\gamma \tag{9-32}$$

Because the variables in the expression for the intensity I_P are the x-ray direction vectors $\mathbf{S}_0$ and $\mathbf{S}$, the form of (9-32) is not yet suitable for direct integration. It is a straightforward matter, however, to express the angular variables in Fig. 9-5 in terms of the vectors required. To help see the necessary geometric relationships, the pertinent vectors have been reproduced in Fig. 9-6.

At the initial setting of the reflecting planes, the reflection condition was satisfied, so that $(\mathbf{S} - \mathbf{S}_0) = \lambda\boldsymbol{\sigma}_{hkl}$. During the time interval dt, this vector has been displaced in direction and by an amount indicated by $\Delta\mathbf{S}$ in Fig. 9-6, so that

$$\begin{aligned}(\mathbf{S}' - \mathbf{S}_0') &= (\mathbf{S} - \mathbf{S}_0) + \Delta\mathbf{S}\\ &= \lambda\boldsymbol{\sigma}_{hkl} + \Delta\mathbf{S}\end{aligned} \tag{9-33}$$

Substituting the instantaneous unit vectors $\mathbf{S}'$ and $\mathbf{S}_0'$ for $\mathbf{S}$ and $\mathbf{S}_0$, respectively, in the original intensity expression (9-14) then leads to the intensity expression (9-24), previously derived.

The displacement $\Delta\mathbf{S}$ is a vector in reciprocal space, so that it can be expressed in terms of the three reciprocal-lattice axes

$$\Delta\mathbf{S} = \lambda(p_1\mathbf{a}^* + p_2\mathbf{b}^* + p_3\mathbf{c}^*) \tag{9-34}$$

where the three coefficients p_1, p_2, p_3 are unspecified, except that it is clear from the preceding section that they are very small quantities, since the actual deviations of the vectors from the correct positions of $\mathbf{S}$ and $\mathbf{S}_0$ in Fig. 9-6 are very small. Substituting (9-34) into (9-23), the sine terms in (9-24) finally assume the form

$$\sin^2\left(\frac{\pi}{\lambda}\,\Delta\mathbf{S}\cdot M\mathbf{a}\right) = \sin^2\left[\frac{\pi}{\lambda}\,\lambda(p_1\mathbf{a}^* + p_2\mathbf{b}^* + p_3\mathbf{c}^*)\cdot M\mathbf{a}\right] = \sin^2 \pi p_1 M \tag{9-35}$$

because of the orthogonality conditions between reciprocal-lattice and crystal axes.

Quite similarly, it can be seen in Fig. 9-6 that the displacement of $\mathbf{S}_0$ during dt is $\Delta\mathbf{S}_\alpha$, while the displacement of $\mathbf{S}$ can be described best by the small vectors $\Delta\mathbf{S}_\beta$ and $\Delta\mathbf{S}_\gamma$ along two mutually orthogonal directions lying in the detector surface such that $(\mathbf{S}' - \mathbf{S}) = \Delta\mathbf{S}_\beta + \Delta\mathbf{S}_\gamma$. The magnitudes of these three vectors clearly are

$$\begin{aligned} |\Delta\mathbf{S}_\alpha| &= |\mathbf{S}_0|\,d\alpha = d\alpha \\ |\Delta\mathbf{S}_\beta| &= |\mathbf{S}|\,d\beta = d\beta \\ |\Delta\mathbf{S}_\gamma| &= |\mathbf{S}|\,d\gamma = d\gamma \end{aligned} \tag{9-36}$$

since $|\mathbf{S}| = 1$ by definition. The vector displacement of $(\mathbf{S} - \mathbf{S}_0)$,

$$\begin{aligned} \Delta\mathbf{S} &= (\mathbf{S}' - \mathbf{S}_0') - (\mathbf{S} - \mathbf{S}_0) \\ &= (\mathbf{S}' - \mathbf{S}) - (\mathbf{S}_0' - \mathbf{S}_0) \\ &= (\Delta\mathbf{S}_\beta + \Delta\mathbf{S}_\gamma) - \Delta\mathbf{S}_\alpha \end{aligned} \tag{9-37}$$

These three vectors also define a small volume element (Fig. 9-7),

$$\begin{aligned} \Delta V &= \Delta\mathbf{S}_\alpha \times \Delta\mathbf{S}_\gamma \cdot \Delta\mathbf{S}_\beta \\ &= |\Delta\mathbf{S}_\alpha|\,|\Delta\mathbf{S}_\beta|\,|\Delta\mathbf{S}_\gamma| \sin 2\theta \end{aligned} \tag{9-38}$$

as can be seen by transposing them to a common origin (at the center of the coordinates Y' and Z' in Fig. 9-6). Upon substituting for the three vector magni-

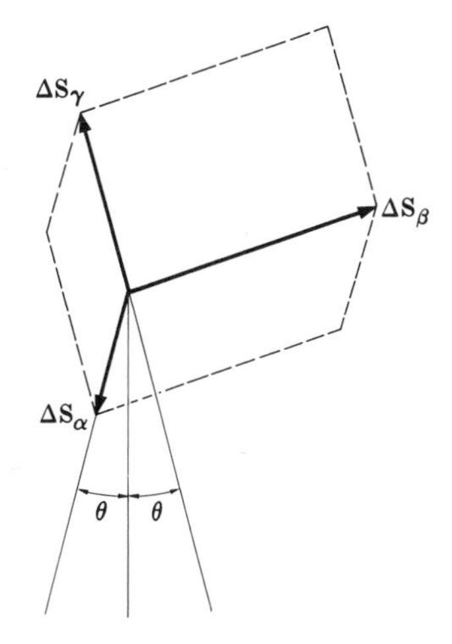

Fig. 9-7. Volume element swept out by the terminal point of $\Delta\mathbf{S}$ in Fig. 9-6. The angular relations among the three vectors can be discerned by noting that Z' is inclined to Z in Fig. 9-6 by the angle θ.

tudes above their value determined in (9-36), the volume element becomes

$$\Delta V = \sin 2\theta \, d\alpha \, d\beta \, d\gamma \tag{9-39}$$

An examination of Fig. 9-6 shows that the small angular changes $d\alpha$, $d\beta$, and $d\gamma$ thus have the effect of causing the terminal point of the displacement vector $\Delta\mathbf{S}$ to sweep out this volume element ΔV in reciprocal space.

Making use of the reciprocal-lattice representation of the displacement vector (9-34), it is possible to express such a motion of $\Delta\mathbf{S}$ in terms of small changes in the three coefficients of amount dp_1, dp_2, dp_3, so that

$$\begin{aligned} \Delta V &= \lambda \, dp_1 \, \mathbf{a}^* \times \lambda \, dp_2 \, \mathbf{b}^* \cdot \lambda \, dp_3 \, \mathbf{c}^* \\ &= \lambda^3 V^* \, dp_1 \, dp_2 \, dp_3 \end{aligned} \tag{9-40}$$

Substituting the unit-cell volume $V = 1/V^*$ (see Table 7-1) for the cell volume in reciprocal space and equating (9-40) to (9-39) gives

$$d\alpha \, d\beta \, d\gamma = \frac{\lambda^3}{V \sin 2\theta} \, dp_1 \, dp_2 \, dp_3 \tag{9-41}$$

Now it is finally possible to make all the necessary transformations that will permit carrying out the integration in (9-32). First, substitute (9-24) for I_P, after expressing the sine terms according to (9-35), and then substitute (9-41) for the three angular variables, so that the expression for the integrated intensity becomes

$$\mathcal{I}_{hkl} = \frac{I_e R^2 \lambda^3}{\omega V \sin 2\theta} F^2_{hkl} \iiint \frac{\sin^2 \pi M_1 p_1}{\sin^2 \pi p_1} \, \frac{\sin^2 \pi M_2 p_2}{\sin^2 \pi p_2} \, \frac{\sin^2 \pi M_3 p_3}{\sin^2 \pi p_3} \, dp_1 \, dp_2 \, dp_3 \tag{9-42}$$

As demonstrated in the preceding section, the diffracted intensity has appreciable values only very close to the ideal peak position, namely, the reciprocal-lattice point specified by hkl, so that the three quotients in (9-42) have appreciable values only for very small values of πp. This, of course, also follows directly from the fact that, for large values of M, the quotients can have appreciable values only when the arguments of the sine terms in the denominator are very nearly integer multiples of π or zero. In the present case, therefore, the assumption that $p\pi \cong 0$ means that only the maximum occurring around $p = 0$ is being considered, so that $\sin \pi p \cong \pi p$. In view of this limitation, it is possible to extend the integration from minus to plus infinity, since each quotient in (9-42) will have only one maximum in this range, namely, at $p = 0$. This leads to the definite integral

$$\int_{-\infty}^{+\infty} \frac{\sin^2 \pi M_1 p_1}{(\pi p_1)^2} \, dp_1 = M_1 \tag{9-43}$$

so that the integrated intensity (9-42), when the three Laue equations are very closely satisfied, becomes

$$\mathcal{I}_{hkl} = \frac{I_e R^2 F^2 \lambda^3}{\omega V \sin 2\theta} M_1 M_2 M_3 \tag{9-44}$$

Upon multiplying (9-44) by V/V and setting $MV = \delta V$, the irradiated crystal volume, and writing out the Thomson intensity,

$$\mathcal{I}_{hkl} = \frac{I_0}{\omega} \frac{e^4}{m^2c^4} F^2 \frac{\lambda^3}{V^2} \left(\frac{1 + \cos^2 2\theta}{2 \sin 2\theta} \right) \delta V \tag{9-45}$$

Defining a reflecting power per irradiated volume element,

$$Q \equiv \frac{e^4}{m^2c^4} F^2 \frac{\lambda^3}{V^2} \left(\frac{1 + \cos^2 2\theta}{2 \sin 2\theta} \right) \tag{9-46}$$

it is possible to rearrange the terms in (9-45) to give a slightly more useful quantity, called the *integrated reflecting power* of a crystal,

$$R_{hkl} = \frac{\mathcal{I}_{hkl}\,\omega}{I_0} = Q\,\delta V \tag{9-47}$$

The integrated reflecting power depends only on the volume of the irradiated crystal δV and its reflecting power, a most important point, because it was previously assumed that the crystal had the shape of a parallelepipedon. According to (9-45), the integrated intensity (unlike the peak intensity) is independent of the crystal shape, since, if necessary, an arbitrarily shaped crystal can be considered to consist of a number of irradiated volume elements, each having the shape of a parallelepipedon. The integrated intensity then can be considered to be the sum of the contributions from such individual blocks. Note also that the integrated reflecting power does not depend directly on either the actual angular speed of the crystal or on the incident-beam intensity, since doubling the speed halves the number of quanta of energy that have an opportunity to be reflected. One important omission has been made in deriving (9-47), however; namely, absorption of the x-ray beam during its passage through the crystal has been completely ignored. Thus the above expression is valid only when absorption is so small a factor that it can be neglected, a condition that is very, very rarely fulfilled by real single crystals in practice. One way to minimize absorption, of course, is to reduce the size of the crystal. The disadvantage of this is that it is then not possible to estimate accurately what portion of the incident beam is actually intercepted by the diffracting planes, and it is possible to measure the reflecting powers of different planes in the same crystal only on a relative basis. This limitation can be overcome, however, without increasing the size of individual crystallites by combining them in a polycrystalline aggregate. The intensities of x rays diffracted by a powder of crystals therefore are considered next.

Diffraction by a polycrystalline aggregate. In deriving relation (9-47) it was assumed that the reflecting planes in a crystal were given every opportunity to diffract x rays by the rotation of the crystal. As already demonstrated, it is possible to attain a similar result by collecting together a large group of randomly oriented crystallites. Suppose a total of N randomly oriented crystallites are present, each of volume δV, and that the total sample is still sufficiently small so that absorption can be neglected. Recalling the reciprocal-lattice construction for the powder method from Chap. 7, it is a simple matter to calculate the integrated reflecting power for such a polycrystalline aggregate. Consider a randomly oriented set of

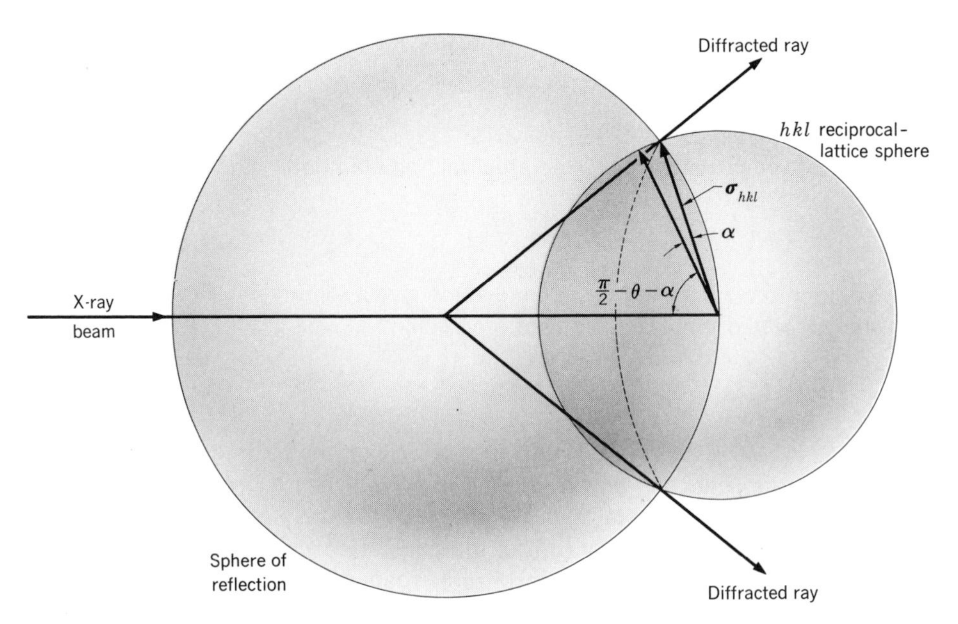

Fig. 9-8

planes (hkl) whose reciprocal-lattice sphere is shown in Fig. 9-8. All planes whose reciprocal-lattice vector $\boldsymbol{\sigma}_{hkl}$ forms the angle $\pi/2 - \theta$ with the incident-beam direction satisfy the reflection condition (7-35). The terminal points of these vectors lie along the circle forming the intersection of the hkl sphere with the sphere of reflection, along the dotted curve in Fig. 9-8. For a slightly diverging incident beam, the reciprocal-lattice vectors may satisfy the diffraction condition even when they deviate from this orientation by a small angle α. Accordingly, the fraction of crystallites oriented to reflect the incident monochromatic beam, dN/N, can be determined by considering the fraction of reciprocal-lattice vectors between $(\pi/2 - \theta - \alpha)$ and $(\pi/2 - \theta - \alpha + d\alpha)$. This is obviously the ratio of the area of the annular ring in Fig. 9-9 to the total area of the surface of the hkl reciprocal-

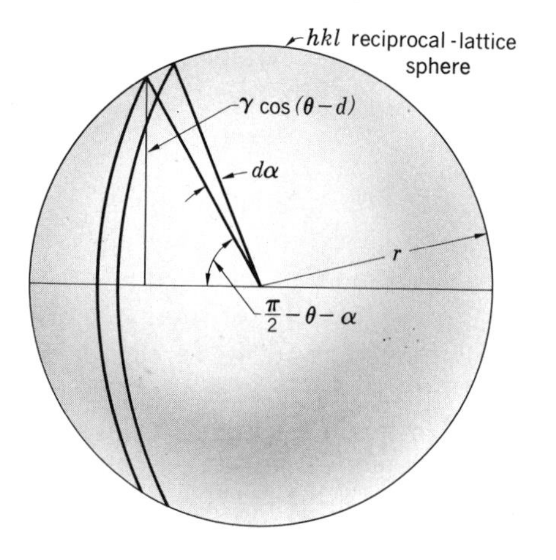

Fig. 9-9

lattice sphere. For convenience, let this sphere have a radius r; then

$$\frac{dN}{N} = \frac{2\pi r^2 \cos(\theta - \alpha)\, d\alpha}{4\pi r^2} = \tfrac{1}{2} \cos\theta\, d\alpha \tag{9-48}$$

as verified in Exercise 9-6. Actually, each crystallite may have more than one plane whose reciprocal-lattice vector terminates on this hkl sphere; for example, this is always true for the two planes (hkl) and $(\bar{h}\bar{k}\bar{l})$. Other planes also may be related by symmetry, as discussed in the introductory chapters, so that the true fraction of planes satisfying the diffraction condition is obtained by multiplying (9-48) by the number of such equivalent planes present j, called the *multiplicity* of that set. The multiplicities for several planes are listed in Table 9-1 for the six crystal systems. Note that they differ for each system, as the symmetry changes, and that, particularly for certain planes in the cubic system, chance coincidences are also possible in the powder method for planes not related by symmetry. Thus the correct number of planes whose reciprocal-lattice vectors lie in the region $(\pi/2 - \theta)\, d\alpha$ is

$$dN = \frac{jN}{2} \cos\theta\, d\alpha \tag{9-49}$$

To obtain the total power of the diffracted beams from all the crystals it is necessary to evaluate the triple integral

$$\iiint I_p\, d\alpha\, d\beta\, d\gamma \tag{9-50}$$

Table 9-1 Multiplicities j for the powder method

hkl	*Cubic*	*Tetragonal*	*Hexagonal*†	*Orthorhombic*	*Monoclinic*‡	*Triclinic*
$h00$		4	6	2	2	2
$0k0$	6		6	2	2	2
$00l$		2	2	2	2	2
$hh0$		4	6	4	2	2
$h0h$	12	8	12	4	4	2
$0hh$				4	4	2
hhh	8	8	12	8	4	2
$hk0$		8	12	4	2	2
$h0l$	24	16	12	4	4	2
$0kl$			12	4	4	2
hhl		8	12	8	4	2
hlh	24	16	24	8	4	2
lhh				8	4	2
hkl	48	16	24	8	4	2

† The (hkl) symbol for the hexagonal system is to be assumed to read $(hk{\cdot}l)$.
‡ The c-unique orientation is assumed; that is, 2 is parallel to c and m is perpendicular to c.

over all possible orientations, after multiplying it by the number of crystals in position to diffract the incident beam, so that

$$\begin{aligned} P &= \tfrac{1}{2}jN \cos\theta \iiint I_p \, d\alpha \, d\beta \, d\gamma \\ &= \tfrac{1}{2}jN \cos\theta \times I_0 Q \, \delta V \end{aligned} \tag{9-51}$$

since I_p is given by (9-14), and the triple integration is exactly the same as that leading to (9-44). Finally, let $N \, \delta V = V_s$, the total sample volume, so that

$$\frac{P}{I_0} = \tfrac{1}{2}jQV_s \cos\theta \tag{9-52}$$

The power P in (9-52) is the energy per second diffracted by the polycrystalline aggregate into the diffraction cone indicated in Fig. 9-8. Normally, one measures only a portion P' of this total power. Suppose an ionization detector having a slit of length l parallel to the circumference of the cone is used to measure the energy per second diffracted along the cone (alternatively, this could be the slit length of a densitometer measuring a photographically recorded diffraction ring). If the specimen-to-detector distance is R, then the energy per second entering the detector is

$$\begin{aligned} \frac{P'}{I_0} &= \frac{1}{2} \frac{jlQV_s \cos\theta}{2\pi R \sin 2\theta} \\ &= \frac{jl}{8\pi R \sin 2\theta} QV_s \end{aligned} \tag{9-53}$$

as derived in Exercise 9-7.

The full meaning of this expression can be understood better by substituting (9-46) for Q and combining like terms.

$$\frac{P'}{I_0} = \frac{e^4\lambda^3 l}{16\pi R m^2 c^4} \times \frac{V_s}{V^2} jF^2 \times \left(\frac{1 + \cos^2 2\theta}{\sin 2\theta \sin\theta}\right) \tag{9-54}$$

Thus it is seen that the energy per second reaching the detector is equal to a term that is constant for a particular experimental arrangement, multiplied by a collection of terms characteristic of the diffracting sample, multiplied by a term that is a function of the reflection angle only. The numerator in this last term obviously expresses the Thomson polarization, while the denominator,

$$\frac{1}{\sin 2\theta \sin\theta} \tag{9-55}$$

is usually called the *Lorentz factor* for the powder method. It will be noted that this factor is in no way connected with the integration required to obtain the integrated reflecting power, so that it applies equally well to peak intensities. As first demonstrated by Lorentz during one of his classroom lectures, the form of this factor depends on the experimental arrangement used. Since both the Lorentz and the polarization factors depend on θ, it is common practice to combine them into a single expression, called the *Lorentz polarization factor*, or simply, the Lp

factor. Complete tables of the Lp factor for a range of diffraction angles and different experimental arrangements are given in *International tables for x-ray crystallography.*

Effect of absorption. The absorption of monochromatic x radiation by a material has been discussed in detail in Chap. 6, where it was shown that the intensity of the transmitted beam I is attenuated by an exponential factor $e^{-\mu x}$, where μ is the linear absorption coefficient of the material, and x is the x-ray pathlength. If a small single crystal is totally immersed in the incident x-ray beam, the attenuation of the diffracted rays obviously depends on the pathlengths of the various incident and diffracted rays. When the crystal has a simple shape such as a sphere, a cylinder, or a prism, the calculation of the attenuation factors is straightforward, if not easy. In particular, tables listing the attenuation as a function of the linear absorption coefficient and the radius are available for spherical and cylindrical crystals. The two parameters are usually combined to give the product μr, and the tables usually list a *transmission factor* $T = I/I_0$ for various values of μr and θ values ranging from 0 to 90°. As an illustration, the variation of T with glancing angle is shown in Fig. 9-10 for $\mu r = 5.0$ for a cylindrically and a spherically shaped crystal. In practice, graphs like this are used to determine the appropriate corrections to apply to measured intensity values. Although such corrections are straightforward, it should be kept in mind that crystals, even when artificially ground, rarely conform exactly to spherical or cylindrical geometries. Moreover, the tabulated values are derived by assuming a parallel incident beam because absorption corrections for nonparallel beams are excessively complex. Thus there remains a residual error in the measured intensity even after a "correction" for absorption has been made. As also discussed elsewhere in this book, absorption is virtually never truly negligible, so that this residual error is one of the principal factors

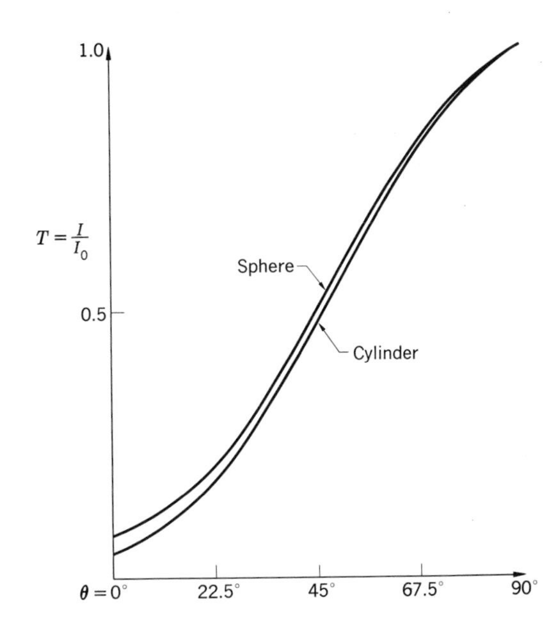

Fig. 9-10. Transmission factor for a spherical and a cylindrical crystal for which $\mu r = 5.0$.

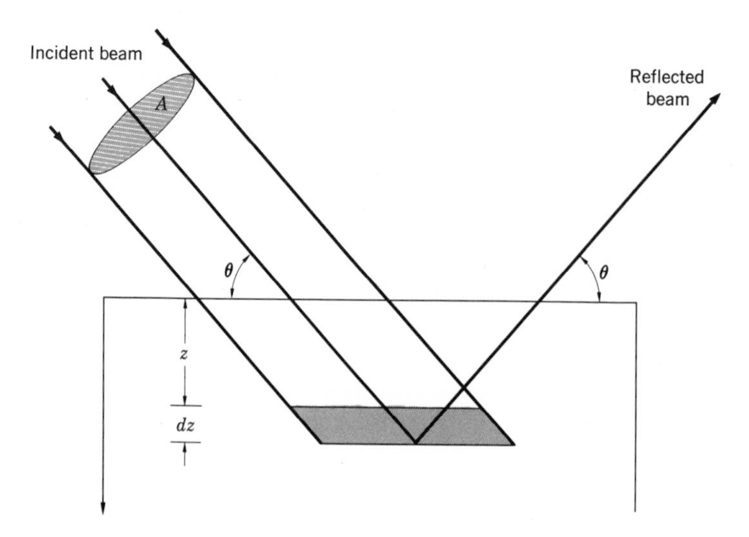

Fig. 9-11

limiting the accuracy of experimentally measured intensities when small single crystals are used.

The problem is considerably simplified when the crystal is larger than the incident beam, so that the crystal intercepts the entire beam. Even here, however, a parallel incident beam will be assumed for simplicity in analysis. Two arrangements must be distinguished. In what is called the Bragg arrangement, the incident and reflected beams leave the same surface of the crystal, as shown in Fig. 9-11. In the symmetrical Bragg arrangement, the reflecting planes are assumed to be parallel to this surface and, in practice, are so arranged when the crystal is cut. Let the cross-sectional area of the incident beam be A. Then the shaded volume element inside the crystal shown in Fig. 9-11 has a volume

$$dV = A \csc \theta \, dz \tag{9-56}$$

If it lies at a depth z below the crystal's surface, the total path traveled by radiation diffracted by this block is $2z \csc \theta$, so that its intensity has been attenuated by an amount $e^{-2\mu z \csc \theta}$. The integrated reflecting power of this volume element, as given by relation (9-47), can be expressed by

$$\frac{\omega \, d\mathcal{I}}{I_0} = Qe^{-2\mu z \csc \theta} \, dV \tag{9-57}$$

If the crystal is assumed to be infinitely thick, so that no radiation can pass through it, it is possible to integrate the right side of (9-57) from the surface, where $z = 0$, out to infinity. The total integrated reflecting power from an infinitely thick crystal slab then becomes

$$\frac{\omega \mathcal{G}}{I_0} = Q \int_0^\infty A e^{-2\mu z \csc\theta} \csc\theta \, dz$$

$$= \frac{QA}{2\mu} \tag{9-58}$$

by direct integration. Dividing both sides of (9-58) by the cross-sectional area of the incident beam, and noting that $I_0 A = \mathcal{G}_0$ is the total energy per second incident on the crystal,

$$\frac{\mathcal{G}\omega}{\mathcal{G}_0} = \frac{Q}{2\mu} \tag{9-59}$$

It is sometimes convenient to designate by $R(\theta)$ the fraction of the total incident beam that is reflected at the angle θ. Then the integrated intensity from a large crystal slab rotated at a constant angular velocity ω about a direction normal to the diffraction plane can be expressed by

$$\mathcal{G}_{hkl} = \mathcal{G}_0 \int R(\theta) \frac{d\theta}{\omega}$$

so that $$R^\theta_{hkl} = \frac{\mathcal{G}_{hkl}\,\omega}{\mathcal{G}_0} = \int R(\theta)\, d\theta \tag{9-60}$$

becomes the integrated reflecting power from an extended crystal face. This expression is directly applicable to intensity measurements from large single crystals and is the one normally used in experimental tests of intensity formulas employing two-crystal diffractometers.

The other experimental arrangement of interest is the transmission, or Laue, arrangement, in which the incident and reflected beams cross opposite faces of a large single crystal slab. In the symmetrical version of this arrangement the reflecting planes are transverse to the slab, as shown in Fig. 9-12; that is, they are exactly normal to the two parallel surfaces of the crystal. In this case, the incident and reflected beams are symmetrically disposed, so that the total pathlength for radiation reflected by any volume element is $t \sec\theta$, where t is the crystal's thickness. The total irradiated volume is $At \sec\theta$, so that the integrated reflecting power, making use of (9-47), becomes

$$\frac{\mathcal{G}\omega}{I_0} = Q e^{-\mu t \sec\theta} A t \sec\theta \tag{9-61}$$

Again, expressing the total energy incident on the crystal per unit time by $\mathcal{G}_0 = AI_0$, the integrated reflecting power for transmission through an extended crystal plate,

$$R^\theta_{hkl} = \frac{\mathcal{G}_{hkl}\,\omega}{\mathcal{G}_0} = \int R(\theta)\, d\theta = Qt' e^{-\mu t'} \tag{9-62}$$

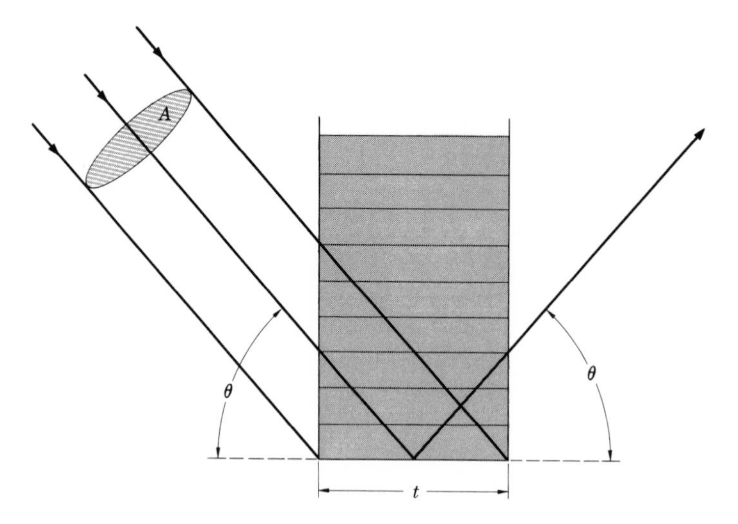

Fig. 9-12

where $t' = t \sec \theta$. This expression also is useful in testing the intensity formulas, as discussed in later chapters. Note that the expression (9-62) has a maximum when $t' = 1/\mu$ (Exercise 9-9), so that relatively thin crystal slabs must be employed in the experiments. Also remember that this expression is strictly valid only for the symmetrical case shown in Fig. 9-12, so that, if the reflecting planes are not normal to the surface, a somewhat more complex absorption correction must be made.

A similar analysis can be carried out to determine the absorption correction for large polycrystalline slabs. Suppose that the polycrystalline aggregate has a density ρ', while the true density of the crystals is ρ. Next assume that the incident and reflected beams are symmetrically disposed to a polycrystalline slab examined in transmission, similar to the arrangement in Fig. 9-12, except that, in the polycrystalline slab, only some of the crystallites have their planes (hkl) oriented to reflect the incident beam. This means that relation (9-52) must be employed. Because the arrangement is symmetric, the incident and reflected beams travel the same distance $t \sec \theta$. The total crystalline volume irradiated is $(\rho'/\rho)At \sec \theta$. Thus, making use of (9-53), the intensity entering a detector having a slit of length l parallel to the diffraction cone is

$$P' = \frac{jlI_0}{8\pi R \sin \theta} Q \frac{\rho'}{\rho} At \sec \theta e^{-\mu t \sec \theta} \tag{9-63}$$

where all the terms have their previously defined meanings. Note that the intensity transmitted in the forward direction is

$$I_t = I_0 e^{-\mu t \sec \theta} \tag{9-64}$$

so that after rearranging the terms,

$$\frac{P'}{I_t} = \frac{jlAt}{4\pi R \sin 2\theta} \frac{\rho'}{\rho} Q \tag{9-65}$$

This expression is quite useful for measuring intensities on an absolute scale. The detector first is placed in position to record the diffracted beam, and then is moved in position to detect the transmitted beam. The ratio then can be used to calculate Q from (9-65) directly on an absolute basis.

A corresponding expression for the case where the incident beam forms the angle α with the surface of a polycrystalline slab and the reflected beam forms the angle β with the same surface $(\alpha + \beta = 2\theta)$ is derived in Exercise 9-10. If $\mathcal{I}_0$ is the total energy per unit time incident on the slab, and P is the power in that part of the reflected beam that enters the detector through a slit of length l at a distance R from the polycrystalline slab, the resulting expression is

$$\frac{P}{\mathcal{I}_0} = \frac{jlQ}{8\pi\mu R \sin\theta}\left(1 + \frac{\sin\alpha}{\sin\beta}\right)^{-1} \tag{9-66}$$

DYNAMICAL THEORY

Abnormal absorption. The theory of x-ray diffraction described above is actually a part of the theory developed by C. G. Darwin immediately following Laue's discovery of the phenomenon, and is contained in two papers published by Darwin in 1914. During that year several investigators set out to test these theories. Generally, a fairly good agreement was found between the predictions of (9-59) or (9-60) and the integrated reflecting powers of such crystals as NaCl, which were readily available in the form of large single crystals. On the other hand, W. H. Bragg observed that the integrated intensities for the strong reflections from tiny diamond crystals did not agree too well with the theory, the observed intensities being considerably smaller. This led Bragg to suspect that the diffracted beams were abnormally absorbed by the diamond crystals. About seven years later, in a very careful study of the integrated reflecting powers of NaCl crystals, measured on an absolute scale, W. L. Bragg, R. W. James, and C. H. Bosanquet found that, although there was a good quantitative agreement between their measurements and the predictions of (9-60), the stronger reflections showed an abnormally high absorption effect. Moreover, it appeared that the excess absorption was directly proportional to the intensity of the reflection, becoming more pronounced as the intensity increased. A complete explanation of the origins of this abnormal absorption effect was presented by Darwin in 1922.

Empirically, the effect can be expressed by an attenuation factor

$$\mu' = \mu + gQ \tag{9-67}$$

where μ is the ordinary linear absorption coefficient due to photoelectric absorption and incoherent scattering, while g is a proportionality constant, and Q is the scattering power of a small (absorptionless) crystallite. The need for this second term becomes clear when all the processes taking place in the crystal are considered. Picture the crystal as an infinite stack of parallel atomic planes (Fig. 9-13). When these planes form an arbitrary angle with the incident beam, its intensity is attenuated by $e^{-\mu x}$, as described in Chap. 6. Suppose that the crystal is then rotated into

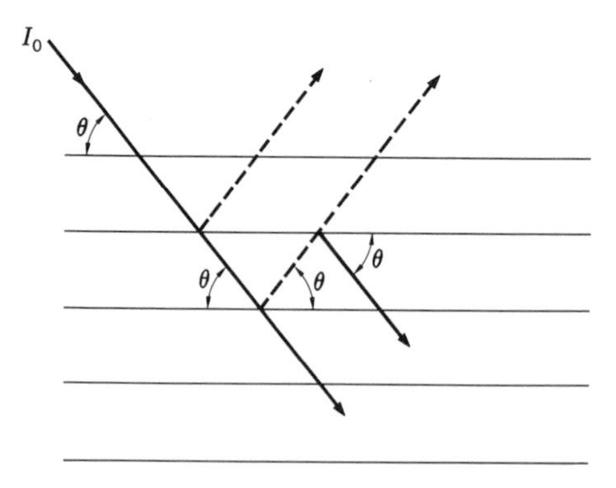

Fig. 9-13. The intensity I_0 of the incident beam reaching successive parallel planes in a perfect crystal is attenuated by the deflection of energy into the reflected beam ($\propto Q$) and by multiple reflections within the crystal.

position to reflect the beam so that the intensity of the incident beam reaching the nth plane below the surface is attenuated, not by the factors comprising μ alone, but also by another factor, namely, the coherent-scattering process whose intensity is proportional to Q.†

Actually, there is still another condition in this crystal that must be recognized; that is, when an incident parallel beam makes the proper angle for reflection on its way into the crystal, the reflected beam also forms the proper angle for reflection by the "backs" of the same planes on its way out of the crystal. Thus a part of the emerging beam is reflected back into the crystal. This twice-reflected beam, of course, is parallel to the incident beam and should reinforce it. Recalling from Chap. 8 that, at each reflection, the phase of the reflected beam changes by $\pi/2$, the twice-reflected beam is $180°$ out of phase with the incident beam and interferes destructively with it. Both of these processes were clearly recognized by Darwin, who called them *extinction.*‡ Without exploring them in depth, some of the more important aspects and consequences of extinction will be discussed in the next sections.

The problem of the interaction between an incident and reflected beam can be formulated in a more direct manner. It is usual to call a theory taking such interactions into account a *dynamical theory.* In a series of papers beginning in 1916, Ewald developed a dynamical theory for x-ray diffraction in which, in place of analyzing the exchange of energy between individual incident and diffracted beams, the wave field set up in the crystal is analyzed as a whole. This is equivalent to analyzing the propagation of any electromagnetic field through a medium having a

† Equation (9-67) is the one first derived by Darwin and later rederived by W. H. Zachariasen. More recently, Zachariasen has shown that, when the incident beam is polarized, the last term in (9-67) should be multiplied by

$$\frac{2(1+\cos^4 2\theta)}{(1+\cos^2 2\theta)^2} \tag{9-67a}$$

‡ Darwin's choice of the term extinction is an unfortunate one because it implies total disappearance. Such total disappearance is encountered when the presence of subtranslations in a crystal causes certain reflections to have zero intensity. This is the case of space-group extinctions, discussed in more detail in Chap. 11. Because the term extinction in Darwin's sense also has gained a regular place in crystallographic nomenclature, it will be so used in this book, but the reader is alerted to keep the above distinctions in mind.

periodically varying, complex dielectric constant, as was pointed out by Laue in 1931. Because the concepts of the dynamical theory are more advanced than those used in this book, only some of its features and predictions are considered below, omitting detailed derivations. This is done in keeping with the general aim of this book, which is to present the essential elements of x-ray diffraction, rather than a treatment in full depth. It should be noted here that most experimental observations can be explained quite successfully by the kinematical theory as modified by Darwin and others to include dynamical interactions. This fact, combined with the mathematical complexity of the dynamical theory, probably is most responsible for its relative neglect until recent years. The currently growing utilization of large perfect single crystals in numerous devices has led to a resurgence of interest in the dynamical theory which is more general than the kinematical one, and hence better suited for explaining what happens in highly perfect crystals. Before considering this theory, however, it is of practical interest to examine how Darwin explained the abnormal intensities referred to above.

Darwin's theory. In his original papers, published in 1914, Darwin had already considered what effect the small deviation from unity in the crystal's index of refraction would have on the diffracted beam. In fact, in these papers, he pointed out that a small deviation from the Bragg reflection angle should be produced, as was later rediscovered experimentally by Stenström in 1919. (See discussion of refraction in Chap. 6.) Darwin analyzed the diffraction of x rays in terms of successive reflections by perfectly parallel planes, and concluded that essentially total reflection of the incident radiation takes place within just a few planes below the crystal's surface. He also showed that the angular range over which reflection takes place is extremely narrow and proportional to the amplitude of the reflected beam. Consequently, the integrated reflecting power of the crystal is proportional to $|F|$, and not to $|F|^2$, as predicted by the kinematical theory. Without repeating the original arguments in detail, a "derivation" indicating the reasons for this difference is presented below. It should be noted that, in what follows, as in Darwin's original papers, ordinary absorption by the crystal is assumed to be so small as to be negligible. In 1930, J. A. Prins modified this theory to include the presence of absorption effects in crystals.

To begin, consider a crystal comprised of a stack of parallel planes set to reflect an incident monochromatic x-ray beam, as indicated in Fig. 9-13. According to the Bragg law, the phase difference between x rays scattered by the uppermost plane and the mth plane below the surface is

$$m\varphi = m \times 2\pi \frac{2d \sin \theta}{\lambda} \tag{9-68}$$

where φ is the phase difference between successive planes in the crystal. Suppose that the ratio of the amplitude of the wave reflected by each plane to that incident on it is correctly given by q. Then a crystal containing M planes (crystal thickness $D = Md$), neglecting absorption and any dynamical interactions, reflects a wave whose amplitude ratio is

$$q(1 + e^{i\varphi} + e^{i2\varphi} + \cdots + e^{iM\varphi}) = q\left(\frac{e^{iM\varphi} - 1}{e^{i\varphi} - 1}\right) \tag{9-69}$$

The intensity of the reflected wave is obtained by multiplying the amplitude by its complex conjugate and, as previously demonstrated in full detail, the peak intensity is equal to $(qM)^2$. By neglecting any attenuation of the primary beam, it is assumed that the intensity reaching each plane is the same, so that the total power is reflected over an angular range, which, by analogy to (9-28), can be expressed

$$\Delta\theta = \left(\frac{\ln 2}{\pi}\right)^{\frac{1}{2}} \frac{\lambda}{2D \cos \theta} \tag{9-70}$$

where D is the crystal's thickness. The integrated intensity then is given by

$$\mathcal{I} = (qM)^2 \, \Delta\theta \tag{9-71}$$

In effect, it has been assumed above that the reflected intensity has the form of a "square" peak, which is not a bad approximation to the actual shape predicted by the more rigorous theory.

In order to recast the above expression into a more familiar form, use can be made of the Fresnel zone theory of optics. According to this theory, the scattering from a plane can be analyzed as a scattering from concentric zones within the plane, the zone boundaries being so chosen that all the scattered rays emanating from points within a single zone are in phase with each other when they reach an observer a distance R away. All zones have the same area, which is approximately equal to λR. In the case of the planes considered in Fig. 9-13, the effective area of a unit cell of volume V parallel to the reflecting planes is $(V/d) \sin \theta$, so that the number of unit cells included in a zone is

$$n = \frac{\lambda R}{(V/d) \sin \theta} \tag{9-72}$$

The amplitude of the wave scattered by a single zone is directly proportional to the amplitude of scattering by one cell, F, times the number of cells in a zone, and inversely proportional to the distance from the zone. Consequently, the ratio of the reflected to incident amplitudes for one zone is

$$q = \frac{nF}{R} = \frac{\lambda R d}{V \sin \theta} \frac{F}{R} = \frac{\lambda \, dF}{V \sin \theta} \tag{9-73}$$

Substituting (9-73) and (9-70) into (9-71) then gives an expression for the integrated intensity.

$$\begin{aligned} \mathcal{I} &= \left[\frac{\lambda(Nd)F}{V \sin \theta}\right]^2 \left(\frac{\lambda}{2D \cos \theta}\right) \left(\frac{\ln 2}{\pi}\right)^{\frac{1}{2}} \\ &= \frac{\lambda^3 F^2 D \sqrt{(\ln 2)/\pi}}{V^2 \sin 2\theta \sin \theta} \end{aligned} \tag{9-74}$$

which differs from that derived previously, primarily by some physical constants that have been omitted from the present discussion. The expression in (9-74) represents the scattering emanating from the first Fresnel zones in each plane in the direction defined by the Bragg angle θ.

When dynamical interactions are not neglected, each plane no longer can be assumed to make an equal contribution q to the total reflected beam. It is quite safe to assume, however, that the maximum intensity of the scattered beam should not exceed that of the incident beam. For simplicity, suppose the incident beam has unit amplitude. As before, the scattering power of each plane can be expressed by q, which now is equal to the amplitude scattered by the plane at the Bragg angle. If a total of M planes can make a contribution, it is still true that the reflected peak intensity is

$$(qM)^2 \cong 1 \tag{9-75}$$

for this case. The angular range over which reflection is possible is still correctly given by (9-70), except that now D is not the crystal thickness, but the effective depth below the surface from which nonzero contributions to the reflected beam are possible. (The physical factors responsible for the attenuation of the beam, called *primary extinction* by Darwin, are the multiply reflected beams shown in Fig. 9-13.) It turns out that the effective depth is given by

$$D_{\text{eff}} = \frac{V \sin \theta}{\lambda |F|} \tag{9-76}$$

and is inversely proportional to the amplitude of scattering by a unit cell of volume V, as might be expected, intuitively. The angular range of the reflection then becomes

$$\Delta\theta = \left(\frac{\ln 2}{\pi}\right)^{\frac{1}{2}} \frac{\lambda^2 |F|}{2V \sin \theta \cos \theta} \tag{9-77}$$

Assuming again, for simplicity, that the shape of the reflection is very nearly a "square" peak, (9-77) and (9-75) can be substituted into (9-71) to give

$$\mathcal{I} = 1 \times \frac{\lambda^2 |F| \sqrt{(\ln 2)/\pi}}{V \sin 2\theta} \tag{9-78}$$

which has the correct form for the integrated intensity reflected by a perfect crystal, including primary extinction effects but neglecting absorption. The complete expression derived by Darwin, including all the physical constants omitted in the above discussion, is

$$\mathcal{I}_{hkl} = \frac{I_0}{\omega} \frac{8}{3\pi} \frac{e^2}{mc^2} |F_{hkl}| \frac{\lambda^2}{V} \left(\frac{1 + |\cos 2\theta|}{2 \sin 2\theta}\right) \tag{9-79}$$

Comparing this expression with the one derived previously by neglecting primary extinction, (9-45), it is clear that they differ primarily in that the quantity $(e^2/mc^2V)|F|$ appears to the first power in (9-79) for a perfect crystal but is raised to the second power for the small crystallite in which primary extinction is negligible.

Extinction correction. It follows from the foregoing discussion that the integrated reflecting power of a crystal will depend rather markedly on whether or not dynamical interactions can be neglected. Darwin realized this clearly himself and, together with Moseley, carried out some measurements on sodium chloride crystals in 1913. Like the observations of later investigators, the integrated intensities from these crystals agreed more closely with those predicted by the kinematic equation (9-45). Nevertheless, others observed considerably decreased intensities, particularly when highly "perfect" crystals like optically pure calcite or diamond were examined. As already mentioned, these observations were at first attributed to anomalous-absorption effects, until Darwin recast the problem in 1922 and showed how the kinematical equation can be corrected to agree with the experimentally observed intensities.

It should be obvious that real crystals, which reflect x rays over a wider angular range than is allowed by (9-77), with a larger intensity than that predicted by (9-79), cannot be made up of a perfectly parallel stack of planes like that depicted in Fig. 9-13. Darwin realized this and reasoned, therefore, that the planes in actual crystals are not perfectly parallel. His personal belief was that the planes were somewhat warped, but he found this difficult to express in a suitable geometrical model. Instead, he postulated the following purely artificial model, which is directly amenable to the formulations already given. Suppose the crystal consists of small perfect regions, or blocks, each of volume δV, so small that primary extinction is negligible within each perfect block. Next, suppose that these blocks are tilted very slightly relative to each other, as illustrated in a greatly exaggerated manner in Fig. 9-14. Darwin called this an *ideally imperfect crystal*, and subsequently Ewald dubbed it a *crystal mosaic*, for reasons that are self-evident in Fig. 9-14. It should be clearly recognized that Darwin did not believe that crystals actually have such structures, but found this particular collection of small blocks, with perfect coherence within each domain but no phasal coherence between adjacent blocks, mathematically most useful in accounting for the observed intensities. For example, if the blocks are so small that a large number of them contribute to each observed reflection, their angular deviations from the mean can be expressed by some continuous distribution function such as the error function

$$W(\Delta) = \frac{1}{\sqrt{2\pi}\,\eta} e^{-\Delta^2/2\eta^2} \qquad (9\text{-}80)$$

Fig. 9-14. Mosaic model of crystal. Note that the blocks comprising the crystal are assumed to be tilted sufficiently so that when one block forms the angle θ with the incident beam, its immediate neighbors do not. In view of the narrow reflecting range, these tilts need not be large, and are shown greatly exaggerated.

where Δ represents the magnitude of the angular deviation from the mean, and η is the standard deviation.

The agreement between the predictions based on the model in Fig. 9-14 and experimental observations such as those discussed in the next section has tended to give this model an aura of reality. In fact, it has prompted various investigators to concoct theories and experiments designed to prove the existence of such a mosaic structure in real crystals. As further discussed in Chap. 10, there exist certain crystals composed of relatively homogeneous blocks not unlike those depicted in Fig. 9-14. These blocks, however, are almost always considerably larger than those invented by Darwin and, more importantly, have not been observed to exist in most crystals, despite the vastly increased observational powers of modern instruments. It must be concluded, therefore, that a segregation of the optically independent regions into physically distinguishable blocks like those in Fig. 9-14 is not essential for the successful application of the kinematical diffraction theory developed by Darwin.

Needless to say, the kinematical theory derived in the early parts of this chapter is especially suited to describe x-ray diffraction by an ideally imperfect crystal. When the size of each block in Fig. 9-14 is increased until primary extinction no longer is negligible, Darwin showed that a correction can be applied to the diffracting power of a single perfect block such that a new quantity is obtained.

$$Q' = Q\frac{\tanh A}{A} \tag{9-81}$$

where Q is the diffracting power of a small block as given by (9-46) and

$$A = \frac{e^2}{mc^2}\frac{|F|}{V}\frac{\lambda t_0}{\gamma} \tag{9-82}$$

where t_0 is the thickness of the block (normal to the reflecting planes), and γ is the direction cosine between the incident (or reflected) beam's direction and the crystal's surface (for a symmetrical arrangement). The integrated reflecting power then is given by analogy to (9-47) by

$$R_{hkl} = Q'V_c \tag{9-83}$$

where V_c is the total irradiated crystal volume.

In a somewhat more elegant treatment of dynamical interactions, presented in his book, Zachariasen showed that it is necessary to distinguish the possible experimental arrangements, namely, the Bragg, or reflection, arrangement from the Laue, or transmission, arrangement. Defining the integrated reflecting power for one block,

$$R = Q'\frac{t_0}{\gamma} = Qf(A)\frac{t_0}{\gamma} \tag{9-84}$$

he has shown that, for the Bragg arrangement,

$$f(A) = \frac{\tanh A + |\cos 2\theta| \tanh |A \cos 2\theta|}{A(1 + \cos^2 2\theta)} \tag{9-85}$$

while, for the Laue case, the primary extinction correction involving Bessel functions is

$$f(A) = \frac{\Sigma J_{2n+1}(2A) + |\cos 2\theta| \Sigma J_{2n+1}(2A|\cos 2\theta|)}{A(1 + \cos^2 2\theta)} \tag{9-86}$$

When absorption is included in the analysis, as shown in detail by Prins, it is possible to utilize the previously derived expressions, except that the attenuation factor in (9-67) should be used in place of the linear absorption coefficient. Thus the integrated reflecting power of the crystal for the symmetrical Bragg case becomes, by analogy with (9-59),

$$R^{\theta}_{hkl} = \frac{Q}{2(\mu + gQp)} \tag{9-87}$$

while for the symmetric Laue case, according to (9-62),

$$R^{\theta}_{hkl} = Q \frac{t}{\gamma} e^{-(\mu + gQp)(t/\gamma)} \tag{9-88}$$

where p is the polarization correction defined in (9-67*a*), and t is the crystal thickness in the transmission arrangement.

The term gQ expresses the attenuation of the incident beam reaching the lower-lying blocks, and was named by Darwin *secondary extinction*. It is possible to relate the coefficient g to the relative block tilts by assuming that their distribution is known. Suppose it is given by the error function (9-80); then it turns out that

$$g = \frac{1}{\sqrt{2\pi}\, \eta} \tag{9-89}$$

and the secondary extinction is inversely proportional to the standard deviation η of the block tilts. As the deviation increases, g decreases, a result consistent with the transition from the perfect to the ideally imperfect case.

The above expressions have been given for the case when either primary or secondary extinction is present but the other one is negligibly small. Usually, both are sufficiently large so that they must be considered jointly. In this case, either (9-87) or (9-88) can be used by simply substituting $Q' = Qf(A)$ for Q everywhere it appears. Note that Q is part of the secondary extinction correction, so that it is not possible to separate the two effects analytically for individual examination experimentally. Suppose, however, that one or the other effect is, in fact, negligibly small. In that case it becomes a relatively straightforward matter to test the above equations. Before proceeding to a discussion of experimental observations of extinction, it is worthwhile to consider under what conditions either type of extinction can be considered to be negligible.

According to the above discussion, primary extinction disappears when $f(A) = 1.0$, in which case $Q' = Q$. An examination of (9-85) and (9-86) shows that $f(A)$ tends toward unity as A goes to zero. This can be expressed by the inequality

$$\frac{e}{mc^2} \frac{|F|}{V} \frac{\lambda t_0}{\gamma} \ll 1.0 \tag{9-90}$$

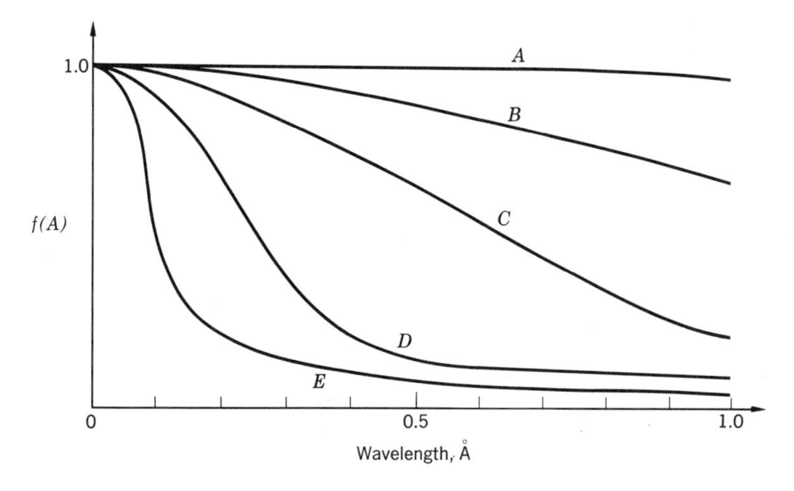

Fig. 9-15. Dependence of primary-extinction coefficient for the 111 reflection of silicon on x-ray wavelength. Curve A is calculated for $t_0 = 10^{-4}$ cm, B for 10^{-3} cm, C for 2×10^{-3} cm, D for 5×10^{-3} cm, and E for 10^{-2} cm. *(After Bragg and Azároff.)*

from which it follows that the block thickness should be

$$t_0 \ll \frac{\gamma V}{\lambda |F|} \frac{mc^2}{e^2} \tag{9-91}$$

It is clear that the critical thickness below which primary extinction is negligible is inversely proportional to the wavelength of x rays used in the experiment. (Note that in the symmetric Bragg arrangement, $\gamma = \sin \theta$, so that $\lambda/\sin \theta = 2d$ and is an invariant quantity. This is not true in the symmetric Laue arrangement, where $\gamma = \cos \theta$.) The critical thickness is also inversely proportional to the structure-factor magnitude, from which it follows that primary extinction is much more prominent for intense reflections than for weaker ones. To see the order of magnitudes involved, Fig. 9-15 shows a plot of the primary-extinction coefficient $f(A)$ calculated for the 111 reflection from a silicon crystal. These curves were calculated for the Laue case according to (9-86) and assuming different values of block thickness ranging from 10^{-4} cm to 10^{-2} cm. It is interesting to note that, for this crystal, primary extinction barely becomes significant for wavelengths longer than 1 Å when $t_0 = 10^{-4}$ cm (1 micron). By comparison, when $t_0 = 5 \times 10^{-3}$ cm, primary extinction reaches its maximum value at approximately the $K\alpha$ line of silver ($\lambda = 0.56$ Å). In order to illustrate the effect of the structure-factor magnitude, the primary-extinction coefficient, calculated using the expression in (9-81), is shown in Fig. 9-16 for three orders of the 100 reflection from sodium chloride. In this case, the wavelength was held constant, while the thickness t_0 was varied as indicated. The conclusion that can be drawn from the above is that the same crystal may appear to be ideally imperfect at very short wavelengths or for relatively weak reflections. By the same token, the magnitude of the primary-extinction coefficient increases markedly with increasing wavelengths and for strong

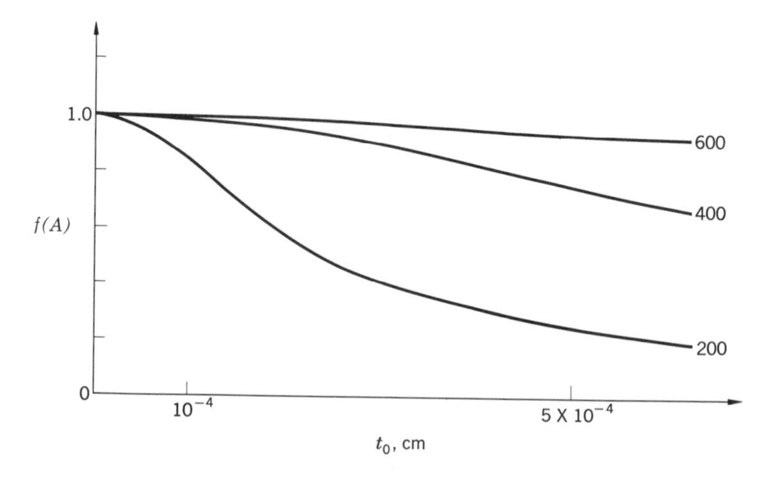

Fig. 9-16. Dependence of primary-extinction coefficient for several orders of the pinacoid reflections of sodium chloride on the block thickness assumed in the mosaic model (Fig. 9-14). *(After James.)*

reflections. This is a nontrivial observation, because one and the same crystal may appear to be an ideal (perfect) crystal when long wavelengths are used to measure a reflection, its truly imperfect nature becoming evident only when short wavelengths are employed.

Secondary extinction is a function of the scattering power of the individual blocks in Fig. 9-14, whereas primary extinction is a function of the scattering amplitudes of the individual planes within a block. Clearly, both are most important when strong reflections are measured. According to (9-67), secondary extinction becomes negligible when

$$\mu \gg gQ \tag{9-92}$$

Recalling that secondary extinction is caused by the deflection of part of the incident beam's intensity reaching a lower-lying block, it is of interest to estimate how much adjacent blocks must be tilted before one block no longer "casts a shadow" on the blocks below it. This can be done by assuming a distribution of blocks as given by the error function (9-80). Then, according to (9-89) and (9-92), the standard deviation of the blocks must satisfy the condition

$$\eta \gg \frac{Q}{\sqrt{2\pi}\,\mu} \tag{9-93}$$

for secondary extinction to be negligible. It is left to Exercise 9-13 to show that this means that η must be of the order of minutes of arc. Finally, it should be apparent from the model in Fig. 9-14 that, as the block size decreases, primary extinction decreases. What may not be as apparent is that the secondary extinction tends to increase in that case. To see how this comes about, consider the reflection curve of a crystal in which the optically independent regions are so small that primary extinction can be neglected. Next suppose that the halfwidth of the reflection by each block is correctly given by (9-70), where

$$D = \frac{t_0 \sin \theta}{\gamma} \tag{9-94}$$

correctly expresses the effective block thickness for either the symmetrical Bragg or Laue arrangement, provided γ is the direction cosine of the x-ray beams relative to the crystal face's normal. If secondary extinction also is to be negligible in this case, the standard deviation must be considerably larger than the halfwidth of each block, so that

$$\eta \gg \frac{\lambda \sqrt{\ln 2/\pi}}{(t_0/\gamma) \sin 2\theta} \tag{9-95}$$

Clearly, as t_0 decreases, it becomes increasingly more "difficult" to keep η large enough for secondary extinction to be negligible also.

The above result serves to demonstrate the physical limitations of Darwin's mosaic model. If crystals literally consisted of blocks, one would expect the smaller blocks to fit together with less misalignment than that required to join together large blocks. If, on the other hand, the "blocks" are in reality merely sets of planes tilted, by randomly distributed imperfections like dislocations, in such ways as to form parallel and optically coherent sets, one would expect that a larger number of such dislocations would be needed to break up the crystal into many small regions. It is easily demonstrable that stress fields surrounding imperfections in crystals are "accommodated" most easily by a "bending" of the atomic "planes" in their vicinity, so that it is natural that an increased number of imperfections present in a crystal serves to increase the magnitudes of the standard deviation from parallelism rather than to decrease it. This expectation forms the basis for several x-ray methods devised to study the perfection of the single crystals, as described in the next chapter.

Experimental observation of extinction. Probably the first really careful attempt to investigate the adherence of experimentally measured intensities to the above theories, on an absolute basis, was the previously mentioned investigation of NaCl by Bragg, James, and Bosanquet. In these measurements, use was made of relation (9-88), which can be written, after rearranging terms and taking the natural logarithms of both sides,

$$\ln \frac{R\gamma}{Qt} = -(\mu + gQ)\frac{t}{\gamma} \tag{9-96}$$

where t is thickness of the crystal set to reflect x rays in the symmetrical-transmission (Laue) arrangement. The reflecting power of several reflections was measured using crystals of different thickness, produced by grinding down the salt crystal after each set of measurements was completed. A plot of the left side of (9-96) for each thickness used, against the thickness t, yielded the straight lines shown in Fig. 9-17. The conclusion that can be drawn from this experiment is that the crystals investigated displayed secondary-extinction effects in a form

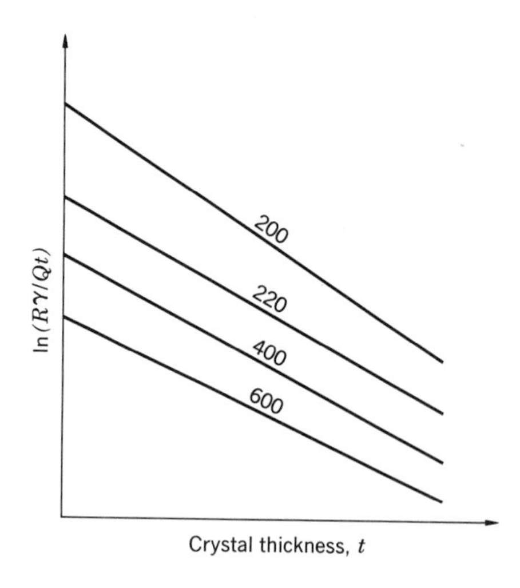

Fig. 9-17. Effect of crystal thickness on the reflecting power of several planes of sodium chloride. *(After Bragg, James, and Bosanquet.)*

predicted by the above equation, so that the slope of the straight lines can be used to determine the secondary-extinction coefficients. The agreement of the experimental curves with (9-96) also suggests that primary extinction was negligible in these crystals, although the same type of plot results if Q', including primary extinction, is substituted for Q in (9-96). In fact, this procedure suggests a way for measuring g and Q' experimentally, provided that the reflecting powers are measured on an absolute scale. The difficulty is that, by grinding the crystal, its perfection may be affected so that the extinction may actually vary with thickness in an unknown way.

A variation of the above procedure was employed by R. H. Bragg and L. V. Azároff to devise a method for studying the perfection of crystals. (See also discussion in Chap. 10.) Instead of changing the thickness of a crystal, the 111 integrated reflecting power of a silicon crystal was measured on an absolute basis in transmission, using x rays having five different wavelengths: $\lambda = 0.39$, 0.57, 0.71, 0.82, and 0.93 Å. Rewriting the terms in (9-96) to include primary extinction, $Q' = Qf(A)$,

$$\ln\left(\frac{R^\theta_{111}\gamma/t}{Qf(A)}\right) + \mu\frac{t}{\gamma} = -gQf(A)\frac{t}{\gamma} \tag{9-97}$$

It is clear that $Qf(A)$ can be calculated for each of the wavelengths employed, using the appropriate equations given above. After inserting the experimentally determined values of R^θ_{111}, the left side of (9-97) can be evaluated and plotted against $Qf(A)t/\gamma$. If the correct value of A, as given by (9-82), is inserted in (9-86) to calculate $f(A)$, then the resultant plot should be a straight line whose negative slope yields g. Referring back to (9-82), it is seen that everything required to calculate A is known except t_0, the block thickness in Darwin's model. The dependence of $f(A)$ on t_0 at different wavelengths was displayed in Fig. 9-15, and quite obviously, the correct value of t_0 to be used can be established by trial and error since the correct value assures a linear dependence on wavelength. The efficacy of this procedure is illustrated in Fig. 9-18, which shows the relation between the two sides of (9-97) for the 111 reflection of silicon for four different values of t_0. It is interesting to note that a clearly detectable deviation from linearity occurs when the block thickness assumed differs from the correct value by as little as 14%.

As a final illustration of both the wavelength dependence of extinction and the kind of variations that can be expected even within the same single crystal, Fig. 9-19 compares the integrated reflecting powers from a thin slice of silicon recorded under virtually identical conditions. The irradiated area on the crystal was approximately 1×10 mm, and the crystal was successively translated parallel to itself in 2-mm increments (at right angles to the long direction of the beam). The intensities recorded at each point are shown for two different wavelengths in Fig. 9-19, beginning near one edge of the crystal, denoted -10 mm in the figure, and ending at the opposite edge of a plate cut transversely from a single-crystal boule. Two aspects are particularly noteworthy: the intensity is reduced in the central portion, which is more perfect according to other evidence, and the difference between the

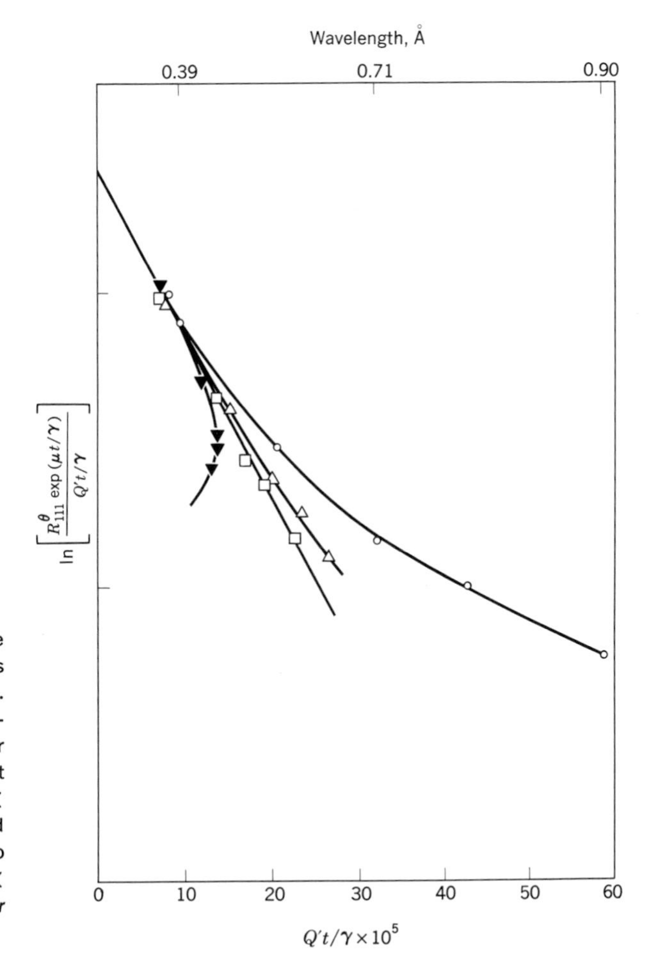

Fig. 9-18. Determination of the block thickness for the fictitious mosaic model (Fig. 9-14) of silicon. The circles represent the measured values before correction for extinction, the squares (straight line) correspond to $t_0 = 1.75 \times 10^{-3}$ cm, while the open and closed triangles correspond to $t_0 = 1.5 \times 10^{-3}$ cm and 2.0×10^{-3} cm, respectively. *(After Bragg and Azároff.)*

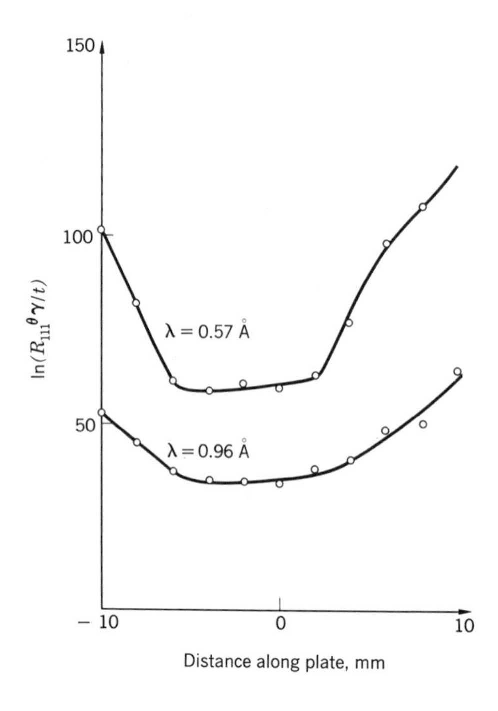

Fig. 9-19. Variation of 111 reflecting power at adjacent positions along a central line in a silicon wafer.

central and less perfect boundary regions is exaggerated by the shorter-wavelength radiation.

It should be noted that the experimental tests of extinction described in this section did not include the polarization correction (9-67*a*) derived subsequently by Zachariasen. An examination of its dependence on the diffraction angle shows that p does not begin to deviate from unity significantly until $\theta \approx 20°$. Thus the preceding results are not appreciably affected because the diffraction angles employed were small. Contemporaneously with the experiments of Bragg and Azároff, a somewhat different procedure for evaluating the extinction coefficients was proposed by S. Chandrasekhar. When a crystal-monochromatized x-ray beam is used to irradiate a crystal, the magnitude of each structure factor is altered by the degree of polarization of the primary beam. In fact, by changing the orientation of the reflecting planes relative to the two distinguishable polarization directions lying within the diffraction plane and normal to it, the magnitude of F_{hkl} can be altered in a known way. The ratio of two such reflecting powers then permits the estimation of the extinction, which is also affected by the degree of polarization in a known way. It is possible to use pairs of such measurements to estimate the total extinction for a particular reflection, but not to evaluate each kind of extinction separately.

Ewald's dynamical theory. In his doctoral thesis published in 1912, Ewald set forth a theory describing the propagation of light through a crystal, which was treated, however, as a discrete lattice array rather than as the conventionally used continuum. Four years later he published an extension of this treatment to include x radiation, and set up therein the elegant formalism of the dynamical theory of x-ray diffraction. In its simplest form, the crystal in Ewald's model is considered to be equivalent to an infinite three-dimensional periodic array of dipoles. An external field then induces oscillations in these dipoles, which can be represented by suitably formulated plane waves propagating through the lattice array with certain velocities. The oscillating dipoles, in turn, generate a new field, and the entire system is maintained in dynamical equilibrium. In 1931, Laue modified this model somewhat by considering a periodically varying distribution of negative electrons and positive nuclei in a crystal so arrayed that their charges normally just neutralize each other. An incident electric field then polarizes this distribution at each point in proportion to its field strength at that point. This kind of a distribution can be described by a three-dimensional periodic dielectric constant (relatable to the three-dimensional electron density), and the x-ray diffraction problem devolves to the solution of Maxwell's equations in such a medium. The solutions are relatively straightforward for an infinitely large crystal, but become considerably more complex when the boundary conditions of finite crystals must be considered. It turns out that the problem is directly soluble for the transmission, or Laue, case, but as first shown by Zachariasen, only approximately for the reflection, or Bragg, case. Incidentally, it should be noted that this treatment is not limited to x rays, and in fact, the parallel problem in the case of electrons has been solved by using Schrödinger's equation (quantum mechanics) in place of Maxwell's classical equations.

The formulation of the dynamical theory is entirely omitted in the present volume. The interested reader is referred to two excellent recent reviews by R. W. James and by B. W. Batterman and H. Cole, as well as texts by James, Laue, and Zachariasen, cited in the literature list at the end of this chapter. Further references to important recent contributions by P. B. Hirsch, N. Kato, and others can be found in the two review articles cited. The present discussion is limited to the results and predictions of the dynamical theory for the Laue arrangement (Fig. 9-20) in order to illustrate its nature. Some of the experimental verifications are given in the next section.

An electric field incident on an ideally perfect crystal, under the conditions indicated in Fig. 9-20, induces an electric field which can be described by the sum of

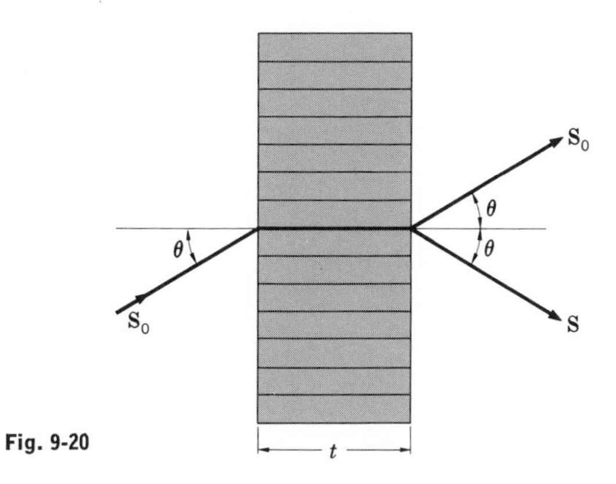

Fig. 9-20

two coherent plane waves like

$$\mathcal{E} = e^{2\pi i\nu t}\mathcal{E}_0 e^{(2\pi i/\lambda)\,\mathbf{S}_0\cdot\mathbf{r}} + e^{2\pi i\nu t}\mathcal{E}_{hkl} e^{(2\pi i/\lambda)\,\mathbf{S}\cdot\mathbf{r}} \tag{9-98}$$

where $\mathcal{E}_0$ and $\mathcal{E}_{hkl}$ are the electric field vectors of the waves traveling in the forward direction, signified by the unit vector $\mathbf{S}_0$ and the reflection direction indicated by $\mathbf{S}$, respectively, and r is a position vector within the crystal. These two waves combine to produce a traveling wave which moves along the direction of the bisector of the angle between $\mathbf{S}$ and $\mathbf{S}_0$, which, for the symmetrical Laue arrangement considered in Fig. 9-20, means that it moves parallel to the reflecting planes. Concurrently, they produce a standing wave at right angles to these planes. (For the reader familiar with the Brillouin-zone theory of crystals, this is equivalent to the case of the two solutions, according to Floquet's theorem, that coexist at the zone boundary between two zones separated by a forbidden-energy gap.) In the case of x rays, it is also necessary to consider the effect of polarization in these fields, so that, in fact, four waves must be considered, two having electric field vectors lying in the diffraction plane, and two having field vectors normal to this plane. This can be expressed by a polarization factor P, which takes on the value $P = 1$ for the normal case and $P = \cos 2\theta$ for the parallel case. Setting $\mathcal{E}_0 \cdot \mathcal{E}_{hkl} = P|\mathcal{E}_0|^2$ and letting $(\mathbf{S} - \mathbf{S}_0) = \lambda\boldsymbol{\sigma}_{hkl}$ at the Bragg angle, the field intensity becomes

$$\begin{aligned}|\mathcal{E}|^2 = \mathcal{E}\cdot\mathcal{E}^* &= |\mathcal{E}_0|^2 + |\mathcal{E}_{hkl}|^2 + 2\mathcal{E}_0\cdot\mathcal{E}_{hkl}\cos(2\pi\boldsymbol{\sigma}_{hkl}\cdot\mathbf{r}) \\ &= 2|\mathcal{E}_0|^2[1 \pm P\cos(2\pi\boldsymbol{\sigma}_{hkl}\cdot\mathbf{r})]\end{aligned} \tag{9-99}$$

where $\pm$ distinguishes between the two cases when the two plane waves are exactly in phase or exactly out of phase with each other.

According to (9-99), it is seen that a standing-wave pattern is set up when $\boldsymbol{\sigma}_{hkl}\cdot\mathbf{r} = \text{const}$, so that the planes of constant intensity are parallel to the reflecting planes and are spaced $|1/\boldsymbol{\sigma}_{hkl}| = d_{hkl}$ apart. The four possibilities are indicated in Fig. 9-21. The two waves whose vectors are perpendicular to the diffraction plane $(P = 1)$ are shown in A and B, while the other two waves $(P = \cos 2\theta)$

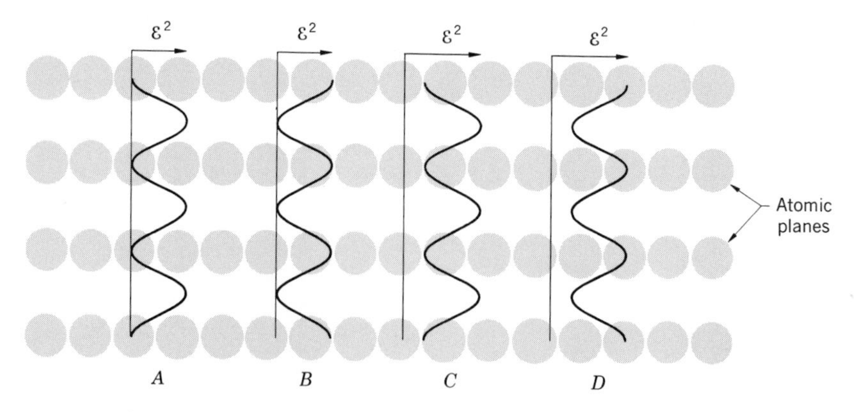

Fig. 9-21. The four possible standing-wave patterns in a perfect crystal set at the reflection angle. (*After Batterman and Cole.*)

are shown in C and D. Note that the field intensity does not go quite to zero in these last two cases because of the term $1 - P$ in (9-99). The four waves recombine upon exit from the crystal (Fig. 9-20), the waves denoted A and C in Fig. 9-21 forming the diffracted beam, while B and D combine in the forward beam direction. Other things being equal, for the symmetric case here considered, the intensity of both beams should be the same. If the crystal has a simple structure, such that all the atoms lie along the reflecting planes shown in Fig. 9-20, then the waves having nodes at the atomic planes (minimum field intensity) are absorbed less than the waves having antinodes or maxima at these planes, since the absorption of the field by atoms is directly proportional to the intensity of the field at the atoms. This leads to the *anomalous-transmission effect* in perfect crystals, further discussed in the next section.

It is possible to express the effective linear absorption coefficient in the forward beam direction,

$$\mu_{\text{eff}} = \mu[1 \pm |P|\epsilon(1 - p^2)^{\frac{1}{2}}] \tag{9-100}$$

where
$$p = \frac{|\mathcal{E}_{hkl}|^2 - |\mathcal{E}_0|^2}{|\mathcal{E}_{hkl}|^2 + |\mathcal{E}_0|^2} \tag{9-101}$$

and
$$\epsilon = \frac{F''_{hkl}}{F''_0} \tag{9-102}$$

relates the real causes of absorption (see Chap. 6), namely, the imaginary components of the atomic scattering factors expressed as crystal scattering factor F''_{hkl} in the diffraction direction and F''_0 in the forward direction. (It should be noted that, since the dynamical theory "converts" the incident field into a crystal field which reemerges as a diffracted field, it makes more sense to speak of a *forward-diffracted* beam in the direction of $\mathbf{S}_0$.)

In considering the detailed propagation of four waves like those in Fig. 9-21, it is generally necessary to take into account explicitly the directions of the unit vectors $\mathbf{S}$ and $\mathbf{S}_0$ relative to the reflecting planes. It turns out that both diffracted beams actually consist of four waves having polarizations and distributions like those shown in Fig. 9-21. Although this further complicates the analysis and changes the simple picture shown in Fig. 9-20, some interesting consequences are discovered. As can be seen in the above equations, a change in the absorption of one of the waves leads to an increase in the *relative* intensity of the other one. This turns out to be a function of the crystal's thickness and the relative inclinations of $\mathbf{S}$ and $\mathbf{S}_0$ to the crystal surface, so that, for a wedge-shaped crystal, thicknesses will occur at which waves that are mutually in phase (or out of phase) emerge from the crystal. The solution of the wave equations describing such an oscillation of energy from wave to wave was called *pendellösung* (pendulum solution) by Ewald, and leads to the appearance of interference bands, usually visible at the lenticular edges of perfect crystals. More recently, Kato has shown how the spacing of such fringes are inversely proportional to the magnitude of the reflecting plane's structure factor, and has demonstrated a possible way for determining structure factors without having to measure the reflecting powers of the planes.

Anomalous transmission. An important stimulus toward reviving interest in the dynamical theory was given indirectly by H. N. Campbell in 1950 when he reported at a meeting of the American Crystallographic Association a rather unusual phenomenon he had observed while examining the reflections from a relatively perfect crystal of calcite. While rotating the crystal into reflecting position, Campbell had a stationary detector monitor the intensity transmitted in the forward direction. He observed to his surprise that the forward beam's intensity increased when the crystal was in reflecting position, and a portion of the incident beam's energy was clearly being deviated by diffraction to a second detector set at the appropriate Bragg angle. As news of this observation spread, it soon became established that similar observations had been made and duly published by G. Borrmann in Germany in 1941. Because of the war then in progress, this received little notice until Campbell rediscovered the effect ten years later. A number of papers appeared in the years 1951–1952, explaining the anomalous transmission, or *Borrmann effect,* in terms of the dynamical theory. Its origin becomes evident when relation (9-100) for the forward-diffracted beam is examined. For incident directions not forming the Bragg angle with a set of planes in the crystal, there is no diffracted beam ($\mathcal{E}_{hkl} = 0$), so that $p^2 = 1$, and the effective absorption coefficient is just the usual linear absorption coefficient μ. When the crystal is set exactly at the reflecting angle, conversely, the energy flow into the Bragg-diffracted and forward-diffracted beams is equal, so that $p^2 = 0$, and the effective absorption coefficient reduces to

$$\mu_{\text{eff}} = \mu(1 \pm |P|\epsilon) \tag{9-103}$$

where the minus sign refers to waves like A and C in Fig. 9-21 (low absorption, $\mu \to 0$), and the plus sign to the other two waves. Normally, only part of the incident beam participates in the diffraction process, so that the forward-diffracted beam is superimposed on the transmitted beam, and its presence is not clearly noticeable. This is illustrated in Fig. 9-22, where the intensity of the diffracted beam I_{hkl} divided by that incident on it, I, is compared with the similarly normalized

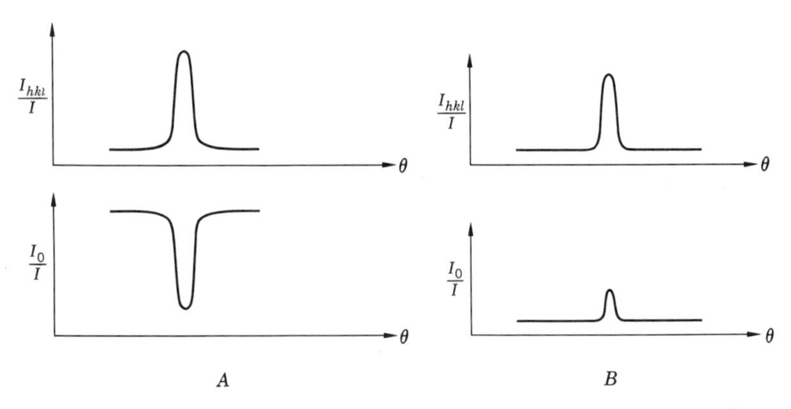

Fig. 9-22. Relative intensities of diffracted and transmitted beams in a thin (A) and a thick (B) perfect crystal.

transmitted intensity, I_0/I. In the case of a thin crystal, shown to the left in Fig. 9-22*A*, the transmitted beam is quite intense, except when some of its energy is being reflected by the (hkl) planes. In a thick crystal, however, very little intensity is transmitted unless the anomalous-transmission effect literally causes additional power to be diffracted in the forward direction. Since I_0/I is considerably attenuated in the thick crystal case, shown to the right in Fig. 9-22, the anomalous-transmission effect causes an actual rise in the intensity of the transmitted (forward-diffracted) beam at the same time that the diffracted beam's intensity I_{hkl}/I goes through its maximum.

The phenomenon has a very interesting corollary. The value of ϵ given by (9-102) is usually slightly less than unity, while the polarization factor P in (9-100) is either unity or $\sin 2\theta$. For one of the four waves in Fig. 9-21, therefore, $\mu_{\text{eff}} \ll \mu$, while for the two for which the plus sign applies (B and D), $\mu_{\text{eff}} > \mu$, and for the fourth, $\mu_{\text{eff}} \approx \mu$ (Exercise 9-15). This means that, while three of the waves are strongly attenuated, the fourth is hardly absorbed at all. (This is an example of dichroism in the case of x-ray transmission.) Specifically, the wave designated A suffers very small effective absorption, so that, if a suitably thick crystal is used the forward-diffracted beam that emerges is not only monochromatic, but also plane-polarized. Such a polarizer-monochromator was first constructed by Cole and his collaborators using the 220 reflection from a germanium crystal about one millimeter thick. An analysis of the angular dependence of the intensity of scattering of this polarized beam clearly demonstrated that it was truly plane-polarized. Incidentally, an analogous analysis for the Bragg-diffracted beam shows that it similarly becomes plane-polarized under these conditions.

A somewhat different experiment was carried out by Borrmann in order to demonstrate that the energy flow in a crystal set at the appropriate Bragg angle does in fact take place along the diffracting planes. This experimental arrangement is illustrated in Fig. 9-23, which shows the incident beam of intensity I forming the Bragg angle with a set of transverse planes in the symmetrical Laue arrangement. According to the dynamical theory, the energy flow through the crystal proceeds parallel to these planes and, upon emergence at the opposite face, breaks up into a forward-diffracted beam of intensity I_0 and a diffracted beam of intensity I_{hkl}. Not all the incident beam's energy is thus diverted, particularly if the incident beam is not perfectly monochromatic. As a result, part of the incident beam continues through, suffering only normal absorption losses. Thus the directions of the two beams, labeled according to their intensities I and I_0 in Fig. 9-23, are parallel, while the diffracted beams (I_0 and I_{hkl}) diverge by an amount 2θ. This can be readily verified by placing a photographic film so as to intercept the three beams at different distances R from the crystal. The expected phenomenon was actually observed by Borrmann, and clearly confirms the predictions of the theory. Note that the kinematical theory is not able to explain this effect, because, according to its predictions, diffracted beams should originate at points lying along the path of the incident beam (shown dashed in Fig. 9-23), not at the point of emergence actually observed.

In concluding the discussion of the dynamical theory in this chapter, it should be emphasized that a rather simplified description has been presented above.

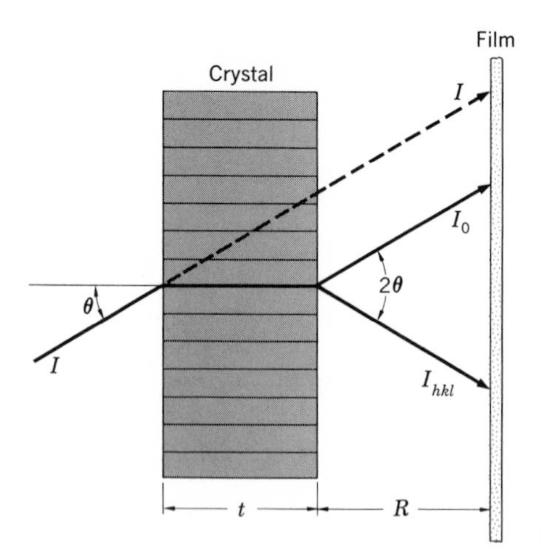

Fig. 9-23

A number of interesting complications actually arise when nonidealized conditions exist. For example, when a narrow parallel incident beam deviates very slightly from the ideal Bragg angle, the four waves in the two diffracted beams do not travel along identical directions parallel to the reflecting planes, but split up into two rays, each of the type A and C and B and D, which diverge slightly, of the order of several degrees. Because it is very difficult to obtain perfectly parallel beams in practice, the divergence present in an incident beam causes a literal fanning out of the diffracted-beam directions. Since the absorption of the diffracted beams is a critical function of their state of polarization and the angular relations discussed above, the beams emerging from a crystal may exhibit rather marked intensity variations from one side of the diffracted beams to the other under such conditions. The full evaluation of the consequences of these and similar variations, as well as a rigorous solution of the diffraction equations for the Bragg, or reflection, arrangement, are some of the problems occupying investigators at the present time.

SUGGESTIONS FOR SUPPLEMENTARY READING

Boris W. Batterman and Henderson Cole, Dynamical diffraction of x rays by perfect crystals, *Reviews of Modern Physics,* vol. 36 (1964), pp. 681–717.

G. Borrmann, Beugung am Idealkristall, *Zeitschrift f. Kristallographie,* vol. 120 (1964), pp. 143–181.

Robert H. Bragg and Leonid V. Azároff, Direct study of imperfections in nearly perfect crystals, in J. B. Newkirk and J. H. Wernick (eds.), *Direct observation of imperfections in crystals* (Interscience Publishers, Inc., New York, 1962), pp. 415–429.

S. Chandrasekhar, Extinction in x-ray crystallography, *Advances in Physics,* vol. 9 (1960), pp. 363–386.

R. W. James, *The optical principles of the diffraction of x-rays* (G. Bell & Sons, Ltd., London, 1948).

——— The dynamical theory of x-ray diffraction, *Solid State Physics,* vol. 15, pp. 53–220.

Max von Laue, *Röntgenstrahl-Interferenzen* (Akademische Verlagsgesellschaft, Frankfurt, 1960).

Kathleen Lonsdale, Extinction in x-ray crystallography, *Mineral. Mag.,* vol. 28 (1947), pp. 14–25.

William H. Zachariasen, *Theory of x-ray diffraction in crystals* (John Wiley & Sons, Inc., New York, 1945).

——— The secondary extinction correction, *Acta Crystallographica,* vol. 16 (1963), pp. 1139–1144.

EXERCISES

9-1. Following the procedure suggested in the text, derive expression (9-19) for the structure factor of a crystal.

9-2. A crystal having a face-centered cubic structure must contain sets of four atoms in each unit cell such that three atoms are related to the fourth atom by the vectors $\frac{1}{2}\mathbf{a} + \frac{1}{2}\mathbf{b}$, $\frac{1}{2}\mathbf{a} + \frac{1}{2}\mathbf{c}$, and $\frac{1}{2}\mathbf{b} + \frac{1}{2}\mathbf{c}$. Consider the simplest case possible, namely, one atom at 000 and three equivalent atoms at $\frac{1}{2}\frac{1}{2}0$, $\frac{1}{2}0\frac{1}{2}$, $0\frac{1}{2}\frac{1}{2}$. (This is the actual structure of many simple metals like Al, Cu, Ni, etc.) Derive the structure-factor expression for this case, and determine the combinations of h, k, and l for which $F = 0$. (Compare this with the result deduced in Exercise 3-12.)

9-3. Following the procedure outlined in Exercise 9-2, derive the structure factor for body-centered cubic cesium having two Cs atoms per cell at 000 and $\frac{1}{2}\frac{1}{2}\frac{1}{2}$, respectively. What combinations of h, k, l give nonzero values in this crystal? How does this result compare with that predicted by a transformation matrix?

9-4. Suppose that the two atoms in Exercise 9-3 are different, for example, one atom is Cs and the other Cl; then one gets the CsCl structure. What combinations of h, k, l give nonzero structure-factor values for this structure? What is the lattice type in this case?

9-5. Making use of the construction shown in Fig. 9-4, derive Eq. (9-26) following the procedure indicated in the text. Alternatively, derive the final expression by rearranging the terms and noting that $|\mathbf{S}' - \mathbf{S}| = \epsilon$, provided it is a small quantity.

9-6. Making use of the construction shown in Fig. 9-9, derive relation (9-48) in the text. Exactly why is it necessary to multiply the fraction deduced by the multiplicity j?

9-7. Using a construction similar to that shown in Fig. 9-9, derive relation (9-53) in the text for the diffraction power entering the detector of a powder diffractometer.

9-8. Equation (9-54) is directly useful in calculating the diffraction intensities to be expected from a polycrystalline specimen. Combining all the terms that are constants for a given experimental arrangement without evaluating them, calculate the relative diffraction powers of the first six reflections to be observed from a CsCl powder using Cu $K\alpha$ radiation ($a_{CsCl} = 4.11$ Å, $\lambda = 1.539$ Å).

9-9. Making use of relation (9-62), show that the optimum thickness for diffraction through a crystalline slab occurs when the x-ray pathlength equals $1/\mu$.

9-10. Suppose an x-ray beam incident on an infinitely thick polycrystalline slab forms the angle α with its surface, and the diffracted beam forms the angle β with this surface. Consider a layer of thickness dz parallel to this surface and a depth z below it. For a beam cross section A, the irradiated volume element in this layer is $(\rho'/\rho)A \csc \alpha \, dz$. By properly determining the total pathlength of x rays diffracted by this volume element and noting that the effective linear absorption coefficient is $(\rho'/\rho)\mu$, derive relation (9-66).

9-11. Making use of the discussion of absorption in Chap. 6 and the discussion of diffraction in this chapter, what quantitative predictions can you make about the relative intensities to be expected from two elements having identical crystal structures but different atomic numbers? Suppose that both crystals have identical cell sizes and contain atoms whose atomic numbers differ by a factor of 2: Specifically, consider x-ray wavelengths that are much shorter than the K absorption edges of either atom, and the relative integrated intensities for the same hkl reflection obtained in the symmetrical Bragg arrangement. What is the ratio between their respective integrated intensities if the crystals are ideally imperfect? Ideally perfect?

9-12. For NaCl, $F_{200} = 84$ and $F_{111} = 16.8$. For Cu $K\alpha$ radiation ($\lambda = 1.54$ Å) and W $K\alpha$ radiation ($\lambda = 0.21$ Å) calculate the critical block thickness t_0/γ for these two reflections at which the crystal becomes ideally perfect ($a_{NaCl} = 5.63$ Å).

9-13. Calculate for sodium chloride the magnitude of Q for the 200 reflection, using the information supplied in Exercise 9-12 for Cu $K\alpha$ radiation. Next, calculate the standard deviation for the block tilts, given that the linear absorption coefficient for this wavelength is 160 cm^{-1}. This calculation is useful for an estimate of the reflection halfwidth to be expected when secondary extinction is negligible. Compare this with the critical value given by (9-104).

9-14. From the experimental curves plotted in Fig. 9-17, it is possible to determine the value or $g = 300$ for the 200 reflection in NaCl. Using the values for NaCl given in the two preceding exercises, compare the secondary-extinction coefficient with the linear absorption coefficient for

Cu $K\alpha$. What can one deduce about the relative block tilts from this value, and how does it compare with the result of Exercise 9-13?

9-15. Consider a perfect germanium crystal so arranged that Cu $K\alpha$ radiation is reflected from its (220) planes at a glancing angle $\theta = 22.5°$ in the symmetrical Laue arrangement. If $\epsilon = F''_{hkl}/F'_0 = 0.95$ for germanium, what are the four possible values that μ_{eff} can have for four waves like those in Fig. 9-20? (This is an example of the principle underlying the *Cole polarizer-monochromator* described in the text.)

TEN
X-ray diffraction by real crystals

IMPERFECTIONS IN CRYSTALS

General classification. The success of the so-called mosaic model of an ideal crystal in accounting for the x-ray diffraction by real crystals, recounted in the preceding chapter, led to a number of postulations regarding its nature, origins, and energetic bases. In fact, very elaborate calculations were made to demonstrate that real crystals would not be stable unless they had such a secondary structure. Most of these ideas have been discredited as the nature of imperfections present in real crystals became better understood. Probably the most durable of the early models is one put forward by Buerger and shown in Fig. 10-1. Buerger concluded, after a microscopic examination of a variety of crystals, that they appear to grow from an ideally perfect nucleus along branches, or *lineages*, which do not maintain strict parallelism with each other. The misalignment between adjacent lineages can be likened to that between Darwin's hypothetical blocks. Buerger's model differs from Darwin's in two important respects, however: the lineages are continuous throughout the crystal, and their existence was deduced from the visual examination of real crystals rather than being postulated merely for theoretical convenience. It should be noted, however, that not all crystals display such lineage structure when examined microscopically.

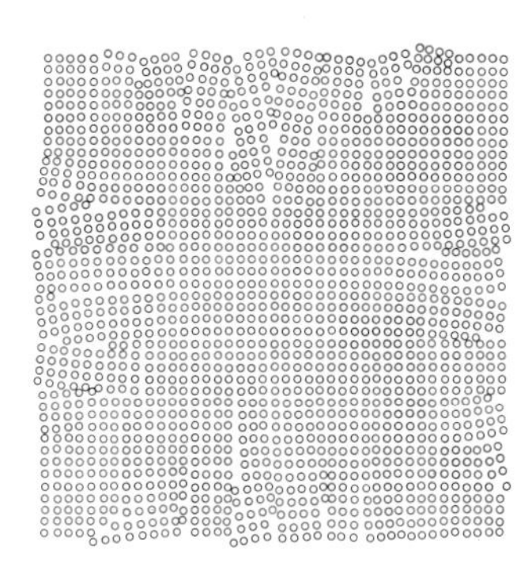

Fig. 10-1. Lineage model of crystal proposed by Buerger after observation of numerous minerals occurring in nature.

In the early 1930's it had already become apparent that a mere knowledge of the atomic arrays in crystals was not sufficient to explain all their mechanical, electrical, and other properties. In fact, it had become evident that many properties could be explained only if deviations from ideal crystal structures were assumed. This led to the postulation of a number of different kinds of imperfections or defects that can exist in crystals.† The confirmation of the actual presence of most of these defects in real crystals continues to occupy solid-state scientists to the present day.

To understand how imperfections present in crystals may affect the diffraction of x rays, it is convenient to classify them into the following groups:

Transient defects, which have lifetimes measured in microseconds. Of the variety of such defects normally present, only phonons affect x-ray diffraction. As shown below, it is more convenient to use their elastic-wave description in subsequent discussions.

Point defects, which can be missing atoms called vacancies, interstitial atoms, substituted, or vicarious, atoms, or combinations of these, usually occurring in pairs.

Line defects, which extend along straight or curved lines in a crystal and are usually composed of edge dislocations and screw dislocations.

Plane defects, which may extend along truly plane or along curved surfaces within a crystal. These include small-angle boundaries and other surfaces marking discontinuities in a crystal's periodicity.

Volume defects, which extend throughout small volumes in a crystal, for example, occluded voids, large clusters of point defects, exsolved precipitates, and other kinds of inclusions.

Not all the above-listed defects are concurrently present in crystals, although even the most perfect crystals are believed to contain some defects at ordinary temperatures. This is a direct consequence of the thermodynamic requirement that the free energy of a system tends toward a minimum by increasing its entropy through disorder.

Effect on crystal reflections. In addition to stimulating the creation of various defects in crystals, thermal energy in the form of phonons propagating through a crystal also causes the individual atoms to undergo small oscillations about their ideal position in the crystal. The reason for this becomes clear when the familiar energy curve as a function of interatomic separation in a crystal is considered. As can be seen in Fig. 10-2, the solid curve reaches a minimum at the equilibrium separation, r_0, usually calculated for the absolute zero of temperature. As the crystal's temperature increases, the energy increases, say, to V_1. At this energy

† For readers not already familiar with imperfections encountered in crystals, an elementary discussion is given in Leonid V. Azároff, *Introduction to solids* (McGraw-Hill Book Company, New York, 1960), chap. 5.

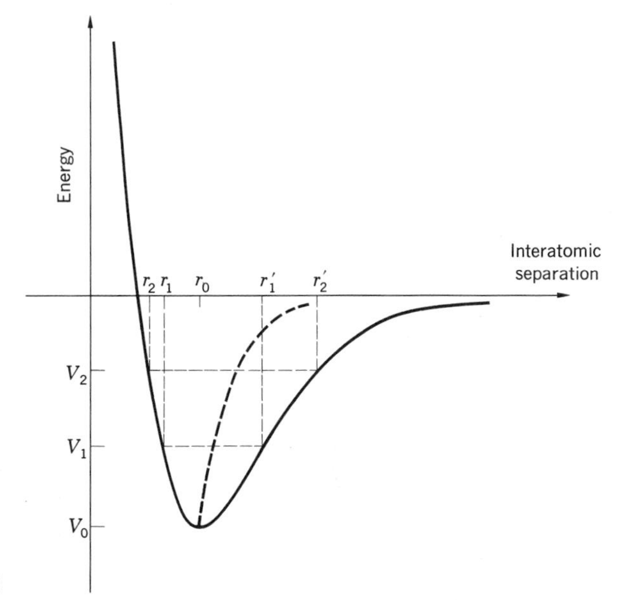

Fig. 10-2

there are two equally stable interatomic separations, r_1 and r_1', so that the atoms can oscillate between these extremes. At a still higher temperature, indicated by the energy V_2, the two stable interatomic separations are r_2 and r_2', and so forth. Thus it is clear that raising the thermal energy of the crystal causes the atoms to undergo vibrations of increasing amplitude. Also note that the mean of the two extreme positions shifts to larger and larger values as the temperature increases, as indicated by the dashed curve in Fig. 10-2. This is the reason for the thermal expansion of crystals whose magnitude and nature are determined by the asymmetry of the energy curve, which, in turn, is a function of the nature of the interatomic bonds in crystals.

The usual effect that heating a crystal has on its diffraction diagram is that the reflection angles decrease as the temperature increases. It is possible, therefore, to study the thermally induced expansion in crystals by careful measurements of lattice constants at several temperatures. It is usually observed that the shifting of diffraction maxima is accompanied by a reduction in their intensity. This decline in intensity becomes increasingly pronounced as the temperature increases. Most of the "lost" intensity reappears in the form of a diffuse background intensity, which has the effect of broadening the base of each diffraction peak, as described in a subsequent section. The magnitude of the background intensity is directly relatable to the thermally induced atomic vibrations, whose magnitude along specific directions depends on the interatomic binding forces along those same directions. Consequently, it is possible to make use of *temperature-diffuse scattering,* as this phenomenon has been named, to study the elastic properties of crystals.

Point defects, as their name implies, occur at individual atomic sites in crystals. When they are present in very small amounts, they produce no detectable effect on x-ray diffraction maxima. As their density increases, they tend to affect the lattice constants in measurable ways. Only in very favorable cases, however, is it possible to determine the actual defect density from lattice-constant changes. At the same time, point defects have an effect on the structure-factor magnitudes, which, when sufficiently large, may be detected by careful measurement of diffraction intensities. In such cases, as briefly described in a later section, it is actually possible to observe their presence in a crystal. When the number of point defects approaches the number of atoms present in a crystal, as is the case in order-disorder transitions, discussed in a later section, their effect on the structure-factor magnitudes is quite large, and it is possible to deduce their distribution directly.

Theoretically, individual dislocations have a very small effect on x-ray diffraction intensities, as several calculations have demonstrated. Nevertheless, they do produce stress fields which cause small displacements of atoms to take place over distances ranging from 50 to 100 Å. These stress fields, therefore, have the effect of disturbing the lattice periodicity in a crystal and, consequently, the coherency of x-ray scattering. Whether this can be described best by including the effect in a refined treatment of extinction or whether it is best treated as a separate phenomenon has not been clearly established as yet. It has been amply demonstrated, however, that the intensity of a reflection from a purportedly parallel set of planes shows local deviations from uniformity in the vicinity of dislocations. This is the basis for the so-called *topographic x-ray methods* used to study the perfection of single crystals. The different experimental arrangements devised for such studies are considered in this chapter.

The effect that plane defects have on x-ray diffraction depends not only on their density in a crystal, but also on the nature of the defects present. For example, it is possible to cause dislocations to array themselves along individual planes in a relatively perfect crystal, as shown by the etch pits formed at their emergence in Fig. 10-3. This array forms a small-angle boundary separating the two relatively perfect regions on each side of it. The two regions are tilted relative

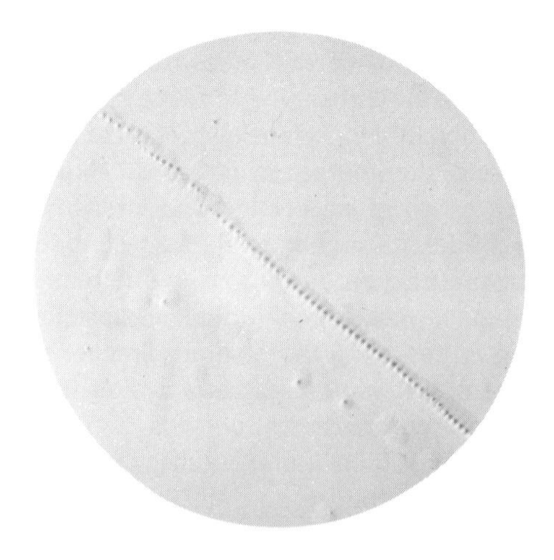

Fig. 10-3. Small-angle boundary formed by parallel array of edge dislocations made visible by etching a germanium single crystal. (*Photograph by F. L. Vogel; shown magnified about 750 times.*)

to each other by about one minute of arc, as careful x-ray diffraction measurements clearly indicate. Such small-angle boundaries, the boundaries between lineages present in crystals, and similar plane defects causing small relative tilts of adjacent undisturbed regions in crystals cause a small broadening of the diffraction maxima in accord with the kinematical theory discussed in the previous chapter. Other kinds of plane defects, for example, stacking faults, also cause a broadening of the diffraction maxima, and may cause a concurrent change in the diffraction angles. Still other types of plane defects may cause asymmetrical broadenings of diffraction maxima. Some of these effects are discussed in this chapter, and a number of others are discussed in the books and review articles in the accompanying literature list.

Just as dislocations tend to gather in planar arrays, point defects also tend to cluster together. Thus vacancies may agglomerate to form larger voids in crystals, which may collapse into dislocations or form voids or cracks in a crystal. Interstitial atoms may tend to cluster along certain planes, forming, in effect, two-dimensional precipitates. Finally, substituted atoms may combine with their host atoms to form ordered arrays differing from that of the host crystal. This ultimately leads to a phase segregation, and a special kind of volume defect is formed. The one thing that distinguishes these volume defects from the host crystal is the difference in their electron density. When the volume defects are small (about 50 to 100 Å in cross section), they scatter x rays in such a way that interference effects are produced at angles close to the direct beam. This *small-angle scattering* is quite similar to that of large gas molecules, discussed in Chap. 8. It is possible, therefore, to deduce approximate sizes and shapes of the volume defects by suitable interpretive procedures. In crystalline media, however, there often occurs a multiple scattering process which greatly complicates such interpretations. Nevertheless, small-angle scattering procedures may be used to characterize volume defects. It also has been shown that the methods used to observe the presence of line defects in single crystals may be used to detect volume defects. Thus precipitates of copper and of oxygen in silicon crystals have been studied by the topographic methods described in later sections.

Because x-ray diffraction experiments normally are conducted at room temperature, the effect that temperature has on diffraction intensities is most important. For this reason it is discussed in some detail next. The theoretical treatment of atomic vibrations in crystals can become very complex, however, so that the present discussion is limited to somewhat idealized cases. The conclusions drawn from these simple examples turn out to be applicable in a satisfactorily approximate way to crystals having more complex structures, so that the procedures for making corrections for the temperature effect also are described below. Next, the imperfections whose presence can be observed in single crystals is considered. The reason for segregating certain experimental methods in a separate classification is that they can be applied most successfully to single crystals only. Imperfections whose presence can be studied equally well in polycrystalline aggregates are considered separately in the final sections of this chapter. The detailed changes that are observed when the single-crystal method or the powder method is employed are described in full.

TEMPERATURE EFFECT

Original Debye theory. The effect that thermally induced vibrations of atoms have on x-ray diffraction maxima was first analyzed by Debye in 1913. In this analysis he assumed that each atom oscillates about a mean position in the crystal like a simple harmonic oscillator and that each atom can vibrate independently of all other atoms. This is clearly an oversimplification, so that, in 1914, Debye linked his x-ray theory to his well-known specific-heat theory, in which the thermally induced atomic displacements in a crystal are described by elastic waves (phonons) propagating through the crystal. The phonon spectrum derived from the specific-heat theory then can be used to calculate the mean-square displacements of atoms in simple cubic structures. This theory is presently limited to cubic crystals having one atom per lattice point, that is, to the structures of elements. An equally rigorous theory for polyatomic crystals or lower-symmetry elements is not available at present, so that approximations based on the theory derived for cubic crystals must be employed. The simplest form of this theory, that is, the original Debye theory, is presented in this section. Although it is founded on the incorrect assumption that the atoms vibrate independently of each other, its qualitative predictions regarding the effect of temperature on the intensity of diffraction maxima are unchanged by more sophisticated analyses.

Consider a simple cubic structure comprised of one atom per lattice point of a primitive cubic lattice. The vector to any atom in this crystal then is, simply, the vector to the lattice point $m_1m_2m_3$.

$$\mathbf{R}_m = m_1\mathbf{a}_1 + m_2\mathbf{a}_2 + m_3\mathbf{a}_3 \tag{10-1}$$

where the three cell edges all have the same length a. Suppose that the instantaneous displacement of an atom from its lattice point due to thermal agitation is denoted $\mathbf{\Delta}_m$; then its position in the crystal is given by

$$\begin{aligned}\mathbf{R}'_m &= m_1\mathbf{a}_1 + m_2\mathbf{a}_2 + m_3\mathbf{a}_3 + \mathbf{\Delta}_m \\ &= \mathbf{R}_m + \mathbf{\Delta}_m\end{aligned} \tag{10-2}$$

The intensity of x rays scattered by such a structure, expressed in electron units as previously done in Chap. 8, is given by

$$I_{\text{eu}} = \sum_m f_m e^{(2\pi i/\lambda)(\mathbf{S}-\mathbf{S}_0)\cdot\mathbf{R}_m'} \sum_n f_n e^{(-2\pi i/\lambda)(\mathbf{S}-\mathbf{S}_0)\cdot\mathbf{R}_n'} \tag{10-3}$$

where m and n are dummy indices representing the various triple indices of the lattice points in the crystal.

Since this simple structure contains only one kind of atom, $f_m = f_n$ and (10-3) can be simplified.

$$\begin{aligned}I_{\text{eu}} &= \sum_m \sum_n f^2 e^{(2\pi i/\lambda)(\mathbf{S}-\mathbf{S}_0)\cdot(\mathbf{R}_m'-\mathbf{R}_n')} \\ &= \sum_m \sum_n f^2 e^{(2\pi i/\lambda)(\mathbf{S}-\mathbf{S}_0)\cdot(\mathbf{R}_m-\mathbf{R}_n)} e^{(2\pi i/\lambda)(\mathbf{S}-\mathbf{S}_0)\cdot(\mathbf{\Delta}_m-\mathbf{\Delta}_n)}\end{aligned} \tag{10-4}$$

Note that, when the atomic displacements $\mathbf{\Delta}$ are vanishingly small, (10-4) reduces

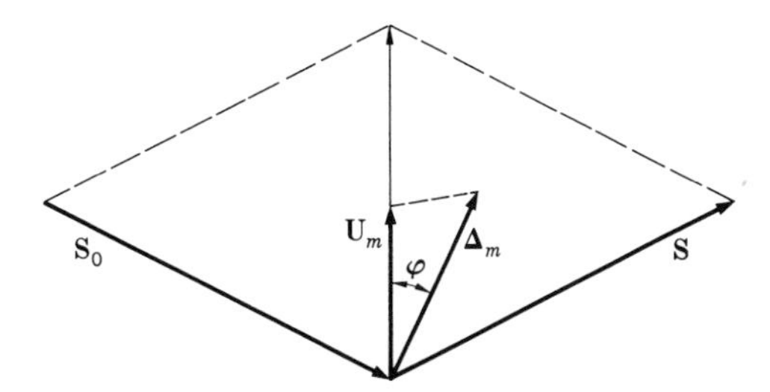

Fig. 10-4

to the usual expression for a crystal-reflection maximum. Also note that no assumption has been made so far regarding the nature or cause of the atomic displacements. This illustrates an important point, namely, that all random atomic displacements, no matter how produced, have an analogous effect on the diffraction maxima to that of the temperature effect here discussed.

It is convenient to relate the instantaneous displacements of the vibrating atoms to the diffraction vector $(\mathbf{S} - \mathbf{S}_0)$ by noting the angle φ that it forms with the displacement vector $\boldsymbol{\Delta}_m$ (Fig. 10-4). Accordingly, the dot products in the last exponent in (10-4) are of the type

$$\begin{aligned}(\mathbf{S} - \mathbf{S}_0) \cdot \boldsymbol{\Delta}_m &= |\mathbf{S} - \mathbf{S}_0||\boldsymbol{\Delta}_m| \cos \varphi \\ &= 2(\sin \theta) u_m \end{aligned} \tag{10-5}$$

where $u_m = |\boldsymbol{\Delta}_m| \cos \varphi$ is the projection of the displacement of the mth atom in the diffraction-vector direction. Using the previously defined constant

$$k = \frac{4\pi \sin \theta}{\lambda} \tag{8-7}$$

the intensity equation (10-4) can be written

$$I_{\text{eu}} = \sum_m \sum_n f^2 e^{(2\pi i/\lambda)(\mathbf{S}-\mathbf{S}_0)\cdot(\mathbf{R}_m - \mathbf{R}_n)} e^{ik(u_m - u_n)} \tag{10-6}$$

where u_m and u_n are the projected instantaneous displacements of atoms m and n, respectively. Note that these displacements are measured along a direction normal to the reflecting planes because parallel displacements do not affect the phase coherency for x-ray scattering by successive parallel planes.

The first part of (10-6), the regular crystal reflection, is independent of time, while the last term changes as the individual atomic displacements change with time. Actual intensity measurements are carried out over relatively long time intervals as compared with temperature-induced oscillations (phonon lifetimes), so that it is only meaningful to talk of a time average of this term. Let $\eta = k(u_m - u_n)$, so that

$$\begin{aligned} e^{ik\overline{(u_m - u_n)}^t} &= e^{i\overline{\eta}^t} = 1 + i\overline{\eta}^t - \frac{\overline{\eta^2}^t}{2} - \frac{i\overline{\eta^3}^t}{6} + \cdots \\ &= 1 - \frac{\overline{\eta^2}^t}{2} + \cdots \end{aligned} \tag{10-7}$$

since the odd powers average out to zero because plus and minus η values are equally probable. Note that the remaining even-power terms in (10-7) constitute the series expansion of $e^{-\frac{1}{2}\overline{\eta^2}^t}$. This suggests that the probability governing atomic displacements goes as $e^{-\eta^2}$, which is rigorously correct provided that the displacements are random. Accordingly,

$$\begin{aligned} \overline{e^{ik(u_m-u_n)}}^t &= e^{-\frac{1}{2}k^2\overline{(u_m-u_n)^2}^t} \\ &= e^{-\frac{1}{2}k^2(\overline{u_m^2}^t+\overline{u_n^2}^t-2\overline{u_m u_n}^t)} \\ &= e^{-k^2(\overline{u^2}^t-\overline{u_m u_n}^t)} \end{aligned} \qquad (10\text{-}8)$$

since the atoms at lattice points denoted by m and n are equivalent and have the same mean-square displacement $\overline{u^2}^t$.

Substituting (10-8) into the original expression for the diffraction intensity,

$$\overline{I_{eu}}^t = \sum_m \sum_n f^2 e^{(2\pi i/\lambda)(\mathbf{S}-\mathbf{S}_0)\cdot(\mathbf{R}_m-\mathbf{R}_n)} e^{-k^2(\overline{u^2}^t-\overline{u_m u_n}^t)} \qquad (10\text{-}9)$$

Taking all terms for which $m = n$ outside the double sum, one obtains

$$\overline{I_{eu}}^t = Nf^2 + \sum_{m \neq n} \sum{}' f^2 e^{(2\pi i/\lambda)(\mathbf{S}-\mathbf{S}_0)\cdot(\mathbf{R}_m-\mathbf{R}_n)} e^{-k^2(\overline{u^2}^t-\overline{u_m u_n}^t)} \qquad (10\text{-}10)$$

where the prime following the summation sign is a conventional way to indicate that the summation excludes terms having like dummy indices m and n. This expression is quite general, and has no assumptions underlying it, except that the crystal has a primitive cubic structure.

The evaluation of the temperature effect clearly depends on how the projected atomic displacements are treated. In Debye's original analysis, the assumption was made that the atoms vibrated independently of each other, so that $\overline{u_m u_n}^t = 0$ provided that $m \neq n$. Physically, this means that each atom vibrates with the same frequency, an assumption also made by Einstein in a formulation of the specific heat of crystals which preceded Debye's more sophisticated treatment. The time-averaged intensity becomes, in this case,

$$\overline{I_{eu}}^t = Nf^2 + e^{-k^2\overline{u^2}^t} \sum_m \sum_n{}' f^2 e^{(2\pi i/\lambda)(\mathbf{S}-\mathbf{S}_0)\cdot(\mathbf{R}_m-\mathbf{R}_n)} \qquad (10\text{-}11)$$

An examination of (10-11) shows that the double sum can be made complete again by restoring the N terms for which $m = n$, that is, by adding and subtracting the quantity $Nf^2e^{-k^2\overline{u^2}^t}$. Making this change in (10-11),

$$\begin{aligned} \overline{I_{eu}}^t &= Nf^2(1 - e^{-k^2\overline{u^2}^t}) + e^{-k^2\overline{u^2}^t} \sum_m \sum_n f^2 e^{(-2\pi i/\lambda)(\mathbf{S}-\mathbf{S}_0)\cdot(\mathbf{R}_m-\mathbf{R}_n)} \\ &= Nf^2(1 - e^{-k^2\overline{u^2}^t}) + e^{-k^2\overline{u^2}^t}[\text{CR}] \end{aligned} \qquad (10\text{-}12)$$

since the restored double sum above is the usual expression for a crystal reflection [CR].

The meaning of (10-12) is that temperature affects the diffraction of x rays by crystals in two ways: The crystal-reflection intensity is attenuated by an exponential temperature factor. Concurrently, an apparently monotonically varying intensity distribution is produced. This is the so-called temperature-diffuse scattering, usually abbreviated TDS. Defining a temperature factor

$$2B \equiv 16\pi^2\overline{u^2}^t \tag{10-13}$$

or alternatively,

$$2M \equiv 2B\frac{\sin^2\theta}{\lambda^2} = k^2\overline{u^2}^t \tag{10-14}$$

the time-averaged intensity, expressed in electron units, becomes

$$\overline{I_{eu}}^t = Nf^2(1 - e^{-2M}) + [\mathrm{CR}]e^{-2M} \tag{10-15}$$

The separate terms in (10-15) are plotted in Fig. 10-5, after dividing both sides by f^2. Note that the temperature factor e^{-2M} declines rapidly with increasing scattering angle and causes the reflection peaks to decline correspondingly. (The dashed peaks indicate I_{eu}/f^2 in the absence of the temperature effect.) Otherwise the peaks remain unchanged. The temperature-diffuse scattering varies monotonically according to (10-15), increasing as θ increases.

Correction for temperature effect. The temperature factor $2B$ can be evaluated according to the specific-heat theory of Debye from the following expression. For a simple cubic structure having one atom per lattice point of a primitive lattice, this factor, first derived by Debye and later corrected by Waller,

$$2B = \frac{12h^2T}{m_Ak\Theta^2}\,Q\left(\frac{\Theta}{T}\right) \tag{10-16}$$

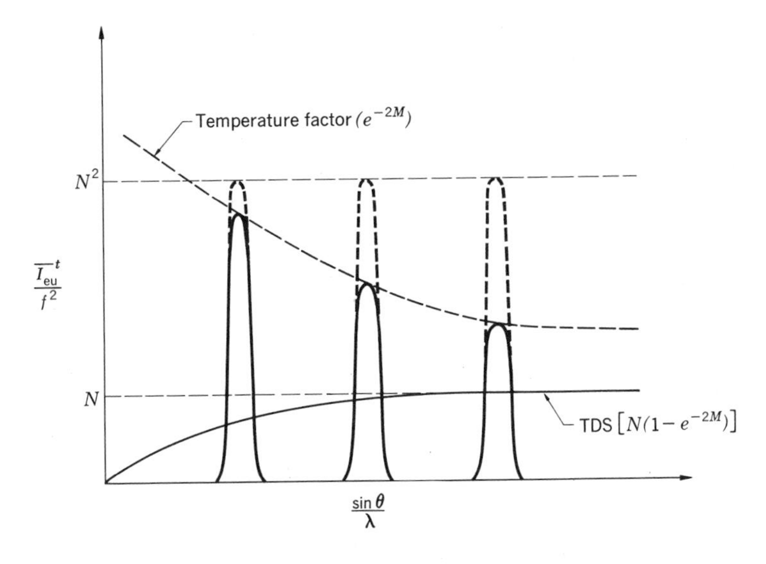

Fig. 10-5

where m_A is the atomic mass; k is Boltzmann's and h is Planck's constant; T is the absolute temperature; and Θ is the Debye characteristic temperature for the crystal. The function $Q(\Theta/T)$ was tabulated by Debye, and is very nearly equal to unity for $T > \Theta$, which is the case at room temperature for most crystals. It declines rapidly below the characteristic temperature, similarly to the decline of the specific heat. Since all the quantities in (10-16) are known, including the characteristic temperature for many crystals, it is possible to calculate the temperature factor and to determine the amount by which crystal reflections are attenuated (Exercise 10-3). By combining (10-16) with (10-13), it is also possible to estimate the magnitude of the mean-square displacements of the atoms.

It is equally easy to determine the Debye-Waller factor experimentally. The integrated reflecting power of each reflection is attenuated by the appropriate temperature factor, so that when (9-47) applies,

$$R_{hkl} = Q_{hkl}\,\delta V\,e^{-2M} \tag{10-17}$$

Rearranging the terms, inserting (10-14) for $2M$, and taking natural logarithms of both sides,

$$\ln\frac{R_{hkl}}{Q_{hkl}\,\delta V} = -2B\,\frac{\sin^2\theta}{\lambda^2} \tag{10-18}$$

When the crystal structure is known, it is possible to calculate all the terms in the denominator on the left side, in (10-18), so that a plot of the logarithm of the ratio of experimental to calculated reflecting powers, evaluated at different glancing angles, that is, different $(\sin^2\theta)/\lambda^2$ values, should be a straight line of slope $-2B$ (Fig. 10-6). In practice, a straight line results when cubic crystals having simple structures are thus examined. Usually, several orders of the same reflection are measured at different temperatures, so that a suitably cut large crystal can be used in the experiments. By this means, one can obtain an x-ray value for the Debye characteristic temperature which compares reasonably well with corresponding values obtained from specific-heat or related measurements. Similarly, it is possible to deduce the mean-square displacement induced by heating the crystal. Note that the mean-square vibration amplitude deduced from (10-13) gives the component of the average displacement normal to the reflecting planes. If the displacements are isotropic, as assumed in the above theory, the actual atomic

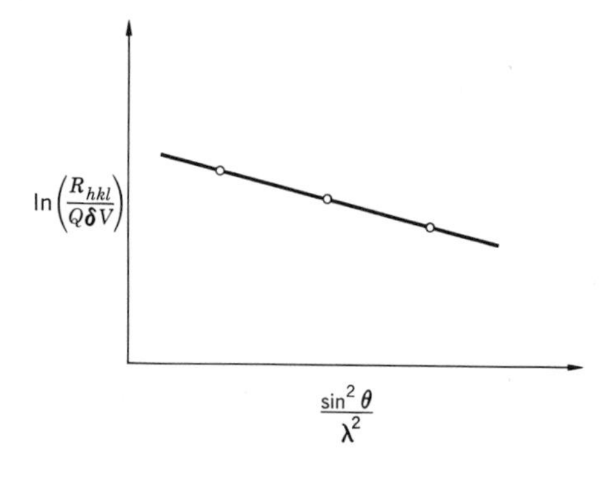

Fig. 10-6

displacements Δ_m can be deduced from their projected values u_m, since

$$\begin{aligned}\overline{u_m{}^2}^t &= \overline{|\Delta_m|^2}^t\,\overline{\cos^2\varphi}^t \\ &= \tfrac{1}{3}|\Delta_m|^2 \end{aligned} \tag{10-19}$$

It should be clearly realized, however, that the foregoing procedure is based on an assumption of independent isotropic atomic vibrations, which is known to be incorrect. Thus the above analysis is really limited only to order-of-magnitude estimates. A more involved analysis, based on the theory developed to explain temperature-diffuse scattering, yields physically more meaningful values for the atomic displacements. Even this theory, however, is limited at the present time to cubic crystals having simple structures.

In most measurements of x-ray diffraction by crystals, the integrated intensities are rarely measured on an absolute basis. Thus it is not possible to make use of relation (10-18) directly. It is possible, however, to convert the integrated intensities measured on a relative basis to structure-factor amplitudes, also on a relative basis, using the appropriate formula from Chap. 9. The measured, or observed, structure factors F_o are related to those calculated on an absolute basis (for atoms that are at rest) F_c by

$$F_o{}^2 = KF_c{}^2 e^{-2B\frac{\sin^2\theta}{\lambda^2}} \tag{10-20}$$

where K is a scale constant. Rearranging the terms and taking natural logarithms of both sides, an expression similar to (10-18) results.

$$\ln\frac{F_o{}^2}{F_c{}^2} = \ln K - 2B\frac{\sin^2\theta}{\lambda^2} \tag{10-21}$$

A plot of the left side of (10-21) as a function of $(\sin^2\theta)/\lambda^2$ (Fig. 10-7) yields a straight line of slope $-2B$ whose intercept on the ordinate at $\theta = 0°$ determines the scale factor K. This procedure, based on one first devised by A. J. C. Wilson, is commonly used to scale experimentally measured intensities before comparing them with calculated values. When (10-21) is plotted using experimental data obtained from a crystal containing different kinds of atoms, the plotted points often fail to lie on a straight line. One physical reason for such deviations is the fact that

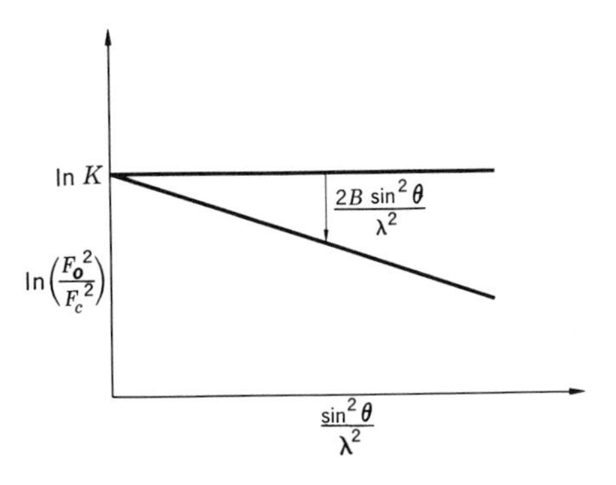

Fig. 10-7

unlike atoms in a crystal do not have identical vibration amplitudes since these are determined by the atomic masses, interatomic forces, and so forth. Thus, not only is it necessary to apply different corrections to each kind of atom present, but the necessary corrections are not known in advance. Assuming that the temperature effect causes a decrease in each atom's scattering amplitude by an amount correctly expressed by e^{-M}, it is possible to make an allowance for this effect by including such terms directly in the structure-factor calculations; that is,

$$F_c = f_1 e^{\varphi_1} e^{-M_1} + f_2 e^{\varphi_2} e^{-M_2} + \cdots + f_N e^{\varphi_N} e^{-M_N} \tag{10-22}$$

where φ_n represents the phase factor $2\pi(hx_n + ky_n + lz_n)$, and M_n is the appropriate Debye-Waller factor for the nth atom in the unit cell. The values for these temperature factors needed to provide the best agreement between observed and calculated structure-factor values can be determined by iteration, most commonly making use of least-squares procedures. (See Chap. 11.) In a further refinement of this procedure, three temperature factors are assigned to each atom in (10-22), one for each of three mutually orthogonal vibration directions. This then allows correction for anisotropic thermally induced vibrations. Although the physical principles on which these refinements are based are quite sound, it is uncertain to what degree the accuracy with which x-ray diffraction intensities are usually measured justifies their utilization in determining actual atomic displacements. It is undoubtedly valid to establish relative atomic displacements in this way, both along different crystallographic directions and for the same atomic species in different environments. Such analyses can be particularly useful when repeated at several different temperatures.

Temperature-diffuse scattering. According to the original Debye theory, the TDS intensity increases monotonically with the scattering angle, as depicted in Fig. 10-5. Such a monotonic increase is not observed in practice, and is a direct consequence of the wrong assumption made in dropping the cross term $\overline{u_m u_n}^t$ in the exponent of the temperature factor in (10-9). To see what effect restoring this term has, suppose that it is possible to relate this cross term to the mean-square displacement $\overline{u^2}^t$ by means of an interaction coefficient α_{mn} which properly represents the interaction between atoms m and n. Without specifying its actual functional dependence, $\alpha_{mn} = 1$ corresponds to zero interatomic separation, that is, when $m = n$ and the temperature factor vanishes (exponential equals unity) in (10-9), and $\alpha_{mn} \rightarrow 0$ as the distance separating atom m from n increases. With this equivalence, the omitted exponential term

$$e^{k^2 \overline{u_m u_n}^t} = e^{k^2 \alpha_{mn} \overline{u^2}^t} = 1 + k^2 \alpha_{mn} \overline{u^2}^t = 1 + 2M\alpha_{mn} \tag{10-23}$$

recalling the definition of the Debye-Waller factor in (10-14) and assuming that the mean-square displacement is a small quantity. Substituting this term in the original intensity expression (10-9),

$$\overline{I_{\text{eu}}}^t = \sum_m \sum_n f^2 e^{(2\pi i/\lambda)(\mathbf{S}-\mathbf{S}_0)\cdot(\mathbf{R}_m - \mathbf{R}_n)} e^{-2M}(1 + 2M\alpha_{mn}) \tag{10-24}$$

As before, the intensity expression can be divided into two parts, an ordinary crystal reflection and the TDS.

$$I_{\text{eu}} = [\text{CR}]e^{-2M} + 2Me^{-2M} \sum_m \sum_n f^2 \alpha_{mn} e^{(2\pi i/\lambda)(\mathbf{S}-\mathbf{S}_0)\cdot(\mathbf{R}_m - \mathbf{R}_n)} \tag{10-25}$$

Note that the crystal reflection is attenuated by a temperature coefficient e^{-2M}, just as predicted by the "incorrect" theory developed in an earlier section. Thus the preceding discussion of how to correct ordinary crystal reflections for the temperature effect needs no modification. This is not true of the temperature-diffuse scattering. To facilitate comparison of the TDS intensity, note that the coefficient of the second term in (10-25) has the form $xe^{-x} = (e^x - 1)e^{-x} = 1 - e^{-x}$ for small values of $x = 2M$. Making this substitution, the TDS intensity

$$I_{\text{TDS}} = f^2(1 - e^{-2M}) \sum_m \sum_n \alpha_{mn} e^{(2\pi i/\lambda)(\mathbf{S}-\mathbf{S}_0)\cdot(\mathbf{R}_m - \mathbf{R}_n)} \tag{10-26}$$

It is clear from the above expression that I_{TDS} is not a monotonically varying function when interactions between the atoms are included. Moreover, it is evident that the TDS intensity has maxima at the same reciprocal-lattice points as the regular crystal reflections. In fact, except for the interaction coefficient α_{mn}, the double sum is exactly like that in the [CR] term. The nature of α_{mn} therefore becomes important in predicting the actual shape of the TDS curve. It is considered to be outside the scope of this book to describe the details of theories that have been developed to date. They are largely limited to simple cubic crystals, and take account of anisotropies of atomic vibrations, as indicated by the elastic constants of such crystals. A theory developed for molecular crystals by W. Cochran and G. S. Pawley makes use of elastic constants and Raman frequencies to determine the molecular-interaction constants. (See the literature list at the end of this chapter.) A few general observations can be made, however, without recourse to these theories. In the case of simple metals, the coefficient α_{mn} decreases rapidly as the separation between atoms m and n increases, since it has been established that interatomic forces become vanishingly small by the time third-nearest neighbors are considered. It is a simple matter to prove graphically that it is these higher-index terms in the double sum of (10-26) that contribute most to the sharpening of the maxima (Exercise 10-5). In their absence, the TDS peaks are broad, and exemplify the kind of broadening that results whenever atoms are randomly displaced from their correct sites in a crystal. If the vibration is anisotropic, for example, when the restoring forces are weaker along certain directions, the TDS intensity distribution about the reciprocal-lattice points becomes anisotropic also. It is possible to deduce the nature of this anisotropy, therefore, by studying the detailed variation of the TDS intensity.

The two terms in (10-25) are plotted in Fig. 10-8. It is clear that the crystal reflections decrease in intensity with increasing scattering angle θ. Concurrently, the diffuse scattering increases, leading to a kind of conservation "law" for the total intensity of scattering (Exercise 10-6). Although the crystal reflection is not actually broadened by the temperature effect, it is superimposed on a diffuse-scattering maximum, which causes it to have a very broad base (Fig. 10-9). When

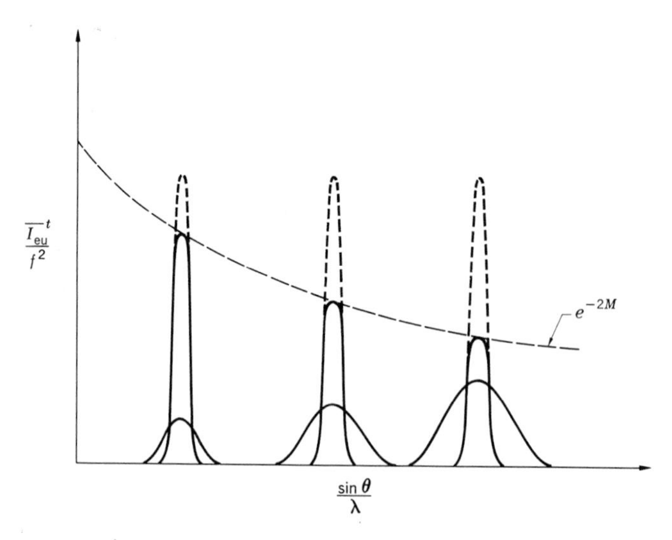

Fig. 10-8. The sharp maxima represent the regular crystal reflections, and the broad maxima are the result of TDS.

the temperature effect is appreciable, therefore, the difficulties inherent in separating [CR] from TDS intensities become magnified. This does not affect TDS studies because they can be carried out using intensities sufficiently removed from the reciprocal-lattice points to contain no crystal-reflection contributions. The reverse does not hold, however, because the true peak shape of the TDS at the reciprocal-lattice point is rarely known exactly. Still another interesting consequence of TDS is illustrated in Fig. 10-10, which shows a symmetric distribution of intensity about each point in the reciprocal lattice. Because the intensity is now spread out over a considerable volume, the sphere of reflection intersects far more "points" than it does in the absence of TDS. This is often the reason why

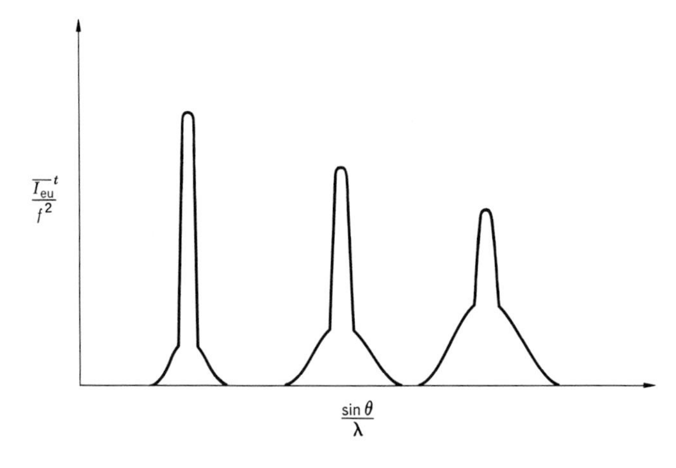

Fig. 10-9

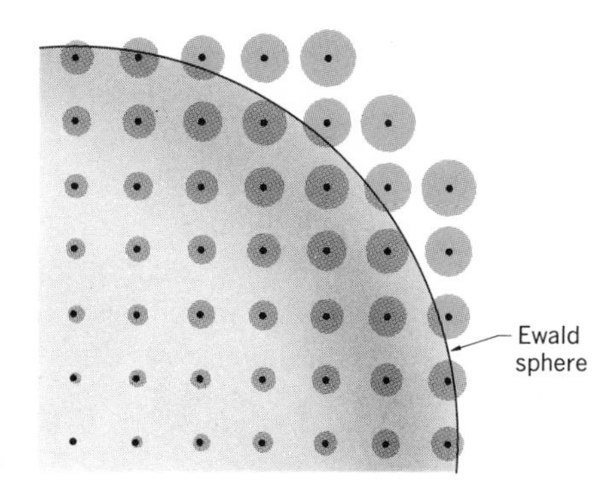

Fig. 10-10

diffuse reflections appear in x-ray photographs in locations where crystal reflections are not ordinarily expected. As shown in the concluding sections of this chapter, however, TDS is not the only cause of diffuse maxima in reciprocal space.

NEARLY PERFECT CRYSTALS

The study of "static" imperfections present in crystals can be divided into two parts; an examination of relatively perfect crystals and of crystals containing large defect densities. Experimental procedures have been developed for detecting the presence of defects in large, relatively perfect single crystals. Some of these methods actually are capable of recording images of individual defects of a crystal surface, and are sometimes called *topographic methods* for that reason, while others rely on more or less rigorous models for their interpretations. Because the theory of x-ray diffraction is far more complex for very perfect crystals, a rigorous theoretical basis for most of these methods is not yet available. Far greater advances have been made in developing rigorous theories to explain defects present in relatively imperfect crystals because these can be treated by the simpler kinematical theory. A representative sampling of some of these theories is presented in later sections of this chapter.

The methods for observing defects in nearly perfect crystals can be classified, according to the experimental arrangement or the physical principles utilized, into the following six groups:

1. Two-crystal diffractometry
2. X-ray reflection microscopy
3. X-ray reflection microradiography
4. Diverging-beam transmission arrangements
5. Anomalous-transmission studies
6. Electron-density variations

Categories 2 and 3 contain the topographic methods most commonly employed. Some of the procedures described in categories 1, 4, and 6 are also applicable to polycrystalline aggregates, and such extensions are noted below.

Two-crystal diffractometry. After the advantages of a two-crystal instrument were first pointed out by Compton in 1917, its utilization in the study of x-ray spectra and x-ray diffraction intensities (Chap. 6) grew in popularity. Probably its main advantage is that the beam incident on the sample crystal has a spectral and geometrical width accurately defined by the first crystal. This requires, however, that a relatively perfect crystal be selected for the monochromator, so that, historically, the first studies of crystal perfection took place in the course of seeking suitable monochromators. The usual procedure employed was to examine the halfwidth of the selected reflection, since it was desired to have as parallel a beam and as narrow a spectral distribution as possible. According to the kinematical theory discussed in the preceding chapter, the halfwidth can be related to the relative misalignments of the optically independent regions, or blocks, in the crystal by a relation like (9-89). To bring this model up to date, suppose that the blocks are divided from each other by small-angle boundaries spaced t_0 apart and relatively tilted by an amount Δ. Then, if the boundaries are comprised of parallel-edge dislocations having a Burger's vector b, their density D is given by

$$D = \frac{\Delta}{bt_0} \tag{10-27}$$

as first shown by P. Gay, P. B. Hirsch, and A. Kelly in 1953. Three years later, A. D. Kurtz, S. A. Kulin, and B. L. Averbach made use of this relation to interpret, similarly, the halfwidths of 111 reflections from germanium crystals. When the true halfwidth $\epsilon_{\frac{1}{2}}$ of a reflection measured on a two-crystal diffractometer is used, the dislocation density, according to (10-27), is given by

$$D = \frac{\epsilon_{\frac{1}{2}}{}^2}{9b^2} \tag{10-28}$$

assuming that the block-tilt distribution (9-89) is, in fact, Gaussian. The actual germanium crystals studied were found to contain 10^6 to 10^7 cm^{-2} dislocations, in agreement with etch-pit counts on the same crystals. The limitation of this interpre-

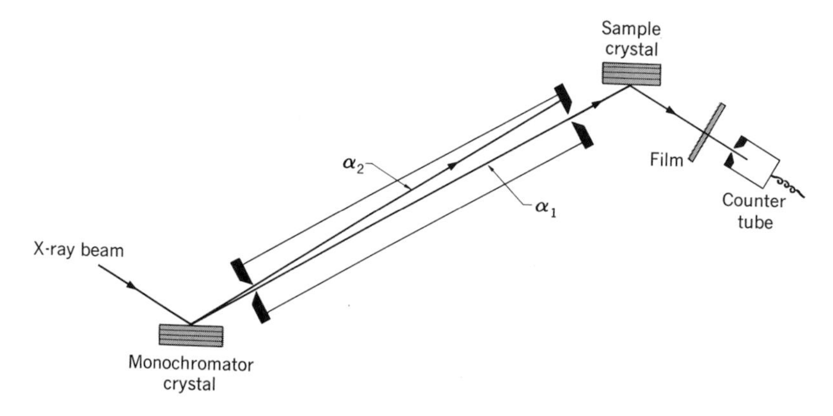

Fig. 10-11. Two-crystal diffractometer arrangement used by Weissman.

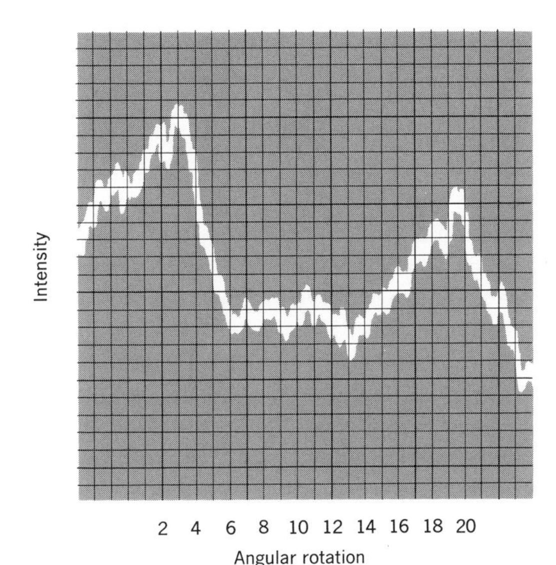

Fig. 10-12. Intensity variation of 200 reflection from aluminum. [*From J. Intrater and S. Weissman, Acta Crystallographica, vol.* 7 (1954), *pp.* 729–732.]

tation is that it presupposed a block model not actually present in the crystal, as clearly demonstrated by the very same etch pits which typically do not show the formations characteristic of a network of small-angle boundaries. Moreover, it does not take into account other defects that may be present and modify the halfwidth. In a later study of more perfect germanium crystals, Batterman concluded that dislocation densities less than 10^5 cm^{-2} do not affect either the widths or the magnitudes of the reflections in a two-crystal diffractometer.

Occasionally, a crystal may consist of fairly large blocks, in which case its reflection peak consists of several slightly displaced maxima instead of a smooth

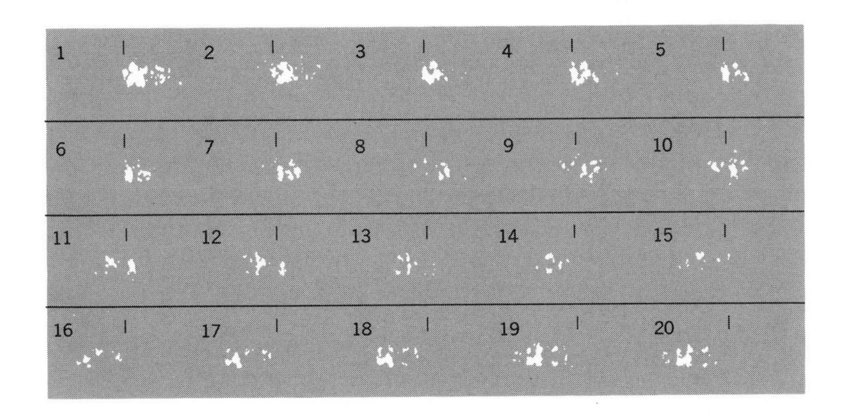

Fig. 10-13. Correlation of intensity variations of 200 reflection from an aluminum crystal (Fig. 10-12) to the reflecting regions in the crystal, recorded photographically. The numbers refer to angular rotations in multiples of 1 minute of arc. The small vertical line is a fiduciary marker, identifying a fixed point on the crystal. [*From J. Intrater and S. Weissman, Acta Crystallographica, vol.* 7 (1954), *pp.* 729–732.]

continuous curve. The prominence of such structure depends on the relative number of blocks in the irradiated area. Suppose a pair of slits are interposed between the two crystals so that the α_2 component of the x-ray beam diffracted by the first crystal is prevented from reaching the second crystal. When the dispersion in the monochromator crystal is sufficiently large, it is possible thereby to reduce the cross section of the beam incident on the second crystal (Fig. 10-11), while concurrently sharpening its spectral width. The increased resolving power of such an arrangement has been used by Weissman and his coworkers to correlate the structure observed in the reflection peak (Fig. 10-12) to the reflecting region on the crystal's surface. As the glancing angle of the second crystal is changed by uniform angular increments, the reflection is first recorded on a film and then by the counter (Fig. 10-11). Between rotation, the film is advanced so that the various portions in the crystal, whose reflection intensity is recorded by the counter, are photographed side by side, as indicated in Fig. 10-13. The combined intensity profile and photographic record provide a complete visual description of the distribution of reflecting planes throughout the irradiated-crystal volume. An examination of the photographs actually obtained by this method (Fig. 10-13) clearly shows that even relatively imperfect crystals do not contain a regular block structure, but that the coherently diffracting regions have very irregular shapes and extend through all parts of the irradiated crystal. Incidentally, it should be noted that, because of the high resolving power of the arrangement shown in Fig. 10-11, the sample crystal can be replaced by a polycrystalline aggregate, and the reflection from individual crystallites similarly can be analyzed. Quite recently, the possibility of adding a third crystal to this arrangement has been suggested in order to analyze the beam diffracted by the sample crystal. The use of such a three-crystal diffractometer offers a number of interesting possibilities.

X-ray reflection microscopy. Reflections from different parts of a single crystal also may be observed without using a two-crystal diffractometer. Suppose polychromatic x rays emanating from a point source fall on the face of a large crystal, as shown in Fig. 10-14. If the reflecting planes are truly parallel to the surface, the reflected rays are dispersed uniformly. If the crystal contains imperfections which cause adjacent regions to become tilted, the reflected rays tend to bunch together, producing an uneven intensity distribution on the film. Such an arrangement was first used by W. Berg in 1931 to study the cleavage surfaces of natural crystals, in an attempt to correlate the recorded intensity distributions with changes produced in the crystal by a previous mechanical deformation.

The use of polychromatic radiation has the disadvantage that more than one

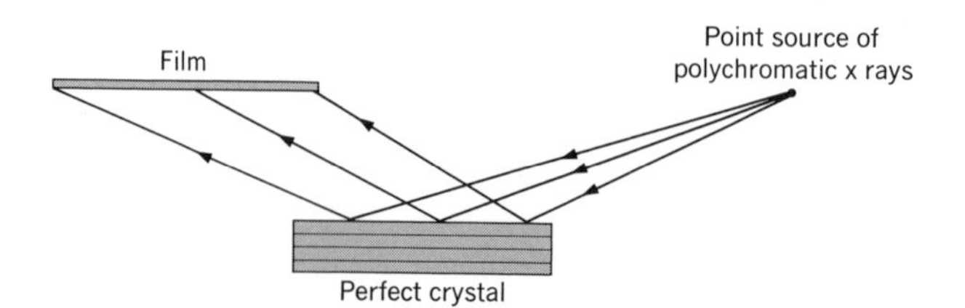

Fig. 10-14. Point-source arrangement of Berg.

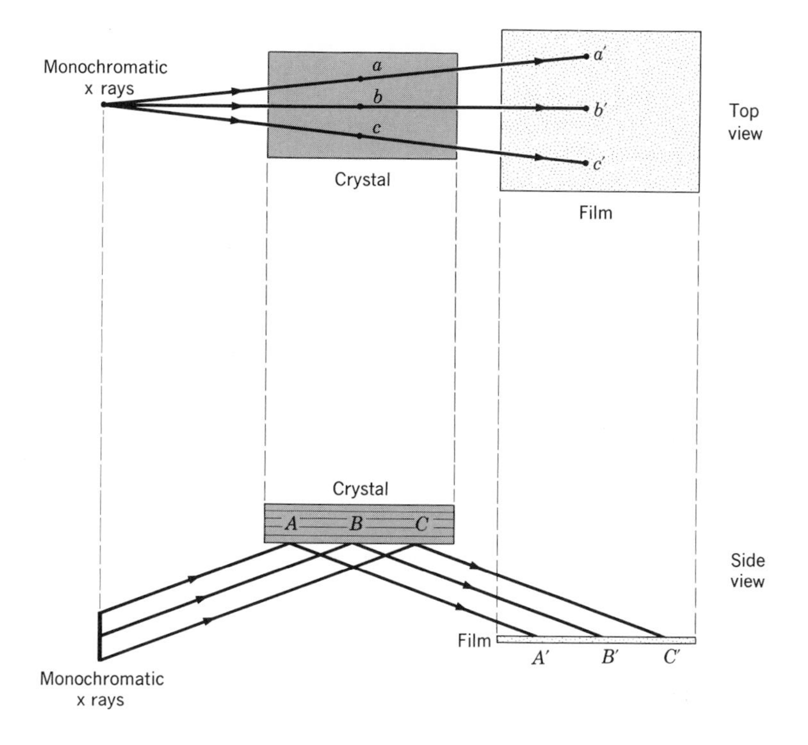

Fig. 10-15. Parallel-ray (monochromatic) arrangement of Berg.

set of planes can give rise to reflected rays. To overcome the interpretive difficulty caused by superimposed reflections, Berg described another arrangement, employing a line source of monochromatic radiation. A ray diagram of this arrangement is shown in Fig. 10-15. Although the rays are drawn parallel to each other, a slight amount of divergence, defined by the source size and slits used, is nevertheless present in the direct beam. As in Fig. 10-14, slight deviations from parallelism cause the reflected rays to become distributed unevenly. This arrangement was rediscovered in 1945 by C. S. Barrett, who suggested that the crystal and the film may be oscillated in synchronism to permit more of the surface layers to reflect x rays. He concluded that optimum resolution can be obtained if the x-ray source-to-crystal distance is large (over 1 ft) and the crystal-to-film distance is very small (about 1 mm) and if long-wavelength radiation is employed. The Berg-Barrett arrangement has become fairly popular because it does not require specialized x-ray diffraction equipment and permits an accurate correlation of substructure and crystal topography. As shown next, individual dislocations can be seen by this means.

After a very careful examination of the Berg-Barrett arrangement, J. B. Newkirk concluded that the principal cause of the unequal intensity distribution in an x-ray reflection micrograph obtained with a slightly diverging beam was a variation in the primary extinction in different parts of the crystal, rather than relative tilts of the reflecting planes. The decrease in extinction caused by strains surrounding dislocations thus renders them visible in the micrograph. It can be shown, furthermore, that when the Burger's vector of a dislocation is perpendicular to the reflecting planes, it produces a maximum distortion normal to these planes. This results in a decrease in the extinction effect for such planes. By carefully comparing the x-ray reflection micrographs from different planes belonging to the same family, for example, (110), $(1\bar{1}0)$, (101), etc., it is possible to detect reflection contrasts, which, in turn, allow one to determine the distribution of imperfections throughout the crystal. Such procedures have been employed to study various crystals by numerous investigators.

An arrangement similar to the one shown in Fig. 10-11 was used by U. Bonse to observe individual dislocations in reflection micrographs and to study the concomitant reflection-intensity profiles. He observed that the contrast produced by a dislocation was different for different portions of the reflection curve. Making use of elasticity theory, Bonse calculated the effect that the strain field surrounding a dislocation should have. According to these calculations, the twists of the reflecting planes and small variations in their interplanar spacings should displace the reflection maximum slightly. As borne out by his measurements, such displacements are actually observed. Analyses such as these, hopefully, will lead to the development of a more complete understanding of the physical basis for the hypothetical Darwin model for a crystal.

X-ray reflection microradiography. Suppose that, instead of reflecting a parallel beam from the surface of a crystal, a set of planes inside the crystal is similarly examined. This is, of course, the Laue, or transmission, case, except that, when a high-resolution arrangement like one of those described above is used, it is apt

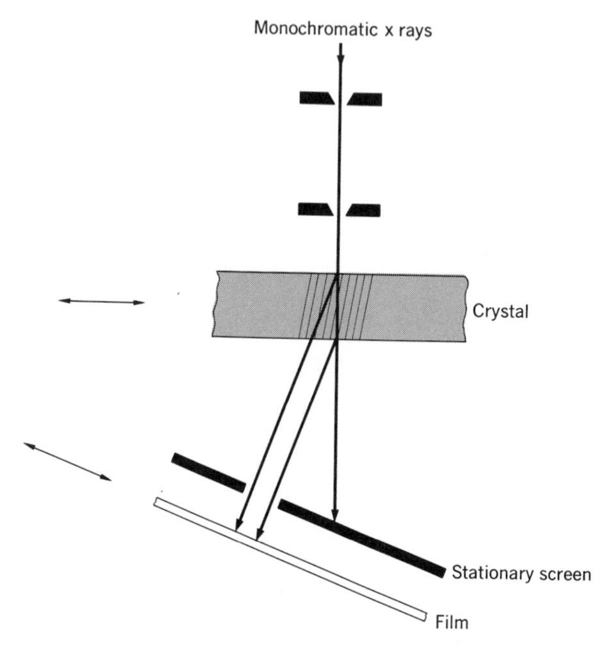

Fig. 10-16. Transmission arrangement of Lang.

Fig. 10-17. Dislocations in a silicon single crystal. Fringes visible on the lenticular edge of the crystal are caused by the pendulum effect described in Chap. 9. [*From A. R. Lang, Journal of Applied Physics, vol.* 30 (1959), *pp.* 1748–1755.]

to call the arrangement reflection microradiography, to distinguish it from reflection microscopy. Such a transmission arrangement was developed by A. R. Lang in 1958, and is shown in Fig. 10-16. A narrow monochromatic beam of x rays irradiates a set of more or less transverse planes lying in the irradiated section of the crystal. A high-resolution photographic film or plate is placed close to the crystal and at right angles to the diffracted beam. An absorbing screen is interposed to prevent the direct beam from striking the film. In order to examine a larger area of the crystal, it is oscillated back and forth, transversely to the incident beam, as indicated by the arrows on the left in Fig. 10-16. The film is coupled to the crystal so that they oscillate in synchronism, with the result that the defect distribution throughout a reasonably large crystal slab can be examined. As can be seen in the representative photograph shown in Fig. 10-17, the "shadows" of individual dislocations are clearly revealed. This is a fairly potent topographic method whose principal interpretive difficulty arises from the superposition of large numbers of dislocations. Lang has suggested that stereoscopic pictures can be prepared by recording pairs of x-ray reflection microradiographs from (hkl) and $(\bar{h}\bar{k}\bar{l})$ planes. Unfortunately, the stereoscopic angle is fixed at 4θ, which usually does not coincide with the proper viewing angle for humans.

Diverging-beam transmission arrangements. Suppose that polychromatic x rays emanating from a point source impinge on a thin crystal slab in the Laue arrangement. As can be seen in Fig. 10-18, a focusing condition exists, since each reflecting plane selects the appropriate wavelength component and reflects it at the Bragg angle to the same point on the focusing circle. Provided all the planes are parallel, the reflected beam should reproduce an accurate image of the point source at the focal point. If the planes are not perfectly parallel, for example, if the crystal contains groups of planes that are slightly tilted with respect to each other, each group of parallel planes will cause a focus at slightly displaced points along the circle in Fig. 10-18. Such an arrangement was first used by A. Guinier and J. Tennevin in 1949 to study the result of heat-treating metal crystals. They observed the formation of small coherent domains separated from each other by polygonization boundaries. According to dislocation theory, such plane defects

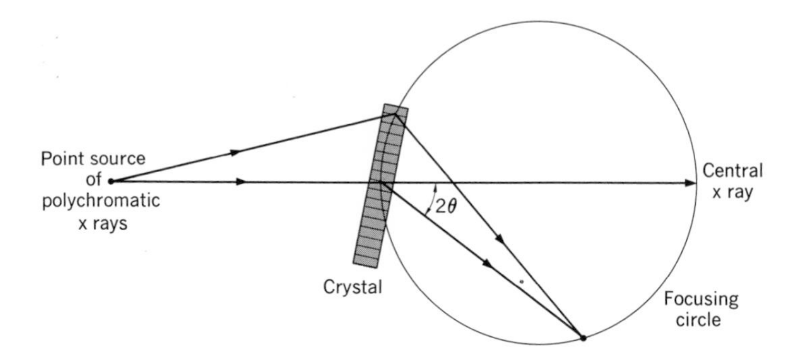

Fig. 10-18

are formed when thermally activated dislocations move together and decrease the crystal's strain energy by forming small-angle boundaries. When the individual grains are large enough, similar effects can be observed in polycrystalline metals also.

An interesting feature of the arrangement shown in Fig. 10-18 is that, when the reflecting planes in the crystal are relatively parallel, the spectral width of the reflected beam is extremely narrow. In fact, for the 111 reflection from a highly perfect silicon crystal, the effective width may be less than the natural linewidth of characteristic radiation. It becomes a simple matter, in this case, to select any wavelength component from the continuous spectrum by simply changing the inclination of the crystal slab relative to the central ray. This procedure was used to study the wavelength dependence of extinction, described in Chap. 9. It also constitutes an easy procedure for obtaining the integrated intensity from small regions in a crystal slab, and by comparing the intensity reflected by adjacent portions of the crystal (Fig. 9-18), to deduce a relative "figure of merit" for each. Such an analysis of a silicon ingot was carried out by Azároff and Bragg in 1957. The resulting distribution is displayed in Fig. 10-19, which shows a sketch of the way

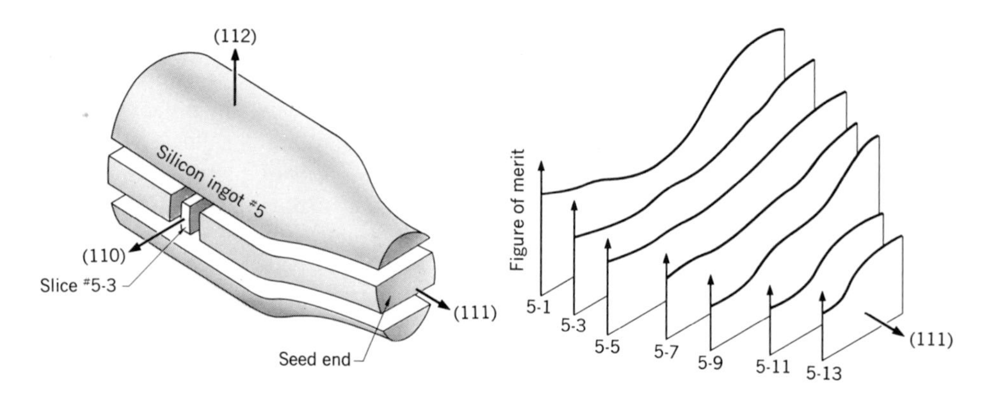

Fig. 10-19. Figure of merit related to the perfection of a silicon boule. *(After Azároff and Bragg.)*

Fig. 10-20. Pseudo-Kossel lines in a back-reflection photograph of a nearly perfect corundum crystal. Discontinuities in the ellipses appear when localized stresses are present in a crystal. *(Photograph by S. Weissman.)*

the ingot was sectioned and the figure of merit (integrated intensity) determined throughout each crystal slice. It should be noted that it is possible to make use of (10-27) and the linewidths measured by this arrangement to deduce dislocation densities. As in the case of the two-crystal arrangement, however, it is questionable whether such estimates have a physical meaning beyond that of a figure of merit.

There is still another way in which diverging x radiation can be utilized in crystal-perfection studies, particularly for crystals of "intermediate" perfection. Suppose a point source of x rays is placed in contact with the surface of a crystal. This can be done, for example, by bombarding a thin metal foil, placed in contact with the crystal, with a focused beam of electrons or x rays so that the characteristic x rays diverge from a point on the crystal's surface. A film placed against the opposite side of such a crystal then should be uniformly irradiated by the diverging rays. The crystal subtends a solid angle of π so that x rays traveling at certain angles are diffracted by the oblique planes present in the crystal. This produces a "deficiency" of x rays in the original cone formed by these rays, so that a film placed behind the crystal records light conical sections wherever such rays have been removed from the diverging beam. If the x rays are diffracted in the direction of the film, a more intense conical section is also recorded. [These conical sections are similar in appearance (Fig. 10-20) to the Kossel lines observed in electron diffraction.] This interpretation of the origin of the x-ray analog of Kossel lines was originally proposed by K. Lonsdale, who pointed out that, when the crystal is highly perfect, primary extinction limits the reflecting region to a very thin layer near the crystal's surface. Conversely, when the crystal is extremely imperfect, the diffracted beams do not form sufficiently sharp cones and disappear in the background irradiation of the film. The appearance of such lines therefore serves as an indication of the relative perfection of a crystal. It turns out, moreover, that small elastic stresses present in the crystal cause small changes in the lattice constants of certain planes and produce corresponding shifts in the line positions on the film. It thus becomes possible to make use of this fact to study the magnitude and distribution of such stresses in single crystals.

Fig. 10-21. Anomalous-transmission photograph of a silicon single crystal. [*From G. Borrmann, W. Hartwig, and H. Irmler, Zeitschrift für Naturforschung, vol.* 13 (1958), *pp.* 423–425.]

Anomalous-transmission studies. The discovery of anomalous transmission of x rays by crystals and the basic conditions required for this phenomenon to take place were briefly outlined in Chap. 9. It will be recalled that when a crystal is set to reflect a nearly parallel beam of monochromatic x rays in the Laue arrangement, the x-ray energy can be pictured as flowing along the atomic planes. If any atoms are displaced from these planes, say, by dislocations present in the crystal, they become displaced from the nodes (or antinodes) of the traveling-wave fields (Fig. 9-21), and discontinuities are observed in the anomalously transmitted beams. If the dislocation density in the crystal is quite low, that is, if it approximates an ideally perfect crystal, the individual dislocations become clearly visible, as can be seen in Fig. 10-21.

It can be shown that the optimum thickness of crystals examined by the normal transmission methods is proportional to $1/\mu$. Preparation of reflection microradiographs in heavily absorbing crystals, using the Lang arrangement, thus requires very thin crystals. The dislocation arrays in such crystals may not be typical of larger crystals, and their handling during the experiment sometimes causes the imperfections present to change in unknown ways. To study the effect of neutron irradiation in copper crystals, F. W. Young, Jr., and coworkers decided to combine the Lang arrangement with the Borrmann effect. Provided that the crystal has the requisite perfection, the reflected-beam component suffering slight attenuation can be used to observe any dislocations lying in the reflecting planes, even in relatively thick crystals. A reflection microradiograph of a highly perfect copper crystal is shown in Fig. 10-22. As in all topographic methods, the presence of large numbers of dislocations complicates the interpretation, but the ability to examine relatively thick crystals is an important virtue of this particular arrangement.

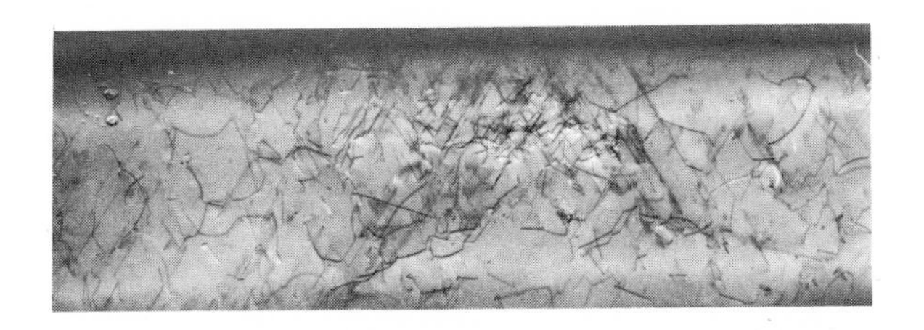

Fig. 10-22. X-ray topograph (enlarged about nine times) of copper crystal 0.4 mm thick, using the diffracted Ag K_α beam in an anomalous-transmission arrangement. *(Photograph by F. W. Young, Jr.)*

The two rays emerging after anomalous transmission through a perfect crystal bear a definite phase relationship to each other, as discussed in Chap. 9. Suppose that an identical second crystal is placed parallel to the first, as shown in Fig. 10-23. The two rays diffracted by the first crystal, called a beam splitter by U. Bonse and M. Hart, who first successfully tested this arrangement, meet the second crystal at essentially the correct angle for diffraction by it. Both rays, therefore, are split into two rays in turn. Note that this has the effect of bending two of the rays back in such a way that their paths must intersect. In effect, the second crystal plays the role of a dual mirror that reflects the two rays toward each other. If a third parallel crystal then is placed at their mutual intersection, both rays set up superimposing fields in this crystal. Keeping in mind that the original phase relation between the two rays emerging from the first crystal has not been destroyed by the second crystal, the two rays recombining in the third crystal can interfere with each other in a predictable way. Thus the arrangement in Fig. 10-23 acts like an interferometer for x rays, and the third crystal is properly called an analyzer. As Bonse and Hart have demonstrated, insertion of an optical wedge in either ray path causes a phase shift that can be clearly detected by the analyzer crystal. This makes it possible to analyze the perfection of crystals by observing how they influence the coherent beams passing through them. It also enables the study of a variety of mechanical deformations, e.g., the effect of force couples and other stress-inducing processes.

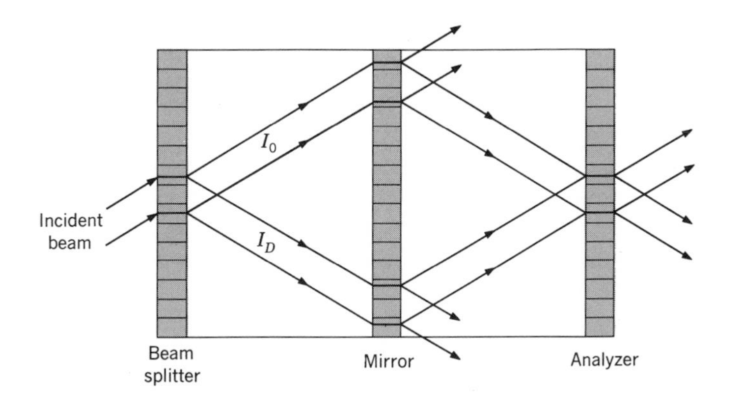

Fig. 10-23. X-ray interferometer devised by Bonse and Hart from a single crystal of silicon. The ray paths at each parallel crystal section are indicated. A more complete description of this instrument can be found in *Applied Physics Letters, vol.* 6 (1965), *pp.* 155–156.

Electron-density variations. As stated in the introductory parts of this chapter, point defects in crystals generally affect the intensities of x-ray reflections because they alter the structure-factor values. For example, when one kind of atom is substituted for another in an isomorphous series, the change in the scattering-factor magnitude causes a corresponding change in the structure factor. In this context, a vacancy can be thought of as an atom with zero scattering power. An interstitial atom, on the other hand, introduces a completely new term in the structure-factor expression. Some of these situations are considered in the next chapter.

As discussed further in Chap. 11, the electron-density distribution in the unit cell of a crystal can be displayed by summing a three-dimensional Fourier series having the structure factors as coefficients. The result is a distribution function which can be contoured to indicate maxima in the electron density at atomic sites. If the electron density is computed using structure factors placed on an absolute scale, it is possible to measure the volume under a peak, which should equal the total number of electrons at that site in the unit cell, averaged over the entire crystal. Suppose a crystal contains a large number of interstitial atoms, all of which tend to occupy the same kind of site in the crystal. An electron-density map of such a crystal should then show a corresponding maximum at the appropriate site, whose volume is directly proportional to the density of interstitial atoms present. An example of such an analysis is given toward the end of Chap. 12, in which a zinc oxide crystal heavily doped with zinc is found to contain the extra zinc atoms in specific interstitial sites. As might be expected, the lattice constants of the doped crystals show an expected increase over similar undoped ones. There is a serious limitation in such studies, however, and that is the relative insensitivity of these methods. For example, electrical-resistivity measurements in insulators or semiconductors are sensitive to point-defect concentrations of the order of parts per million. In view of the inherent errors in x-ray intensity measurements, particularly when small crystals are employed, the concentration of point defects present has to be quite high before their presence can be detected unambiguously in electron-density maps. The small maxima detected in ZnO (Fig. 11-20) correspond to approximately 1 zinc atom per $1{,}000$ unit cells. Just for comparison, a parallel study carried out, using doped CdS crystals, which have the same structure as ZnO, failed to show any evidence of interstitial atoms present, even though there is other x-ray evidence (lattice-constant changes), suggesting that interstitial atoms are present. It should be noted that other methods, such as electrical-conductivity measurements, although far more sensitive to small-defect concentrations, are unable, by themselves, to disclose the nature of the defects present.

An interesting example of when electron-density studies can be used successfully is afforded by indium antimonide. The atomic numbers of indium and antimony are 49 and 51, respectively, and there is reason to believe that, on the average, about half of one electron is transferred from one atom to the other, according to infrared-absorption and electrical-conductivity measurements. The natural question arises, can x-ray diffraction methods determine the electron density at each atomic site in the unit cell with sufficient accuracy to enable verifying the amount of charge transferred? Not only is such a corroboration useful, but it concurrently would establish the direction of the transfer, since other measurements are unable to tell on which kind of atom the excess charge resides. An x-ray study was based on extremely carefully measured intensities; the statistical accuracy of the data showed a reproducibility between equivalent reflections within $\pm 2\%$. Both kinds of atoms in the cubic crystal structure of InSb have very nearly the same mass, and each atom is equally surrounded by four tetrahedrally disposed unlike atoms. Consequently, an isotropic temperature factor proved quite adequate for placing the measured intensities on an absolute basis. More importantly, the two kinds of atoms occupy special positions in the unit cells such that, whatever contributions possible errors in the final F's make at one atomic site, identical errors are introduced at the site of the other atom. This is indeed fortuitous, because such errors cannot possibly affect the difference between the two electron-density maxima located at each site. In order to facilitate the final volume estimates, one other step was employed. The electron-density distribution of a fictitious crystal having two identical palladium atoms respectively occupying the indium and antimony sites was calculated and subtracted at each point in the unit cell from the experimentally determined electron density of InSb. This procedure has two advantages: it reduces the number of electrons localized near the atomic sites from an average of 50 to an average of 50 less 46 (the

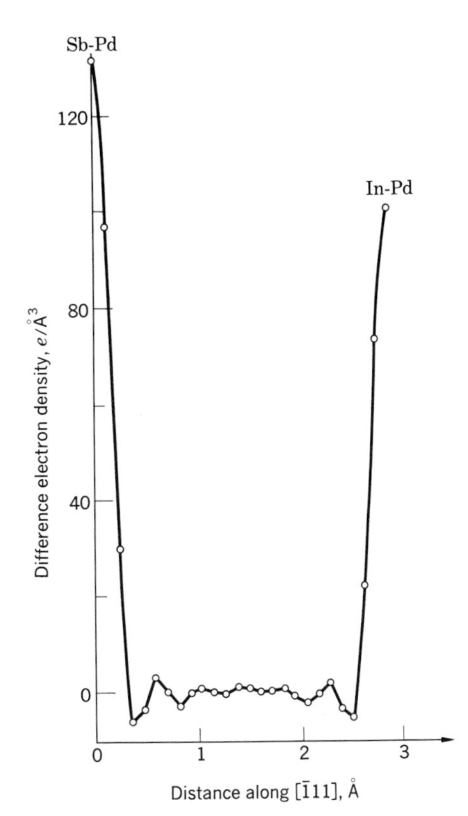

Fig. 10-24. Difference electron density for indium antimonide. [*From A. E. Attard and L. V. Azároff, Journal of Applied Physics, vol.* 34 (1963), *pp.* 774–776.]

atomic number of palladium), and it minimizes a source of error present in all Fourier series because finite rather than infinite series are summed in practice. The resulting "difference" electron density is shown in Fig. 10-24, evaluated along a line passing through the two kinds of atoms. By measuring the volumes of the two peaks shown in Fig. 10-24, it was found that the peak at the In site contains 4.51 ± 0.05 electrons, and the other peak contains 3.61 ± 0.04 electrons. Their sum of $8.12 \pm 0.09e$ differs from the expected number of eight electrons by about the standard deviation. Their difference corresponds to a charge of $0.45 \pm 0.05e$ transferred on the average to an antimony atom, and can be compared with $0.34e$ and $0.43e$ deduced from infrared and electron-mobility measurements by two different investigators.

The surprisingly good agreement between the x-ray diffraction value and the two independent measurements cited above should not be construed as proof of the general applicability of such methods. Not only was the crystal structure conducive to accurate evaluation of such sources of error as absorption (virtually perfect spheres of InSb could be ground), temperature effect (the x-ray value of Θ was determined to be 180°K as compared with 200 and 208°K determined by other methods), extinction (only secondary extinction was detected), and so forth, but the accumulation of errors could not affect the final outcome, which was to determine the difference between the electron densities at two equivalent sites in the same unit cell. On the other hand, this analysis clearly demonstrates that such x-ray methods can be valuable, when used wih discretion.

DISORDERED CRYSTALS

Geometry of order-disorder. The kind of disorder considered in the present section involves a rearrangement of the atoms among the atomic sites available in a particular crystal structure. The structure is said to be *ordered* when all the atoms occupy their correct sites, and it is said to be *disordered* when some of the atoms occupy incorrect sites. As a simple example of this, consider the array of open and shaded circles in Fig. 10-25, symbolizing an ordered structure of a binary compound. When this structure is disordered (Fig. 10-26), the two kinds of atoms may occupy the two kinds of sites completely at random, so that the sites become statistically equivalent. Since their occupation by either atom is equally likely, the scattering factor for such a statistically averaged atom is a weighted average of the scattering factors of each of the two atoms. A comparison of the two drawings shows that the point-group symmetry of the two structures is the same, $4mm$, but the plane-group symmetry has changed from $p4mm$ for the ordered array to $c4mm$ for the disordered one. Note that the primitive cell in Fig. 10-26 is, of course, smaller than the centered one. Thus it appears as if the ordered crystal structure has a superlattice compared with the disordered one. Actually, this is simply one example of the relation between what Buerger has called the *basic structure* and the *derivative structure* which results when certain translations or symmetry elements, or both, are suppressed in a crystal structure.

When the diffraction of x rays by two such structures is considered, it is clear

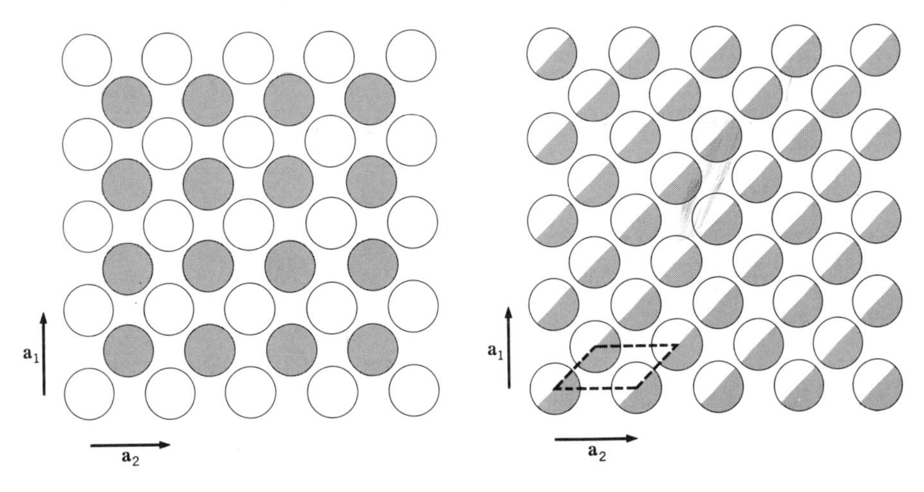

Fig. 10-25 **Fig. 10-26**

that reflections from the ordered structure, having a primitive lattice, can have all possible hkl values, while those from the disordered structure can have only those indices that are allowed by the extinction rules for the appropriate centered lattice. This means that both structures will have certain hkl reflections in common, called *fundamental reflections.* The ordered structure, however, gives rise to an additional set of reflections, called *superlattice reflections.* It will be shown below that the intensity of these superlattice reflections can be related to the degree of order present in a crystal.

Most commonly, crystals become disordered when heated to some temperature above what is called the transformation temperature. Upon cooling such a crystal, it often happens that different parts of the crystal start to order simultaneously. When a twofold choice exists for the correct occupation of a site, it is possible that the crystal starts to order in two different ways at once. The way that the two ordered regions, or *domains,* finally meet at a *domain boundary* depends on their structure and relative orientations. For the two-dimensional array considered above, at least two kinds of domain boundaries can be distinguished. The two domains in Fig. 10-27 are separated by a domain wall having glide symmetry, while the two domains in Fig. 10-28 are related to each other by reflection across the domain boundary. Normally, a crystal contains many such domains, and they give rise to characteristic diffuse scattering effects. Because the scattering by parallel planes in two neighboring regions are exactly out of phase with each other, they have been called *antiphase* domains. It should be noted, however, that they are actually related to each other by ordinary twin symmetry elements, so that the domains are, in reality, twins. This point is sometimes overlooked, and has led to the invention of unnecessarily complicated relationships to describe this phenomenon, particularly in crystals having complex crystal structures.† Some

† The concepts of derivative structures, twinning, and transformations in crystals are described in an elementary way in L. V. Azároff, *Introduction to solids* (McGraw-Hill Book Company, New York, 1960), chaps. 4, 7, and 8.

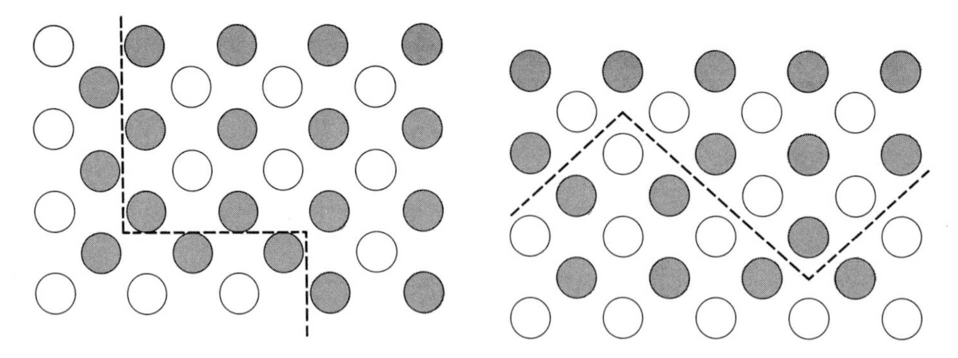

Fig. 10-27

Fig. 10-28

examples of the effect that such order-disorder phenomena can have on x-ray diffraction are considered next.

Long-range order. Whether the atoms in a crystal should order or not is obviously a problem best suited to thermodynamical analysis. The effect of such order on x-ray diffraction, however, is more of a geometrical problem. A formulation was devised by W. L. Bragg and E. J. Williams for this purpose, and is presented below. Consider a binary compound containing two kinds of atoms A and B whose correct sites in the unit cell are denoted by α and β, respectively. Let there be N_A atoms of type A, and N_B atoms of type B, so that the crystals contain a total of $N = N_A + N_B$ atoms. At the same time, note that then there must be N_A sites α and N_B sites β in the crystal. At any temperature some fraction of these sites will be correctly occupied. This state of affairs can be summarized by defining the fraction:

r_α of rightly occupied α sites
w_α of wrongly occupied α sites
r_β of rightly occupied β sites
w_β of wrongly occupied β sites

Here a rightly occupied α site contains an A atom, whereas a wrongly occupied α site contains a B atom, and similarly for the β sites. It is clear from these definitions that

$$r_\alpha + w_\alpha = r_\beta + w_\beta = 1.0 \tag{10-29}$$

Similarly, when there is complete order, $r_\alpha = r_\beta = 1.0$. By comparison, when there is complete disorder, the fraction of rightly occupied α sites is just equal to the fraction of A atoms present. Let this fraction be denoted $m_A = N_A/N$, and let $m_B = N_B/N$. The number of wrongly occupied sites then obeys the condition

$$N_A w_\alpha = N_B w_\beta$$

or

$$m_A w_\alpha = m_B w_\beta \tag{10-30}$$

The degree of order present can be expressed by an order parameter, usually called the *long-range-order parameter* S, to distinguish it from a short-range-order parameter considered later. Bragg and Williams chose a linear relation to express this parameter (Exercise 10-9), according to which

$$S = r_\alpha - w_\beta = r_\beta - w_\alpha \tag{10-31}$$

To see how this parameter can be utilized in practice, consider the binary alloy phase Au_3Cu. The ordered crystal structure of Au_3Cu is shown in Fig. 10-29. In the primitive cubic cell, the α sites are located at the centers of the faces $\frac{1}{2}\frac{1}{2}0$, $\frac{1}{2}0\frac{1}{2}$, and $0\frac{1}{2}\frac{1}{2}$, while the β sites occupy the corners 000. Since this is an A_3B type of compound, $m_A = \frac{3}{4}$ and $m_B = \frac{1}{4}$, so that it follows from (10-30) that $3w_\alpha = w_\beta$. Denoting the scattering factors of A atoms f_A and of B atoms f_B, the scattering factor for the equipoints can be correctly expressed for any degree of long-range order by

$$\begin{aligned} f_\alpha &= r_\alpha f_A + w_\alpha f_B \qquad \text{for an } \alpha \text{ site} \\ f_\beta &= r_\beta f_B + w_\beta f_A \qquad \text{for a } \beta \text{ site} \end{aligned} \tag{10-32}$$

The structure factor for this crystal then can be expressed according to (9-20).

$$F_{hkl} = (r_\beta f_B + w_\beta f_A)e^0 + (r_\alpha f_A + w_\alpha f_B)(e^{2\pi i(h/2+k/2)} + e^{2\pi i(h/2+l/2)} + e^{2\pi i(k/2+l/2)}) \tag{10-33}$$

It is most convenient to consider this expression for two sets of possible hkl values. When the indices are unmixed, that is, all are odd or even, then

$$\begin{aligned} F_{hkl} &= (r_\beta f_B + w_\beta f_A) + (r_\alpha f_A + w_\alpha f_B)(3) \\ &= 3f_A + f_B \end{aligned} \tag{10-34}$$

since $(w_\beta + 3r_\alpha) = (3w_\alpha + 3r_\alpha) = 3$, and so forth. Since these reflections do not depend on the degree of order present, they are called the fundamental reflections for this structure.

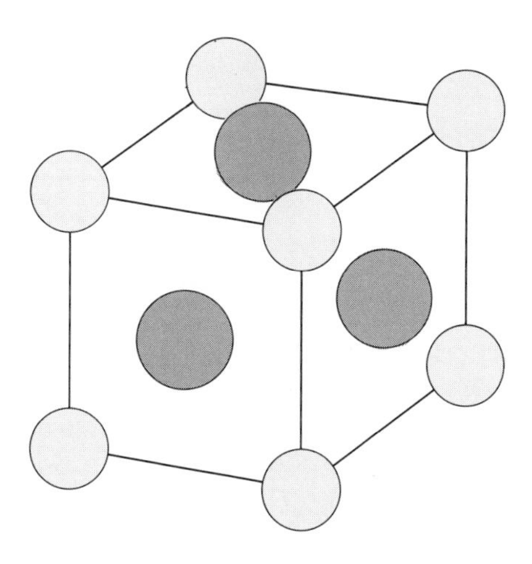

Fig. 10-29. Crystal structure of ordered Au_3Cu.

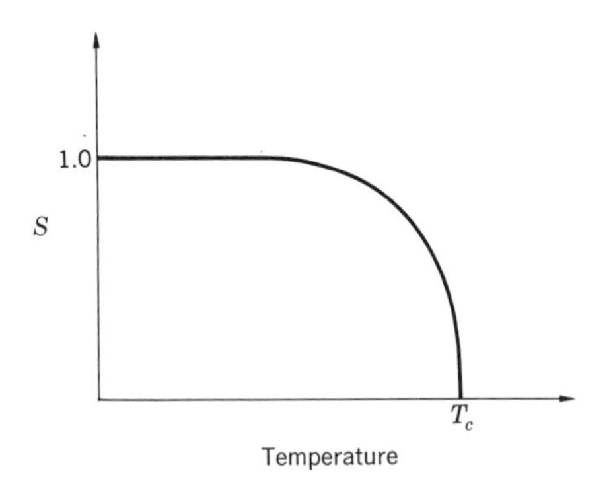

Fig. 10-30. Long-range-order parameter for CuZn.

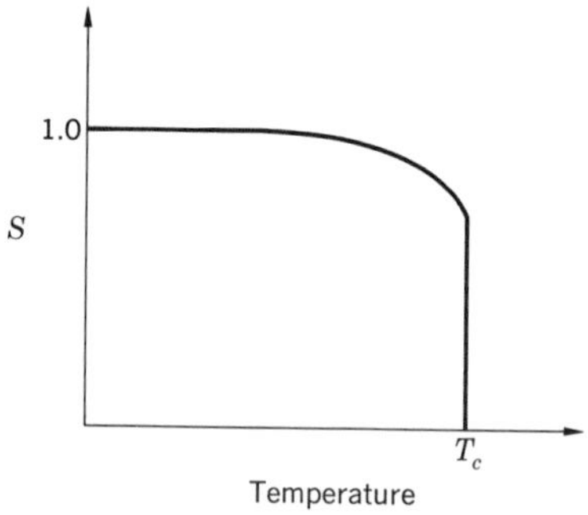

Fig. 10-31. Long-range-order parameter for Au_3Cu.

Next consider the case when the hkl indices are mixed. Then

$$
\begin{aligned}
F_{hkl} &= (r_\beta f_B + w_\beta f_A) + (r_\alpha f_A + w_\alpha f_B)(-1) \\
&= (f_B - f_A)S
\end{aligned}
\qquad (10\text{-}35)
$$

since $(w_\beta - r_\alpha) = -S$, and so forth. These are the so-called superlattice reflections which are present in full force when the crystal is ordered and $S = 1$ but tend to disappear as $S \rightarrow 0$. It is thus possible to observe the amount of disorder present in a crystal of Au_3Cu by noting the intensities of superstructure reflections relative to those of the fundamental reflections. Making use of the long-range-order parameter to indicate the degree of order present, one would expect it to vary from unity at $T = 0°K$ to zero at the transition, or critical, temperature T_c, something like the curve shown in Fig. 10-30. Actually, it turns out that such a continuous variation is observed in the alloy CuZn (β brass), whereas in Au_3Cu, the decline of S with temperature is more slight, but becomes discontinuous at T_c, as shown by the curve in Fig. 10-31. Because the variation of S with temperature is directly related to the interatomic forces in crystals, studies of order-disorder phenomena are quite useful in elucidating the nature of such interactions.

Antiphase domains. The disorder described above affects only the superstructure-reflection intensities because only their structure-factor magnitudes are altered. When ordering is accompanied by the formation of twinned, or antiphase, domains in a crystal, an additional effect takes place. Consider the collection of circles in Fig. 10-32, which could be, for example, depicting the atomic array in a (100) face of the Au_3Cu structure (Fig. 10-29). Next consider a set of horizontal planes of the type (001), which would be seen edge on as horizontal rows in Fig. 10-32. Assuming that the total volumes for the two kinds of twin orientations shown in that figure are equal, it is clearly evident that the 001 reflection has zero intensity, since the atoms in one twin scatter exactly out of phase with those in the other twin. (This is, of course, the origin of the name antiphase domains.) The above condition is strictly true, however, only if the reflecting planes are

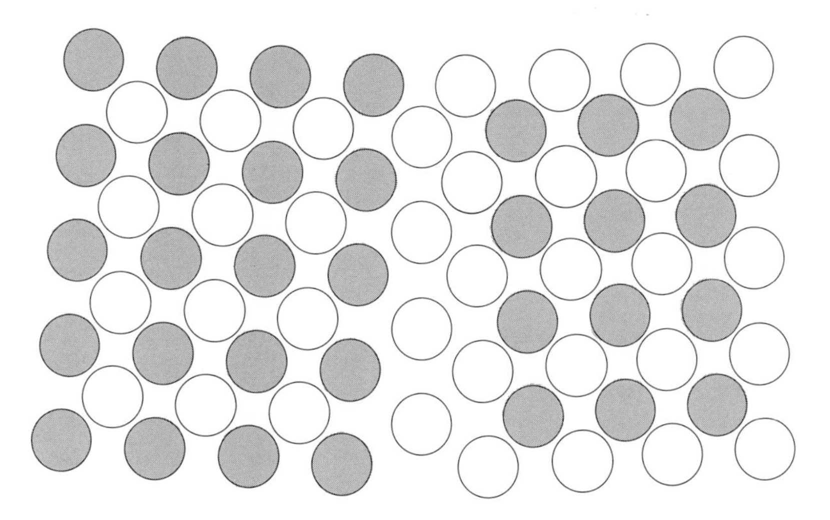

Fig. 10-32

perfectly parallel to each other throughout the crystal. Since real crystals are not so perfect, the same kinds of planes can be expected to be slightly tilted with respect to each other, as shown in an exaggerated manner in Fig. 10-32. This, in turn, makes constructive interference possible at the two extremes of the reflection curve, as shown in Fig. 10-33. In fact, it turns out that the total intensity of the reflections is exactly the same whether the crystal is twinned in this manner or not. This is demonstrated next.

Consider a crystal divided into a total of N_b blocks whose average dimensions

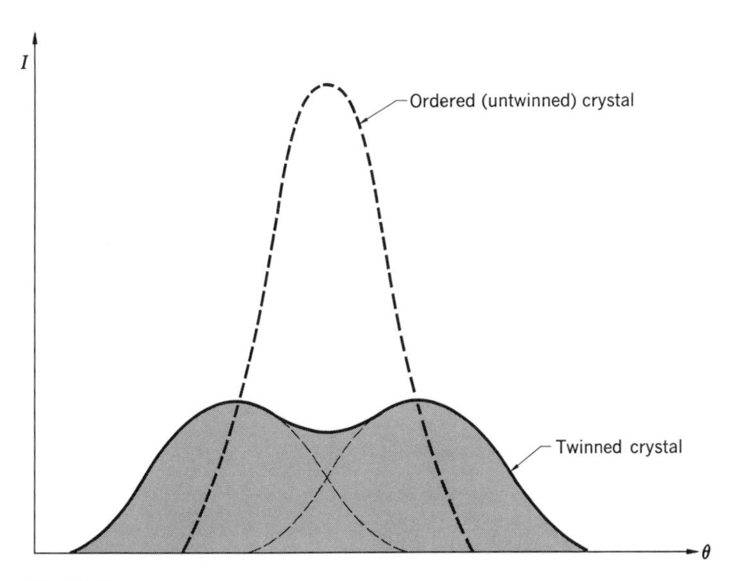

Fig. 10-33

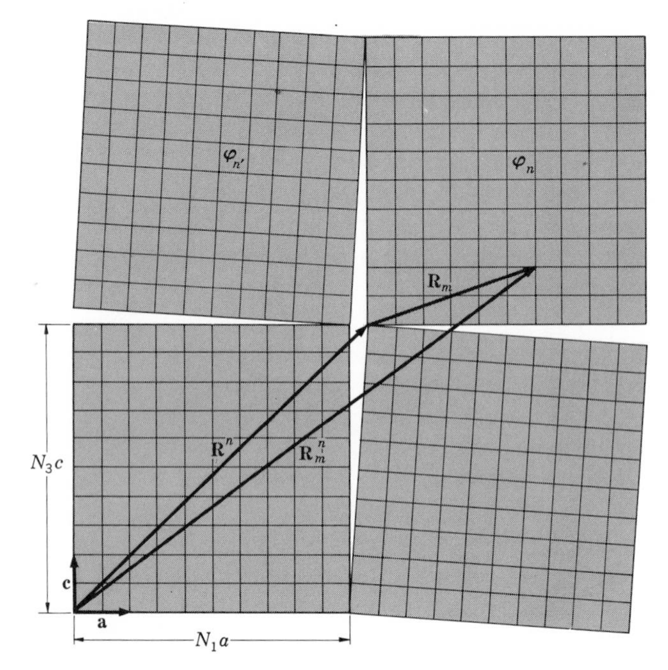

Fig. 10-34

are $N_1a \times N_2b \times N_3c$, where N_1, N_2, and N_3 enumerate the number of cells contained in each block parallel to the cell edges of length a, b, and c, respectively, and are typically of the order of 20 to 30. Labeling the unit cells in each block in the usual way by three integers $m_1m_2m_3$, it is possible to distinguish two vectors, as illustrated in the two-dimensional representation in Fig. 10-34. One vector, $\mathbf{R}_m = m_1\mathbf{a} + m_2\mathbf{b} + m_3\mathbf{c}$, locates a particular cell within each block, whereas $\mathbf{R}_n = n_1N_1\mathbf{a} + n_2N_2\mathbf{b} + n_3N_3\mathbf{c}$ locates the origin of the $n_1n_2n_3$ block. To locate a particular unit cell, therefore, the usual vector

$$\begin{aligned}\mathbf{R}_m{}^n &= \mathbf{R}_m + \mathbf{R}^n \\ &= m_1\mathbf{a} + m_2\mathbf{b} + m_3\mathbf{c} + n_1N_1\mathbf{a} + n_2N_2\mathbf{b} + n_3N_3\mathbf{c} \qquad (10\text{-}36)\end{aligned}$$

The relative phase of each block is either zero (in phase) or π (out of phase) and can be denoted by the phase angles φ_n or $\varphi_{n'}$, as indicated in Fig. 10-34. (The subscript n is a contraction for the three integers $n_1n_2n_3$.) Letting the scattering power of each cell be represented by its structure factor F, the amplitude of scattering by the crystal can be expressed in the usual way.

$$\begin{aligned}A &\propto Fe^{(2\pi i/\lambda)(\mathbf{S}-\mathbf{S}_0)\cdot\mathbf{R}_m{}^n} \\ &\propto \sum_{m_1}\sum_{m_2}\sum_{m_3} Fe^{\frac{2\pi i}{\lambda}(\mathbf{S}-\mathbf{S}_0)\cdot(m_1\mathbf{a}+m_2\mathbf{b}+m_3\mathbf{c})} \sum_{n_1}\sum_{n_2}\sum_{n_3} e^{i\varphi_n}e^{\frac{2\pi i}{\lambda}(\mathbf{S}-\mathbf{S}_0)\cdot(n_1N_1\mathbf{a}+n_2N_2\mathbf{b}+n_3N_3\mathbf{c})}\end{aligned} \qquad (10\text{-}37)$$

Clearly, the first sum in (10-37) represents the scattering by one block, and the second sum then adds up the contributions from N_b such blocks. The relative phases of the blocks are related to each other by the exponential $e^{i\varphi_n}$, with $\varphi_n = 0$ or π.

To calculate the intensity, it is necessary to multiply (10-37) by its complex conjugate. Expressing the intensity in electron units and assuming that the three Laue conditions are satisfied,

$$I_{\text{eu}} = [\text{CR}](N_b + \sum_{n_1}\sum_{n_2}\sum_{n_3}\sum_{n_1'}\sum_{n_2'}\sum_{n_3'}{}' e^{i(\varphi_n - \varphi_{n'})} e^{2\pi i[hN_1(n_1-n_1')+kN_2(n_2-n_2')+lN_3(n_3-n_3')]}) \tag{10-38}$$

where $$[\text{CR}] = F^2 \sum_{m_1}\sum_{m_2}\sum_{m_3}\sum_{m_1'}\sum_{m_2'}\sum_{m_3'} e^{2\pi i[h(m_1-m_1')+k(m_2-m_2')+l(m_3-m_3')]}$$

is the usual crystal-reflection expression. Note that the term in parentheses has been written in two parts: the first term represents the N_b term, for which the differences in the exponents vanish, that is, the cases when $n = n'$, while the remaining summation includes all the unlike pairs of terms, and is primed for that reason.

To see what the nature of the contributions made by each of these terms is, it is convenient to examine the intensity distribution in reciprocal space. As indicated, but not stressed, in earlier discussions, the intensity distribution in the vicinity of each reciprocal-lattice point is produced by the corresponding set of planes in the crystal. Hence it follows that an integration over the unit volume surrounding a reciprocal-lattice point must yield the integrated intensity of the reflection. This is known as the *Laue theorem*, and is most useful in analyzing many x-ray diffraction effects. Note that the intensity expression (10-38) really is a function of the three indices h, k, and l, which also constitute the coordinates of individual points in the reciprocal lattice. For present purposes, it is convenient to divide the reciprocal lattice into equivalent unit volumes having sides of length $1h$, $1k$, and $1l$ (Fig. 10-35). Treating h, k, and l as continuous variables, then, it is convenient to think of an *interference function* $I(hkl)$ which can be integrated

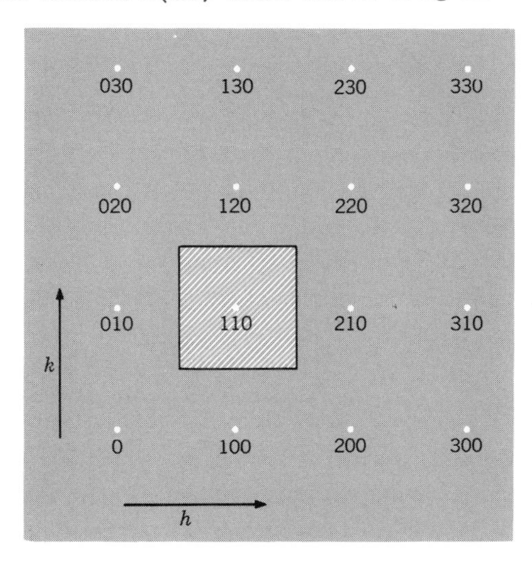

Fig. 10-35. $hk0$ net of reciprocal lattice showing the cross section of the unit volume surrounding the 110 reciprocal-lattice point.

throughout this unit volume to yield the integrated reflecting power

$$R_H = K \int_{-\frac{1}{2}}^{\frac{1}{2}} \int_{-\frac{1}{2}}^{\frac{1}{2}} \int_{-\frac{1}{2}}^{\frac{1}{2}} I(hkl)\, dh\, dk\, dl \tag{10-39}$$

where the constant K converts the intensity to absolute units and takes into account the correction factors for the particular experimental arrangement employed.

Returning to the discussion of scattering by a crystal containing antiphase domains, the intensity expression (10-38) can be substituted directly for the interference function in (10-39). After this has been done, the second term will consist of summations of individual terms such as

$$\begin{aligned} \int_{-\frac{1}{2}}^{\frac{1}{2}} e^{2\pi i h[(m_1-m_1')+N_1(n_1-n_1')]}\, dh &= \int_{-\frac{1}{2}}^{\frac{1}{2}} e^{2\pi i M h}\, dh \\ &= \frac{1}{2\pi i M}\, e^{2\pi i M h}\Big|_{-\frac{1}{2}}^{\frac{1}{2}} \\ &= \frac{1}{\pi M} \sin \pi M \\ &= 0 \end{aligned} \tag{10-40}$$

since $M = [(m_1 - m_1') + N_1(n_1 - n_1')]$ must be an integer. This is a most interesting result, because it not only eliminates the cross terms from further consideration, but it shows that they equal zero independently of what the phases φ_n of the different blocks are.

Returning to the interference function in (10-38), it is clear that

$$\begin{aligned} I(hkl) &= I_{eu} = [\text{CR}]N_b \\ &= F^2 N_b \sum_{m_1} \sum_{m_2} \sum_{m_3} \sum_{m_1'} \sum_{m_2'} \sum_{m_3'} e^{2\pi i[h(m_1-m_1')+k(m_2-m_2')+l(m_3-m_3')]} \end{aligned} \tag{10-41}$$

which is exactly what is obtained for a perfectly ordered crystal containing N_b "blocks," each of whose contribution to total scattering was correctly expressed by the above summation. The above analysis shows that the slight tilting of adjacent blocks does not decrease the integrated reflecting power of the crystal, even though it may serve to smear it out in reciprocal space so that the recorded reflection may appear as shown in Fig. 10-33.

It should be noted that not all crystals undergoing order-disorder transitions necessarily form antiphase domains upon ordering. Domains have been observed in Au_3Cu, but not in $CuZn$, for example. Also, it should be noted that the correctness of the above prediction regarding the integrated reflecting power is very difficult to test experimentally. This is so because the broad diffraction maxima must be separated quantitatively from the general background scattering, which receives contributions from incoherent scattering, TDS, and diffuse scattering due to various other kinds of imperfections that may be present.

Random disorder. According to the foregoing discussion, all vestiges of long-range order disappear above the transition temperature, as indicated in Figs. 10-30 and 10-31. Suppose that the crystal is heated to a temperature well above T_c, so that one can assume that the atoms are distributed quite randomly and the occupation of a given site is statistical; that is, for x-ray diffraction purposes, all sites can be considered to be occupied by indistinguishable atoms. The vector from the origin to any such atom then is

$$\mathbf{R}_l = l_1\mathbf{a} + l_2\mathbf{b} + l_3\mathbf{c} \tag{10-42}$$

where **a**, **b**, and **c** are lattice vectors defining a primitive cell in the disordered crystal containing one statistically equivalent atom. The diffraction intensity, in electron units, then is simply

$$I_{\text{eu}} = \sum_l \sum_{l'} f_l f_{l'} e^{(2\pi i/\lambda)(\mathbf{S}-\mathbf{S}_0)\cdot(\mathbf{R}_l-\mathbf{R}_{l'})} \tag{10-43}$$

Limiting this analysis to a binary crystal containing $N_A = m_A N$ atoms of type A and $N_B = m_B N$ atoms B, the fractional ratios $m_A + m_B = 1$, so that

$$N_A + N_B = N$$

Denoting the scattering factors of the two kinds of atoms by f_A and f_B, it is possible to distinguish the following unique combinations of l and l' atoms in (10-43): First of all, consider the case when $l = l'$. This will happen N_A times when the l atom is an A atom and N_B times when it is a B atom. It therefore follows that

$$I_{\text{eu}} = N_A f_A^2 + N_B f_B^2 + \sum_{l \neq l'}\sum{}' f_l f_{l'} e^{(2\pi i/\lambda)(\mathbf{S}-\mathbf{S}_0)\cdot(\mathbf{R}_l-\mathbf{R}_{l'})} \tag{10-44}$$

Next, consider the case when $l \neq l'$. Only four possible cases can be distinguished: both atoms are A atoms (m_A^2 times); both atoms are B atoms (m_B^2 times); the first is a B atom and the second is an A atom ($m_B m_A$ times); or the reverse ($m_A m_B$ times). Thus

$$\begin{aligned} I_{\text{eu}} &= N_A f_A^2 + N_B f_B^2 + \sum_{l \neq l'}\sum{}' (m_A^2 f_A^2 + m_B^2 f_B^2 + 2m_A m_B f_A f_B) \\ &\qquad\qquad \times e^{(2\pi i/\lambda)(\mathbf{S}-\mathbf{S}_0)\cdot(\mathbf{R}_l-\mathbf{R}_{l'})} \\ &= N(m_A f_A^2 + m_B f_B^2) + \sum_{l \neq l'}\sum{}' (m_A f_A + m_B f_B)^2 e^{(2\pi i/\lambda)(\mathbf{S}-\mathbf{S}_0)\cdot(\mathbf{R}_l-\mathbf{R}_{l'})} \end{aligned} \tag{10-45}$$

Adding and subtracting N terms $(m_A f_A + m_B f_B)^2 e^0$ from (10-45), the terms missing from the double sum are restored, and

$$\begin{aligned} I_{\text{eu}} &= N(m_A f_A^2 + m_B f_B^2) - N(m_A f_A + m_B f_B)^2 \\ &\qquad + \sum_l \sum_{l'} (m_A f_A + m_B f_B)^2 e^{(2\pi i/\lambda)(\mathbf{S}-\mathbf{S}_0)\cdot(\mathbf{R}_l-\mathbf{R}_{l'})} \\ &= m_A m_B N (f_A - f_B)^2 + [\text{CR}] \end{aligned} \tag{10-46}$$

after expanding the squared parentheses and combining like terms.

The second term in (10-46) is the ordinary crystal-reflection term for a crystal containing equivalent atoms whose scattering factor is $m_A f_A + m_B f_B$. This is, of course, the statistically weighted average value of the scattering factor for a randomly disordered AB crystal. The first term in (10-46) is a small, very slowly varying function, representing a continuous diffuse scattering.† It is called the *Laue monotonic scattering* and, because of its virtual indistinguishability from general background scattering, it has never been quantitatively evaluated in an experiment.

Short-range order. Long-range order begins to set in when a disordered crystal is cooled below the transition temperature. What about temperatures just slightly above this critical temperature—does a crystal structure disorder completely or not? There is a growing amount of evidence indicating that a truly randomly disordered structure is almost never attained. The reasons for this are quite simple. Either the interaction forces between atoms are such that they prefer to be surrounded by like atoms or they prefer unlike neighbors. The first case leads to the formation of clusters of like atoms, while the second case is conducive to ordering. At temperatures at which long-range order is not thermodynamically stable, it is therefore possible for a short-range order to persist. The actual atomic array in this case is very similar to that in a disordered crystal, except that an A atom tends to become surrounded primarily by B atoms, and conversely, each B atom tends to be surrounded by A atoms. This ordering effect persists only over a few neighboring atomic coordinations, so that no long-range order develops. Nevertheless, it results in an atomic array that is not truly random.

To see what effect short-range order has on x-ray diffraction by crystals, consider a binary crystal of composition A_xB_y. Let the total number of atoms present be N, so that the number of A atoms present N_A is determined by the fraction m_A; that is, $N_A = m_AN$. Similarly, the number of B atoms present is $N_B = m_BN$, while the sum of the fractions $m_A + m_B = 1$. In selecting the most convenient method for describing the location of atoms in an essentially disordered crystal, suppose that the statistically distributed atoms are located at the ends of vectors, such as

$$\mathbf{R}_l = l_1\mathbf{a}_1 + l_2\mathbf{a}_2 + l_3\mathbf{a}_3 \tag{10-47}$$

where the $\mathbf{a}_i$ vectors are chosen along atom rows but are not necessarily true translations of the lattice. Let the distance between two such vectors,

$$|\mathbf{R}_l - \mathbf{R}_{l'}| = |\mathbf{r}_i| \tag{10-48}$$

be an interatomic vector from atom l to atom l'. The subscript i in (10-48) identifies the coordination shell surrounding the atom placed at its center (Fig. 10-36), so that $\mathbf{r}_1$ is the vector to nearest neighbors (radius of first coordination shell), $\mathbf{r}_2$ is

† It has been pointed out by T. Ishibachi and M. L. Rudee [*Journal of Applied Physics,* vol. 37 (1966), pp. 3359–3361] that the assumption made by Laue that

$$\langle f_lf_{l'}\rangle_{\text{av}} = \langle f_l\rangle_{\text{av}}\langle f_{l'}\rangle_{\text{av}} = \langle f_l\rangle_{\text{av}}^2 = \langle f\rangle_{\text{av}}^2$$

is not correct for the formulation of random disorder in alloys since the removal of atoms for which $l = l'$ from the double sum in (10-45) affects the probability distribution (randomness) of atoms for which $l \neq l'$. As shown by these authors, the correct formulation of these equations has the consequence that the Laue monotonic scattering vanishes at reciprocal-lattice points, although it is unaffected (except for scale) in the rest of the reciprocal lattice.

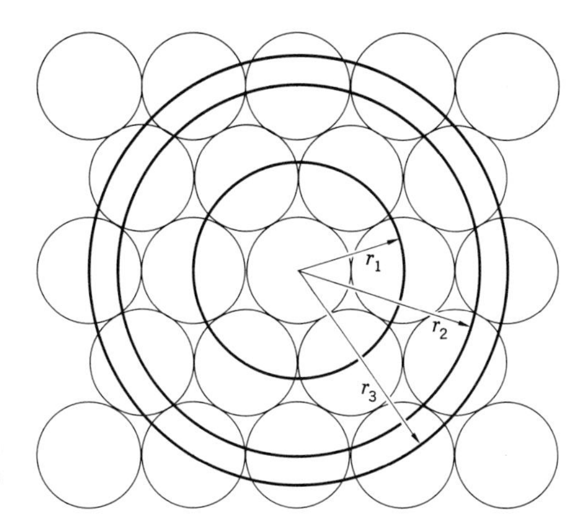

Fig. 10-36. Coordination shells surrounding an atom in a closest packing. Note that, in two dimensions, $c_1 = 6$, $c_2 = 6$, $c_3 = 6$, etc.

the vector to second-nearest neighbors (radius of second shell), and so forth. The number of atoms at each of these distances from the origin atom, that is, the number of atoms in the ith shell about the origin, c_i, depends on the crystal structure. For example, in a cubic closest packing $c_1 = 12$, $c_2 = 6$, etc.

With the foregoing definitions, it is possible to express the intensity of x-ray diffraction in electron units as follows:

$$I_{\text{eu}} = \sum_l \sum_{l'} f_l f_{l'} e^{\frac{2\pi i}{\lambda}(\mathbf{S}-\mathbf{S}_0)\cdot(\mathbf{R}_l-\mathbf{R}_{l'})}$$

$$= \sum_l f_l^2 + \sum_l \sum_{\substack{l' \\ i=1}}^{i=p}{}' f_l f_{l'} e^{\frac{2\pi i}{\lambda}(\mathbf{S}-\mathbf{S}_0)\cdot(\mathbf{R}_l-\mathbf{R}_{l'})} + \sum_l \sum_{\substack{l' \\ i>p}}^{\infty}{}' f_l f_{l'} e^{\frac{2\pi i}{\lambda}(\mathbf{S}-\mathbf{S}_0)\cdot(\mathbf{R}_l-\mathbf{R}_{l'})} \quad \textbf{(10-49)}$$

where the first sum represents each atom "interacting" with itself, while the primed sum over atoms at sites for which $l \neq l'$ is divided into two sums: one for atoms lying in concentric shells up to $i = p$, and one for all the other, more distant atoms.

Suppose all the atoms in shells $i \leqq p$ are in an ordered array and there are n_i atoms of type A in shell i surrounding a B atom at a distance r_i from it. Similarly, when an A atom is considered, there are n_i' atoms of type B surrounding it in the ith shell, a distance r_i from the origin A atom. With these definitions one can conclude that, for an interatomic separation r_i, the number of bonds from A to B atoms must equal the number of bonds from B to A atoms, so that

$$m_B N n_i = m_A N n_i'$$

$$m_B n_i = m_A n_i' \quad \textbf{(10-50)}$$

It is now possible to express the three terms in the intensity expression (10-49) more explicitly for the two kinds of atoms present in the crystal. Considering each term separately,

First term $\quad \sum_l f_l^2 = (m_A N f_A^2 + m_B N f_B^2) \quad$ **(10-51)**

The summation in the second term is carried out most conveniently by considering the atoms in sites denoted by l' to be arrayed in concentric shells of radius r_i, like Fig. 10-36, with the number of atoms in the ith shell c_i being counted by a sum over j, so that

Second term
$$\sum_{l} \sum_{\substack{l' \\ i=1}}^{i=p}{}' f_l f_{l'} e^{(2\pi i/\lambda)(\mathbf{S}-\mathbf{S}_0)\cdot(\mathbf{R}_l - \mathbf{R}_{l'})}$$
$$= \sum_{l} \sum_{i=1}^{p} \sum_{j} \left[m_B f_B \left(\frac{n_i}{c_i} f_A + \frac{c_i - n_i}{c_i} f_B \right) + m_A f_A \left(\frac{n_i'}{c_i} f_B + \frac{c_i - n_i'}{c_i} f_A \right) \right] e^{(2\pi i/\lambda)(\mathbf{S}-\mathbf{S}_0)\cdot \mathbf{r}_i} \qquad (10\text{-}52)$$

where the first term in brackets is obtained by examining the environment of a B atom placed at the origin, and the second term is obtained by placing an A atom at the origin. Finally,

Third term
$$\sum_{l} \sum_{\substack{l' \\ i>p}}^{\infty}{}' f_l f_{l'} e^{(2\pi i/\lambda)(\mathbf{S}-\mathbf{S}_0)\cdot(\mathbf{R}_l - \mathbf{R}_{l'})}$$
$$= \sum_{l} \sum_{\substack{l' \\ i>p}}^{\infty}{}' (m_A f_A + m_B f_B)^2 e^{(2\pi i/\lambda)(\mathbf{S}-\mathbf{S}_0)\cdot(\mathbf{R}_l - \mathbf{R}_{l'})} \qquad (10\text{-}53)$$

since, when $r_i > r_p$, random disorder prevails.

Note that this third term is an incomplete sum, similar to the crystal-reflection term in (10-46), except that the terms for which $l = l'$ and $i \leq p$ are missing. To restore these terms, add and subtract N terms $(m_A f_A \pm m_A f_A)^2 e^0$ and $\sum_{i=1}^{p} \sum_{j} (m_A f_A + m_B f_B)^2 e^{(2\pi i/\lambda)(\mathbf{S}-\mathbf{S}_0)\cdot(\mathbf{R}_l - \mathbf{R}_{l'})}$ to the original intensity expression (10-49). Again proceeding term by term, subtract $N(m_A f_A + m_B f_B)$ from (10-51). After expanding the squares and combining like terms,

First term
$$N[m_A f_A{}^2(1 - m_A) + m_B f_B{}^2(1 - m_B) - 2m_A m_B f_A f_B] = m_A m_B N(f_A - f_B)^2 \qquad (10\text{-}54)$$

expresses the Laue monotonic scattering. Subtracting $(m_A f_A + m_B f_B)^2 c_i/c_i$ from the quantity in the brackets in (10-52), it is easily demonstrated (Exercise 10-14) that

Second term
$$\sum_{l} \sum_{i=1}^{p} \sum_{j} \left(\frac{m_A m_B c_i - m_B n_i}{c_i} \right) (f_A - f_B)^2 e^{(2\pi i/\lambda)(\mathbf{S}-\mathbf{S}_0)\cdot \mathbf{r}_i}$$
$$= m_A m_B N (f_A - f_B)^2 \sum_{i=1}^{p} \sum_{j} \left(1 - \frac{n_i}{m_A c_i} \right) e^{(2\pi i/\lambda)(\mathbf{S}-\mathbf{S}_0)\cdot \mathbf{r}_i} \qquad (10\text{-}55)$$

since the sum over l merely repeats the remaining quantity N times. Finally, adding the terms that have been subtracted above, to the third term, (10-53), simply converts it to the ordinary crystal-reflection term (10-46) for a randomly disordered crystal. The total diffracted intensity in electron units, (10-49), then becomes

$$I_{\mathrm{eu}} = I_D + [\mathrm{CR}] \tag{10-56}$$

where the diffuse intensity expressed in electron units I_D is the sum of (10-54) and (10-55).

$$I_D = m_A m_B N (f_A - f_B)^2 \left(1 + \sum_{i=1}^{p} \sum_{j} \alpha_i e^{(2\pi i/\lambda)(\mathbf{S} - \mathbf{S}_0)\cdot \mathbf{r}_i}\right) \tag{10-57}$$

where $\alpha_i = 1 - \dfrac{n_i}{m_A c_i}$ (10-58)

is called a *short-range-order parameter* because it is possible to express the degree of ordering present in terms of these quantities (Exercise 10-15). Note that the summation over j, intended to add up the contributions of the atoms in each shell i, cannot be carried out formally in (10-57) because none of the factors depend on j directly. The actual number of atoms in the ith shell is, of course, c_i, and the kinds of atoms they are (A or B) is determined by α_i in (10-58). Hence it is possible to replace $\sum_j$ by c_i in (10-57). Also, since $\alpha_i = 0$ when total disorder sets in, that is, when $i > p$, the summation over i can be extended up to infinity. Noting that, experimentally, one measures the time-averaged intensity

$$\begin{aligned} \overline{I_D}^t &= m_A m_B N (f_A - f_B)^2 \left(1 + \sum_{i=1}^{\infty} c_i \alpha_i \overline{e^{(2\pi i/\lambda)(\mathbf{S} - \mathbf{S}_0)\cdot \mathbf{r}_i}}^t\right) \\ &= m_A m_B N (f_A - f_B)^2 \left(1 + \sum_{i=1}^{\infty} c_i \alpha_i \frac{\sin kr_i}{kr_i}\right) \end{aligned} \tag{10-59}$$

where $k = (4\pi \sin \theta)/\lambda$ as before.

Rearranging the terms in (10-59),

$$\begin{aligned} \frac{\overline{I_D}^t}{m_A m_B N (f_A - f_B)^2} - 1 &= \sum_{i=1}^{\infty} c_i \alpha_i \frac{\sin kr_i}{kr_i} \\ &= \int_0^{\infty} f(r) \frac{\sin kr}{kr}\, dr \end{aligned} \tag{10-60}$$

Here a continuous function of r, namely, $f(r)$, is chosen to represent the quantities $c_i \alpha_i$. This changeover is legitimate provided $f(r)$ is suitably selected. Since $c_i \alpha_i$ have real values only when $r = r_i$, the proper choice of $f(r)$ must be like the function shown in Fig. 10-37, where the areas of the peaks at the points r_1, r_2, r_3, etc., are equivalent to the magnitudes of the products $c_i \alpha_i$ for the respective values of i.

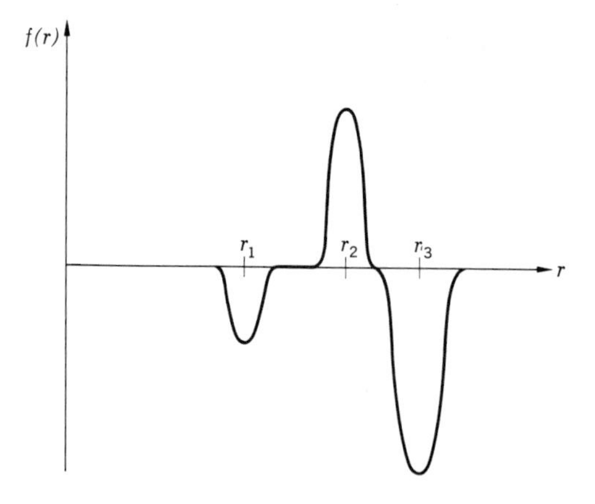

Fig. 10-37. Expected form of $f(r) = c_i\alpha_i$ for a nearly ordered binary compound. The magnitude of $c_i\alpha_i$ at each value of r_i is given by the area under the peak.

Note that the peaks may be positive or negative, depending on whether α_i is positive or negative.

The left side of (10-60) is a known function of $(\sin \theta)/\lambda$, so that it can be written

$$k\varphi(k) = \int_0^\infty \frac{f(r)}{r} \sin kr\, dr \tag{10-61}$$

which can be inverted by making use of the Fourier integral theorem (Appendix 1) to

$$\begin{aligned} f(r) &= \frac{2r}{\pi} \int_0^\infty k\varphi(k) \sin rk\, dk \\ &= \frac{2r}{\pi} \int_0^\infty k \left(\frac{\overline{I_D}^t}{m_A m_B N (f_A - f_B)^2} - 1 \right) \sin rk\, dk \end{aligned} \tag{10-62}$$

All the quantities on the right of (10-62) are known or can be measured experimentally. It is thus possible to construct the function $f(r)$ from a numerical integration of (10-62) after the measured diffuse intensity has been suitably corrected and expressed in electron units. Note that I_D is an oscillating function of k in reciprocal space according to (10-59), the oscillations taking place about the value of the Laue monotonic scattering intensity. Thus the integral in (10-62) will take on positive and negative values, as suggested by Fig. 10-37, and the area under each resulting peak is equal to $c_i\alpha_i$. Since the number of atoms in the ith shell c_i is known, it thus becomes possible to determine the short-range-order parameters experimentally for any temperature at which the diffuse intensity can be measured.

It is possible to express the intensity in a somewhat different way, counting the individual atoms at each point $l_1l_2l_3$ separately (10-47). In that case, one obtains a short-range-order parameter $\alpha_{l_1l_2l_3}$ which relates the ordering at each such site. Some of the consequences of such a formulation, as well as practical procedures for evaluating the integral in (10-62), measuring the diffuse intensity, and correcting it appropriately, are described in some of the books listed at the conclusion of this chapter.

DEFECTIVE CRYSTALS

Closest packings. The crystal structures of many elements, including most metals, can be likened to the closest packings of hard impenetrable spheres. These structures have the highest packing density possible for simple three-dimensional periodic arrays and, by permitting each atom to form bonds with 12 nearest neighbors, result in greater structural stability. An even larger group of inorganic solids, including most alloy phases, have structures that are based on closest packings of one kind of atom, with the other atoms present distributed among the interstitial voids.† The basis of a closest packing is a *closest-packed layer* like the one shown in Fig. 10-38. The spheres are in a hexagonal array, plane group $p6mm$, and represent the most efficient utilization of space in one layer (Exercise 10-16). Focusing attention on the outlined unit cell, note that it contains two triangular voids, which have been labeled B and C in Fig. 10-39. The reason for distinguishing them is that, when a sphere occupies the site B, it overlaps the other site, so that it is not possible for another sphere to occupy site C at the same time. Denoting the sites already occupied by spheres A, it is clearly seen that identical hexagonal closest-packed layers result regardless of whether the spheres occupy all the A sites, all the B sites, or all the C sites.

In constructing a three-dimensional structure by stacking such closest-packed layers on top of each other, it is possible to distinguish them by noting the sequence in which the various sites are occupied. To begin, suppose that the first layer is an A layer like the one in Fig. 10-38. Let the next layer above be formed by placing the spheres above the triangular interstices labeled B, so that the second layer is a B layer, as shown in Fig. 10-40. For each successive layer, obviously, a twofold choice exists. For the third layer in Fig. 10-40, the choice is between another A

† An introduction to the geometry and structural aspects of closest packings can be found in Leonid V. Azároff, *Introduction to solids* (McGraw-Hill Book Company, New York, 1960), chap. 3, with examples of actual crystal structures in chaps. 11, 13, and 15.

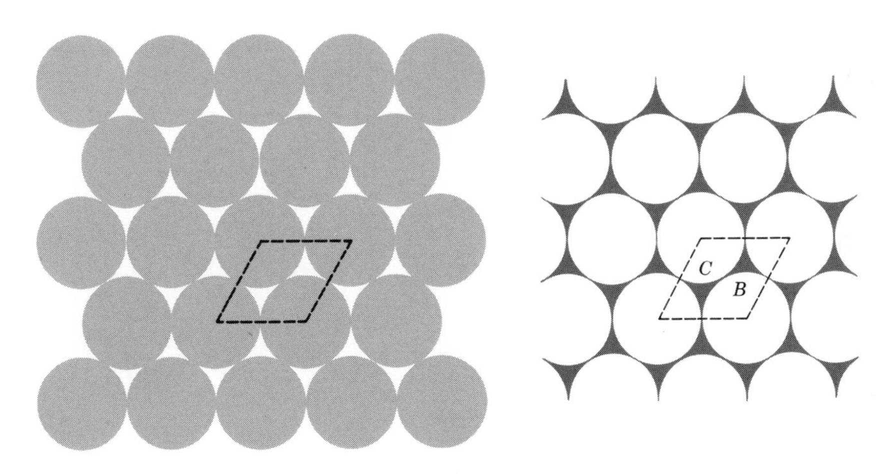

Fig. 10-38

Fig. 10-39

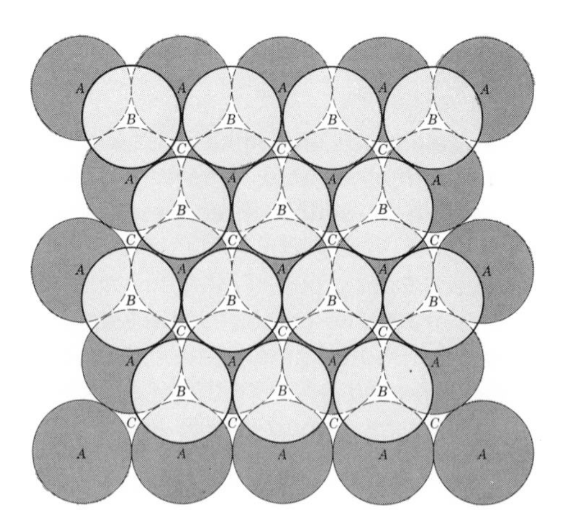

Fig. 10-40

layer or a C layer. (The placing of a B layer on top of a B is not as stable as placing the spheres within the triangular valleys between spheres.) Suppose the third layer is an A layer, so that the stacking sequence is given by the formula . . . ABA so far. If the distance separating the two A layers is to be a translation in this two-layer structure, it is possible to represent this closest packing by the stacking formula . . . AB . . . (= . . . $ABABAB$ . . .), where the three dots, called ellipsis, indicate that the sequence is periodically repeated in both directions. This two-layer sequence (Fig. 10-41) is called the *hexagonal closest packing*, often abbreviated hcp, because it is the simplest stacking possible that has hexagonal symmetry (Exercise 10-17), and is one of the two most common packings encountered. The other one is obtained by placing the third layer in the C sites, so that the stacking formula becomes . . . ABC . . . (. . . $ABCABCABC$ . . .). The stacking direction in this three-layer packing is parallel to the body diagonal of a cube in what turns out to be a face-centered cubic array of the atoms (Fig. 10-42). This is naturally called the cubic closest packing, abbreviated ccp, or the face-centered cubic packing, abbreviated fcc.

Other closest packings can be formed by varying the stacking sequence and the stacking period from two-layer to infinite-layer sequences. Moreover, it is

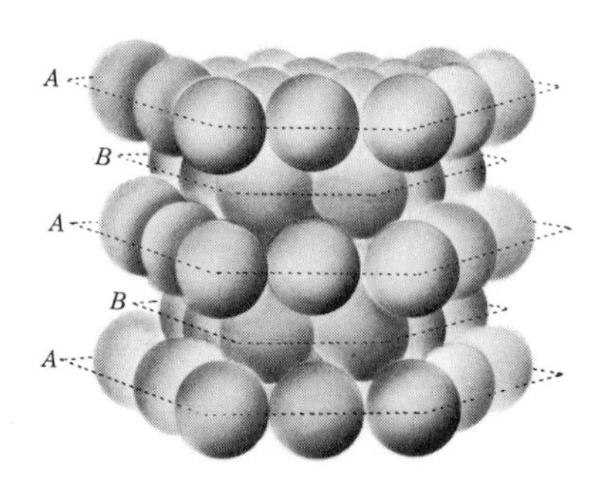

Fig. 10-41. Hexagonal closest packing of like spheres.

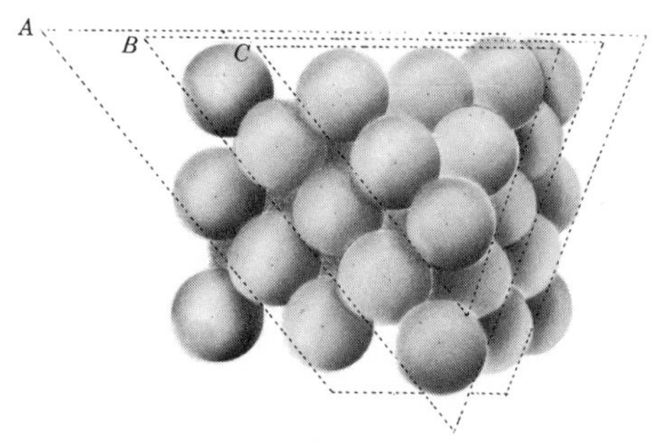

Fig. 10-42. Cubic closest packing of like spheres.

possible to form composite layers of two or more kinds of atoms. For example, the Laves phases encountered in certain alloy systems consist of double layers in which some atoms have been removed in order to make room for larger atoms also included in the double layer. The resulting double layers then stack on top of each other quite similarly to the way simple hexagonal closest-packed layers stack. Micaceous minerals or clay minerals offer examples of even more complex layers, consisting of two or three closest-packed oxygen layers bound to each other by cations occupying some of the available voids. Similarly, such compounds as ZnS, for example, form closest packings of anions, with the cations occupying tetrahedral voids. Since there are different ways possible for forming closest packing, the resulting structures of such binary compounds can differ also. Probably the best-known example is SiC, which can crystallize with many different stacking sequences, ranging from the simple hexagonal closest packing to some which contain over one hundred layers in each repeat unit. It is not too surprising to discover, therefore, that when such layers are stacked on top of each other, it often happens that the correct stacking sequence is not preserved, and a *stacking fault* results. Before considering the effect that such plane defects have on x-ray diffraction, it is instructive to consider what happens when the layers can be stacked above each other quite randomly.

Random-layer structure. In order to keep this discussion perfectly general, consider a two-dimensional layer whose plane lattice is defined by two vector translations $\mathbf{a}$ and $\mathbf{b}$. The location of the nth atom within the cell defined by these two vectors is given by the vector $\mathbf{r}_n$ drawn from the cell's origin. Next, suppose that such layers are stacked on top of each other so that the $\mathbf{a}$ and $\mathbf{b}$ vectors in each layer remain respectively parallel but the origin of each layer bears no fixed relation to that of its neighboring layers. Literally, this means that there is no periodic repetition along a third noncoplanar direction, so that a random-layer structure results. Thus it is not possible to speak of a third axis $\mathbf{c}$, but it is possible to consider a perpendicular distance to the next layer, that is, the interplanar distance d. The origin in an adjacent layer can be considered to be displaced by an amount δ along $\mathbf{a}$ and ϵ along $\mathbf{b}$, so that, in order to reach the origin in a layer that is m_3 layers away, it is necessary to travel the distance $m_3\mathbf{d} + \delta_{m_3}\mathbf{a} + \epsilon_{m_3}\mathbf{b}$, where the subscripts identify the displacements within the layer. Any atom in this random-layer structure now can be reached from a common origin by following the vector

$$\mathbf{R}_m{}^n = \mathbf{R}^n_{m_1m_2m_3} = \mathbf{r}_n + m_1\mathbf{a} + m_2\mathbf{b} + m_3\mathbf{d} + \delta_{m_3}\mathbf{a} + \epsilon_{m_3}\mathbf{b} \tag{10-63}$$

The intensity of x-ray scattering by this array, expressed in electron units, then can be written

$$I_{\text{eu}} = \sum_n \sum_{m_1} \sum_{m_2} \sum_{m_3} f_n e^{(2\pi i/\lambda)(\mathbf{S}-\mathbf{S}_0)\cdot\mathbf{R}_m{}^n} \sum_{n'} \sum_{m_1'} \sum_{m_2'} \sum_{m_3'} f_{n'} e^{(-2\pi i/\lambda)(\mathbf{S}-\mathbf{S}_0)\cdot\mathbf{R}_{m'}{}^{n'}} \tag{10-64}$$

where the primed indices in the second term indicate complex conjugates of the unprimed first term.

The summations over n, m_1, and m_2 lead to the same expressions as those encountered previously in this book, so that it is possible to rewrite (10-64) by inspection to give

$$I_{eu} = FF^* \frac{\sin^2 [(\pi/\lambda)(\mathbf{S} - \mathbf{S}_0) \cdot N_1\mathbf{a}]}{\sin^2 [(\pi/\lambda)(\mathbf{S} - \mathbf{S}_0) \cdot \mathbf{a}]} \frac{\sin^2 [(\pi/\lambda)(\mathbf{S} - \mathbf{S}_0) \cdot N_2\mathbf{b}]}{\sin^2 [(\pi/\lambda)(\mathbf{S} - \mathbf{S}_0) \cdot \mathbf{b}]}$$
$$\times \sum_{m_3} \sum_{m_3'} e^{(2\pi i/\lambda)(\mathbf{S}-\mathbf{S}_0)\cdot[(m_3 - m_{3'})\mathbf{d} + (\delta_{m_3} - \delta_{m_{3'}})\mathbf{a} + (\epsilon_{m_3} - \epsilon_{m_{3'}})\mathbf{b}]} \tag{10-65}$$

where N_1 and N_2 are the number of cells in each layer parallel to the a and b axes.

From the nature of the terms preceding the double sum in (10-65), it is clear that the intensity will be zero unless the two Laue conditions

$$(\mathbf{S} - \mathbf{S}_0) \cdot \mathbf{a} = h\lambda$$
$$(\mathbf{S} - \mathbf{S}_0) \cdot \mathbf{b} = k\lambda$$

are very nearly satisfied. When this is the case, (10-65) becomes

$$I_{eu} = F^2 \frac{\sin^2 \pi h N_1}{\sin^2 \pi h} \frac{\sin^2 \pi k N_2}{\sin^2 \pi k} \sum_{m_3} \sum_{m_3'} e^{2\pi i\left[\frac{(\mathbf{S}-\mathbf{S}_0)}{\lambda}\cdot(m_3 - m_{3'})\mathbf{d} + h(\delta_{m_3} - \delta_{m_{3'}}) + k(\epsilon_{m_3} - \epsilon_{m_{3'}})\right]} \tag{10-66}$$

In examining the intensity expression (10-66) it is worthwhile to distinguish between $00l$ reflections and all others for which h and k are both not zero. When $h = k = 0$, it is clear that the remaining sum in (10-66) becomes $\sin^2 [(\pi/\lambda)(\mathbf{S} - \mathbf{S}_0) \cdot N_3\mathbf{d}]/\sin^2 [(\pi/\lambda)(\mathbf{S} - \mathbf{S}_0) \cdot \mathbf{d}]$, which reduces to N_3^2 when the third Laue condition is satisfied. Thus $00l$ reflections are normal crystal reflections. Another way of stating this is that planes parallel to the layers diffract x rays in the usual way, and such reflections are independent of random displacements of the layers parallel to the reflecting planes. This is no longer true when the reflecting planes are not parallel to these layers, that is, when h or k differs from zero. It is easy to show (Exercise 10-18) that the randomness of the atomic positions contributes a randomness to the phases of their scattering, so that, loosely speaking, the scattering becomes incoherent.

To see under what conditions coherent scattering is possible, consider the reciprocal-lattice construction for a random-layer structure. The simplest way to derive it is to begin with some reciprocal-lattice vector,

$$\mathbf{r} = p_1\mathbf{a}^* + p_2\mathbf{b}^* + p_3\mathbf{c}^* \tag{10-67}$$

and to determine its three coefficients, p_1, p_2, and p_3, by making use of the orthogonality relations

$$\mathbf{a} \cdot \mathbf{r} = \mathbf{a} \cdot (p_1\mathbf{a}^* + p_2\mathbf{b}^* + p_3\mathbf{c}^*) = p_1$$

so that $\quad \mathbf{b} \cdot \mathbf{r} = p_2$

and $\quad \mathbf{c} \cdot \mathbf{r} = p_3$

Substituting this in (10-67) and letting $\mathbf{r} = (\mathbf{S} - \mathbf{S}_0)$, it follows that

$$(\mathbf{S} - \mathbf{S}_0) = [(\mathbf{S} - \mathbf{S}_0) \cdot \mathbf{a}]a^* + [(\mathbf{S} - \mathbf{S}_0) \cdot \mathbf{b}]\mathbf{b}^* + [(\mathbf{S} - \mathbf{S}_0) \cdot \mathbf{c}]\mathbf{c}^* \tag{10-68}$$

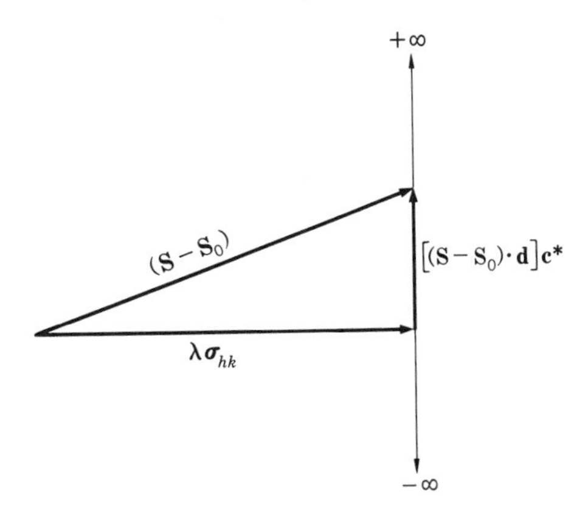

Fig. 10-43

Replacing $\mathbf{c}$ by $\mathbf{d}$ for the random-layer structure, when the first two Laue conditions are satisfied,

$$(\mathbf{S} - \mathbf{S}_0) = \lambda h\mathbf{a}^* + \lambda k\mathbf{b}^* + [(\mathbf{S} - \mathbf{S}_0) \cdot \mathbf{d}]\mathbf{c}^*$$
$$= \lambda \boldsymbol{\sigma}_{hk} + [(\mathbf{S} - \mathbf{S}_0) \cdot \mathbf{d}]\mathbf{c}^* \tag{10-69}$$

where $\boldsymbol{\sigma}_{hk}$ is a vector in the $hk0$ net of the reciprocal lattice.

In order to evaluate (10-69), it is necessary to know the magnitude of the term in brackets. Note that $|\mathbf{c}^*| = |1/\mathbf{d}|$ and that, since $\mathbf{d}$ is perpendicular to $\mathbf{a}$ and $\mathbf{b}$ by choice, it must be parallel to $\mathbf{c}^*$. Thus it is possible to diagram the relation in (10-69) as shown in Fig. 10-43. It is clear that the vector $(\mathbf{S} - \mathbf{S}_0)$ can range from a minimum value of $\lambda\boldsymbol{\sigma}_{hk}$ to an unspecified upper limit. Thus reflections for which h or k is not zero constitute continuous rods in reciprocal space rather than discrete spots. One can speak, therefore, of two-dimensional layer reflections hk.

Another way to appreciate the origin of these reciprocal-lattice rods is to consider the nature of the function $\sin^2 Nx/\sin^2 x$. As N, the number of periodic repetitions along x, decreases, the maximum about the point $x = 0$ becomes broader, until, in the limit, when $N \rightarrow 0$, the peak becomes infinitely broad. For the three-dimensional peaks constituting the normal reciprocal-lattice points, this means that when one of the three N's decreases, the peak broadens along the corresponding direction in reciprocal space. This viewpoint will be pursued later. Returning to the random-layer structure, it follows from the above discussion that its reciprocal lattice consists of discrete $00l$ points and reciprocal-lattice rods parallel to the $\mathbf{c}^*$ axis, passing through the $hk0$ points and extending from plus to minus infinity (Fig. 10-44).

Stacking faults. Ideally, the stacking sequence of a simple closest packing follows a regular sequence. During the growth of the crystal, however, it is possible that the atomic layer being formed on a previously established one begins to form incorrectly; that is, the atoms adhere to the wrong set of sites on the crystal's surface. If growth then proceeds rapidly, it often happens that the crystal does not have time to "correct" this mistake, and a *growth fault* is produced. For

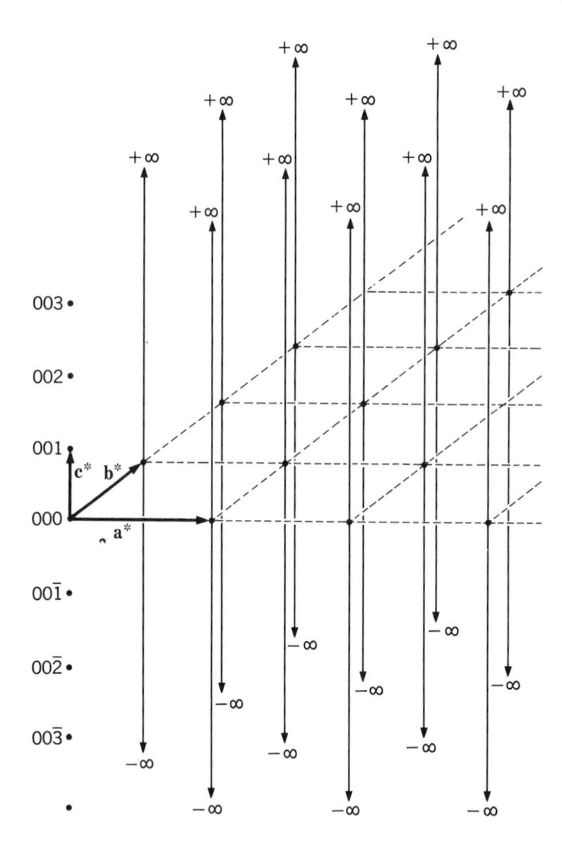

Fig. 10-44. Portion of reciprocal-lattice construction for a random-layer structure.

example, consider the cubic closest packing. Suppose that after some time the sequence . . . $ABCAB$ is disrupted by the formation of another A layer. The thermodynamically more stable cubic three-layer sequence then requires that the next layer differ from the preceding one, so that the sequence becomes

$$. . . ABCABACBAC . . . \tag{10-70}$$

with the growth fault lying at the center of the stacking sequence in (10-70).

Suppose that an otherwise perfect crystal is deformed in shear by sliding one half of it past the other half along a closest-packed plane. Since the stacking sequence in both halves is unchanged, the resulting sequence, assuming that the crystal has a cubic closest packing, then becomes

$$. . . ABCABABCABC . . . \tag{10-71}$$

The stacking fault in (10-71) again follows the B layer at the center of the sequence, and is called a *deformation fault.* Note that the growth fault in (10-70) actually is a twin boundary relating the twins by a mirror reflection, whereas no such relation exists between the mechanically displaced halves in (10-71). It is also possible to distinguish a double deformation fault in a ccp, in which case the stacking sequence becomes . . . $ABCACBCABC$ . . . . Similar faults can be distinguished in hcp (Exercise 10-19).

Regardless of the nature of the stacking fault, it is clear that the structure in the immediate vicinity of this plane defect differs from that encountered elsewhere in the crystal. Thus a kind of two-dimensional structure is produced which tends to elongate the reciprocal-lattice point into a rod normal to the stacking plane. The degree and exact nature of the broadening obviously depend on the structure of the fault, that is, whether it is a growth fault or a deformation fault and whether it occurs in a ccp or hcp, and on the frequency of occurrence of the faults. The mathematical formulation is carried out by introducing probability functions to express the occurrence frequency and by defining a displacement vector which describes the shifts of the successive layers, that is, the structure of the fault. Generally, experimental studies are carried out using polycrystalline aggregates, notably metals, which introduces a further complication, briefly considered in the closing sections of this chapter. A big stimulus was given to such studies by M. S. Paterson in 1952 when he showed that it is possible to distinguish growth faults from stacking faults by noting their effects in a powder photograph or diffractometer tracing. It turns out that deformation faults in cubic closest packings produce a symmetrical broadening of the reflection peaks, accompanied by a small change in the reflection angle. Growth or twin faults in ccp, on the other hand, produce a symmetric broadening. Others have shown how one can set up equations for separating these effects from each other when both kinds of stacking faults are present in the same closest packing.

The occurrence of stacking faults is not limited to simple metals or even to relatively simple structures like the SiC structures based on closest packings. More complicated structures, such as many silicates, also exhibit systematic plane defects that can be classified as faults. In all cases, the presence of such defects causes an elongation or broadening of the reciprocal-lattice points into rods, platelets, or other characteristic shapes. A relatively simple example of such faults in wollastonite, $CaSiO_3$, is discussed in Exercise 10-20. Further examples and more detailed mathematical expositions can be found in some of the references at the end of this chapter. It should be noted here, however, that similar distortions are produced by any two-dimensional defects in crystals. For example, it sometimes happens that a new phase is exsolved by the parent crystal. This new phase frequently precipitates along certain preferred planes in a crystal, attaining a fairly large size in two dimensions, but remaining limited to just a few atomic layers in the third. As should be expected, the reciprocal-lattice construction for the new precipitate consists of rods. Because the new phase generally has two translations that tend to coincide with those of the parallel plane in the parent crystal, the reciprocal-lattice rods of the new phase pass through the reciprocal-lattice points of the parent crystal. An examination of the reciprocal lattice of the crystal therefore enables a determination of the orientation relationships and a deduction of the shape and size of the precipitates. Such analyses are considerably simplified when the undistorted reciprocal lattice is recorded using some of the procedures described in Chap. 16.

POLYCRYSTALLINE AGGREGATES

Laue theorem. It has been shown previously in this chapter that it is possible to determine the integrated reflecting power of a crystal by integrating the interference function $I(hkl)$ over one unit cell in reciprocal space. It is the object of the present discussion to combine this procedure with the relations derived in Chap. 9 for the diffraction power of polycrystalline aggregates. Recall from that chapter that the fraction of N irradiated crystallites that are oriented to reflect x rays at a glancing angle θ is given by

$$\frac{dN}{N} = \tfrac{1}{2} \cos \theta \, d\alpha \tag{9-48}$$

where $d\alpha$ measures the small deviations of the reflecting planes from the glancing angle.

The total power reflected by one crystal toward a detector a distance R away is

$$R_{hkl} = \iint I(hkl) R^2 \, d\beta \, d\gamma \tag{10-72}$$

where $R\,d\beta \times R\,d\gamma$ is an infinitesimal area on the detector's surface. Multiplying (10-72) by the total number dN of crystallites in position to reflect x rays at the angle θ, one gets the total diffracted power P of the polycrystalline aggregate

$$\begin{aligned} P &= \int R_H \, dN \\ &= \frac{R^2 N \cos \theta}{2} \iiint I(hkl) \, d\alpha \, d\beta \, d\gamma \end{aligned} \tag{10-73}$$

where both $I(hkl)$ and P are expressed in electron units. As already shown in detail in the preceding chapter, the small angular changes $d\alpha$, $d\beta$, and $d\gamma$ produce small vector displacements $\mathbf{d\alpha}/\lambda$ perpendicular to $\mathbf{S}_0/\lambda$ and $\mathbf{d\beta}/\lambda$ and $\mathbf{d\gamma}/\lambda$ normal to $\mathbf{S}/\lambda$. The resultant volume swept out by the vector $(\mathbf{S} - \mathbf{S}_0)/\lambda$ then is

$$\frac{\mathbf{d\alpha}}{\lambda} \cdot \frac{\mathbf{d\beta}}{\lambda} \times \frac{\mathbf{d\gamma}}{\lambda} = \frac{\sin 2\theta}{\lambda^3} \, d\alpha \, d\beta \, d\gamma \tag{10-74}$$

But this same volume element can be expressed in terms of small changes in the reciprocal-lattice vectors

$$dh \, \mathbf{a}^* \cdot dk \, \mathbf{b}^* \times dl \, \mathbf{c}^* = V^* \, dh \, dk \, dl \tag{10-75}$$

Combining relations (10-75) with (10-74) and substituting in (10-73), the final form of the relation becomes

$$P = \frac{R^2 N \lambda^3}{4V \sin \theta} \iiint I(hkl) \, dh \, dk \, dl \tag{10-76}$$

after substituting $V = 1/V^*$ and $\sin 2\theta = 2 \sin \theta \cos \theta$ and factoring out like terms. Relation (10-76) is the Laue theorem expressed for a polycrystalline aggregate. It is the analytical expression of the statement that, regardless of the nature of the interference function, the total diffracted power is given by integrating

$I(hkl)$ over the appropriate volume in reciprocal space. Its chief usefulness arises from the fact that, usually, one can construct an interference function for a particular structure model. This function is then inserted in (10-76), and the Laue theorem used to convert it to the experimentally measured diffraction power. As an example of this procedure, we consider next the case of crystals having layered structures.

Layered structures. When the structure of a crystal consists of layers of atoms stacked on top of each other, it is convenient to think of the diffraction by a two-dimensional layer. Without specifying the nature of the stacking, it is clear from (10-66) that the corresponding two-dimensional interference function has the form

$$I(hkl) = F^2 \frac{\sin^2 \pi h N_1}{\sin^2 \pi h} \frac{\sin^2 \pi k N_2}{\sin^2 \pi k} \tag{10-77}$$

and expresses the intensity of scattering by one layer.

For simplicity, suppose the stacking sequence is quite random, so that the reciprocal-lattice construction is as shown in Fig. 10-44. It is clear that the interference function in this case has appreciable values only close to the reciprocal-lattice rods. Let $h = n_1 + h'$ and $k = n_2 + k'$, where n_1 and n_2 are the integral indices of the $hk0$ reciprocal-lattice point through which the rods pass, while h' and k' are very small quantities. Substituting in (10-77) and noting that

$$\sin^2 \pi(n_1 + h') = \sin^2 \pi h'$$

$$\begin{aligned} I(hkl) &= F^2 \frac{\sin^2 \pi h' N_1}{\sin^2 \pi h'} \frac{\sin^2 \pi k' N_2}{\sin^2 \pi k'} \\ &= F^2 \frac{\sin^2 \pi h' N_1}{(\pi h')^2} \frac{\sin^2 \pi k' N_2}{(\pi k')^2} \end{aligned} \tag{10-78}$$

since $\pi h'$ and $\pi k'$ are very small angles.

Substituting the interference function (10-78) in the Laue theorem (10-76) and noting that $dh = 0 + dh'$ and, similarly, $dk = dk'$, the total power diffracted is

$$\begin{aligned} P &= \frac{R^2 N \lambda^3}{4V} \iiint_{-\infty}^{+\infty} \frac{F^2}{\sin\theta} \frac{\sin^2 \pi h' N_1}{(\pi h')^2} \frac{\sin^2 \pi k' N_2}{(\pi k')^2} \, dh' \, dk' \, dl \\ &= \frac{R^2 N \lambda^3}{4V} N_1 N_2 \int_{-\infty}^{+\infty} \frac{F^2 \, dl}{\sin\theta} \end{aligned} \tag{10-79}$$

Experimentally, one does not measure the total diffracted power of a polycrystalline aggregate, but rather the diffracted power per unit angle, usually expressed in units of 2θ. The total power then is given by

$$P = \int P(2\theta) \, d(2\theta) \tag{10-80}$$

To express (10-79) in such units, consider one reciprocal-lattice rod stretching from minus to plus infinity in Fig. 10-44. Next, consider the vector $(\mathbf{S} - \mathbf{S}_0)/\lambda$ from the reciprocal-lattice origin to this particular rod. When it has moved a distance $dl\,\mathbf{c}^*$ along the rod, its length changes by an amount $d[(\mathbf{S} - \mathbf{S}_0)/\lambda]$. According to the construction shown in Fig. 10-45,

$$
\begin{aligned}
dl|\mathbf{c}^*| \sin \varphi &= d\left(\frac{\mathbf{S} - \mathbf{S}_0}{\lambda}\right) \\
\lambda|\mathbf{c}^*|\, dl \sin \varphi &= d(2 \sin \theta) \\
&= 2 \cos \theta \, d\theta
\end{aligned}
\tag{10-81}
$$

Since $d\theta$ is very small, $2d\theta = d(2\theta)$, so that, rearranging the terms in (10-81),

$$dl = \frac{\cos \theta \, d(2\theta)}{\lambda|\mathbf{c}^*| \sin \varphi} \tag{10-82}$$

From the construction in Fig. 10-45,

$$\cos \varphi = \frac{|\boldsymbol{\sigma}_{hk}|}{|(\mathbf{S} - \mathbf{S}_0)/\lambda|} = \frac{\lambda|\boldsymbol{\sigma}_{hk}|}{|(\mathbf{S} - \mathbf{S}_0)|} = \frac{2 \sin \theta_{\min}}{2 \sin \theta} \tag{10-83}$$

where $\theta_{\min}$ is the smallest glancing angle at which this reciprocal-lattice rod intersects the sphere of reflection. Substituting $(1 - \cos^2 \varphi)^{\frac{1}{2}}$ for $\sin \varphi$ in (10-81), it is possible to express the total power (10-79) in units of the diffraction angle θ. To complete the expression, however, it is necessary to multiply the right side of (10-79) by 2, to take account of identical contributions from the top and bottom halves of the reciprocal-lattice rod, and by a factor j', expressing the multiplicities between permutations of h and k, dependent on the crystal symmetry. Finally, the total diffracted power, in electron units,

$$P = \frac{j'NR^2\lambda^2N_1N_2}{2V|\mathbf{c}^*|} \int \frac{F^2 \cos \theta \, d(2\theta)}{(\sin^2 \theta - \sin^2 \theta_{\min})^{\frac{1}{2}}} \tag{10-84}$$

This expression is identical with (10-80), so that differentiating both sides with

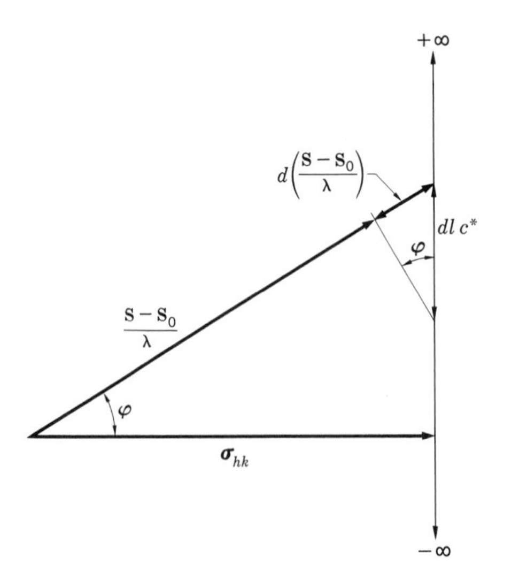

Fig. 10-45

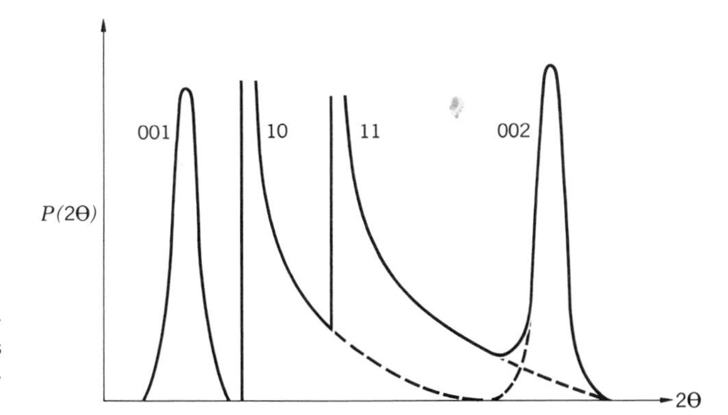

Fig. 10-46. Idealized diffractometer tracing for crystallites having random-layer structures.

respect to (2θ),

$$\frac{dP}{d(2\theta)} = P(2\theta) = \frac{j'NR^2\lambda^2N_1N_2F^2\cos\theta}{2A(\sin^2\theta - \sin^2\theta_{\min})^{\frac{1}{2}}} \tag{10-85}$$

where $A = Vc^* = V/d$ is the area of a cell in the layer.

The appearance of a powder photograph of crystallites having a random layered structure can be predicted by considering the reciprocal-lattice construction in Fig. 10-44 and expression (10-85) for the diffraction power of hk reflections. The $00l$ reflections, obviously, are ordinary crystal reflections yielding sharp peaks. The hk reflections, on the other hand, begin at an angle $\theta = \theta_{\min}$, at which point $P(2\theta)$ is infinite, and decrease rapidly as θ increases, because $\cos\theta$ tends to zero. Thus a diffractometer tracing of such a sample has the general characteristics shown in Fig. 10-46.

The foregoing analysis illustrates the main features of the procedure used in interpreting the experimentally measured diffraction power from imperfect crystals in a polycrystalline aggregate. First, the interference function for the particular defect model is formulated and inserted into the Laue theorem (10-76). By judicious choice of variables, the power is then made a function of the glancing angle. This is recognized to be the experimentally measured power, so that it can be compared with the measured intensity forthwith. In practice, it is first necessary to express this as a power per unit length of the Debye ring and to convert it into absolute units. The procedure for doing this has been discussed in detail in Chap. 9.

SUGGESTIONS FOR SUPPLEMENTARY READING

Leonid V. Azároff, X-ray diffraction studies of crystal perfection, *Progr. Solid State Chem.*, vol. 1 (1964), pp. 347–379.

——— Imperfections in crystals and their effect on diffraction by crystals, in G. N. Ramachandran (ed.), *Advanced methods of crystallography* (Academic Press Inc., New York, 1964), pp. 251–269.

W. W. Beeman, P. Kaesberg, J. W. Andregg, and M. B. Webb, Size of particles and lattice defects, *Handbuch der Physik*, vol. 32 (1957), pp. 321–442.

W. Cochran and G. S. Pawley, The theory of diffuse scattering by a molecular crystal, *Proc. Royal Soc. of London,* ser. A, vol. 280 (1964), pp. 1–22.
A. Guinier, *X-ray diffraction* (W. H. Freeman and Company, San Francisco, 1963).
A. Guinier and J. Tennevin, Researches on the polygonization of metals, *Progress in Metal Physics,* vol. 2 (1950), pp. 177–192.
P. B. Hirsch, Mosaic structure, *Progress in Metal Physics,* vol. 6 (1955), pp. 236–339.
H. Jagodzinski, Diffuse disorder scattering by crystals, in G. N. Ramachandran (ed.), *Advanced methods of crystallography* (Academic Press Inc., New York, 1964), pp. 181–219.
R. W. James, *The optical principles of the diffraction of x-rays* (G. Bell & Sons, Ltd., London, 1950).
H. Lipson, Order-disorder changes in alloys, *Progress in Metal Physics,* vol. 2 (1950), pp. 1–52.
J. B. Newkirk and J. H. Wernick (eds.), *Direct observation of imperfections in crystals* (Interscience Publishers, Inc., New York, 1962).
G. N. Ramachandran (ed.), *Crystallography and crystal perfection* (Academic Press Inc., New York, 1963).
B. E. Warren, X-ray studies of deformed metals, *Progress in Metal Physics,* vol. 8 (1957), pp. 147–202.
A. J. C. Wilson, *X-Ray optics,* 2d ed. (Methuen & Co. Ltd., London, 1962).
W. A. Wooster, *Diffuse x-ray reflections from crystals* (Oxford University Press, Fair Lawn, N.J., 1962).
William H. Zachariasen, *Theory of x-ray diffraction in crystals* (John Wiley & Sons, Inc., New York, 1945).

EXERCISES

10-1. The presence of point defects in crystals can be described as a solid solution of such defects in the host crystal. Three types of solid solutions can be distinguished:

1. Interstitial solid solution
2. Substitutional solid solution
3. Omission solid solution

If the third classification is applied to a crystal "dissolving" vacancies in its structure, prepare a graph showing how the density of a crystal should vary with defect concentration for each of three kinds of solid solutions.

10-2. Zinc oxide has a hexagonal structure, $a = 3.243$ Å and $c = 5.195$ Å, with two formula weights per cell. Consider a ZnO crystal containing a stoichiometric excess of zinc atoms. What is the possible nature of the defects that may be present? If the measured density of the crystal is 5.606 g/cm³, what is the concentration for each type of defect possible?

10-3. The diffraction by a silver crystal is being studied at a temperature of $500°$C, using $\mathrm{Cu}\ K\alpha$ radiation ($\lambda = 1.54$ Å). If the Debye characteristic temperature for silver is $215°$K, evaluate the magnitude of the temperature factor $2B$ at large scattering angles. By how much is the intensity of the 444 reflection attenuated if the unit-cell edge of the face-centered cubic cell $a = 4.086$ Å? What would the attenuation be at room temperature ($20°$C)?

10-4. Making use of the $2B$ values in Exercise 10-3 for a silver crystal, what is the order of magnitude of the mean-square displacements at room temperature and at $500°$C?

10-5. The double summation in the expression for the crystal reflection leads to the quotients of the type $\sin^2(Nx/2)/\sin^2(x/2)$, where N is the total number of unit cells in the crystal, as described in Chap. 10. Considering the value of this quotient near $x = 0$, it can be represented by a sum of the type

$$N + 2(N-1)\cos x + 2(N-2)\cos 2x + 2(N-3)\cos 3x + \cdots$$

Plot each of the first 10 terms in the above series, for x ranging from 0 to π, on the same graph paper. Then compare the curve resulting when only the first five terms are summed with that obtained when all 10 terms are added together. This illustrates an important aspect of x-ray diffraction, namely, the mathematical basis of the formation of sharp diffraction maxima.

10-6. Under what conditions does the TDS intensity given by (10-26) reduce to the monotonically varying function given in (10-12)? What does this tell you about the physical meaning of the temperature effect? Show, by simple algebra, that there exists a conservation "law" for the total intensity of coherent scattering by a crystal.

10-7. In studying x-ray diffraction by a copper crystal using Cu $K\alpha$ radiation, assuming the original Debye theory, at what temperature does the diffuse scattering at large angles consist of equal contributions from TDS and Compton modified radiation? ($\lambda = 1.54$ Å, $\Sigma f_e^2 = 8$, and $f = 10.95$.)

10-8. (*a*) What is the difference between the kinds of crystals that can be studied best by the three kinds of transmission arrangements described in this chapter? (*b*) Could the Guinier-Tennevin arrangement be used with a very perfect crystal to observe the anomalous-transmission effect? (*c*) Could it be used for the examination of crystals by the procedure developed by Lang?

10-9. Derive the Bragg and Williams long-range-order parameter as given by (10-31) in the text. To obtain such a linear expression, assume $S = a + br_\alpha$ and solve for a and b, remembering that $S = 1$ for perfect order and $S = 0$ for random disorder.

10-10. Ordered beta brass, CuZn, has the CsCl structure in which one kind of atom occupies the corners and the other the center of a cubic cell. Derive the expressions for the intensity of fundamental and superstructure reflections for β brass. If Mo $K\alpha$ radiation is used, would you expect to be able to detect any superstructure reflections? What means for observing them in CuZn can you suggest?

10-11. Consider a fully disordered crystal of CuZn. Can any of the methods proposed for observing superstructure reflections in this crystal be applied to detecting the Laue monotonic scattering? How is the diffuse scattering related to the intensity-conservation principle evinced by TDS?

10-12. Determine the values of c_i for a cubic closest packing and for a body-centered cubic structure for i values up to 5. Express r_i in terms of the cubic cell edge $a = 1.0$.

10-13. Plot the intensity of x-ray diffraction by an A_3B crystal (having the Au_3Cu structure) when it is fully ordered and when it is randomly disordered. Include the crystal reflections from (100), (110), (111), (200), (210), and (211) planes, referred to the ordered cell. For simplicity, assume that $f_A = f_B = f$ and plot I/f^2.

10-14. Derive expression (10-55) in the text. Note that this is a simple exercise in algebra, except that it is necessary to recognize simplifying relations like $m_A n_i' = m_B n_i$.

10-15. Evaluate the short-range-order parameter α_i defined by (10-59) in the text for Au_3Cu and CuZn types of structures. Do this for the first five values of i. What is the physical significance of a negative α_i value?

10-16. Consider the hexagonal closest-packed layer in Fig. 10-38. By dividing the area of the plane into equal hexagons surrounding each circle, prove that 90.7% of the available space in a plane is occupied by the circles. For comparison, consider a square array of circles in the plane group $p4mm$. Show that only 78.5% of the area is occupied by the circles in this case.

10-17. Prepare a scaled drawing of the stacking sequence . . . $ABAB$ . . . by drawing the A layer superimposed by a B layer and another A layer. Show that the hexagonal unit cell contains two atoms at 0,0,0 and $\frac{2}{3},\frac{1}{3},\frac{1}{2}$, respectively. Also determine the space group and the c/a ratio of this hexagonal closest packing. How many atoms and how many lattice points are there in each unit cell?

10-18. Consider the scattering by a random-layer structure when h or k is not zero. In this case, the amplitude of scattering by the nth atom is proportional to $E_n e^{i\varphi_n}$, where E_n is the amplitude of the electric vector. Show that, if the phases φ_n are random, the intensity obtained by multiplying the amplitude by its complex conjugate is analogous to the incoherent scattering intensity discussed in Chap. 8.

10-19. Show the nature of a growth fault and a deformation fault in a hexagonal closest packing by listing the stacking sequences for each case. Is a double-deformation fault possible in hcp?

10-20. An examination of the reciprocal lattice of wollastonite shows that the hkl reflections are sharp when k is an even number, but are drawn out into streaks parallel to $\mathbf{a}^*$ when k is odd. If the broadening is due to stacking faults, what is the stacking-fault plane in the crystal? What is the nature of the relative displacements of the two layers on each side of the stacking fault? **Hint:** Consider the significance of the fact that only when k is odd do the reflections broaden.

10-21. Consider a closest packing of like spheres, and locate the origin of each hexagonal closest-packed layer by the vector $\eta(\mathbf{a}_1 - \mathbf{a}_2)/3$, where

$$\eta = \begin{cases} 0 & \text{for an } A \text{ layer} \\ +1 & \text{for a } B \text{ layer} \\ -1 & \text{for a } C \text{ layer} \end{cases}$$

Within the layer there is only one atom per lattice point, so that $R_m = m_1\mathbf{a}_1 + m_2\mathbf{a}_2 + m_3\mathbf{c} + \eta(\mathbf{a}_1 + \mathbf{a}_2)/3$, where η depends on the layer considered. Show that this leads to the usual crystal reflections for all hkl values as long as η changes in a periodic fashion. Show that it leads to layer reflections hk when η can change in random manner from layer to layer. Note that, in a closest packing, adjacent layers must have different η values.

ELEVEN
Crystal-structure analysis

STRUCTURE-FACTOR CALCULATION

Simple structures. The way a structure factor combines the amplitudes and relative phases of x rays scattered by individual atoms in a unit cell has been discussed in Chaps. 8 and 9. To calculate the structure factor from the relation

$$F_{hkl} = \sum_{n=1}^{N} f_n e^{2\pi i(hx_n+ky_n+lz_n)} \tag{9-19}$$

it is only necessary to know what the scattering-factor magnitudes and cell coordinates of the N different atoms are. (Before proceeding with the subject matter of this chapter, the reader is advised to review carefully the contents of Chap. 4.) As an example, consider the simple face-centered cubic structure, say, that of copper, in which the Cu atoms occupy the equipoint set $0,0,0$; $\frac{1}{2},\frac{1}{2},0$; $\frac{1}{2},0,\frac{1}{2}$; and $0,\frac{1}{2},\frac{1}{2}$. The structure-factor expression then becomes

$$\begin{aligned} F_{hkl} &= f_{\mathrm{Cu}}(e^0 + e^{\pi i(h+k)} + e^{\pi i(h+l)} + e^{\pi i(k+l)}) \\ &= \begin{cases} 4f_{\mathrm{Cu}} & \text{when } h,k,l \text{ unmixed} \\ 0 & \text{when } h,k,l \text{ mixed} \end{cases} \end{aligned} \tag{11-1}$$

Next consider another structure based on a face-centered cubic lattice, NaCl. The two kinds of atoms in NaCl occupy the equipoints (in space group $F\frac{4}{m}\bar{3}\frac{2}{m}$)

$$\begin{aligned} &\text{Na in } 4a\text{:} \quad 0,0,0;\ \tfrac{1}{2},\tfrac{1}{2},0;\ \tfrac{1}{2},0,\tfrac{1}{2};\ 0,\tfrac{1}{2},\tfrac{1}{2} \\ &\text{Cl in } 4b\text{:} \quad \tfrac{1}{2},\tfrac{1}{2},\tfrac{1}{2};\ 0,0,\tfrac{1}{2};\ 0,\tfrac{1}{2},0;\ \tfrac{1}{2},0,0 \end{aligned} \tag{11-2}$$

The structure-factor expression for NaCl written out term by term is

$$\begin{aligned} F_{hkl} &= f_{\mathrm{Na}}(e^0 + e^{\pi i(h+k)} + e^{\pi i(h+l)} + e^{\pi i(k+l)}) + f_{\mathrm{Cl}}(e^{\pi i(h+k+l)} + e^{\pi il} + e^{\pi ik} + e^{\pi ih}) \\ &= \begin{cases} 4(f_{\mathrm{Na}} + f_{\mathrm{Cl}}) & \text{when } h,k,l \text{ all even} \\ 4(f_{\mathrm{Na}} - f_{\mathrm{Cl}}) & \text{when } h,k,l \text{ all odd} \\ 0 & \text{when } h,k,l \text{ mixed} \end{cases} \end{aligned} \tag{11-3}$$

Note that, after writing out the structure-factor expression term by term, it is usually possible to simplify the expression considerably. As discussed below, and illustrated in Exercises 11-1 and 11-2, such simplifications arise even in the case of more complex structures.

The foregoing discussion also illustrates another important point, that is, the consequences of the presence of subtranslations in the lattice array. When a centered cell is chosen to describe an atomic array, this means that any atom placed at some point xyz in the unit cell is repeated by the subtranslations of the chosen cell. For example, in a face-centered cell the four lattice points per cell have the coordinates listed in the equipoint set $4a$ in (11-2). An atom placed at some point xyz then must be repeated to the translation-equivalent points: $x + \frac{1}{2}, y + \frac{1}{2}, z + 0$; $x + \frac{1}{2}, y + 0, z + \frac{1}{2}$; and $x + 0, y + \frac{1}{2}, z + \frac{1}{2}$. This is the case, for example, for the four chlorine atoms in (11-2). The presence of additional symmetry may cause the atom at xyz to be repeated even more times (Chap. 4), but the above translation-equivalent sites represent the minimum number of repetitions required by the centered cell. This is a very important consequence of selecting a nonprimitive cell in a lattice.

In order to emphasize this, consider one final example. The crystal structure of cesium has a body-centered cubic lattice with one atom per lattice point. Thus the two Cs atoms per cell occupy the sites $0,0,0$ and $\frac{1}{2},\frac{1}{2},\frac{1}{2}$ so that

$$\begin{aligned} F_{hkl} &= f_{\mathrm{Cs}}(e^0 + e^{\pi i(h+k+l)}) \\ &= f_{\mathrm{Cs}}[1 + (-1)^{h+k+l}] \\ &= \begin{cases} 2f_{\mathrm{Cs}} & \text{when } h + k + l = 2n \\ 0 & \text{when } h + k + l = 2n + 1 \end{cases} \end{aligned} \tag{11-4}$$

Next consider the crystal structure of CsCl, which has a primitive cubic lattice with two atoms (one formula unit) per cell. The atomic coordinates in space group $P\frac{4}{m}\bar{3}\frac{2}{m}$ are

$$\begin{aligned} &\text{Cs in } 1a\text{:} \quad 0,0,0 \\ &\text{Cl in } 1b\text{:} \quad \tfrac{1}{2},\tfrac{1}{2},\tfrac{1}{2} \end{aligned} \tag{11-5}$$

so that

$$\begin{aligned} F_{hkl} &= f_{\mathrm{Cs}}e^0 + f_{\mathrm{Cl}}e^{\pi i(h+k+l)} \\ &= \begin{cases} f_{\mathrm{Cs}} + f_{\mathrm{Cl}} & \text{when } h + k + l = 2n \\ f_{\mathrm{Cs}} - f_{\mathrm{Cl}} & \text{when } h + k + l = 2n + 1 \end{cases} \end{aligned} \tag{11-6}$$

Note that the primitive lattice does not require a repetition of the atoms by any other than the full cell translations. Thus the *systematic absences* of reflections when $h + k + l$ is an odd number, caused by the choice of a body-centered cell in (11-4), do not occur when a primitive cell is chosen. As discussed below, the detection of such systematic absences among the observed reflections can be used to determine the correct lattice type of the crystal examined.

Symmetry factors. When a crystal structure is such that the unit cell contains many atoms, the calculation of structure factors can become quite tedious. Fortunately, the presence of symmetry in the crystal makes it possible to reduce this labor because the contributions to a structure factor by atoms occupying equivalent positions in the cell can be considered in sets. Let the symmetry factor of a set be

$$S = \sum_{j=1}^{m} e^{2\pi i(hx_j + ky_j + lz_j)} \tag{11-7}$$

where the sum over j includes all the coordinates of the m equivalent positions in one equipoint set. Then the structure-factor expression (9-19) can be written

$$\begin{aligned} F_{hkl} &= \sum_{n'} f_{n'} S_{n'} \\ &= f_A S_A + f_B S_B + f_C S_C + \cdots \end{aligned} \tag{11-8}$$

because the A atoms distributed among their equipoint set, whose trigonometric contribution is expressed by the symmetry factor S_A, all scatter with the same amplitude f_A, and so forth, so that the summation in (11-8) is carried out over the atoms in each set as a group rather than individually.

The symmetry factor in (11-7) can be expanded into a real part A and an imaginary part B by Euler's relation

$$\begin{aligned} S &= A + iB \\ &= \sum_j \cos 2\pi(hx_j + ky_j + lz_j) + i \sum_j \sin 2\pi(hx_j + ky_j + lz_j) \end{aligned} \tag{11-9}$$

where $i = \sqrt{-1}$. When a crystal has a centrosymmetric structure and the symmetry center is chosen as the origin of coordinates, then for every atom at a point xyz there is another at $\bar{x}\bar{y}\bar{z}$ in the same equipoint set. In this case $B = 0$ in (11-9) because $\sin(-x) = -\sin x$, and the sine terms sum to zero in pairs. When the structure lacks an inversion center or when the origin is not placed at a symmetry center, both A and B have nonzero values.

Experimentally, one measures the magnitude of F_{hkl}, and it is possible to calculate it from (11-8) and (11-9) since

$$\begin{aligned} |F_{hkl}| &= \left| \sum_{n'} f_{n'} S_{n'} \right| \\ &= \left| \sum_{n'} f_{n'} (A + iB)_{n'} \right| \\ &= \left[\left(\sum_{n'} f_{n'} A_{n'} \right)^2 + \left(\sum_{n'} f_{n'} B_{n'} \right)^2 \right]^{\frac{1}{2}} \end{aligned} \tag{11-10}$$

It is also possible to calculate the phase angle of the complex structure factor

$$F_{hkl} = |F_{hkl}| e^{i\alpha(hkl)} \tag{11-11}$$

where $$\alpha(hkl) = \tan^{-1} \frac{\sum_{n'} f_{n'} B_{n'}}{\sum_{n'} f_{n'} A_{n'}} \tag{11-12}$$

Note that, for centrosymmetric structures, $B = 0$ and $\alpha(hkl) = \tan^{-1} 0 = 0$ or π, so that $F_{hkl} = \pm|F_{hkl}|$ according to (11-11).

It is obviously possible to substitute the atomic coordinates of the equipoint set of a general position in (11-9) and to calculate the real and imaginary components of the symmetry factor for any space group. This has been done for all 230 space groups, and the results are tabulated in Volume 1 of *International tables for x-ray crystallography*. The way this information can be used in structure-factor calculations is considered next.

Complex structures. When a unit cell contains a large number of different kinds of atoms but the atoms occupy special positions, it is usually more convenient to calculate the structure-factor expressions directly from (9-19) without recourse to tabulated symmetry-factor expressions (Exercise 11-2). When the atoms occupy more generalized positions, however, it is much easier to use the symmetry factors and relation (11-10). As an illustration of this procedure, consider the crystal structure of cubanite, $CuFe_2S_3$. The space group of cubanite is $Pnma$, No. 62 in *International tables,* and because it is centrosymmetric, $B = 0$ when the symmetry center is chosen as the origin of coordinates. For this choice, Volume 1 lists the following symmetry-factor expression:

$$A = 8 \cos 2\pi \left(hx - \frac{h+k+l}{4}\right) \cos 2\pi \left(ky + \frac{k}{4}\right) \cos 2\pi \left(lz + \frac{h+l}{4}\right) \tag{11-13}$$

It simplifies considerably, however, when special combinations of h, k, and l are considered. Specifically, when

$h + l$ is even, k is even:

$$A = 8 \cos 2\pi hx \cos 2\pi ky \cos 2\pi lz \tag{11-14}$$

$h + l$ is even, k is odd:

$$A = -8 \sin 2\pi hx \sin 2\pi ky \cos 2\pi lz \tag{11-15}$$

$h + l$ is odd, k is even:

$$A = -8 \sin 2\pi hx \cos 2\pi ky \sin 2\pi lz \tag{11-16}$$

$h + l$ is odd, k is odd:

$$A = -8 \cos 2\pi hx \sin 2\pi ky \sin 2\pi lz \tag{11-17}$$

The unit cell of cubanite contains four formula weights, so that the following distribution of atoms among the available equipoints in $Pnma$ takes place:

$$\begin{array}{ll} 4\text{Cu in } c: & x,\tfrac{1}{4},z;\ \bar{x},\tfrac{3}{4},\bar{z};\ \tfrac{1}{2}-x,\tfrac{3}{4},\tfrac{1}{2}+z;\ \tfrac{1}{2}+x,\tfrac{1}{4},\tfrac{1}{2}-z \\ 8\text{Fe in } d: & x,y,z;\ \tfrac{1}{2}+x,\tfrac{1}{2}-y,\tfrac{1}{2}-z;\ \bar{x},\tfrac{1}{2}+y,\bar{z};\ \tfrac{1}{2}-x,\bar{y},\tfrac{1}{2}+z \\ & \bar{x},\bar{y},\bar{z};\ \tfrac{1}{2}-x,\tfrac{1}{2}+y,\tfrac{1}{2}+z;\ x,\tfrac{1}{2}-y,z;\ \tfrac{1}{2}+x,y,\tfrac{1}{2}-z \\ 4\text{S}_\text{I} \text{ in } c: & x,\tfrac{1}{4},z;\ \text{etc.} \\ 8\text{S}_\text{II} \text{ in } d: & x,y,z;\ \text{etc.} \end{array} \tag{11-18}$$

The values of the unspecified coordinates are

$$\begin{array}{llll} \text{Copper in } c: & x = 0.127 & y \text{ (special)} & z = 0.583 \\ \text{Iron in } d: & x = 0.134 & y = 0.088 & z = 0.088 \\ \text{Sulfur I in } c: & x = 0.262 & y \text{ (special)} & z = 0.913 \\ \text{Sulfur II in } d: & x = 0.274 & y = 0.413 & z = 0.084 \end{array} \tag{11-19}$$

The above structure illustrates several aspects of atomic distributions among the equipoints of a space group. Because the general position (d) in $Pnma$ has a rank of 8, the 12 sulfur atoms must be distributed among at least two crystallographically distinct sets of points. Similarly, different kinds of atoms may occupy the same sets of general positions, provided that their respective coordinates are suitably chosen. Also note that the y coordinate of the atoms in position c need not be specified, since it lies on the mirror planes at $y = \frac{1}{4}$ and $y = \frac{3}{4}$ when the cell origin is placed at the symmetry center.

Proceeding to the calculation of $h0l$ structure factors, the same symmetry-factor expression is used for each of the atomic sets whose coordinates are given in (11-19). Note that the symmetry-factor expressions do not change in value, regardless of which coordinate triplet in an equipoint set is chosen (Exercise 11-8). Thus the structure-factor expression for cubanite can be written

$$F_{hkl} = 4f_{\mathrm{Cu}}A_{\mathrm{Cu}} + 8f_{\mathrm{Fe}}A_{\mathrm{Fe}} + f_{\mathrm{S}}(4A_{\mathrm{S_I}} + 8A_{\mathrm{S_{II}}}) \tag{11-20}$$

where the numerical coefficients take account of the equipoint ranks. As a specific example, consider hkl reflections, for which $h + l$ and k are both even. Then, from (11-14) and (11-19),

$$\begin{aligned} A_{\mathrm{Cu}} &= 8 \cos 2\pi h(0.127) \cos 2\pi k(\tfrac{1}{4}) \cos 2\pi l(0.583) \\ A_{\mathrm{Fe}} &= 8 \cos 2\pi h(0.134) \cos 2\pi k(0.088) \cos 2\pi l(0.088) \\ A_{\mathrm{S_I}} &= 8 \cos 2\pi h(0.262) \cos 2\pi k(\tfrac{1}{4}) \cos 2\pi l(0.913) \\ A_{\mathrm{S_{II}}} &= 8 \cos 2\pi h(0.274) \cos 2\pi k(0.413) \cos 2\pi l(0.084) \end{aligned}$$

Similar expressions are obtained for the other reflections.

The actual structure-factor calculation thus consists of three parts: (*1*) The symmetry factors are derived using the known atomic coordinates for each equipoint set separately. (*2*) The appropriate scattering-factor values are determined. (*3*) Each term is weighted by the rank of the equipoint (number of symmetry-related atoms in set), and the terms are added together, as in (11-20). When many structure factors are to be calculated, it is usually worthwhile to systematize the calculations. Ways for doing this, and computational aids, ranging from tables of $\cos 2\pi hx$ values to high-speed digital computers, are described in some of the books listed at the end of this chapter.

Space-group extinctions. When a nonprimitive cell is chosen to describe the lattice, and the hkl indices of a reflection are based on this cell, the systematic absences that occur are usually called *space-group extinctions* in x-ray crystal-

Table 11-1 Space-group extinctions†

Class of reflection	Condition for nonextinction (n = an integer)	Interpretation of extinction	Symbol of symmetry element
hkl	$h + k + l = 2n$	Body-centered lattice	I
	$h + k = 2n$	C-centered lattice	C
	$h + l = 2n$	B-centered lattice	B
	$k + l = 2n$	A-centered lattice	A
	$\begin{cases} h + k = 2n \\ h + l = 2n \\ k + l = 2n \end{cases}$ $\approx h, k, l$, all even or all odd	Face-centered lattice	F
	$-h + k + l = 3n$	Rhombohedral lattice indexed on hexagonal reference system	R
	$h + k + l = 3n$	Hexagonal lattice indexed on rhombohedral reference system	H
$0kl$	$k = 2n$	(100) glide plane, component $\frac{\mathbf{b}}{2}$	$b(P,B,C)$
	$l = 2n$	$\frac{\mathbf{c}}{2}$	$c(P,C,I)$
	$k + l = 2n$	$\frac{\mathbf{b}}{2} + \frac{\mathbf{c}}{2}$	$n(P)$
	$k + l = 4n$	$\frac{\mathbf{b}}{4} + \frac{\mathbf{c}}{4}$	$d(F)$
$h0l$	$h = 2n$	(010) glide plane, component $\frac{\mathbf{a}}{2}$	$a(P,A,I)$
	$l = 2n$	$\frac{\mathbf{c}}{2}$	$c(P,A,C)$
	$h + l = 2n$	$\frac{\mathbf{a}}{2} + \frac{\mathbf{c}}{2}$	$n(P)$
	$h + l = 4n$	$\frac{\mathbf{a}}{4} + \frac{\mathbf{c}}{4}$	$d(F)$, (B)
$hk0$	$h = 2n$	(001) glide plane, component $\frac{\mathbf{a}}{2}$	$a(P,B,I)$
	$k = 2n$	$\frac{\mathbf{b}}{2}$	$b(P,A,B)$
	$h + k = 2n$	$\frac{\mathbf{a}}{2} + \frac{\mathbf{b}}{2}$	$n(P)$
	$h + k = 4n$	$\frac{\mathbf{a}}{4} + \frac{\mathbf{b}}{4}$	$d(F)$

Table 11-1 Space-group extinctions (continued)†

Class of reflection	*Condition for nonextinction (n = an integer)*	*Interpretation of extinction*	*Symbol of symmetry element*
hhl	$l = 2n$	(110) glide plane, component $\frac{\mathbf{c}}{2}$	$c(P,C,F)$
	$h = 2n$	$\frac{\mathbf{a}}{2} + \frac{\mathbf{b}}{2}$	$b(C)$
	$h + l = 2n$	$\frac{\mathbf{a}}{4} + \frac{\mathbf{b}}{4} + \frac{\mathbf{c}}{4}$	$n(C)$
	$2h + l = 4n$	$\frac{\mathbf{a}}{2} + \frac{\mathbf{b}}{4} + \frac{\mathbf{c}}{4}$	$d(I)$
$h00$	$h = 2n$	[100] screw axis, component $\frac{\mathbf{a}}{2}$	2_1, 4_2
	$h = 4n$	$\frac{\mathbf{a}}{4}$	4_1, 4_3
$0k0$	$k = 2n$	[010] screw axis, component $\frac{\mathbf{b}}{2}$	2_1, 4_2
	$k = 4n$	$\frac{\mathbf{b}}{4}$	4_1, 4_3
$00l$	$l = 2n$	[001] screw axis, component $\frac{\mathbf{c}}{2}$	2_1, 4_2, 6_3
	$l = 3n$	$\frac{\mathbf{c}}{3}$	3_1, 3_2, 6_2, 6_4
	$l = 4n$	$\frac{\mathbf{c}}{4}$	4_1, 4_3
	$l = 6n$	$\frac{\mathbf{c}}{6}$	6_1, 6_5
$hh0$	$h = 2n$	[110] screw axis, component $\frac{\mathbf{a}}{2} + \frac{\mathbf{b}}{2}$	2_1

† From M. J. Buerger, *X-ray crystallography* (John Wiley & Sons, Inc., New York, 1942).

lography. Such extinctions also arise whenever a symmetry operation in a space group includes a subtranslation that differs from the lattice translations. For example, the operation of a screw axis combines a rotation with a subtranslation into a single hybrid operation. Similarly, it will be recalled from Chap. 4 that a glide plane combines a reflection with a subtranslation. To see what effect this has on the structure-factor expression, simply insert the symmetry equivalent-point coordinates into (9-19). For example, a 2_1 passing through the origin parallel to c repeats a point xyz at $\bar{x},\bar{y},z + \frac{1}{2}$. Focusing attention on the subtranslation

$c/2$, it is more convenient to express the relative coordinates of two such points by $x,y,z + \frac{1}{4}$ and $\bar{x},\bar{y},z - \frac{1}{4}$. Substituting in (9-19),

$$F_{hkl} = \sum_m f_m(e^{2\pi i(hx_m+ky_m+lz_m)}e^{\pi il/2} + e^{2\pi i(-hx_m-ky_m+lz_m)}e^{-\pi il/2}) \tag{11-21}$$

where the summation over m implies a summation over pairs of like atoms related by the 2_1 screw axis. A critical examination of (11-21) reveals that the two-fold screw axis parallel to c does not affect the general hkl reflections. Consider, however, the $00l$ reflections, for which

$$F_{00l} = \sum_m f_m e^{2\pi i(lz_m)}(e^{\pi il/2} + e^{-\pi il/2})$$

$$= 2\cos\pi\left(\frac{l}{2}\right)\sum_m f_m e^{2\pi ilz_m} \tag{11-22}$$

It is immediately apparent that $F_{00l} = 0$ unless l is even.

It is possible to arrive at the same conclusion directly by examining the repetition of the $00l$ planes in a crystal containing a 2_1 (Exercise 11-7). This was done systematically by Buerger, who compiled a listing of space-group extinctions, reproduced in Table 11-1. Its use in interpreting x-ray diffraction diagrams is considered next.

DETERMINATION OF SPACE GROUP

Systematic absences. It should be clear from an examination of Table 11-1 that it is possible to deduce the lattice type and the possible presence of screw axes or glide planes from a systematic examination of the x-ray reflections recorded, using some of the methods discussed in later chapters. In this connection it should be noted that a correct analysis is possible only when the full set of hkl reflections is examined, and that a mistake frequently made is to base an analysis on fragmentary data, for example, an examination of $hk0$, $h0l$, or $0kl$ reflections only. Two other rules must be kept in mind when using Table 11-1:

1. Systematic absences in a more general class of reflections automatically causes certain less general classes of reflections to be absent also. An example characteristic of this is the systematic absence of hkl reflections when $h + k + l$ is odd, corresponding to a body-centered lattice. Note that this automatically implies that $h00$ reflections are absent unless h is even, so that it is not possible to utilize this criterion in deciding whether a 2_1 parallel to a is present or not.
2. The presence of reflections other than those listed in Table 11-1 merely means that the corresponding symmetry elements are not present in the crystal. This fact cannot be used to deduce the presence (or absence) of other kinds of symmetry. For example, when $0kl$ reflections having odd k values are

observed, it simply means that a b glide is absent; no inferences can be drawn, however, about the presence or absence of other symmetry, such as a parallel mirror plane, for example.

Diffraction symmetry. It is possible to detect the presence of certain symmetry elements in other ways, however. An examination of the intensity distribution in the reciprocal lattice of a single crystal clearly displays the presence or absence of specific symmetry elements in the crystal. For example, a mirror plane parallel to the a and b axes of a crystal relates all points at xyz to symmetry-equivalent points at $xy\bar{z}$ in the unit cell. This has the effect on the structure factor that $|F_{hkl}| = |F_{hk\bar{l}}|$, so that pairs of reflections hkl and $hk\bar{l}$ have the same intensity. The same applies to rotation axes; for example, a four-fold rotation axis parallel to c causes $|F_{hkl}| = |F_{\bar{k}hl}| = |F_{\bar{h}\bar{k}l}| = |F_{k\bar{h}l}|$. Similarly, a center of symmetry in the crystal causes $|F_{hkl}| = |F_{\bar{h}\bar{k}\bar{l}}|$. Such equivalences between the structure-factor magnitudes of symmetry-related reflections are tabulated for each space group in *International tables.* Note that when anomalous dispersion is negligibly small, it is not possible to distinguish $|F_{hkl}|$ from $|F_{\bar{h}\bar{k}\bar{l}}|$ also in noncentrosymmetric crystals. This automatic introduction of a center of symmetry in most diffraction experiments is known as Friedel's law in x-ray crystallography.

The ability to detect the presence of rotation axes or reflection planes from the intensity distribution in reciprocal space is actually not as helpful as may appear at first glance. First of all, only the reciprocal-lattice symmetry about its origin can be determined, and is called its *Laue symmetry*, or *Laue group*. Second, the automatic presence of an inversion center means that only the 11 centrosymmetric crystal classes can be distinguished. Thus x-ray diffraction enables only the determination of the lattice type, the Laue group, and the possible presence of screw axes and glide planes. Nevertheless, with this information it is possible to determine what is called the *diffraction symbol* of a crystal. This consists, first of all, of a listing of the Laue symmetry. Thus, if a crystal is orthorhombic, its Laue group automatically is mmm, since this is the only centrosymmetric point group possible in this system. In the tetragonal system, by comparison, two Laue classes can be distinguished, namely, $\frac{4}{m}$ and $\frac{4}{m}mm$. In forming the diffraction symbol, the Laue class is followed next by the lattice-type symbol, which is deduced, of course, from a scrutiny of all hkl reflections for any systematic absences. Finally, the presence of screw axes or glide planes parallel to the appropriate crystallographic directions is noted by following the same sequence as used in writing space-group symbols (Chap. 4). For example, consider an orthorhombic crystal which is determined to have a body-centered lattice. Suppose that no other systematic absences, not included in the requirement that $h + k + l$ be even, can be detected. Then the diffraction symbol is $mmmI$---, where the three dashes signify the absence of symmetry detectable by extinctions parallel to a, b, and c. Alternatively, suppose that $h0l$ reflections are present only when h is even. This means that an a glide lying in a plane perpendicular to b is present. The corresponding diffraction symbol becomes $mmmI$-a-. Finally, suppose that $hk0$ reflections are present only when $h = 2n$; $h0l$ reflections are present only when

$l = 2n$; and $0kl$ reflections, when $k = 2n$. An examination of Table 11-1 then discloses that the appropriate diffraction symbol is $mmmIbca$.

In order to deduce what the space group of the crystal is from its diffraction symbol, it is necessary to know what the diffraction symbols are for the 230 possible space groups. This can be done by adding an inversion center to each space group and then grouping them accordingly. This was first done by Buerger, who showed that 121 diffraction symbols can be distinguished. His table is reproduced in Volume 1 of *International tables.* Consulting that table to find which space groups belong to the three diffraction symbols described above reveals that

$mmmI$--- includes $Imm2$, $I222$, $I2_12_12_1$, and $Immm$.
$mmmI$-a- includes $Ima2$ and $Imam$.
$mmmIbca$ includes only $Ibca$.

The above example demonstrates why 230 space groups require only 122 diffraction symbols.† Although most of the diffraction symbols represent more than one space group, 58 diffraction symbols represent only one space group, so that they can be recognized directly from the diffraction data, by inspection. The remaining 172 space groups, however, cannot be identified uniquely quite as easily, and in some cases the correct space group is not determined finally until the very last stages of the crystal-structure determination.

Detection of inversion symmetry. The problem of space-group determination would be considerably eased if it were possible to establish whether a center of symmetry is present or not, despite Friedel's law. Sometimes this can be done by utilizing crystal properties that depend on symmetry. The most obvious one is to deduce the correct point-group symmetry from the crystal habit. Unfortunately, this is not possible when a crystal lacks sufficiently well developed faces. Alternatively, tests can be carried out to determine whether the crystal is piezoelectric or pyroelectric, both properties being characteristic of noncentric structures. Such physical tests, however, are conclusive only when they are positive; that is, the presence of, say, piezoelectricity definitely indicates a noncentric structure. The absence of detectable piezoelectricity does not establish conclusively the presence of an inversion center, since it may simply indicate that the effect was too small for the limited sensitivity of the instrument used.

Fortunately, it is possible to make use of x-ray diffraction intensities to help resolve this dilemma. As discussed in Chap. 8, when x rays having energies close to an absorption edge of an atom are used, the resulting anomalous-dispersion effect produces a phase lag between the anomalous scatterer and the rest of the atoms. In a centrosymmetric structure this simply serves to change the magnitude of that atom's contribution. (This also can be quite helpful in determining the phases of x-ray reflections, as discussed in a later section.) In a noncentric structure, however, the relative magnitudes of the reflection intensity of pairs of reflections hkl and $\bar{h}\bar{k}\bar{l}$ are changed by dispersion. Thus a fairly straightforward

† In his original derivation, Buerger obtained 121 diffraction symbols, but a reexamination of enantiomorphous pairs led to the corrected number.

test for the presence of inversion symmetry in a crystal is to irradiate it with an appropriate x-ray beam and to note whether Friedel's law is violated. Although the applicability of this test is limited by readily available x-ray tube targets, it is actually much simpler and easier to use than is generally realized. This is particularly true when a crystal monochromator is available to select x-ray wavelengths, and an ionization detector to measure small intensity differences.

Actually, the x-ray diffraction intensities themselves should reflect the presence of symmetry in a crystal. This is so because their magnitude is determined by the crystal structure whose atomic array is constrained in definite ways by any symmetry present. This was first stressed by Buerger, whose ideas are described more fully in a later section, dealing with Fourier methods. The same viewpoint was formulated more rigorously some time later by Wilson. He derived two expressions for the average value of the structure-factor amplitude, based on the assumption that it is noncentric: The ratio between the square of the mean magnitude of structure factors and the average intensity turns out to be 0.785 for a centrosymmetric crystal, as opposed to 0.637 for a noncentric one. In practice, the calculation of this ratio is hampered by inadequate data, and the resulting number often lies midway between the two ideal values, so that a certain amount of ambiguity remains. Although extensions of this approach have been carried out and, in principle, can be used to establish the presence of a symmetry center as well as other symmetry elements, the application of these statistical treatments often fails to give clear-cut indications. As is the case with all statistical analyses, these methods work best when the number of intensities used is large and systematic constraints are minimal.

Multiple reflections. The identification of space-group extinctions is sometimes confounded by the presence of reflections seemingly violating one of the extinction rules in Table 11-1. Whenever this occurs, of course, it must be concluded that the corresponding symmetry element is absent. If in an entire $hk0$ zone only one reflection violates the condition that $h + k$ must be even, however, there is a strong temptation to assume that the presence of this one reflection rather than the systematic absence of all others for which $h + k$ is odd should be suspected. One way that a "forbidden" reflection may be produced was pointed out by M. Renninger, who had observed that the 222 reflection from diamond could exhibit anomalously high intensities when the crystal was rotated about its $[111]$ axis. He called this phenomenon *umweganregung*, translated "indirect reflection" in English, and attributed it to double reflections by two planes, such as (331) and $(\bar{1}\bar{1}\bar{1})$. Using the reciprocal-lattice construction, it is a simple matter to establish that such double reflections may occur whenever two reciprocal-lattice points happen to intersect the sphere of reflection simultaneously. Consider the construction in Fig. 11-1. Suppose the reciprocal lattice is so oriented relative to the incident x-ray beam direction that the two points P_1 and P_3 lie on the sphere of reflection at the same time. The diffracted beam traveling in the direction C_1P_1 can be thought of as an x-ray beam moving through the crystal (like another incident beam), in which case the Ewald construction is that shown by the broken lines in Fig. 11-1. By construction, the dashed sphere of reflection must intersect a

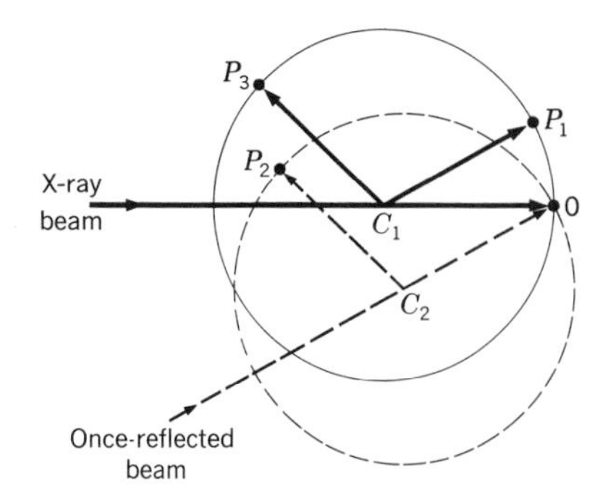

Fig. 11-1

reciprocal-lattice point, so that the once-reflected beam C_2O is diffracted in that direction. In the example illustrated in Fig. 11-1 this is the case for P_2. Note that C_2P_2 is parallel to C_1P_3, so that the twice-diffracted beam C_2P_2 travels in the same direction as the ordinary reflection C_1P_3, and the two beams superimpose in the x-ray diffraction experiment. The relation between the indices of the points P_1, P_2, and P_3 must obey the condition that

$$h_3 = h_1 + h_2 \qquad k_3 = k_1 + k_2 \qquad l_3 = l_1 + l_2 \tag{11-23}$$

Under these conditions, the intensity of the x-ray beam diffracted by the plane $(h_3k_3l_3)$ will be anomalously high, as noted by Renninger. Because of the more regular shape of such double reflections, they can be identified when photographic recording of intensities is employed, so that they should not cause confusion in space-group determinations. (This is one of the reasons why film methods always should be used to examine a crystal before beginning quantitative intensity measurements with counters.) On the other hand, when the double reflection superimposes on a normally present reflection, it merely changes its intensity, which is not clearly recognizable. This occurs much more often than is generally realized, and accounts for at least part of the residual experimental errors in intensity measurements. As pointed out by Zachariasen, multiple reflections involving three or more planes are also possible. One way to test for the presence of multiple reflections is to rotate the crystal about the reflecting plane's normal and to repeat the measurement. Such a rotation removes the additional reciprocal-lattice points from their positions on the Ewald sphere in the case of multiple reflection, causing it to vanish.

ITERATIVE METHODS

Symmetry-fixed structures. To begin, consider one of the simplest possible cases for a binary compound, epitomized by cesium chloride. Following the procedure suggested in Chap. 4, it is determined that the cubic unit cell contains one formula weight of CsCl. Following the discussion in the preceding sections, the diffraction symbol for CsCl is either $m3P$--- or $m3mP$---, which include the space groups $P23$, $Pm3$, $P432$, $P\bar{4}3m$, and $Pm3m$. An examination of the equipoints tabulated for these space groups discloses that all of them contain

$$1a: \quad 0,0,0 \qquad \text{and} \qquad 1b: \quad \tfrac{1}{2},\tfrac{1}{2},\tfrac{1}{2} \tag{11-24}$$

The next one ($3c$) has a rank of 3 and need not be considered because there is only one cesium and one chlorine atom to be placed in this cell. Thus the crystal structure of CsCl is completely fixed by symmetry, there being no choice whatever in what positions the atoms occupy, since $1a$ and $1b$ are equivalent, except for choice of the cell origin. This leads to the atomic array shown in Fig. 11-2, which, clearly, has the highest symmetry of the above listing, namely, $Pm3m$.

As another simple example, consider sodium chloride. The diffraction symbol for this crystal is either $m3F$--- or $m3mF$---, which include the space groups $F23$, $Fm3$, $F432$, $F\bar{4}3m$, and $Fm3m$. As in the discussion of CsCl, no attempt is made to establish the correct Laue class, although it is assumed that the presence of glide planes or screw axes would have been detected if present. This is an important point, because one often uses powder data or rotating-crystal photographs to study such simple structures. As discussed in later chapters, these two methods do not allow the determination of any symmetry that may be present, although their systematic absences can be used to determine the correct lattice type. This means that, for the case of sodium chloride, all other cubic space groups having a face-centered lattice would have to be added to the above list in any systematic attempt to deduce the correct space group. An examination of the equipoint lists of these other space groups (for example, $Fd3$ or $F4_132$) discloses, however, that the lowest-ranking equipoint must contain eight identical atoms. Since it turns out that the face-centered unit cell contains only four NaCl units, these other space groups can be eliminated from further consideration. (Note that this assumes an ordered array of Na and Cl atoms, and not a random array like those discussed in the section on order-disorder, in the preceding chapter.)

Returning to the space groups belonging to the diffraction symbols $m3F$--- and $m3mF$---, an examination of their equipoint lists shows the following possible sites of rank 4 (without regard to which space groups contain which):

$$\begin{aligned} 4a&: \quad 0,0,0 \\ 4b&: \quad \tfrac{1}{2},\tfrac{1}{2},\tfrac{1}{2} \\ 4c&: \quad \tfrac{1}{4},\tfrac{1}{4},\tfrac{1}{4} \\ 4d&: \quad \tfrac{3}{4},\tfrac{3}{4},\tfrac{3}{4} \end{aligned} \tag{11-25}$$

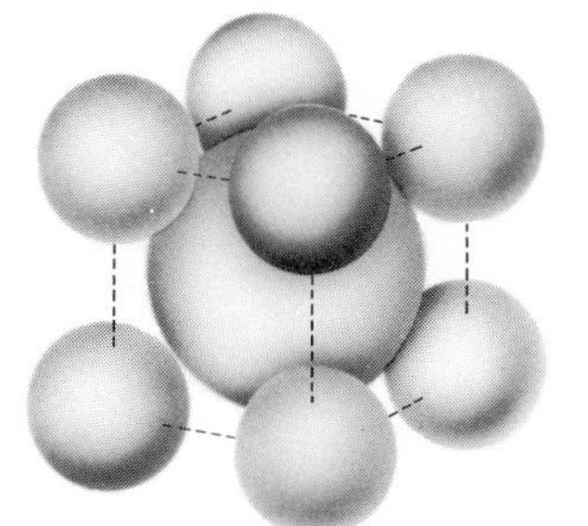

Fig. 11-2. CsCl structure.

where the components of the face-centering translations $(0,0,0;\ \frac{1}{2},\frac{1}{2},0;\ \frac{1}{2},0,\frac{1}{2};\ 0,\frac{1}{2},\frac{1}{2})$ must be added to each point listed in (11-25) to obtain the complete set of equivalent positions. An examination of the various permutations of pairs of points in (11-25) shows that two unique sets can be distinguished, namely,

$$4a + 4b \qquad \text{and} \qquad 4a + 4c \tag{11-26}$$

the others being equivalent to these two, except for shift in origin. (Shifting the origin from $0,0,0$ to, say, $\frac{3}{4},\frac{3}{4},\frac{3}{4}$ obviously cannot affect a crystal structure.) The correct choice of space group reduces, therefore, to the problem of finding the highest symmetry group encompassing the two possibilities in (11-26). It turns out that $F\bar{4}3m$ contains all four sets in (11-25), whereas the higher-symmetry space group $Fm3m$ contains only $4a$ and $4b$. Thus one can distinguish two possible space groups, for which the atomic arrays are given below.

$$\begin{array}{lll} Fm3m & \text{Na in } 4a\text{:} & 0,0,0;\ \frac{1}{2},\frac{1}{2},0;\ \frac{1}{2},0,\frac{1}{2};\ 0,\frac{1}{2},\frac{1}{2} \\ & \text{Cl in } 4b\text{:} & \frac{1}{2},\frac{1}{2},\frac{1}{2};\ 0,0,\frac{1}{2};\ 0,\frac{1}{2},0;\ \frac{1}{2},0,0 \\ F\bar{4}3m & \text{Na in } 4a\text{:} & 0,0,0;\ \frac{1}{2},\frac{1}{2},0;\ \frac{1}{2},0,\frac{1}{2};\ 0,\frac{1}{2},\frac{1}{2} \\ & \text{Cl in } 4c\text{:} & \frac{1}{4},\frac{1}{4},\frac{1}{4};\ \frac{3}{4},\frac{3}{4},\frac{1}{4};\ \frac{3}{4},\frac{1}{4},\frac{3}{4};\ \frac{1}{4},\frac{3}{4},\frac{3}{4} \end{array} \tag{11-27}$$

In order to determine which space group, and therefore which structure, is the correct one for sodium chloride, it is necessary to calculate the structure-factor magnitudes, using both sets of coordinates in (11-27). The set giving the best agreement with experimentally determined values then must be the correct one for $NaCl$. As shown from such an analysis in Exercise 11-10, the correct structure for sodium chloride is that belonging to space group $Fm3m$ above, as can be seen from the model in Fig. 11-3. The other set of coordinates in (11-27) corresponds to what is called the sphalerite structure (Fig. 11-4), which is one of the two common polymorphs of zinc sulfide. Note that, except for the selection of correct pairs of special positions, the structures considered so far have been completely fixed by the symmetry of the space groups. In more complicated structures, some of the atoms may occupy positions for which one or more coordinate values must be established. Examples of this are considered next.

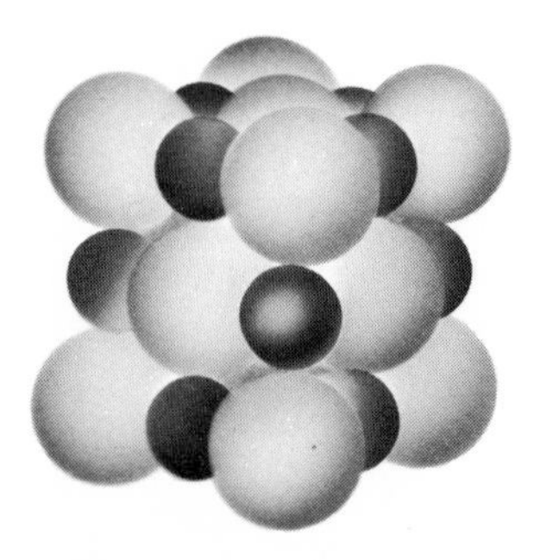

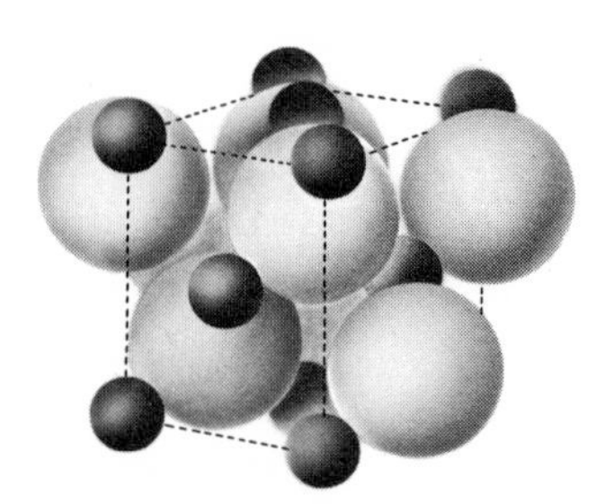

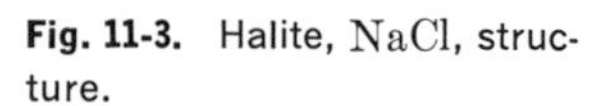

Fig. 11-3. Halite, $NaCl$, structure.

Fig. 11-4. Sphalerite, ZnS, structure.

One-parameter structures. As an example of a (commercially) most important crystal structure having one parameter to be specified, consider the spinel structure, which is adopted by numerous transition-metal oxides having the general formula AB_2O_4, where $A = \mathrm{Mg}$ and $B = \mathrm{Al}$ in the mineral spinel. The diffraction symbol of $\mathrm{MgAl_2O_4}$ is $m3mFd$--, which, fortunately, includes only one space group, $Fd3m$. (If the Laue class were not known, $Fd3$ also would have to be considered, but a comparison of equipoint lists would quickly eliminate the need for its further consideration, since the lower-ranking sets are identical in both space groups.) The face-centered unit cell contains eight formula weights, so that it is necessary to consider all equivalent positions in $Fd3m$ of rank 32 or less. When the listing in *International tables* is consulted, it is found that two sets of coordinates are given, depending on whether the origin is placed at the center of symmetry or at the point in the unit cell having the higher symmetry $\bar{4}3m$, which is $\bar{\frac{1}{8}},\bar{\frac{1}{8}},\bar{\frac{1}{8}}$ away from the location of the inversion center. Although the final analysis clearly does not depend on the location of the origin, a choice of the symmetry center for the origin usually simplifies the ensuing calculations of the structure factor, since $B = 0$ for all reflections in that case. With this choice of origin, the equivalent positions of interest are

$$\begin{array}{ll}
8a: & \frac{1}{8},\frac{1}{8},\frac{1}{8};\ \frac{7}{8},\frac{7}{8},\frac{7}{8} \\
8b: & \frac{3}{8},\frac{3}{8},\frac{3}{8};\ \frac{5}{8},\frac{5}{8},\frac{5}{8} \\
16c: & 0,0,0;\ 0,\frac{1}{4},\frac{1}{4};\ \frac{1}{4},0,\frac{1}{4};\ \frac{1}{4},\frac{1}{4},0 \\
16d: & \frac{1}{2},\frac{1}{2},\frac{1}{2};\ \frac{1}{2},\frac{1}{4},\frac{1}{4};\ \frac{1}{4},\frac{1}{2},\frac{1}{4};\ \frac{1}{4},\frac{1}{4},\frac{1}{2} \\
32e: & x,x,x;\ x,\frac{1}{4}-x,\frac{1}{4}-x;\ \frac{1}{4}-x,x,\frac{1}{4}-x;\ \frac{1}{4}-x,\frac{1}{4}-x,x; \\
 & \bar{x},\bar{x},\bar{x};\ \bar{x},\frac{3}{4}+x,\frac{3}{4}+x;\ \frac{3}{4}+x,\bar{x},\frac{3}{4}+x;\ \frac{3}{4}+x,\frac{3}{4}+x,\bar{x}
\end{array} \tag{11-28}$$

to which the components of the face-centering translations must be added to complete the listings.

The space group having been uniquely determined, the structure-analysis reduces to selecting the correct special positions for the 8 A atoms and 16 B atoms in the spinel structure, plus the correct determination of the x parameter for the 32 oxygen atoms. It is clear that if the A atoms occupy either of the two 8-fold positions, the B atoms can occupy only one of the two 16-fold positions, after which the 32 oxygen atoms must be in $32e$. Even though this seemingly reduces the number of permutations to an extremely small number, the labor required to deduce the correct structure is by no means small, as will become clear presently. The choice of the placement of the 8 A atoms is arbitrary, since it merely serves to determine which inversion center is chosen as the origin. Suppose that $8a$ is chosen. Then the choice for the 16 B atoms must be made between $16c$ and $16d$, by calculating structure-factor magnitudes, similarly to the procedure described in the preceding section for NaCl. Before it is possible to calculate F, however, it is necessary to determine the value of x in $32e$, since the oxygen-atom contribution must be included also. Here use is made of crystal chemistry. As early as the beginning of the twentieth century, it had been postulated that the atomic arrays in simple crystals consisted of closest packings† of large anions, with the smaller cations distributed among the voids existing in their interstitial sites. It since has been shown that this is indeed the case for virtually all ionic crystals and many metallic phases. The commonest cubic closest packing is that leading to an atomic array described by a face-centered cubic lattice with one atom at each lattice point. To form such an array of oxygen atoms in the unit cell of the spinel structure, it is only necessary to double the length of each a axis of the cubic closest packing, so that the spinel cell contains $2 \times 2 \times 2 = 8$ unit cells of the closest packing (Fig. 10-42), or a total of $8 \times 4 = 32$ atoms. If the oxygen atoms are in a perfect closest packing, their location in the spinel cell, whose origin is at the inversion center chosen above, is given by the coordinates in $32e$ with $x = 0.250$. With this choice of x, the structure can be considered to be fixed by symmetry, so that the calculation of just a few structure-factor magnitudes reveals that the 16 B atoms occupy the octahedral voids in the cubic closest packing given by $16d$ in (11-28).

It should be noted that the oxygen atoms in most compounds crystalizing with the spinel structure do not form a perfect closest packing. This means that the value of x in $32e$ may deviate from 0.250 by some fraction, usually less than ± 0.020. Once the atomic distribution of the cations has been established, the oxygen parameter can be refined by calculating the variation of the structure-factor magnitudes for several values of x, and plotting F_{hkl} as a function of x, for selected reflections, on the same graph paper. It is usual in such refinements to use higher-order reflections, since they are more sensitive to small changes in the atomic coordinates. The x value at which the best agreement with experimentally observed magnitudes is obtained then is chosen as the correct value for that structure. It should be noted here, however, that the correct distribution of the cations in transition-metal oxides crystalizing with the spinel structure cannot be assumed to be that of 8 A atoms in $8a$ and 16 B atoms in $16d$. As first pointed out by T. F. W. Barth and E. Posnjak, it is possible to form what is called an *inverse spinel* in which 8 B atoms occupy the sites in $8a$ while the other 8 B atoms and the 8 A atoms are distributed at random among the $16d$ sites. In fact, later studies have shown that the division need not be even as regular as this.

Because this is a frequently encountered problem, the most direct way to establish the atomic distribution in a spinel is considered in more detail in Exercise 11-4. As shown there, the symmetry factor for the atoms in $8a$ is zero unless $h + k + l = 2n + 1$ or $4n$. Similarly, $S = 0$ for the atoms in $16d$ unless either h, k, l are all odd or all even, in which latter case all three must be odd multiples of 2 or 4, while the oxygen atoms make a contribution to all reflections not excluded by the diffraction-group symmetry. This means that the atoms in $16d$ do not contribute to reflections such as 422, so that such reflections can be used to determine the atomic occupation of the $8a$ position. Alternatively, reflections 200, 222, 420, 600, etc., can be used to determine the occupation of the $16d$ sites, since the other cations make no contribution when $h + k + l \neq 4n$. This means that it is a rela-

† A fairly thorough introduction to closest packings and the elements of packing theory can be found in Leonid V. Azároff, *Introduction to solids* (McGraw-Hill Book Company, New York, 1960), chaps. 3 and 4.

tively straightforward matter to determine the crystal structure of ferrites and similar compounds crystalizing with the spinel structure, because the contributions of atoms in different sites can be clearly distinguished. It should be noted in passing that the reflections receiving contributions from atoms in specific equivalent position are tabulated next to their coordinates in *International tables.* Thus the conclusions reached in Exercise 11-4 can be verified by checking the equipoint listing for space group $Fd3m$ (No. 227) in the tables. It also should be realized by now that, in any structure analysis, the correctness of the final atomic array is determined by the agreement between the structure-factor magnitudes calculated for that array and the experimentally observed ones.

Multiparameter structures. Although the increasing availability of high-speed computers makes it more likely that trial-and-error methods may be utilized in the determination of complex structures, without such aids iterative procedures can become excessively burdensome when the number of undetermined parameters is very large. It is of interest, therefore, to examine briefly the kinds of approaches that one can use in deducing likely structure models, at least for fairly simple crystal structures. Stereochemical considerations, such as the closest-packing theories mentioned above or the beforehand knowledge of the configuration of an organic molecule, and so forth, all can be utilized to help initiate a suitable trial model. Mechanical or physical properties, such as cleavage, optical properties, and others, also often are most helpful in suggesting possible trial arrays. Sometimes similarities in the cell dimensions or chemical formulas may suggest correspondences to crystal structures already known. It is here where familiarity with such concepts of crystal chemistry as derivative structures or an intimate knowledge of geometrical crystallography becomes most useful.

In attempting to deduce a crystal's structure from comparisons with previously known structures, it is essential to compare the intensity distributions in reciprocal space of both crystals. When they show definite similarities, it is safe to make certain deductions. As an example, consider the compound potassium ozonide, KO_3, whose powder photograph closely resembles that of potassium azide, KN_3. A careful scrutiny of the two shows that the intensity distribution for many of the reflections is quite similar for both compounds, except that KO_3 has more reflections than KN_3. Although the identification of the correct reciprocal-lattice net from powder photographs is more involved than that from single-crystal studies (Chap. 18), it is nevertheless possible to show quite clearly that KO_3 and KN_3 both have tetragonal lattices which are related in that $a_{KO_3} \cong \sqrt{2}\, a_{KN_3}$ and $c_{KO_3} \cong c_{KN_3}$. It is known from spectroscopic and other data that whereas the N_3^- ion is linear, the O_3^- is bent. An examination of the projection of the known KN_3 structure along its c axis, shown on the left in Fig. 11-5, suggests that, unless the O_3^- ion is bent to form a right angle, the "fit" of the bent ion in the voids between the K^+ ions is such as to require a suppression of certain four-fold screw axes and a corresponding increase in the translations as indicated on the right in Fig. 11-5. With this as a basis, it is possible to make comparisons between structure-factor magnitudes calculated for the trial structure and those observed experimentally.

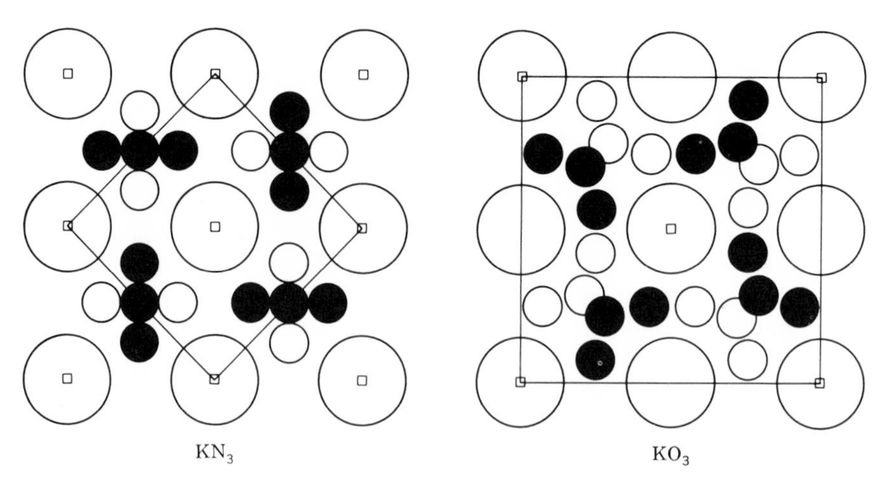

Fig. 11-5. Comparison of c-axis projections of KN_3 and KO_3 structures. The small open and closed circles represent the coplanar anions lying halfway between the larger potassium ions. [*From L. V. Azároff and I. Corvin, Proceedings of the National Academy of Sciences, vol.* 49 (1963), *pp.* 1–5.]

In such analyses it is most convenient to work with projections in which only two coordinates of each atom need be considered at a time. Similarly, only the reflections of the corresponding zone are required, so that the analysis of the projection of a structure requires relatively much less effort than a direct analysis of the three-dimensional structure. Projections also often have higher symmetries than the full structure, which again serves to simplify the analyses by restricting the degrees of freedom in the placement of the various atoms. Finally, careful analyses of certain projections very often permit the deduction of the correct space group, and sometimes of the correct structure as well. This is particularly true when the corresponding zone in the reciprocal lattice has an unusual intensity distribution, for example, only a few very strong reflections, indicating that most of the atoms contribute to those reflections with nearly identical phases; that is, the atoms tend to lie in arrays parallel to such reflecting planes. It is, of course, not possible to consider even a few of the myriad of actual examples of such iterative analyses here, so that the reader is referred to the literature and to his own ingenuity.

FOURIER METHODS

Electron-density functions. It is possible to represent any well-behaved function by means of a suitable series of trigonometric terms, called a Fourier series. As an example of how this can be done, consider the one-dimensional array of four maxima in Fig. 11-6. Suppose each maximum represents the electron density ρ_A of an atom in a one-dimensional "crystal." Then the array of maxima in Fig. 11-6 represents the electron-density distribution $\rho(x)$ of the one-dimensional "crystal" whose translation period is a. It is possible to express this electron density by a Fourier series of cosines, because the array in Fig. 11-6 has appropriate symmetry. This means that one can write

$$\rho(x) = \sum_{n=-\infty}^{\infty} A_n \cos 2\pi n \frac{x}{a} \tag{11-29}$$

where the dummy variable n is any positive or negative integer including zero; x is the actual distance measured along a; and the coefficients must be properly chosen to reproduce the distribution of maxima in Fig. 11-6.

The coefficients can be determined analytically whenever the function $\rho(x)$ is known by making use of the orthogonality of Fourier series. Multiply both sides of (11-29) by $\cos 2\pi m \frac{x}{a}\, dx$ and integrate over one translation period a.

$$\int_{-a/2}^{a/2} \rho(x) \cos 2\pi m \frac{x}{a}\, dx = \sum_n A_n \int_{-a/2}^{a/2} \cos 2\pi n x \cos 2\pi m x\, dx \tag{11-30}$$

Direct integration of the integral on the right side of (11-30) is possible after expanding it as follows:

$$\begin{aligned} \int_{-a/2}^{a/2} \cos 2\pi n \frac{x}{a} \cos 2\pi m \frac{x}{a}\, dx &= \frac{1}{2} \int_{-a/2}^{a/2} \cos 2\pi (n+m) \frac{x}{a}\, dx \\ &\quad + \frac{1}{2} \int_{-a/2}^{a/2} \cos 2\pi (n-m) \frac{x}{a}\, dx \\ &= \frac{1}{2} \left\{ \left[\sin 2\pi (n+m) \frac{x}{a} \right]_{-a/2}^{a/2} + \left[\sin 2\pi (n-m) \frac{x}{a} \right]_{-a/2}^{a/2} \right\} \\ &= \frac{1}{2} \begin{cases} a & \text{if } n = m \\ 0 & \text{if } n \neq m \end{cases} \end{aligned} \tag{11-31}$$

This means that it is possible to determine the coefficients of the Fourier series (11-29) for each value of $n = m$ by means of relation (11-30), since, according to (11-31),

$$A_n = \frac{2}{a} \int_{-a/2}^{a/2} \rho(x) \cos 2\pi n \frac{x}{a}\, dx \tag{11-32}$$

For the more general case of any symmetry, a series of exponentials is used in place of cosine (or sine) series. Thus the three-dimensional electron density of a

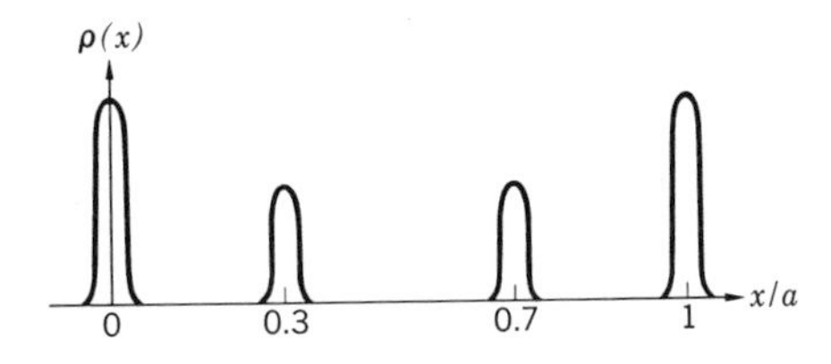

Fig. 11-6. Symmetric one-dimensional electron density.

real crystal can be expressed

$$\rho(xyz) = \sum_{n} \sum_{m=-\infty}^{\infty} \sum_{p} A_{nmp}\, e^{-2\pi i\left(n\frac{x}{a}+m\frac{y}{b}+p\frac{z}{c}\right)} \tag{11-33}$$

provided that the coefficients A_{nmp} are suitably chosen. As derived in Exercise 11-12, the coefficients can be found analogously to the above procedure, so that when the integer dummy indices $n = h$, $m = k$, $p = l$, then

$$A_{nmp} = A_{hkl} = \frac{1}{V} \int_{-a/2}^{a/2} \int_{-b/2}^{b/2} \int_{-c/2}^{c/2} \rho(xyz) e^{2\pi i\left(h\frac{x}{a}+k\frac{y}{b}+l\frac{z}{c}\right)} \frac{V}{abc}\, dx\, dy\, dz \tag{11-34}$$

where the geometrical factor V/abc serves to normalize the coefficients for the case when the unit cell is not orthogonal.

If the electron density $\rho(xyz)$ of a real crystal is considered to be the sum of the electron densities of the individual atoms, the structure factor

$$\begin{aligned} F_{hkl} &= \sum_{n} f_n e^{2\pi i\left(h\frac{x}{a}+k\frac{y}{b}+l\frac{z}{c}\right)_n} \\ &= \iiint_{\substack{\text{unit}\\ \text{cell}}} \rho(xyz) e^{2\pi i\left(h\frac{x}{a}+k\frac{y}{b}+l\frac{z}{c}\right)} \frac{V}{abc}\, dx\, dy\, dz \end{aligned} \tag{11-35}$$

according to the concepts underlying the scattering factor discussed in Chap. 8. This means that, combining (11-34) with (11-35),

$$A_{nmp} = A_{hkl} = \frac{1}{V} F_{hkl} \tag{11-36}$$

so that the proper form of the Fourier series representing the electron density of a crystal is

$$\rho(xyz) = \frac{1}{V} \sum_{h} \sum_{k=-\infty}^{\infty} \sum_{l} F_{hkl}\, e^{-2\pi i\left(h\frac{x}{a}+k\frac{y}{b}+l\frac{z}{c}\right)} \tag{11-37}$$

Note that, in the present discussion, the coordinates are expressed by the fractions x/a, y/b, z/c, so that x, y, and z represent actual distances measured, respectively, along the three crystallographic axes. The ratios are, of course, the same as the fractional coordinates used elsewhere in this book. As demonstrated by the equations in this section, the use of the ratios makes it easier to maintain dimensional accuracy, and should not cause any real confusion in the reader's mind.

Since the series in (11-37) must be summed over all possible positive and negative hkl values, it follows that the intensity distribution in the entire three-dimensional reciprocal lattice must be measured. Not only does this require a great deal of work, but the consequent summation is also lengthened by the inclusion of all these terms. As demonstrated in the preceding section, it is sometimes possible to gain a good idea of the crystal structure even when it is projected onto a plane, say, along the c axis. The corresponding electron-density projection can be determined from (11-37) in the following way: Let ζ be the total number of electrons in one unit cell, so that

$$\zeta = \iiint_{\substack{\text{unit}\\\text{cell}}} \rho(xyz) \frac{V}{abc} \, dx \, dy \, dz \tag{11-38}$$

If the electron density projected along c and denoted $\rho(xy)$ contains the same number of electrons as the full cell, it follows that

$$\zeta = \int_a \int_b \rho(xy) \frac{A}{ab} \, dx \, dy \tag{11-39}$$

where A is the cell area in the ab plane, and A/ab is a normalizing factor. Equating (11-39) with (11-38) makes it possible to discard the integrations over a and b on both sides, so that after dividing both sides by dx and dy,

$$\begin{aligned}\rho(xy) &= \frac{ab}{A} \cdot \frac{V}{abc} \int_{-c/2}^{c/2} \rho(xyz) \, dz \\ &= \frac{V}{Ac} \int_{-c/2}^{c/2} \frac{1}{V} \sum_h \sum_k \sum_l F_{hkl} \, e^{-2\pi i \left(h\frac{x}{a} + k\frac{y}{b} + l\frac{z}{c}\right)} \, dz \\ &= \frac{1}{Ac} \sum_h \sum_k e^{-2\pi i \left(h\frac{x}{a} + k\frac{y}{b}\right)} \sum_l \int_{-c/2}^{c/2} F_{hkl} \, e^{-2\pi i l \frac{z}{c}} \, dz \\ &= \frac{1}{A} \sum_h \sum_k F_{hk0} \, e^{-2\pi i \left(h\frac{x}{a} + k\frac{y}{b}\right)}\end{aligned} \tag{11-40}$$

since the integral equals c if $l = 0$ but vanishes if $l \neq 0$. Note that not only does the projected electron density (11-40) require only two summations, but also only the structure factors of the $hk0$ reflections. This means that only the intensities of one zone need be measured.

It is current practice, however, to measure full three-dimensional sets of reflections in crystal-structure analyses. This is done partly because the remaining unsolved structures have unit cells that are so large that it is usually not possible to resolve individual atoms clearly in a projection, and partly because the atomic coordinates, and hence the interatomic-bond distances, can be determined much more precisely from three-dimensional analyses. The three-dimensional electron density (11-37) is not, however, summed all at once. Instead, two-dimensional summations are carried out at regularly spaced intervals along one of the axes.

These are called *sections*, and as an example, consider the section at $z = 0$. According to (11-37), when $z = 0$,

$$\rho(xy0) = \frac{1}{V} \sum_h \sum_k \left(\sum_l F_{hkl} \right) e^{-2\pi i \left(h\frac{x}{a} + k\frac{y}{b} \right)} \tag{11-41}$$

Note that only the structure factors need be summed over l, since the exponential term does not depend on l when $z = 0$. Nevertheless, the full three-dimensional set of F's is needed so that a section should not be confused with a projection. Further examples of projections and sections are considered in Exercises 11-13 and 11-14.

Utilization of electron densities. The final electron density represents, of course, the distribution of atoms in the unit cell of a crystal. Thus the utilization of the electron densities and the reason for their calculation are self-evident. In practice however, it is not a simple matter to carry out the summations discussed above.

Fig. 11-7. $\rho(xy)$ of KO_3. *(After Azároff and Corvin.)*

This is so, not because the calculations are tedious—modern high-speed computers have diminished that difficulty—but because the coefficients necessary for the series summation are incompletely known. As previously stressed, only the structure-factor magnitudes are determined experimentally, so that it is possible to calculate electron densities only after the individual phase factors have been determined by some other means. Thus electron densities are more commonly calculated as a check to see whether a postulated structure is indeed the correct one. As discussed in a later section, it is possible to refine the atomic coordinates by successive electron-density syntheses. As an example of the appearance of a simple electron-density projection, Fig. 11-7 shows $\rho(xy)$ for KO_3, which should be compared with the model structure in Fig. 11-5.

Patterson functions. The inability to synthesize an electron density straightforwardly from the measured intensities naturally leads to the question, what useful information, if any, can be obtained from a Fourier series employing the intensities F^2 directly? This question was first successfully answered by A. L. Patterson, who devised the series that now bears his name. The Patterson series stems from a well-known function in Fourier integral theory, called the correlation operation. Considering first only one dimension, the function can be expressed by

$$P(X) = \frac{1}{a}\int_{-a/2}^{a/2} \rho(x)\rho(x+X)\,dx \tag{11-42}$$

where X is some particular value of x measured along a. Substituting the one-dimensional version of (11-37) for the electron densities in (11-42),

$$\begin{aligned} P(X) &= \frac{1}{a}\int_{-a/2}^{a/2} \frac{1}{a}\sum_n F_n\, e^{-2\pi i n\frac{x}{a}} \frac{1}{a}\sum_{n'} F_{n'} e^{-2\pi i n'\frac{x+X}{a}}\, dx \\ &= \frac{1}{a^3}\sum_n\sum_{n'} F_n F_{n'}\, e^{-2\pi i n'\frac{X}{a}} \int_{-a/2}^{a/2} e^{-2\pi i(n+n')x}\, dx \end{aligned} \tag{11-43}$$

The integral in (11-43) equals zero unless $n = -n'$, and it equals a when $n = -n'$. Note, however, that when this is true, $F_{n'} = F_{-n} = F_n^*$; that is, the two structure factors in (11-43) are complex conjugates of each other, so that, when the integral does not vanish,

$$\begin{aligned} P(X) &= \frac{1}{a^2}\sum_n F_n{}^2\, e^{2\pi i n(X/a)} \\ &= \frac{1}{a^2}\sum_n F_n{}^2 \cos 2\pi n(X/a) \end{aligned} \tag{11-44}$$

since, in the summation from minus to plus infinity, the sine terms in the series cancel each other in pairs.

The important feature of the Fourier series in (11-44) is that it utilizes the

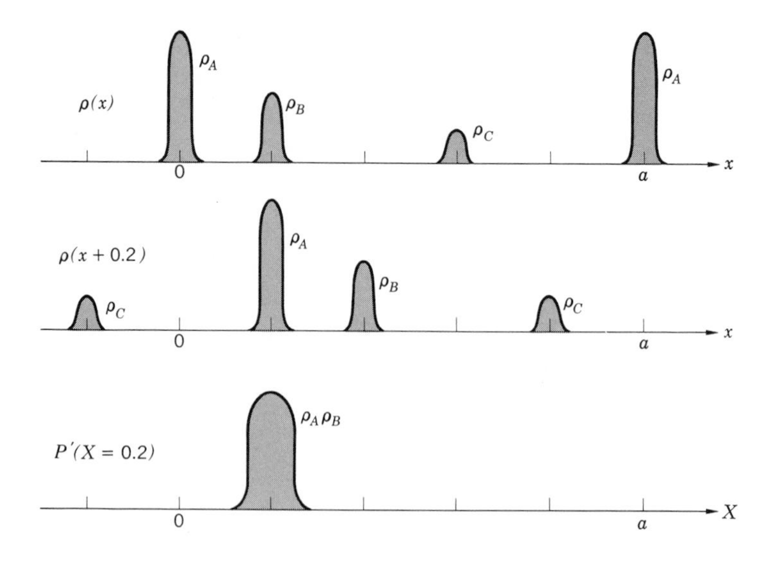

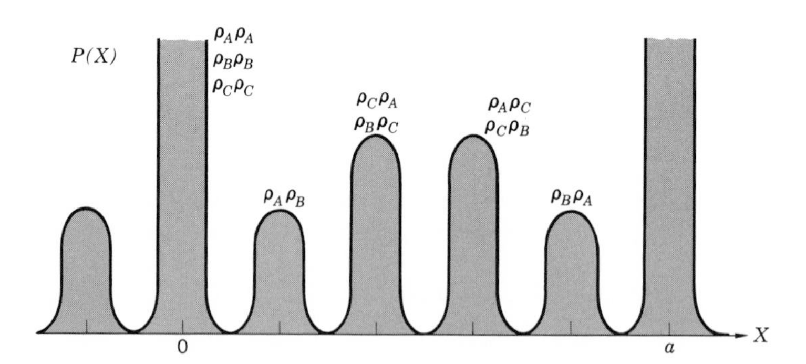

Fig. 11-8. Evolution of a one-dimensional Patterson function. The top two plots show $\rho(x)$ and its displacement by $X = 0.2$, namely, $\rho(x + 0.2)$. The third plot is that of the product $\rho(x)\rho(x + 0.2) = P'(X = 0.2)$, actually shown over the range of $X = 0.2 \pm 0.1$, since the width of Patterson peaks equals the sum of the electron-density peaks. Finally, the bottom plot shows the complete one-dimensional Patterson function.

experimentally measurable structure-factor amplitude squared. To see what physical meaning it has, consider Fig. 11-8. At the top, an asymmetric one-dimensional electron-density distribution contains three maxima, respectively, of density ρ_A, ρ_B, and ρ_C. Below it, the same electron-density distribution is shown shifted by an amount $X = 0.2$. This is the meaning of $\rho(x + X)$ in (11-42). When the product between these two functions is formed at each point x along a, either one or the other of the two functions has zero magnitude, except near $x = x + X = 0.2$, so that $P'(X) \propto \rho(x)\rho(x + .2)$ contains only one maximum at $X = 0.2$, and its total density is equal to the product of the two overlapping electron densities $\rho_A\rho_B$. Similarly, when X is allowed to take on all values from 0 to 1, a peak in $P(X)$ is produced whenever the X value corresponds to an interpeak distance in the one-dimensional density distribution at the top of Fig. 11-8. Since the original function contains three peaks, the Patterson function contains $3 \times 3 = 9$ peaks, three of which, representing an atom overlapping with itself ($X = 0$), lie at the origin, as shown at the bottom of Fig. 11-8. Thus the Patterson function displays all the interatomic distances in the electron-density function quite accurately. Because these distances are measured from a common origin in the Patterson function, however, their correct location in the electron density is not obvious. Fortunately, it is often possible to unravel the Patterson function by a systematic analysis of its peak positions, as discussed further in a later section.

The full three-dimensional Patterson function

$$P(XYZ) = \frac{1}{V^2} \sum_h \sum_{k=-\infty}^{\infty} \sum_l F_{hkl}^2 \cos 2\pi(hX + kY + lZ) \tag{11-45}$$

where $F_{hkl}^2 = |F_{hkl}|^2 = F_{hkl} \cdot F_{\bar{h}\bar{k}\bar{l}}$ and X, Y, and Z are expressed as fractions of a, b, and c, respectively, in keeping with the notation used elsewhere in this book. Quite frequently, one desires a projection of the Patterson function such as

$$P(XY) = \frac{1}{AV} \sum_h \sum_k F_{hk0}^2 \cos 2\pi(hX + kY) \tag{11-46}$$

where A is the area of the projected cell. Similarly, to the synthesis of electron-density projections, the reflections of a single zone are sufficient for (11-46).

It is not possible in this book to consider all the aspects of Patterson functions nor how they can be utilized in crystal-structure analysis. To see some of the more important ones, however, consider an idealized electron density whose projection contains three atoms consistent with the plane group $p2$. Four adjacent cells are shown in Fig. 11-9 in order to illustrate how the Patterson function can be

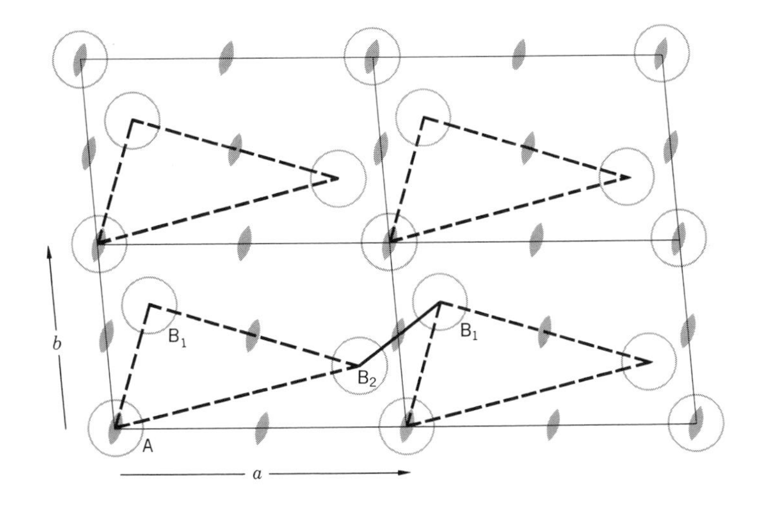

Fig. 11-9. $\rho(xy)$.

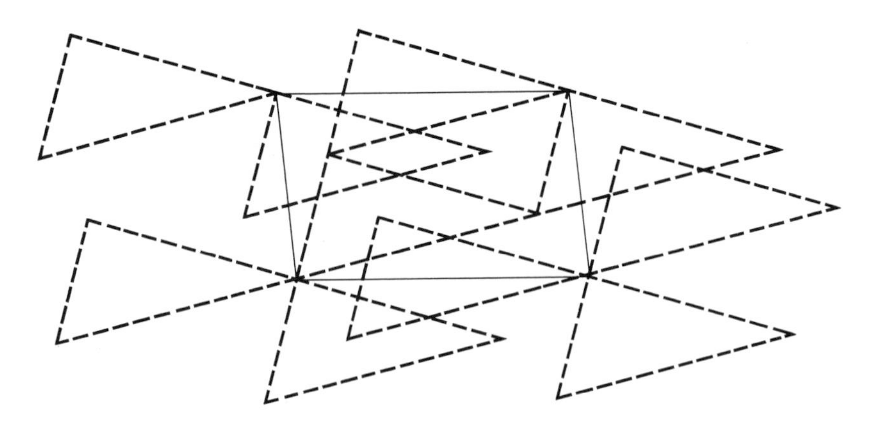

Fig. 11-10

derived graphically from the electron density. Since the Patterson map contains peaks at the terminal points of interatomic vectors, drawn from a common origin in the Patterson map, one way to locate the maxima in $P(XY)$ is to place the origin of the unit cell successively at each of the atomic sites in the electron density, and note the positions of the other atoms in the cell. For the triangular array in Fig. 11-9, the result is shown in Fig. 11-10. The reason for drawing four unit cells in Fig. 11-9 is simply so that all the peaks lying in the displaced cell, during the construction of the Patterson map, can be seen at a glance. Note that some of the electron-density peaks recur in the same place more than once, so that the final Patterson map looks like Fig. 11-11, where each circular contour represents the number of electron-density peaks superimposing at that point. It is now possible to distinguish the Patterson peaks as being single, double, or multiple peaks. Note that, for N peaks in the electron density, there will be N peaks superimposed at the origin, and $N^2 - N$ peaks distributed throughout the rest of the cell. This is one of the chief limitations of Patterson maps in practice: the same unit-cell volume must accommodate N^2 rather than N peaks, and moreover, the peaks in the Patterson function have volumes proportional to the products of pairs of corresponding volumes in the electron density. This problem, obviously, is further aggravated when projections are used.

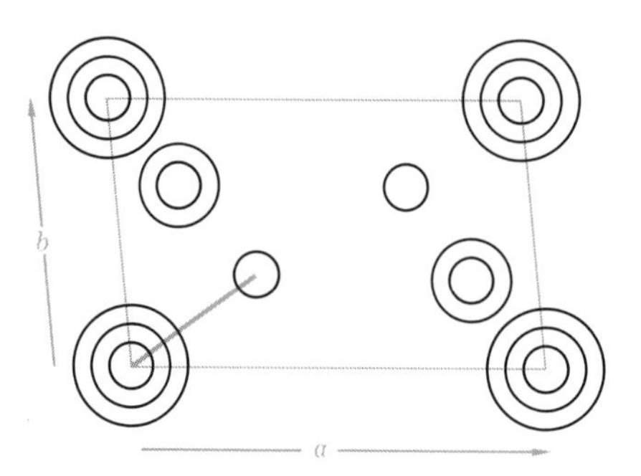

Fig. 11-11. $P(XY)$. Solid line joins atoms B_2 and B_1 in Fig. 11-9.

The weights of the peaks in a Patterson synthesis depend on the symmetry present. Consider a pair of atoms related by a two-fold axis, for example, the pair joined by a solid line in the bottom half of Fig. 11-9. Since this is an interatomic vector, it must appear in the Patterson map, and it is indicated by the solid line in Fig. 11-11. Note that this peak is of single weight; that is, it represents a single pair of atoms; whereas the other nonorigin peaks have double weights; that is, they represent the superposition of two peaks. This relation between the weights of peaks was systematically studied by Buerger. He suggested that Patterson peaks which result from pairs of symmetry-equivalent atoms be distinguished by calling them *rotation peaks* if the two atoms are related by an operation of rotation, and *reflection peaks* if related by mirrors. These peaks characteristically have single weight. When more than two equivalent atoms are related by the symmetry at a point, additional peaks systematically accompany the rotation peaks. These peaks characteristically have double weights, and Buerger named them *satellites*. The satellites accompanying rotation peaks are particularly helpful in distinguishing such peaks from other Patterson peaks caused by nonequivalent atoms, and in determining the correct symmetry.

As an example of these different classes of peaks, consider the array of peaks in Fig. 11-12. The projected electron density on the left has the symmetry pmm. Consider the set of four equivalent atoms in its central region. The two pairs related by the operation of rotation give rise to rotation peaks of single weight in the Patterson map shown to the right in Fig. 11-12. The reflection-related atom pairs give rise to four pairs of identical reflection peaks in the Patterson map, so that they superimpose in pairs. Consequently, these peaks all have double weights. Note that two superposed reflection peaks having coordinates X,0 and $0,Y$, respectively, accompany each rotation peak at X,Y in the Patterson map. This is the reason for calling them satellite peaks. Next, consider the eight Patterson peaks produced by pairing the origin atom in the electron-density map with each of the four symmetry-equivalent atoms in its central region. These interactions give rise to four additional peaks having double weight. It is clearly seen on the right in Fig. 11-12 that these peaks can be distinguished from the rotation peaks by the absence of accompanying satellites. This is a considerable aid in the interpretation of Patterson maps in practice, because the presence of a

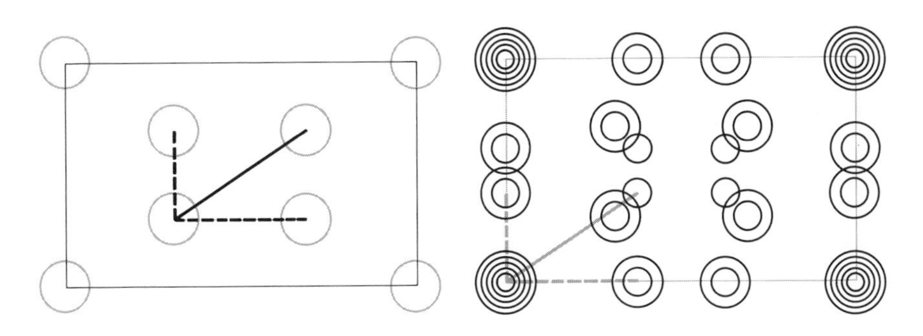

Fig. 11-12

certain amount of "background" makes the determination of relative peak heights imprecise. In concluding this section it should be noted that the relative disposition of satellites is determined by the symmetry present (Exercise 11-15), and, in three dimensions, it is necessary to consider also *inversion peaks* resulting from pairs of atoms related by an inversion center.

Harker sections. Despite the problems of N^2 overlapping peaks, for many years Patterson functions were used in projections because the computational labor attendant on calculating three-dimensional sections was deemed to be prohibitive. (The ready availability of high-speed computers has changed that, so that nowadays it is fairly common practice to calculate a sufficient number of sections to obtain an accurate model of the peak distribution throughout the three-dimensional Patterson space.) Within a couple of years of Patterson's publication of his function, however, D. Harker pointed out that in certain cases the additional labor of calculating a three-dimensional section was quite justifiable. For example, in the space group $P2_1$, for each atom at x,y,z, there is another atom at $\bar{x},\bar{y},z+\frac{1}{2}$. This means that the rotation vectors all occur in symmetric pairs about 2_1 at $2x,2y,\frac{1}{2}$ and $2\bar{x},2\bar{y},\frac{1}{2}$, while the satellites have variously spaced z coordinates. It is clear from this that the section

$$P(XY\tfrac{1}{2}) = \frac{1}{V^2}\sum_h \sum_k \left[\sum_l F^2_{hkl}(-1)^l\right] \cos 2\pi(hX + kY) \tag{11-47}$$

contains only rotation peaks, barring chance differences of $z = \frac{1}{2}$ between a pair of atoms not related by the 2_1. This means that this one section, called the Harker section for 2_1, correctly gives the x and y coordinates of the various atoms. Actually, this is not quite correct. Since the Patterson function places all the interatomic vectors at the same origin, an ambiguity exists in its interpretation. As can be seen in Fig. 4-5, there are four distinct 2_1 axes in the unit cell, so that the pairs of atoms identified in $P(XY\frac{1}{2})$ must be correctly placed about these screw axes, and finally, their z coordinates determined. Further discussions of the interpretation of Patterson functions, the use of Harker sections, and many specific examples can be found in some of the books listed at the end of this chapter, particularly Buerger's *Vector space*.

DIRECT METHODS

Heavy-atom method. When a crystal contains one kind of atom having many more electrons than the other atoms, the heavy atom dominates the structure-factor magnitudes. Very often this means that the relative phases of the structure factors can be determined unambiguously once the heavy atom or atoms are correctly located in the unit cell. In such cases, it is possible to synthesize the electron density at once, using measured F's and calculated phases. Even when some of the phases initially are incorrectly determined, the other atomic positions revealed in the electron density permit the correct determination of doubtful phases, so that an iterative structure analysis becomes possible. The literature abounds with practical illustrations of this, because the presence of heavy atoms, for example in most metal oxides, greatly facilitates the structure determination.

Another way to utilize heavy atoms is possible whenever isomorphous compounds exist in nature or can be synthesized artificially. This is the case, most fortunately, for many more complex organic molecules, including those having considerable biological importance. Recalling that the structure factor can be expressed by the sum of the contributions of the component atoms, let F, for one isomorphous crystal containing a light metal whose scattering factor is f_L, be written

$$F_{hkl} = f_L S_L + \sum_{\substack{\text{other}\\ \text{atoms}}} f_n S_n \tag{11-48}$$

while for the crystal containing the heavy atom represented by f_H,

$$F_{hkl} = f_H S_H + \sum_{\substack{\text{other}\\ \text{atoms}}} f_n S_n \tag{11-49}$$

If the two crystals are truly isomorphous, then the heavy and light atoms occupy identical sites, so that $S_L = S_H$. Similarly, the two remaining summations on the right in (11-48) and (11-49) also equal each other. Assuming for simplicity that both crystals are centrosymmetric and that the replacement is for the atom occupying the $0,0,0$ site in the crystal, it is possible to deduce the phases (signs) of most reflections directly from a comparison of the magnitudes of corresponding pairs of reflection intensities. In a centrosymmetric crystal the phases can be $0°$ or $180°$ only, so that $e^{i\varphi} = \pm 1$. Suppose that

$$(I_{hkl})_H > (I_{hkl})_L \tag{11-50}$$

Since $f_H > f_L$, and because the metal atom at the origin makes a positive contribution to F_{hkl}, it is clear that the sign of the hkl reflection in (11-50) is positive, so that the magnitude of I_{hkl} is increased by increasing the magnitude of the positive contribution of the origin atom. Quite similarly, when

$$(I_{hkl})_H < (I_{hkl})_L \tag{11-51}$$

the sign of the hkl reflection is negative, because increasing the positive contribution has served to decrease the magnitude of (11-48). The signs thus determined can be used with measured F values to synthesize the electron density, which, under the simplifying condition assumed above, should yield the correct final structure.

When the substituted atoms do not occupy a site at the origin, it is not known in advance whether they make positive or negative contributions to the structure factors in (11-48) and (11-49). Provided the structure is centrosymmetric, however, this merely introduces an ambiguity, but does not prevent utilization of the above principles in the structure analysis (Exercise 11-17). In this connection it should be noted that, when a set of trial phases are used in synthesizing an electron density, it is better to omit doubtful terms than to include incorrect phases. It should be intuitively obvious, and can be proved more rigorously, that the effect on a Fourier series of including a negative term when the correct term is positive is more pronounced than if its amplitude is incorrectly assumed to be zero. Thus it is better to begin with a few certainly correct terms in the initial syntheses, gradually increasing their number as more is learned about the structure.

There is still another way to utilize isomorphous substitution in structure analysis. Consider a crystal whose atomic composition is $MABCD$. Its Patterson function contains the following peaks:

$$\begin{matrix} MM & MA & MB & MC & MD \\ AM & AA & AB & AC & AD \\ BM & BA & BB & BC & BD \\ CM & CA & CB & CC & CD \\ DM & DA & DB & DC & DD \end{matrix} \tag{11-52}$$

An isomorphous crystal then must have the atomic composition $NABCD$, and a Patterson containing the same matrix of peaks as (11-52), with M replaced by N. If the Patterson functions of both crystals are synthesized and subtracted from each other, a *difference Patterson* is obtained containing only the peaks

$$\begin{aligned} &(MM - NN)(MA - NA)(MB - NB)(MC - NC)(MD - ND) \\ \text{and} \qquad &(AM - AN)(BM - BN)(CM - CN)(DM - DN) \end{aligned} \tag{11-53}$$

The relative disposition of these peaks is shown in Fig. 11-13, from which it is clear that the difference Patterson represents the actual structure, with the $(M - N)$ atom placed at the origin. Note that the peaks have the same relative positions as they would have in the electron density, except that their magnitudes are equally "scaled" by the factor $(M - N)$. The preceding observation is literally true when the structure is centrosymmetric. If it is acentric, the difference Patterson contains the correct structure and its centrosymmetric inversion, so that it is necessary to distinguish the two superimposed sets of atoms. Even when a center of symmetry is present, the presence of additional symmetry may seriously complicate the analysis unless the atom being replaced is at the origin.

In practice, it is not necessary to synthesize the two Patterson functions prior

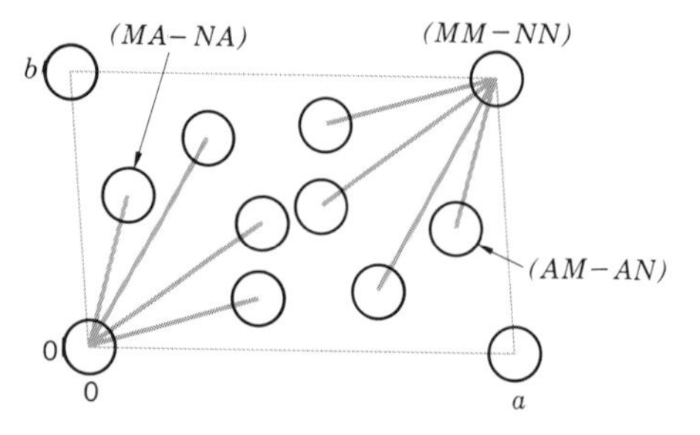

Fig. 11-13. $P_M(XY) - P_N(XY)$.

to subtracting one from the other because the difference Patterson

$$P(XYZ) = P_M(XYZ) - P_N(XYZ)$$
$$= \frac{1}{V^2} \sum_h \sum_k \sum_l (F_M{}^2 - F_N{}^2)_{hkl} \cos 2\pi(hX + kY + lZ) \tag{11-54}$$

Thus only one Fourier series has to be computed after pairs of suitably scaled structure-factor amplitudes squared are subtracted from each other.

Anomalous-scattering method. The above straightforward procedures for structure analysis are very convenient to use because they reduce the structure determination to the relatively routine process of gathering x-ray data and suitably synthesizing it. Unfortunately, it is often not possible to obtain suitable isomorphous crystals. It is nevertheless sometimes possible to circumvent this difficulty by making use of the anomalous scattering of x rays, discussed in Chaps. 6 and 8. Recalling that the use of x rays whose energies lie close to an absorption edge of an atom causes them to be scattered anomalously, it is possible to alter the magnitude (and phase) of an atom's contribution to F_{hkl} in this way. The details of the quantitative effect produced on the structure factor depend on whether the crystal is centrosymmetric or not, and on whether the x-ray wavelength used is shorter or longer than that of the absorption edge. Whenever possible, two wavelengths are chosen so as to produce the maximum difference possible in the scattering-factor magnitudes. In certain crystals containing atoms that are neighbors to each other in the periodic table, for example, some of the elements in the first transition series, it is often possible to alter the scattering-factor magnitudes of several atoms successively by choosing three or more x-ray wavelengths.

Although in principle this is a most appealing method for producing "isomorphous" pairs without needing more than one crystal, in practice one is limited by the x-ray wavelengths readily available in commercial tubes. Nevertheless, the employment of crystal monochromators and highly sensitive ionization detectors and the possibility of utilizing L or M edges as well as K edges should serve to increase considerably the applications of this direct approach to structure analysis. The chief remaining difficulty is the purely experimental one of measuring small intensity differences with adequate precision, a situation that is aggravated by the increased absorption by the anomalous scatterer. Here the utilization of the same crystal, however, simplifies estimation of the absorption correction.

Image-seeking methods. A little consideration of what has been discussed so far should make it clear that a crystal containing N atoms repeats the structure N times in the Patterson function to yield the N^2 total peaks present. This observation led Buerger to call the Patterson function the Fourier representation of a "squared crystal." In view of the preceding statement, it follows that, if one can recognize an array of N peaks that represent the correct atomic array, one should be able to find it repeated N times in the Patterson function. Suppose the atoms in this array are joined by lines (interatomic vectors) to form a polygon of N sides, called an *N-gon*. Then this N-gon (representing the peaks in the electron density) is repeated by each of the N atoms in the Patterson, obviously, N times. Suppose that each time this N-gon is found to come into coincidence with the appropriate peaks in Patterson space, a fiducial marker is placed at some point, say, at the midpoint along one of its sides. It follows from the preceding discussion that, since N such coincidences must occur in the Patterson, N such fiducial markers will have been noted, and these must be arrayed in a similar manner to the array of peaks (atoms) in the initial N-gon. It turns out that this is indeed the case, except that the array of markers represents an inverted array. To see this, consider Fig. 11-10, which was constructed by repeating the 3-gon to show how the "squared crystal" is formed. Suppose that a fiducial marker is placed at the midpoint of the short legs of each of the dashed 3-gons. Clearly, when these midpoints are joined to each other, the same 3-gon, or triangle (structure), is reproduced, except that it is the centrosymmetric inverse of the original triangle.

In practice, of course, one lacks a priori knowledge of what the structure, or N-gon, looks like. An important aspect of the above procedure comes to one's assistance, however. Note that the self-coincidences were recorded *by placing the marker at the center of one edge of the N-gon.* This means that all one really needs to identify is one edge, that is, one interatomic vector in the Patterson function, and then to seek N coincidences or images of that vector while maintaining it parallel to itself! Using the previous example, four unit cells of the Patterson function in Fig. 11-11 are reproduced in Fig. 11-14. Suppose the rotation vector previously identified is chosen as the *image-seeking vector,* and its coincidences or images are noted by placing a cross at its center, as indicated in Fig. 11-14. Joining up these crosses to form triangles (3-gons) clearly shows a reconstruction of the electron density in Fig. 11-9, except that a different triangle, or 3-gon, has been chosen to represent the structure.

The actual process of carrying out such image seeking is illustrated in Fig.

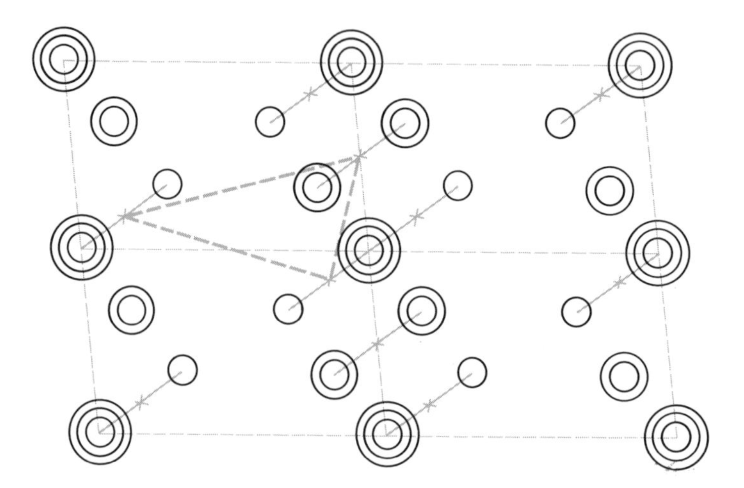

Fig. 11-14. Self-coincidences of the rotation vector joining circles B_2 and B_1 in Fig. 11-9. Note that the 3-gon defined is the centrosymmetric inverse of the original triangle.

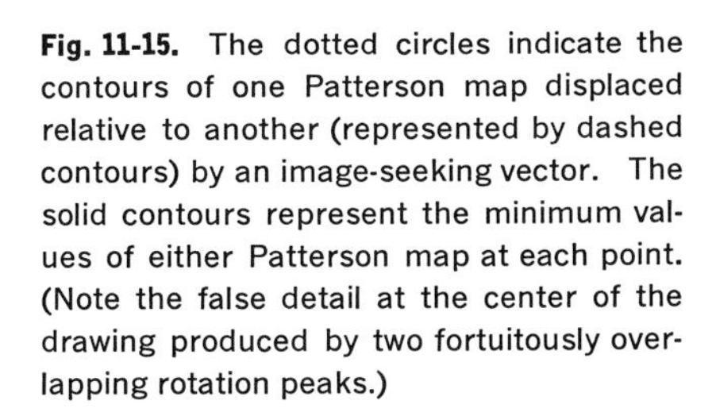

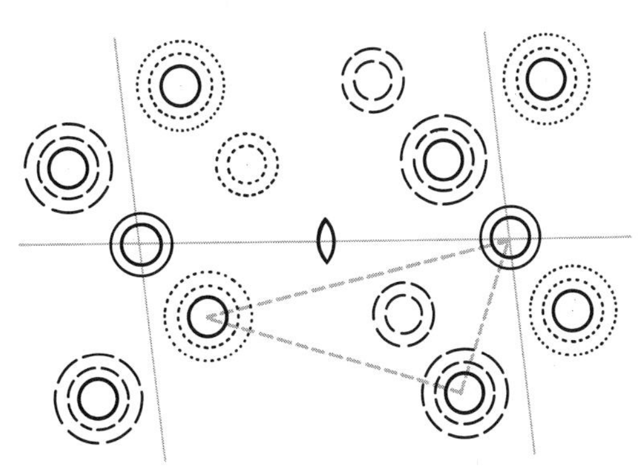

Fig. 11-15. The dotted circles indicate the contours of one Patterson map displaced relative to another (represented by dashed contours) by an image-seeking vector. The solid contours represent the minimum values of either Patterson map at each point. (Note the false detail at the center of the drawing produced by two fortuitously overlapping rotation peaks.)

11-15. The peaks in the four unit cells of Patterson space in Fig. 11-14 are drawn on one sheet of paper (indicated by dashed circles in Fig. 11-15). An identical drawing on a transparent sheet (dotted circles in Fig. 11-15) is overlain on the first drawing, except that its origin is shifted so that it superimposes the peak identified as a rotation vector. The two relatively displaced Patterson maps are then examined at each point X,Y, and the minimum value of either Patterson is noted on a third sheet overlying the two maps. Thus, when a peak in one map lies above background in the other map, the background level is recorded on this third sheet. When two peaks superimpose, the lowest values at each point are recorded on this third sheet. The solid circles in Fig. 11-15 indicate where such superpositions occur, and the resulting *minimum function* is shown in Fig. 11-16. A comparison of this figure with the original electron density in Fig. 11-9 clearly shows that the correct atomic positions have been recovered. This way of analyzing the superimposed Patterson maps, first proposed by Buerger, has the tremendous advantage over alternative ways of examining such superpositions, in that most of the unwanted "false" peaks are removed at once, even in the minimum function M_2, based on identifying the correct relative position of only two atoms in the structure. Patterson maps of complex crystal structures, however, particularly when projections are used, contain so much detail that the simple M_2 function may contain numerous spurious peaks. In that case, the identification of one or more additional atom positions permits the synthesis of a higher-ranking minimum function, for example, M_4, obtained by superimposing two relatively displaced M_2 maps. This operation is equivalent, of course, to examining the original Patterson map with a 4-gon. In principle, this procedure can be continued as many times as deemed necessary,

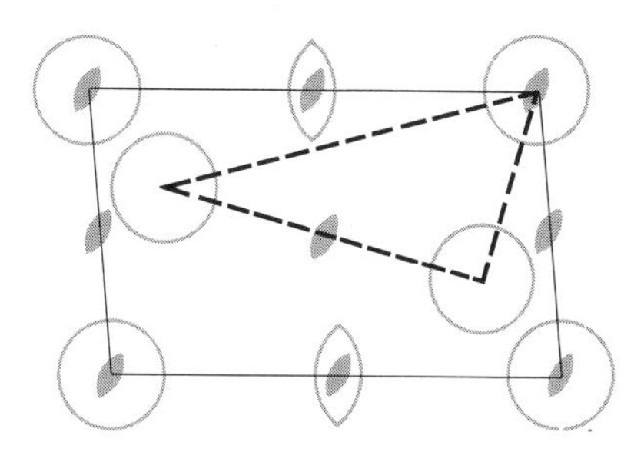

Fig. 11-16. Minimum function obtained from superposition in Fig. 11-15.

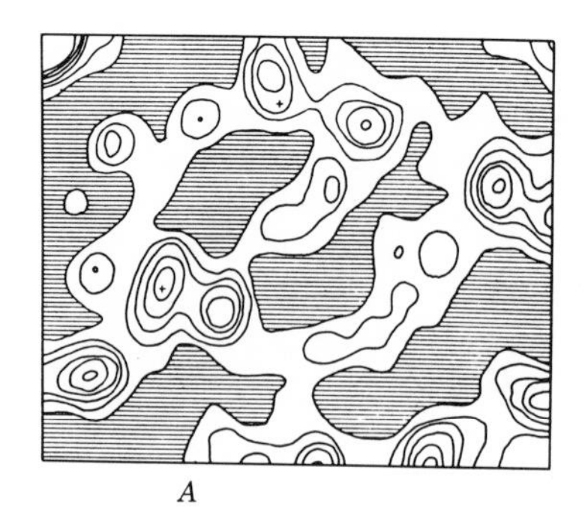

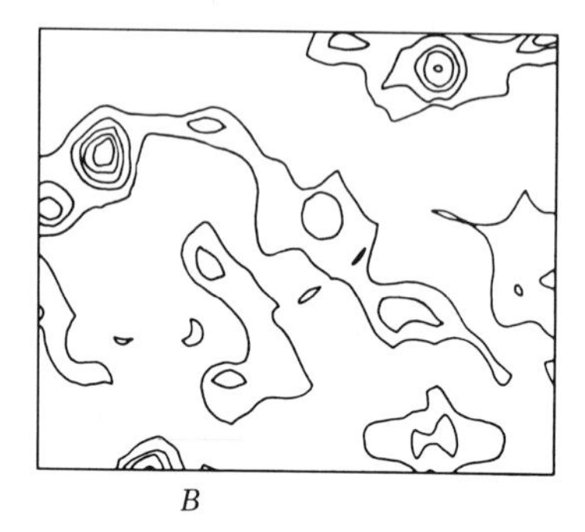

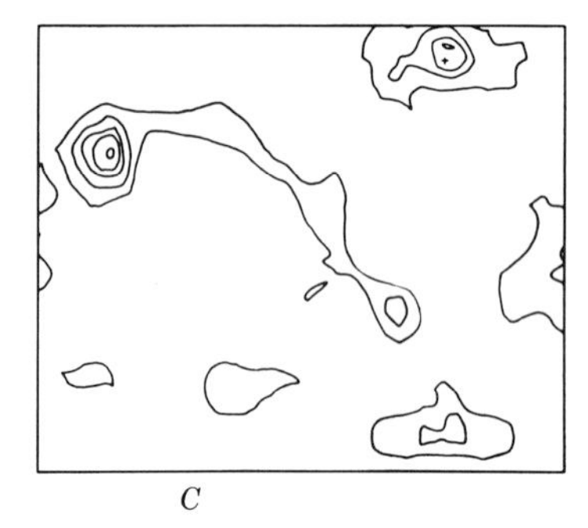

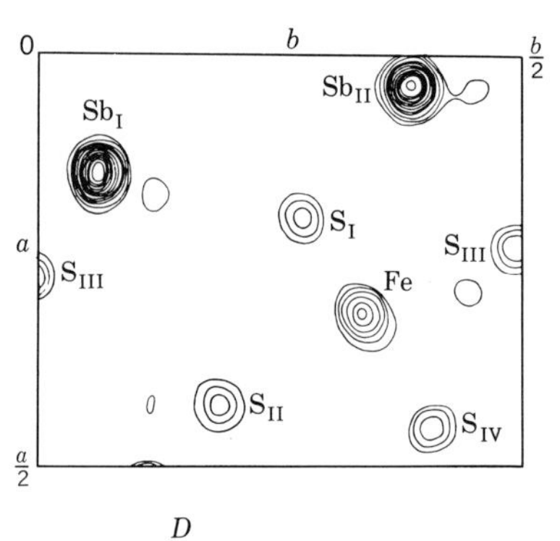

Fig. 11-17. Structure analysis of berthierite, $FeSb_2S_4$. (*A*) Patterson projection $P(XY)$; (*B*) one of two minimum-function $M_2(XY)$ maps; (*C*) $M_4(XY)$ obtained by superposing two $M_2(XY)$ maps; (*D*) electron-density projection $\rho(xy)$. [*From M. J. Buerger, Acta Crystallographica, vol.* 4 (1951), *pp.* 531–544.]

although in practice it is rarely necessary to go to higher ranks than M_6. The successive stages of applying this image-seeking process are illustrated in Fig. 11-17. It should be noted that projections cannot be used when there are too many fortuitous superpositions of peaks, since even M_2 tends to reduce the number of peaks quite drastically. It is possible, in such cases, to use three-dimensional sections and conduct the image-seeking process throughout the volume of a Patterson cell. This analysis is greatly facilitated by the adaptation of computer methods, so that it is possible nowadays to feed the observed intensity values to a computer which converts them to F^2 values; calculates the Patterson function and, after being instructed regarding the vector shift to be tried, analyzes this function for possible images of the trial vector; and finally, synthesizes the minimum-function maps. At this point, the investigator can ask the computer to read off the peak coordinates and use them to calculate structure-factor magnitudes and phases, to verify the correctness of the structure deduced.

Inequalities. The foregoing discussion of the relations between relative structure-factor magnitudes and their phases has focused attention on the fact that the observed intensities do contain information regarding phases, even though it is not directly apparent. The information regarding the phases is contained, of course, in the relative disposition of the atoms in the unit cell. In other words, if all the atoms could occupy the origin site, they would all scatter in phase, and the magnitudes of successive reflection intensities would be a systematic function of the scattering angle. Because they cannot, destructive interference between x rays scattered by different atoms decreases the magnitude of I_{hkl}. The presence of symmetry in a crystal, however, controls the way that the magnitudes of certain sets of reflection intensities are affected, so that their relative magnitudes clearly contain the necessary information to determine their phases. The problem is, how to extract this information directly. By considering the way certain symmetry present in a crystal affects the relative peak positions in a Patterson map, Buerger concluded that it is possible to recognize certain constraints imposed on the atomic arrays, and from them to deduce certain relations between sets of structure-factor magnitudes. He called this approach *implication theory*.

A somewhat different approach was taken by Harker and J. S. Kasper, who borrowed from mathematical theory two relations known as the Cauchy inequality and the Schwartz inequality and examined the relations between structure-factor magnitudes that could be established with their aid. As an illustration, consider Schwartz's inequality

$$|\int fg\,d\tau|^2 \leq (\int |f|^2\,d\tau)(\int |g|^2\,d\tau) \tag{11-55}$$

and let the variables take on the values

$$f = [V\rho(xyz)]^{\frac{1}{2}}$$
$$g = [V\rho(xyz)]^{\frac{1}{2}}\,e^{2\pi i(hx+ky+lz)}$$
$$d\tau = dx\,dy\,dz$$

Substituting these values on the left of (11-55),

$$\left| \int fg \, d\tau \right|^2 = \left| V \iiint\limits_{\substack{\text{unit}\\ \text{cell}}} \rho(xyz) \, e^{2\pi i(hx+ky+lz)} \, dx \, dy \, dz \right|^2$$

$$= |F_{hkl}|^2 \tag{11-56}$$

according to (11-35). Substituting the same values on the right of (11-55) and noting that $|e^{2\pi ix}|^2 = 1$, the inequality becomes

$$|F_{hkl}|^2 \leq V^2 \left[\iiint\limits_{\substack{\text{unit}\\ \text{cell}}} \rho(xyz) \, dx \, dy \, dz \right]^2$$

$$\leq \zeta^2 \tag{11-57}$$

where ζ is the total number of electrons contained in the unit cell. This is an important, but not original, result, which states that the magnitude of a structure factor cannot exceed ζ.

To see the influence that symmetry has on structure factors, consider what happens when the crystal contains a symmetry center. In this case

$$\rho(xyz) = \rho(\bar{x}\bar{y}\bar{z})$$

and

$$F_{hkl} = V \iiint\limits_{\substack{\text{unit}\\ \text{cell}}} \rho(xyz) \cos 2\pi(hx + ky + lz) \, dx \, dy \, dz \tag{11-58}$$

Choosing the values for the variables in (11-55) as before, Schwartz's inequality becomes

$$|F_{hkl}|^2 \leq [V \textstyle\int\int\int \rho(xyz) \, dx \, dy \, dz][V \int\int\int \rho(xyz) \cos^2 2\pi(hx + ky + lz) \, dx \, dy \, dz] \tag{11-59}$$

Making use of the relation $\cos^2 x = \frac{1}{2}(1 + \cos 2x)$ and noting that the first integral in (11-59) equals ζ

$$|F_{hkl}|^2 \leq \frac{\zeta}{2} [V \textstyle\int\int\int \rho(xyz) \, dx \, dy \, dz + V \int\int\int \rho(xyz) \cos 2\pi(2hx + 2ky + 2lz) \, dx \, dy \, dz]$$

$$\leq \frac{\zeta}{2} (\zeta + F_{2h,2k,2l}) \tag{11-60}$$

Making use of (11-57) to define a *unitary structure factor* $U_{hkl} = F_{hkl}/\zeta$, the inequality in (11-60) can be written after dividing both sides by ζ.

$$|U_{hkl}|^2 \leq \tfrac{1}{2} + \tfrac{1}{2} U_{2h,2k,2l} \tag{11-61}$$

the more conventional form of this relation.

The importance of this Harker-Kasper inequality is that it makes it possible to compare pairs of reflections hkl and $2h,2k,2l$ and to deduce their phases. Recalling that the phases in centrosymmetric crystals serve to make the structure factor either positive or negative, it follows that the sign of $U_{2h,2k,2l}$ may be deduced from (11-60), even when the sign of U_{hkl} is not known. For example, suppose $|U_{hkl}| = 0.9$. Clearly, $U_{2h,2k,2l}$ must be positive if the inequality in (11-61) is to be maintained. (See Exercise 11-19.) Other examples can be presented in which an unambiguous relation is not established. It does become possible, however, by using many reflections, to group the reflections into sets whose signs can be determined once the signs of a select small number of reflections is known. This, of course, greatly reduces the amount of iteration that otherwise would be required. Moreover, the presence of additional symmetry permits further relations to be deduced. For example, when a two-fold axis is present,

$$|U_{hkl}|^2 \leq \tfrac{1}{2} + \tfrac{1}{2}U_{2h,2k,0} \tag{11-62}$$

as derived in Exercise 11-20.

The above discovery by Harker and Kasper, that it is possible to deduce phases directly from the magnitudes of structure factors, inaugurated a new era in structure analysis. Although it has subsequently turned out that these procedures, including many additional relations based on inequalities, are limited to solving structures containing a relatively small number of atoms in the unit cell, the inauguration in 1948 of the concept that crystal structures can be solved directly from the manipulation of measured intensities served to stimulate major efforts in developing more powerful direct methods. One of these, representing the direct interpretation of Patterson syntheses based on the observed intensities, has already been discussed. Another, which utilizes statistical analyses of relative structure-factor magnitudes and their sign relations, is briefly reviewed below.

Statistical methods. The need to examine a number of relations between a much larger number of observed structure-factor magnitudes has led several investigators to seek more fundamental relations between them. By considering what relations are produced when an electron-density function is multiplied by itself (squared), D. Sayre discovered a fundamental relation that must be obeyed by a crystal composed of equal atoms.

$$F_H = \frac{1}{g_H V} \sum_{H'} F_{H'} F_{H+H'} \tag{11-63}$$

where g_H is a scaling factor; H is shorthand for hkl; and H' is a particular value of H. When the sum of the products on the right of (11-63) is examined, it is clear that, for a very large product, that is, when a particular $F_{H'}$ and $F_{H+H'}$ are both large, this product dominates the sum, and hence the sign, of F_H. Expressing the sign of F_H in a centrosymmetric crystal by S_H, relation (11-63) means that

$$S_H = S_{H'} S_{H+H'}$$

or

$$S_{H+H'} = S_H S_{H'} \tag{11-64}$$

Clearly, the sign relation in (11-64) should hold when the three structure factors considered all have large magnitudes.

By considering the equation from which Schwartz's inequality is derived, Zachariasen derived not only Sayre's relation (11-64), but a more powerful one that can be written

$$S_H = S(\overline{S_{H_i}S_{H+H_i}}) \tag{11-65}$$

The first S on the right of (11-65) represents the sign of the averaged products. It should be noted that the application of these sign relations requires that the intensities be grouped into appropriate sets and their products examined, noted, and stored, while other sets are compiled. Finally, cross correlations between such sets are established, so that the entire procedure is quite laborious unless high-speed computers are utilized.

A purely statistical approach was adopted by H. Hauptman and J. Karle, who sought to establish the probabilities that govern the magnitudes of structure factors in the presence of certain symmetries in the crystal. The equations they obtained for centrosymmetric crystals can be epitomized by the relation

$$\Sigma_H = \Sigma_1 + \Sigma_2 + \Sigma_3 + \Sigma_4 \tag{11-66}$$

which fixes the sign of F_H in terms of four sums of terms. Σ_1 and Σ_4 in their nomenclature involve only structure-factor magnitudes, so that they can be summed directly. Σ_2 and Σ_3 involve structure factors directly, so that their phases (signs) must be included in carrying out the summation. This means that this procedure, like the others described in this section, is essentially iterative in character. First, the various compilations are expressed in terms of a number of structure factors whose phases remain unspecified. Subsequently, signs are assigned to some of them, and the permutations continued until a self-consistent set of signs is obtained.

The foregoing discussion naturally raises the following question: With the ready availability of large-storage high-speed computers, why not simply use direct iterative procedures? The difficulty is, of course, that for N independent reflections there are 2^N possible combinations of signs (restricting the analysis to centrosymmetric crystals). Suppose that 21 $hk0$ reflections have been observed. The signs of two structure factors can be chosen arbitrarily, thereby fixing the origin of $\rho(xy)$. This leaves $2^{21-2} = 2^{19} = 489{,}088$ possible sign combinations to be explored! As suggested by M. M. Woolfson, however, suppose a set of seven of these structure factors are considered separately. Initially, one might assume that $2^7 = 128$ combinations would have to be considered. Recalling that two of the signs can be fixed arbitrarily, and assuming that one wrong sign may be tolerated without seriously affecting the structure analysis, the number of possible combinations is reduced to $128 \div 2^3 = 16$, a much more reasonable number of permutations to explore. It is possible to extend this method to include the remaining structure factors by forming another set of seven in a like manner. Care must be exercised in doing this to include the two reflections whose signs were arbitrarily fixed, so that all signs are related to the same choice of origin for the unit cell. The extent to which such a "frontal assault" on the phase problem in crystal-structure analysis will be successful remains to be seen. It has the considerable appeal of turning over the task of determining phases to a computer, thereby freeing the scientist for more rewarding pursuits.

REFINEMENT PROCEDURES

Correctness of structure. Having considered some of the methods developed for the determination of atomic coordinates in a unit cell, it is important to establish a criterion for assessing the correctness of the resulting structure. Clearly, if the atomic coordinates in the proposed structure model are the same as those in the crystal, the calculated and observed structure-factor magnitudes should be the same. Denoting them, respectively, $|F_c|$ and $|F_o|$, the differences $|F_c| - |F_o|$ represent the order of the disagreement, and hence the incorrectness of the proposed model. The magnitude of this difference, however, is strongly influenced by the magnitudes of the two terms being subtracted, so that a means for "averaging" such terms over all reflections is needed to establish a useful figure of the lack of reliability for the proposed model. Several different functions have been proposed, but the following has proved to be most popular:

$$R = \frac{\Sigma||F_o| - |F_c||}{\Sigma|F_o|} \tag{11-67}$$

where R is called the *residual* for a set of structure-factor magnitudes. As shown by Wilson, $R = 0.828$ for a centrosymmetric crystal and a totally wrong set of atomic coordinates. (The corresponding value is $R = 0.586$ for an acentric structure.) Thus, in principle, $R < 0.50$ suggests a correct structure, although considerably smaller R values are usually deemed necessary to lend credibility to the proposed structure.

A number of problems are associated with the use of R as defined in (11-67). A practical problem arises when small but nonzero F_c values are calculated but the corresponding F_o values are not observed. Clearly, there is some lower limit below which x-ray intensities cannot be measured by the particular instrument employed. Thus there are several possibilities for treating these differences. The simplest one is to omit them entirely from the summations in (11-67), so that only actually observed reflections are included. This obviously leads to an exaggeratedly smaller value of R. A more satisfactory procedure is to assess the minimum intensity that can be recorded, and then to assume that the unobserved reflection had an intensity less than this. This procedure not only requires the calculation of F_o values from this minimum-intensity value at each reciprocal-lattice point concerned, but it still leaves the question of how much less than this minimum value was the actual unobserved intensity. The most probable value is one-half the detection limit, and this is the value frequently used in accurate assessments of R. Whatever scheme is in fact adopted, it is clearly necessary to state it when reporting the residual, so that a meaningful comparison with other values can be made.

Successive Fourier syntheses. Once a set of phases have been determined, it is possible to calculate the electron density, as discussed earlier in this chapter. If the structure model used to calculate the phases is entirely correct, the electron-density calculation should reproduce the proposed atomic coordinates exactly. Any errors in the initial assignation of phases cause the maxima in the electron density to deviate from the postulated locations. Provided that the Fourier series employed for this purpose contains the observed structure-factor amplitudes, it is meaningful to use the newly discerned atomic coordinates to recalculate the phases. If any phases are found to have changed, the entire process is repeated until no further change in phases results. This procedure is particularly useful when the phases have been determined by considering only some of the atoms having especially strong scattering powers. Similarly, it provides a means for establishing the phases for reflections that may have been omitted from an initial synthesis because of uncertainties as to their correct value. Thus successive Fourier syntheses really represent a means for determining the correct structure. The procedure is quite straightforward and fairly rapid when the structure is centrosymmetric, since the phase angles are either 0 or $180°$. In acentric structures the phases can have all values between these two limits, so that the convergence to the finally correct synthesis is considerably slower.

Even when the above procedure for establishing the correct phases is concluded, it is possible that small additional changes in atomic coordinates may reduce any disparities remaining between observed and calculated structure-factor magnitudes. This is the real purpose of refining a crystal structure, so that the final atomic coordinates derived constitute the very best fit possible between the proposed atomic coordinates and the true ones represented by the experimentally measured reflection powers. Ideally, the differences $|F_o| - |F_c|$ should vanish for a correct structure, so that the residual (11-67) for a refined structure reflects the extent of the unavoidable experimental errors and incorrectness of assumptions made in converting measured intensities to F values and in calculating theoretical atomic scattering factors.

Difference synthesis. One reason why a regular Fourier series utilizing observed structure-factor amplitudes and calculated phases fails to locate the atomic peaks quite correctly is that the true electron-density synthesis requires that an infinite series be employed. Since one observes a finite number of terms in practice, small ripples caused by *series-termination errors* are introduced in the resulting maps. Whenever very accurate electron densities are required, therefore, it is necessary to supply the missing terms. This can be done, for example, by calculating the unobservable structure-factor amplitudes and including them in the series. Although their magnitude declines fairly rapidly with increasing $(\sin\theta)/\lambda$, it is necessary to increase the computational labor considerably thereby. A method first proposed by A. D. Booth virtually eliminates series-termination errors and has another, most important advantage of displaying graphically any errors remaining in the atomic locations deduced from the final electron density, as well as other errors made in calculating the structure-factor magnitudes.

Booth proposed that a *difference synthesis* be formed by subtracting an electron density, calculated using F_c from the usual one containing F_o. For a one-dimensional synthesis, this can be expressed

$$\Delta\rho = \rho_o - \rho_c = \frac{1}{a}\sum_h (F_o - F_c)_h e^{2\pi i h x} \qquad (11\text{-}68)$$

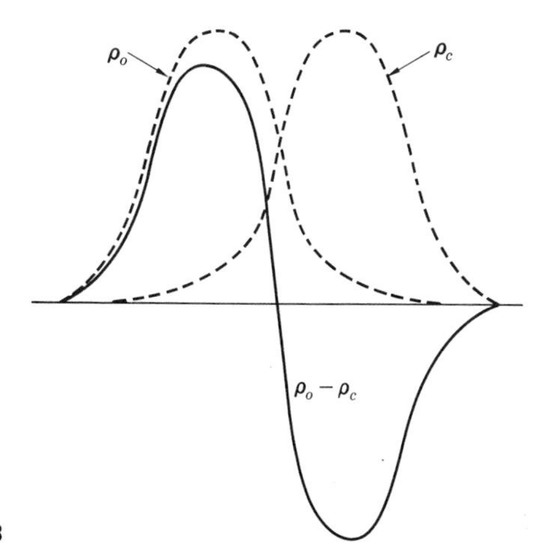

Fig. 11-18

so that, actually, only one synthesis is required after the differences $F_o - F_c$ are calculated. Assuming that the magnitudes of the unobserved terms are the same as those omitted from ρ_c in (11-68), it is clear that series-termination errors are automatically eliminated, since the coefficients of such terms in the final series in (11-68) are zero.

When peaks representing atomic locations in ρ_o do not coincide with those in ρ_c, then the difference synthesis shows a characteristic structure. This is illustrated for a one-dimensional synthesis in Fig. 11-18. The gradient of the curve representing $\rho_o - \rho_c$ at the atomic site indicates the direction and magnitude of the shift necessary to cause the two peaks to coincide. Although such small shifts do not affect any of the phases, they serve to improve the agreement between the magnitudes of observed and calculated structure factors. In one of the early applications of this procedure to three-dimensional syntheses, the computational labor was reduced by synthesizing three mutually orthogonal line sections through each atom in the structure of cubanite, $CuFe_2S_3$. The three difference curves for the general position occupied by iron (11-18) are shown for two consecutive difference syntheses in Fig. 11-19. The residual given by (11-67) declined from 14 to 12 % as a result of the indicated shifts.

A somewhat specialized application of difference syntheses arises when it is desired to detect very small differences in the electron densities of two virtually

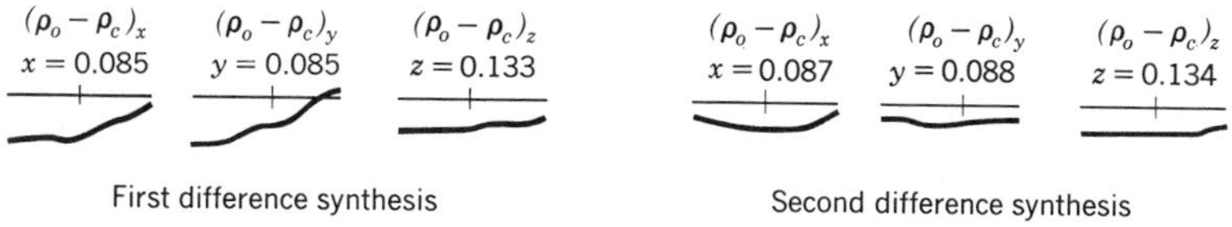

$(\rho_o - \rho_c)_x$ $x = 0.085$ $(\rho_o - \rho_c)_y$ $y = 0.085$ $(\rho_o - \rho_c)_z$ $z = 0.133$

First difference synthesis

$(\rho_o - \rho_c)_x$ $x = 0.087$ $(\rho_o - \rho_c)_y$ $y = 0.088$ $(\rho_o - \rho_c)_z$ $z = 0.134$

Second difference synthesis

Fig. 11-19. Three-dimensional difference syntheses calculated for the iron atoms in cubanite. The xyz values of the iron atoms prior to each calculation are indicated. Note the absence of a gradient in the second synthesis at the indicated positions. [*From L. V. Azároff and M. J. Buerger, Amer. Mineralogist, vol.* 40 (1953), *pp.* 213–225.]

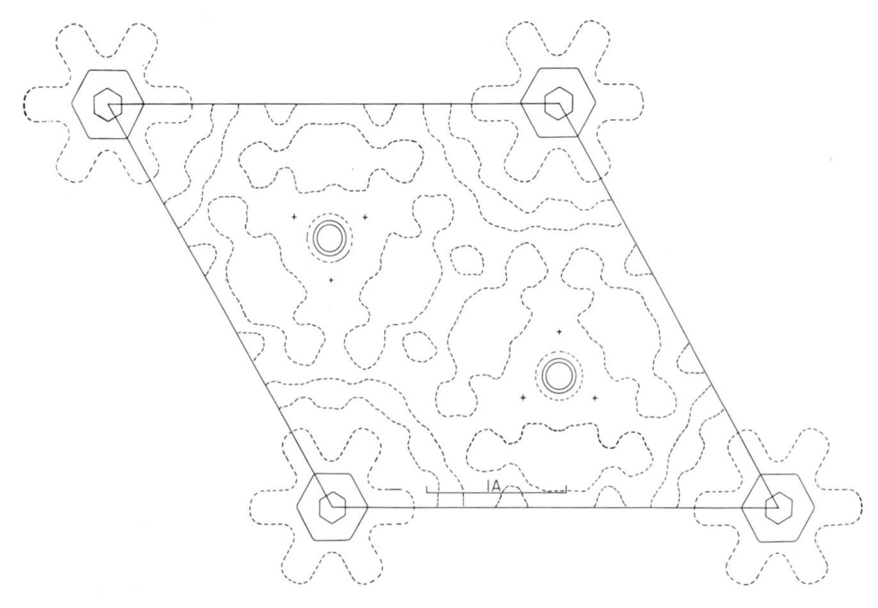

Fig. 11-20. Two-dimensional difference synthesis obtained by subtracting the electron-density projection of a "pure" ZnO crystal from that of another ZnO crystal doped to saturation with excess zinc atoms. [*From G. P. Mohanty and L. V. Azároff, J. Chemical Physics, vol.* 35 (1961), *pp.* 1268–1270.]

identical crystals. This is the case, for example, in studies of crystals containing certain imperfections in their structures, as discussed in the previous chapter. As an example of such an application, consider the electron densities of two crystals of zinc oxide, one of which contains a stoichiometric excess of zinc atoms, presumably in interstitial sites. The difference synthesis obtained by subtracting the electron density of a "pure" crystal from that of the doped crystal is shown in Fig. 11-20. The dashed contour represents the zero line separating the regions of negative and positive values. After the positive and negative densities throughout the unit cell are summed, they are found to just cancel, except for the small peaks occurring at the origin. These peaks lie at the projections of the most probable interstitial sites in the crystal, so that it seems reasonable to assume that they represent the concentration of the excess zinc atoms in the doped crystal. Further discussions of such applications have been presented in Chap. 10.

Least-squares procedures. Of the several other procedures proposed for refining crystal-structure models, the least-squares procedure first employed by E. W. Hughes has become most popular in recent times. The mathematical basis for this procedure goes back to the beginning of the nineteenth century, and consists of the proposition that the best fit between a large number of experimental quantities and a finite set of functional equations properly related to these quantities is obtained when the sum of the squares of the discrepancies between observed and calculated values is a minimum. To apply this principle to structure refinement, it is necessary to express each structure factor as the sum of a calculated value, plus an error in calculation that is proportional to the magnitude of the error in correctly locating the atom in the unit cell. Since this means that for N atoms in the structure there are $3N$ unknown errors, the calculation is quite laborious. On the other hand, it is a purely mathematical operation, so that it lends itself particularly well to digital computers. In fact, modern-day programs for least-squares refinements not only include corrections to atomic coordinates, but also estimate three-dimensionally anisotropic temperature factors, and sometimes even attempt to fit scattering-factor curves in order to improve the agreement. It is not unusual, therefore, to find R values of just a few percent reported at the conclusion of a refinement procedure. In view of the many errors inherent in the measurement of x-ray intensities discussed in previous chapters, such very small residuals are probably more indicative of the proficiency of this refinement procedure than of the true accuracy or physical validity of some of the resulting parameters. Nevertheless, these small R values are good indications of the degree of refinement attainable.

SUGGESTIONS FOR SUPPLEMENTARY READING

J. M. Bijvoet, N. H. Kolkmeyer, and C. H. MacGillavry, *Röntgenanalyse von Kristallen* (D. B. Genten's Uitgevers-Maatschappij N.V., Amsterdam, 1948).

R. Brill (ed.), *Advances in structure research by diffraction methods* (Interscience Publishers, Inc., New York, 1964).

Martin J. Buerger, *Crystal-structure analysis* (John Wiley & Sons, Inc., New York, 1960).

——— *Vector space* (John Wiley & Sons, Inc., New York, 1959).

A. I. Kitaigorodskii, *The theory of crystal structure analysis,* translated by David and Katherine Harker (Consultants Bureau, New York, 1961).

H. Lipson and W. Cochran, *The determination of crystal structures* (G. Bell & Sons, Ltd., London, 1953).

Dan McLachlan, Jr., *X-ray crystal structure* (McGraw-Hill Book Company, New York, 1957).

Werner Nowacki, *Fouriersynthese von Kristallen* (Verlag Birkhauser, Basel, 1952).

EXERCISES

11-1. Calculate the structure-factor expression for a crystal having a C-centered lattice. What is the space-group extinction rule for this case? To what group of reflections does it apply?

11-2. Calculate the structure-factor expression for the sphalerite (cubic ZnS) structure belonging to space group $F\bar{4}3m$ with

Zn in $4a$: 0,0,0; etc.

S in $4c$: $\frac{1}{4},\frac{1}{4},\frac{1}{4}$; etc.

where "etc." means that the face-centering translations must be added to each of the points listed.

11-3. Look up the symmetry-factor expression for $F\bar{4}3m$ (No. 216) in *International tables for x-ray crystallography.* Using the atomic coordinates for ZnS given above, show that it reduces to the same expression derived in Exercise 11-2. Which procedure was faster?

11-4. Look up the symmetry-factor expressions for space group $Fd3m$ to which the spinel structure adopted by ferrites and other transition-metal oxides belongs. In spinel, $MgAl_2O_4$, the atoms occupy the following equipoints:

Mg in $8a$: $\frac{1}{8},\frac{1}{8},\frac{1}{8}$; $\frac{7}{8},\frac{7}{8},\frac{7}{8}$; etc.

Al in $16d$: $\frac{1}{2},\frac{1}{2},\frac{1}{2}$; $\frac{1}{2},\frac{1}{4},\frac{1}{4}$; $\frac{1}{4},\frac{1}{2},\frac{1}{4}$; $\frac{1}{4},\frac{1}{4},\frac{1}{2}$; etc.

O in $32e$: x,x,x; etc.

The above coordinates occur when the origin is placed at a symmetry center, in which case $x = 0.25$ (very nearly). Derive the symmetry factors for each kind of site, and indicate the reflections to which the various kinds of atoms do or do not make contributions. Clearly, such reflections can be used to establish the actual distribution of metal atoms among sites $8a$ and $16d$.

11-5. Prove that the presence of a 4_2 parallel to the a axis of a crystal leads to the systematic absence of all $h00$ reflections for which $h \neq 2n$.

11-6. Prove that the presence of an n glide having the subtranslations $a/2 + b/2$ leads to the systematic absence of all $hk0$ reflections for which $h + k \neq 2n$.

11-7. By means of a scale drawing, show that the presence of a 2_1 parallel to c makes it impossible for the crystal to contain any $00l$ planes for which l is not an even number.

11-8. The coordinates of a copper atom in cubanite are given in (11-19) in the text. Calculate the symmetry factors for the 101 and 111 reflections, using any two of the coordinate triplets listed there. Prove in a general way that the values of S do not change when different points of the same equipoint set are used. **Hint:** Utilize the symmetry of trigonometric functions.

11-9. It is particularly important in the orthorhombic system to keep track of symmetry elements and their respective orientation in the unit cell. For example, the space groups $Pbcm$ and $Pmcb$ are actually two different space groups. (Using the conventional abc sequence for listing the symmetry elements, $Pmcb$ becomes $Pbam$.) Can these two space groups be distinguished by their diffraction symbols?

11-10. Using the two sets of coordinates listed in (11-27), derive your own structure-factor expressions and calculate the $hk0$ structure-factor magnitudes for both structure types. If the observed relative intensities on a powder photograph of NaCl are

Reflection	200	220	400	420	440	600	620
Intensity	80	100	15	50	15	10	35

which is the correct structure for NaCl? (In carrying out this analysis, ignore Lp factors and other corrections needed to convert intensities to F^2 magnitudes, except for powder multiplicities. Also, ignore the $\sin\theta$ dependence of the scattering factors, but assume that $f_{Cl} = 1.5f_{Na}$.) It is interesting to note that, despite the large quantitative errors introduced by these simplifications, it is nevertheless possible to determine the correct structure unambiguously.

11-11. The diffraction symbol deduced for KO_3 is $4/mmmI/\text{-}c\text{-}$, which includes space groups $I4/mcm$, $I4cm$, and $I\bar{4}c2$. The eight potassium atoms present in a cell have to be distributed among an appropriate combination of equipoints of rank 2, 4, or 8. Consider a projection of the structure along c, and deduce the only acceptable projected potassium atom positions consistent with the cell size $a = 8.597$ Å and $c = 7.080$ Å. (The radius of a potassium ion is 1.33 Å.)

11-12. Following the one-dimensional example in the text, derive the general three-dimensional equation for determining Fourier coefficients given in (11-34).

11-13. In order to minimize the superposition of atoms in a regular projection, it is possible to project a portion of the cell. Consider the projection of one-half a cell along the c axis. By following a procedure like that used in deriving (11-40), derive the correct expression for the Fourier series representing such a projection.

11-14. When a structure is very nearly determined, considerable computational labor can be saved by calculating electron-density distributions along lines passing through atoms parallel to the crystallographic axes. Derive the correct expression for such a line section parallel to a and passing through the origin. Compare this with the expression for a Fourier series representing a projection on [100].

11-15. Consider the four symmetry-equivalent atoms in plane group pgg. Derive the Patterson map for this group of atoms. Where are the satellites located relative to the rotation peaks in this case? (Compare it with Fig. 11-12 in the text.) Repeat the procedure for $p4$.

11-16. Suppose that the electron-density projection in Fig. 11-9 actually represents the symmetry $p2_1$ (instead of the $p2$ symmetry shown), with the origin atom at $z = 0$ and the other two atoms at $z = 0.200$ and $z = 0.700$, respectively. Construct the Harker section $P(XY\frac{1}{2})$ for this structure, and compare it with the projection $P(XY)$ in Fig. 11-11. Is this the correct $P(XY)$ for the structure assumed in the present exercise? Can you determine the structure uniquely from the Harker section in this case?

11-17. Make a list of the possibilities for pairs of intensities recorded for two isomorphous crystals when one contains a much stronger scatterer in a nonorigin site. How many phases can be fixed by this comparison if both crystals are centrosymmetric?

11-18. Making use of the complex scattering factors, including dispersion terms discussed in Chap. 8, show how anomalous dispersion affects the magnitude of the structure factors for a simple body-centered cubic structure. Next consider the cesium chloride structure, and determine how the structure-factor amplitude would be altered when x rays having wavelengths just shorter and just longer than the K edge of Cs are employed.

11-19. Consider a centrosymmetric crystal and one of its strong reflections, $|U_{hkl}| > 0.7$. Consider a wave scattered by these planes (hkl), and compare it with the wave scattered by the planes $(2h,2k,2l)$, assuming that $F_{2h,2k,2l}$ is positive and negative. Demonstrate that, regardless of the sign of F_{hkl}, the Harker-Kasper inequality in (11-61) correctly predicts the sign of $F_{2h,2k,2l}$.

11-20. Making use of Schwartz's inequality, derive the Harker-Kasper inequality given in (11-62) for a two-fold axis parallel to c. **Hint:** $\rho(xyz) = \rho(\bar{x}\bar{y}z)$ in this case.

11-21. The statement is made in the text that small changes in x,y,z for certain atoms may decrease the residual R without altering the phases of the structure factors. What are some of the physical sources of error in the measured values of F_{hkl} that may account for this? Prepare a list, and underscore the three that you think are most likely to cause large errors.

11-22. Consider a difference synthesis at an atomic site. Suppose the atom in ρ_c is at the correct point but the electron density has been calculated using an incorrect temperature factor when scaling the structure factors. How is this manifested in the difference synthesis?

PART FOUR
Elements of Experimental Methods

TWELVE
Production and detection of x rays

X-RAY TUBES

Gas tubes. The original tube used by Röntgen consisted of a glass envelope which had been evacuated to a residual-gas pressure of the order of 10^{-3} to 10^{-4} mm Hg. (Gas pressure is expressed in terms of the height of a column of mercury it can support; atmospheric pressure at sea level is 760 mm.) Consequently, when an electric field is produced inside the tube, the negatively charged electrons are attracted to the positive end of the field while the positively charged gas ions are attracted to the negative end. As already described in Chap. 5, Röntgen realized soon after his initial discovery that the quantity of x rays produced could be increased by placing a metal target in the path of the onrushing electrons, and a schematic view of such an early x-ray tube is shown in Fig. 12-1. The tube, an evacuated and sealed-off glass envelope, usually had three lobes containing electrodes. (Sometimes a fourth lobe was used to admit or expel gas.) One lobe contained the cathode, or negative electrode, usually made of aluminum and with a parabolic shape, at whose focal point was placed the anticathode, or positive electrode, usually made from platinum or tungsten. The third lobe contained an auxiliary anode, which, however, was utilized only during the manufacture of the tube to collect negative ions during its evacuation.

The actual generation of x rays in such a tube proceeds essentially as follows: When an electric field is applied to the tube, the few ionized gas atoms and free electrons are accelerated, respectively, to the cathode and anticathode (anode). Upon striking the cathode, the positive ions tend to knock out some electrons from it, which are then accelerated toward the positive anode. Aluminum makes a good cathode material because metals of low atomic number disintegrate less under bombardment by positive ions. The electrons, or cathode rays, tend to collide with the gas ions on their way to the anode, thus producing more ion-electron pairs, and so forth. Upon reaching the anode, of course, the electrons produce x rays. As first observed by Röntgen, the larger the atomic number (or weight) of the atoms struck, the more intense the x-ray beam produced, which was the reason for choosing metals like tungsten. Another factor, to be discussed further in a

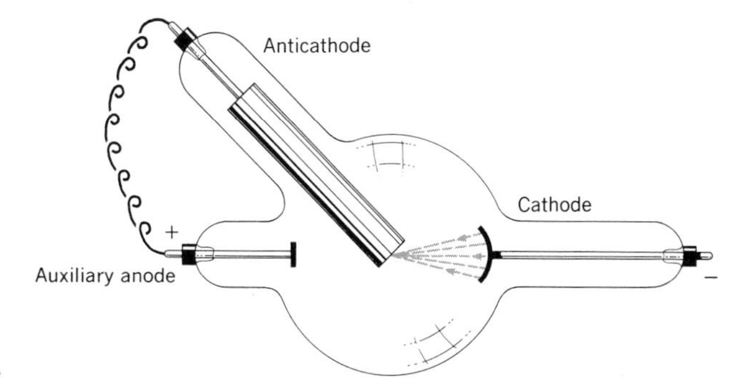

Fig. 12-1. Gas x-ray tube.

subsequent section, was the heating of the target by the electrons striking it, so that the relatively higher melting point of tungsten favored its choice over the slightly heavier platinum.

The potential difference applied to the gas x-ray tube determines the kinetic energy of the electrons, or cathode rays. If the tube is to sustain a current of some magnitude, this energy must be sufficient to ionize the gas atoms present at each collision. Note that, if the gas pressure is too small, insufficient collisions will occur and an adequate current of electrons reaching the anticathode cannot be maintained. Thus, regardless of how large the potential difference applied to the tube, very little current is produced when the gas pressure drops below 10^{-5} mm Hg. Conversely, when the gas pressure exceeds about 10^{-3} mm Hg, a current of about 100 mA can be produced by just a few thousand volts. Any attempt to increase the applied voltage leads to an increase in the tube current, with resultant damage to the electrodes. A working range of about 5 to 50 kV and 5 to 30 mA can be obtained at a gas pressure of about 10^{-4} mm Hg. Although they are relatively inexpensive to construct and operate, the difficulty in maintaining a constant x-ray output and in changing the operating conditions has largely diminished the popularity of gas x-ray tubes, although there were several still in operation as late as the 1940's.

Coolidge tubes. An important improvement over the gas-filled tube was made by W. D. Coolidge, who replaced the aluminum cathode by a tungsten filament, which was heated by passing an electric current through it. The heating of the filament serves to increase the kinetic energy of the valence-conduction electrons present in the metal, until they can escape from the metal via a process called thermionic emission. The thermionically emitted electrons then are accelerated toward the positive electrode, as in the gas tube, except that now the cathode-ray current is controlled by the number of electrons "boiled off" from the filament rather than by collisions with gas atoms. In fact, such collisions tend to disrupt the efficiency of the process, so that a Coolidge x-ray tube is normally evacuated to a pressure of less than 10^{-5} mm Hg. The filament commonly is surrounded by a metal cathode, which is maintained at the same negative potential and whose function is to focus the thermionically emitted electrons to a small area of the target, called the *focal spot* of the tube, as indicated in Fig. 12-2.

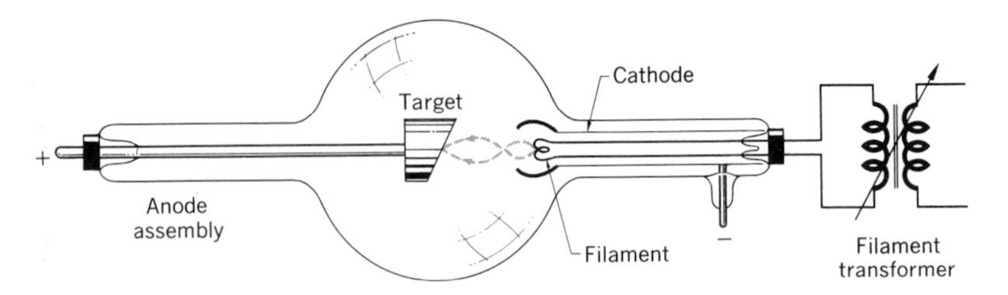

Fig. 12-2. Coolidge tube.

The operation of the Coolidge tube is more easily controlled because the number of electrons reaching the target is directly related to the temperature of the filament, as illustrated in Fig. 12-3. The temperature of the filament, of course, is directly variable by means of the filament transformer, shown in Fig. 12-2. At a given temperature, the maximum number of electrons is fixed, so that increasing the tube potential merely increases their velocity, but does not change the tube current. This is illustrated in Fig. 12-4, where it can be seen that, after an initial rapid rise, the tube current levels off, or reaches saturation, upon further increase in the applied voltage. This is the important advantage of the Coolidge tube over the gas tube; that is, the voltage used to produce x rays is controlled independently of the tube current, which, in turn, is determined by the filament temperature. Because the proper operation of the tube requires thermionically emitted electrons, it is necessary to make sure that the filament is heated before any high voltage is applied to the tube. Otherwise cold emission may take place, with resultant damage to the filament wire, although it should be noted that some field emission takes place from a heated filament also. Similarly, any stray gas atoms left in the tube or its components that may be emitted after it is placed in operation cause undesirable fluctuations in the tube current. Modern x-ray tubes are carefully

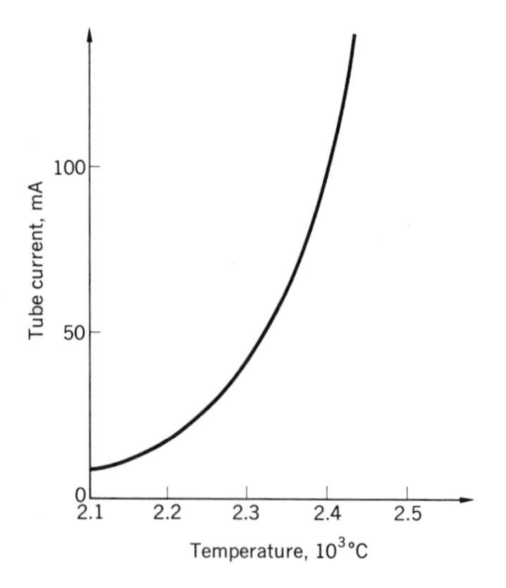

Fig. 12-3. Dependence of tube current on filament temperature. *(After Schall.)*

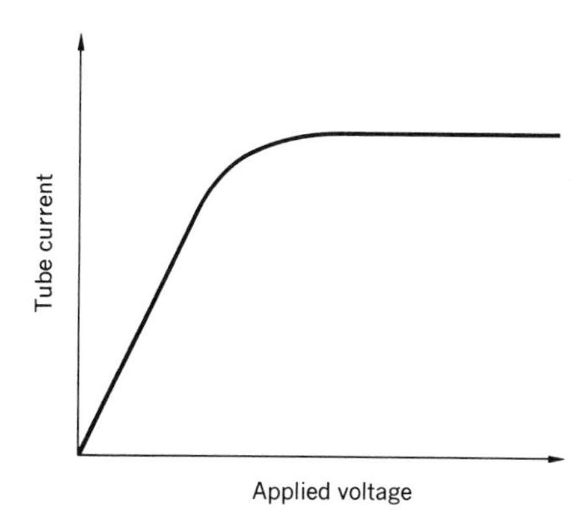

Fig. 12-4

"baked out" during evacuation and before they are sealed, to remove completely all gases absorbed in the metal components.

Demountable tubes. In order to control the gas pressure in an x-ray tube, it is possible, of course, to connect it permanently to a vacuum pump. In the case of a gas tube, suitable leak valves can be used to admit room air whenever an increased gas pressure is desired. In the case of Coolidge tubes, the vacuum can be maintained at any desired level in a continuous manner. It should be realized that mechanical pumps are insufficient for this purpose, but either oil or mercury diffusion pumps, or the more recently perfected ion-exchange pumps, are capable of attaining and maintaining pressure levels much lower than those necessary for x-ray tubes over very long periods of time. The incorporation of a vacuum pump as an integral component in an x-ray tube offers certain advantages, because no longer need the various components be permanently sealed up inside a glass envelope. This means that the target metal in the anode assembly (Fig. 12-2) can be readily replaced, so that a single unit takes the place of several sealed-off tubes having specific target materials. Similarly, when a filament is burned out, it can be replaced, so that a demountable tube has virtually an infinite lifetime.

There are still other advantages to such a tube. Because most of the cathode-ray energy absorbed by a target metal goes into heating it, the maximum wattage that can be readily dissipated by the anode assembly is severely limited. This imposes practical limitations, which experience has shown to be of the order of 200 W of electrical power per square millimeter of focal-spot area for a tungsten target. Exceeding this limit for prolonged time periods causes erosion of the target surfaces by volatilization and melting of the metal. The number 200 W/mm^2 is just an order of magnitude, and can be exceeded by a factor of 2 or more, provided that the total illuminated area, that is, the focal spot, is made very small. Such a small focal spot, of the order of 40×60 microns (1 micron $= 10^{-6}$ m), can dissipate the heat much more readily than a more conventional focal spot of the order of 1×10 mm in size, because the heat flow lines diverge more rapidly from the smaller spot. To produce such microbeams requires the ability to focus the electrons emanating from the filament, and such focusing can be carried out by external adjustments in a demountable tube.

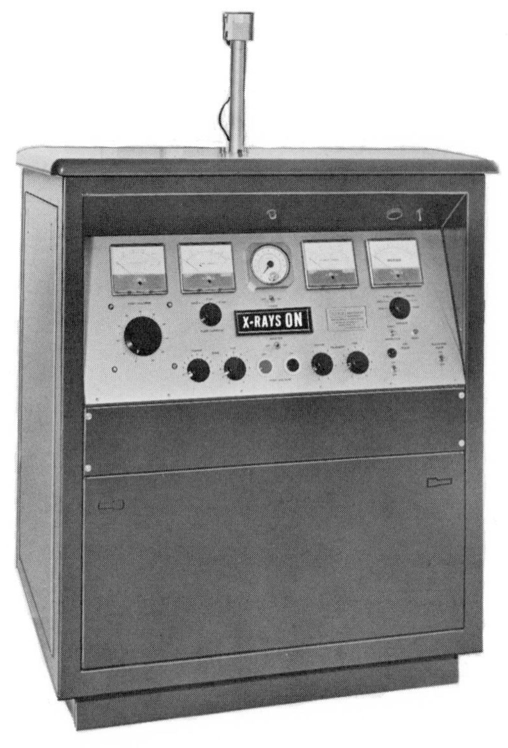

Fig. 12-5. Hilger microfocus x-ray unit. *(Courtesy of Engis Equipment Co.)*

Another way to increase the tolerable power loading on a tube target is to move the target during its bombardment by the cathode rays, for example, by suitably rotating the anode so that a new target surface is constantly presented to the electron beam. The difficulties in the mechanical construction of temporary vacuum seals and the maintenance of truly gastight enclosures are the chief difficulties in the utilization of demountable tubes in practice. Because of the current variations produced by changes in the pressure, demountable tubes have been limited chiefly to special-purpose apparatus, for example, microfocus x-ray units or high-power rotating-anode tubes. A commercial microfocus x-ray unit is shown in Fig. 12-5. The x-ray tube is the small prismatic enclosure sitting atop the vertical column which connects the tube to a vacuum pump, and a horizontally projecting cylinder has a micrometer adjustment for positioning the cathode assembly at the appropriate distance from the target. The size of the focal spot on the target can be varied also, by inserting differently shaped filaments in the cathode assembly.

Sealed-off tubes. The long-term stability of a demountable tube under specific loading conditions depends on how well the vacuum can be maintained at a constant value. When quantitative intensity measurements are made, it is essential that the x-ray output of the tube remain virtually constant during the entire experiment. It is, of course, possible to monitor independently the x-ray tube output, but such monitoring should be limited chiefly to telling when the tube output changes, and should not be used as a means for correcting such deviations. After a tube is properly constructed and degased, its vacuum can be maintained almost indefinitely by simply sealing off all air passages. Although it is no longer possible to change the target metal in the anode assembly or replace a burnt filament, the operational stability of sealed-off tubes is generally superior to demountable ones. Another advantage of sealed-off tubes is that they can be used as rectifiers in the electrical circuit supplying power to the tube. As discussed in a subsequent section, however, it is common practice nowadays to use external rectification.

Practical considerations. A commercial sealed-off x-ray tube is shown in Fig. 12-6. A schematic cross section of such a tube is shown in Fig. 12-7. The anode assembly is encased in a metal block, and consists primarily of a means for cooling the target proper, which is either the same metal from which the anode assembly is made or a metal plated onto or inserted into the assembly, as shown. The importance of cooling the target becomes clear when it is realized that the efficiency with which x rays are generated in the target is given by the empirical relation

$$\begin{aligned}\text{Efficiency} &= \frac{\text{x-ray energy emitted}}{\text{electron energy absorbed}} \\ &= 1.4 \times 10^{-9} ZV \qquad (12\text{-}1)\end{aligned}$$

where Z is the atomic number of the target metal, and V is the voltage applied to the tube in volts. At 50 kV, the efficiency of a tungsten-target tube is about one-half of one percent, while that of a lighter target is even less. Thus over 99% of the energy incident on the target is converted into heat which must be carried away. To fully appreciate the problem, consider a small space heater rated at 1,200 W and used to supply additional heat in a seldom-used room. The heating element in such a heater typically has a radiant surface of about 10,000 mm^2, so that its power dissipation in heating a room is about 0.12 W/mm^2. By comparison, a copper-target x-ray tube normally is operated at 50 kV and 20 mA, or at 1,000 W, all concentrated into a focal spot typically 1×10 mm^2, so that the specific loading is 100 W/mm^2, which is virtually all used up in heating the target. Obviously, it is essential to remove the heat as rapidly as possible. In fact, the principal limitation in loading an x-ray tube is the ability to remove the heat generated in the target. The anode assembly normally is cooled by water circulating through it, which may be supplied directly from the city mains. Because rather complex pathways are devised for the cooling water in order to increase its cooling efficiency, it is often not desirable to pass city water directly through the assembly because the concentration of silt and dirt in it may cause the cooling system to fail. One way to mitigate this problem is to pass the water through a filter, although this is really satisfactory only when the water is reasonably clean initially. A far simpler and more maintenance-free system is to use a self-circulating distilled-water system such as the one illustrated in Fig. 12-8. A small reservoir having an approximately 5-gal capacity is placed in series with a circulating water pump and an ordinary heat exchanger about one foot long. The pump must be able to deliver as much water as the sum of the requirements of all the units operated from this system [about $\frac{1}{2}$ gal/(min)(unit)] at a sufficient head pressure to reach the most remote unit. The water pressure actually is not important, provided the required flow rate through the anode assembly is maintained. Thus it is good practice to provide a flow switch (not pressure switch) in the return line from each unit so that the electric power to the x-ray unit is cut off the moment the water flow rate drops below some prescribed value. Most modern commercial x-ray units have such flow switches directly incorporated in their cooling systems.

In similarly cooled x-ray tubes, the choice of target metal dictates the total power input to the x-ray tube that can be dissipated without damaging the target surface. Because of its high melting point and relatively good heat conductivity and capacity, tungsten makes an ideal target whenever noncharacteristic x rays are desired, because the total intensity of the continuous spectrum produced increases with the atomic number of the target metal. When characteristic radiation of specific wavelength is desired, however, the thermal properties of the metal required must be examined. Thus copper has a much lower melting point than tungsten, but it is a much better conductor of heat, so that it is capable of dissipating the heat about as efficiently. In fact, the anode assembly in x-ray tubes

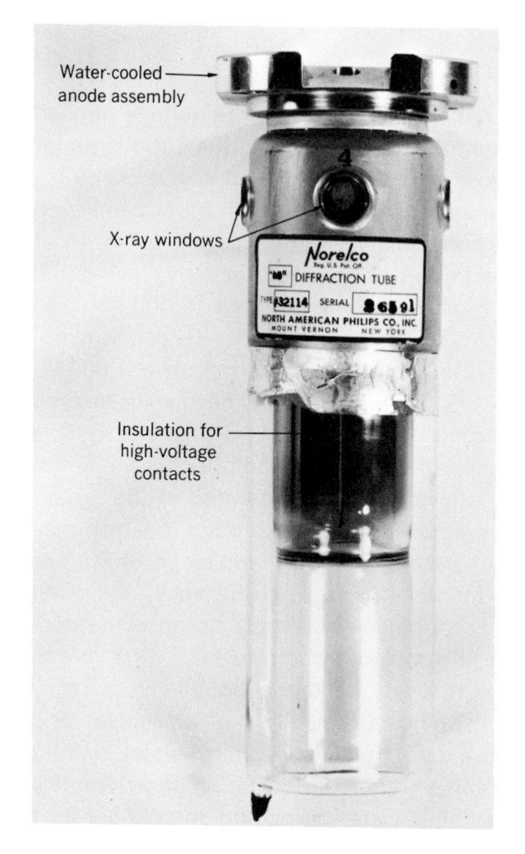

Fig. 12-6. Norelco x-ray tube. *(Courtesy of Philips Electronic Instruments.)*

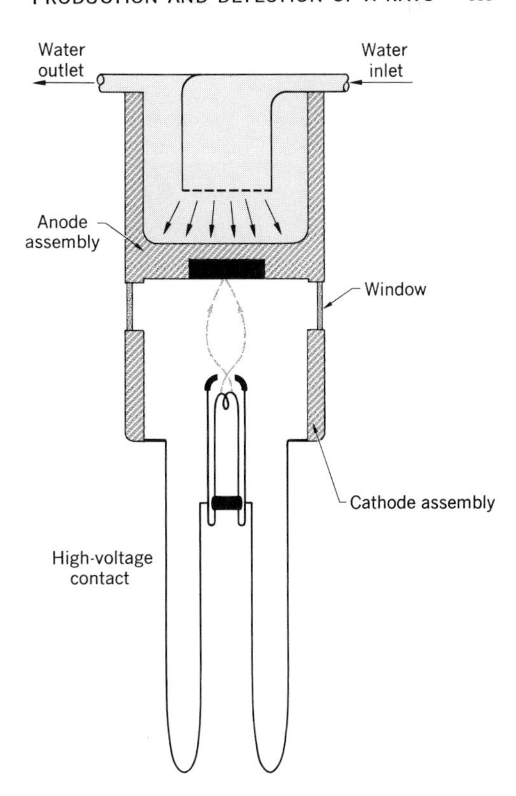

Fig. 12-7. Schematic cross section of a sealed-off tube like that in Fig. 12-6.

is normally made out of copper, and the other metals are either plated onto its surface or mechanically inserted into it. Other elements, such as chromium or iron, have much poorer thermal properties, so that tubes containing such metals as targets must be operated at considerably lower power ratings.

The rest of the x-ray tube (Fig. 12-7) is usually a glass envelope supporting the cathode assembly, joined to the metal housing, which, in addition to the anode assembly, also supports three or four x-ray windows, or ports. Since an x-ray port must transmit the x rays generated in the target, it is made of the most transparent (thin and relatively nonabsorbing) material that can be affixed to the metal housing without producing an air leak. Nonporous beryllium, mica, aluminum, and very thin glass are sometimes used, their relative transmissivity or general suitability decreasing in the order in which they are listed; that is, beryllium is the best window material for sealed-off x-ray diffraction tubes. (X-ray tubes in which much higher energy x rays are produced can use other window materials without significant intensity loss.)

Because city water is used directly or indirectly to cool the anode assembly, it is convenient to place the target housing at ground potential. The cathode assembly therefore must be placed at a correspondingly large negative potential, so that the desired potential difference between cathode and anode is produced. At the same time, the filament located inside the cathode assembly must be energized by its transformer. Both power requirements are supplied to the tube

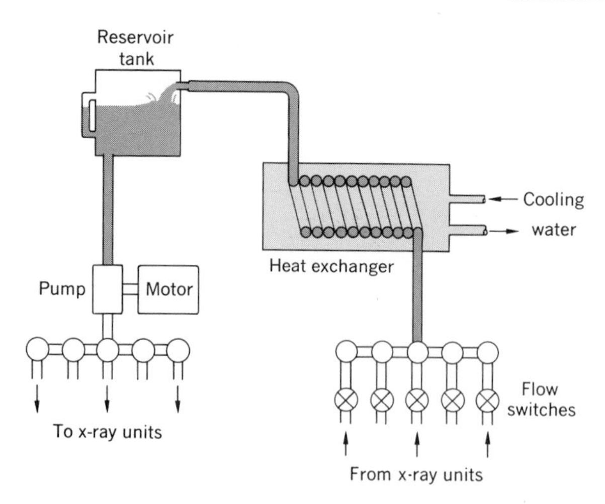

Fig. 12-8. Recirculating cooling system.

via a long insulated rod at whose end are located two concentric, mutually insulated springs bearing the electrodes that contact correspondingly positioned metal electrodes in the tube. The entire assembly is enclosed by a glass cylinder, which is completely surrounded by a suitably grounded shield to eliminate any personnel shock hazards. This enclosure must be provided with appropriate electrical interlocks so that removal of the protective shield immediately disconnects the high-voltage supply. It should be borne in mind that during continuous operation, the entire x-ray tube becomes heated, so that the metal springs ensuring electrical contact tend to lose some of their springiness with time. One of the commonest reasons why x-ray units fail to start up after a shutdown period is the slight contraction of these springs on cooling and their failure to make proper contact inside the tube. When this happens, the high-voltage rod should be removed (after carefully disconnecting and grounding the high-voltage transformer) and the springs elongated slightly until they again make intimate contact with the tube.

It was stated in a preceding section that one way to increase the specific loading capacity of the target is to decrease the focal spot to such a small size that it becomes easier to remove the heat generated, that is, by increasing its circumference-to-area ratio. As a matter of fact, the reverse procedure can be used to produce an increased *effective* intensity output from a target. To see how this is done, consider the two views of a target shown in Fig. 12-9. The focal spot shown has a rectangular shape, such as is produced by using a line filament, which may be made from a flat ribbon or a length of coiled filament wire. When the line-shaped focal spot is viewed from its side, the effective size of the focal spot is unchanged in length, but is narrowed, as shown. When viewed from one of its two ends, the focal spot becomes foreshortened, and by choosing the viewing, or takeoff, angle properly, the effective focal spot can have the shape of a square. The advantage of such an arrangement is that the intensity of the x-ray beam is increased without increasing the effective size of the focal spot, which, as will become clear later, is important in determining the resolution of adjacent diffraction maxima. In commercial x-ray diffraction tubes (Fig. 12-7), the target is normal

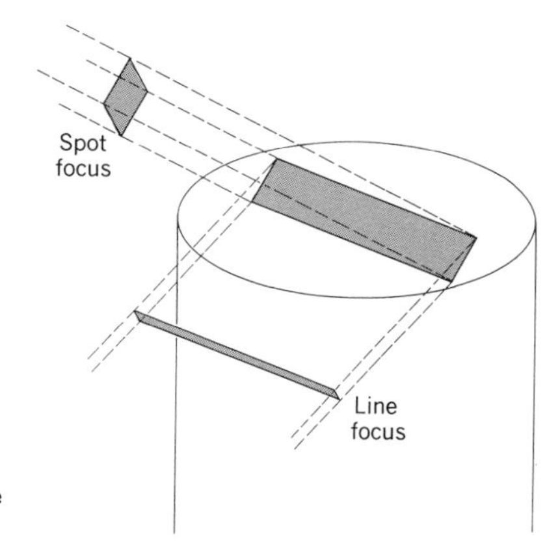

Fig. 12-9. Two commonly used views of the x-ray tube focal spot.

to the cathode-ray beam, so that the x rays can be viewed conveniently only at a takeoff angle of 0 to 10°. The increase in the intensity with takeoff angle for a side view of a 1- by 10-mm focal spot is shown in Fig. 12-10. As can be seen therein, increasing the takeoff angle significantly above 6 to 7° does not produce a corresponding increase in the intensity, so that this is the takeoff angle selected for most diffraction experiments. When considerably improved resolution is desired, a takeoff angle as small as 2 to 3° may be used.

It should be noted in passing that most commercial x-ray tubes for x-ray diffraction employ linear focal spots, except that the rectangle is not uniformly bombarded by electrons. Instead, the electrons emitted primarily from the two sides of a filament coil are so focused by the cathode cup that they strike the target in two slightly separated parallel lines, forming the sides of the rectangular line focus. This is done, of course, to increase the specific loading of the target without

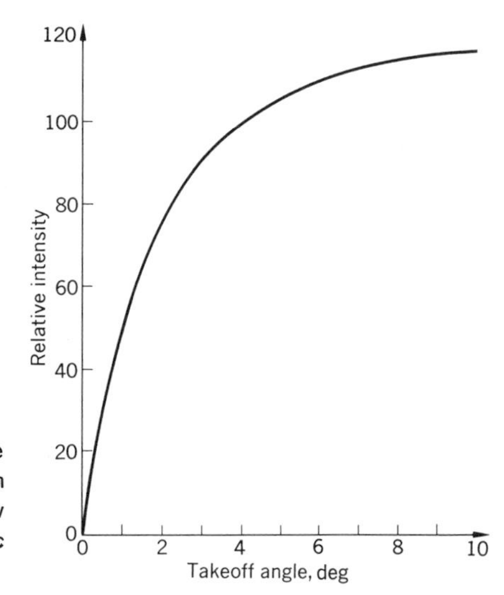

Fig. 12-10. Variation of intensity with takeoff angle measured through a slit 30 in. from a line focus, with an aperture at the tube window allowing a full view of the focal spot. *(Courtesy of General Electric Company.)*

eroding it, and is of little importance in all but a very few x-ray diffraction experiments. To see this effect one can prepare a pinhole photograph of the target by placing a lead foil, through which an extremely fine hole has been pricked, in front of the x-ray port, and a photographic film about two feet behind it. By noting the target-to-pinhole and pinhole-to-film distances, it is possible to determine the exact size and shape of the effective focal spot (Exercise 12-4). The exact size of the pinhole is not important provided only it is considerably smaller than the focal spot.

X-RAY GENERATORS

Self-rectification. A sealed-off x-ray tube has the same basic components as an ordinary diode tube. It can be used, therefore, in the role of a rectifier for an alternating current, as indicated in the simple circuit shown in Fig. 12-11. The filament transformer supplies the current necessary to heat the filament, one side of which is connected to one side of the output winding of the high-voltage transformer. The other side of this winding is connected to the anode. When the ac direction is such that the anode is positive relative to the cathode, electrons are accelerated toward it, and x rays are produced. When the other half cycle passes through the transformer, the negative potential on the anode repels the emitted electrons, and x rays are not produced. In fact, excluding the possibility of thermionic emission of electrons from an overheated target, no current at all flows through the tube during this half cycle, and the result is a half-wave-rectified current (Fig. 12-12). This is, obviously, a very simple method for obtaining x rays and is quite suitable for intermittent x-ray exposures and wherever relatively trouble-free compact x-ray equipment is needed, for example, in dental radiography. It has the disadvantage for continuous operation that the tube target becomes excessively heated and a reverse current may be produced, with resultant serious damage to the tube.

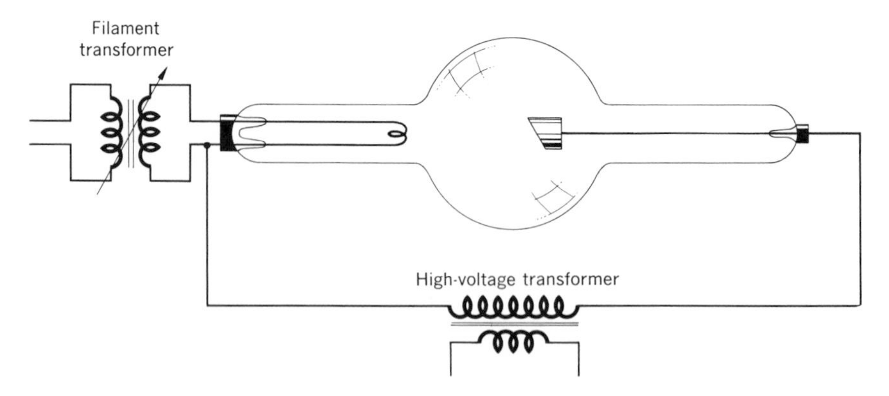

Fig. 12-11. Self-rectified circuit.

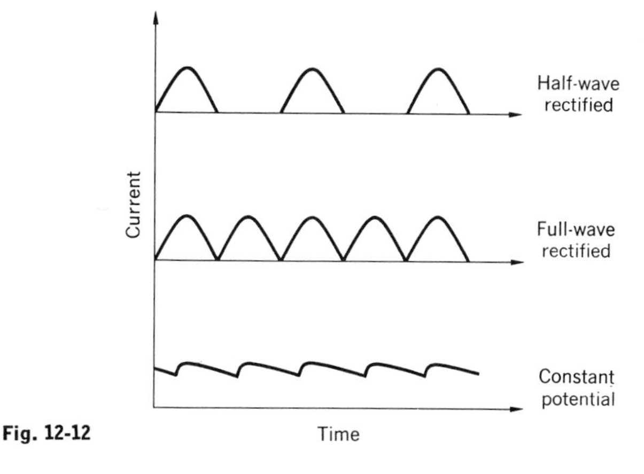

Fig. 12-12

Full-wave rectification. The potential danger to the tube from a reverse current can be removed by placing a diode rectifier in the circuit. In fact, two such rectifiers can be combined to ensure that a positive potential is applied to the anode during each half cycle. As can be seen in the circuit diagram of Fig. 12-13, by grounding the center tap of a transformer and the anode of the tube, either one rectifier or the other blocks a current flowing in that half of the high-voltage winding that would make the cathode positive. Thus one-half of the voltage produced in the transformer secondary appears across the tube during both half cycles, so that a current passes through the x-ray tube, as shown by the central curves in Fig. 12-12. This is called a full-wave-rectified current. Note that the anode of the x-ray tube can be grounded in the circuits shown in Fig. 12-13. This makes such full-wave-rectified circuits particularly suitable for use with water-cooled x-ray tubes such as are commonly employed in x-ray diffraction units. Incidentally, it should be noted that the effective current flowing through the x-ray tube is not the peak

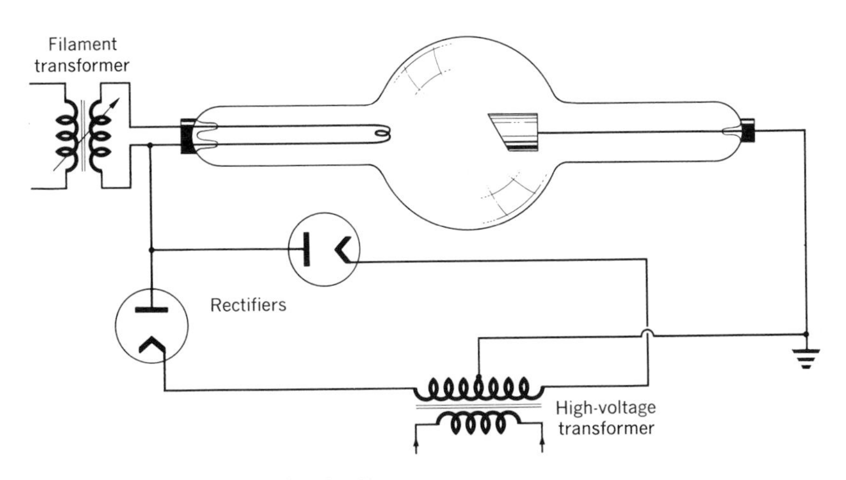

Fig. 12-13. Full-wave-rectifier circuit.

current shown in Fig. 12-12, but what is called the root-mean-square (rms) current, which is equal to the maximum current divided by the square root of 2.

Constant-potential generation. By suitably inserting two capacitors into the circuit shown in Fig. 12-13, its characteristics can be modified to increase the effective current flowing through the x-ray tube and to minimize the voltage changes on the anode. Consider the circuit shown in Fig. 12-14. During one-half cycle, the high-voltage secondary winding passes a current through the diode rectifier R_1 and charges the lower capacitor C_1 to the full voltage on the secondary. In the next half cycle, both the secondary and capacitor C_1 in series with it discharge through the x-ray tube. Concurrently, the second capacitor, C_2, becomes charged during this half cycle, since the polarity is such that no current can flow through rectifier R_1. The capacitor C_1 becomes again charged during the next half cycle, and is prevented from interfering by the blocking rectifier R_2. Thus a virtually constant potential is maintained on the x-ray tube anode, although the current flowing through the tube has an asymmetric ripple like the one shown by the bottom curve in Fig. 12-12. The ripple can be minimized, of course, by suitable filtering circuits, which are not considered in detail here. It should be noted that the current flowing through the x-ray tube in a constant-potential generator exceeds the rms current in a similarly rated full-wave-rectified generator, so that the x-ray tube must be operated at reduced ratings. Despite this, the big advantage in having a constant potential is that the same energy x radiation is produced throughout the entire operation as compared with the pulsating distribution produced by a rectified generator. This factor is very important when x-ray spectra are studied, for example, in x-ray fluorescence analysis, but less so in x-ray diffraction studies, where only the characteristic spectra are utilized.

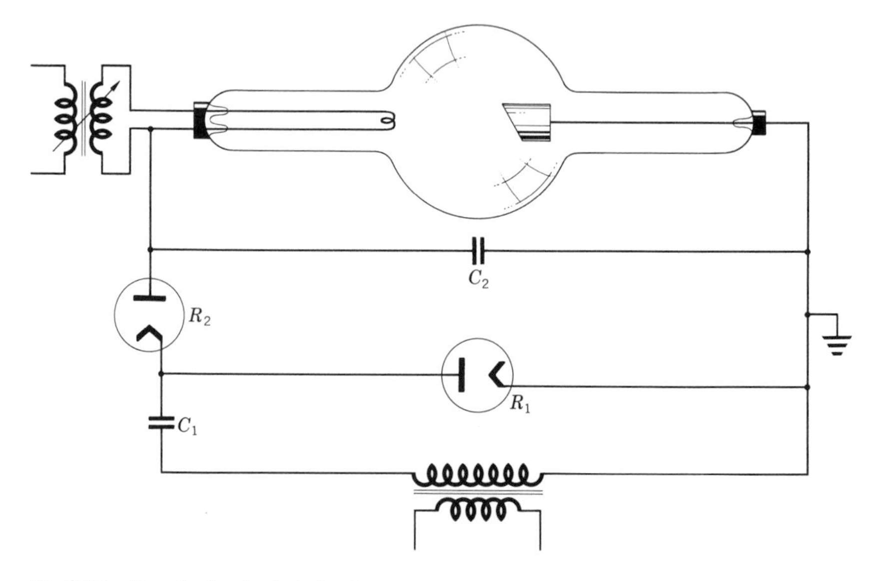

Fig. 12-14. Constant-potential circuit.

Fig. 12-15. Norelco table-top x-ray unit. *(Courtesy of Philips Electronic Instruments.)*

A similarly undesired variation in the x-ray tube output comes from irregularities in the input line voltage to the x-ray generator. It is a well-known phenomenon that, during certain peak periods, the voltage in the mains may drop by as much as 5 to 10%. This is caused partly by a drop in the city mains, but more usually by other equipment connected to the same line as the x-ray unit. Such input-voltage fluctuations have two undesired effects of varying importance: A 5 to 10% variation in the high-voltage transformer is undesirable, but has an effect similar to the ripple present in all but the most thoroughly filtered circuits. Relatively simple and inexpensive line-voltage regulators, comprised of transformers having an extra secondary winding, are adequate to eliminate this problem. A variation of 5% in the filament transformer, however, produces a far more serious change in the tube current (Fig. 12-3). Maintaining the filament current at the same value, regardless of such line-voltage variations, requires a somewhat more elaborate stabilizer. Modern x-ray units can be obtained commercially with varying degrees of rectification and stability to meet almost any need the user may envision.

To illustrate the range of commercially available x-ray diffraction generators, Fig. 12-15 shows a compact "table-top" x-ray generator that has full-wave rectification but no stabilizers or special filters. It is thus best suited for photographic recording of x-ray intensities, so that any transient irregularities are averaged out during the relatively long exposures. Figure 12-16 shows a power supply that includes line-voltage stabilizers and current stabilizers, as well as the filtering necessary for constant-potential operation of the x-ray tube. This particular power supply has another interesting feature. A comparison of Figs. 12-11 and 12-13 suggests that it should be possible to buck two x-ray tubes, not unlike the arrangement of the rectifiers in Fig. 12-13, so that current flows through one tube during one half cycle and through the other tube during the second half cycle. A means for doing this is provided on the console shown in Fig. 12-16, so that this power supply can be used to operate one tube at full-wave rectification or two tubes at half-wave rectification.

Fig. 12-16. X-ray generator designed to operate one or two tubes concurrently. *(Courtesy of General Electric Company.)*

X-RAY DETECTION

Detector types. Historically, the first x-ray detector was a screen coated with barium platinocyanide lying near Röntgen's cathode-ray tube. Röntgen also was the first to use photographic films for detecting and recording x rays and to study the effect of ionization of a gas through which x rays had passed. It is possible to classify the various kinds of detectors that evolved from these initial studies into three groups, as shown in Table 12-1.

Of the three kinds of detectors listed in Table 12-1, only photographic films produce a permanent record. Other detectors produce pulses of light, electricity, etc., when suitably energized by x rays, but these pulses then disappear with the passage of the x rays. It is, of course, possible to record the pulses produced for later examination, and some of the ways of doing this are considered later in this chapter. Compared with such recordings, however, photographic films have two very important advantages in x-ray diffraction studies. The most important one is that a photographic record not only shows the number of x rays crossing a particular point on the film (beam intensity), but also the relative disposition in space of various diffracted beams. The importance of being able to discern at a glance the spatial disposition of diffracted beams will become clearer when actual diffraction experiments are discussed in later chapters, but it should be realized here that a film can intercept all the x-ray beams diffracted at the same time or over a prolonged time period, whereas ionization detectors, for example, can be used only to detect one diffracted beam at a time.

The other advantage of films is that the permanent record produced is small and compact and therefore easy to examine and store. Photographic recording equipment for x-ray diffraction experiments is also about an order of magnitude less costly than other types of recorders commonly used. Its chief limitation is in its relatively lower accuracy and sensitivity in the measurement of x-ray intensities. Also, it is not possible to measure x-ray intensities on an absolute scale with films. The advantages of films far outweigh their disadvantages for many applications, however, so that their use remains widespread. In fact, it is good practice to utilize film-recording arrangements even in laboratories equipped with more elaborate facilities. The reason for this is the ability of films to show the disposition of the diffracted beams. It is this use of films that requires the utilization of the two kinds of films listed in Table 12-1. As is well known, an image is produced in the $AgBr$ grains in the photographic emulsion when (*1*) an x-ray photon passing through a grain changes the ionization of one or more of the silver atoms, and (*2*) the silver atoms are subsequently reduced to free silver by suitable chemical developers. By coating two sides of a thin supporting membrane with silver bromide grains it is possible to double the probability that the x-ray photon will interact in the above manner, so that a *double-coated film* requires shorter exposures and is most commonly used in diffraction studies. Double-coated films have one disadvantage, however, which is illustrated in Fig. 12-17. An x-ray beam incident on the film produces a more intense blackening in the first layer than in the second layer because the absorption decreases with penetration. When the x-ray beam is incident at an oblique angle (Fig. 12-17), the image formed does not have a uniform

Table 12-1 X-ray detectors

Photographic films	Double-coated film
	Single-emulsion film
	Polaroid prints
Ionization detectors	Ionization chamber
	Proportional counter
	Geiger-Müller counter
Solid-state detectors	Fluorescent screens
	Scintillation counter
	Photoelectric detectors

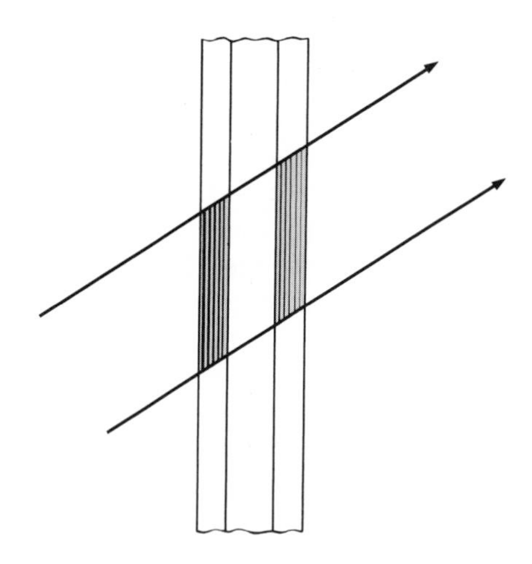

Fig. 12-17

blackening when subsequently viewed with visible light at right angles to the film. This has the effect of producing a small shift of the apparent center of blackening on the film. Whenever the positions of x-ray beams incident at oblique angles have to be measured with great accuracy, therefore, it is not advisable to use double-coated films, and single-emulsion films must be used instead, despite the increased exposure times.

Recently, a very rapid x-ray photographic process has been developed by E. H. Land. This process, quite similar to the one employed in regular Polaroid cameras, combines the chemical-processing materials into a single package with the photographic emulsion. Thus an actual print of the x-ray diffraction spectrum can be obtained 10 sec after the exposure is completed. The mechanical arrangement of the process currently limits it to flat-plate casettes (Fig. 12-18); however, the reduction in exposure times by a factor of 8 to 10 times, combined with the rapid development of the final print, makes this a very convenient procedure for many applications.

Ionization detectors. When an x-ray beam passes through a gas, some of the x-ray photons collide with gas atoms. The usual result of such collisions is that an electron-ion pair is formed; that is, the gas becomes ionized. Suppose that x rays are constrained to enter a gas-filled cylindrically shaped detector through an end window, as shown in Fig. 12-19. Next, suppose that a metal wire, centrally placed in the metal cylinder, is made positive relative to the outer cylinder by an externally applied voltage. When an electron-ion pair is now produced inside the gas, the negatively charged electron will be accelerated toward the positive wire anode, while the positive ion will move (more slowly because of its much larger mass) toward the negative case. Whenever an electron reaches the anode, it serves to charge the condenser connected to it, so that an electrical pulse can be recorded as this charge leaks off through the resistor connected in series. The schematic appearance of such a circuit is also indicated in Fig. 12-19. It is clear that such an

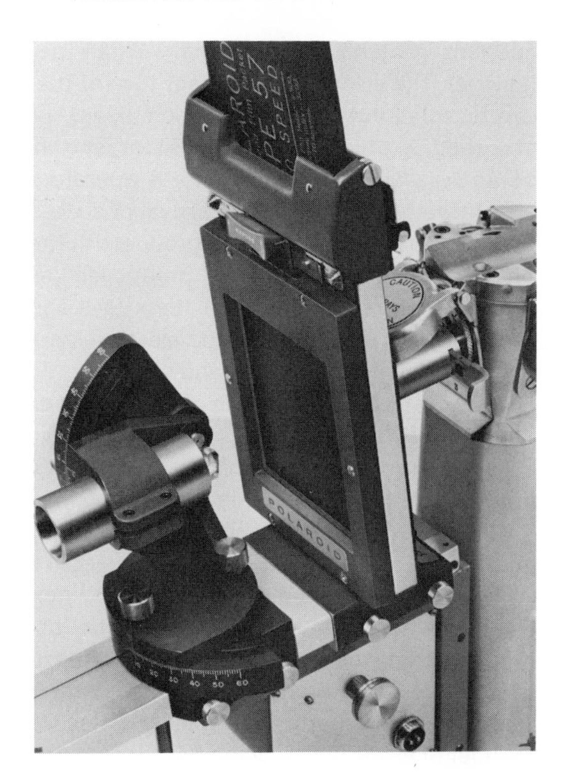

Fig. 12-18. Land diffraction cassette in back-reflection arrangement. *(Courtesy of Polaroid Corporation.)*

instrument can be used to detect x-ray photons and was, in fact, first employed in x-ray diffraction studies by Bragg in 1913.

To better understand the operating characteristics of the instrument in Fig. 12-19, it should be realized that the efficiency of absorbing an x-ray photon increases with the atomic number of the absorbing atoms. Also, increasing the density of atoms by increasing the gas pressure or lengthening the tube in Fig. 12-19 serves to increase the probability of interaction. Another, most important factor in determining the response of the detector is the magnitude of the voltage difference between the outer cylinder and the central wire. To understand its role, assume than an argon-filled detector is used to detect x-ray photons whose energy is 6,000 eV ($\lambda \approx 2.0$ Å). It takes about 30 eV to produce an electron-ion pair in argon, so that, for each x-ray photon wholly absorbed by the gas, about 200 electron-ion pairs are produced. If there is no applied voltage or if it is too low, the electrons

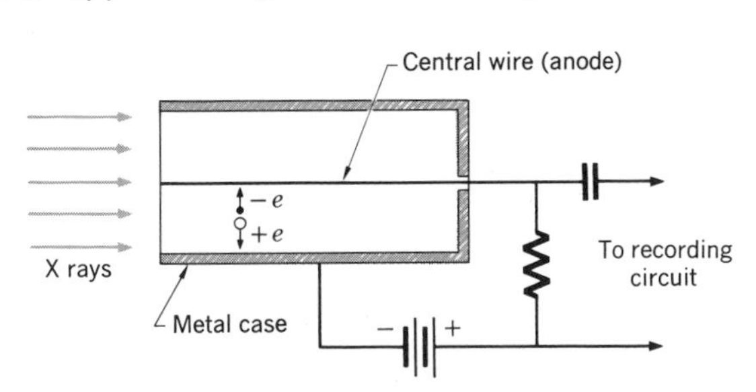

Fig. 12-19

and ions will tend to recombine. When the applied voltage exceeds about 100 V, however, the electrons are swept toward the anode so rapidly that recombinations can be ruled out, and an electric current, proportional to the number of photons absorbed, is produced in the resistor, shown in Fig. 12-19. The detector is said to be an *ionization chamber* when it is operated in this manner.

If the time between the arrival of successive x-ray photons is sufficiently long, the individual electrical pulses produced in the ionization chamber can be measured. The magnitude of the pulse produced, usually related to its height, is directly proportional to the energy of the x rays absorbed; that is, each 6,000-eV photon produces, on the average, 200 electron-ion pairs, whereas a 12,000-eV (1 Å) x-ray photon produces 400 pairs, and so forth. This fact makes it possible to use pulse heights for the determination of the photon energies, in ways to be discussed below. The number of electrons produced by a single photon is quite small, and early investigators employing ionization chambers in x-ray diffraction experiments had great difficulty in measuring x-ray beam intensities very accurately.

By 1908, however, E. Rutherford and H. Geiger had demonstrated that it was possible to increase the number of electron-ion pairs formed by increasing the voltage applied to the central wire. When the voltage exceeds approximately 300 V, each electron liberated by an x-ray photon receives a sufficient acceleration (energy) from the field, so that it can, in turn, form new electron-ion pairs upon colliding with gas atoms. It turns out that this amplification effect A is almost linearly proportional to the applied voltage, up to about 800 to 900 V. Thus the number of electrons reaching the central wire is equal to An, where n is the number of electron-ion pairs formed by each photon absorbed. When the detector in Fig. 12-19 is operated in this voltage range, it is called a *proportional counter*, because the size of the electrical pulse detected for each x-ray photon absorbed is proportional to the energy of the x-ray phonon. Although this is also true of the ionization chamber, the much larger amplification of the proportional counter makes the detection of differences in pulse heights much easier. In fact, as discussed later, it is possible to make use of this proportionality to measure the energy, as well as the number, of the x rays absorbed.

As the voltage is increased still further, the number of secondary electron-ion pairs produced for each photon absorbed increases, until at about 1,000 V or more, an avalanche effect is triggered off, resulting in a much increased amplification factor A. In fact, the acceleration that each freshly liberated electron undergoes is so large that ionization of the gas rapidly spreads from the point where the photon is first absorbed until electrons are arriving at the central wire along its entire length. The pulses produced are very large indeed, and require much less amplification by the recording circuitry than do those emanating from proportional counters. Because the electron avalanche spreads throughout the counter so rapidly (about one microsecond) and so completely, virtually the same size pulse is produced by a photon regardless of its energy. A practical version of such a detector was first built by Geiger and perfected in 1928 by Geiger and Müller. Because of its relative simplicity, the *Geiger-Müller counter* has been primarily responsible for popularizing the quantitative measurement of x-ray intensities in the second half of the twentieth century.

Solid-state detectors. Even though they were the earliest x-ray detectors used, phosphor-coated screens are still commonly used in medical and industrial fluoroscopy. The chief improvements have been in their efficiency to convert the x-ray energy absorbed into visible light. The process of fluorescence taking place inside the phosphor is simply an excitation of an atom to a higher energy state by the incident photon, and the atom's subsequent return to its ground state, accompanied by the emission of a visible-light photon. The process occurs almost instantaneously, so that the phosphor glows when x rays impinge on it. Because the intensity of diffracted beams is relatively small, fluorescent screens are used in x-ray diffraction, primarily to trace the direct-beam path during the alignment of diffraction apparatus. It is, of course, possible to use some highly photosensitive device to measure the luminosity produced in the screen and to attempt to measure x-ray diffraction intensities by this means. Although several such instruments have actually been devised and tried, their overall usefulness is limited by relatively poor energy resolution. Such devices have found use, nevertheless, in measuring x-ray exposures in x-ray radiographic applications.

Suppose the pulses of visible light or scintillations are produced by x rays in a single crystal that is virtually transparent to such light. Then it is possible to amplify the light pulses produced by means of a *photomultiplier*, as shown in Fig. 12-20. A photomultiplier tube contains a photocathode which emits electrons when struck with visible-light photons by the ordinary photoelectric-emission process. The electrons are then accelerated toward a positively charged electrode, called a dynode. Upon striking the dynode, each electron liberates several other electrons, which are then accelerated toward the next dynode, where the process is repeated. It is clear that each successive dynode serves to increase the number of electrons, so that the gain by the time the last dynode is reached may be sufficient to produce an amplification of the order of 10^7. The total amplification in the *scintillation counter* shown in Fig. 12-20 depends on the number of light photons or scintillations produced by the incident x-ray photon. It is usually as large as the highest attainable by proportional counters, and can be made to rival that of Geiger-Müller

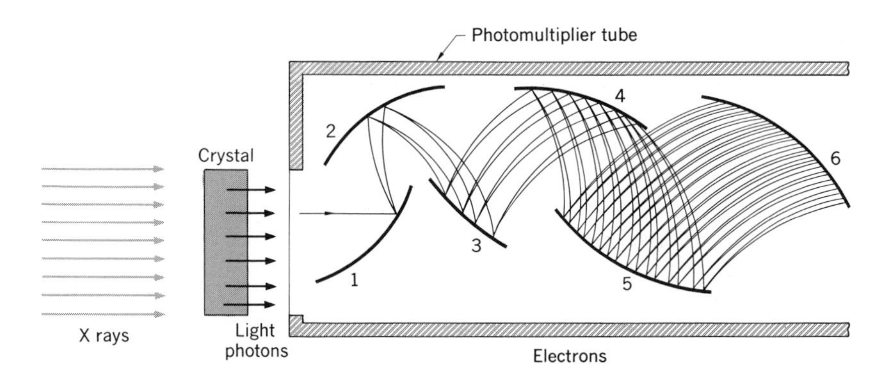

Fig. 12-20. Schematic representation of a scintillation counter showing what happens when one of the light photons excited in the crystal enters the photomultiplier tube. (Only the first six dynodes are indicated.)

counters. The scintillation counter must be light-tight, so that only the light photons produced in the crystal enter the photomultiplier. Even when such precautions are taken, there may be a considerable amount of noise or background present in a photomultiplier, largely due to thermionically emitted electrons at room temperature.

There is still another way that radiation can be detected by crystals. When photons of suitable energy are absorbed by certain insulators, such as cadmium sulfide or silicon, for example, some of the atoms become ionized within the crystal, and the liberated electrons can contribute to electrical conductivity. Since the crystals are nonconducting in the absence of such external excitation, it is possible to measure the photoinduced current fairly easily and to use it as a means for detecting radiation incident on the crystal. Although *photoconducting detectors* find widespread application for the detection of visible and near-visible radiation in photoelectric eyes, photometers, and similar devices, it has not yet proved practical to use them to measure the very low intensities of x-ray beams in diffraction experiments. Relatively simple devices can be devised, however, to monitor the direct beam or to aid in the alignment of diffraction apparatus.

Practical considerations. The principal difference between the processes taking place in a proportional counter and a Geiger-Müller counter is that the electron-ion-pair generation is localized to a region of about one millimeter in the proportional counter, and does not extend along its full length as in the avalanche process. This is important, because the electrons reach the central wire within times less than one microsecond, whereas a time of the order of one hundred microseconds is required for the ion to reach the negative metal sheath. This means that, following each avalanche in the Geiger-Müller counter, a space charge of positive ions extends the full length of the counter and moves toward the negative sheath at a much slower pace than is required to trigger the electron avalanche. During its presence in the detector, this space charge decreases the effective electric field in the counter and inhibits the triggering of the next avalanche by an incoming x-ray photon. Accordingly, a minimum time must elapse between two successive photons entering the counter, if both are to be detected. The actual *dead time* of the counter thus limits its usefulness at very high rates of photon incidence, that is, at large intensities. This is an important limitation in the use of Geiger-Müller counters. With increasing x-ray intensity, it manifests itself in practice by the increase in what are called coincidence losses, so that, at some maximum rate, of the order of several thousand counts per second, the response of the counter ceases to increase regularly with the number of incident x-ray photons. Although this can be circumvented by placing suitably calibrated absorbers in front of the detector, it remains the chief limitation of Geiger-Müller counters. In this connection it should be noted that, if the rate of x-ray incidence is sufficiently large, the counter fails to respond to successive photons entirely, and it is then said to have become blocked. Because the signals produced by this counter are several volts high, however, little amplification is needed for their detection and recording, so that the attendant instrumentation is much simpler. This accounts for the continued popularity of such counters in x-ray diffraction applications.

The inability to distinguish the relative energies of x rays triggering the avalanches renders Geiger-Müller counters virtually useless in x-ray spectroscopy. Here either proportional or scintillation counters are commonly employed, since the size of the pulses produced in each type of detector is proportional to the energy of the incident photon. Also, the dead time in both kinds of counters is less than one microsecond, so that they remain linear for all but the most intense x-ray beams encountered in practice. Nevertheless, it is good practice to check their linearity by interposing known absorbers in a manner discussed in Exercise 12-6. Furthermore, because the pulse heights produced are proportional to x-ray energy, it is possible to use an electronic circuit that transmits only pulses above a certain magnitude and blocks lower-height pulses. Such a *pulse-height discriminator* can be combined with another circuit that transmits only pulses that are smaller than a certain maximum height to form a single-channel *pulse-height analyzer*. Thus it is possible to transmit pulses that are limited to a very narrow range of pulse heights and to detect only pulses triggered by x rays having a narrow range of energies. A practical limitation is placed on how monochromatic this x-ray beam can be rendered electronically by the actual pulse-height distribution that may be produced in each type of counter by an x-ray photon. It turns out that it is easier to manufacture proportional counters having narrow pulse-height distributions, so that such counters are preferred in x-ray spectroscopy. On the other hand, because proportional counters are gas-filled, they are more sensitive (efficient) in the detection of x rays having certain energies. For example, argon-filled counters are most efficient for detecting relatively low energy (long wavelength) x rays, and should not be used for x rays whose wavelengths are less than 1.0 Å. (This pro-

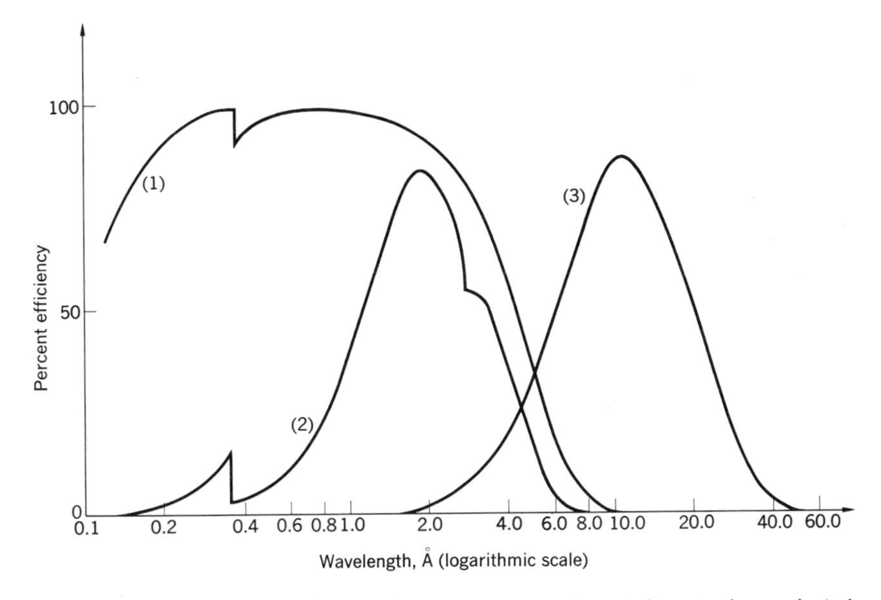

Fig. 12-21. Relative responses of (1) a scintillation counter equipped with a thallium-activated NaI crystal; (2) a xenon-filled proportional counter; (3) a proportional counter containing N_2HeCH_4. *(Courtesy of General Electric Company.)*

vides a partial means for minimizing the background intensity caused by scattering of short-wavelength radiation present in the continuous spectrum.) By comparison, scintillation counters normally are equally efficient for all energies. Their chief limitation is their wider pulse-height distributions and their electronic noise, which makes them inefficient for very low energy x rays whose pulse heights may not be much larger than those in the background noise. A comparison of the efficiencies of several counters is shown in Fig. 12-21.

It should be noted that at very long wavelengths (low energies) another factor limits detector efficiency, namely, the absorption of such less penetrating x rays by the detector window. To minimize this problem, ever-thinner windows have been employed; however, such thin windows are unable to support the pressure at which the compressed gases are maintained in conventional counters. A gas-flow proportional counter was first employed around the middle of the present century to overcome this limitation. A very thin window is sufficient to contain the counter gas which is continuously moving through the counter. This motion does not affect the rapidly accelerated electrons, but tends to sweep out some of the slower-moving ions, and provides a continuously fresh supply of unionized gas atoms, so that an improved efficiency is obtained over a stationary-gas counter under equal pressure.

INTENSITY RECORDING

Photographic intensities. As already noted, the photographic process consists of the photoactivation of silver bromide grains in the emulsion, followed by their subsequent conversion to free-silver grains in the chemical development process. The optical density of the freed silver therefore is a measure of the amount of radiation received by that spot on the film. Because the final film is viewed with visible light, it is convenient to define density D or blackening B in terms of this light. This is defined as

$$B = D = \log \frac{I_0}{I} \tag{12-2}$$

where I_0 and I are, respectively, the intensities of light incident upon and transmitted through the film. Thus a blackening of unity means that one-tenth of the incident light is transmitted. To the eye, a blackening of two ($I/I_0 = \frac{1}{100}$) appears very dark and constitutes an upper limit. The double-coated film normally used, however, is capable of recording blackenings up to $B = 3.0$. To prevent overexposure of the film, one can make use of the *reciprocity law* of optics, which states that

$$B = \text{intensity} \times \text{exposure time} \tag{12-3}$$

The meaning of (12-3) is that the same blackening can be obtained by doubling the exposure time while halving the x-ray intensity. It is thus possible to prepare films having linearly related blackenings by exposing them for different time periods under otherwise identical conditions.

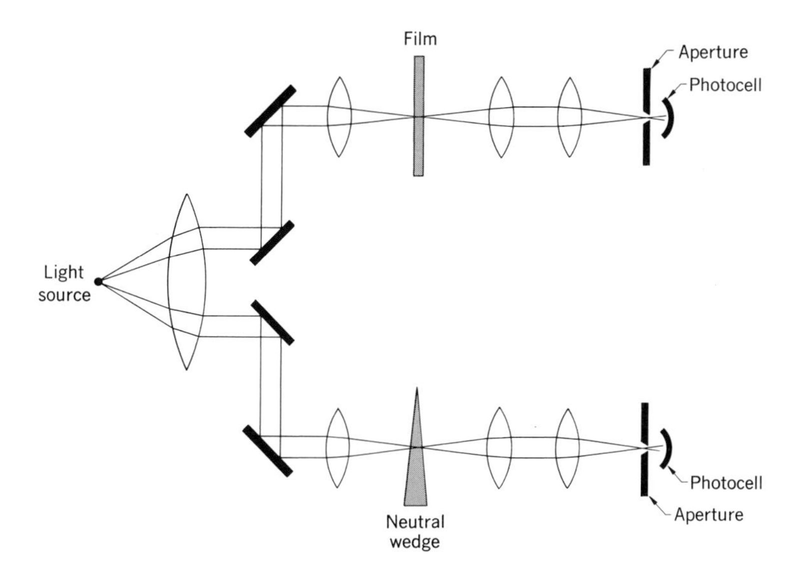

Fig. 12-22. Schematic representation of a neutral-wedge photometer.

Although the human eye is a most effective evaluator of x-ray diffraction intensities, it has definite limitations. Whenever most accurate film intensities are desired, therefore, a densitometer, or more correctly, a photometer, must be employed. Figure 12-22 shows a double photometer in which light emanating from a single source passes through identical focusing lenses and falls on the film and on a neutral wedge having a continuously varying optical density. The light passed through both can be recorded either by two similar photocells, or alternatively, by the same one. The wedge is adjusted until the readings in the film and the calibrated wedge agree. This null method can be automated, and affords one of the most accurate means possible for measuring film densities.

Pulse counters. The pulses produced in an ionization detector or a scintillation counter are irregularly spaced because the absorption of incident x-ray photons is a random process. The intensity of an x-ray beam is defined as the number of photons crossing a unit area per unit time. The laws which govern the precision attainable in the measurement of intensities, therefore, are the laws of probabilities of random events, and a large number of events must be recorded to attain adequate precision in the measurements. In order to record the large number of pulses produced in a counter at irregularly spaced intervals, a device called a *scaler* is employed. Its function is to count the pulses arriving in specific groups, usually by 2's or by 10's, in what is called a binary or a decade scaler. In essence, a scaler consists of several stages, each of which is capable of recording the prescribed number of pulses. Thus, in a decade scaler, the first stage records up to 10 pulses. The tenth pulse wipes out the record in the first stage and transmits one pulse to the second stage. When 10 pulses have been received by the second stage, it in turn transmits a pulse to the third stage, and so forth. By this means, a very large number of pulses can be accumulated, because n stages can store 10^n pulses in a decade scaler (or 2^n pulses in a binary scaler). Suppose five decade stages are used. The first stage counts individual pulses, the second counts tens of pulses, the third counts hundreds, the fourth counts thousands, and the last stage counts tens of thousands. If the numbers 5, 4, 0, 2, 0 are, respectively, recorded by each stage, the total number of pulses counted is

$$5(1) + 4(10) + 0(100) + 2(1{,}000) + 0(10{,}000) = 2{,}045 \text{ pulses} \tag{12-4}$$

The comparable interpretation of a binary scaler is left to the exercises at the end of this chapter.

Precision of measurements. The need for accumulating a large number of pulses becomes evident when the precision with which intensities can be measured is considered. A detailed consideration of the laws of probabilities governing random processes lies outside the scope of this book. Nevertheless, use must be made of some of the results to evaluate the oft-recurring problem: Suppose that the above-noted 2,045 pulses have been recorded, what is the precision of this measurement? Or putting it another way, what is the probability that the same number of pulses will be counted the next time this same measurement is repeated? The answer to this problem is given by a simple formula. Exactly one half of the times, the error ϵ_N associated with counting a total of N pulses will be less than

$$\epsilon_N = \frac{67}{\sqrt{N}} \quad \% \tag{12-5}$$

The other half of the times that this measurement is made, the error will exceed the value given in (12-5), or in other words, there is a probability of 0.50 that the experimental value N differs from the true value by ϵ_N.

If this is not an acceptable precision, one can use another expression from probability theory, which states that there is a probability of 0.96 that the experimental value N differs from the true value by ϵ_N, where now

$$\epsilon_N = \frac{201}{\sqrt{N}} \quad \% \tag{12-6}$$

In order to make use of these relations, consider the experimental value in (12-4). According to (12-5), half of the time the precision of measuring 2,045 pulses is 1.5%, so that the true value will be 2045 ± 31 pulses. Alternatively, 96 times out of 100, the true value will be 2045 ± 93 pulses, according to (12-6). Suppose, however, it is essential that the measurement have a precision of 1.5%, with a probability of 0.96. How many pulses must be measured to attain this precision? According to (12-6), a precision of 1.5% is attained when approximately 18,000 pulses have been counted. Thus it becomes clear that a high precision accompanied by a high probability that the measured value resembles the true value within the stated precision requires a very large number of pulses to be counted. The numbers of pulses, or counts, needed to attain certain precisions with the probabilities stated are given in Table 12-2.

Table 12-2 Counting statistics

Number of counts	*Percent error for probabilities of:* 0.50	0.96	0.99
1,000	4.23	6.32	9.48
3,200	2.37	3.54	5.31
4,000	2.12	3.16	4.74
6,400	1.68	2.50	3.75
8,000	1.50	2.24	3.36
10,000	1.34	2.00	3.00
16,000	1.06	1.58	2.37
25,600	0.84	1.26	1.89
40,000	0.67	1.00	1.50
64,000	0.53	0.80	1.20
256,000	0.27	0.40	0.60
512,000	0.19	0.28	0.42

There are two methods of accumulating large numbers of pulses during an intensity measurement. In one, the intensity is counted for a fixed time period during which the pulses are accumulated in the scaler. This is called the *fixed-time method*, and the time period is selected to give a number of counts commensurate with the precision desired. It is particularly useful in certain experiments in which a crystal is in position to reflect x rays for only a fixed time interval. It has the disadvantage that the intensities of different reflections will be recorded with unequal precision unless the fixed-time period is varied. This disadvantage can be overcome by using the *fixed-count method*, in which the precision desired is predetermined by selecting the number of pulses to be counted. At the beginning of a measurement, the scaler and a synchronous clock are jointly activated. When the scaler has recorded the preselected number of pulses, it shuts itself and the clock off. The time recorded on the clock then indicates the time period during which the counts were accumulated. In either method, the actual counting rate is determined by dividing the total number of pulses by the total elapsed time.

Radiation dosimetry. It should be clear to the reader by now that when x rays are absorbed by atoms they produce changes in the absorbers. In the case of human tissues such changes can be therapeutic, as in certain controlled exposures to high-energy radiations, or very harmful, as in direct exposures to the low-energy x rays used in x-ray diffraction studies. With our present knowledge of the hazards involved, there is absolutely no excuse for anyone suffering from undue exposure. In this connection it should be realized that the intensity of x-ray beams diffracted by most crystals is too small to represent any danger even during prolonged exposures. This is not true, however, of the direct beam emanating from an x-ray tube or any portion of that beam accidentally deviated from its intended path by inadequate shielding. The matter of properly enclosing the direct-beam path in x-ray diffraction experiments is discussed in a recent report of the Apparatus and Standards Committee of the American Crystallographic Association,† from whom copies can be secured.

By comparison, the question of what constitutes a tolerable exposure is far more difficult to answer. The whole matter of measuring exposures is complicated by differences in penetrating power of various radiations, by the various reactions such radiations can produce, and by differing sensitivities of human organs. Moreover, as more is learned about such matters, previous estimates have to be revised constantly. Thus the only safe procedure to follow in an x-ray diffraction laboratory is to avoid undue exposure entirely. Once exposed, one can consult a physician for palliatives, but little is known yet about treatments that actually cure radiation burns. An account of recent developments in the field of health physics is given in the first reference at the end of this chapter. It is germane to draw attention here to one factor. It is good practice to use a portable ionization detector to check periodically whether there is any stray radiation emanating from any part of an x-ray diffraction arrangement. In doing this, however, care should be exercised to use a detector that is properly sensitive to the x-ray energies employed. This is particularly true when the x-ray intensity is converted to a radiation dose, which is measured in *rontgens*, if an exposure dose is desired, or in radians (*rads*), if the dose absorbed per unit volume of irradiated material is desired. A rad is a relatively new unit in dosimetry, and equals 100 ergs of absorbed energy per gram of absorber, and thus depends on the nature of the absorber when the cross-sectional area of irradiation is considered.

SUGGESTIONS FOR SUPPLEMENTARY READING

G. W. Dolphin, W. J. Megaw, and J. Rundo, Health physics, *Reports on Progress in Physics,* vol. 25 (1962), pp. 337–384.

William Parrish (ed.), *X-ray analysis papers* (Centrex Publishing Co., Eindhoven, The Netherlands, 1965).

W. E. Schall, *X-rays,* 8th ed. (John Wright and Sons, Ltd., Bristol, England, 1961).

Wayne T. Sproull, *X-rays in practice* (McGraw-Hill Book Company, New York, 1946).

EXERCISES

12-1. In view of the importance of not exceeding tube ratings, it is essential to observe the manufacturer's operating instructions. Suppose an iron-target tube can be operated at a maximum of 10 mA and 40 kV. What is the total power input to the tube? Under what conditions could the tube be operated with a tube current of 15 mA?

12-2. Assuming that 100% of the energy striking a copper target in a tube operated at 40 kV and 25 mA is converted into heat, how long would it take to melt a 200-g copper target after the cooling water (20°C) has been shut off? (For Cu, melting point = 1083°C, specific heat = 0.094 cal/g, and the latent heat of fusion is 49 cal/g mole.)

12-3. The thermionic emission from a filament is controlled by the Richardson-Dushman equation $I = AT^2e^{-b/T}$, where A and b are constants and T is the absolute temperature. Suppose a tungsten filament for which $b = 50{,}000°$ emits 1 mA at 2000°K; neglecting the small change in T^2, calculate the current emitted at 2100 and at 2200°K.

† Karl E. Beu, *Safety Considerations in the design of x-ray tube and collimator couplings on x-ray diffraction equipment* (Apparatus and Standards Committee Report 1, 1962).

12-4. The target-to-pinhole distance is about 3 cm when a lead foil is placed flush against a tube face. Suppose that a 1-by-10-mm focal spot is viewed at a takeoff angle of $6°$ parallel to its long axis. What is the size of the image formed by this focal spot on a film placed 30 cm from the pinhole?

12-5. If a tube is operated at 30 kV and 20 mA in a constant-potential unit, what is the maximum voltage (peak) that should be applied to it in a full-wave-rectified unit to obtain the same rms current?

12-6. The absorption of x rays obeys the well-known relation $I/I_0 = e^{-\mu x}$ where I and I_0 are the intensities of the transmitted and incident beams, μ is the linear absorption coefficient, and x the thickness of the absorber. To test the linearity of a detector, it is usual practice to monochromatize the x-ray beam by reflection from a crystal and then to interpose successively increasing numbers of foils, each having the same thickness. The incident and transmitted beams are measured, and $\ln(I/I_0)$ is then plotted directly against the number of foils placed in the beam, since μx is constant for each foil. In a typical experiment, the successive experimental ratios I/I_0 for the first 10 foils were 10.0, 8.0, 6.0, 4.7, 3.4, 2.5, 1.7, 1.05, 0.71, 0.49. Plot these values on semilogarithmic graph paper. What are the possible reasons for the observed deviations from a straight line? What experimental precautions can be taken to assure that only detector nonlinearity is being measured?

12-7. In a binary scaler, each stage counts two pulses; that is, the first pulse turns the stage on, while the second pulse shuts off the stage and serves to activate the next stage. Consider a bank of eight lights arrayed in a row. When a light is on, it means that it has received the first of two pulses. If it is off, it either has not been activated or else it has just passed on a pulse to the next light. What are the total numbers of pulses recorded by these eight lights if they are, in sequence, ON, ON, OFF, OFF, ON, OFF, ON, OFF, and for another sequence, OFF, ON, OFF, ON, ON, ON, OFF, ON?

12-8. Suppose a mechanical counter records the number of pulses transmitted by the last of the eight stages in Exercise 12-7; that is, it registers one count each time the eighth bulb is shut off by the second pulse reaching it. What are the total counts recorded when the mechanical counter reads 17 and the lights are as in the first sequence above? When the counter reads 120 and the lights are in the second sequence in Exercise 12-8?

12-9. The probable errors in relations (12-5) or (12-6) express only the errors inherent in counting random processes. In the case of x-ray diffraction, other sources of error may invalidate these simple relations. Suppose that the background intensity in the laboratory is high, for example, because of the presence of nearby radioactive sources. The background then contributes a total count of N_B, whose randomness may differ from that in the diffracted beam. This is taken into account in probability theory by changing relation (12-6) to

$$\epsilon = 201 \frac{\sqrt{N + N_B}}{N - N_B} \quad \% \tag{12-7}$$

Suppose the background counting rate is about 50% of that in the diffracted beam. By how much must the total count be increased to give a precision 1.5%, with a probability of 0.96, in this case, over the number of counts needed to attain the same precision with negligible background? By how much must the counting time be increased if a probability of 0.50 is deemed adequate?

12-10. Another source of error in x-ray intensity measurements arises from the instability in the x-ray source. When several sources of error are combined, the total error is given by a relation such as the following:

$$\% \text{ total error} = [(\text{statistical error})^2 \times (\text{instability error})^2]^{\frac{1}{2}} \tag{12-8}$$

Most modern commercial x-ray units claim to have instabilities less than 0.5% during a period of about four hours. Assuming that this is a valid estimate, how does this instability affect the precision discussed in Exercise 12-9?

THIRTEEN
X-ray diffractometers

DIFFRACTOMETER GEOMETRY

Reciprocal-lattice construction. The relationship between an individual crystal and its associated reciprocal-lattice construction has been examined in Chap. 7 for various possible experimental arrangements. A diffractometer measures the intensity of x rays reflected from one stack of planes at a time, so that it is of interest to examine what kind of spatial arrangements are best suited for such measurements. Suppose a single crystal has one of its pinacoid faces set parallel to the incident x-ray beam; then the reciprocal-lattice construction appears as shown to the left in Fig. 13-1. (Only a small part of the zero level of the reciprocal lattice is shown, for clarity.) Next, suppose that the crystal has been rotated by an amount θ, so that the reciprocal-lattice construction now appears as shown to the right in Fig. 13-1. As previously demonstrated, the arc from the reciprocal-lattice origin O to the reciprocal-lattice point P subtends an angle 2θ at the center of the Ewald sphere. As can be easily demonstrated (Exercise 13-1), this relationship between the angle θ by which the crystal is rotated and the diffraction angle 2θ holds true for any reciprocal-lattice row passing through the origin. It follows from Fig. 13-1, therefore, that if the crystal is initially set with its pinacoid face parallel to the x-ray beam (zero-order reflection) and an ionization detector is set to intercept the direct beam, a subsequent rotation of the crystal by an amount θ and of the detector by an amount 2θ will automatically assure that the detector is in the correct position to detect the diffracted beam. This principle was utilized by the Braggs (Fig. 5-2) to construct the first such instrument in 1913. Note that an automatic coupling of the crystal and detector rotations in a $1:2$ ratio allows the measurement of the first- and higher-order reflections from one stack of planes simply because only one central reciprocal-lattice row maintains the geometric condition indicated in Fig. 13-1 (Exercise 13-1). This is one of the chief limitations of the original Bragg spectrometer; namely, it requires that the crystal be reset each time so that a different central row of the reciprocal lattice is perpendicular to the x-ray beam in the zero-order setting. Also, only one zone of reflections can be examined with a single mounting of the crystal, because only one layer of the reciprocal lattice can

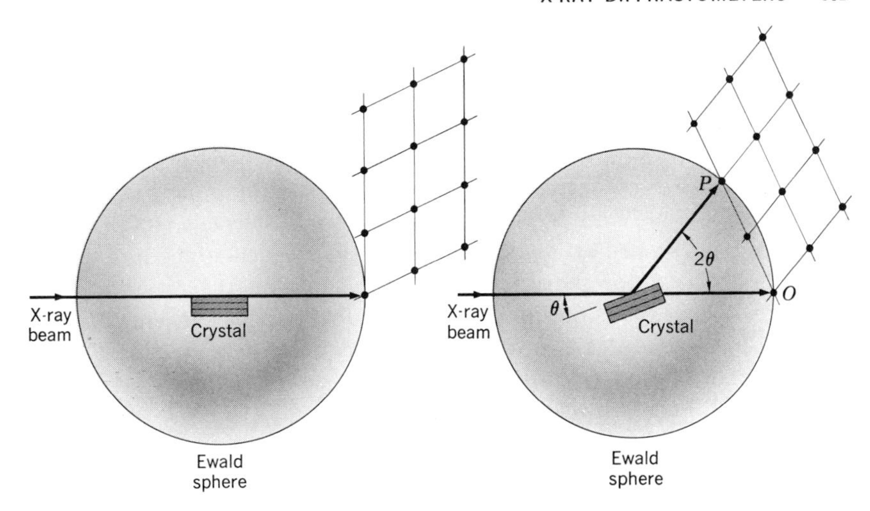

Fig. 13-1

be made to coincide with the equatorial plane within which the incident and diffracted beams lie. This is similar to the restrictions imposed in a one-circle optical goniometer, and ways to ameliorate this condition are discussed in later sections.

When a polycrystalline aggregate is placed at the center of the Ewald sphere, the reciprocal-lattice construction has the appearance shown to the left in Fig. 13-2. The concentric reciprocal-lattice spheres intersect the Ewald sphere and give rise to diffraction cones, as described in Chap. 7. In the equatorial plane shown in Fig. 13-2, this means that there are two diffracted beams (for each reciprocal-lattice sphere) forming the appropriate angle 2θ with the incident x-ray beam, as shown to the right in Fig. 13-2. Provided that a sufficiently large number of crystal-

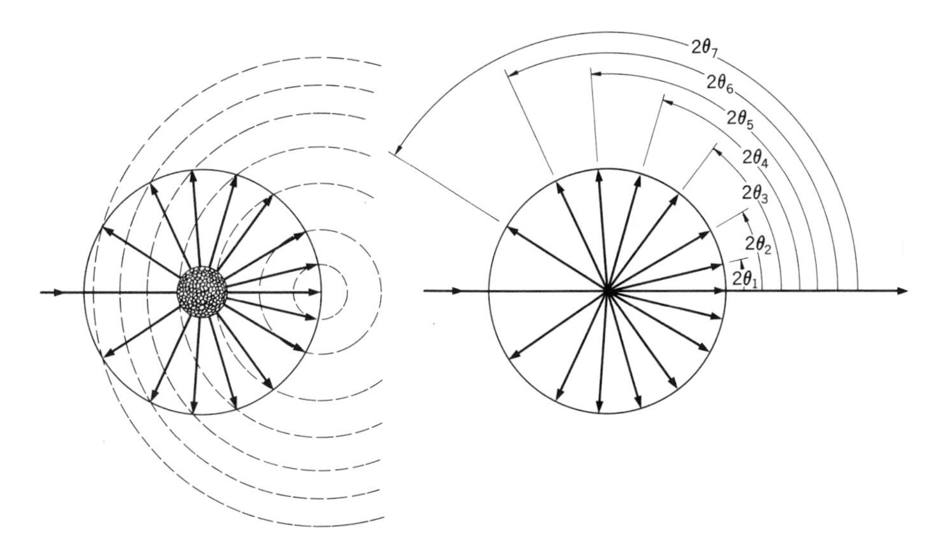

Fig. 13-2

lites are in position to reflect the incident beam, therefore, a detector moving in a circle about the polycrystalline sample will intersect, and thereby detect successively each reflection cone. Thus a much simpler instrument can record both the intensity and the diffraction angle 2θ from all possible reflections of a polycrystalline substance. This was realized by Bragg, but when he attempted to use for this purpose the instrument he had previously built for single crystals, the intensities of the reflected beams were too feeble to make this a practical procedure. The construction of a practical powder diffractometer had to wait till 1945, when the availability of Geiger-Müller tubes having higher quantum-counting efficiencies was combined with special geometries to produce a highly efficient system by H. Friedman. The development of modern-day instrumentation is largely the result of the continued efforts of W. Parrish and his collaborators.

Parafocusing. When a large single crystal is irradiated by a diverging beam of x rays, it can be shown (Exercise 13-2) that small angular displacements δ cause it to reflect slightly different wavelength components to the same point on a circle, having the crystal at its center and the x-ray source on its circumference. This is illustrated in an exaggerated manner in Fig. 13-3. As realized quite early by Bragg, this has the effect of "focusing" the reflected rays when a slightly diverging and imperfectly monochromatic x-ray beam is used. As suggested by J. C. M. Brentano, this is an ever-present consequence of the finite spectral widths of x-ray sources, so that it should be distinguished from true optical focusing by employing the term *parafocusing*. The inadvertent parafocusing effect when large single crystals are used in the Bragg arrangement (Fig. 13-1) is responsible for the relatively high intensity obtainable with this type of instrument. A similar increase in the reflected-beam intensity can be obtained with polycrystalline samples by utilizing parafocusing geometries.

The need to cut, or cleave, large crystals along many different directions in order to utilize a Bragg-type diffractometer for measuring their reflection intensities severely limited its utility in x-ray diffraction studies. Instead, it became the practice to examine small single crystals with photographic recording of the relative reflection intensities. As demonstrated in subsequent chapters, photographic methods are convenient to use for many purposes, but the accuracy of intensity

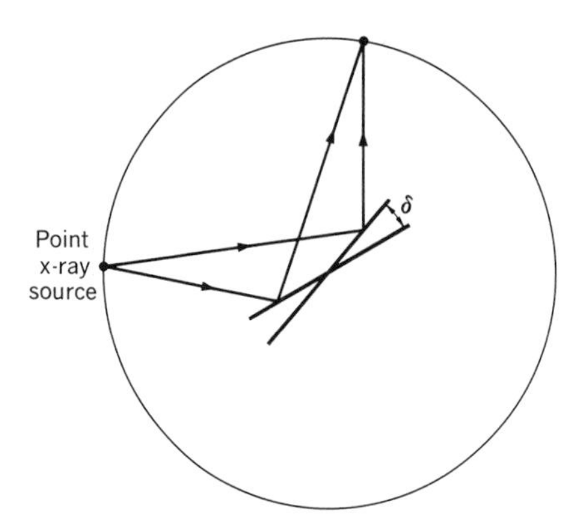

Fig. 13-3

measurement by this means is limited by the response of the film and the ability to convert film exposures to x-ray intensities. The development of ever more sensitive direct ionization detectors (Chap. 12) naturally stimulated a resurgence of interest in single-crystal diffractometry. When small single crystals are examined, the parafocusing condition illustrated in Fig. 13-3 cannot be utilized. It is also necessary to provide a means for reorienting the crystal so that many different planes can be positioned to reflect the incident beam. This can be done most easily by mounting the crystal on a device, called a *goniometer*, on which the necessary angles can be laid off and measured. The manipulation of such instruments is considered briefly next.

Goniometry. In order to help the early crystallographers measure interfacial angles on small crystals, W. H. Wollaston devised an optical goniometer in 1809. One narrow parallel light beam was directed at a fixed mirror, and another onto the crystal, which was mounted on a rotatable spindle. When one face of the crystal became parallel to the mirror, it reflected the second light signal to the viewer, who manipulated the crystal until the two images coincided. It was thus possible to measure the angles through which the crystal had to be rotated to afford successive faces an opportunity to reflect the second light beam. The operating principle of such a one-circle goniometer is illustrated in Fig. 13-4. In more recent versions, a carefully collimated light beam, having, say, a cross section shaped like a Maltese cross, is directed at the central rotation axis of the goniometer. A telescope equipped with cross hairs is similarly aligned to the same rotation axis.

It is clearly seen in Fig. 13-4 that only the reflections from planes belonging to a single zone can be detected on a one-circle instrument. In order to eliminate the need for remounting the crystal, a second rotation can be provided, as shown in Fig. 13-5. With the aid of the two circular motions, ω and ϕ, it is actually possible to observe reflections from all faces except those facing the crystal holder. It

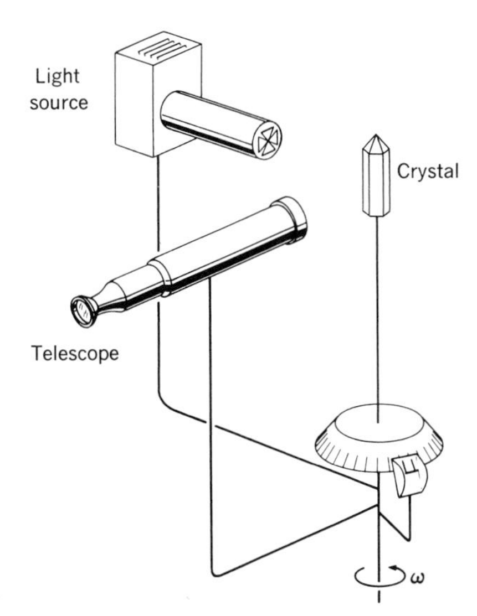

Fig. 13-4. Schematic view of a one-circle goniometer.

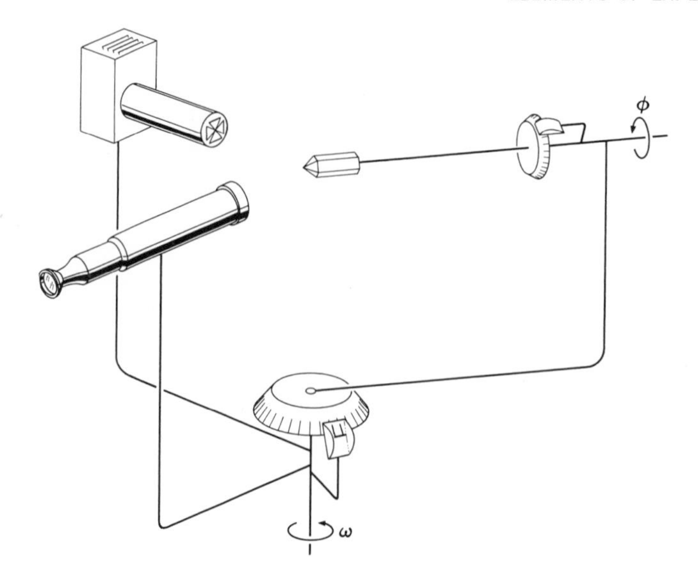

Fig. 13-5. Schematic view of a two-circle goniometer.

is somewhat more convenient, however, to provide still a third rotation, χ, as shown in Fig. 13-6, by the use of which successive settings can be made more rapidly. The three rotations shown in Fig. 13-6 represent the basic three angles with whose aid any orientation can be described uniquely, as first pointed out by L. Euler in 1776.

Although the two- and three-circle goniometers were devised originally to aid the measurement of interfacial angles on crystals, they can be used equally well for orienting crystals before beginning x-ray diffraction studies. When the sole object is the measurement of crystal angles, the crystal can be affixed to the rotation spindle by means of soft wax, so that, subsequently, it can be prodded into

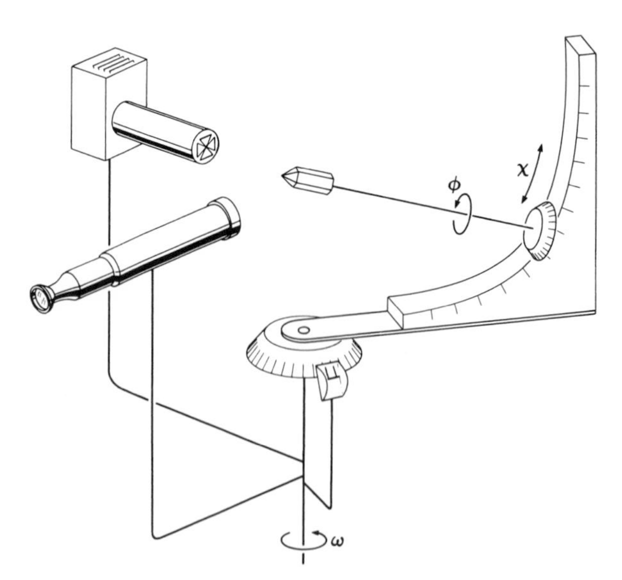

Fig. 13-6. Schematic view of a three-circle goniometer.

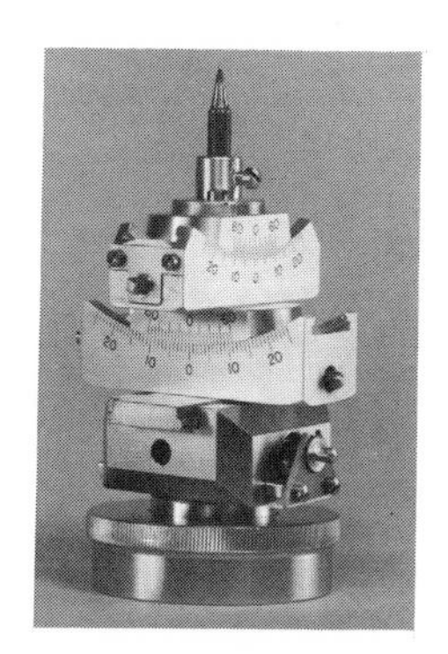

Fig. 13-7. Goniometer head. *(Courtesy of Charles Supper Company.)*

alignment. When a small crystal is being readied for an x-ray diffraction study, however, it is advisable to bond it more rigidly to a suitable holder, which can then be moved from one instrument to another without disturbing the alignment. This is done most easily by affixing the crystal to a glass fiber bonded to a metal pin, as described further in Chap. 15. Although not essential, it is usually more convenient to align the crystal on a *goniometer head*, like the one shown in Fig. 13-7, which contains two mutually perpendicular arcs for small tilt adjustments and two translation screws for centering the crystal on its rotation axis. The goniometer head is provided with a keyed slot in its base, so that it can be repositioned on various instruments in nearly the same way.

POWDER DIFFRACTOMETER

X-ray optics. The parafocusing geometry governing a polycrystalline sample is illustrated in Fig. 13-8. An x-ray beam diverging from a point source meets the curved crystal in such a way that each ray is reflected to the same point on a circle passing through the source, sample, and focus. (The focusing is exact only for

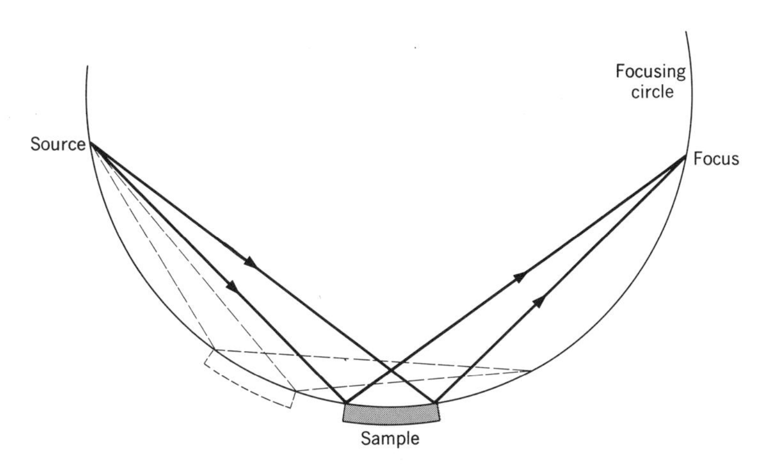

Fig. 13-8

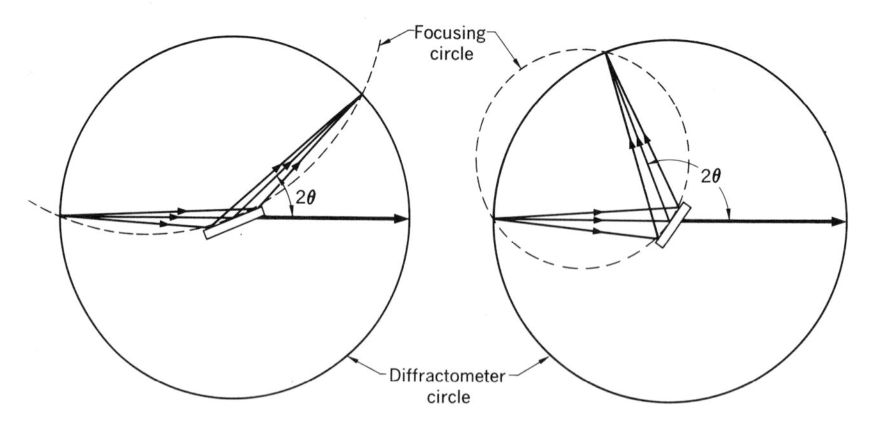

Fig. 13-9

the layer in the sample immediately touching the focusing circle.) When a different incident angle is desired, the sample must be moved along the circle, as indicated in Fig. 13-8. This has the effect of displacing the focal point of the reflected rays. Because x rays penetrate a powder sample to some finite depth, the focusing is not exact. It turns out that approximating the curvature of the focusing circle by a tangent plane introduces a defocusing of comparable magnitude, provided that the radius of the circle is not too small. With this criterion in mind, it is possible to construct a diffractometer in which a plane sample is placed at the center of the diffractometer circle, on which lie the x-ray source and the focal point of the reflected x rays. The attendant focusing circles are indicated in Fig. 13-9 for two possible diffraction angles 2θ. Note that the sample-to-detector distance remains unchanged, so that a relatively simple mechanical linkage between the sample and detector is adequate to maintain the parafocusing geometry. Note also that the sample surface always forms the angle θ with the central incident ray, while the

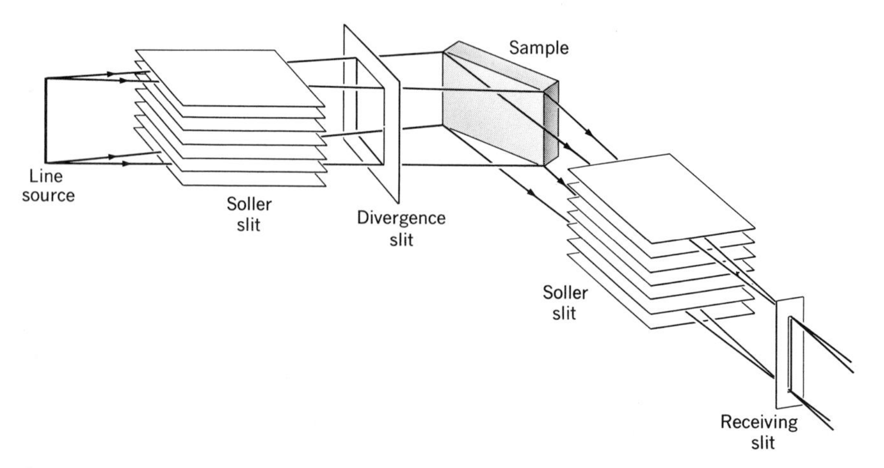

Fig. 13-10

diffracted beam forms the angle 2θ. The linkage between the detector and the sample, therefore, must maintain a $2:1$ rotational ratio.

A typical x-ray system is illustrated in Fig. 13-10. X rays emitted from a line-shaped source pass through a stack of parallel metal foils, called a *Soller slit*, and fall upon a flat sample. A *divergence slit* preceding the sample serves to limit the divergence of the incident beam so that it does not strike other parts of the instrument. The diffracted beam passes through another stack of parallel foils and converges onto a *receiving slit*. As previously pointed out in Chap. 6, the actual resolution of such an arrangement is determined by the effective width of the line source, as determined by the takeoff-angle defining slit, in this case the receiving slit. For various reasons, other slits may be inserted in the x-ray system, but their function is to limit parasitic scattering, not to define the ray paths.

The function of the parallel metal foils in the Soller slit is to limit the angular divergence of the x-ray beam at right angles to the focusing plane. It has become customary to think of the focusing plane as being horizontal (regardless of its actual orientation in space), so that the Soller slit limits the *vertical divergence* of the beam. The ray diagrams in the horizontal (focusing) and vertical planes are shown at the top and bottom, respectively, in Fig. 13-11. The relative dimensions of the various slits employed are considered in the next section.

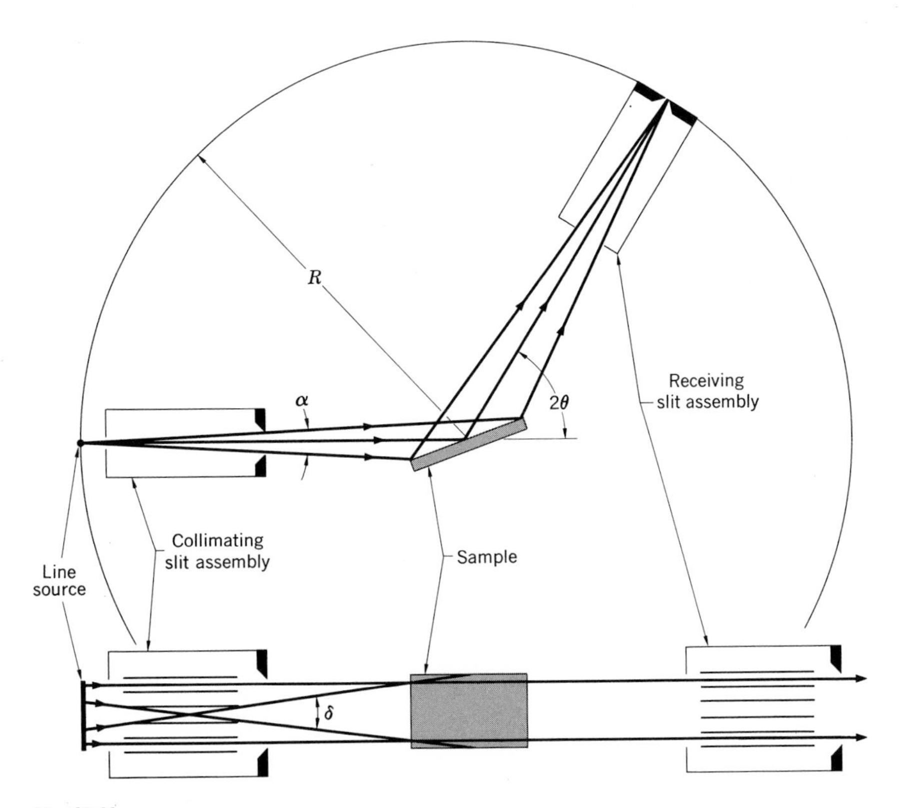

Fig. 13-11

Instrumental factors. The angular aperture α of the incident beam is selected so that a maximum portion of the sample is irradiated. As can be seen directly from Fig. 13-11, if α is expressed in radians, the length l of the irradiated sample portion is

$$l = \frac{\alpha R}{\sin \theta} \tag{13-1}$$

Thus l decreases as the diffraction angle 2θ increases, so that increasingly larger divergence slits are required to assure that comparable areas are irradiated. At the smaller angles, it is necessary to utilize a sufficiently narrow divergence slit to assure that the sample holder is not also irradiated. Some of the limitations that this imposes in practice are considered in Exercise 13-3.

It is fortunate that the absorption factor in a symmetric reflection arrangement like the one depicted in Fig. 13-11 is independent of the angle θ (Exercise 13-4). This means that the intensities of diffracted beams can be compared directly, provided the same size divergence slits are employed. Thus there are competing practical factors to be considered in the selection of the divergence angle α. It should be noted, however, that the constancy of the absorption factor depends on the powder sample having an effective "infinite" thickness. Since $I \propto 1/\mu$, it is possible to estimate how closely this criterion is met in any given case. When a sample is relatively transparent to x rays, not only does the absorption factor have an angular dependence, but the deeper penetration of the incident beam causes an increasing deviation from the parafocusing geometry of Fig. 13-7. This causes the reflection to become broadened and asymmetric. These observations assume importance when x-ray diffraction studies dependent on reflection profiles are attempted.

The finite penetration of the sample produces an aberration similar to that caused by using a flat specimen in the first place. The detailed effect that this and other factors have on the shape and position of the reflections is discussed in several of the books listed at the end of this chapter, and will not be considered here. It turns out that in most cases the angular displacement of the reflection peak is less than $0.01°$ (2θ), so that correction for it is not required. At small angles, however, the angular resolution of a powder diffractometer is inherently poor, so that the additional broadening caused by using a flat specimen becomes more serious. In order not to sacrifice intensity by drastically reducing the divergence angle, R. E. Ogilvie devised a specimen holder (Fig. 13-12) that automatically curves a metal (or plastic) strip, to which the powdered sample is affixed, so as to conform to the parafocusing conditions illustrated in Fig. 13-9.

Vertical divergence causes asymmetric broadening and a peak shift toward smaller angles when $2\theta < 90°$ and toward larger angles when $2\theta > 90°$. The amount of the shift is proportional to $\cot \theta$, so that it is particularly important for small diffraction angles. If the length of the parallel plates in Fig. 13-11 is L and their separations s, then the vertical divergence angle determined by a Soller slit is

$$\delta = 2 \tan^{-1} \frac{s}{L} \tag{13-2}$$

Fig. 13-12. Automatically focusing specimen holder devised by R. E. Ogilvie. *(Courtesy of Philips Electronic Instruments.)*

Convenient dimensions for a Soller slit are $s = 0.5$ mm and $L = 15$ mm, giving $\delta = 4.5°$. For $2\theta > 10°$, this amount of divergence causes a negligible peak shift, but such a Soller slit decreases the beam intensity by a factor of 2. Thus the improved performance of the diffractometer can be procured only at the sacrifice of some intensity. Value judgments have to be made in practice as to when the aberrations introduced by increased vertical divergence are better tolerated than weak intensities.

The receiving slit in the diffractometer serves to define the angular resolution of the instrument. Normally, the width of the receiving slit should be comparable with the halfwidth of the source. For the relatively sharp line sources used in most diffractometer applications ($\approx 0.02°$), the various aberrations cited above introduce a broadening such that receiving slits two to three times wider can be employed without significant loss of resolving power. Opening the receiving slit further of course broadens the reflection peaks, but also increases their total intensity. In this connection it should be noted that an increase in peak intensity caused by relaxing some of the geometric restrictions also increases the background intensity. The efficiency of an instrument is usually judged by the peak-to-background ratio, although sometimes high efficiency must be sacrificed to obtain discernible peaks. Ways to improve the peak-to-background ratio will be discussed in later sections.

The preparation of diffractometer samples is considered in Chap. 17. It should be noted here, however, that irregularities in the sample's surface, such as displacement from the true center of rotation, for example, also introduce errors in peak shapes and positions. Similarly, it is essential that the detector, and the receiving slit, obey very accurately the $2:1$ ratio between their rotation angles. Deviations by as little as $1°$ cause asymmetric reflection broadening and significant intensity variations. It is most important, therefore, that the diffractometer be properly aligned before any meaningful measurements are attempted.

Instrumental alignment. At the present time there are about a dozen versions of powder diffractometers available commercially. All are designed to conform with the x-ray optics described above, but differ in how they are positioned relative to the x-ray tube, in the orientation of their focusing plane (horizontal or vertical), in how many slits they employ, in what the sequential array of slits is, the kinds of drive mechanisms available, and so forth. They also differ in the kinds of alignment aids that the manufacturer supplies and in the facility with which various adjustments can be made. For this reason it is advisable to consult the manufacturer's instructions for detailed guidelines in making actual adjustments. It is nevertheless meaningful to review briefly the way a diffractometer should be aligned to obtain optimum resolution and precise angular readings at a maximum intensity.

One can distinguish between alignment of the diffractometer proper and its positioning relative to the x-ray tube. The former is partly the responsibility of the manufacturer. The interchange of components during ordinary use, however, occasions a need for frequent checking of instrumental adjustments and actual realignment whenever optimum results are desired. The intrinsic alignment of the diffractometer is completed when the central rotation shaft of the instrument is parallel to all the slits employed; when they are all centered on the same straight line, with the diffractometer dials set to zero degrees; and when the rotational ratio between the sample holder and detector arm remains exactly $2:1$ at all angles. Ideally, all alignments should be checked optically, using visible light, not x rays. This minimizes a very real hazard due to accidental exposure while adjustments are being made, and assures correct alignment, rather than a mere "best fit," with respect to the incident x-ray beam.

The way the alignment can be checked optically is illustrated in Fig. 13-13. With the diffractometer dials set to zero, a plane mirror is placed at $45°$ to the diametral line passing through the various slits. An adjustable-focus telescope equipped with cross hairs is then mounted at right angles to this line, as shown.

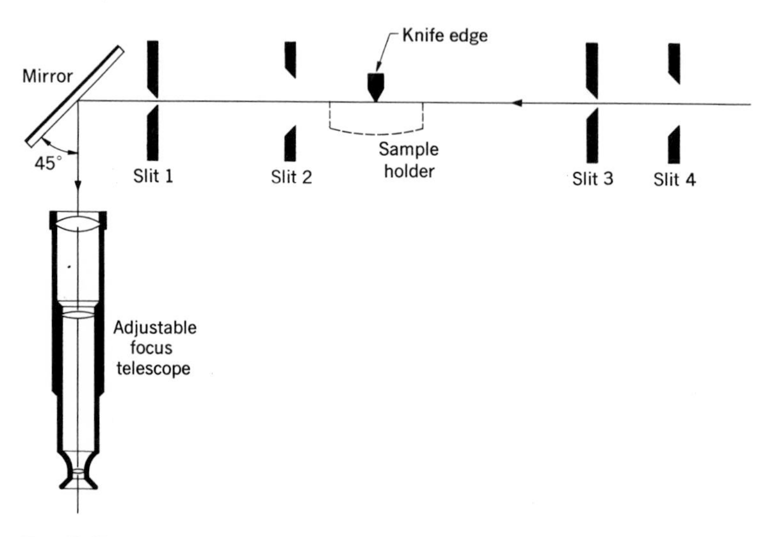

Fig. 13-13. Outline of an optical arrangement for aligning any number of slits parallel to a knife-edge.

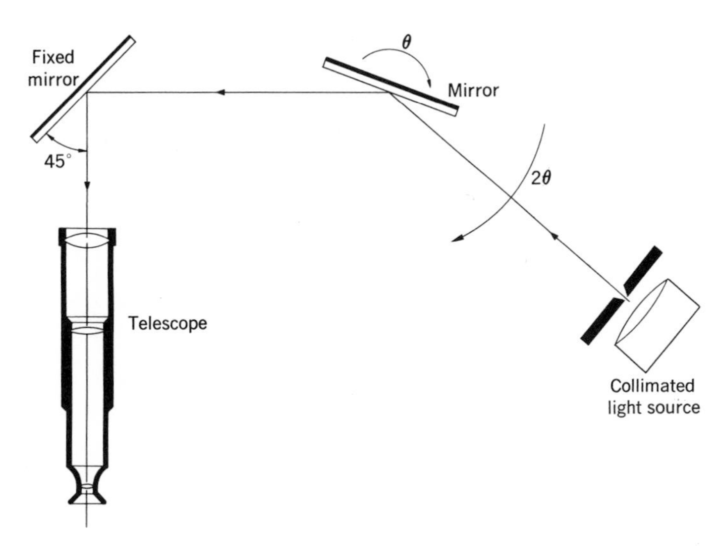

Fig. 13-14. Outline of an optical arrangement for verifying the $2:1$ adjustment of a diffractometer.

Since the sample holder normally does not project into the optic (x-ray) path, a reference edge is placed flush against it to indicate the vertical direction of the diffractometer's rotation axis. The telescope is first focused on this reference edge, and its vertical cross hair is made parallel to it. The reference edge is then removed, and the telescope is focused on the most remote slit (slit 4 in Fig. 13-13), and its parallelism and centering checked (adjusted). The other slits are then successively inserted, and their alignment verified in a like manner.

As shown in Fig. 13-14, the same optical arrangement can be employed to adjust the $2:1$ ratio. With the diffractometer dials set to zero, the slits are removed, and a collimated light source is affixed to the detector arm. The image of its aperture is then centered on the telescope's cross hairs. Following the insertion of a plane mirror in the sample holder, the detector arm is rotated to any convenient angle, say, $2\theta = 80°$. The reflected image should be centered when the sample holder is set at $\theta = 40°$. If necessary, either the setting of the holders or the readings of corresponding dials now can be adjusted. Note that the tracking accuracy of the detector can be checked at various angles in the same way. In fact, the stability of the diffractometer during continuous operation can be verified by means of this simple one-circle optical goniometer. A commercial instrument that makes use of an optical system similar to that described above is shown in Fig. 13-15.

After the intrinsic alignment of the diffractometer has been verified, it is only necessary to align it parallel to the desired x-ray beam. Here it will be recalled from the preceding chapter that the intensity of the beam is proportional to the takeoff angle (Fig. 12-10), increasing to an optimum value at about 6 to 7°. The effective source size also increases with takeoff angle, so that the resolving power decreases. When the line focus (Fig. 12-9) is used with a diffractometer, it turns

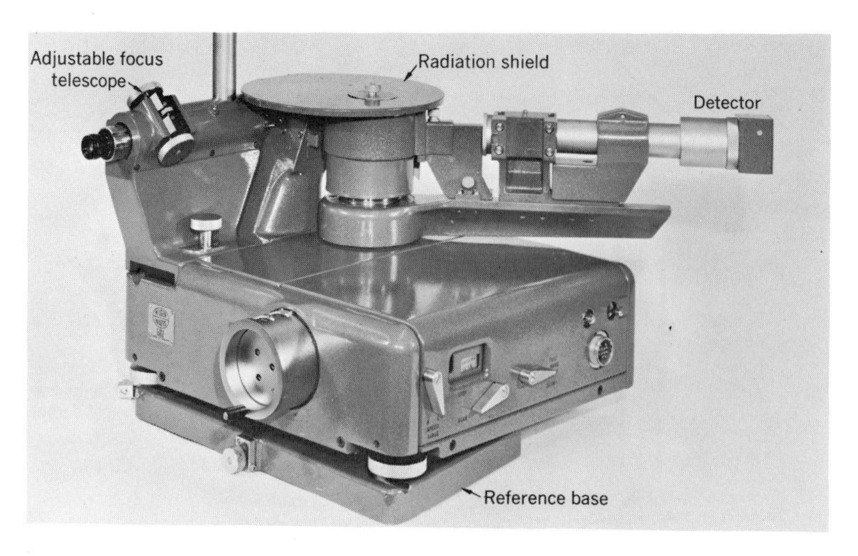

Fig. 13-15. Hilger powder diffractometer mounted on a reference base and equipped with a telescope allowing accurate alignment of essential components optically. *(Courtesy of Engis Equipment Company.)*

out that a takeoff angle of $3°$ gives optimum resolution without too great an intensity loss. (At $3°$, a 1.6-mm-wide source has an effective width of about 0.008 mm.) Obviously, larger angles may be preferred when maximum intensity is desired, whereas takeoff angles smaller than $2°$ are not practical. The adjustment of the diffractometer to the correct takeoff angle is best carried out by having a mechanical means for adjusting the diffractometer's position relative to the x-ray tube. The irradiated portion of a fluorescent screen placed in the sample holder can be used to check the takeoff angle, once the appropriate limits for a particular divergence slit and diffractometer radius have been established. The same screen also can be used to check whether the source-to-sample distance is correctly set. (According to Fig. 13-9, it is clear that the distances between source to sample and sample to receiving slit must be equal.) Assuming that the correct takeoff angle has been established, the diffractometer can be translated parallel to itself, until the most intense portion of the direct beam passes through the center of the sample. This can be checked using the fluorescent screen, or more accurately, by placing a limited aperture (slits or pinhole) at the sample center and monitoring the beam intensity on the ionization detector when it is set at $2\theta = 0°$. (Note that, because the x-ray beam is diverging, the geometric center of the direct beam, visible on a fluorescent screen at the sample position, will change with the rotation angle.) The correctness of centering the diffractometer then can be checked by rotating the aperture $180°$ and confirming that the intensity transmitted is unchanged. A scan of the direct beam, using very fine slits and attenuating foils (if necessary), for several degrees to both sides of zero, should verify the correctness of the chosen alignment position. Finally, one other aspect of the alignment must be checked, namely, the parallelism of the line focus and the vertical axis of the diffractometer. This is best done by blocking out the upper half of the receiving slit, and then the lower half of the receiving slit, with the detector slightly removed from zero. If the line focus is tilted relative to the diffractometer axis, the intensity coming through one half will exceed that passing through the other and indicate the direction in which the diffractometer must be tilted. Since x rays are employed in making these adjustments, great care should be exercised to avoid undue exposure, not only to the direct beam, but also to radiation scattered by various components in the path of the direct beam.

Intensity measurements. The nature and operation of x-ray detectors has been discussed briefly in Chap. 12. It is of interest here to consider how such detectors can be utilized in measuring the shapes and intensities of x-ray reflections from polycrystalline materials. Suppose the total number of counts N entering a detector during a preselected time period t is counted at different angles 2θ. The counting rate

$$n = \frac{N}{t} \tag{13-3}$$

then can be calculated and, if desired, plotted against the diffraction angle. The time involved in such point-by-point counting is considerable, so that a more rapid measure of the counting rate is desired. This can be done by feeding the incoming pulses directly to a special circuit, called a *counting-rate meter.* In such a circuit, the current flowing in a resistor R and smoothed by a parallel condenser C is proportional to the sum of the pulse currents received during the RC-time of the circuit. The voltage across the resistor, averaged over the RC-time, or *time constant*, therefore constitutes a continuous measure of the counting rate, similar to fixed-time point-by-point counting. This voltage can be displayed on a meter or used to deflect a pen in a chart recorder.

According to the discussion in the preceding chapter, the fractional error having a 50% probability can be expressed by

$$\epsilon_N = \frac{\Delta n}{n} = \frac{\Delta N/t}{N/t} = \frac{0.67}{\sqrt{N}} \tag{13-4}$$

and by comparable expressions for higher-confidence values. When a rate meter having a time constant RC is used to measure a counting rate n, it turns out that the significant time interval is equal to $2RC$. Provided that the rate meter is operating for times longer than this value, the fractional error of an individual rate-meter reading having a 50% probability is

$$\epsilon = \frac{0.67}{\sqrt{2nRC}} \tag{13-5}$$

Note that the accuracy of the total intensity recorded does not depend on the counting rate n according to (13-4), but only on the total number of counts N accumulated. This quantity can be controlled in practice by varying the scanning speed and the receiving-slit size.

When a counting-rate meter is used in conjunction with a chart recorder, it should be realized that the time constant influences the shape of the reflection profile, even though it has very little influence on the measured intensity. The amount of the distortion is closely coupled to the scanning speed, and also depends on the receiving-slit width. Because the RC-time causes a time lag between the input signal and the rate-meter output, this effect is most pronounced when the rate of change of the counting rate is largest. This is illustrated in Figs. 13-16 and 13-17. Note that the peak positions are actually shifted and the relative heights of the $\alpha_1\alpha_2$ peaks distorted when too large a time constant or a too rapid speed is employed. In fact, it is evident from these figures that the actual distortion produced is proportional to the product of the speed with the time constant.

This leads to the matter of what strategy one should employ in actual experiments. Apparently, the "best" results are obtained by slow scanning times and small time constants. Actually, this is only partly true because a short time constant, by allowing the recorder pen to respond more rapidly to changes in the counting rate, causes a "jitter" in the recorded trace. (A similar jitter can be caused by coarse-grained samples, in which case some relief may be obtained by rotating the sample rapidly in its own plane, using a special spinner for this purpose.) Thus one must decide on the end result desired, and select the operating parameters accordingly. For a rapid survey of the diffraction spectrum of a sample, a

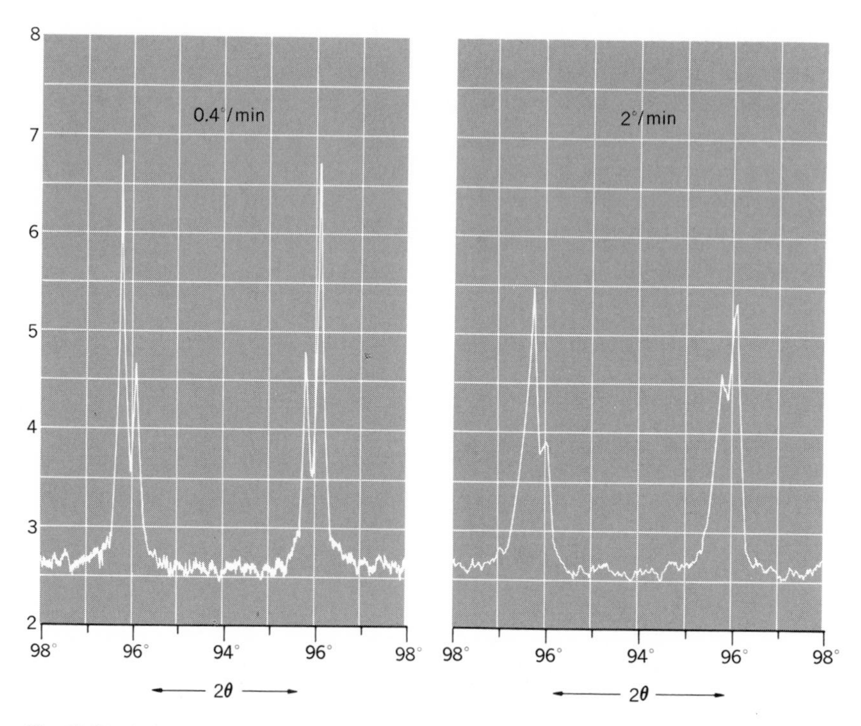

Fig. 13-16. Effect of scanning speed on diffractometer tracings when the time constant of the recorder is held fixed at 4 sec.

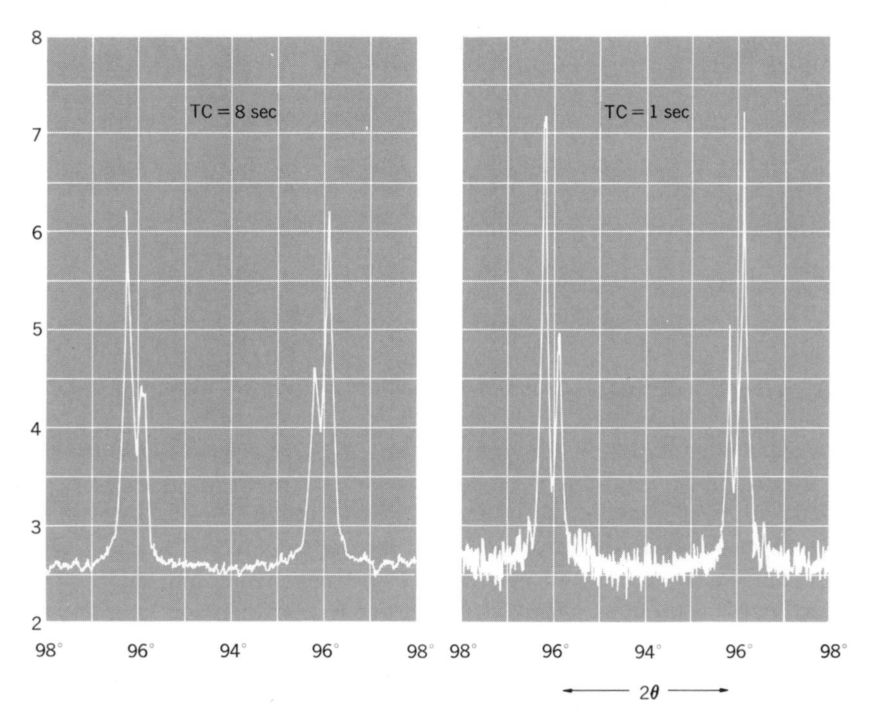

Fig. 13-17. Effect of time constant on diffractometer tracings when the scanning speed is held fixed at $0.4°$/min.

relatively rapid scanning speed, wide receiving slit, and small time constant should be employed. For more accurate measurements, more time must be expended.

When seeking to compare total intensities of reflections (quantitative analysis), or when the detailed variation of intensity with reflection angle is required, it is important that the relative error of all comparable measurements be kept constant. According to (13-4), this error is inversely proportional to $\sqrt{N}$, so that a fixed number of counts N should be obtained at each comparable point. A special device, called a *step scanner*, can be used for this purpose. Although the manufacturing details may vary, in essence, a step scanner "sits" at each of successive predetermined angular intervals until a preselected number of counts has been accumulated. Since the counting rate at each angular setting is different, the time between steps varies, and can be noted on a chart recorder by a proportional displacement of the pen. The resulting trace of a reflection consists of successive steps, instead of the usual smooth line; whence the name of this process. It is very useful when accurate intensities or profiles are sought, because the automatic point-by-point counting relieves an operator of the attendant drudgery.

When accurate intensities are sought, it is possible to achieve the desired goal by integrating the reflection mechanically. This is done by scanning a reflection beginning at some angular value sufficiently removed from the peak position to assure that true background intensity is being measured, and continuing past the reflection an equivalent distance. By suitably adjusting the scanning speed (or by repeating the scan) any desired number of counts can be accumulated. The background intensity is measured similarly on both sides of the reflection, and the suitably normalized counting rates are subtracted from each other. The result is an integrated intensity value that is completely independent of the instrumental factors discussed in this section. It is obviously much more reliable than measurements of peak heights, which, as clearly seen in Figs. 13-16 and 13-17, depend markedly on the operating conditions.

Beam monochromatization. As discussed in previous chapters, there are a number of ways in which a nearly monochromatic x-ray beam can be obtained in practice. A rather straightforward way is to employ filters, in particular balanced filters (Fig. 6-13), that produce a relatively narrow band transmitted by difference. Here the availability of a direct ionization counter is essential, so that this method is particularly well suited to diffractometers. In addition to experimental problems of accurate balancing, this method requires dual measurements of each intensity. It also transmits a finite bandwidth determined by the positions of the two absorption edges, so that special background corrections are required.

An alternative procedure for producing an x-ray band is to use electronic discrimination in conjunction with proportional, or scintillation, counters. Although developed primarily to aid the energy discrimination in spectral analyses, it is possible to use the same instrumentation in x-ray diffraction studies as well. In a typical pulse-height analyzer, it is possible to set a "window" that transmits, primarily, pulses of only one height (wavelength) and rejects those of larger or lesser height. As demonstrated by Parrish and his collaborators, a scintillation counter used with a suitable metal foil to filter the β component can virtually eliminate the short-wavelength continuum. This instrumentation also reduces markedly the long-wavelength continuum for $\mathrm{Cu}\ K\alpha$ and longer-wavelength radiation. In the case of $\mathrm{Mo}\ K\alpha$, a considerable portion of the long-wavelength continuum is transmitted, although the overall peak-to-background ratio is markedly increased in this case also.

The best way to monochromatize an x-ray beam, of course, is by reflection from a crystal. Either plane- or bent-crystal monochromators can be employed, and they can be positioned in the path of the incident or the reflected beam. Bent-crystal monochromators have the advantage that their parafocusing action improves the resolution and increases the intensity over identical plane monochromators by almost ten-fold. The placement of a monochromator between the receiving slit and the detector has one further advantage. Since only the selected wavelength radiation is transmitted, fluorescence radiation originating in the sample is eliminated. This means that a single x-ray tube can be employed in a variety of studies, with considerably improved overall results.

SINGLE-CRYSTAL DIFFRACTOMETERS

Manual instruments. The rapid growth in popularity of powder diffractometers in the late 1940's served to stimulate a revival of interest in single-crystal diffrac-

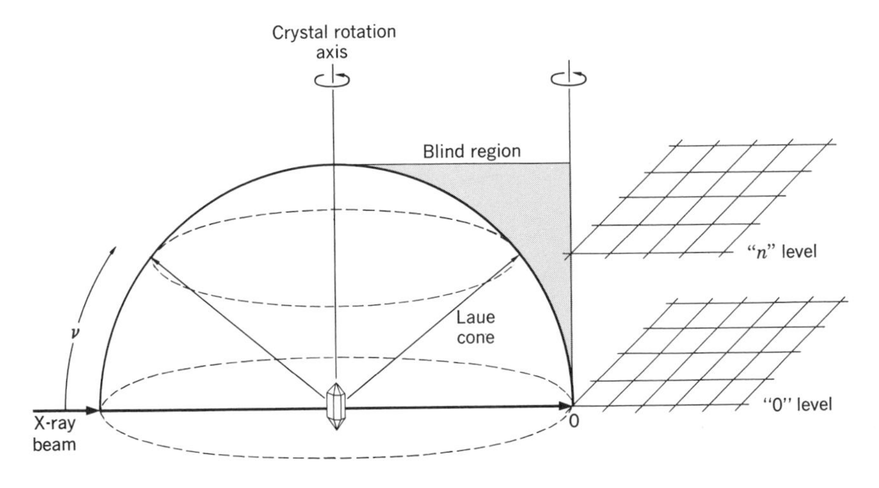

Fig. 13-18

tometry. Initial trials were limited to feasibility tests in which the one-circle geometry of powder diffractometers was employed to demonstrate that small single crystals would yield adequately large intensities. Because reflections belonging to a single zone only can be measured this way, attempts next were made to extend the instrumentation to include other reflections also. The first and most obvious extension follows the reciprocal-lattice construction shown in Fig. 13-18. Suppose the incident x-ray beam is perpendicular to some rational direction of the mounting or rotation axis of a crystal. The reciprocal lattice can be considered to be made up of parallel nets orthogonal to the rotation axis, as indicated by the two grid portions shown on the right in Fig. 13-18. Rotation of the crystal then causes each of the nets to intersect the Ewald sphere, giving rise to Laue cones, as previously described in Chap. 7. Denoting the angle measured from the horizontal plane to the cone generators forming the n-layer cone by ν, it is clear that an arc having the crystal at its center can be used to support a detector at the appropriate angle ν, so that reflections forming the n-layer cone can be intercepted. This is illustrated in Fig. 13-19, where it can be seen that the crystal must be rotated by an amount ω to bring a reciprocal-lattice point to intersect the sphere of reflection, while the detector arc is then rotated by an amount $2\theta'$, where this angle is the projection onto the horizontal plane of the true diffraction angle. Returning to Fig. 13-18, note that, in the normal-beam arrangement illustrated, no amount of rotation can bring into reflection position those reciprocal-lattice points that are in the so-called blind region. Only in the zero level does every reciprocal-lattice point get a chance to intersect the sphere (out to some limiting value σ). Thus, when a normal-beam arrangement is employed, it is necessary to remount the crystal one or more times to record all possible reflections.

A way to eliminate the blind region was suggested by Buerger in 1936, and is illustrated in Fig. 13-20. Remembering that the origin O of the reciprocal lattice lies at the point where the direct beam leaves the Ewald sphere, incline the beam with respect to the crystal rotation axis by an amount μ equal to the angle ν meas-

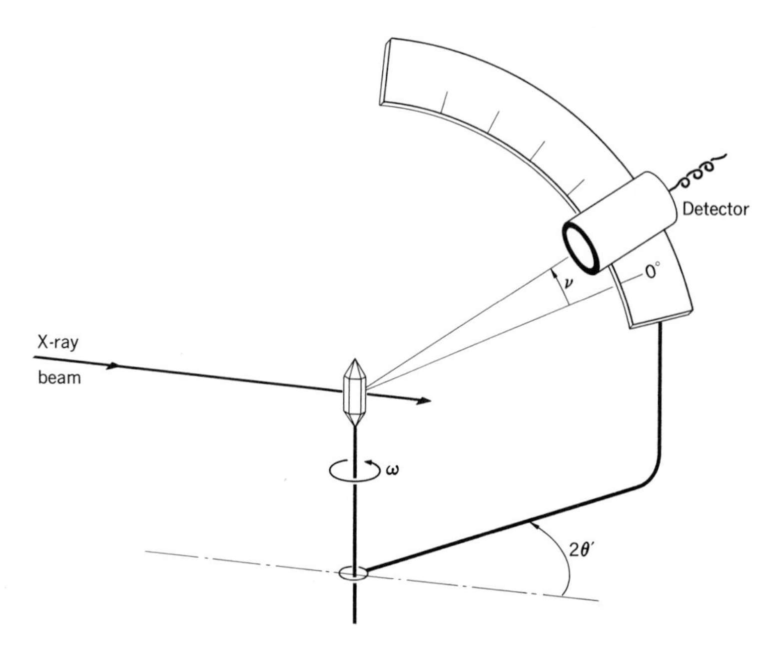

Fig. 13-19. Normal-beam arrangement of single-crystal diffractometer.

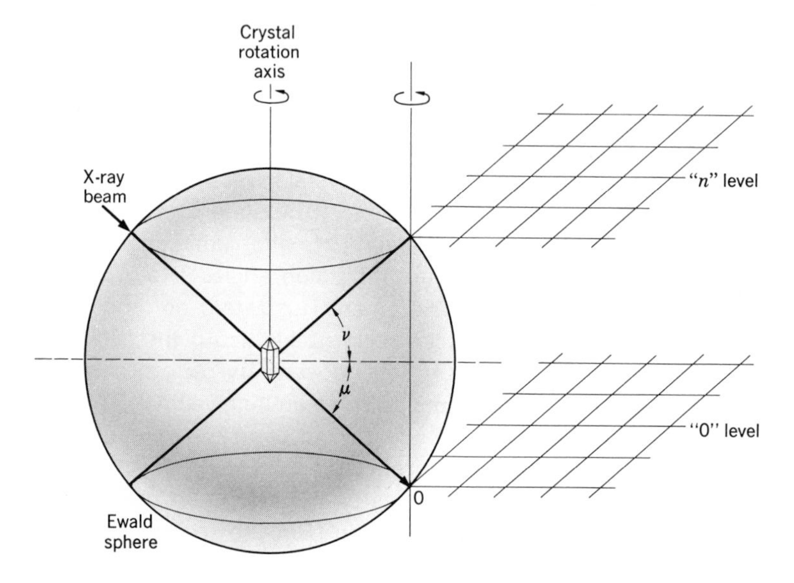

Fig. 13-20. Equi-inclination geometry.

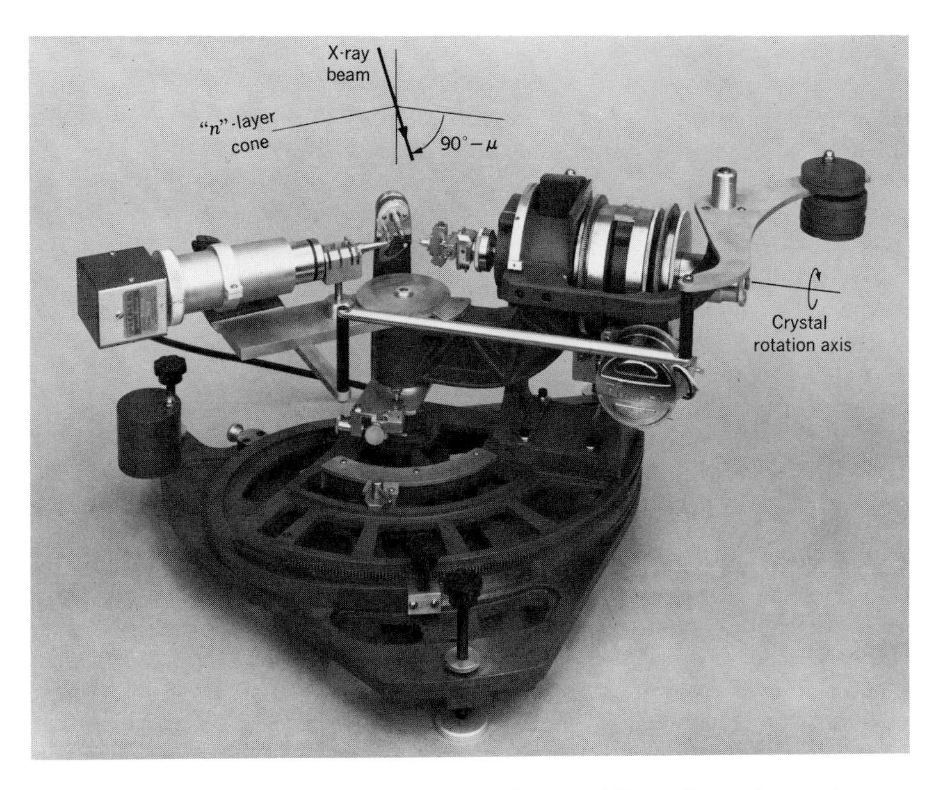

Fig. 13-21. Buerger single-crystal diffractometer. *(Courtesy of Charles Supper Company.)*

ured to the Laue cone of the same upper level n. When this *equi-inclination condition* is met, the origin of the n-level also lies on the sphere of reflection. Rotation of the crystal about its rotation axis then has the effect of rotating the n-level exactly the same way as the zero-level, without a blind region. A commercially manufactured version of an instrument designed by Buerger and embodying this principle is shown in Fig. 13-21. The crystal is mounted on a rotating spindle which can be inclined relative to the x-ray beam by moving its support along the base plate. The detector arm is then inclined so that the counter is made coaxial with the generators of the Laue cone (it always points at the crystal). A means for rotating the detector arm about the crystal rotation axis is also provided. Thus one sets the instrument to record one reciprocal-lattice level at a time (Fig. 13-20), and then, by rotating the crystal by an amount ω and the detector by an amount $2\theta'$ (the projection of 2θ), individual reflections in each level can be detected.

A somewhat different approach was adopted by T. C. Furnas and D. Harker. Seeking a way to adapt a commercial one-circle diffractometer, they made use of the Eulerian geometry of a three-circle goniometer (Fig. 13-6) to bring any reflection into the equatorial plane of the Ewald sphere. In effect, this is a normal-beam geometry, in which the crystallographic direction parallel to the rotation axis is changed successively. The strategy employed in examining the reciprocal lattice

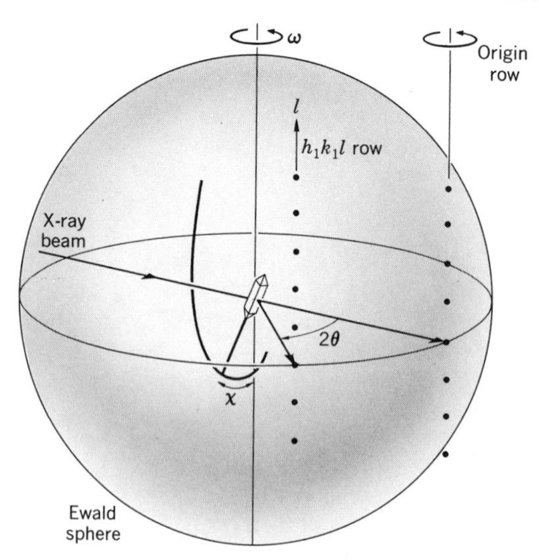

Fig. 13-22

with such an arrangement is illustrated in Fig. 13-22. The crystal is initially oriented so that densely populated rows of the reciprocal lattice are vertical (perpendicular to the equatorial plane). Considering only the upper part of the Ewald sphere in Fig. 13-22, suppose the rows are parallel to c^*, so that successive points along each row have the same h and k indices and differ only in l. The crystal then can be rotated by an amount ω about a vertical axis, until some row (h_1k_1) intersects the Ewald sphere at the equator. By inclining the crystal along the arc indicated by an appropriate amount χ, successive points h_1k_1l along this row can then be brought to lie on the equator as well. The basic motions of such an instrument are indicated in Fig. 13-23. The detector is moved into the appropriate 2θ position along the equatorial plane, and remains there while all the points along one row are examined. Because two angular motions, χ and ω, are required to orient the crystal and one, 2θ, to position the detector, instruments making use

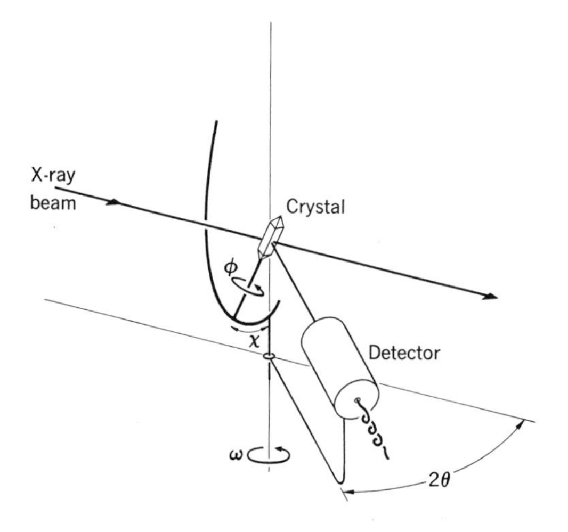

Fig. 13-23

Fig. 13-24. Furnas-Harker three-circle diffractometer. *(Courtesy of General Electric Company.)*

of such an arrangement are commonly referred to as *three-circle diffractometers.* Quite often, the ϕ circle is also utilized, so that such instruments are called *four-circle diffractometers.* The first commercially manufactured instrument embodying this principle is shown in Fig. 13-24.

Semiautomatic instruments. Suppose an orthorhombic crystal having a "typical" cell dimension of 10 Å is to be examined on a diffractometer, using molybdenum radiation. Even though the reflections lying in one octant only need be considered, approximately 1,400 reflections have to be measured. Assuming a minimum of 5 to 10 min to measure each reflection, one can calculate the time needed to complete the measurements. When crystals having much larger cells and lower symmetry are being studied, the time required increases further. It is natural that means to automate the procedure were sought as soon as the practicality of single-crystal diffractometry became apparent. It is interesting to note that automation of the original Bragg spectrometer was attempted by W. A. Wooster and A. J. P. Martin as early as 1936. The inefficiency of the ionization detectors available at that time, however, precluded further developments.

An examination of the reciprocal-lattice construction of a single crystal suggests immediately that a geometrically simple instrument should be able to analogize the motions necessary to place a crystal in various reflection positions and a detector so as to receive them. U. W. Arndt and D. C. Phillips devised an instrument in which the reciprocal-lattice coordinates a^*, b^*, and c^* can be set on three suitably linked slides, after which their interplay automatically generates all necessary crystal-setting coordinates, ha^*, kb^*, lc^*. In its most efficient operating mode, the instrument is set to scan, continuously or at reciprocal-lattice points only, along a densely populated reciprocal-lattice row. Upon completing one row, it shifts over to a parallel row automatically, thereby examining a two-dimensional layer. The dimensions of the mechanical components used in a commercially manufactured version of this instrument limit its scanning range to reflections having diffraction

Fig. 13-25. Norelco single-crystal diffractometer system. The diffractometer, shown on the right, incorporates equi-inclination geometry (arcs) and linear geometry (vertical and horizontal bars on extreme right). The electronic consoles on the left encompass all programming and intensity-recording controls. *(Courtesy of Philips Electronic Instruments.)*

angles $2\theta < 65°$, but this is usually not too serious a limitation in practice. The instrument must be reset manually for each successive layer, but this is virtually the only attention it requires from an operator, except for periodic tests of its continuing performance. These tests consist in checking the crystal alignment and the stability of electronic circuitry, usually by measuring certain reference reflections.

When developing the equi-inclination instrument shown in Fig. 13-21, Buerger envisaged the desirability of semiautomatic operation and incorporated external drive shafts for both the crystal rotation axis and the detector arm's rotation (about the same axis). At first an analog device was contemplated for setting the two drive shafts by remote control. The rapid concurrent developments in machine-tool automation suggested that similar devices could be used to operate the diffractometer also. Briefly, the necessary angular coordinates for setting a crystal and the detector are precalculated for each reciprocal-lattice level. This information, stored on cards or tapes, is then fed to an electronic device that translates it into electrical signals that operate two drive motors. The operator sets the instrument manually for each reciprocal-lattice level, but the individual reflections within each level are measured automatically.

The equi-inclination geometry can be combined with the linear scanning geometry devised by Arndt and Phillips, since, in both cases, one reciprocal-lattice layer is being examined at a time. An instrument that encompasses both of these aspects is shown in Fig. 13-25. It retains the inherently simpler mechanical linkages of the equi-inclination instrument, so that it can be built more easily to produce very accurate crystal settings. At the same time it utilizes the strategy of scanning along noncentral reciprocal-lattice rows (rather than simply rotating two concentric shafts), so that the scanning strategy for a number of x-ray diffraction problems in which the intensity distribution between reciprocal-lattice points is of interest can be easily programmed. When crystal-monochromatized x radiation is employed, the time spent in going from one reflection to the next can be utilized to measure the background intensity in the normal operating mode. This fact, as well as some of the other advantages of using crystal monochromatization cited in the closing section of this chapter, was recognized by its designer, J. Ladell, and the instrument shown in Fig. 13-25 is normally supplied with a suitable monochromator.

Fully automatic instruments. As the reader has surely realized by now, it is also possible to automate the four circles in an arrangement embodying the motions indicated in Fig. 13-23. Because the detector remains in the horizontal plane for all reflections in this method, the automation of four-circle diffractometers received its first impetus from crystallographers interested in neutron diffraction, where the detectors and shielding required are too cumbersome to permit easy vertical mobility. At present, a relatively large number of different kinds of three- and four-circle diffractometers are available commercially. They are distinguishable primarily by the kinds of motors, drive shafts, etc., that they employ, as well as by the attendant equipment for setting the angles and carrying out the intensity measurements that they encompass. Figure 13-26 shows one such instrument, which is unique in that it uses Moiré fringes to make the angular settings, rather

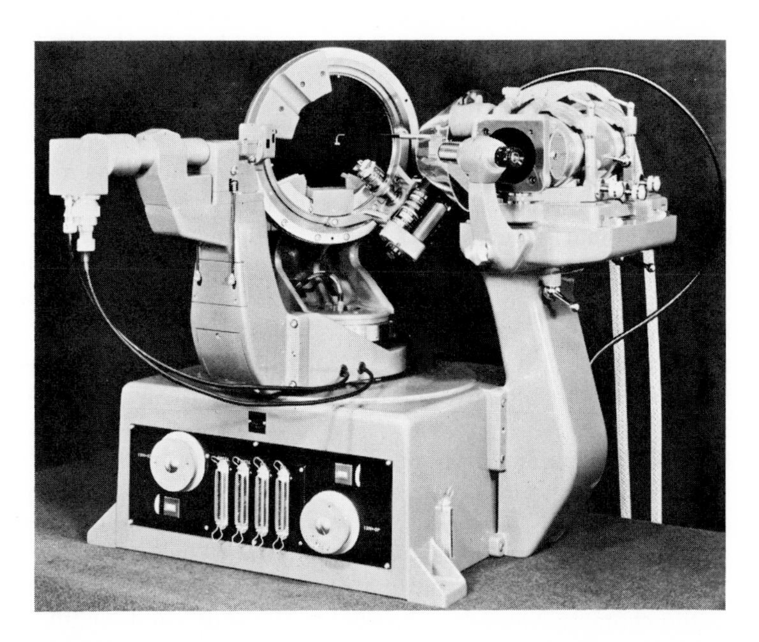

Fig. 13-26. Hilger four-circle diffractometer. *(Courtesy of Engis Equipment Company.)*

than relying on the positioning accuracy of drive shafts or gears. Each circle is engraved with 3,600 regularly spaced lines (the detector circle has only 1,800 lines) and passes by a measuring head containing four silicon photocells behind a suitably engraved grating. As the circular grating passes the fixed one, light directed through the two gratings sets up a Moiré-fringe pattern, which moves across the four photocells. The angular settings are made by counting the appropriate number of fringes passing by the photocells, and are independent of the mechanical components responsible for achieving the settings.

The operation of four-circle diffractometers employs the following strategy: Utilizing high-speed computers, the information necessary to make individual crystal (and detector) settings is precomputed for each crystal. This information, along with operating instructions and identifying labels (usually the hkl indices), is stored on cards or tape and then fed through a suitable "interface" to the electronic control mechanism of the diffractometer. Using one of the intensity-measuring procedures described in the next section, the intensity of each reflection is obtained (along with background measurements) and recorded on cards or tape, ready to be processed. It has become common practice, in fact, to incorporate, in the output, instructions for carrying out various operations on the intensities, so that, when the output information is processed by a computer, it ultimately provides the operator with structure-factor magnitudes suitable for direct interpretation. This procedure is commonly called an open-loop operation, because the operator must initiate it by supplying the input data, and he terminates it by accepting the output. If any part of the final data requires redoing, the operator then can recycle those operations.

Since the entire operation of the diffractometer is controlled automatically by the input program, it is possible and highly advisable to incorporate into the program a number of periodic tests of its performance. With the proliferation of very-high-speed computers, which can be used on a time-sharing basis, it also is becoming practical to consider closed-loop operation of diffractometers. In this operating mode, each reflection is treated essentially as a separate event. The measured intensity (output) is processed in a prescribed manner, and a decision is made (input) whether to go on with the next measurement or to repeat the previous one after making some indicated adjustment. The extent to which automatic diffractometry can be made impervious to instrumental slips or other mishaps then is limited only by the investigator's ingenuity in devising suitable tests and corrective procedures. Closed-loop operation has been used successfully for several years already, and its increased application can be forecast readily.

Intensity measurements. Each stack of parallel planes (hkl) in a single crystal reflects the x rays incident on it at appropriate angles. If a polychromatic beam (filtered or not) is employed, the intensity distribution in reciprocal space can be visualized as radial streaks of finite extent, directed along central reciprocal-lattice rows and having maxima near the reciprocal-lattice points corresponding to the characteristic peaks present. The width and length of these streaks are determined by the spectral range of the tube target, by beam divergence, and by crystal perfection, as discussed in earlier chapters. In order to establish the correct intensity of the "background" onto which the characteristic reflection is superimposed, it is necessary to consider how much of it is recorded by the detector during each measurement. This means that the way a reflection is measured determines the way the appropriate background intensity should be sought. One can recognize essentially three kinds of measurement strategies:

1. *Fixed crystal and fixed detector:* The crystal is set to reflect the maximum intensity and the detector to receive it.
2. *Moving crystal and fixed detector:* The crystal is rotated slowly through its reflecting range about the diffractometer axis (ω scan), while the detector is stationary at the appropriate 2θ angle.
3. *Moving crystal and moving detector:* The crystal and detector are suitably coupled for each reflection, and both rotate about the diffractometer axis in unison (ω, 2θ scan).

Fixing the crystal and detector at the appropriate settings places the most stringent demands on any diffractometer's operation. It is a reliable method for measuring single-crystal intensities, provided that the crystal is irradiated by a convergent beam from a very uniform source. If the crystal "sees" the entire source at all reflecting angles, the peak height is directly proportional to total intensity, obtained by integrating over the entire reflection. A fairly uniform "plateau" in the reflection profile can be achieved also by using a bent-crystal monochromator, as shown in Fig. 13-27. A suitably ground and bent crystal aids in uniformalizing the convergent beam, so that fairly uniform "flat-topped" reflections can be obtained in practice. As already stated, the accuracy of this mode of operation is closely linked to the x-ray source uniformity.

The two operating modes involving a rotating crystal can be compared by considering their respective reciprocal-lattice constructions, shown in Figs. 13-28 and 13-29. The ω, 2θ scan is practical only in the normal-beam arrangements, so that the intensity distribution along a central reciprocal-lattice row in the zero level is portrayed in these figures. Considering Fig. 13-28 first, it is seen that rotation of the crystal about the diffractometer axis (ω rotation) causes the intensity distribution associated with a reciprocal-lattice point to intersect the Ewald sphere first at some point indicated by the ray I_1. As the rotation continues, the entire intensity distribution, up to I_2, can be detected by a stationary detector, provided it has a

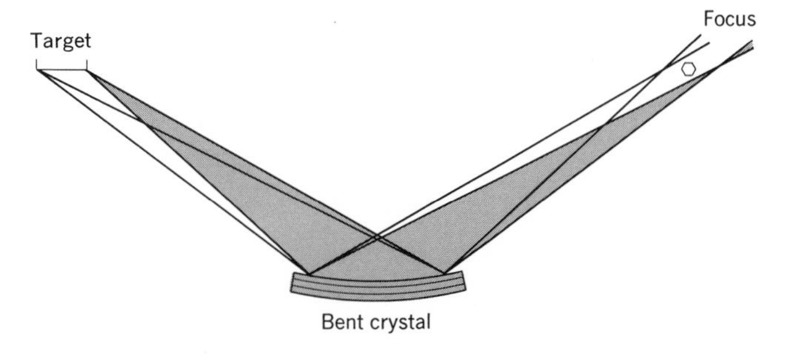

Fig. 13-27. Bent-crystal parafocusing of target on a specimen crystal. Provided that the crystal is irradiated by a uniformly intense beam, the reflected beam has a uniform plateau-shaped top.

sufficiently large receiving surface (slit). Alternatively, a much narrower slit can be placed at the 2θ value of ray I_1, and, by tracking the crystal with the usual $2:1$ coupling, it will receive each reflected ray in turn, until the position of I_2 is reached. Note that in this case the detector always measures the intensity of some spectral component of the reflection, so that, if the true reflection range lies between I_1 and I_2, intensity measurements just past the end points can be used conveniently to establish the correct background intensity. In the arrangement where the detector is stationary (Fig. 13-28), rotating the crystal out of the reflecting range also removes from reflecting position any portion of the central reciprocal-lattice row. Since this row contains contributions from higher-order (and lower-order) reflections and other sources, an error in estimating the background intensity is introduced thereby. To illustrate the nature of the problem, Fig. 13-30 compares the intensity distribution in the zero level of a reciprocal lattice when a crystal-monochromated beam is used and in the absence of a monochromator.

When the equi-inclination arrangement is used, the ω, 2θ scan is not practical, since the correct relative motions are possible only in the zero level (Exercise 13-8). This means that the use of a crystal monochromator is highly desirable, so that the reflected intensity distribution is confined to a small volume about each reciprocal-lattice point. The use of a monochromator also serves to uniformalize the background intensity distribution (Fig. 13-30) and, by improving the relative peak-to-background ratio, generally facilitates intensity measurements.

The correct assessment of the background intensity is as important as the measurement of the reflection intensity at the Bragg angle, because the actual value sought is the difference between them. This is largely the reason for the

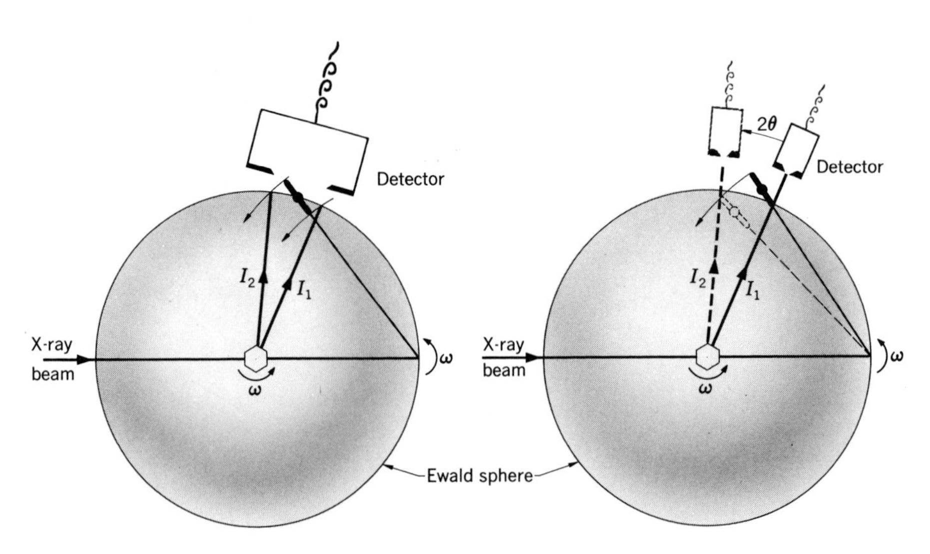

Fig. 13-28. ω scan.

Fig. 13-29. ω, 2θ scan.

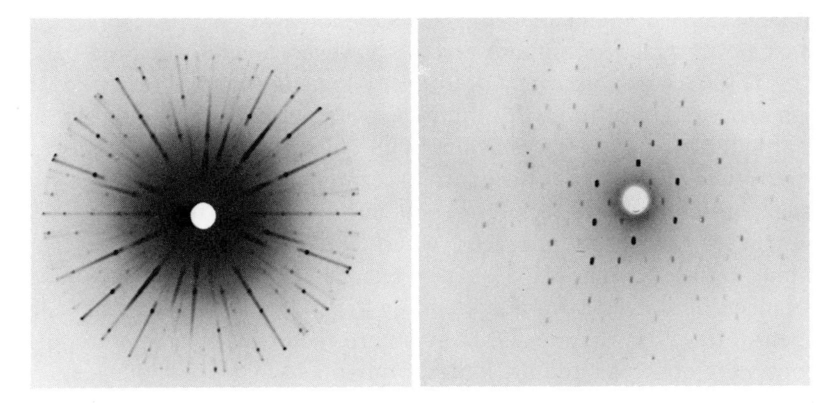

Fig. 13-30. Photographs of the zero level of the reciprocal lattice of a triclinic crystal recorded with a Buerger precession camera (Chap. 16). The appearance of the intensity distribution when a filtered beam is used is shown on the left. The effect of a monochromator arrangement like Fig. 13-27 is shown on the right. [*From L. V. Azároff, Acta Crystallographica, vol.* 10 (1957), *pp.* 413–417.]

popularity of balanced filters in single-crystal diffractometry. Combined with appropriate pulse-height discriminators, a fairly high degree of spectral purity can be attained. Since the desired reflection is the difference between the intensities transmitted through two different filters, separate background corrections are not normally necessary. For good accuracy it is essential, however, that exactly the same scanning ranges be used for both measurements. On most diffractometers this means that the same scanning direction should be employed to minimize missettings due to gear backlash, etc., a particularly important consideration when the ω, 2θ scanning mode is used.

By far the most satisfactory method for measuring intensities, in principle, is one employing a crystal monochromator. (An additional polarization correction is required for a twice-reflected beam, but this is a well-known function of the diffraction angle.) By eliminating virtually all background not due to the scattering of characteristic radiation, it eliminates the possibility of mistaking a reflection of the more intense portion of the continuous spectrum for a real reflection, a potential source of difficulty with fully automatic instruments. The presence of a monochromator does require that the sample crystal remain accurately centered, so that its "view" of the x-ray source is not changed by the goniometer's motions. For maximum accuracy, the sample crystal should have a simple and reasonably uniform shape, so that the volume of the crystal contributing to various reflections is the same. This is really a matter of relative absorption, and suitable corrections can be incorporated into the data-reduction programs. Basically, the same requirements must be met when filtered radiation is employed, except that the presence of a monochromator allows one to evaluate any flaws present more precisely because of the enhanced peak-to-background ratio. It is of course possible to place the monochromator following the crystal and ahead of the detector. In this case some of the mechanical stringencies are relaxed, but a possible difficulty may arise because of the increased $\alpha_1\alpha_2$ dispersion at large diffraction angles. Suitable choice of the monochromator crystal, however, should eliminate this objection.

The improved accuracy of intensity measurement inherent in diffractometry has aroused concern about a number of factors affecting intensity measurement that had been largely ignored when film detection was employed exclusively. The occurrence of multiple reflections, for example, invariably present in many upper-level measurements, whether equi-inclination or normal-beam geometry is employed, has been observed when crystallographically equivalent reflection intensities are compared. (See also Exercise 13-10.) In a comparison study using balanced filters and several spherically ground sodium chloride crystals, errors of about 3 to 4% in the measured structure factor magnitudes of equivalent reflections were observed and attributed primarily to instrumental causes. In another comparison of symmetry-related reflections, the intensities were found to vary by less than 2% on the same SiC crystal, but as much as 9% when the values obtained, using two crystals, were compared. This points out what is probably the principal contribution that automatic diffractometry can make to accurate intensity measurement, namely, the gathering of comparison data from several presumably equivalent crystals. In the past, such studies had not been attempted, because of the prohibitive attendant labor. The relatively short times in which an automatic diffractometer can complete the examination of each crystal (now reckoned in days instead of weeks) and the rapidity with which the raw intensities can be converted into suitably corrected structure-factor magnitudes will make it possible to base crystallographic studies on several sets of equivalent, but independently determined, sets of data. Only in this way will it be possible to establish the true physical significance of the results and to achieve accuracy estimates having a validity comparable with that taken for granted in other kinds of physical measurements.

SUGGESTIONS FOR SUPPLEMENTARY READING

U. W. Arndt, Analogue and digital single-crystal diffractometers, *Acta Crystallographica*, vol. 17 (1964), pp. 1183–1190.

——— and B. T. M. Willis, *Single crystal diffractometry* (Cambridge University Press, New York, 1966).

Martin J. Buerger, *Crystal-structure analysis* (John Wiley & Sons, Inc., New York, 1960), pp. 112–146.

Harold P. Klug and Leroy E. Alexander, *X-ray diffraction procedures* (John Wiley & Sons, Inc., New York, 1954), pp. 235–318.

William Parrish (ed.), *Advances in x-ray diffractometry and x-ray spectroscopy* (Centrex Publishing Co., Eindhoven, The Netherlands, 1962).

———, *X-ray analysis papers* (Centrex Publishing Co., Eindhoven, The Netherlands, 1965).

EXERCISES

13-1. By reference to Fig. 13-1 and using plane geometry, prove that rotation of a central row of the reciprocal lattice (one passing through the origin) by an amount θ causes its point of intersection with the Ewald sphere to move an amount 2θ about the center of the sphere.

13-2. By reference to Fig. 13-3 prove that the Bragg-Brentano parafocusing condition is obeyed, provided that the angular displacement of the reflecting plane δ has a magnitude comparable with the beam divergence angle.

13-3. In a typical diffractometer arrangement (Fig. 13-11), the value of R is chosen to be 170 mm and the specimen length $l = 20$ mm. For divergence angles of 0.5, 1.0, and 4.0°, calculate the minimum glancing angle ($2\theta_{\min}$) possible without exceeding the sample's dimensions. For each of these three cases also calculate the maximum d value that a plane can have and still reflect x rays when Mo $K\alpha$ radiation ($\lambda = 0.71$ Å) and Cu $K\alpha$ radiation ($\lambda = 1.54$ Å) are employed.

13-4. Consider a polycrystalline slab of infinite thickness and incident and reflected rays each of which forms the angle θ with its top surface. Derive an expression for the total diffracted intensity and for a parallel incident beam with a cross section 1 cm², and show that it is inversely proportional to the linear absorption coefficient μ and independent of the diffraction angle.

13-5. Suppose an orthorhombic crystal having $a = 7.46$ Å, $b = 9.68$ Å, and $c = 5.42$ Å is to be examined on a four-circle single-crystal diffractometer with Cu $K\alpha$ radiation ($\lambda = 1.54$ Å). If the crystal has a face-centered lattice, how many reflections would you have to measure to detect all reflections that are observable with copper radiation? With Mo $K\alpha$ ($\lambda = 0.71$ Å)? With Ag $K\alpha$ ($\lambda = 0.56$ Å)?

13-6. How will you mount the crystal in Exercise 13-5 if you are going to measure all the reflections manually, using a three-circle goniometer to orient the crystal? Assuming that, when all dials are set to zero, the crystal is oriented so that one of its cell edges coincides with the diffractometer's rotation axis while another is parallel to the x-ray beam, by how much must you rotate the crystal along the Eulerian angles (Fig. 13-23) in order to bring the (222) plane into reflection position? At what angle 2θ must the diffractometer be set to record this reflection?

13-7. How will you mount the crystal in Exercise 13-5 if you are planning to use the equi-inclination arrangement (Fig. 13-20) for measuring upper levels? At what angle μ will you incline the diffractometer axis to examine the first upper level if the crystal has a primitive lattice? A body-centered lattice?

13-8. Show by suitable cross sections of the Ewald sphere why it is not possible to employ the ω, 2θ scan for measuring upper-level reflections in the equi-inclination arrangement.

13-9. In the equi-inclination arrangement, it is possible to arrange the inclination angle so that two levels of the reciprocal lattice intersect the Ewald sphere concurrently. Using a suitable drawing, portray the reciprocal-lattice construction for this case, and suggest how a pair of detectors might be disposed to measure two reflections at the same time.

13-10. By drawing the reciprocal-lattice construction, illustrate how multiple reflections may arise in the equi-inclination and in the normal-beam arrangements.

FOURTEEN
Laue method

RECIPROCAL-LATTICE CONSTRUCTION

In the discussion of x-ray diffraction so far, it has been convenient to construct the reciprocal lattice on a module determined by (8-1) and to vary the radius of the sphere of reflection as a function of the wavelength of the incident x-ray beam. In many practical applications of x-ray diffraction theory, however, it is often more convenient to fix the radius of the sphere of reflection and to vary the size of the reciprocal lattice in proportion to the x-ray wavelength. As will become evident in the ensuing chapters, this has the advantage of displaying more clearly the effect of using radiations having different wavelengths, and also facilitates the comparisons between the x-ray diffraction effects obtained from different crystals on the same instrument. Returning to the important diffraction condition (8-35), it is clear that the dependence on wavelength can be transferred to the reciprocal-lattice vector by multiplying both sides of the expression by λ.

$$(\mathbf{S} - \mathbf{S}_0) = \lambda \boldsymbol{\sigma}_{hkl} \tag{14-1}$$

It should be noted that, with the reciprocal-lattice vector $\boldsymbol{\sigma}_{hkl}$ defined in (8-1) as being proportional in magnitude to $1/d_{hkl}$, it is possible to incorporate the wavelength directly into the reciprocal-lattice vector by setting the proportionality constant in that equation $K = \lambda$. The reciprocal-lattice construction for a plane in reflecting position now appears as shown in Fig. 14-1, and can be compared directly with Fig. 8-8.

Unlike the case depicted in Fig. 14-1, the wavelength of the x-ray beam employed in the Laue method does not have a single value since the entire polychromatic beam emanating from an x-ray target is used. As already pointed out in Chap. 8, the most important portion of the continuous spectrum is the large hump extending from λ_{swl} to approximately twice that value. [The characteristic line, although far more intense, is of no practical interest in a stationary crystal because it is unlikely that more than one plane can satisfy the reflection condition (Fig. 14-1) for a fixed wavelength value.] Incorporating the wavelength in the reciprocal lattice itself as in (14-1), each reciprocal-lattice point becomes a line ranging from

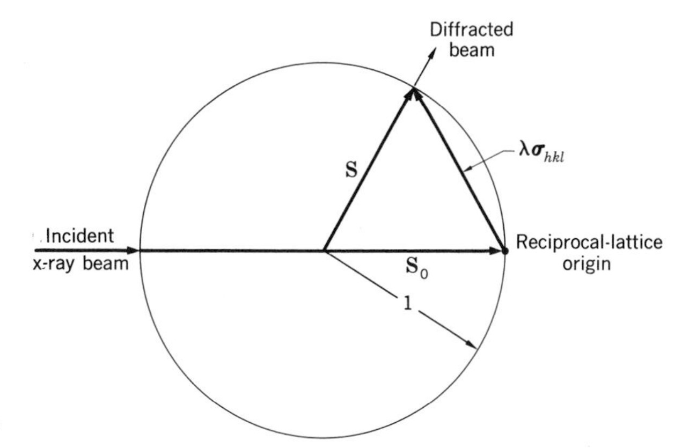

Fig. 14-1

$\lambda_{swl}\boldsymbol{\sigma}_{hkl}$ to $\lambda_{max}\boldsymbol{\sigma}_{hkl}$, as shown in Fig. 14-2. Note that all the resulting reciprocal-lattice lines point radially outward from the reciprocal-lattice origin, as do the vectors determining their limits.

As demonstrated in Fig. 14-1, the diffraction condition is satisfied whenever a reciprocal-lattice point lies on the sphere of reflection, so that a construction such

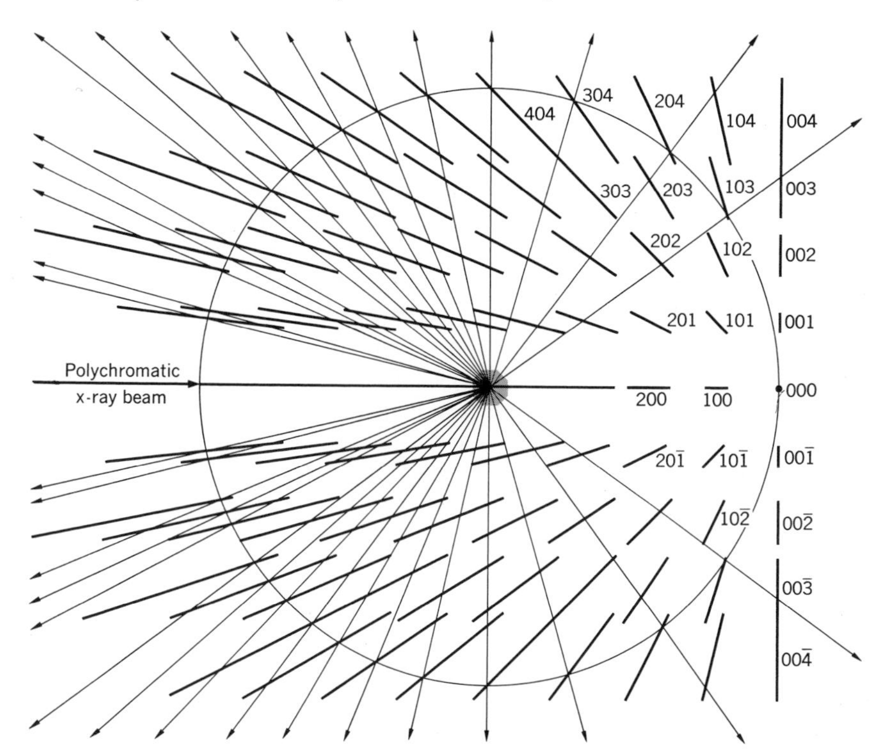

Fig. 14-2. Reciprocal-lattice construction for the Laue method with some of the reciprocal-lattice streaks indexed. Note that the third and fourth orders begin to overlap for such indices as $h00$, $00l$, $h0l$, etc. Whenever the Ewald sphere intersects a reciprocal-lattice streak, a reflected ray emanates from the crystal, as indicated by the arrows.

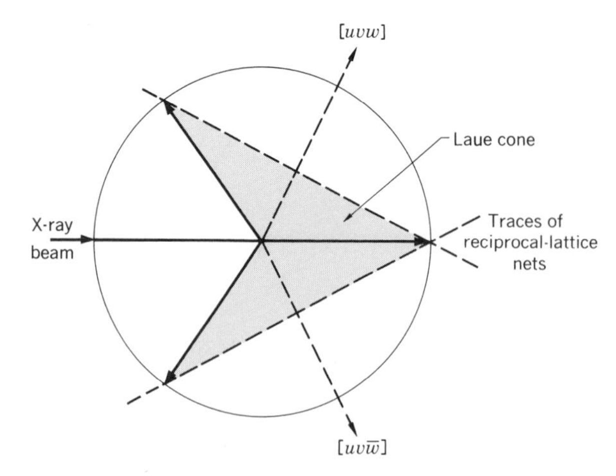

Fig. 14-3. Two symmetrically disposed Laue cones produced when two symmetry-equivalent zone axes form the same angle with the direct-beam direction.

as Fig. 14-2 can be used to determine directly which planes in the crystal are in reflecting position for some component of the incident x-ray spectrum. (The diffracted beams, of course, all originate from the center of the sphere of reflection, where the actual crystal is located.) The reciprocal-lattice construction shown in Fig. 14-2 also illustrates that the reciprocal-lattice points of a set of planes belonging to a common zone lie on a plane that passes through the reciprocal-lattice origin, so that it cuts the sphere of reflection along some circle. The diffracted beams passing through the reciprocal-lattice points along this circle (on the sphere) form a cone, as shown in Fig. 14-3. The axis of the cone is normal to the reciprocal-lattice plane, and therefore it is parallel to the corresponding zone axis $[uvw]$.

When a symmetry element in the crystals is parallel to the incident x-ray beam, the diffraction cones surrounding the beam are symmetric, and their spatial disposition corresponds to the symmetry element. On the other hand, if the crystal is tilted slightly so that symmetry-equivalent zone axes in the crystal no longer form equal angles with the x-ray beam, the cones are no longer equivalent, as can be

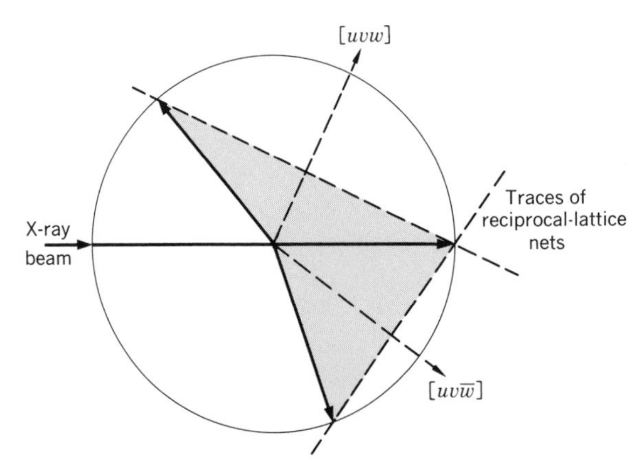

Fig. 14-4. Two asymmetrically disposed Laue cones.

seen in Fig. 14-4. It is clear from this discussion that the Laue method can be used to determine the symmetry of a crystal and its orientation relative to the x-ray beam. Since it is desirable for these purposes to record a large number of reflections about the direct-beam direction, it is usual practice to place a sheet of film at right angles to the beam either in the front- or in the back-reflection region.

INSTRUMENTATION

Flat-plate cameras. All that is necessary to record the diffraction cones produced in the Laue method is an x-ray beam, a single crystal, and a flat film. Figure 14-5 shows a commercial instrument suitable for recording both the front- and back-reflection regions. The x-ray beam should include an intense continuous radiation spectrum, so that heavy-metal targets (W, Pt, Ag, Mo) are preferable and maximum operating voltages are recommended. The crystal-to-film distance is made quite small and is usually fixed at 3 or 5 cm, so that published charts can be used in the indexing procedures discussed below. The shorter distance has two advantages; namely, it decreases the exposure time and increases the number of reflections intercepted by the film. Note that it is not easy to measure the crystal-to-film distance precisely, because the film is inside a light-tight cassette, because it is usually difficult to place a ruler directly up against the crystal, and so forth. When reproducibly accurate distances are required, therefore, it is advisable to establish the desired distance first by some auxiliary procedure. Subsequently, a feeler gauge of appropriate length can be placed between two machined surfaces, respectively, on the film holder and on the crystal holder, and the desired distance is reproduced by pushing them in contact with each other along the camera track.

The film is placed in a flat cassette, which should be light-tight. A proper design for the cassette is shown in cross section in Fig. 14-6. Note that the cassette frame and the back plate have indentations around their edges which permit them to fit together in such a way that light cannot leak in around the edges. (The cassette should be painted a dull black to minimize light scattering.) An

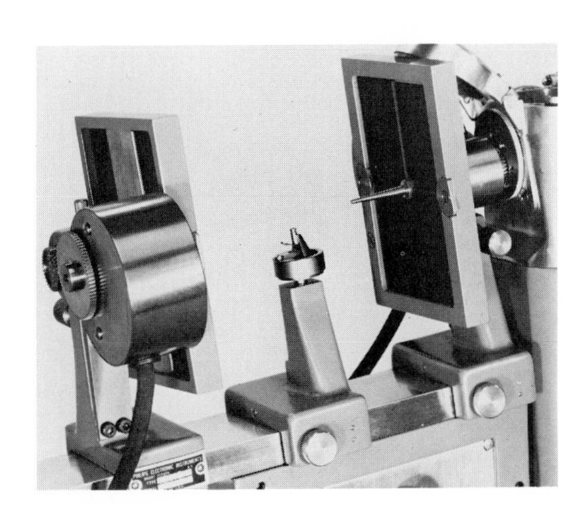

Fig. 14-5. Front- and back-reflection cassettes for the Laue method. *(Courtesy of Philips Electronics Instruments.)*

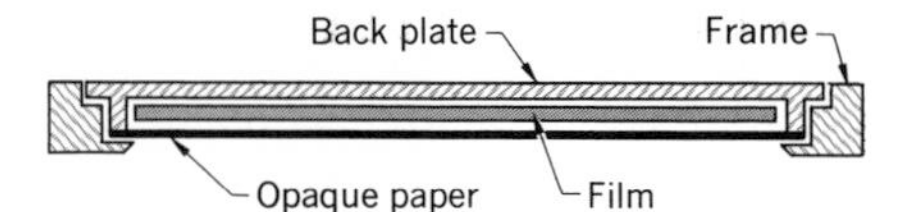

Fig. 14-6

opaque sheet of paper is placed in front of the film, and it may be held in place by the pressure of the back plate or it may be cemented in place permanently. Both the front- and back-reflection cassettes are constructed this way. They differ in one important respect, however; a small lead cup is attached centrally on the front-reflection cassette to intercept the incident beam, whereas a central opening must be provided in the back-reflection cassette to permit the x-ray beam to pass through. Alternatively, the Land cassette (Fig. 12-18) and Polaroid film can be used to great advantage in the Laue method. The diffraction diagrams presented in this chapter actually were prepared in this way.

Collimating system. The main function of the collimating system is to define a narrow pencil of x rays and to prevent stray radiation from striking the film or leaking into the room. It is common practice to use a series of circular apertures to form a *pinhole system* or a combination with rectangular openings to form a *slit system*. The principle of such systems, popularly called *collimators*, is illustrated in Fig. 14-7. The radiation emanates from the x-ray tube target equally in all directions, so that two apertures, 1 and 2, commonly in the form of holes drilled in lead, serve to define the main pencil of x rays. Because the focal spot on the target usually is larger than the pinholes in the apertures, rays coming from other parts of the target (dotted lines) also can pass through. These rays do not strike the crystal, however, so that the effective beam diameter is determined by the two defining apertures and the crystal size. Note that the x-ray beam is diffracted by aperture 2, and becomes superimposed on the diffraction effects to be studied. In order to prevent these diffracted rays from reaching the film, a third aperture, 3, is added to the sequence. This hole is designed to just pass through the x-ray beam defined by apertures 1 and 2, but to intercept the smallest cone of diffracted radiation arising from aperture 2. The relative sizes and shapes of the x-ray

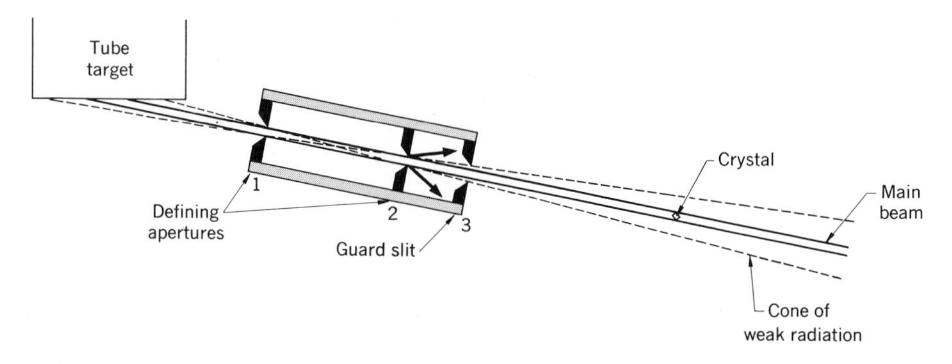

Fig. 14-7

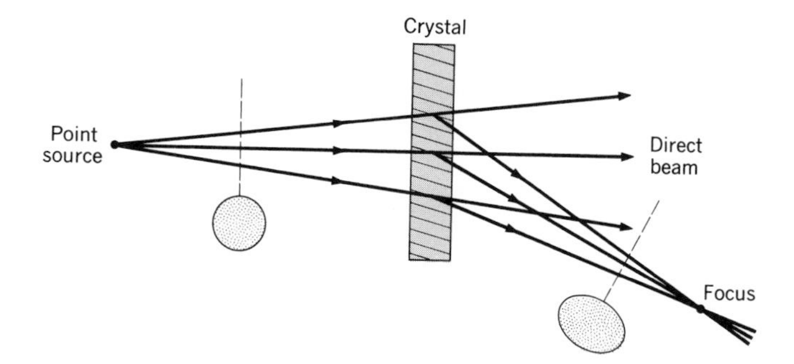

Fig. 14-8

source, collimator apertures, and crystal serve to determine the size and shape of the diffracted-beam cross sections, as described next.

Shape of reflections. It is left to the reader to prove that a truly parallel beam of x rays having a cross section smaller than the crystal is reflected at all angles without change in its cross-sectional size or shape (Exercise 14-4). As can be seen in Fig. 14-7, however, it is not possible to obtain a parallel beam of x rays by simple collimation because the original source emits x rays uniformly in all directions. Consequently, there always exists a certain amount of cross fire. Reducing the defining apertures decreases the divergence while simultaneously decreasing the intensity, so that this approach to parallelism is self-defeating. To see exactly what effect the presence of diverging rays has on an x-ray reflection in the Laue method, consider a virtual point source of x rays and a large crystal, as shown in Fig. 14-8. The incident beam diverges uniformly, so that its cross section is circular, as shown to the left of the crystal. After reflection by a set of parallel planes, the rays in the plane of the drawing are focused (see also Fig. 10-19), while at right angles to this plane the rays continue to diverge. The result is that the reflected beam has an elliptical cross section. Note that the semiminor axis of the ellipse points radially outward from the center of the film, that is, from the direct-beam direction (Exercise 14-5).

By comparison, a converging beam of x rays (Fig. 14-9) is caused to diverge on reflection, while, at right angles to the focusing plane, it remains unaffected and continues to converge. This produces an elliptical cross section in the reflected beam whose semimajor axis points radially from the direct-beam direction (Exercise 14-6). In practice, the incident beam usually contains a slightly larger number of

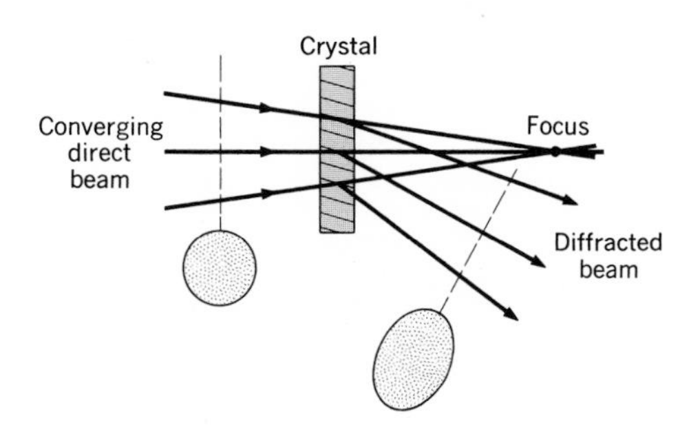

Fig. 14-9

diverging rays, so that the reflected beams deviate somewhat from being circular in cross section. This is less true when the crystal is smaller than the incident-beam cross section, in which case there tends to be an equal number of diverging and converging rays. It should be realized, however, that two other factors also affect the form of the reflected-beam shape as intercepted by the film. At increasing reflection angles, the reflected beam makes a more oblique angle with the film, so that an increasingly larger section is intercepted by the film. The other factor is the perfection of the crystal. Some of these effects have already been discussed in Chap. 10, while the effect of microstresses that may be present in a crystal are considered further in Chap. 20.

The foregoing discussion of the shape of x-ray reflections was limited to the transmission arrangement. The interest in these shapes stems primarily from their use as a guide to the perfection of a crystal. In back reflections, no focusing occurs, regardless of whether the incident beam is predominantly converging or diverging. The result is that back-reflection spots on the film are more or less circular, increasing in size radially outward from the center of the film.

FRONT-REFLECTION REGION

Appearance of photographs. It has been demonstrated in the introductory discussion in this chapter that the reflections from a set of planes belonging to the same zone form a cone about the zone axis (Fig. 14-3). These cones intersect the front-reflection photograph in ellipses (conic sections), as can be seen in Fig. 14-10. When the same crystal is displaced slightly from proper alignment with the direct beam, the ellipses become distorted, as shown in Fig. 14-11. (See also Fig. 14-4.)

Fig. 14-10. Transmission Laue photograph of a silicon crystal oriented with its [111] direction parallel to the incident beam. (Note slightly elliptical shapes of reflections.) (2-cm crystal-to-film distance)

Fig. 14-11. Transmission Laue photograph of same silicon crystal as that shown in Fig. 14-10, with [111] direction displaced by $10°$ from that of the incident beam (3-cm crystal-to-film distance).

It is possible, therefore, to use such photographs to orient single crystals. For this purpose, it is helpful to mount the crystal on two mutually perpendicular arcs such as those provided by the *goniometer* shown in Fig. 13-7. One additional degree of freedom can be provided by mounting the goniometer head on a shaft that allows rotation about the vertical goniometer axis. The orientation of the crystal can then be accomplished by successive trial-and-error adjustments, until symmetry such as is evident in Fig. 14-10 is produced, or by making use of the more elaborate projection methods described next.

Use of gnomonic projection. It has been shown above that the x-ray reflections from planes belonging to a common zone lie along the generators of a single cone in the Laue method. It was shown in Chap. 2 that the poles of a set of planes belonging to a common zone lie along a great circle in a spherical projection. Recalling also that all great circles project as straight lines in a gnomonic projection, it is clear that such a projection should prove useful in the interpretation of Laue photographs. In order to see the relationship between the gnomonic projection and a Laue photograph, suppose that the incident x-ray beam is parallel to the c

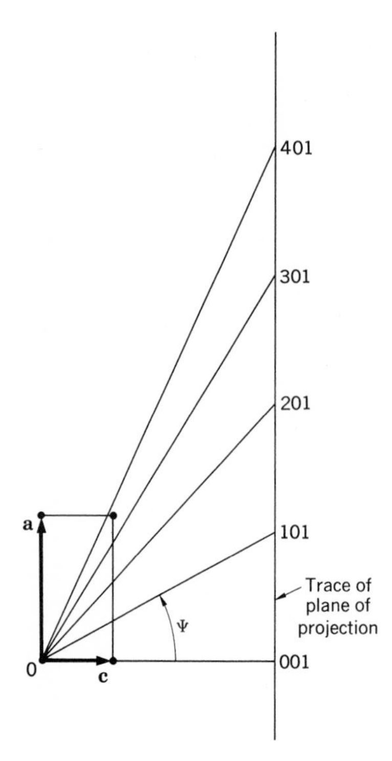

Fig. 14-12

axis of an orthorhombic crystal, while its a axis is vertical. Since the b axis is perpendicular to both, the $(h0l)$ planes belonging to its zone reflect along a vertical line on the film. For example, the $h01$ poles project in a gnomonic projection, as shown in Fig. 14-12. It is easy to see (Exercise 14-8), with the aid of a construction like that in Fig. 14-13, that

$$\cot \psi = \frac{a/h}{c} \tag{14-2}$$

The relationship between the reflecting plane, its normal, and the direction of the reflected x rays is illustrated in Fig. 14-14. If it is assumed that the photographic film and the plane of projection are coincident, it is possible to establish a simple

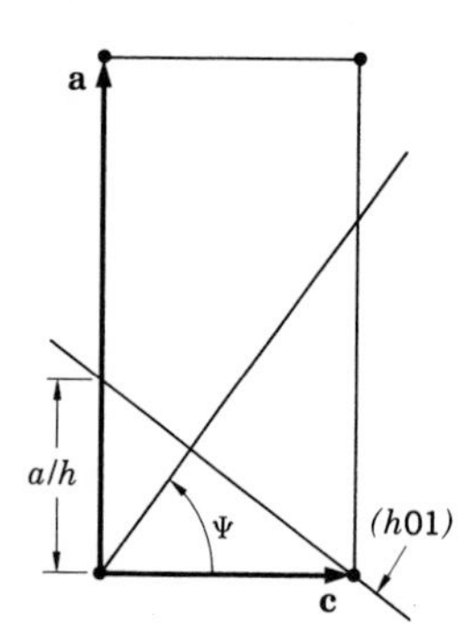

Fig. 14-13

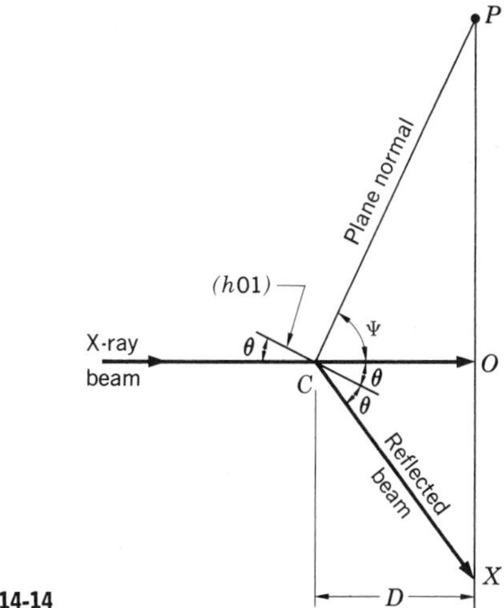

Fig. 14-14

relationship between the radial distance OX to an x-ray reflection spot X, and the radial distance OP to the projection of the corresponding pole P. According to Fig. 14-14,

$$\frac{OP}{CO} = \cot \theta \tag{14-3}$$

while $$\frac{OX}{CO} = \tan 2\theta \tag{14-4}$$

The distance CO is simply the crystal-to-film distance D, so that combining (14-3) and (14-4),

$$OP = D \cot \left(\frac{1}{2} \tan^{-1} \frac{OX}{D} \right) \tag{14-5}$$

On the other hand, noting that

$$\frac{OP}{D} = \cot \theta = \tan \psi = \frac{1}{\cot \psi} = h \frac{c}{a} \tag{14-6}$$

it follows that the distance to successive $h01$ poles, OP, will increase uniformly with a period $D \times (c/a)$. It should be noted that the poles of other $(h0l)$ planes also lie along the same vertical line in Fig. 14-14, but they do not have the above periodicity.

An examination of (14-5) suggests a simple means for constructing a gnomonic

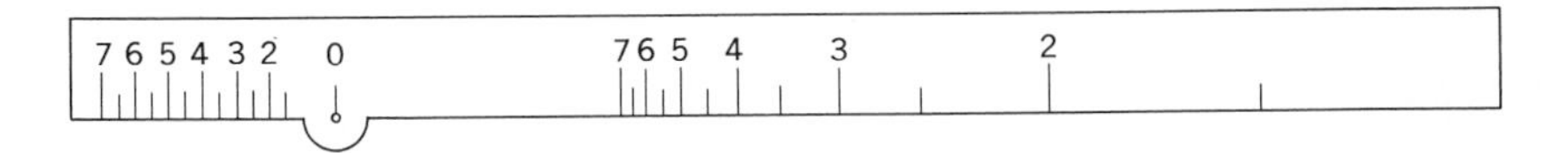

Fig. 14-15. Ruler for construction of gnomonic projections directly from Laue photographs. *(After Wyckoff.)*

projection directly from a Laue photograph. By fixing the specimen-to-film distance D at some convenient value, it is possible to construct a ruler (Fig. 14-15) on one side of which OX is measured while on the other side OP is given according to (14-5). The center of the ruler O is placed at the center of a Laue photograph, and the distance to a reflection is read off on the linear scale (to the left of O). The corresponding pole P (Fig. 14-14) then lies at the same distance marked along the nonlinear scale on the right of O. As an illustration of this procedure, Fig. 14-16 shows the gnomonic projection derived in this way from the Laue photograph of an MgO crystal shown at the center of the figure. It should be clear from the preceding discussion that the poles lie along straight lines. To determine the indices of the poles, it is necessary to know beforehand the orientation of the crystal relative to the x-ray beam. The photograph of the cubic MgO crystal in Fig. 14-16 was prepared with the x-ray beam parallel to a_3, while a_2 was vertical and a_1 horizontal. This means that the l index of the regularly spaced poles can be taken as unity. (See Fig. 14-12 and the accompanying discussion.) Since the a_2 axis is vertical, the $h01$ poles are regularly spaced along the central horizontal line. Similarly, the $0k1$ poles lie along the vertical central line. The indices of other $hk1$ poles can then be read off in a straightforward manner. To determine the indices of the other poles, one proceeds as follows: Consider the pole labeled R. Its h index obviously is 2, and its k index apparently is $\frac{\bar{1}}{2}$. Taking $l = 1$ and clearing fractions by multiplying all indices by 2 gives $4\bar{1}2$.

Although it is clear from Fig. 14-16 that the gnomonic projection can be used in the interpretation of transmission Laue photographs, the foregoing discussion has shown that the indexing of poles having $l \neq 1$ is somewhat laborious. The situation becomes further aggravated if a crystallographic axis is not initially truly paralle

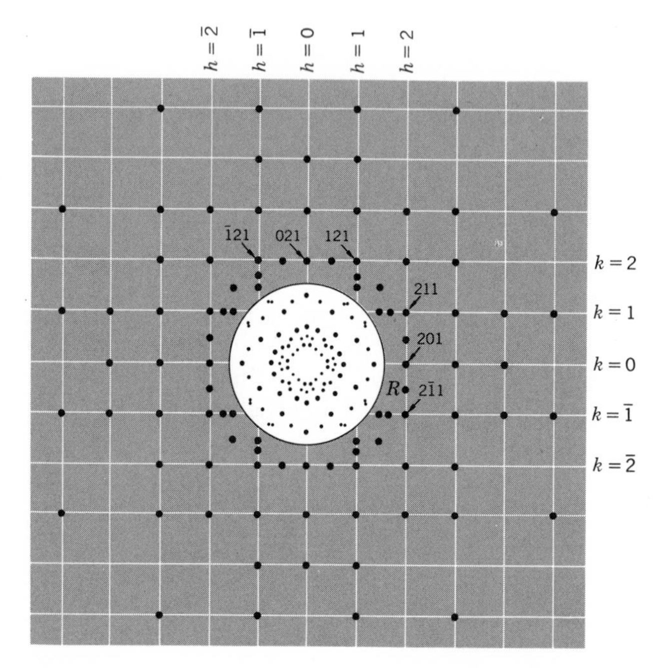

Fig. 14-16. Gnomonic projection constructed from a transmission Laue photograph of a cubic MgO crystal shown in the central circle. *(After Barrett.)*

to the x-ray beam, since then the regularity of the net evident in Fig. 14-16 becomes disrupted. Largely for this reason, the gnomonic projection is rarely used nowadays.

Use of stereographic projection. The reciprocal-lattice points corresponding to a zone of crystal planes lie on the same plane in the reciprocal lattice. Suppose such a reciprocal-lattice plane, tilted by an angle ϕ from the vertical, cuts the Ewald sphere, as shown in Fig. 14-17, thereby producing a set of reflection spots arrayed along an ellipse on the front-reflection photograph. Since the reciprocal-lattice vectors $\lambda\boldsymbol{\sigma}_{hkl}$ satisfying the reflection condition (14-1) must lie in this plane, the orientation of each reciprocal-lattice vector can be specified by the angle ϕ and an angle δ measured from the vertical within the plane. The reciprocal-lattice vectors being parallel to plane normals, the corresponding disposition of the plane normal within a reference sphere is shown in Fig. 14-18. Recalling from Chap. 2 that the normals of planes belonging to the same zone also must lie in a plane, it is clearly seen that this plane forms the angle ϕ with the vertical, while δ is then measured along the great circle which marks the intersection of this plane with the reference sphere. The zone axis of this zone, being perpendicular to the plane in which the normals lie, therefore must form the angle ϕ with the horizontal. The poles corresponding to the normals can be readily transferred to a stereographic net (Fig. 14-18) by noting the angle along the equatorial great circle and the respective values of each normal along the corresponding meridinal great circle, as shown. Similarly, the position of the pole P of the corresponding zone axis can be located next—$90°$ away from the meridinal trace.

A comparison of Figs. 14-17 and 14-18 suggests that it would be desirable to be able to measure the positions of reflections on a front-reflection photograph directly in terms of the two angular coordinates ϕ and δ required for a stereographic projection. As is seen in Fig. 14-17, the exact distance of a reflection spot from the center of the film (direct-beam trace) depends on the crystal-to-film distance. Except for this scale factor, however, the values of ϕ and δ of each point on a film can be readily deduced, so that a chart such as the one first prepared by J. Leonhardt and shown in Fig. 14-19 can be utilized for this purpose. The ϕ values are sketched in by the dashed curves. Note that they start out having the shape of ellipses, but become hyperbolas for $\phi > 45°$. The curves corresponding to positions having the same δ value are shown by solid lines in the Leonhardt chart. The protractor on the bottom half of the chart provides a reference mark to aid in the transfer of differently oriented ellipses (or hyperbolas).

The use of a Leonhardt chart is illustrated in Fig. 14-20. The photographic film, bearing an identifying mark to indicate its original orientation in the front-reflection cassette (letter F in upper left corner), is placed over the chart so that their respective centers superimpose. The film is then rotated through an angle ϵ until the ellipsoidal array of reflections from a diffraction cone coincides with a curve of constant ϕ. Concurrently, a transparent sheet overlaying a Wulff net is rotated by an amount ϵ in the same sense. The ϕ and δ coordinates can now be transferred from the chart to the net, as previously discussed in connection with Figs. 14-17 and 14-18. After this has been done, the film and projection overlay sheet are rotated by equal amounts, and the poles of another zone are plotted similarly. The final stereographic projection can then be used to index the individual poles or the zone axes by measuring the angles between their poles along some great circle on the stereographic net and by consulting a suitable table of interplanar angles. It will be recalled that one such table is sufficient for all cubic crystals (Table 2-1), since their interplanar angles do not depend on cell size. It follows from this that the indexing of noncubic crystals requires the prior knowledge of cell dimensions, so that interplanar angles can be precalculated.

It should be noted that a *stereographic ruler* similar to the one described for gnomonic projections also can be constructed (Exercise 14-9). Its use is analogous to that illustrated by Fig. 14-16, and generally, a more accurate plot is possible using such a ruler than a Leonhardt chart. Because the most common use for the Laue method has become limited to the orientation of single crystals, there is little need for such accurate projections, and the use of the chart is more rapid, particularly since individual poles are rarely plotted. As demonstrated in the next sections, the back-reflection photographs can be interpreted in a manner identical with that described in this section. Back-reflection photographs have the further advantage that crystals too large to yield adequate intensities in the transmission arrangement can be studied, as well as relatively large single crystals within a

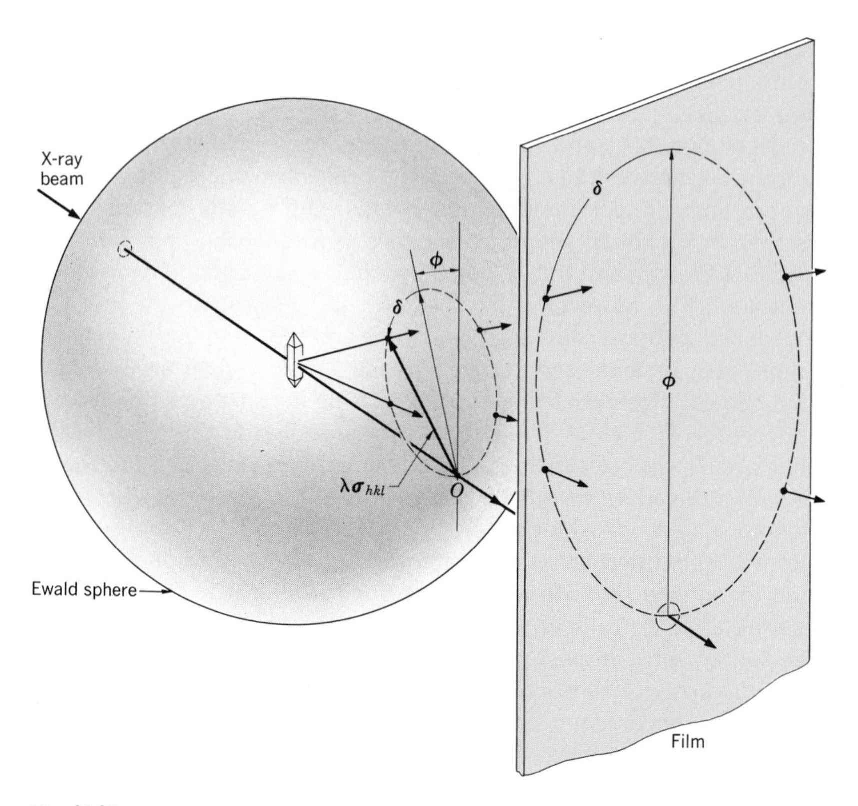

Fig. 14-17

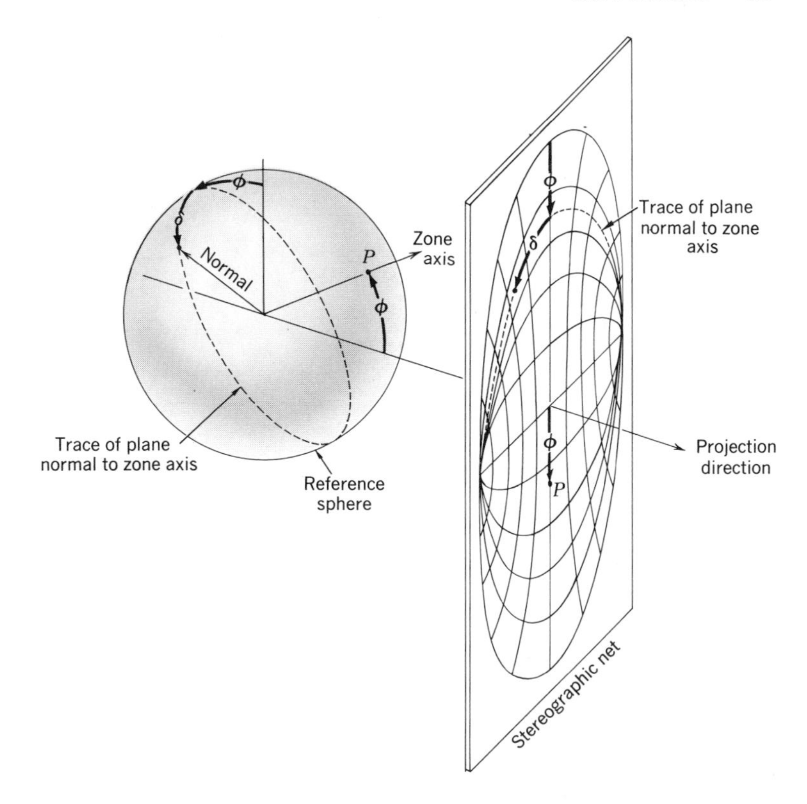

Fig. 14-18. Stereographic projection of normals belonging to the same zone. Note that ϕ for the zone axis and for the plane normal to it is measured in the same sense on the stereographic net.

multicrystal aggregate. For this reason, the parallel details of interpretation are discussed more fully below.

BACK-REFLECTION REGION

Appearance of photographs. When the angle ϕ that a zone axis forms with the incident-beam direction exceeds $45°$, the reciprocal-lattice plane that contains the reciprocal-lattice points for this zone intersects the Ewald sphere in the back-reflection region. As previously demonstrated, the diffraction cone produced intersects a film placed at right angles to the incident beam along a hyperbola (more accurately, a conic) in this case. An example of a back-reflection Laue photograph is shown in Fig. 14-21 in which some of the reflections belonging to a common zone are indicated.

As in the case of transmission Laue photographs, it is possible to relate the positions of individual reflection spots to the orientation of their plane normals (reciprocal-lattice vectors) directly. As shown in Fig. 14-22, the reciprocal-lattice plane forms the angle γ (in a vertical sense) with the incident x-ray beam. This is

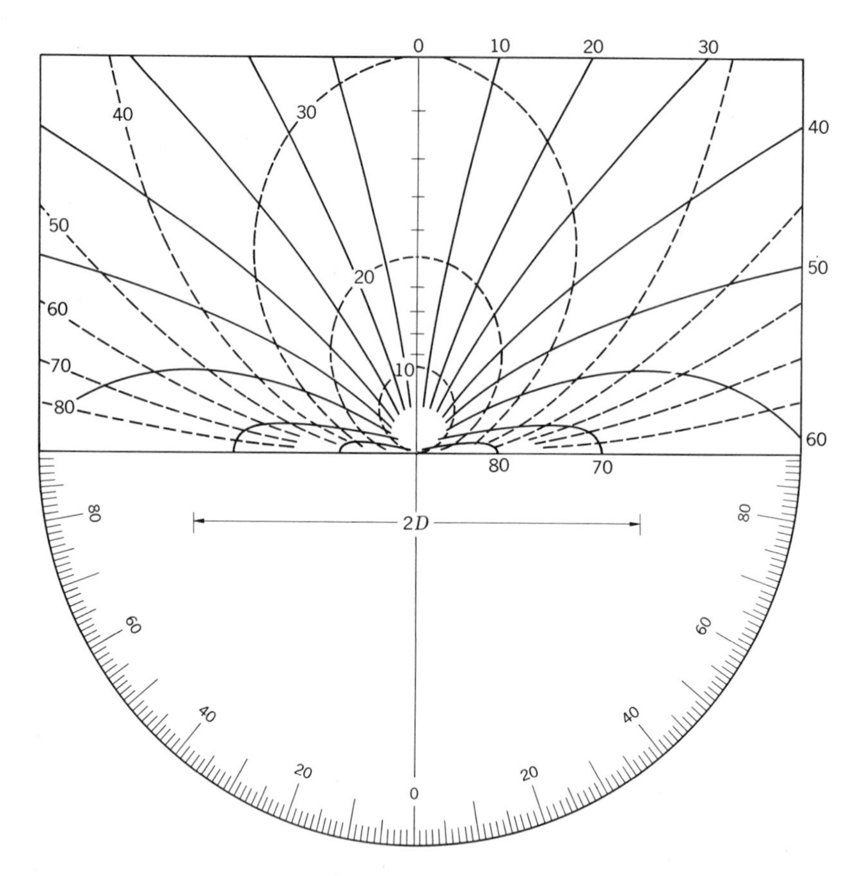

Fig. 14-19. Leonhardt chart for converting the film coordinates of a reflection to angular coordinates of corresponding normals. A scale for establishing the appropriate crystal-to-film distance D is indicated by the horizontal line between arrowheads. [*From C. G. Dunn, Transactions AIME, vol.* 185 (1949), *p.* 421.]

the complement of the angle ϕ used in the transmission arrangement (Fig. 14-18). Within the reciprocal-lattice plane, the reciprocal-lattice vectors form the angle δ with the central vector, as before. Thus the radial distance on the film, from its center to a hyperbola, is proportional to γ, while the distances along a hyperbola are proportional to δ. It follows, therefore, that it is possible to prepare a chart which indicates the γ and δ values of every point on a film for a fixed crystal-to-film distance D. Such a chart was first prepared by A. B. Greninger, and is shown in Fig. 14-23.

Use of stereographic projection. The disposition of the zone axis and the normals to the planes of the zone shown in Fig. 14-22 is indicated relative to a reference sphere in Fig. 14-24. (It should be recognized that the reference sphere bears no relation whatsoever to the Ewald sphere.) A stereographic projection of the hemi-

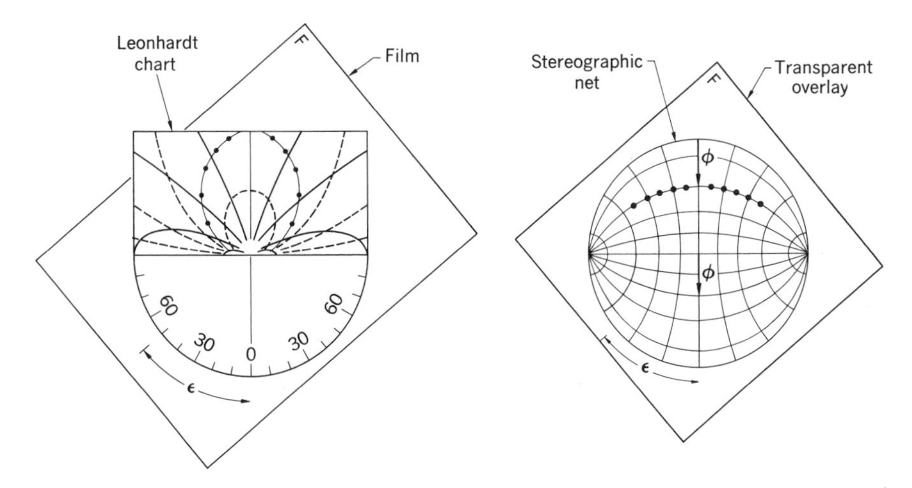

Fig. 14-20. Transferral of angular coordinates from film to stereographic projection with the aid of the Leonhardt chart.

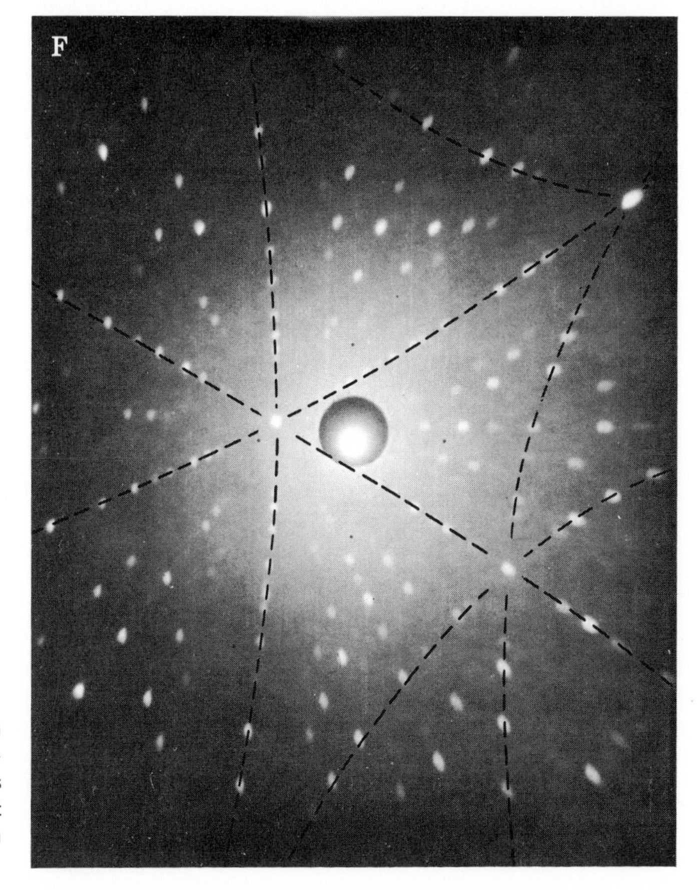

Fig. 14-21. Back-reflection Laue photograph of a silicon crystal. Reflections from several prominent zones are indicated (3-cm crystal-to-film distance).

sphere closest to the projection plane is also indicated in Fig. 14-24. The projection direction is chosen parallel to the x-ray beam direction, so that the plane containing the normals intersects the projected hemisphere above the central meridian, while the pole of the zone axis P lies below. Note that the Wulff net in the projection plane is oriented so that its central meridian is horizontal. This is done so that the great circles on the Wulff net correspond in orientation to those indicated on the reference sphere. The angle γ that the planes (great circles) form with the projection direction is indicated. The angle δ that the individual normals form within a plane can then be marked off directly along the appropriate meridian. In this way, the interpretation of a back-reflection Laue photograph with the aid of a Greninger chart is directly analogous to the interpretation of front-reflection photographs previously discussed.

As an example of this procedure, a tracing of the most prominent conic in the back-reflection photograph previously considered (Fig. 14-21) is shown superimposed on a Greninger net in Fig. 14-25. The fiducial marker $\mathbf{F}$ in the upper left corner of the film serves to relate its orientation to that of the crystal during the exposure. A similar marker then is placed on a transparent sheet superimposed on a stereographic net. The center of the film is pinned to the center of the Greninger chart, and the transparent overlay is similarly pinned to the Wulff net. (The presence of a hole at the center of the film can be overcome by a piece of tape.) Next, the amount of rotation of the film ϵ that is required to line up a hyperbola on the film parallel to the ruled conics on the Greninger chart is duplicated by corresponding rotation of the transparent overlay on the Wulff net, care being taken to rotate both in the same sense. The γ value read off the chart then is measured vertically upward from the center along the equator, and the appropriate great circle is drawn in. Since this is the projection of the trace on the reference sphere of the

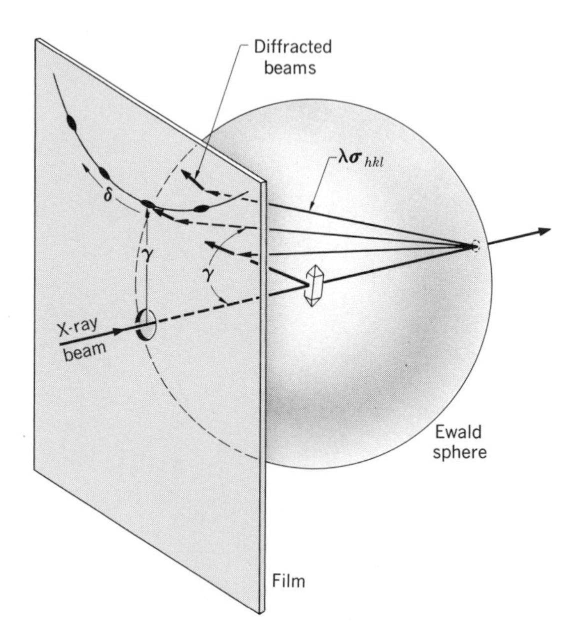

Fig. 14-22. Relation between reciprocal lattice plane containing plane normals σ_{hkl} and diffracted beams in the back-reflection arrangement. Note that the plane containing the normals, if extended, cuts the film in a straight line, while the diffracted beams intersect the film along a conic section.

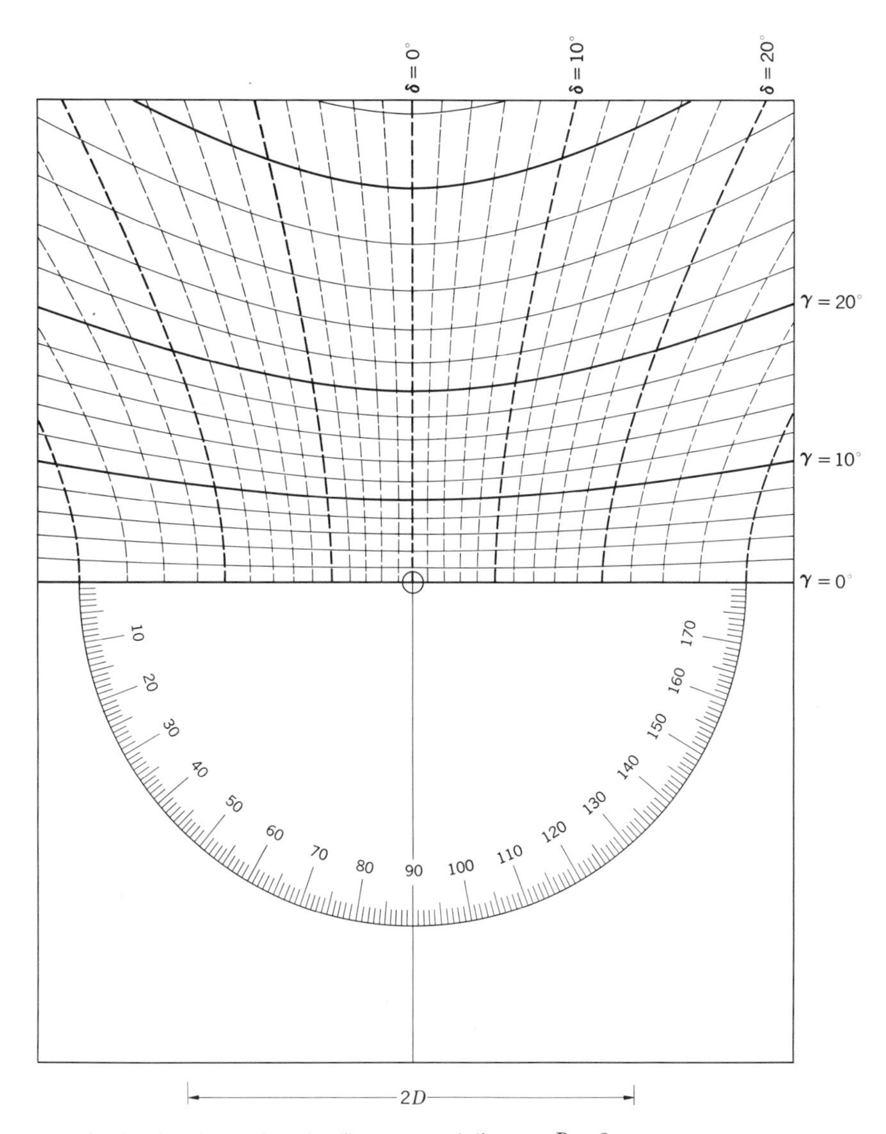

Fig. 14-23. Greninger chart for film-to-crystal distance $D = 3$ cm.

plane containing the normals of planes of one zone, the positions of the individual normals corresponding to individual reflecting planes can be marked off along this great circle by noting the δ values of the reflections on the film. (It should be remembered that the Greninger chart is calibrated to give the angular coordinates of plane normals for each point of the film directly.) Note also that the pole P of the zone axis must lie $90°$ away from the trace of this plane, measured along the vertical great circle (equator) in Fig. 14-25. Alternatively, the pole P can be located by measuring γ along the equator from the bottom of the outer circle. After this

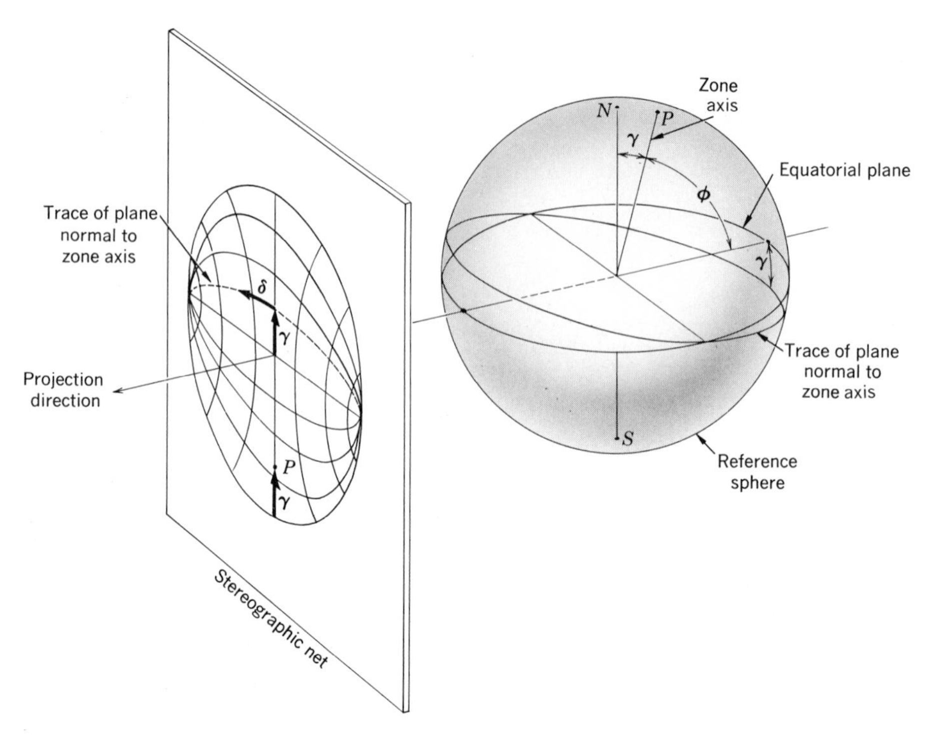

Fig. 14-24. Stereographic projection of normals belonging to the same zone. Note that δ is measured within this plane. (Compare with Fig. 14-18.)

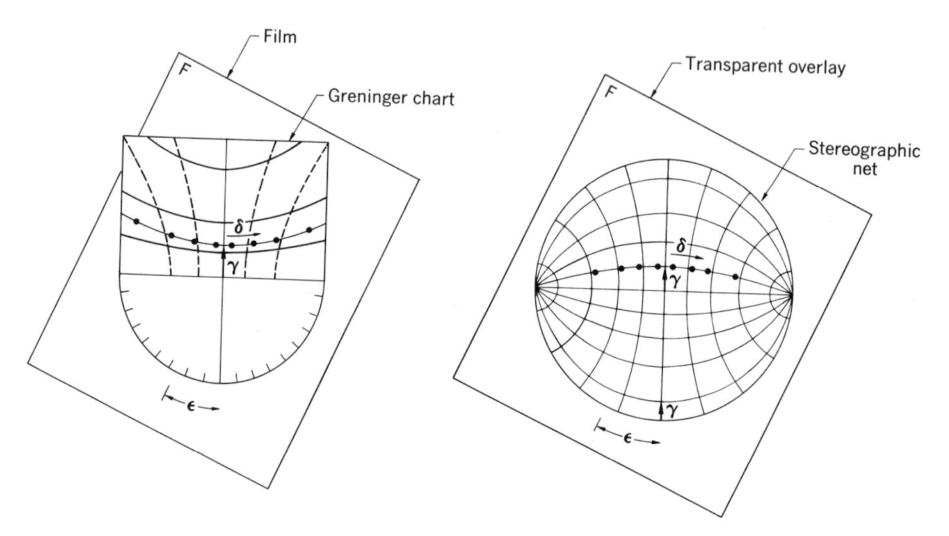

Fig. 14-25. Transferral of angular coordinates from a back-reflection photograph to a stereographic projection with the aid of the Greninger chart.

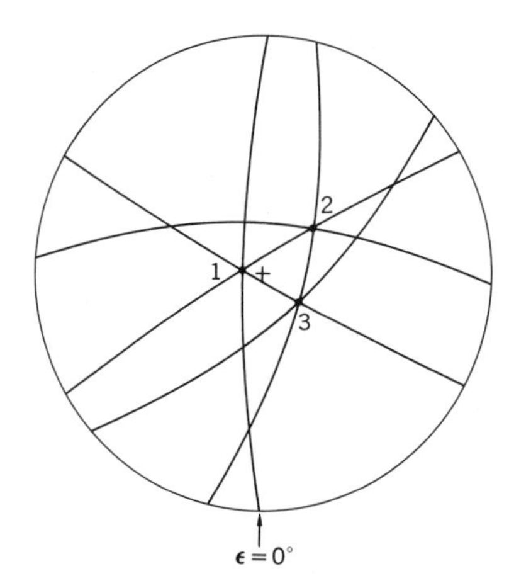

Fig. 14-26. Stereographic projection of the zone traces indicated in Fig. 14-21.

zone is recorded, the film and overlay are both rotated by equal amounts, so that the next zone can be recorded, and so on until the stereographic projection is completed. The stereographic projection of the most prominent zones recorded in Fig. 14-21 is shown in Fig. 14-26.

Crystal orientation. The principal use of the Laue method is to study the orientation of single crystals, so that there is usually little interest in identifying the individual reflections or their normals. For this reason, it is sufficient to indicate the projections of entire zones (traces of planes containing the normals) or the poles of their zone axes. It can be seen in Fig. 14-24 that the zone traces (great circles) intersect each other at poles, some of which lie on a number of such traces. These poles therefore mark the normals of planes that belong to several zones, and such planes must have relatively simple indices, such as (100), (110), (211), etc.† To identify such poles it is necessary to measure the angles between them along the great circles joining them. This means that the overlay is rotated about its center on a Wulff net (Fig. 14-25) until the two poles selected lie on the same meridian, so that the angle between them can be read on the net. The three poles indicated in Fig. 14-26 are separated, respectively, by 35, 30, and 31°. Consulting the list of interplanar axes for cubic crystals (Table 2-1), it is noted that [111] forms an angle of 35.3° with [110] and an angle of 29.5° with [311]. Concurrently, [110] forms an angle of 31.5° with [311]. This suggests that the pole marked 1 is a 110 pole and the pole marked 2 is a 111 pole, while the pole marked 3 must be a 311 pole in order to satisfy its angular relation to both of the other two poles.

As another example of this triangulation procedure, consider the zone traces shown in Fig. 14-27. The angle between poles 1 and 2 and between poles 2 and 3 turns out to be 90°. From Table 2-1 this suggests that these pairs of poles are 100

† Note that the actual plane satisfying the reflection condition (Fig. 14-2) may have higher-order indices. Because reflections from all planes (nh,nk,nl) superimpose in Laue photographs, it is common practice to use the mutually prime indices, that is, those of the first-order reflection.

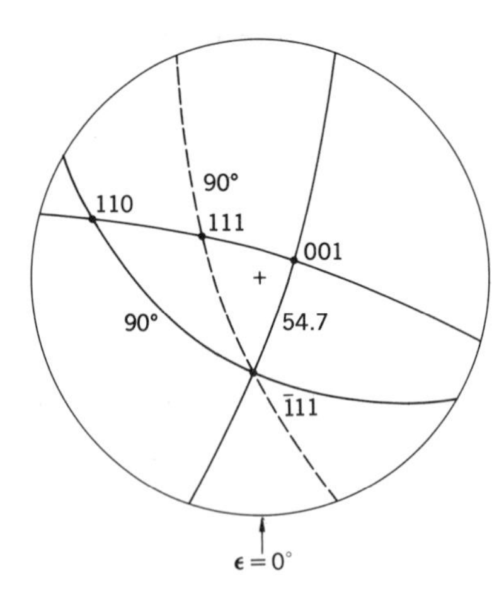

Fig. 14-27. Stereographic projection of the zone tracings of a cubic crystal having nearly correct [[001]] orientation.

and 110, 100 and 100, or 110 and 111. Clearly, the first and last possibility are more reasonable, particularly when it is noted that the angle between 1 and 3 is 55°, which is very close to 54.7°, the angle between [111] and [100]. This fixes pole 2 as a 110 pole, but does not uniquely identify 1 and 3. To resolve this ambiguity, still another zone trace is needed, for example, the dashed one in Fig. 14-27. Pole 4 is found to lie 45° from 2 and from 3, which identifies it as a 111 pole and establishes pole 1 as the 001 pole.

The orientation of the crystal now can be specified directly in several equivalent ways. Noting that the 001 pole lies close to the center (cross) of the projection, the standard [001] projection (Fig. 2-14) can be consulted. A direct comparison with Fig. 14-27 then permits the assignation of the correct indices to all the poles (Fig. 14-28). It is seen that the center of the projection lies in a triangle bounded by 001, $1\bar{1}1$, and $\bar{1}\bar{1}0$. After measuring the angles between these poles and the center (along great circles joining them), it is customary to show the appropriate triangle in its standard orientation (Fig. 2-14) and to locate the center point as shown in Fig. 14-29. More commonly, one wishes to know how far removed the crystal is from some desired orientation or how to bring a particular direction in coincidence

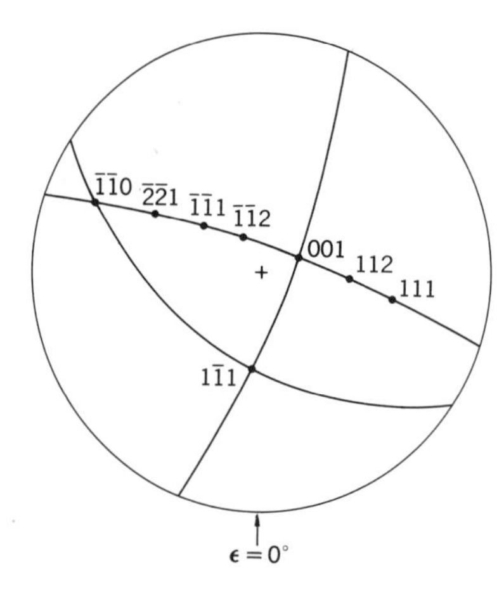

Fig. 14-28. Indices of some of the individual poles lying along the zone traces shown in Fig. 14-27.

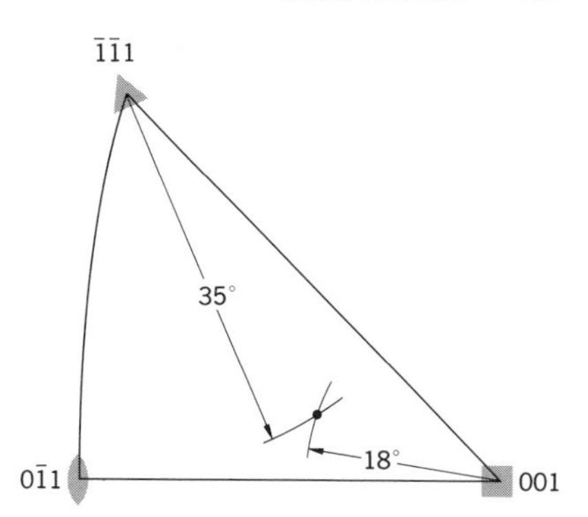

Fig. 14-29. Use of the appropriate stereographic triangle to describe crystal orientation. The dot inside the triangle locates the center of the stereographic projection shown in Fig. 14-27.

with the direct x-ray beam direction. The necessary angular coordinates can be determined directly in terms of the three angles indicated in Fig. 14-25.

It should be noted that the poles of the zone axes in Fig. 14-27 could have been used to determine the crystal's orientation by a triangulation procedure similar to that described above. In fact, it is quite usual to combine the use of poles of zone axes of prominent zones with the poles lying at mutual intersections of several plane tracings, because these poles usually have the simplest indices. Alternatively, the stereographic projection obtained can be compared with several standard projections, and similarities between them can be used to identify the various poles more quickly. It is, of course, not at all essential to prepare stereographic projections. All the necessary information is contained in the back-reflection photograph, and the angle between any two poles can be read directly from a Greninger chart by measuring the angle between the corresponding two reflection spots along an appropriate conic. Thus it is possible to identify the reflections lying at the intersections of several conics on the film, and from this to deduce the crystal orientation.

Where it is desired to orient a series of crystals in a definite way, for example, prior to carrying out some physical measurements, an even faster procedure is to prepare in advance a series of reference photographs in which the crystal has been displaced by known amounts from the desired orientation. Because of the high symmetry of cubic crystals, the unique section of a standard projection is quite small (triangle in Fig. 14-29), and the range that such photographs need to cover is quite limited. Finally, a direct display of the reflection maxima on a fluorescent screen would permit a rapid adjustment of the crystal to any orientation. Although ordinary fluorescent screens can be used for this purpose, the necessity to work in a completely dark room makes such procedures impractical. Several manufacturers are currently developing direct-image intensifier tubes, however, which will be most helpful for such purposes.

USE OF DIFFRACTOMETER

The popularity of diffractometers in x-ray crystallographic studies, primarily because of their superior intensity-measuring ability, raises the question of whether

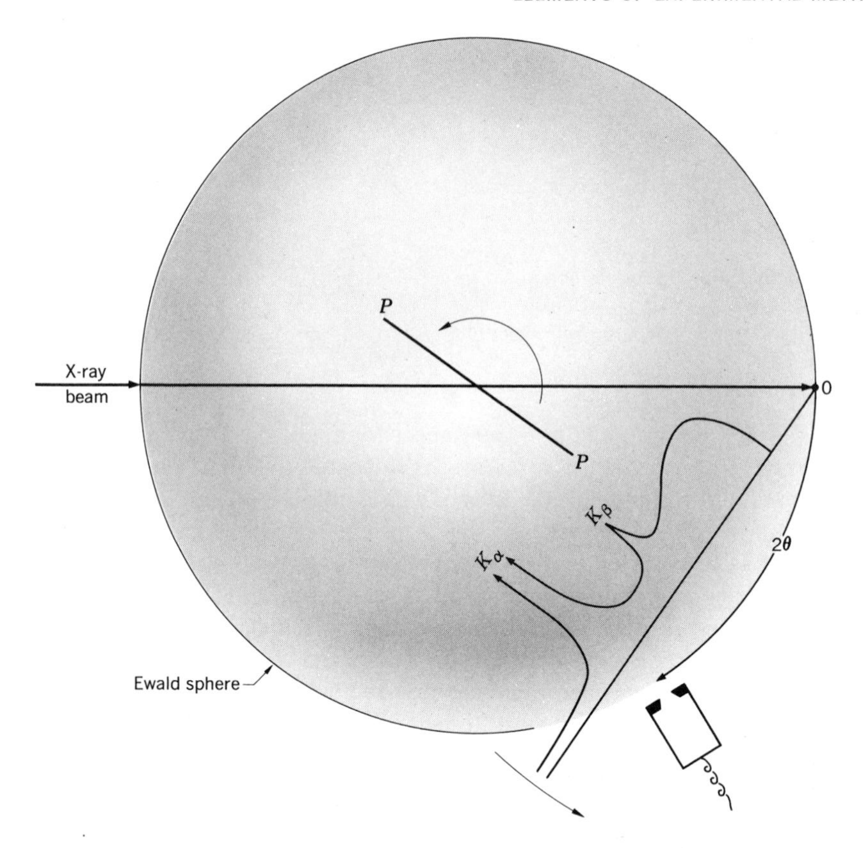

Fig. 14-30. Relation of reciprocal-lattice streak to arbitrary 2θ setting of detector. Rotation of the reflecting plane PP also rotates the intensity streaked out along its normal as shown by the arrows. When a portion of the streak intersects the Ewald sphere in front of the detector, it will record the reflected beam.

they can be utilized for crystal-orientation purposes, and if so, what is the best procedure to follow. An examination of the reciprocal-lattice construction in Fig. 14-2 shows that the use of continuous radiation has the effect of "extending" a single point into a continuous row of points. This suggests that it should be relatively easy to locate a reflection from a single crystal using continuous radiation even when nothing is known in advance about its lattice. To see how this can be done, consider the equivalent of the construction in Fig. 14-2 for just one plane (hkl). The intensity distribution along a line passing through the reciprocal-lattice origin normal to such a plane is shown in Fig. 14-30. If this line lies in the plane of the drawing, clockwise rotation about a vertical line will cause the reflection condition (Fig. 14-1) to be satisfied for different points along this line. The placement of the detector at some small angle of 2θ, say $30°$, will assure that the reflection from this plane of some portion of the x-ray spectrum will be detected. Once such a reflection is detected, the synchronous rotation of the crystal and detector in the usual diffractometer motion can be used to record the complete intensity distribution shown. Note that the exact angle 2θ need not be known in advance for this reflection, because it does not matter which portion of the continuous spectrum is initially detected.

For an arbitrary mounting of the crystal, there is no assurance that such central lines lie in the diffractometer plane. On the other hand, it would be most unusual if some central lines did not, in fact, turn out to lie more or less in that plane. Thus the detector is normally set at some small angle, with entrance slits "wide open." The crystal is rotated slowly about its rotation axis until some signal is detected by the counter. By suitable rotation of the crystal about line PP in Fig. 14-30, parallel to the reflecting plane and the plane of the drawing, the intensity of the reflection is maximized. This has the effect of bringing the reflection line to lie in the diffractometer plane. Subsequent synchronous rotation of the crystal and detector then locates the position of the $K\alpha$ reflection. Once one reciprocal-lattice point thus has been fixed, another one is sought approximately $90°$ away. The detector again is placed at some small angle, while the crystal is slowly rotated as before. The reason for looking for a reflecting plane approximately $90°$ from the initial one is that the two mutual perpendicular arcs on a goniometer head (Fig. 13-17) then can be utilized to best advantage in orienting the crystal. Once two reciprocal-lattice points have been made to lie in the diffractometer plane, some plane of the reciprocal lattice, defined by these two points and the reciprocal-lattice origin, is coplanar with the diffractometer.

The subsequent procedure to be followed is determined by the purpose of the experiment. For example, other points in this plane can be located fairly rapidly now by repeating the above procedure, but without the need to adjust further the goniometer-head arcs. Alternatively, upper levels of the reciprocal lattice can be explored provided a suitable device for tilting the crystal's rotation axis is available,

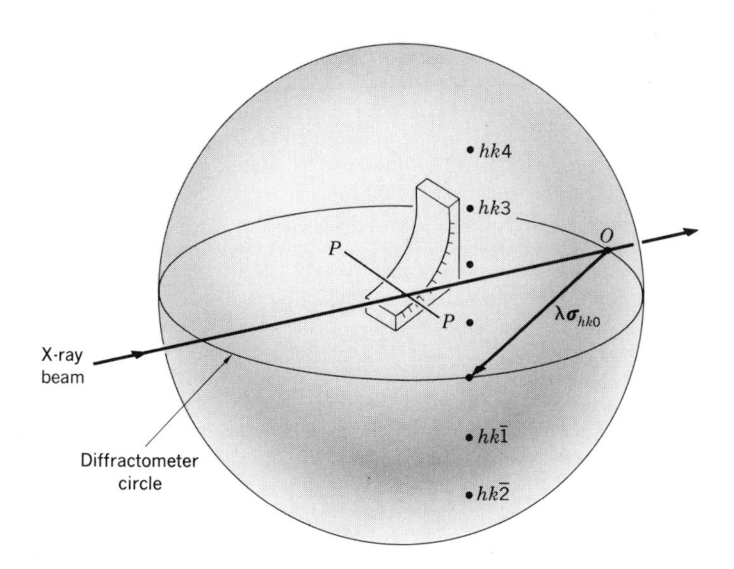

Fig. 14-31. Relation of vertical row in reciprocal lattice to goniostat setting when $hk0$ net is parallel to diffractometer circle.

such as the one shown in Fig. 13-24, for example. If the reciprocal-lattice points lie above each other in rows that are normal to the diffractometer plane, they can be brought to intersect the Ewald sphere in the diffractometer plane by rotation about the axis PP in Fig. 14-31, which is readily accomplished with a goniostat (Fig. 13-24) by motion along its arc. If these reciprocal-lattice rows are not perpendicular to the diffractometer plane, they are more difficult to locate unless the lattice type and its accurate dimensions are known beforehand. This points out one important limitation of diffractometers. Because only one reflection can be detected at a time, only one reciprocal-lattice point can be located at a time, so that the study of a previously unknown crystal becomes relatively difficult and tedious. Even when "completed," there is no way to be absolutely certain that some reflections have not been overlooked. Thus, before attempting any studies on a diffractometer, the crystal should be examined by one of the photographic methods described in this book. Only then is it possible to carry out diffractometer studies in a meaningful way.

SUGGESTIONS FOR SUPPLEMENTARY READING

B. D. Cullity, *Elements of x-ray diffraction* (Addison-Wesley Publishing Company, Inc., Reading, Mass., 1956), especially pp. 215–242.

E. Schiebold, *Die Lauemethode* (Akademische Verlagsgesellschaft Grest & Portig KG, Leipzig, 1932).

R. W. G. Wyckoff, *The structure of crystals*, 2d ed. (Chemical Catalog, New York, 1931).

EXERCISES

14-1. Prepare a reciprocal-lattice construction like that in Fig. 14-2 for the $hk0$ net of an orthorhombic crystal having $a = 4.10$ Å and $b = 5.46$ Å. Assume that a tungsten-target tube is operated at 30 kV and that $\lambda_{\max} = 0.920$ Å, which corresponds to the K absorption edge of bromine in the photographic emulsion. If the x-ray beam is parallel to the a^* axis, which $(hk0)$ planes will satisfy the reflection condition?

14-2. What is the shape of the intersection of the diffraction cones in the front-reflection Laue method when the angle φ that the zone axis forms with the direct-beam direction is $45°$? When $\varphi = 90°$? Can the angle exceed $90°$?

14-3. Suppose a cylindrical film is placed around the crystal, with the cylinder axis perpendicular to the x-ray beam. Will the extra reflections recorded on this film be of any practical value in determining the orientation of the crystal? Does such an arrangement have any advantage over the flat-film methods discussed in this chapter?

14-4. By means of a suitable drawing, show that the incident and reflected beams have the same cross sections, regardless of the reflection angle θ for a parallel pair of rays. **Hint:** This can be proved quite rigorously by considering similar (identical) triangles.

14-5. Consider a front-reflection photograph along the incident-beam direction. Show what the shape of typical reflections is for the case of a diverging incident beam and a fairly large crystal. Do this by drawing the shape of at least two reflections along a vertical and along a horizontal line on the film and several reflections lying in one quadrant.

14-6. Repeat Exercise 14-5 for the case of a converging beam and a crystal larger than the incident-beam cross section.

14-7. With the aid of the construction in Fig. 14-13, prove that the normals of the $(h0l)$ planes shown in cross section in Fig. 14-12 intercept the trace of the plane of projection in equal steps proportional to a/c, thus proving Eq. (14-2).

14-8. Assign suitable indices to all the poles (black dots) plotted in the gnomonic projection in Fig. 14-16.

14-9. Determine the relation necessary to construct a stereographic ruler for a front-reflection Laue photograph. If the radial distance in a projection is OP and the radius of the stereographic net is R, show that the necessary relation is $OP = R \tan(45° - \theta/2)$. Next, relate OP to the radial distance to a reflection on the film ($OX = D \tan 2\theta$).

14-10. Determine the relation necessary to construct a stereographic ruler for a back-reflection photograph. Keep in mind that the angle determined by the radial distance on the film is the supplement of 2θ. How do the necessary relations differ from those in Exercise 14-9?

14-11. Suppose the crystal discussed in Exercise 14-1 is rotated about its c axis. With a detector set at $2\theta = 30°$, which of the $hk0$ reflections can be detected, assuming that the energy sensitivity of the detector is the same as that of a photographic film?

FIFTEEN
Rotating-crystal method

RECIPROCAL-LATTICE CONSTRUCTION

Number of reflections developed by a crystal. The relative disposition of the x-ray beam, the crystal, and the accompanying reciprocal-lattice construction for a rotating crystal have been shown in Fig. 7-10. As indicated there, when a crystal is rotated about an axis oriented normal to the incident-beam direction, usually called the *normal-beam method*, the individual layers of the reciprocal lattice cut the Ewald sphere along parallel circles. Each time a reciprocal-lattice point hkl intersects the sphere of reflection, a reflected beam passing through that point on the sphere is produced, since the crystal at the center of the sphere now satisfies the diffraction condition for the corresponding (hkl) plane. The result is that, during a complete rotation of the crystal, the individual diffracted beams form cones that are coaxial about the rotation axis of the crystal. In order to record the maximum number of the reflections produced, therefore, a film is placed cylindrically about the crystal's rotation axis, as shown in Fig. 15-1. The

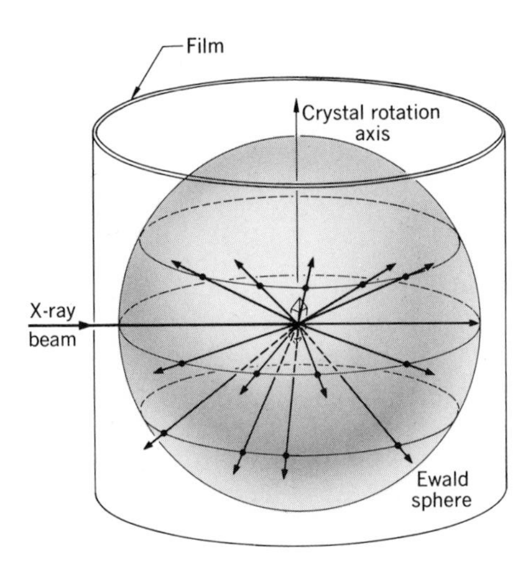

Fig. 15-1. Rotating-crystal arrangement showing some of the diffracted rays coming toward the observer.

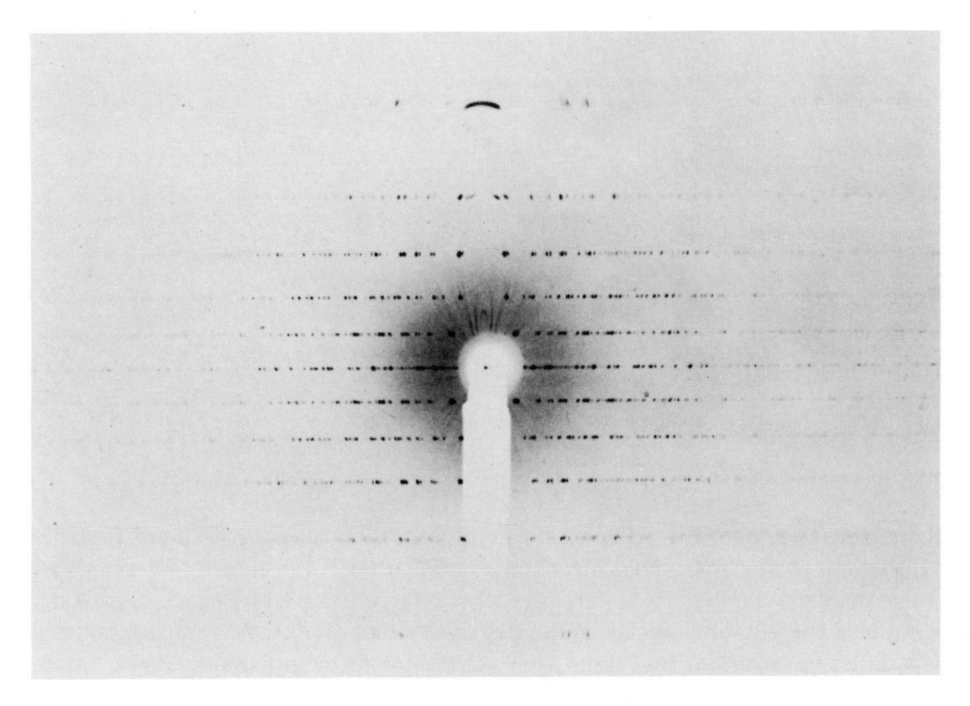

Fig. 15-2. Rotating-crystal photograph of $(NH_4)_2Pt(S_5)_3 \cdot H_20$ rotated about its b axis.

individual reflections thus are recorded as spots lying along horizontal rows on the film (Fig. 15-2), called *layer lines*.

In comparing the rotating-crystal photographs of different crystals, usually prepared employing the same kind of x-ray beam (Cu $K\alpha$), it is convenient to combine λ with the reciprocal-lattice vector σ in the reciprocal-lattice construction. Thus the Ewald sphere in Fig. 15-1 is assigned a radius of unity, and the spacing between layer lines in photographs of different crystals is directly proportional to $\lambda\sigma_{hkl} = \lambda/d_{hkl}$, where d_{hkl} is the interplanar spacing of the planes that are perpendicular to the rotation axis. One of the advantages of this choice for the scale factor is that distances in the reciprocal lattice now are measured in dimensionless units. Note, for example, that the maximum value that $\lambda\sigma$ can have along the rotation axis and still intersect the sphere of reflection is unity, while at right angles to the rotation axis and in the equatorial, or zero, level, the maximum value is 2.

In order to calculate how many reflections can be developed by a single crystal during a complete rotation, it is more convenient to hold the reciprocal lattice stationary and rotate the Ewald sphere about the reciprocal-lattice origin instead. As shown in Fig. 15-3, rotation of the sphere of reflection about a point produces a *tore of reflection* lying within a sphere of radius 2. This is called the *limiting sphere*, because it contains all the reciprocal-lattice points that can possibly be recorded in an x-ray diffraction experiment. As can be seen in Fig. 15-3, the rotating-crystal method can record only the points lying within the tore of reflection, so that, in order to record all possible reflections, the crystal must be remounted to rotate

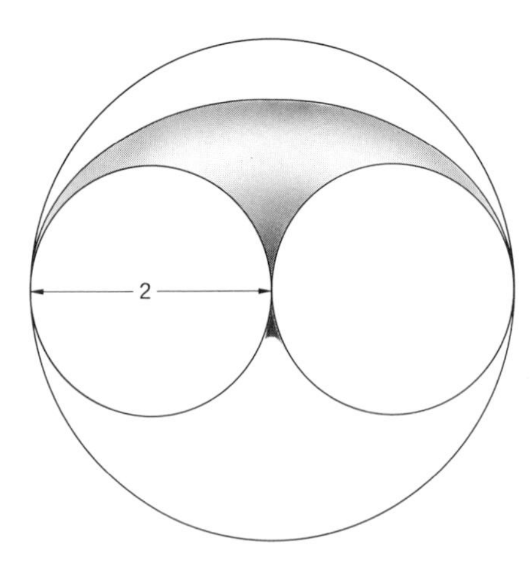

Fig. 15-3

about several other directions also. A moment's consideration shows that the number of crystal mountings necessary to record all the outermost reciprocal-lattice points in the limiting sphere is much too large to be a practical undertaking. Nowadays the rotating-crystal method is used primarily for determining unit-cell constants, however, so that it is usually sufficient to mount the crystal to rotate about two or three different directions only.

The maximum number of reflections that can be recorded during a complete rotation of the crystal depends, of course, on the wavelength of the radiation employed. This is simply the number of reciprocal-lattice points contained within the tore of reflection in Fig. 15-3. To calculate this number, divide the volume of the tore, $2\pi^2r^3$, by the volume of a reciprocal-lattice cell.

$$\text{Maximum number of reflections} = \frac{2\pi^2 r^3}{\lambda^3 V^*} = \frac{2\pi^2 1^3}{\lambda^3 V^*} = 19.74 \frac{V}{\lambda^3} \tag{15-1}$$

The necessary formulas for calculating the cell volume V are given in Table 7-1. Note that if the volume of a nonprimitive unit cell is substituted in (15-1), it must be multiplied by the number of lattice points per cell present.

It is evident from (15-1) that the use of shorter-wavelength radiation serves to increase the total number of reflections recorded. This in turn means that the individual spots become more closely spaced. Similarly, the larger the unit cell of a crystal, the larger the number of reflections recorded and the closer they are spaced. It follows from this that longer-wavelength radiation is required for the study of crystals having large unit cells. Copper $K\alpha$ radiation turns out to be satisfactory for all but the very large unit cells having cell edges longer than 20 Å. On the other hand, the use of shorter-wavelength radiation like Mo $K\alpha$ rarely serves to increase the actual number of reflections recorded because the intensities of x-ray reflections decrease as $(\sin\theta)/\lambda$ $(= 1/2d)$ increases, as discussed in Chap. 11. Thus such short-wavelength radiation may be employed in the rotating-crystal method only when highly absorbing crystals are studied.

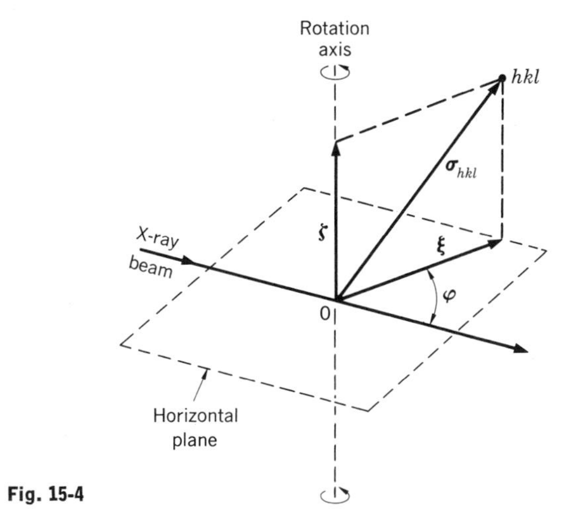

Fig. 15-4

Reciprocal-lattice coordinates. Since it has been demonstrated above that the reciprocal lattice of a crystal is recorded directly in diffraction experiments, it follows that it is most meaningful to discuss such experiments in terms of the reciprocal-lattice coordinates. In the rotating-crystal method, a cylindrical film is used, so that it is convenient to select cylindrical coordinates. Thus the orientation of the reciprocal-lattice vector $\boldsymbol{\sigma}_{hkl}$ is specified by two mutually orthogonal vectors, $\boldsymbol{\zeta}$ along the rotation axis and $\boldsymbol{\xi}$ in plane containing the x-ray beam, and an angle φ formed by the direct beam and the plane containing $\boldsymbol{\zeta}$ and $\boldsymbol{\xi}$, as shown in Fig. 15-4.

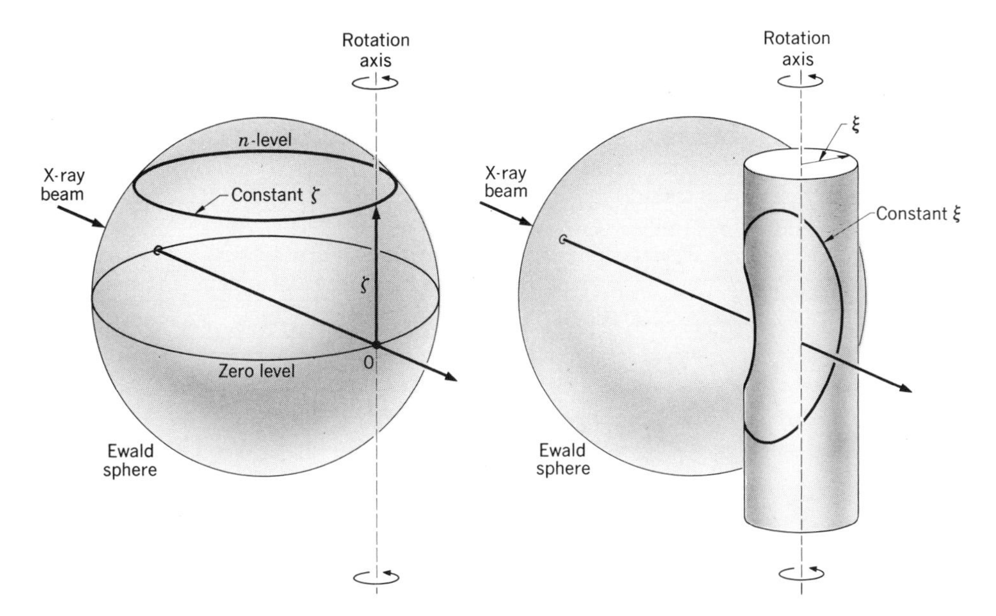

Fig. 15-5. Intersection of planes having constant ζ with Ewald sphere (left side) and cylinders having constant ξ (right side). *(After Buerger.)*

The important vector relation between these coordinates is

$$\boldsymbol{\sigma} = \boldsymbol{\zeta} + \boldsymbol{\xi} \tag{15-2}$$

Similarly, their magnitudes are related by

$$\sigma^2 = \zeta^2 + \xi^2 \tag{15-3}$$

It follows from the above that points in the reciprocal lattice having a constant ζ value lie in a plane normal to the rotation axis. When the crystal is rotated, therefore, these planes cut the sphere of reflection at a constant height ζ above the equator, as shown in Fig. 15-5. Similarly, points having a constant ξ value lie in a cylinder about the rotation axis and cut the Ewald sphere along curves, as also shown in Fig. 15-5. These curves appear on a cylindrical film that surrounds the crystal's rotation axis (Fig. 15-1) in a way shown on the Bernal chart reproduced in Fig. 15-6. It should be apparent that when the scale used in preparing the Bernal chart takes proper account of the crystal-to-film distance, such a chart can be

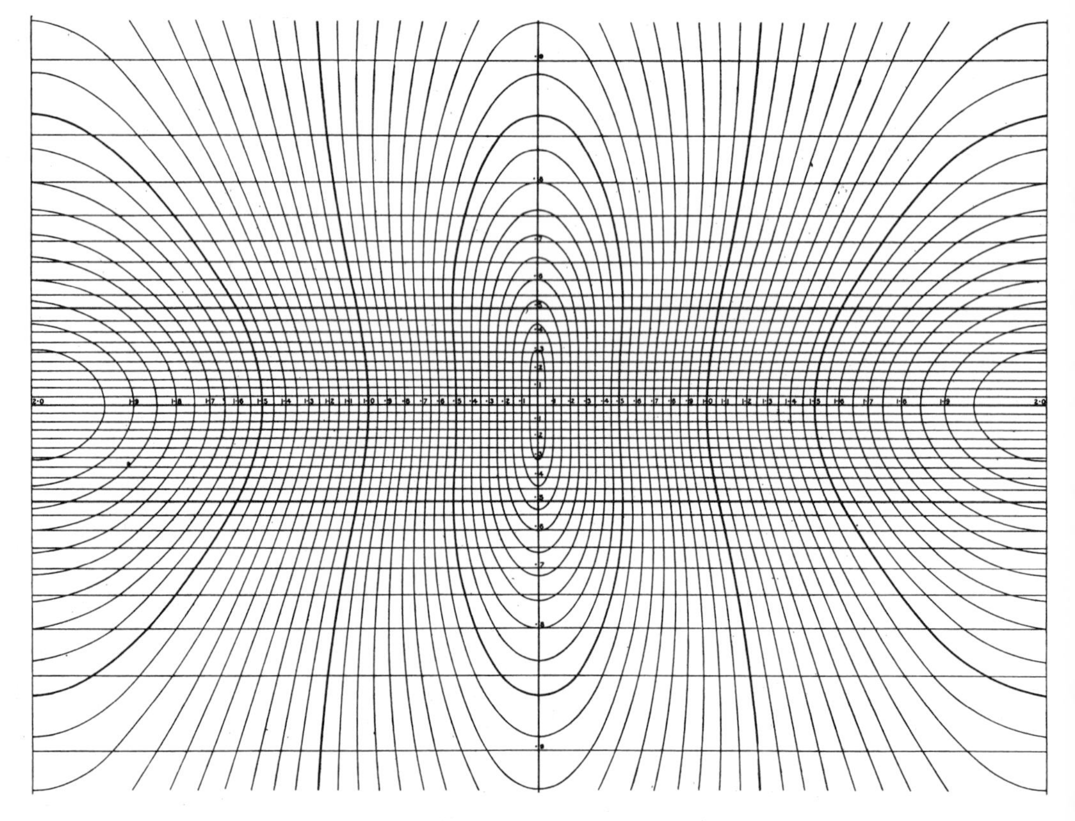

Fig. 15-6. Bernal chart.

Fig. 15-7. Rotating-crystal camera placed before a gas tube x-ray unit. *(Courtesy of M. J. Buerger.)*

superimposed directly over the film (Fig. 15-2), and the ζ and ξ coordinates of each spot can be read directly off the chart. Although not often used because of its limited recording range, a flat-plate camera can be utilized in place of a cylindrical film. Obviously, a chart showing the contours of constant ζ and ξ similarly can be constructed for a flat film, and such a chart also has been prepared by J. D. Bernal.

INSTRUMENTATION

Cylindrical camera. The camera used for recording rotating-crystal photographs consists of a cylindrical film holder, a collimation system, a goniometer head on which the crystal is mounted, and a small motor for rotating the goniometer-head spindle. An assembled view of such a camera is presented in Fig. 15-7, showing the cylindrical film holder directly above a base housing the drive motor. Note that in this instrument the collimator is attached directly to a vertical plate. The goniometer head is shown mounted on the base in Fig. 15-8. The photographic

Fig. 15-8. Components of a rotating-crystal camera. *(Courtesy of M. J. Buerger.)*

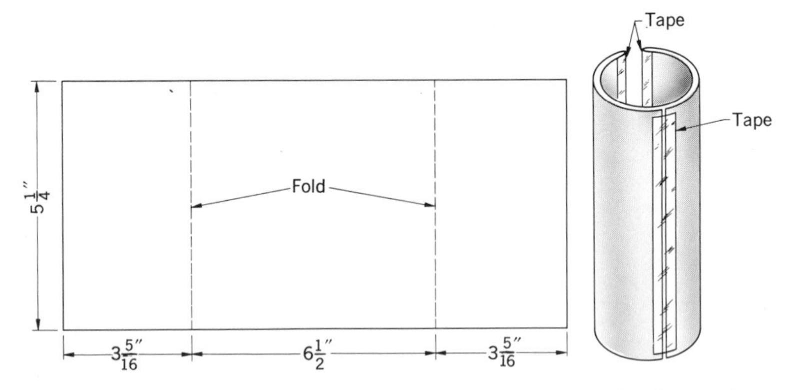

Fig. 15-9. Preparation of film envelope from opaque paper. *(After Buerger.)*

film, normally a sheet 5 by 7 in. in size† is held tightly against the inside walls of the cylindrical film holder by suitable clips. To prevent exposure of the film to visible light, it should be placed in an opaque holder. A simple way to prepare a film envelope is shown in Fig. 15-9. A flat piece of opaque black paper (photographic masking paper) is cut to the size indicated for a film envelope that just fits inside a cylindrical holder having a 57.3-mm diameter. (Note that the film must be trimmed in this case to a length of $6\frac{17}{32}$ in.) Then the two sides are folded back and taped together. Note that the back is deliberately made $\frac{1}{8}$ in. longer than the front (solid piece), to eliminate any buckling of the paper when the envelope is bent into a cylindrical form. The lifetime of the envelope can be extended considerably by placing a strip of reinforcing tape over the two bent edges. A slight gap in the back of the envelope is permissible, since it is pressed against the inner wall of the film holder. The two open ends are made light-tight by slipping recessed rings over each end of the envelope after the film has been inserted in the darkroom.

Collimation system. The basic requirements of a collimator for an x-ray diffraction arrangement have been discussed in the preceding chapter, and are displayed in Fig. 14-7. In order to limit the shadow or blind region produced by the intrusion of the collimator, its end can be tapered, as shown in Fig. 15-10. In fact, the cross-sectional drawing in Fig. 15-10 shows how an actual collimator is constructed from a brass tube, into which two lead disks containing approximately 1 mm-diam holes are centrally pressed, and a cone-shaped cap housing the final, or guard, slit designed to prevent radiation diffracted by the lead aperture from reaching the film.

The direct beam is prevented from striking the film by means of a lead cup, such as the one shown in a cross-sectional view in Fig. 15-10. The reason for using a deep cone-shaped receptacle rather than a flat disk is twofold: The direct beam striking the beam stop is scattered in a backward sense and causes the film to

† Commercial x-ray films labeled 5 by 7 in. are actually 4.94 by 6.94 in. in size. Most cameras are built to accept commercial film sizes, a fact which must be taken into account when trimming oversize films.

become fogged, and the sides of the beam catcher shown serve to eliminate this. Similarly, the direct beam is scattered in a forward direction by the air molecules along its path in the camera, and this scattered radiation also is largely intercepted by the cone. Such a cone, of course, produces a blind spot surrounding the direct beam in the front-reflection region. The use of a cone-shaped catcher, however, serves to minimize this blind region.

Mounting and adjustment of crystal. The optimum size that a crystal should have in a diffraction experiment, that is, one in which the intensity gained from increasing the crystal volume in the beam path is just balanced by the intensity attenuation caused by absorption, is proportional to $1/\mu$, the reciprocal of the linear absorption coefficient for the x radiation selected (Exercise 15-4). In order to assure a uniformly exposed diffraction photograph, however, the crystal size is usually limited to below a half millimeter in diameter. This size limitation assures that the crystal will be surrounded completely by the incident beam during the exposure. Once a suitable crystal has been selected, it is necessary to mount it in such a way that it can be affixed to a goniometer head for proper adjustment in the x-ray beam. Here it should be kept in mind that the amount of extraneous material employed to hold the crystal that is introduced in the x-ray beam path should be kept to a minimum. For this reason, a thin glass capillary is preferred to a solid fiber for holding the crystal in the beam. A simple way to prepare a suitable capillary is to heat a small portion of a Pyrex test tube until the glass is molten, and then, by quickly pulling the two ends apart, reduce the diameter of the tube to the order of tenths of a millimeter. A quick tug can produce several yards of suitable capillaries.

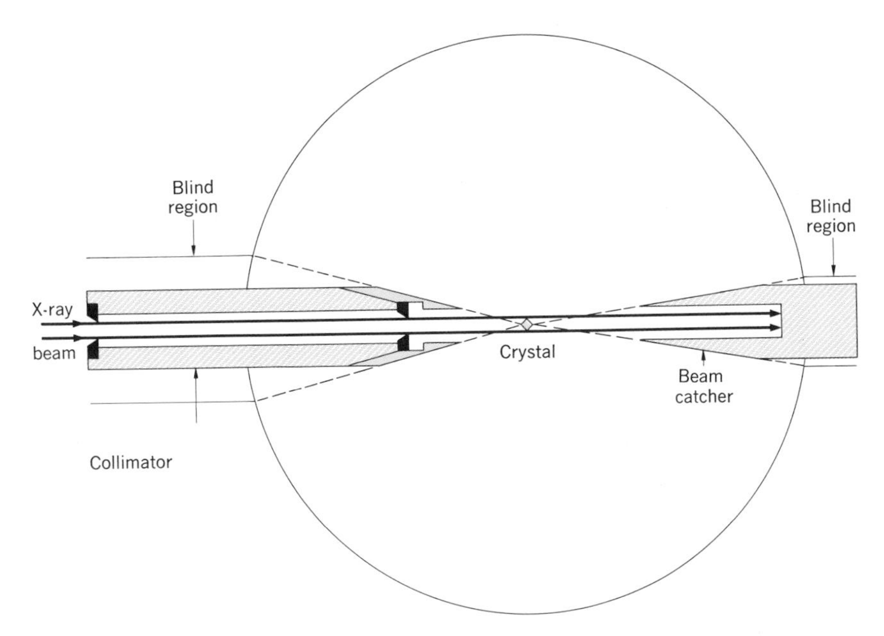

Fig. 15-10

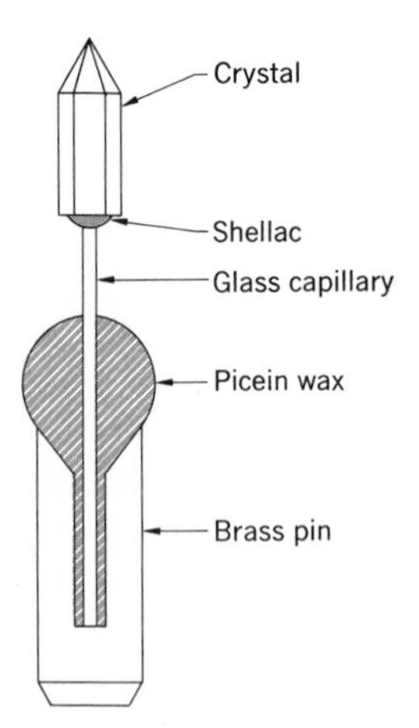

Fig. 15-11

Several kinds of glue then can be used to affix the crystal to the glass. The actual choice should be based on such considerations as, for example, the speed with which the glue dries. Too rapid hardening makes the proper alignment of some desired direction parallel to the glass capillary most difficult. Too slow drying time taxes the experimenter's patience. Finally, the glue should not interact chemically with the crystal. Ordinary shellac and various household cements (usually thinned with amyl acetate or some other suitable diluent) are frequently used for this purpose.

At the other end, the capillary must be attached to a goniometer head. The proper way to do this is illustrated in Fig. 15-11. A brass pin, cut from a $\frac{3}{32}$-in.-diam rod, has a central hole drilled partway through it, as shown. Then a daub of picein wax is attached to it, while the pin is heated at the bottom, so that the wax flows into the hole and forms a meniscus over the countersunk opening. Finally, heating the pin softens the wax, and the glass capillary is introduced into the pin. In actual practice, the glass capillary is first affixed to the pin and, through slight manipulation, is made parallel to the pin. (Both the length of the brass pin and the glass capillary are so adjusted that the crystal is located at the approximate center of curvature of the arcs in the goniometer head.) The brass pin now serves as a handle for manipulating the capillary during the mounting operation described above. The tip of the capillary is covered by the glue and touched to one end of the crystal. The surface tension of the glue holds the crystal in place while it is poked into proper alignment with the capillary with the aid of a pin mounted in a suitable holder. When the glue has dried, the entire assembly (Fig. 15-11) is placed into the opening in the goniometer head. It should be noted that some of the work can be eliminated by sticking the capillary directly into some kind of soft wax affixed to the goniometer head. This small apparent gain in effort sometimes leads to a great deal of extra work later, because such impermanent mounts are too easily disrupted in the course of an experiment. Since usually several days, weeks, or even months of work ensue the mounting of a crystal, the few additional minutes necessary to prepare a permanent mount like the one shown in Fig. 15-11 are well spent.

Once the crystal mount is placed in a goniometer head, a small set screw serves to lock the brass pin in position. Whenever a crystal has one or more well-developed faces, at least one of them should be made parallel to a goniometer arc. This is usually done by rotating the brass pin while the crystal is viewed through a binocular microscope. The presence of such faces then permits the proper alignment of the crystal rotation axis on an optical goniometer. Alternatively, when an anisotropic crystal is transparent, a polarizing microscope can be used to align the rotation axis. In both cases, the two arcs on the goniometer head are adjusted until the rotation axis of the crystal is made absolutely parallel to the goniometer axis. Opaque crystals lacking suitable faces or other identifying markings, such as cleavage planes, striations, etc., have to be oriented using x-ray diffraction methods. Although small corrections from nonparallelism can be made, as discussed in the closing section of this chapter, the rotating-crystal method is ill-suited to the orientation of crystals. Other procedures, such as the Laue method (Chap. 14) or the precession method, discussed in the next chapter, should be used, therefore, to align such a crystal parallel to the goniometer axis. After the crystal is correctly aligned, the goniometer head is mounted on the camera's spindle, and the crystal is viewed directly through the collimator. A small lens having the crystal position at its focus is usually slipped over the collimator to aid in positioning the crystal at the center of the small circular field of view. The centering adjustments are carried out by means of two mutually orthogonal translations built into the goniometer head, while the translation along the rotation axis usually is an integral part of the rotating spindle assembly. When the crystal is properly centered, it appears to remain stationary while the spindle is rotated. If this stationary center does not coincide with the center of the field of view defined by the collimator, the camera is not in proper adjustment and should be adjusted by a qualified person, usually its manufacturer.

INTERPRETATION OF PHOTOGRAPHS

Unit-cell determination. The discussion in the early parts of this chapter has demonstrated that the spacing between layer lines in a rotation photograph is directly proportional to ζ, the spacing between reciprocal-lattice planes normal to the rotation axis. Superimposing a rotation photograph on an appropriately scaled Bernal chart (Fig. 15-12) allows the determination of this value directly. In doing this, the film is so positioned that the ζ values of corresponding layer lines are the same at both extremes of the photograph. The distance between the uppermost and lowermost layers is measured and divided by the number of layers spanned. (In Fig. 15-12, ζ is thus determined to equal 0.36.) Assuming that the crystal was mounted to rotate about the normal to the (001) planes, the value of d_{001} is then given by

$$d_{001} = \frac{\lambda}{\zeta} \tag{15-4}$$

Accordingly, when the symmetry of the crystal is known, the dimensions of the unit

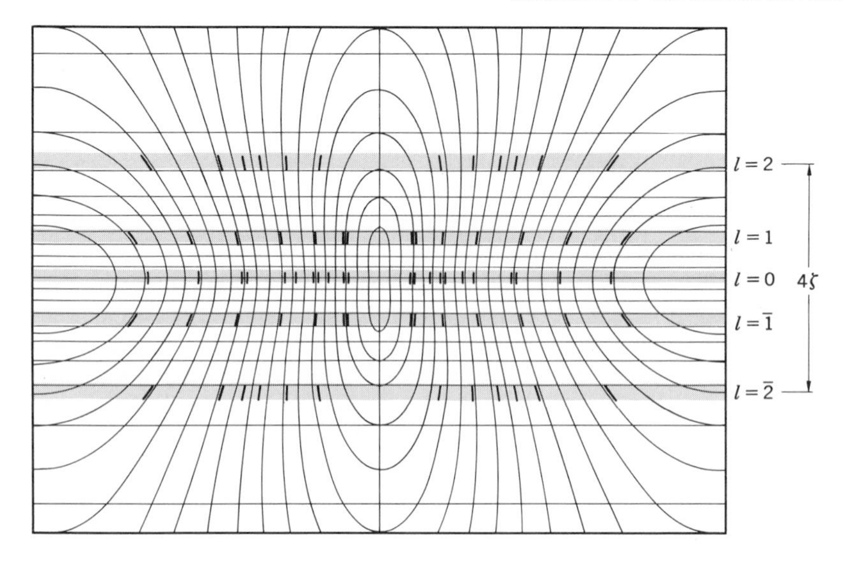

Fig. 15-12. Schematic view of a film overlain on a Bernal chart. Note that the x-ray diffraction spots lie along rows of constant ξ and ζ. (Only a few possible reflections are indicated.)

cell can be determined by rotating the crystal about each of the uniquely different crystallographic axes.

In the cubic system, all three cell edges are the same, so that one rotation photograph about a is all that is needed. In the tetragonal and hexagonal systems, rotations about c and a are necessary. Moreover, it is not possible, normally, to select the correct a axis from an examination of the morphology, so that rotations about a and [110] are necessary to select the smallest (correct) unit cell. (See Exercise 15-6.) Alternatively, by indexing the reciprocal-lattice spots as described in the next section, the correct unit-cell choice can be established directly. In the orthorhombic system, the three cell edges are distinct and mutually orthogonal, so that three rotations are needed to determine their lengths. In the monoclinic and triclinic systems, rotation photographs can be used to determine the length of the cell edges, but the presence of arbitrary angles complicates their interpretation, particularly for the triclinic system, so that moving-film methods described in the next chapter are preferred for such studies.

It should be noted that the use of a Bernal chart and Eq. (15-4) does not produce very accurate lattice-constant values. Although rotating-crystal cameras can be used for their precise determination (see reference to the book by M. J. Buerger at the end of this chapter), an oversize film holder and special film readers are necessary for this purpose. The principles involved in precision lattice-constant measurements are briefly discussed in Chap. 17.

Indexing procedure. Once the unit-cell constants of a crystal have been determined, it is possible to index all the reflections recorded in a rotation photograph rather quickly. To illustrate how this is done, consider an orthorhombic crystal whose a^*, b^*, and c^* values have been determined. Next, suppose that Fig. 15-12 is a rotation photograph of this crystal taken about its c axis. For this crystal, the vector components $\boldsymbol{\xi}$ within each reciprocal-lattice layer are given by

$$\boldsymbol{\xi} = h\mathbf{a}^* + k\mathbf{b}^* \tag{15-5}$$

The ξ values of each reflection can be read from the Bernal chart. Since a^* and b^* have been determined previously, the a^*b^* reciprocal-lattice net is prepared ($\gamma^* = 90°$), and the ξ values in each level (separately) are marked off on a strip of paper, using the same scale as in the drawing. $\xi = 0$ is then pinned to the origin of the reciprocal-lattice net, and the strip of paper is rotated as shown in Fig. 15-13. Whenever a reflection whose ξ value appears on the strip coincides with a point on the net, it can be assumed that the reciprocal-lattice point responsible for that reflection has been identified. By marking in that point (using different colors for different layers), the reciprocal lattice can be reconstructed from a single photograph. The foregoing process is a graphical reconstruction of the actual diffraction process taking place in the rotating-crystal method, so that the indices of each reflection are automatically established.

Note that the left and right sides of the rotation photograph of Fig. 15-2 are identical because each reciprocal-lattice point passes through the sphere of reflection in an identical manner on both sides of the direct beam. Because the reciprocal lattice normally is centrosymmetric, the top and bottom halves of the film are identical also. In other words, all rotation photographs have the symmetry mm about the direct beam, so that they are of very limited value in determining

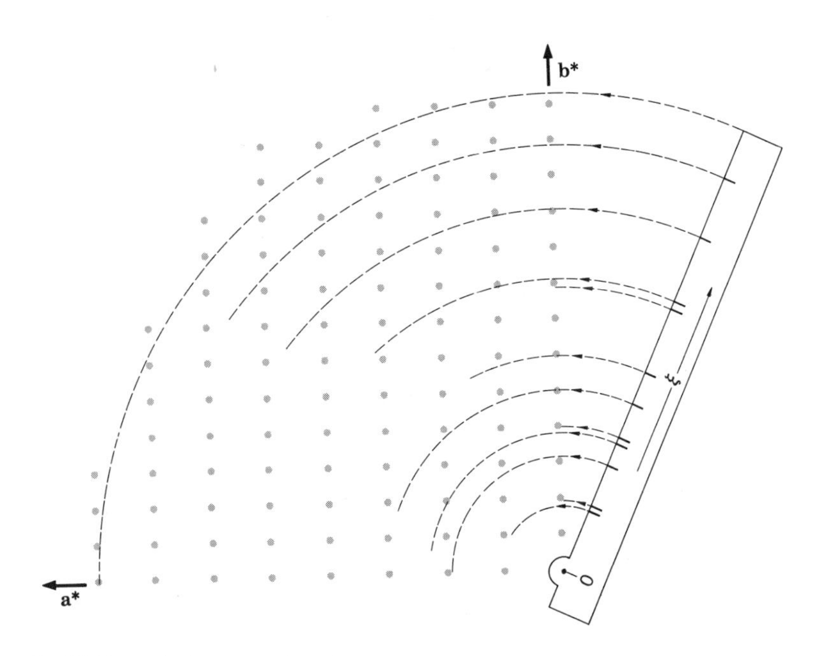

Fig. 15-13. Indexing procedure for a rotating-crystal photograph. The ξ values for the zero level in Fig. 15-12 are shown marked off along the paper strip.

crystal symmetry. Moreover, the really unique information is constrained to only one quadrant of the film, and any actual variations in intensity between reflections having identical or accidentally similar ξ values in any level cannot be discovered, because such reflections obviously superimpose on each other. Thus the rotating-crystal method is really useful only for the determination of the lengths of lattice constants when the crystal's symmetry is known.

OSCILLATING-CRYSTAL ARRANGEMENT

Choice of oscillation range. The limitation of the rotating-crystal method described above also can be described by noting that only two of the three cylindrical coordinates (Fig. 15-4) can be determined because the individual φ values are lost during a complete rotation. Suppose that, instead of a complete rotation, the crystal is rotated through a very small angle only. Then the φ angle of a reciprocal-lattice point producing a reflection is constrained to lie within this limited rotation range. In order to build up a detectable blackening in the film, the crystal is rotated back and forth, that is, oscillated through this limited angular range. Most rotating-crystal cameras have provisions for oscillating a crystal built into the drive mechanism either in the form of cams or angular limit switches (Fig. 15-8).

The reciprocal-lattice construction for this arrangement is illustrated in Fig. 15-14. Instead of rotating the reciprocal lattice, the sphere of reflection is shown rotated about the reciprocal-lattice origin at 0. The two circles of radius R_0 represent the positions of the Ewald sphere at the two extremes of the oscillation range. All the reciprocal-lattice points lying within the shaded region pass through the sphere during the oscillation, so that only the corresponding planes have an opportunity to satisfy the reflection condition. It is easy to see, in Fig. 15-14, that the φ angle of any reciprocal-lattice spot can now be identified, since the $\boldsymbol{\xi}$ vector must terminate within the shaded region of the level. Note, however, that the actual area of the shaded region varies from level to level. As shown in Fig. 15-15, the radius of the circle where the nth level cuts the sphere R_n decreases as ζ_n

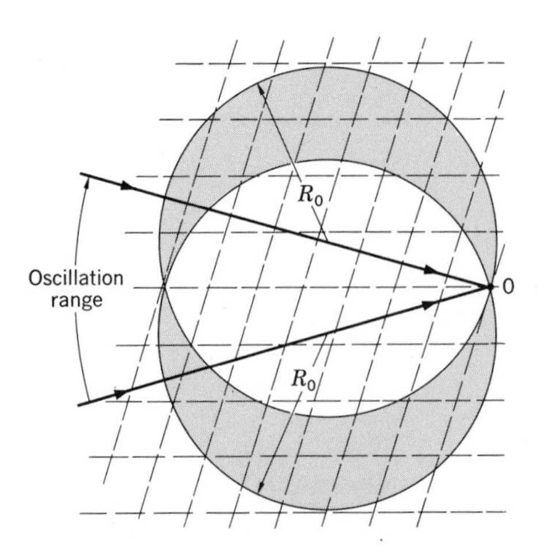

Fig. 15-14

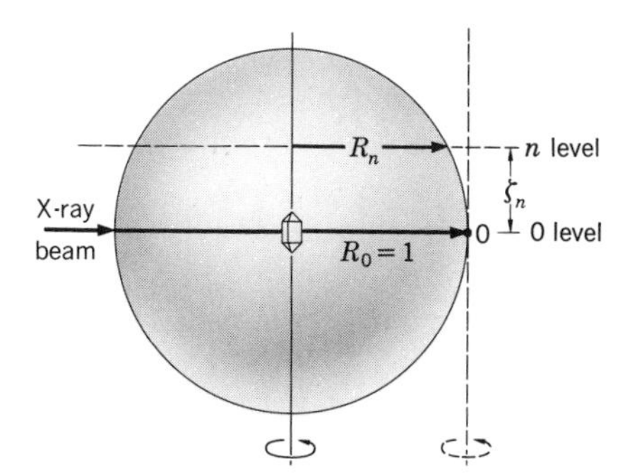

Fig. 15-15

increases. Also note that there is a blind region about the rotation axis of the reciprocal lattice that cannot be recorded in the normal-beam arrangement.

The actual oscillation range to be employed in any given case depends on the density of reciprocal-lattice points in the net. The larger the reciprocal cell (smaller crystal cell), the larger the oscillation range can be. Usually, either 10, 15, or 20° oscillation ranges are selected. In preparing successive photographs intended to map out the entire reciprocal lattice, it is best to overlap successive oscillation ranges by a few degrees. The total number of photographs required, of course, depends on the symmetry of the crystal, as well as on the cell size of its lattice. Referring to Fig. 15-3, it should be kept in mind that only the points lying in the tore of reflection can be recorded with a single mounting of the crystal.

Interpretation of photographs. The indexing of reflections in an oscillation photograph (Fig. 15-16) is carried out quite similarly to the procedure described above for indexing rotation photographs. The reciprocal-lattice coordinates must be known in advance, for example, from previously prepared rotation photographs. A reciprocal-lattice net then is prepared like the one in Fig. 15-13. Since the oscillation range is known from the experiment, by trial-and-error matching, the ξ values are identified in either of two crescent-shaped regions, like those shaded in Fig. 15-14. Once the orientation of the net relative to the x-ray beam in the experimental arrangement used has been established, the subsequent indexing takes place by shifting, mentally, the crescent-shaped regions as successive photographs are analyzed. By this means, the individual reflections recorded can be identified properly with their respective reciprocal-lattice points.

As can be seen in Fig. 15-16, an oscillation photograph is normally not symmetric about the direct beam. The reason for this becomes evident when Fig. 15-14 is studied. Unless the oscillation range happens to bracket a symmetry element lying halfway between the direct-beam directions shown, the reciprocal-lattice points lying in the crescent-shaped sections of each level are different. Thus, by eliminating the overlap of reflections, it is possible to study their individual reflecting powers. As discussed in Chap. 11, this makes it possible to establish systematic absences and to deduce the symmetry of the crystal by formulating its diffraction symbol. This requires, of course, that the entire unique portion of the reciprocal lattice be explored so that any reflections actually present are not overlooked. For an orthorhombic crystal, for example, this requires that a $90°$ sector be recorded. Although this necessitates several photographs, it should be realized that the exposures are proportionately shorter than those employed in a complete rotation.

Crystal-orienting procedure. The parallelism between a rational direction in the crystal and the rotation axis of the camera must be maintained during the entire rotation to within a small fraction of one degree. Unless some auxiliary device was employed to thus align the crystal on the goniometer head, it sometimes becomes necessary to make the final adjustments directly on the instrument. The need for this becomes evident when a complete rotation photograph contains groups of spots (dispersed along the contours of equal ξ) instead of single spots, the dispersal increasing radially outward from the direct-beam spot. Provided that individual layer lines nevertheless can be discerned, it is possible to refine the orientation by the following simple procedure: An oscillation range of about $15°$ is selected, so that, at its midpoint, one of the arcs in the goniometer head is parallel to the x-ray beam, and then an oscillation photograph is prepared. If the zero-level net is tilted as shown in Fig. 15-17, it will intersect one side of the sphere above the equator and the other below. Thus the layer lines in the photograph will appear tilted. Because the amount of the tilt is usually only a few degrees, two oscillations $180°$ apart are recorded, superimposed on the same film (after the

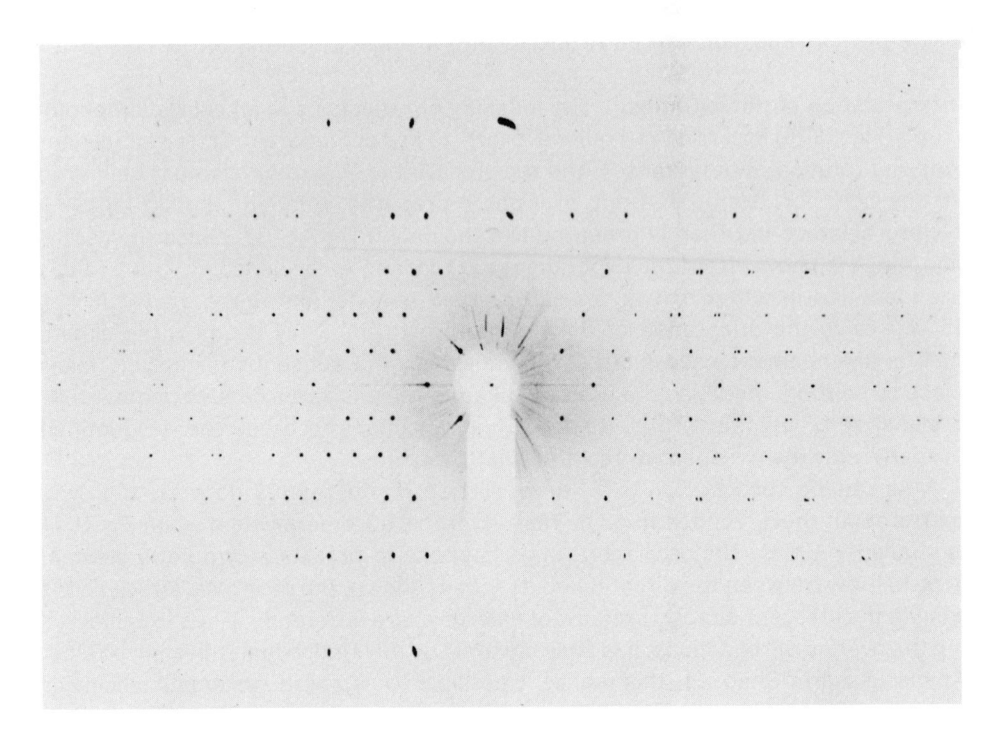

Fig. 15-16. Oscillation photograph of $(NH_4)_2Pt(S_5)_4 \cdot H_2O$ rotated about its b axis. (Compare to Fig. 15-2.)

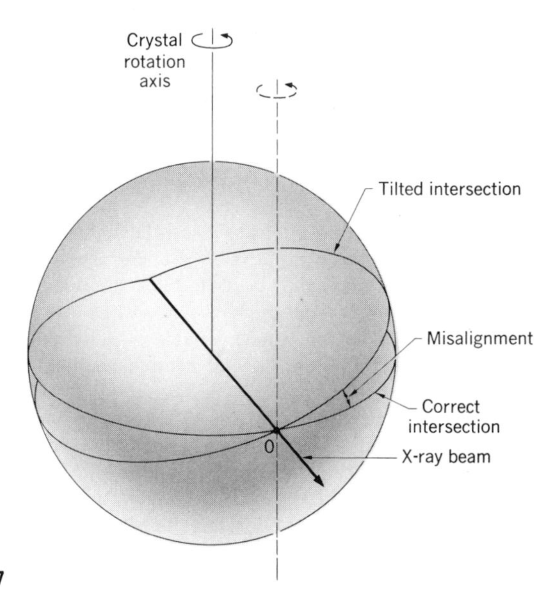

Fig. 15-17

correct sense of the tilt has been established). The amount of angular adjustment required for the arc that is normal to the x-ray beam then can be measured directly on the film. After one arc has been correctly adjusted, the crystal is rotated by $90°$, and the procedure is repeated by oscillating about the second arc. If both arcs are out of adjustment, the correction needed for one arc is partly masked by the tilt about the other. Such interference is minimized by limiting the oscillation range. It usually turns out that one or two repeated cycles are sufficient to obtain the desired parallelism in the alignment.

SUGGESTIONS FOR SUPPLEMENTARY READING

J. D. Bernal, On the interpretation of x-ray, single crystal, rotation photographs, *Proceedings of the Royal Society of London*, ser. A, vol. 113 (1926), pp. 117–160.

M. J. Buerger, *X-ray crystallography* (John Wiley & Sons, Inc., New York, 1942), especially pp. 133–213. (Precise measurement of lattice constants of single crystals is described in chap. 21.)

EXERCISES

15-1. Prepare a drawing like Fig. 8-10 or a cross section like Fig. 15-15, showing the reciprocal-lattice construction when the incident x-ray beam is inclined to the rotation axis of the crystal. **(Hint:** Remember that the origin of the reciprocal lattice lies on the sphere.) Show that by suitable selection of this angle it is possible to bring the point $\xi = 0$ of any level onto the sphere of reflection, eliminating thereby the blind region of the normal-beam method.

15-2. What is the total number of reflections that can be recorded in a complete rotation photograph of a body-centered cubic crystal having $a = 5.2$ Å when Cu $K\alpha$ radiation ($\lambda = 1.54$ Å) is employed?

15-3. Using Fig. 15-10 as a guide, design a collimator and beam catcher for a cylindrical camera having a diameter of 57.3 mm. Assume that the width of the x-ray beam striking the entrance pinhole is much wider than the 1 mm diam of the main pencil of x rays transmitted. Limit the projection of the collimator outside the camera cylinder to 1 cm, and assume that the target-to-crystal distance in this case is 6 cm. **Hint:** Remember to include beam divergence in deciding on the opening in the beam catcher, and try to select the lengths of the collimator, etc., to produce optimum results.

15-4. Suppose that the intensity of a reflected beam increases in proportion to the length x of the x-ray path through the crystal. Concurrently, the beam is attenuated by an amount that increases exponentially with x. Determine the relation between the linear absorption coefficient of the crystal and the optimum pathlength at which the two opposing factors balance. (**Hint:** The maximum of a dependent variable is determined by setting its derivative equal to zero.) What is the relation between μ and x if it is assumed that the intensity increases in proportion to the volume?

15-5. Using the results of Exercise 15-4, what are the optimum sizes that a crystal of aluminum, quartz, and zinc should have for a diffraction experiment employing $\mathrm{Cu}\ K\alpha$ radiation? Look up the necessary physical constants in a handbook.

15-6. Compare the reciprocal-lattice construction of a tetragonal crystal set to rotate about its a axis with that of the same crystal set to rotate about $[110]$. Which will have the shorter ζ value between layer lines? Does this same rule apply to a body-centered tetragonal crystal? **Hint:** The reciprocal lattice is face-centered.

15-7. Suppose that the orthorhombic crystal whose c-axis-rotation photograph is shown in Fig. 15-12 has $a^* = 0.25$, $b^* = 0.15$, and $c^* = 0.36$. Construct the appropriate reciprocal-lattice net to index all the reflections shown, and then proceed to determine the indices, using the values read from the Bernal chart shown. What is the lattice type of this crystal?

15-8. What are the indices of the reflections lying in the first layer line above the zero level in Fig. 15-12? (See Exercise 15-7.)

15-9. Consider a crystal oscillating $5°$ to each side of a coincidence between the direct-beam direction and one of the two arcs in the goniometer head. Next suppose that the arc perpendicular to the beam is correctly adjusted but the parallel arc is off by about $3°$. What effect will this misalignment have on the zero layer line? Suppose now that the other arc is deliberately misadjusted by $3°$ so that the brass pin holding the crystal moves up along the arc and to the right. What effect will this misadjustment have on the zero layer line of the oscillation photograph? **Hint:** Readers having difficulty picturing the reciprocal-lattice net should prepare a stereographic projection along the x-ray beam direction, remembering, as shown in Fig. 15-17, that the zero level must pass through the reciprocal-lattice origin.

SIXTEEN Moving-film methods

WEISSENBERG METHOD

Reciprocal-lattice construction for zero level. It has been demonstrated in the preceding chapter that the degeneration of reflections recorded on a rotating-crystal photograph can be removed by limiting the rotation of the crystal during each exposure to just a few degrees. The oscillating-crystal method then is used to record several levels at a time, but only a small portion of each reciprocal-lattice layer. An alternative arrangement for resolving individual reflections was proposed by K. Weissenberg in 1924, and is illustrated schematically in Fig. 16-1. The rotating crystal is first surrounded by a metal cylinder containing an annular slit through which the diffraction cone of only one reciprocal-lattice layer can pass to the film. The rotating spindle on which the crystal is mounted is then coupled, usually by means of bevel gears, to a worm gear, which causes the film cassette to be translated (parallel to the rotation axis) during the rotation of the crystal. Thus the instrument in Fig. 16-1 performs in the following manner: While the irradiated crystal rotates, causing successive reciprocal-lattice points to intersect the sphere of reflection, the film is continuously moving, so that, as successive reflections emanate from the crystal, the film is displaced by a finite amount. Since the metal cylinder surrounding the crystal, called a *layer-line screen*, allows reflections corresponding to only one layer to reach the film, each Weissenberg photograph records the reflections for one layer of the reciprocal lattice, clearly resolved on different portions of the film. As shown below, the recording of complete levels on a single photograph considerably simplifies the interpretation and indexing procedures.

In order to see the relationship between the reciprocal-lattice construction and the appearance of individual reflections on a film, consider a top view of the crystal shown in Fig. 16-2. Suppose that at time t_1 a central row in the reciprocal lattice is oriented at right angles to the direct beam. As the crystal is rotated in a counterclockwise manner, at a later time t_2, this row has been rotated through an angle ω. Assuming that a reflected beam issues for each point of intersection of this row with the Ewald sphere, the reflected beam at this point forms the angle

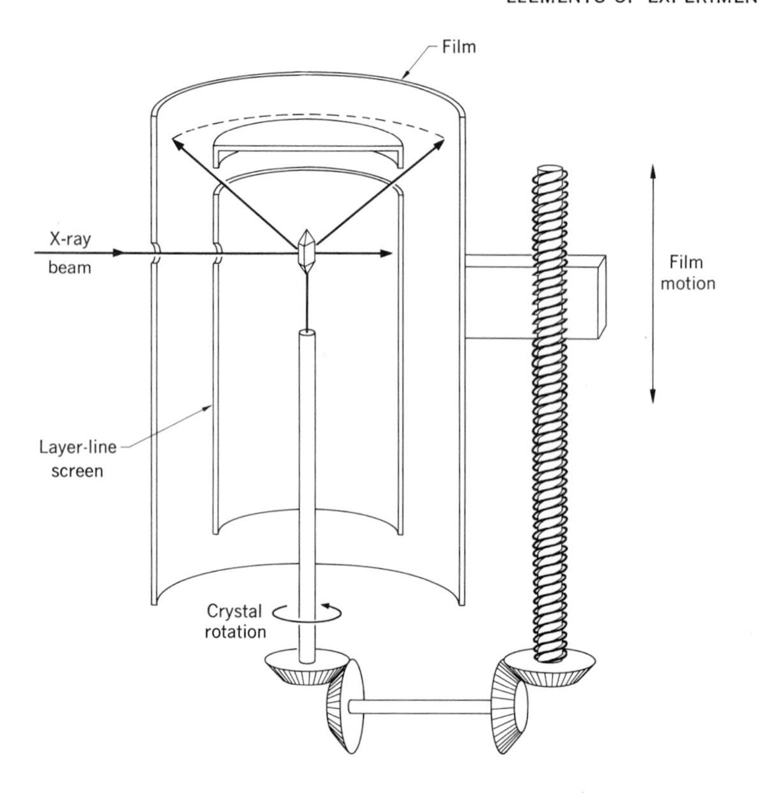

Fig. 16-1

Υ with the direct beam. (It is left to Exercise 16-1 to prove that $\Upsilon = 2\omega$.) Concurrent with this rotation of the crystal, the film is being displaced parallel to the rotation direction (at right angles to the plane of the drawing). Consequently, during the rotation of the crystal and the accompanying central reciprocal-lattice row through an angle ω, the direct beam traces out the middle line from t_1 to t_2 on the film, shown opened up to the right in Fig. 16-2. The reflected beam strikes the film at a distance above this line that is directly proportional to Υ. Since $\Upsilon = 2\omega$, the trace of the reflected beams issued for each point along the central reciprocal-lattice row can be drawn in on the film as shown. Note that as ω approaches $90°$, the central row of the reciprocal lattice cuts the Ewald sphere near the entrance point of the direct beam. This is the position where the film cylinder surrounding the crystal is split, so that the reflections produced when the same central row intersects the other half of the sphere are recorded on the bottom half of the film. Continuous rotation then produces the second line shown on the film in Fig. 16-2. It should be evident that at time t_3 the central reciprocal-lattice row is again normal to the incident x-ray beam direction, so that the distance from t_1 to t_3 along the middle line of the film corresponds to a rotation of $\omega = 180°$.

Consider next a central row in the reciprocal lattice that is $90°$ away from the one just discussed. These two rows could be, for example, the directions of the a^* and b^* axes of an orthogonal crystal. It follows from the preceding discussion

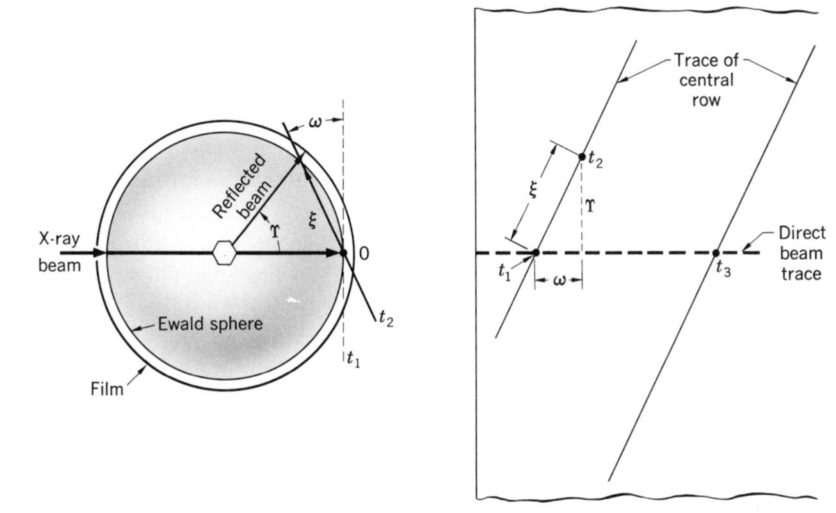

Fig. 16-2

that the two rows produce parallel lines on the film that are separated by $\omega = 90°$ along the direct-beam trace. Next consider a reciprocal-lattice row that is parallel to the a^* axis but does not pass through the origin. The reciprocal-lattice points in this row obviously have a constant k value, say, $k = 1$, as indicated in Fig. 16-3. Now consider the two parallel rows during a rotation of the crystal. Since the central row ($k = 0$) is tangent to the Ewald sphere, its intersection produces a line on the film that originates at the direct-beam trace. The parallel noncentral row intersects the sphere some instant of time later, and since it is not tangent to the sphere

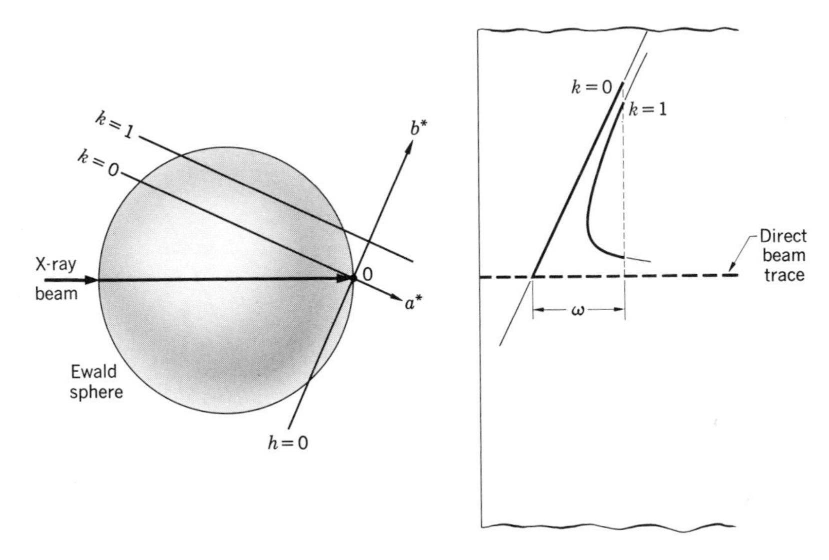

Fig. 16-3

at the origin, its intersection does not produce a line originating at the direct-beam trace on the film. In fact, as can be seen in Fig. 16-3, such a row intersects the sphere of reflection at some point, and then proceeds to cut it on both sides of this point as the crystal is rotated. The appearance of the resulting curve on the film after rotation by an amount ω is shown to the right in Fig. 16-3.

As the crystal's rotation continues, the central reciprocal-lattice row parallel to b^* begins to intersect the Ewald sphere. Some time later, therefore, the relative disposition of the three rows in the reciprocal-lattice, shown in Fig. 16-3, and their traces on the film appear as shown in Fig. 16-4. The solid lines on the film indicate the locations of possible reflections generated by the intersection of these three rows following a rotation of the reciprocal-lattice to the point indicated on the left in Fig. 16-4, assuming that the exposure began with b^* parallel to the x-ray beam. The dotted lines indicate the complete curves produced during a $360°$ rotation. An actual Weissenberg photograph of the zero level of an orthorhombic crystal is shown in Fig. 16-5. Since the crystal was rotated about its b axis in producing this photograph, the reciprocal-lattice points recorded all have $h0l$ indices. The actual indexing procedures are discussed in more detail later, but it should be clear now that the points lying along each reciprocal-lattice row, clearly visible in Fig. 16-5, have one of the two indices constant. The shape of the curves formed by these

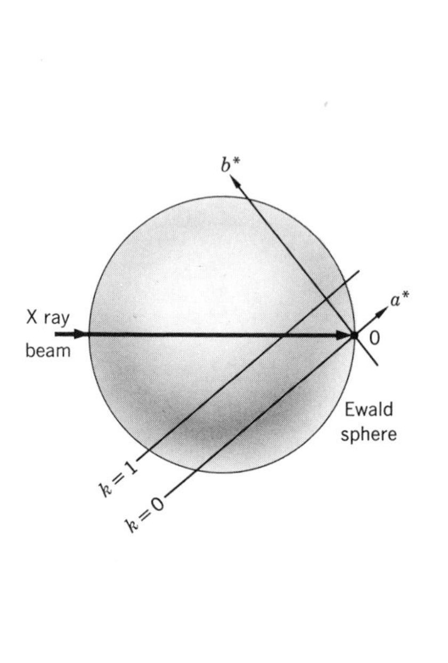

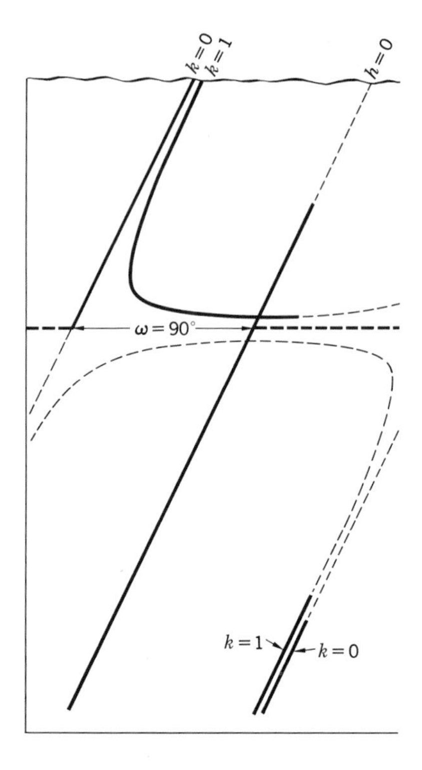

Fig. 16-4

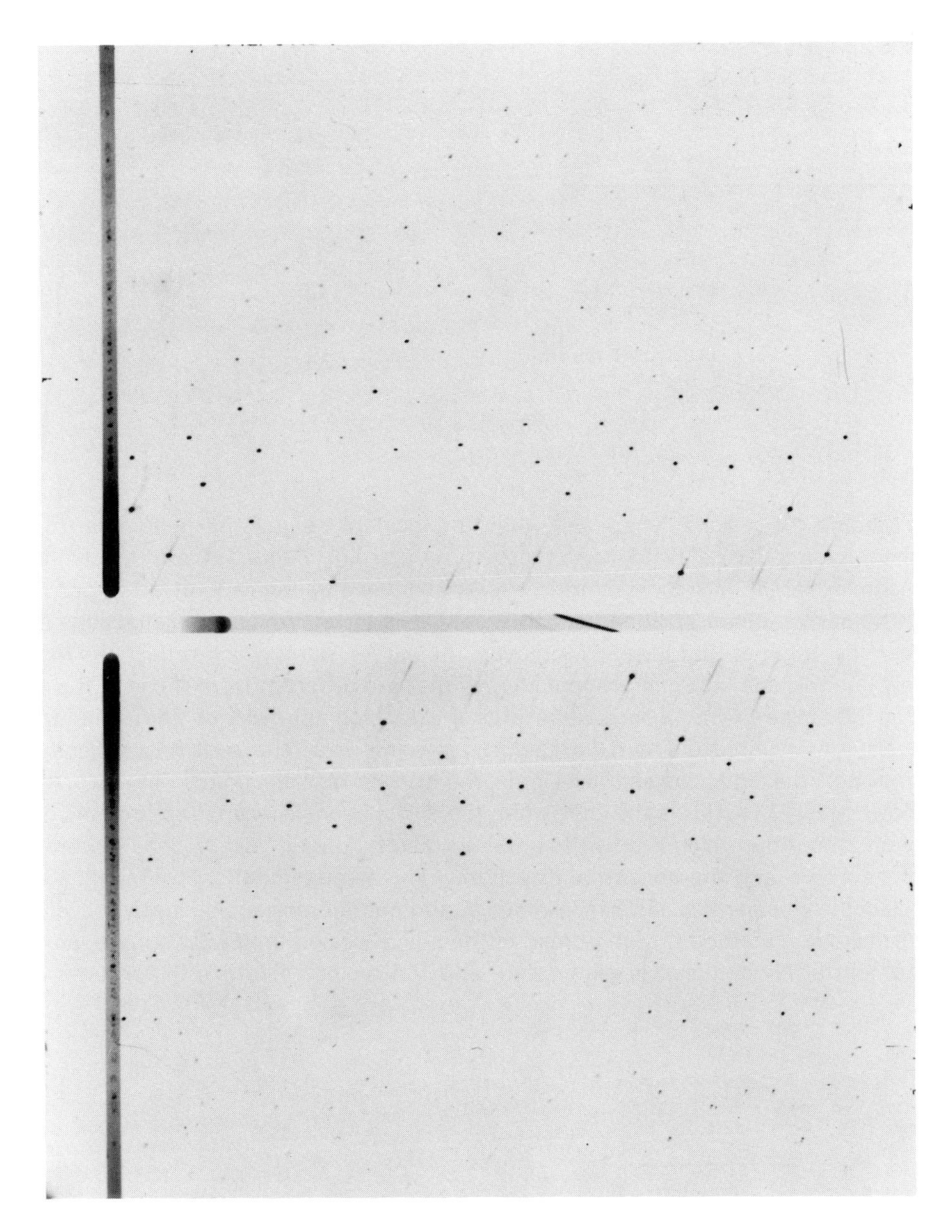

Fig. 16-5. Zero-level Weissenberg photograph of $(NH_4)_2Pt(S_5)_4 \cdot H_2O$ rotated about its b axis.

rows illustrates the way the reciprocal lattice is distorted by the Weissenberg method.

Reciprocal-lattice construction for upper levels. A side view of the rotating crystal surrounded by the Ewald sphere is shown in Fig. 16-6. As already pointed out in

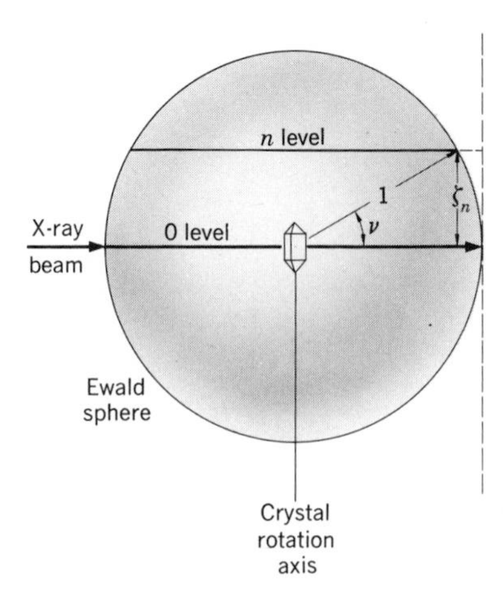

Fig. 16-6

preceding discussions (Chap. 13), when the incident beam is perpendicular to the rotation direction, a blind region develops near the center of any upper level, becoming larger with increasing ζ. A Weissenberg photograph of an upper level in the normal-beam arrangement therefore fails to record certain reflections lying near the origin of that level. Moreover, the curves along which reflections having only one variable index lie assume shapes that are different from those in the zero level (Exercise 16-3). This complicates a direct comparison of zero and upper levels. As pointed out by Buerger, it is possible to circumvent this difficulty by "moving" the origin of any level onto the sphere of reflection. For the n-level, shown in Fig. 16-6, this is the case when the direct beam is inclined to the horizontal by an amount μ, which is also the angle formed by the diffraction-cone generators of the n-level and the horizontal direction. The arrangement shown in Fig. 16-7 is called the *Buerger equi-inclination method*, and has the obvious advantage over the normal-beam method that the origin of the n-level lies on the Ewald sphere, so that the central blind region is eliminated, and the reciprocal-lattice distortion in the

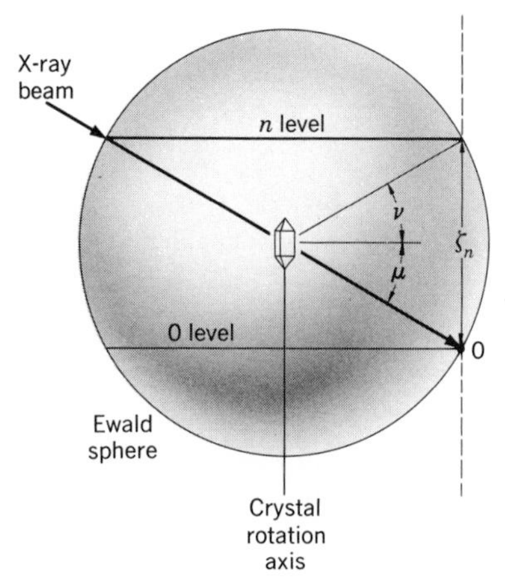

Fig. 16-7

Fig. 16-8. Upper-level Weissenberg photograph of $(NH_4)_2Pt(S_5)_4 \cdot H_2O$ rotated about its b axis.

n-level photograph is identical with that of a zero-level photograph. An actual upper-level photograph is shown in Fig. 16-8.

Camera arrangement. An instrument designed for recording Weissenberg photographs by the equi-inclination method is shown in Fig. 16-9. The uppermost

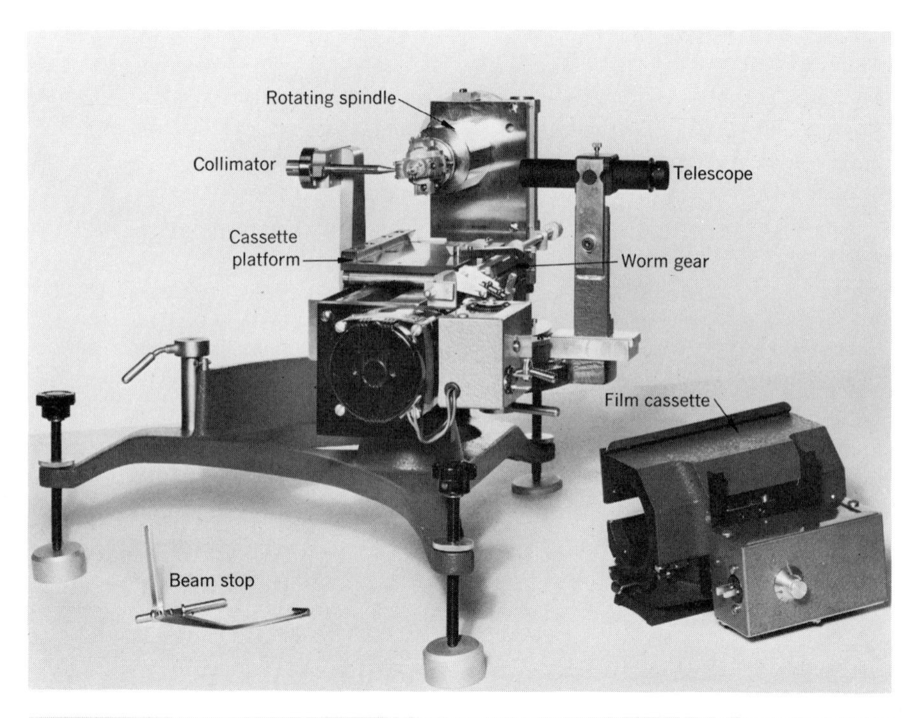

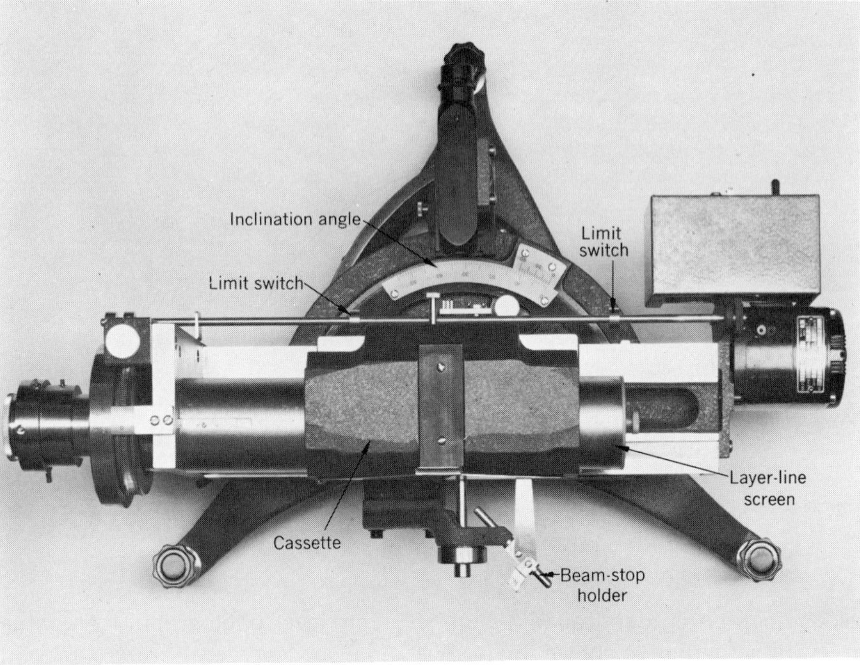

Fig. 16-9. Equi-inclination Weissenberg camera. End view of camera (upper picture) shows the components, and the top view (lower picture) shows the assembled camera. *(Courtesy of Charles Supper Company.)*

photograph shows the crystal mounted in a goniometer head and a telescope, attached in front, which is used to center the crystal and constitutes an alternative arrangement to sighting the crystal through the collimator. The use of the telescope is more convenient when the camera is in position up against the x-ray tube tower; on the other hand, it necessitates the maintenance of the telescope in correct adjustment to the rest of the instrument, whereas sighting the crystal through the collimator verifies the actual position of the crystal in the x-ray beam. In the arrangement shown in Fig. 16-9, the instrument can be used as a rotating-crystal camera by inserting a suitable direct-beam stop and surrounding the crystal with a stationary cylindrical film. This can be done by placing the film holder on the platform (below the crystal) and disengaging the platform from the horizontal worm gear. The camera is then centered opposite the collimator and, when the instrument is turned on, the crystal rotates in the x-ray beam. By means of two suitably placed microswitches, the crystal can be oscillated through a preselected angular range as well.

To convert the instrument to take Weissenberg photographs, a layer-line screen must be slipped over the crystal, as indicated in the lower photograph in Fig. 16-9. The instrument is shown in a top view in this picture to indicate the location of the angular scale which must be adjusted when some upper level is to be recorded. Finally, the film camera is placed on its platform, which is engaged to the horizontal worm gear. The two limit switches are adjusted to limit the camera's travels along the gear, the full range being equal to slightly more than $180°$ rotation of the crystal. The assembly shown in the lower photograph in Fig. 16-9 is ready to start recording a zero-level photograph.

The setting of the instrument to record upper levels requires a knowledge of the layer-line separations parallel to the rotation axis. This is usually determined by first taking a rotating-crystal photograph. The relationship between the inter-layer distance z on the film and the radius of the camera (film radius R) is, according to Fig. 16-6,

$$\tan \nu = \frac{z}{R} \tag{16-1}$$

while from the reciprocal-lattice construction,

$$\sin \nu = \frac{\zeta}{1} \tag{16-2}$$

Consulting Fig. 16-7 shows that the equi-inclination angle μ is related to the interlayer spacing by

$$\sin \mu = \frac{\zeta}{2} \tag{16-3}$$

Combining the above equations yields an expression for the equi-inclination angle in terms of the camera's parameters,

$$\mu = \sin^{-1} \left[\frac{\sin \tan^{-1} (z/R)}{2} \right] \tag{16-4}$$

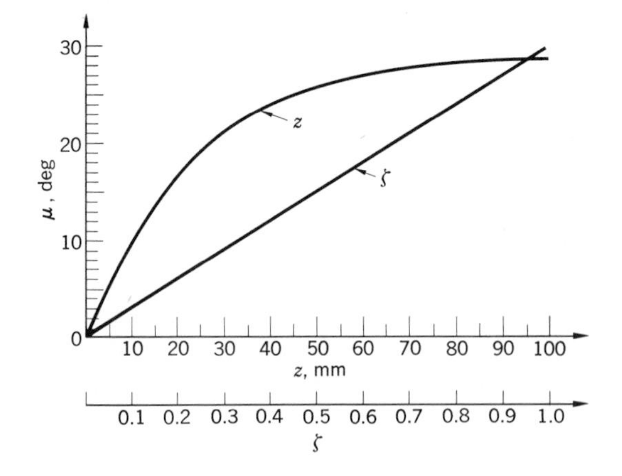

Fig. 16-10. Graph showing equi-inclination angle μ as a function of the reciprocal-lattice coordinate ζ, or the actual height z, for a camera of diameter 57.3 mm. *(After Buerger.)*

Using Eq. (16-3) or (16-4), a chart can be prepared (Fig. 16-10) from which the equi-inclination angles can be read directly.

It is also necessary to advance the layer-line screen opening by an appropriate amount to transmit the desired level cone. The necessary displacement s is determined by the radius of the layer-line screen cylinder r (Exercise 16-4).

$$s = r \tan \mu \tag{16-5}$$

Again, the relation in (16-5) can be plotted for the cylinder radius employed (Fig. 16-11), so that, once the equi-inclination angle μ has been determined from Fig. 16-10, the screen displacement s is quickly established. Note that when the direct-beam catcher is attached to the screen cylinder, it must be displaced by a corresponding amount in the reverse direction. This is not necessary when a separately mounted direct-beam catcher (Fig. 16-9) is employed. The physical dimensions of the instrument set an upper limit to the equi-inclination angle that can be attained at about $25°$.

Practical considerations. A Weissenberg camera is more convenient to use than a rotating-crystal camera when small adjustments to the crystal's orientation in the goniometer head are required. The orienting procedure described in Chap. 15 for the rotating-crystal method can be refined by placing the layer-line screen over the crystal so that only the zero-level reflections can reach the film. Any deviations

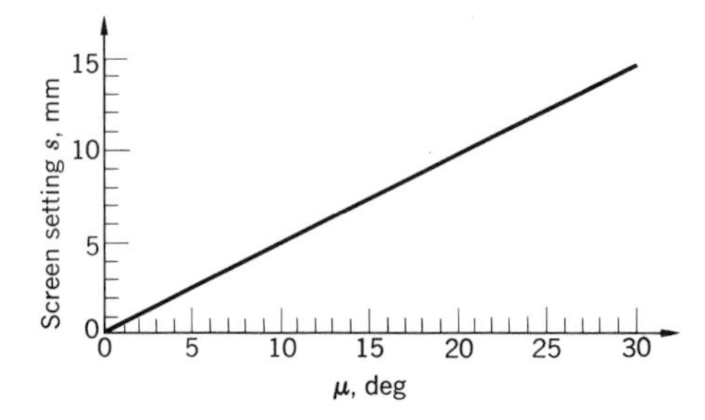

Fig. 16-11. Relation of layer-line screen setting s and equi-inclination angle μ for a layer-line screen of diameter 51 mm. *(After Buerger.)*

from linearity during the oscillations about each arc are now more clearly evident because the shadows cast by the opening in the screen serve as two parallel reference lines. Even when the crystal is properly oriented, it is good practice to take a short rotating-crystal exposure for each level. Usually, this is done by pushing the camera platform to one extreme end so that the rotating-crystal photograph of that level is recorded at one end of the film. When the Weissenberg motion is coupled in, the limiting switches are set so that this end portion of the film is not exposed during the film's subsequent movement. This procedure should be made an integral step in recording each photograph (Figs. 16-5 and 16-8), because it shows at a glance whether the layer-line screen was properly positioned before the exposure was begun.

Intensity measurement. The Weissenberg arrangement is one of the two most commonly used methods for recording diffraction intensities photographically. The other one is the Buerger precession method, described in the second half of this chapter. The practical aspects of measuring the intensities from the blackening of a spot are similar for both methods, so that they can be considered here. Assuming that the crystal being photographed is smaller than the incident beam's cross section, the size and shape of each reflection are proportional to the size and shape of the crystal. Because the intensity distribution across the spot is not uniform, the logarithmic relation between the opacity and the actual blackening of the film (Chap. 12) precludes the use of a densitometer. It should be noted that a procedure wherein a positive film print of the x-ray negative is prepared under suitable processing conditions that render this relationship linear has been developed by R. H. V. M. Dawton. Because the Dawton method encounters considerable difficulties in standardizing the various film-processing steps involved, it has not been widely used. Instead, it is common practice to use eye estimates of the blackening of the spots. A valuable aid in making such estimates is to prepare a graded intensity scale. This can be done by selecting some reflection of the crystal and retaining the crystal at the appropriate setting, while the film carriage is displaced manually, following successive x-ray exposures made in predetermined time intervals. (The reciprocity rule of x-ray exposure states that $I \times t = \text{const.}$) The series of spots constitute a scale of intensities (exposure times) ranging from, say, 1 to 20. Each spot on the film then can be matched with its counterpart on the graduated scale.

Recalling from Chap. 12 that the maximum blackening distinguishable on a film is approximately 2, ideally, the maximum range of intensities that can be measured on one film is from 1 to 100. The diffraction intensities emanating from a single crystal, however, may span a range from 1 to 10,000. In order to record the full intensity range, therefore, more than one film is needed. Two alternative procedures can be used. One makes use of the reciprocity law by preparing successive time exposures in the ratio of $1:2:4:8:16:32$. Only the most intense reflections are recorded on the shortly exposed films in such a *multiple-exposure series*, but their blackening is reduced to well below 2. The repetition of the same reflections on several films also serves as internal calibration for establishing relative intensities. The other scheme, called the *multiple-film method*, simply stacks several films behind each other inside the film holder. Depending on the radiation used, each film reduces the intensity transmitted through it by a known factor. (For Eastman No-Screen film this factor is 4.4 for $\mathrm{Cu}\ K\alpha$ and 1.3 for $\mathrm{Mo}\ K\alpha$.) Thus the intensity reaching the third film exposed with copper radiation is reduced to almost one-twentieth (4.4^2) of the value recorded on the first film. Even larger reductions in intensity can be achieved by interposing thin sheets of metal foil between the films.

The intensities measured by the procedures described above are merely proportional to the true integrated reflecting powers of the crystal planes. This is so because the eye is most conscious of the darkest portions of the reflection spots. Suppose, however, that the intensity going to every part of a spot could be made to strike the film in exactly the same place. This one point would receive an equivalent contribution from each part of the reflection (Fig. 16-12), so that its total blackening would be the sum of the blackening of each part of the reflection, that is, the true integrated intensity of the reflection. The successive displacements depicted in Fig. 16-12 require successive rotation of the film about the crystal's rotation axis, as well as small shifts parallel to it during the exposure. Such *integrating cameras*, based on the Weissenberg arrangement described above and on the Buerger precession method described below, have been devised and are available commercially. Since the optical density of the multiply exposed region in Fig. 16-12 is directly proportional to the integrated intensity, a microdensitometer can be used to measure the intensity of each reflection. The chief limitation of this arrangement is the accumulation of background intensity during the necessarily prolonged exposure.

Interpretation of photographs. Despite the distortion of the reciprocal-lattice lines by the Weissenberg method indicated in Fig. 16-4, it is possible to deduce the diffraction symmetry directly from the photographs. One way of doing this is to construct a device consisting of a triangle sliding along a horizontal ruler so that ξ along one side of the triangle and ω along the ruler can be measured directly (Fig. 16-2). The ξ and ω values of each reflection then can be plotted on a polar

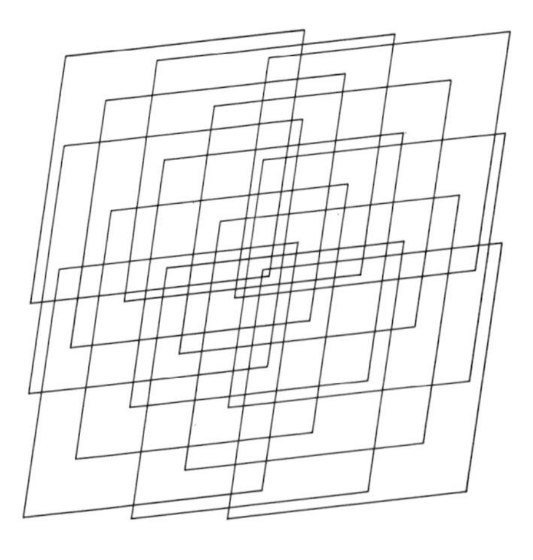

Fig. 16-12. Successive displacements of a reflection, causing central region to receive contributions from each portion of the reflection.

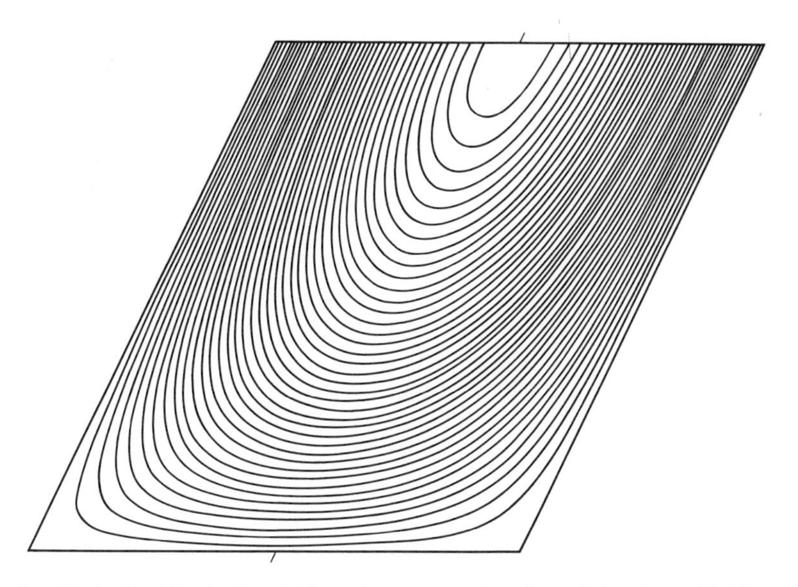

Fig. 16-13. Equi-inclination Weissenberg projection of parallel reciprocal-lattice rows for a film diameter of 57.3 mm. *(After Buerger.)*

net, reconstructing the undistorted reciprocal-lattice thereby. In view of the characteristic distortion that the noncentral reciprocal-lattice lines undergo (Fig. 16-4), it is also possible to construct a template like the one shown in Fig. 16-13. Superposition of such a template on a film then allows the identification of one of the two indices characterizing each spot in the layer. The template is positioned so that the spots on the film coincide with the lines on the template and a densely populated central reciprocal-lattice row lies at its center. (See Fig. 16-4.) The curves joining the reflections of each reciprocal-lattice row then can be drawn in directly on the film. Next the template is displaced until another parallel set of reciprocal-lattice lines having the other index constant is found in a similar way. Finally, the resulting curves appear as shown in Fig. 16-14, and the indices of each reflection can be read off directly.

A little bit of experience in the analysis of Weissenberg films permits the determination of the diffraction symmetry directly from their visual examination. For example, the number of densely populated central reciprocal-lattice rows recorded in a $180°$ rotation interval of the crystal serves to indicate the angles (along the horizontal ω scale) separating them, and hence the nature of the symmetry about the crystal's rotation axis. If the intensity of equivalent reflections on each side of a central row is the same, it can be reasoned that the reciprocal-lattice row lies along a two-fold axis or on a mirror plane. This ambiguity can be resolved by examining an upper-level photograph. If the symmetry element is a mirror plane, the same relation between intensities persists in the upper-level photographs. If it does not, the symmetry element must be a two-fold axis. Further relations between the appearance of Weissenberg photographs and the presence of symmetry elements in a crystal also can be developed. (See Exercises 16-2 and 16-4.)

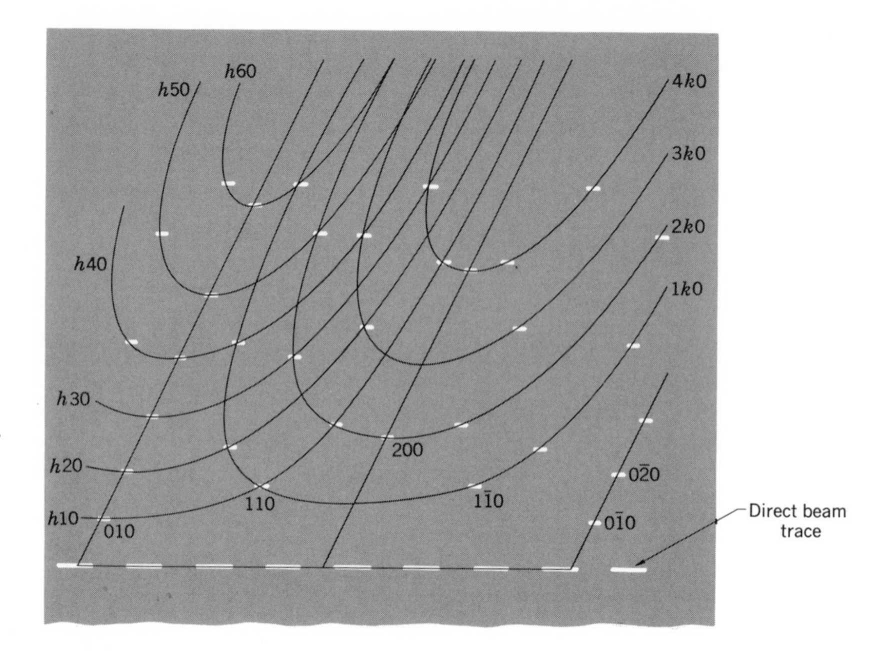

Fig. 16-14. Schematic view of top half of a Weissenberg film, with reciprocal-lattice rows indicated.

A complete discussion of symmetry determination from the Weissenberg method can be found in the book *X-ray crystallography* listed at the end of this chapter. The ease with which the indices of all reflections can be determined by the moving-film methods described in this chapter also simplifies the space-group determination procedure described in Chap. 11.

BUERGER PRECESSION METHOD

Reciprocal-lattice construction for zero level. In the discussion of single-crystal methods so far, it has been convenient to consider the reciprocal lattice as made up of parallel layers stacked along a direction coinciding with the crystal rotation axis. In the ensuing discussion it will be more convenient to consider the reciprocal-lattice layers that are perpendicular to the incident x-ray beam. Enumerating by $0, 1, 2, \ldots, n$ the layers along the direct-beam direction, as indicated in Fig. 16-15, consider first the zero level that contains the origin of the reciprocal lattice. Suppose that, initially, it is perpendicular to the incident x-ray beam, but that subsequently it is inclined by some angle $\bar{\mu}$. The way it will intersect the Ewald sphere is indicated in Fig. 16-16, where the inclination $\bar{\mu}$ is measured by the tilt of the layer's normal $\mathbf{n}$ relative to the x-ray beam direction. Note that the intersection takes place along a circle passing through the reciprocal-lattice origin 0, while its radius clearly depends on the inclination $\bar{\mu}$ (Exercise 16-5). If a reciprocal-lattice point (P in

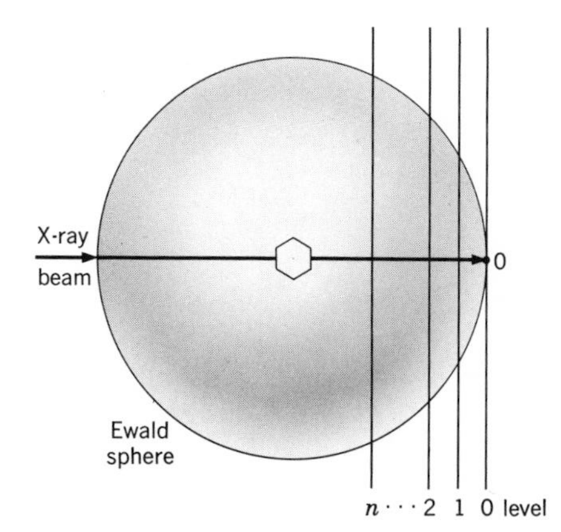

Fig. 16-15

Fig. 16-16) happens to lie along the circle, a diffracted beam will pass through it as shown. Next, suppose that the normal **n** is caused to precess about the x-ray beam without changing the inclination angle $\bar{\mu}$. This will have the effect of generating a sequence of circular intersections like the one shown in Fig. 16-16, so that all the reciprocal-lattice points lying in the zero level will be given a chance to intersect the Ewald sphere. Finally, if a film is similarly inclined and precessed about the x-ray beam direction, without altering its orientation relative to the reciprocal-lattice layer, it will intercept all the diffracted rays thus produced. The way that some of the circles generated during the precession appear when recorded on the film is illustrated in Fig. 16-17.

By keeping the film parallel to the reciprocal-lattice layer throughout the precession motion, the diffracted beams strike the film in exactly the same array as that of the reciprocal-lattice points in the layer. This means that the resulting photograph is simply an "enlargement" of the reciprocal-lattice layer and that such a precession motion allows the recording of the reciprocal lattice without any distortion whatever. The "magnification" factor on the film is directly proportional to the crystal-to-film distance, as indicated in Fig. 16-18. Measuring reciprocal-lattice distances in dimensionless units ($\sigma = \lambda/d$) so that the radius of the Ewald

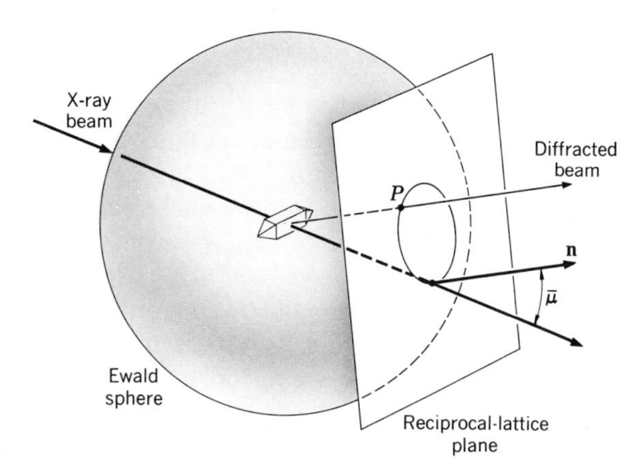

Fig. 16-16

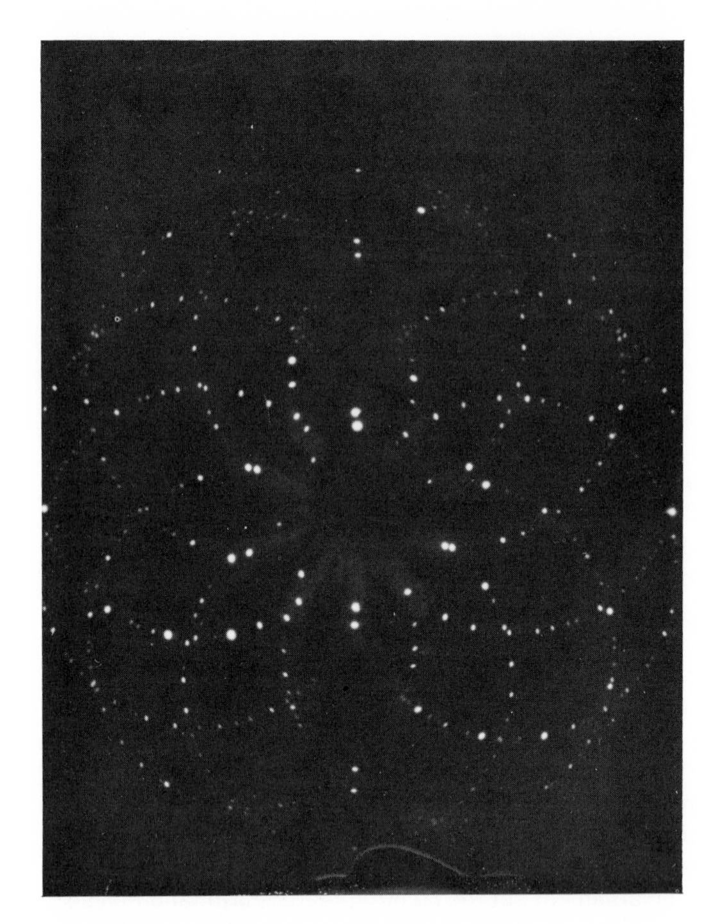

Fig. 16-17. Successive intersections of zero level with Ewald sphere, at regular intervals, during one complete precession.

sphere equals unity, from similar triangles,

$$\frac{P_F O_F}{PO} = \frac{CO_F}{CO} = \frac{M}{1}$$

from which $$P_F O_F = M(PO) \tag{16-6}$$

Thus the reciprocal-lattice spacings measured on the film can be converted into corresponding σ values by dividing them by the known crystal-to-film distance M, expressed in the same units. This fact and the ease of interpreting and indexing an undistorted reciprocal-lattice photograph are the principal reasons for the increasing popularity of the precession method invented by Buerger in 1942.

Reciprocal-lattice construction for upper levels. Consider an upper level inclined to the x-ray beam so that its normal $\mathbf{n}$ forms the angle $\bar{\mu}$ with the direct beam. Since the upper level lies a distance ζ away from the reciprocal-lattice origin, the origin of the upper level, O_u, is displaced from the direct beam, as indicated in Fig. 16-19. In order not to distort the reciprocal-lattice photograph of this upper

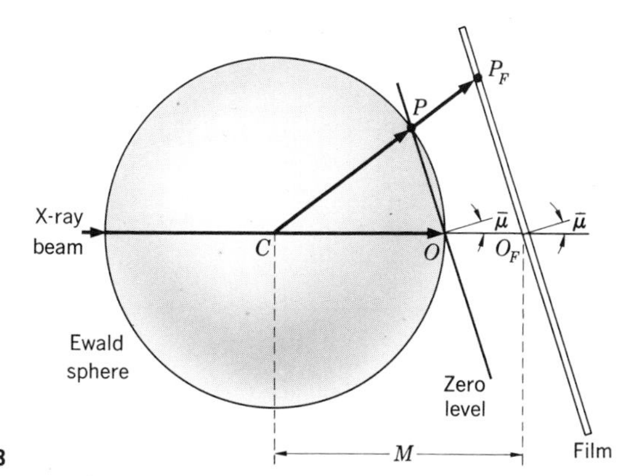

Fig. 16-18

layer, the film must be similarly displaced by an amount

$$\zeta_F = M\zeta \tag{16-7}$$

where M is the crystal-to-film distance, as before. It is left to Exercise 16-6 to prove that distances measured on the film then retain the same proportionality to reciprocal-lattice distance that held true for the zero level (16-6).

Note that a central portion of each upper level is unable to intersect the Ewald sphere during the precession of its normal about the x-ray beam. This causes a blind region to develop in upper-level photographs whose radius increases in proportion to ζ. Physical limitations of the camera similarly limit the magnitude of $\bar{\mu}$ and the uppermost level that can be recorded. The portion of the reciprocal lattice that can be explored with a single setting of the crystal is shown in cross section in Fig. 16-20. As described in the next section, the mounting axis of the crystal is perpendicular to the x-ray beam (and to the plane of Fig. 16-20), so that it

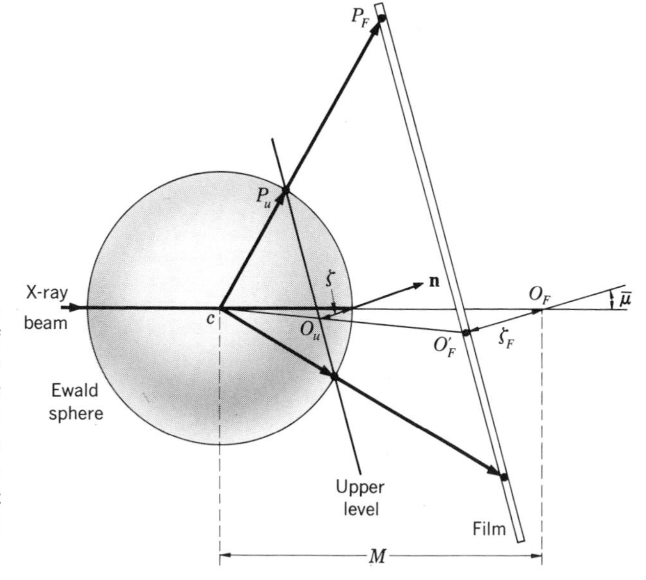

Fig. 16-19. Upper-level geometry. Note that the normal **n** of the upper level forms the precession angle $\bar{\mu}$ with the x-ray beam direction, but its origin O_u is displaced by ζ from the reciprocal-lattice origin. The film, similarly inclined, must be proportionately displaced from its origin O_F.

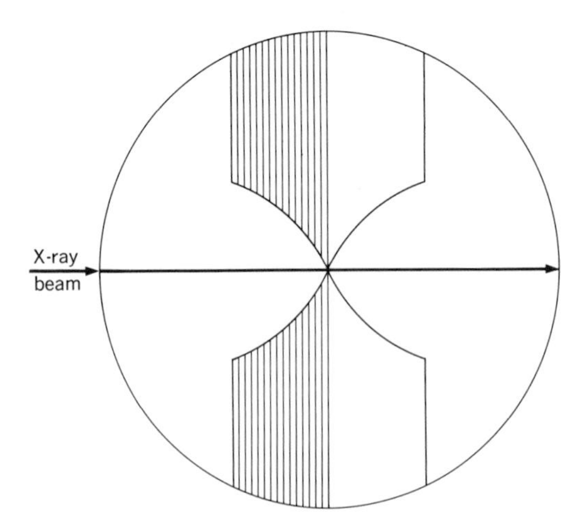

Fig. 16-20. Cross section of reciprocal-lattice sphere showing portion that can be recorded with a single setting of the crystal. The shaded portion is actually recorded, while the centrosymmetrically equivalent portion is shown in outline only. [*From L. V. Azároff, Review of Scientific Instruments, vol.* 25 (1954), *pp.* 928–929.]

is possible to explore the entire reciprocal lattice within the indicated limiting sphere without having to remount the crystal (Fig. 16-21). The number of successive settings that may be required depends on the symmetry of the crystal about the mounting axis, and Table 16-1 lists the symmetry axes that should be chosen as mounting axes to minimize the number of settings required. Note in Table 16-1 that this number never exceeds four, even when the crystal is triclinic.

Camera arrangement. Since all the levels of the reciprocal lattice intersect the sphere of reflection concurrently, it is necessary to isolate each level by means of a suitable layer-line screen, as indicated in Fig. 16-22. As might be expected from the preceding discussion and Fig. 16-17, the layer-line screen is a plane metal sheet, opaque to x rays, except in an annular opening through which the diffraction cone generated by the particular reciprocal-lattice layer desired can pass through.

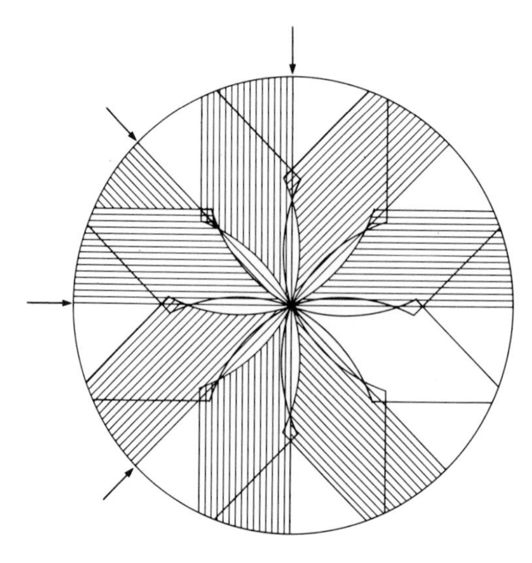

Fig. 16-21. Maximum number of crystal settings (four) required to record entire reciprocal-lattice sphere. The open areas are centrosymmetrically related to the shaded ones (Fig. 16-20).

Table 16-1 Crystal settings for upper-level photography

Crystal system	*Laue symmetry*	*Crystal mounting axis*	*Crystal settings†*							
			Crystal space	*Reciprocal space*	*Crystal space*	*Reciprocal space*	*Crystal space*	*Reciprocal space*	*Crystal space*	*Reciprocal space*
Hexagonal	$6/m$; $6/mmm$	c	a	$\|\|d_{100}^*$	[120]	$\|\|a^*$				
	$\bar{3}$; $\bar{3}m$	c	a	$\|\|d_{100}^*$	[120]	$\|\|a^*$				
Cubic	$m3$; $m3m$	a	a	$\|\|a^*$	[110]	$\|\|d_{110}^*$				
Tetragonal	$4/m$; $4/mmm$	c	a	$\|\|a^*$	[110]	$\|\|d_{110}^*$				
Orthorhombic	mmm	b	a	$\|\|a^*$	c	$\|\|c^*$	$[u0w]$	$\|\|d_{h01}^*$		
Monoclinic	$2/m$	$[uv0w]\|\|d_{100}$	c	$\|\|c^*$	b	$\|\|d_{010}^*$	$[0vw]$	$\|\|d_{0k1}^*$		
Triclinic	$\bar{1}$	$[uvw]\|\|d_{010}$	c	$\|\|d_{001}^*$	$[u_1v_1w_1]$	$\|\|d_{h_1k_1l_1}^*$	$[u_2v_2w_2]$	$\|\|d_{h_2k_2l_2}^*$	$[u_3v_3w_3]$	$\|\|d_{h_3k_3l_3}^*$

† The directions listed in the columns headed *Crystal space* should be placed parallel to the x-ray beam when $\bar{\mu} = 0°$. (The nomenclature $[uvw]||d_{100}$ means the direction $[uvw]$ that is normal to the (100) plane and therefore parallel to d_{100}.)

Source: L. V. Azároff, Crystal settings for upper level photography, precession method, *Review of Scientific Instruments*, vol. 25 (1954), pp. 928–929.

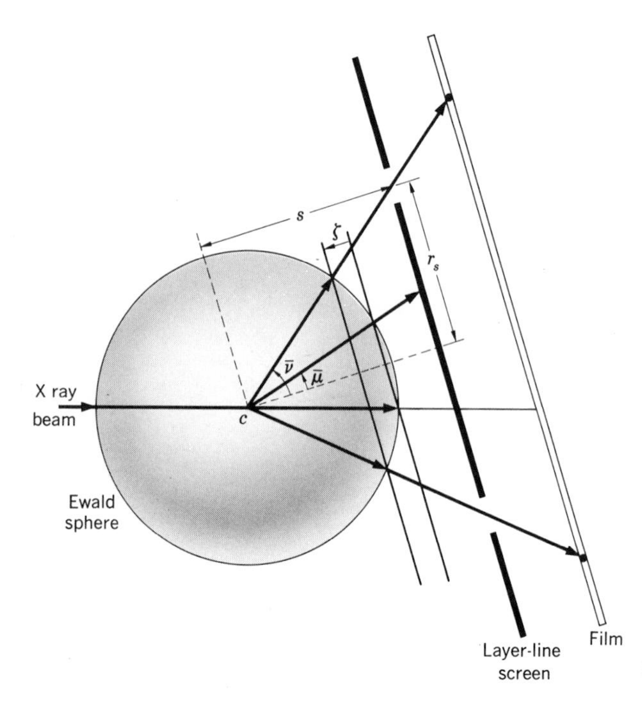

Fig. 16-22. *(After Buerger.)*

The radius of the annular opening in the screen, r_s, is related to the screen-to-crystal distance s, according to Fig. 16-22, by

$$\frac{r_s}{s} = \tan \bar{\nu} \tag{16-8}$$

where $$\bar{\nu} = \cos^{-1} (\cos \bar{\mu} - \zeta) \tag{16-9}$$

as demonstrated in Exercise 16-7.

The magnitude of the spacing ζ separating adjacent reciprocal-lattice levels can be determined by a construction similar to that in Fig. 16-22. The diffraction cones of the zero level and the n-level, a distance ζ_n removed from it, are shown in Fig. 16-23. With the radius of the Ewald sphere chosen as unity, it is easy to prove, considering the two right triangles in Fig. 16-23, that

$$\zeta_n = \cos \bar{\mu} - \cos \bar{\nu} \tag{16-10}$$

where, similarly to (16-8), $\bar{\nu} = \tan^{-1} (r_n/s)$, so that

$$\zeta_n = \cos \bar{\mu} - \cos \tan^{-1} \frac{r_n}{s} \tag{16-11}$$

and s is the crystal-to-screen distance. By inserting a photographic film in the holder normally used for a layer-line screen, it is possible to record the diffraction cones of each upper level as concentric circles. This has been named a *cone-axis photograph* by Buerger, and it is the counterpart of a rotating-crystal photograph used in the Weissenberg method to determine ζ.

Practical considerations. By preselecting the screen distance s and measuring r_n on the film, it is possible to calculate ζ_n with the aid of (16-11). Knowing ζ then enables the determination of the settings necessary to record upper levels. Relation (16-9) can be combined with (16-8) to give the screen distance s for any value of the screen radius r_s.

$$s = r_s \cot \cos^{-1} (\cos \bar{\mu} - \zeta) \tag{16-12}$$

where $\zeta = 0$ for the zero level.

A commercially available Buerger-precession camera is shown in Fig. 16-24. The crystal, mounted on a conventional goniometer head, is affixed to a horizontal spindle at right angles to the x-ray beam. The entire camera is mounted on a rotatable base so that it can be swung aside to allow centering of the crystal by sighting through the collimator. When a crystal has well-developed faces, an autocollimator supplied with the camera can be used to align the crystal in the goniometer head. When it does not, the precession instrument itself can be employed most conveniently for this purpose, as described in the next section. Once the crystal is correctly positioned, the camera is swung back into alignment with the x-ray beam. A suitable layer-line screen is selected from those supplied with the camera ($r_s = 15$, 20, 25, and 30 mm), and the appropriate s value determined with the aid of (16-12). A film is placed in the cassette, the preselected $\bar{\mu}$ value is set on the arc, and the camera is ready to record the zero level of any crystal. The precession motion is attained with the aid of a universal joint on which the crystal's axis and the film cassette are mounted. The two are linked with each other by a parallel-motion bar so that the necessary relationships described above are maintained at all angles. The physical dimensions of the camera in Fig. 16-24 limit the maximum value that $\bar{\mu}$ can have to $30°$.

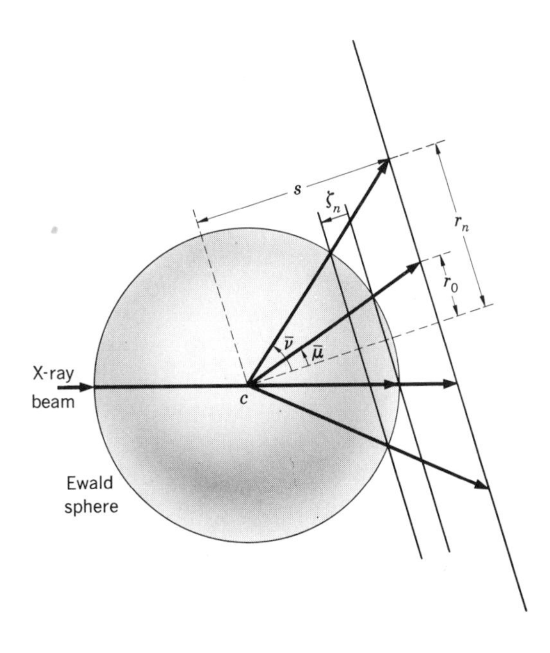

Fig. 16-23. *(After Buerger.)*

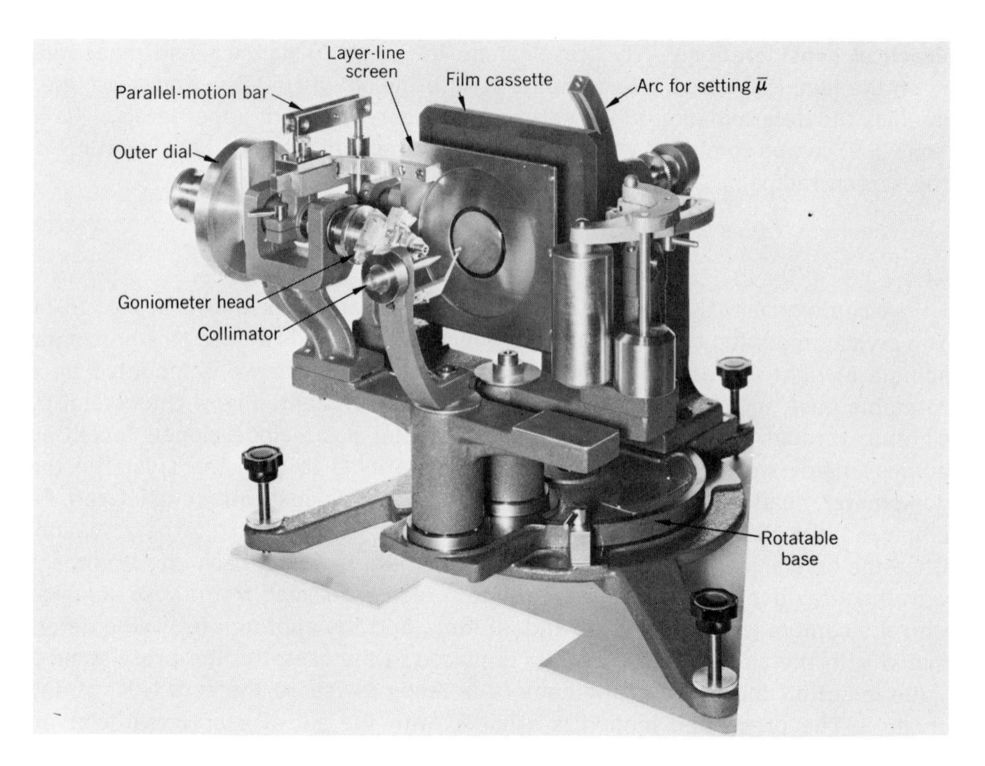

Fig. 16-24. Buerger precession camera. *(Courtesy of Charles Supper Company.)*

When the ζ value is not known in advance, a film is placed in the position of the layer-line screen, and a cone-axis photograph is prepared as already described. A suitably shaped film enclosed in a light-tight paper envelope can be used, or a special cone-axis cassette can be constructed. Once ζ is established, upper levels can be photographed in an analogous manner. The only step that must be added to the selection of r_s and s for the layer-line screen is the advancement of the film cassette by $M\zeta$ so as to assure an undistorted upper-level photograph. Any errors present in making this setting, including, for example, accidental displacement of the film caused by its buckling in the cassette, produces a doubling of the spots recorded. This is a consequence of the fact that each reciprocal-lattice spot intersects the Ewald sphere twice, once when it lies on one side of the ring (Fig. 16-17) and then again on the opposite side.

A correctly prepared zero-level precession photograph is shown in Fig. 16-25. The precession angle $\bar{\mu} = 30°$, so that the recorded region shown extends to $\xi = \sigma = 1/\lambda$. In a camera like the one shown in Fig. 16-24, the value of M is usually fixed at 6.0 cm, so that the maximum radius on the film when $\bar{\mu} = 30°$ is also 6.0 cm (Exercise 16-5). This makes it possible to use a commercial 5- by 7-in. film, trimmed to 5 by 5 in. (actually $4\frac{15}{16}$ by $4\frac{15}{16}$ in.), to record the entire portion of each level that possibly can be recorded. Because the maximum recordable value of σ in a precession instrument is $1/\lambda$, as compared with $2/\lambda$ in a Weissenberg instrument, it is usual to employ Mo $K\alpha$ radiation in precession photography instead of Cu $K\alpha$, normally used in the Weissenberg method. The presence of a more pronounced maximum in the continuous spectrum emanating from a molybdenum target results in its diffraction by the various planes in the crystal. This is the reason why a typical precession photograph prepared with Mo $K\alpha$ radiation contains continuous radiation streaks (Fig. 13-30), particularly notable in connection with the more intense reflections. The streaks can be eliminated by using a crystal monochromator, with the result shown on the right in Fig. 13-30.

Orientation of crystals. The presence of continuous radiation streaks can be extremely helpful when it is necessary to orient a crystal on the precession instrument. Suppose that the correct orientation of the zero level is shown by the solid line in Fig. 16-26, while the actual orientation, indicated by the dashed line, is rotated by an amount ϵ. Such a misalignment causes the circle of intersection with the Ewald sphere to increase in a vertical sense in the top half of the sphere, as shown to the left in Fig. 16-26. (A corresponding decrease occurs in the bottom half.) On the photographic film, this causes an extension of the recorded region at the top of the film and a decline at the bottom. To detect this most clearly, a layer-line

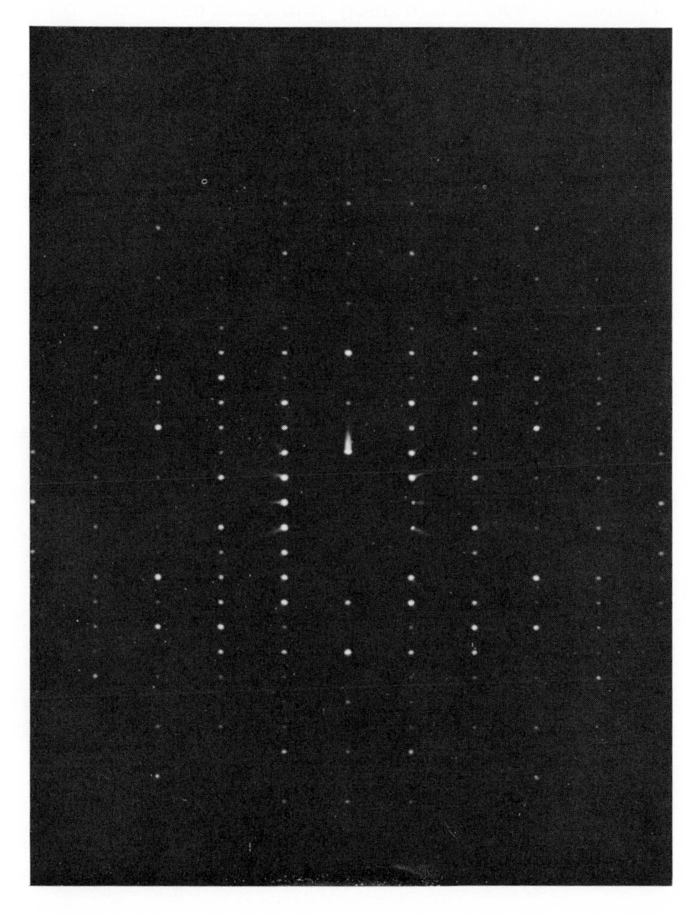

Fig. 16-25. Zero-level photograph of $(NH_4)_2Pt(S_5)_3 \cdot H_2O$.

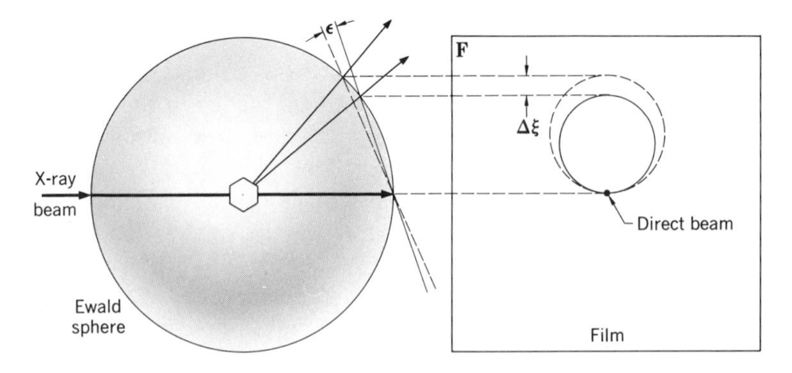

Fig. 16-26

screen should be used to isolate the zero-level diffraction cone, and a relatively small value of $\bar{\mu}$ to decrease the exposure time. Convenient values in practice are $\bar{\mu} = 10°$, $r_s = 5$ mm, and $s = 28.4$ mm. The appearance of the precession photographs when an alignment error of 0, 3, or $6°$ is present can be seen in Fig. 16-27. Such orientation photographs should be prepared with unfiltered radiation. This reduces the exposure times considerably and increases the relative intensity of the continuous radiation reflected, so that the asymmetry produced by misalignment is more readily visible. In this connection it should be noted that a Polaroid film pack can be used instead of conventional films by inserting a suitable cassette in the Buerger precession instrument (Fig. 16-28). The Polaroid films require exposure times roughly one-tenth as long as conventional films. This fact, together with the simplified and brief development process, make such film extremely convenient to use whenever flat-film cassettes can be employed. The x-ray photographs illustrating the Buerger precession method in this chapter were all prepared using Polaroid film.

The geometrical relation between the increment of $\xi_{\max}$ that can be recorded

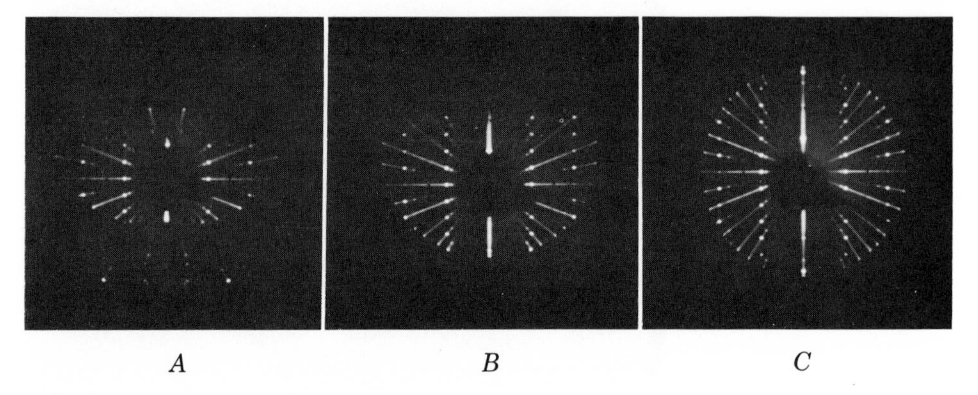

A *B* *C*

Fig. 16-27. Sequence of orientation-photograph ($\bar{\mu} = 10°$) layer-line screen used with $r_s = 5$ mm and $s = 28.4$ mm. (*A*) Vertical displacement $\epsilon = 6°$; (*B*) vertical displacement $\epsilon = 3°$; (*C*) correct orientation. Note that the top of the circle (Fig. 16-26) is not clearly visible when $\epsilon \geq 3°$.

Fig. 16-28. Buerger precession camera equipped with a Polaroid Land cassette. *(Courtesy of Charles Supper Company.)*

when an alignment error is present ($\Delta\xi$ in Fig. 16-26) and the angular error ϵ can be established for various values of $\bar{\mu}$. It is most convenient to measure the distance from the center of the film to the extended part of the circle, x_1, and to the shortened part of the circle, x_2, and to relate this difference on the film to the actual difference in reciprocal-lattice coordinates. As shown in Fig. 16-29,

$$x_1 - x_2 \cong M(2\Delta\xi) \tag{16-13}$$

if it is assumed that the displacements at top and bottom are very nearly the same. (See Exercise 16-8.) The relation between $2\Delta\xi$ and the angular error ϵ turns out to be very nearly linear, so that a simple correction chart can be prepared on ordinary graph paper. The necessary values for $\bar{\mu} = 10°$ are

$\epsilon°$	0	0.5	1.0	1.5	2.0	2.5	3.0
$2\Delta\xi$	0	0.354	0.709	0.106	0.142	0.178	0.213

(16-14)

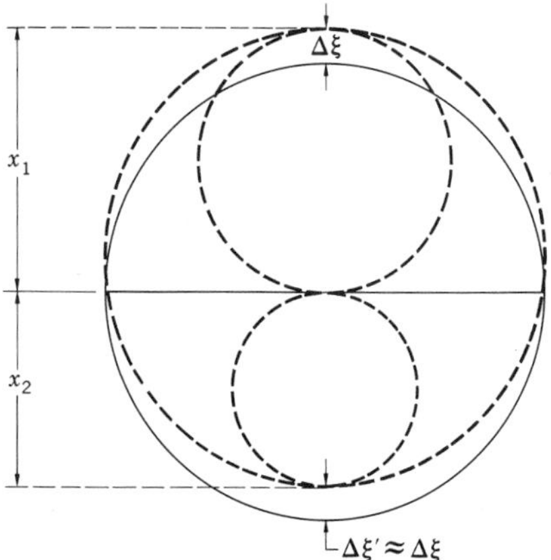

Fig. 16-29

The values in (16-14) can be converted into film units directly by multiplying by M according to (16-13).

The strategy for orienting a crystal on a Buerger precession camera is quite simple. The arcs of the goniometer head, on which the crystal is mounted (Fig. 16-24), are first set so that they are, respectively, parallel and perpendicular to the x-ray beam. Assuming that a preselected row of the reciprocal lattice is nearly parallel to the mounting axis, any rotational error in the alignment of the arc that is parallel to the x-ray beam (horizontal arc) will cause a displacement $\Delta\xi$ either to the left or to the right on the film. By measuring the magnitude of the displacement, it can be used to correct the alignment after determining the angular error with the aid of (16-14). Next, the goniometer head is rotated by $90°$ on the outer dial (Fig. 16-24), and the process is repeated for the second arc. When the preselected reciprocal-lattice row is thus aligned, all the reciprocal-lattice levels containing this row can be located by rotating the crystal about its mounting axis (outer dial). This procedure also serves to orient the reciprocal-lattice layer vertically, as already illustrated in Fig. 16-27.

When the original mounting of the crystal is completely random, as is the case when either irregular fragments or deliberately ground spheres have to be used, the above procedure still is employed, except for one preliminary step. With both arcs on the goniometer head set to zero, the outer dial is rotated in angular amounts commensurate with the magnitude of the precession angle, until a dense reciprocal-lattice layer is encountered. This is easy to recognize because of the characteristic appearance of orientation photographs like those in Fig. 16-27. The crystal, mounted on its brass pin, is then rotated within the goniometer head by the amount necessary to orient this reciprocal-lattice layer parallel to one of the two arcs, and the previously described procedure is commenced. When completed, it guarantees that some densely populated layer in the reciprocal lattice is properly oriented. From there it is a simple matter to record the entire undistorted reciprocal lattice, as described in the preceding sections.

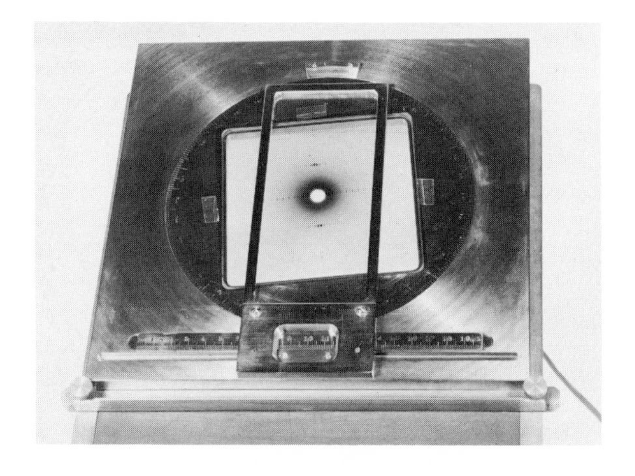

Fig. 16-30. Precession film-measuring device. *(Courtesy of Buerger.)*

Interpretation of photographs. Because the reciprocal lattice is recorded without any distortion, the interpretation of Buerger precession photographs is extremely straightforward. One chooses the appropriate reciprocal-lattice axes and indexes all reflections by inspection. The measurement of distances and angles in the reciprocal lattice is similarly straightforward. It is convenient to mount the film on a reader like the one shown positioned on an illuminated viewer in Fig. 16-30. The distance between parallel reciprocal-lattice rows is measured along the horizontal scale, to which the transparent slider is attached. For maximum reading accuracy, the hairline on the slider is set to pass through rows on opposite sides of the film. The angular rotation necessary to align another set of rows with the hairline is measured along the circumference of the film reader, so that the interaxial angles are also directly determined. Film distances are converted to reciprocal-lattice distances by dividing them by M, expressed in the same units. Because the radius of the Ewald sphere was chosen to equal unity for the relations derived in this chapter,

$$\xi = \frac{\lambda}{d} \tag{16-15}$$

Except for the inclusion of λ in the conversion to crystal coordinates, the relations between the reciprocal lattice and direct lattice are the same as those discussed in Chap. 7.

The Buerger precession method has many advantages over other single-crystal film methods. They almost all derive from the fact that the reciprocal lattice is recorded without distortion. When relatively transparent films are used to record the zero and upper levels, it is possible to stack them above each other in such a way as to reproduce an accurate three-dimensional array. Such models can be extremely helpful in examining the geometry and symmetry of the reciprocal lattice, particularly when truly unknown crystals are first being studied.

SUGGESTIONS FOR SUPPLEMENTARY READING

M. J. Buerger, *X-ray crystallography* (John Wiley & Sons, Inc., New York, 1942).

Crystal-structure analysis (John Wiley & Sons, Inc., New York, 1960), pp. 78–112.

The precession method in x-ray crystallography (John Wiley & Sons, Inc., New York, 1964).

EXERCISES

16-1. Prove that $\Upsilon = 2\omega$ for the zero level of a Weissenberg photograph.

16-2. Consider the central reciprocal-lattice rows corresponding to the a^* directions in a hexagonal crystal rotating about its c axis. Show on a sketch how they would appear in a zero-level Weissenberg photograph. What does this exercise suggest regarding the deduction of symmetry from such photographs? For example, if the space group is $P3$, how would parallel central rows appear in the 1-level equi-inclination photograph? If the space group is $P6mm$?

16-3. By constructions similar to that shown in Fig. 16-2, determine and illustrate the appearance of central and noncentral rows in a normal-beam Weissenberg photograph of an upper level.

16-4. Consider the reciprocal lattice of a tetragonal crystal whose space group is $I4$. Construct the 0- and 1-level Weissenberg photographs for such a crystal rotated about its c axis. (Select arbitrary values for a^* so as to cause intersections out to at least 600.) How can you distinguish a^* from σ_{110} among the central reciprocal-lattice rows recorded on the film?

16-5. Derive the relation between the radius of the limiting circle on a Buerger precession photograph and the precession angle $\bar{\mu}$. For the usual case of a crystal-to-film distance of 6 cm and $\bar{\mu} = 30°$, what is the radial distance from the center to the outermost recordable reflection?

16-6. Derive relation (16-6) by considering the construction in Fig. 16-17.

16-7. Derive relation (16-9). To do this note that the opening angle of the diffraction cone is $2\bar{\mu}$ and the Ewald sphere in Fig. 16-22 has unit radius.

16-8. In deriving relation (16-23), it was assumed that the displacements of the diffraction cones at the top and bottom in Fig. 16-29 are very nearly the same. This approximation is valid for small values of $\bar{\mu}$ and ϵ. Let ξ_0 represent the diameter when the level is correctly oriented, and $\Delta\xi$ and $\Delta\xi'$ the increase and decrease, respectively, caused by ϵ. Derive the relation between the total difference $\Delta\xi - |\Delta\xi'|$ and ϵ for any value of $\bar{\mu}$.

SEVENTEEN
Powder method

THEORETICAL CONSIDERATIONS

Reciprocal-lattice construction. In distinction from the arrangements described in the immediately preceding chapters, the specimen used in the powder method is an aggregate of a large number of tiny crystallites. In all the discussions in this chapter, it will be assumed that their orientations are completely random and the number of crystallites present in the irradiated specimen is very large. When this is not the case, the specimen can be rotated so as to simulate this condition. Consider, then, the reciprocal-lattice construction of a polycrystalline specimen. Each crystallite has associated with it a reciprocal lattice whose origin lies on the Ewald sphere at the point where the incident beam emerges from it. Because there are many such crystallites present and their orientations are completely random, it is no longer possible to distinguish the individual lattices. Instead, they coalesce into concentric spheres whose radii are the various possible reciprocal-lattice vectors. Note that, as a result of symmetry or chance, many of the reciprocal-lattice vectors have the same length, so that all such reciprocal-lattice points come to lie on the same sphere in the reciprocal-lattice construction of a powder. As already indicated in Fig. 7-12, each reciprocal-lattice sphere intersects the sphere of reflection along a circle, so that a cone of diffracted rays passing through this circle emanates from the specimen. The way that three concentric spheres intersect the Ewald sphere is shown in a cross-sectional view in Fig. 17-1. Note that the diffraction cones are concentric about the x-ray beam, so that a cylindrical strip of film having the direct beam as a diameter intercepts each cone in two arc segments, as illustrated in Fig. 17-2.

The half-opening angles of the cones are 2θ, as demonstrated in subsequent discussions, so that it is possible to measure the diffraction angles directly from the arc locations on the film strip in Fig. 17-2. This, in turn, allows the determination of the lengths of the reciprocal-lattice vectors, since $\sigma_{hkl} = 1/d_{hkl}$. Unfortunately, the relative disposition of these vectors cannot be determined from the powder photograph since the reciprocal lattices of individual crystals have coalesced. It is possible, however, to deduce the correct intervector angles indirectly. The

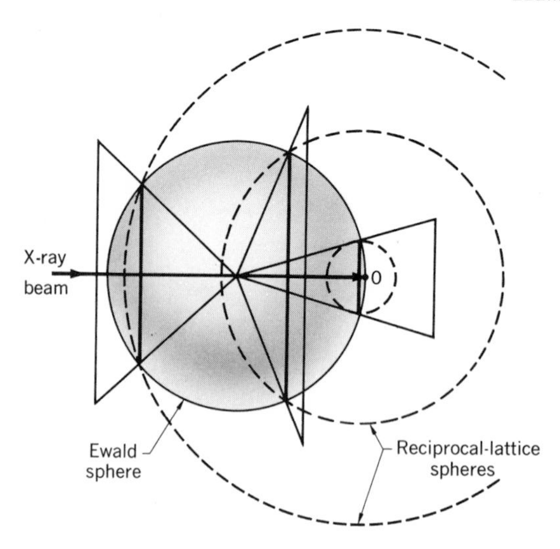

Fig. 17-1

process employed consists of postulating a reciprocal lattice whose vectors σ_{hkl} have magnitudes identical with the experimentally observed vector magnitudes. This essentially iterative procedure can be systematized as described in Chap. 18. The correct choice of a postulated lattice is confirmed by assigning appropriate indices to all the observed reflections. Quite often, the symmetry of the crystallites is known beforehand, so that the entire indexing process becomes considerably simplified.

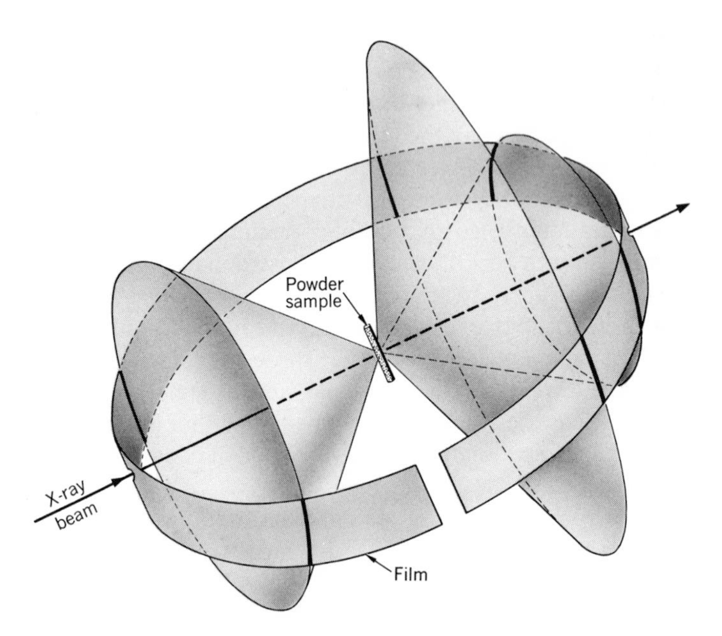

Fig. 17-2. Placement of film strip about a cylindrical sample in the powder method. The three diffraction cones indicated are those shown in Fig. 17-1.

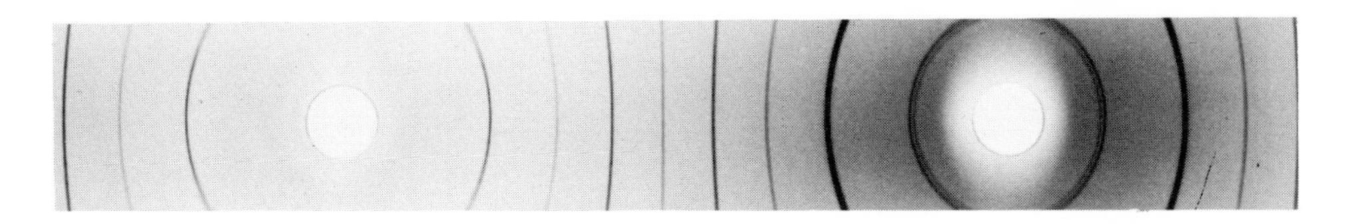

Fig. 17-3. Powder photograph of tungsten. Note $\alpha_1\alpha_2$ resolution in back-reflection region.

Measurement of d values. Only the details of the interpretations of experimental observations in the powder method depend on which kind of experimental arrangement is used. Some of these details are discussed in a later section. For the present discussions it will be assumed that a photographic film is employed to intercept the diffraction cones, as indicated in Fig. 17-2. Note that the film is split halfway between the points where the direct beam enters and leaves the film. This particular film placement was first suggested by M. Straumanis and A. Ieviņš, and has two desirable features. The complete diffraction record from 0 to $90°\theta$ is recorded on the same film strip (Fig. 17-3), and the presence of pairs of arcs about the $\theta = 0°$ and $\theta = 90°$ positions enables their location on the film with utmost precision. It is not essential that the film break occur exactly midway, so that, as pointed out by A. J. C. Wilson, when crystals give rise to reflections only at small theta angles, it may be desirable to locate the break in the front-reflection region instead.

To deduce the interplanar spacing values from a film it is necessary, first, to calculate the theta values of each reflection, after measuring the distance between the corresponding pair of arcs on the film. The measurement of the distance between a pair of arcs is facilitated by the use of a special film-measuring device. Such devices can be obtained from the manufacturers of powder cameras, and a typical film-measuring device is illustrated in Fig. 17-4. When the film is placed in such a device, care should be exercised to ascertain that the long direction of the film is truly parallel to the scale of the measuring device, as shown in Fig. 17-5.

The readings are made in succession on two corresponding arcs (left and right) by placing the cross hair at the center of each arc and recording the value to 0.1 mm. The sequence of operations can be understood best by preparing a data sheet as shown in Fig. 17-6. The larger reading is placed under the column heading x_2, and the smaller under x_1. The next column contains the sum of x_1 and x_2,

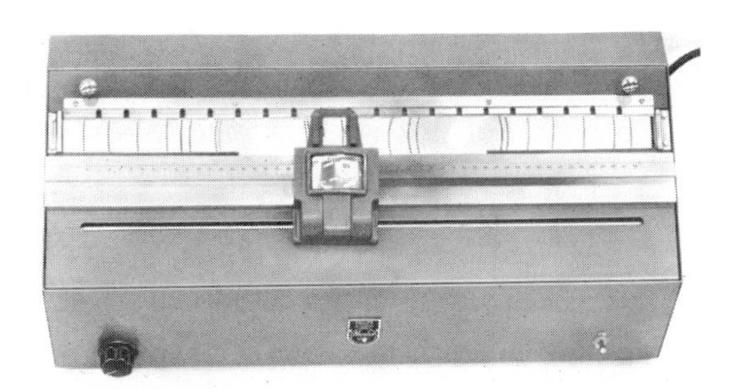

Fig. 17-4. Powder film reader. *(Courtesy of Philips Electronic Instruments.)*

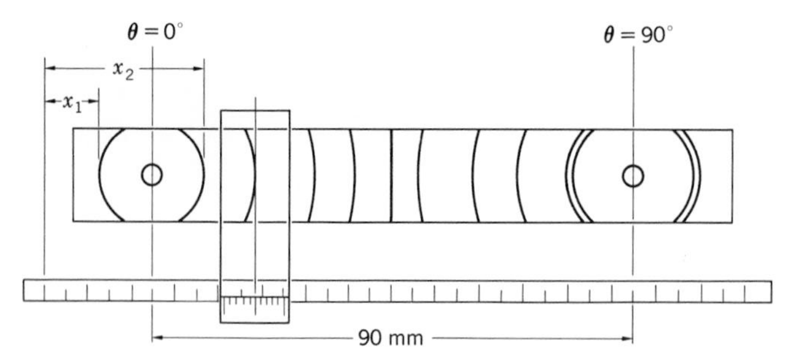

Fig. 17-5

which should remain constant with ± 0.1 mm, and serves as a check that the readings have been correctly made.

The relationship of the diffraction angle θ to the measured arc length S is shown in Fig. 17-7. From this it is seen that

$$4\theta = \frac{S}{R}$$

$$\theta = \left(\frac{1}{4R}\right) S \qquad \text{rad} \tag{17-1}$$

$$= \left(\frac{180}{\pi} \cdot \frac{1}{4R}\right) S \qquad \text{deg} \tag{17-2}$$

The terms in parentheses are constants, depending on the radius of the camera used. S and R are normally measured in millimeters. Under S in Fig. 17-6, write the difference between x_2 and x_1. This is the arc length corresponding to the angle of 4θ.

The locations corresponding to $\theta = 0°$ and $\theta = 90°$ (Fig. 17-5) are next determined by taking a pair of arcs about each hole in the film and halving the sum of the readings for such a pair. The difference between these two averages should be 90 mm for cameras of diameter 57.26 mm (or 180 mm for cameras of diameter 114.6 mm). If it is not, but differs from 90 mm by $p\%$, a corresponding correction of $S \times p/100$ should be added to each value of S recorded. This correction is tabulated in the column headed "correction," and the corrected value is tabulated under S' in Fig. 17-6.

Line no.	x_2	x_1	Check $x_2 + x_1$	Arc length $S = x_2 - x_1$	Correction $\frac{p}{100} \times S$	Corrected arc length S'	$\theta = \frac{S'}{4}$	$\sin\theta$	$d = \frac{\lambda}{2 \sin\theta}$	$Q = \frac{1}{d^2}$	hkl

Fig. 17-6

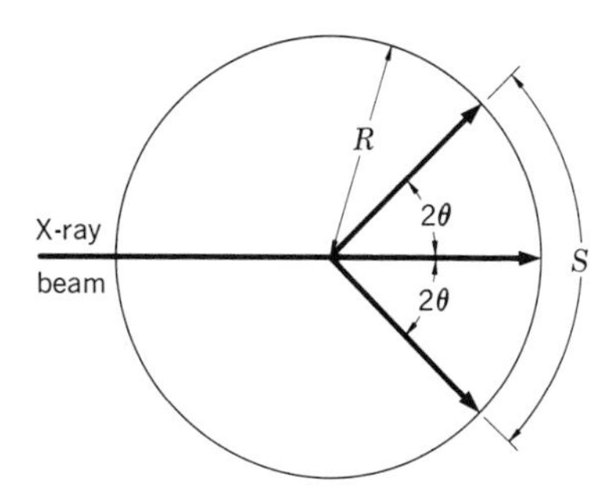

Fig. 17-7

Substituting for R in Eq. (17-2) the assumed radius, $180/\pi$ mm, and for S the corrected arc length S', it follows that

$$\theta = \left(\frac{180}{\pi} \cdot \frac{1}{4} \cdot \frac{\pi}{180}\right) S'$$

$$= \frac{S'}{4} \quad \text{deg} \tag{17-3}$$

From the above it is clear why it is convenient to work with a camera whose radius is a multiple of $180/\pi = 57.3$ (the conversion factor from radians to degrees). This choice makes it possible to measure the diffraction angle directly with a millimeter scale.

The remaining columns in Fig. 17-6 are self-explanatory and lead to a solution of Bragg's equation for the value of the spacing d. These last columns can be eliminated if suitable tables listing θ vs d for different wavelengths are available. The last two columns, headed Q and hkl, will be discussed in the next chapter.

Sources of errors in measured spacings. Two types of errors are associated with any measurements: random errors of observation, and experimental errors which are inherent in the arrangement. The principal observational error in the powder method lies in the location of the center of a line on the film. This reading error can be minimized by repeating the reading several times or by preparing and measuring several films. For the best accuracy, it is better to average the readings of several different investigators since some peculiarity on the film, such as a speck of dirt, may bias a particular investigator into making the same reading several times in succession. It has been shown that reading errors are not appreciably reduced by preparing a microphotometer tracing of the film, since the added detail present tends to obscure the position of the center of the peak. This difficulty remains when a direct measurement of the diffracted intensity profile is carried out with the aid of a diffractometer. In fact, the rather remarkable integrating ability of the human eye cannot be utilized when diffractometer profiles are analyzed, and there exists some controversy regarding the relative advisabilities of using peak heights in preference to centroids or other parameters.

In addition to reading difficulties, there are several situations which cause a shift in the center of an x-ray line on a film from the position it ideally should occupy.

These can be classified into physical and geometrical causes as follows:

Physical errors

1. Absorption of x rays by the specimen
2. Refraction of x rays by the specimen
3. Uneven distribution in the intensity of the background
4. Relative displacement of lines appearing on the two sides of a double-coated film

Geometrical errors

1. Eccentricity of the specimen with respect to the axis of the film cylinder
2. Lack of knowledge of the exact film radius
3. Divergence of the x-ray beam

These errors usually are classified into two groups, according to whether the error is completely random or whether its magnitude depends in a systematic fashion on the θ value of the reflection. The so-called systematic errors tend to decrease as θ increases, vanishing completely when $\theta = 90°$. This tendency serves as the basis for the precision determination of lattice constants, described in a later section of this chapter. The relative importance of the errors depends on the actual experimental arrangement employed. Thus physical error 4 does not apply when the specimen is located at the center of a cylindrical camera (Fig. 17-2). Conversely, a comparable displacement may be produced when a gas-filled detector is used in a diffractometer, since gas absorption usually displays an energy dependence. It is important, therefore, to establish which errors are pertinent before attempting significant work.

Of the four kinds of physical errors listed above, absorption is by far the most troublesome. Refraction of x rays is quite small, and for all but the most precise work it is negligible. Uneven background due to various causes discussed in preceding chapters is more of a nuisance than a serious obstacle, while the double-coated film effect appears only when the diffracted x rays do not intersect the film at perpendicular incidence. This is the case when flat films or asymmetric cameras are used. It can be eliminated quite simply by using a single-emulsion film, even though this increases the exposure time.

Similarly, specimen eccentricity can be eliminated by careful construction of the instruments used, while the exact knowledge of the camera radius (film shrinkage) is unimportant when the Straumanis-Ievinš film-loading arrangement is used. Divergence of the x-ray beam, on the other hand, is an irremedial problem. It is here that absorption by the specimen proves most troublesome. As has been shown by various investigators, it is possible to deduce the effect that absorption by the specimen has on shifting the recorded position of a diffraction maximum, provided one assumes a parallel incident beam. Although divergence by itself does not cause a peak shift (assuming a transparent sample), it so confounds the calculation of the absorption effect that several independent analyses have resulted in different geometrical relationships. This problem is further discussed briefly in a later section.

EXPERIMENTAL PROCEDURES

Choice of radiation. For most applications, the relative monochromatization attained in a filtered x-ray beam (Chap. 6) is adequate. The filter materials and thicknesses recommended for commercially available x-ray targets are listed in Table 6-1. The beam transmitted through these filters is composed predominantly of the $K\alpha_{1,2}$ lines, so that the proper wavelength to use in the interpretation of diffraction diagrams is the intensity-weighted mean of the two wavelength components, namely, $\frac{1}{3}(2K\alpha_1 + K\alpha_2)$. These wavelength values are indicated for most target materials of interest in Table 17-1. Note that the weighted-mean $K\alpha$ wavelength should be used with powder reflections for which α_1 and α_2 are unresolved. At larger diffraction angles, where the two components frequently are resolved (Fig. 17-3), it is necessary to use the respective $K\alpha_1$ and $K\alpha_2$ values with the corresponding maxima in the doublet.

In selecting the most appropriate wavelength to use in a particular experiment, several factors should be considered:

1. Effect of unit-cell size
2. Effect of specific absorption
3. Relative exposure times

The locations of diffraction lines on a powder photograph are determined entirely by the dimensions of the unit cell of the crystals in the sample and by the wavelength of x radiation used. Whether one thinks of the size of the Ewald sphere (radius $= 1/\lambda$) in reciprocal space or considers the Bragg equation (5-3), it is clear that, for a given spacing, the diffraction lines appear at angles directly related to the

Table 17-1 Wavelengths of useful x radiations and elements which highly absorb them

Element	$K\alpha = \frac{1}{3}(K\alpha_2 + 2K\alpha_1)$, Å	$K\alpha_2$, Å	$K\alpha_1$, Å	$K\beta$, Å	$K\alpha$ is highly absorbed by	$K\beta$ is highly absorbed by
Cr	2.2909	2.29352	2.28962	2.08479	Ti, Sc, Ca	V
Mn	2.1031	2.10570	2.10174	1.91016	V, Ti, Sc	Cr
Fe	1.9373	1.93991	1.93597	1.75654	Cr, V, Ti	Mn
Co	1.7902	1.79279	1.78890	1.62073	Mn, Cr, V	Fe
Ni	1.6591	1.66168	1.65783	1.50008	Fe, Mn, Cr	Co
Cu	1.5418	1.54434	1.54050	1.39217	Co, Fe, Mn	Ni
Zn	1.4364	1.43894	1.43510	1.43510	Ni, Co, Fe	Cu
.	.	.	.	.	.	.
.	.	.	.	.	.	.
.	.	.	.	.	.	.
Mo	0.7107	0.71354	0.70926	0.63225	Y, Sr, Ru	Cb, Zr

x-ray wavelength. When extremely short wavelength radiation is employed, all visible diffraction lines are crowded closely together at relatively small angles on the photograph. Conversely, radiation whose wavelength is too long disperses the lines to such an extent that only a limited number of lines is recorded. When the wavelength is fixed, the relative size of the unit cell has an inverse effect, namely, the larger the cell, the more closely spaced the diffraction lines become on the film. Thus, when it is known in advance that the unit cell is large, it is advantageous to use the increased dispersion of long-wavelength radiation.

The foregoing considerations may have to be modified in the light of the sample's composition. As discussed in Chap. 6, the absorption of x rays by an element increases with increasing wavelength up to its absorption edge. Much of the energy thus absorbed is reemitted by the element as fluorescent radiation. Consequently, when the incident radiation has a wavelength lying on the short-wavelength side of the absorption edge of an element, this element will be excited in fluorescence and will strongly absorb that wavelength component. For this reason, Table 17-1 indicates which elements are thus affected most strongly by the main wavelength components emanating from conventional tube targets. The $K\beta$ component is also included because, although strongly minimized by an appropriate filter, it is not suppressed entirely and may contribute to the background by exciting fluorescence excitation. Clearly, whenever possible, x radiation should be so chosen that none of the elements listed in the same row in the last two columns of Table 17-1 are present in the sample.

Fluorescence radiation raises the background intensity and, when excessive, may obscure weak reflections. One way to minimize it is to use a crystal monochromator instead of a filter to limit the wavelength spread in the incident x-ray beam. This is a particularly useful procedure with a diffractometer because the monochromator can be placed between the sample and the detector. Thus, regardless of the amount of fluorescent radiation produced, only the monochromatic component selected is transmitted to the detector. An alternative trick can be utilized when a double-coated film is employed. After the film is exposed in the usual way, but before it is developed, the side of the film nearest to the sample is covered by a strip of ordinary masking tape. The film is then developed in the usual way and rinsed in wash water. Before immersing it in the hypo, however, the masking tape is removed and the film is then fixed in the usual way. The result of not developing one side of the film is that the layer exposed most strongly to the fluorescent radiation is rendered completely transparent by the hypo. In the case of a conventional powder photograph, it is possible to affix the masking tape lengthwise along the film in such a way that one half of the film only is covered. One half of the final photograph will then show the full exposure, while the other half will have only one developed emulsion.

Another factor to consider is the exposure times required by various radiations. The absorption of x rays by a film emulsion (or by a gas in an ionization detector) depends on wavelength. Generally, the longer-wavelength radiations are more readily absorbed ($I \propto \lambda^3$), so that the exposure time should decrease for longer wavelengths quite rapidly. Militating against this is the increased absorption of such radiation by the sample and the air in the camera. Another factor to consider is the maximum power loading of various tubes, which is limited by the tube target's ability to dissipate the heat generated in it. Thus, although Fe $K\alpha$ radiation is absorbed more readily by a photographic film than Cu $K\alpha$, normally, a copper-target tube can be operated to yield a considerably larger intensity and requires a correspondingly reduced exposure. In fact, copper is a commonly used target material for other reasons also. As can be seen in Table 17-1, its characteristic wavelengths lie midway between the other conventional target metals. Thus, when an a priori reason does not exist for choosing a specific target metal, copper affords the best compromise for the competing factors discussed above.

Sample preparation. The final shape of the sample desired depends on the experimental arrangement to be employed. The three common shapes are indicated in Table 17-2. Ideally, the sample should contain only the crystallites to be examined. Sometimes this goal can be attained by suitably fashioning the sample itself, for example, by machining the polycrystalline materials or by extruding them into rods. Care must be exercised in all such fabrication procedures not to destroy the randomness of crystallite orientation that is essential to produce continuous diffraction cones of uniform intensity. As discussed in Chap. 20, this is not normally possible when a material is worked. In fact, the nature of the preferred orientations induced is frequently characterized by x-ray diffraction studies and correlated to the machining process and its effects on the material.

A cylindrical specimen is desirable for a cylindrical powder camera so that the absorption of x rays at all angles is affected in a known manner. Of the different procedures indicated, mixing the powder with a binder like collodion and then rolling it into a cylinder is preferred because collodion does not contribute discernible x-ray diffraction lines, nor does it absorb the x radiation significantly.

To prepare a collodion mount, the specimen is ground to a uniform size (approximately 200- to 300-mesh) and placed on a glass plate. (If the sample is

Table 17-2 Specimen shapes used in powder method

Instrument used	*Specimen shape*	*Fabrication procedure*
Cylindrical camera	Cylinder	Extrusion through cylindrical die
		Rolling with binder
		Filling a capillary
		Coating a fiber
Diffractometer	Plate	Filling a flat holder
		Coating a flat plate
		Forming a flat film
Focusing camera	Thin film	Floating a film
		Coating an adhesive tape

a metal which cannot be formed readily into a rod or wire, it can be filed carefully with a clean jeweler's file. Any cold-working strains induced in the filings must be removed by annealing at suitable temperatures.) A few milligrams of the sample is then mixed with a drop of collodion until a homogeneous paste results. The amount of powder added to the collodion should not be too large, lest the paste become too brittle. The paste is then scooped up on a razor blade and rolled into a rod-shaped specimen with the fingers. A more uniformly shaped cylinder can be obtained by rolling the rod between two pieces of ground glass. The thickness of the rod should be approximately $\frac{1}{2}$ mm, and the length, 10 mm. After allowing about 10 min for the rod to harden, it is affixed to a brass pin by means of ordinary glue. A suitable specimen holder can be made from a $\frac{3}{16}$-in. brass rod, cut to the desired length and predrilled to allow insertion of the specimen rod part way into the pin. If a highly absorbing sample is being prepared, the total absorption of the final mount can be controlled readily by adjusting the powder-to-collodion ratio.

An alternative way of making powder mounts is first to affix a very thin capillary tube, made of lead-free glass, to a brass pin by means of picein wax. When the wax has hardened, the tube is rotated and pulled through some thinned shellac, so that a very thin coating of shellac adheres to the tube walls. The tube thus coated is then rolled in the powder until it is uniformly covered by the specimen.

Instead of coating the outside of a capillary, it is possible to place the powder inside the capillary. This is an essential procedure whenever the substance forming the sample is affected by solvents or is deliquescent. Specially tapered capillary tubes, having one widely flared end, are manufactured commercially from lead-free glass. After sealing the narrow end of the tube, the powder is introduced through the other. By stroking the capillary with the edge of a file, its vibrations cause the powder particles to pack more rapidly and more uniformly. A highly absorbing substance can be mixed with a less absorbing noncrystalline one, such as casein, lampblack, or ground glass. Glass capillaries, of course, introduce extraneous material into the x-ray beam, so that their use is not recommended unless the preferred methods described above are not practical.

A flat plate is placed at the center of rotation in a diffractometer. The irradiated area is controlled by the slits employed, but is usually of the order of $\frac{1}{2}$ by 1 in. Most diffractometers are provided with specimen holders containing flats or grooves machined in them. Two important points must be kept in mind when preparing specimen mounts for a diffractometer: the surface exposed to x rays should be smoothly flat and accurately positioned so that it is tangent and parallel to the diffractometer rotation axis, and the powder should be so arrayed that preferred orientation is absent. When the crystallite shapes in a powder are more or less isotropic, the simplest way to pack the sample holder is to pour the powder into the machined openings, tamp it down lightly with an ordinary glass slide, and then draw the edge of the slide across the holder so as to wipe off any powder extending above it. If the powder is properly packed and compacted, this procedure should leave a uniformly flat surface coincident with the holder's surface. Moreover, the powder should not spill out in that case even when the specimen holder is placed in a vertical position.

Whenever the above procedure does not seem to work, it is safe to assume that only a relatively imperfect specimen can be prepared. If the packing is unsatisfactory because the crystallites are acicular or platy, preferred orientation invariably will be present in the specimen. Of the numerous schemes devised to minimize preferred orientation, none remove it entirely and all disturb the ideal perfection of the specimen. For example, an easy but crude method is to coat a glass slide with a thin layer of petroleum jelly. The powdered material is then sprinkled onto the slide, and the individual crystallites tend to stick in a random way. This, unfortunately, prevents the formation of a uniformly flat surface, since any tamping of the powder induces preferential alignments. Other schemes involve the formation of slurries and filling specially constructed sample holders. In this connection, dispersion of the powder by mixing it with finely ground glass before packing is sometimes effective in randomizing orientation and reducing absorption. Most commonly, however, it is necessary to devise a different procedure for each specific sample type.

As discussed in a later section, focusing cameras, whether they employ monochromators or not, usually require the specimen to conform to some kind of curved surface. Thus it is usually most convenient to fashion the sample in the form of a relatively pliable film. In some instruments x rays must pass through the specimen, so that it is important for it to be relatively transparent also. A convenient procedure for fashioning thin films is to mix the powder with a binder such as collodion, and then to float it on an appropriate liquid (water) or to spray it onto a soluble substrate. This produces very thin films which sometimes have to be stacked into multilayer sandwiches to provide sufficient scattering matter in the beam. Alternatively, thin films can be rolled or pressed by placing the powder in a binder between flat plates. An entire technology of thin-film preparation has evolved in recent years, so that the interested reader can explore it and adapt it to x-ray diffraction. A procedure often used because of its simplicity is to take an adhesive tape and to coat the adhesive side with powder. Since the backing of the tape is relatively transparent to x rays and typically is noncrystalline, it does not affect significantly the diffraction effects produced. Occasionally, a sandwich can be formed by sticking two such tapes together or by coating both sides of a suitably adhesive tape.

Use of powder camera. Of the different experimental arrangements that can be used in the powder method, only cameras are discussed below, because the operation of diffractometers has been discussed in Chap. 13. In essence, all cylindrical cameras resemble the one shown in cross section in Fig. 17-8. A strip of film is placed against a machined cylinder, the specimen is centered in the camera by sliding about a central block to which it is attached, and the collimator and beam catcher are inserted to confine the direct x-ray beam. Note that the collimator and beam catcher reach almost to the specimen. The purpose of this is to enclose the direct beam over most of its path, so as to minimize fogging of the film by air scattering.

By using a pair of specially shaped pliers (Fig. 17-9), the brass pin on which the specimen is mounted is pushed into a small block a (Fig. 17-8), located centrally

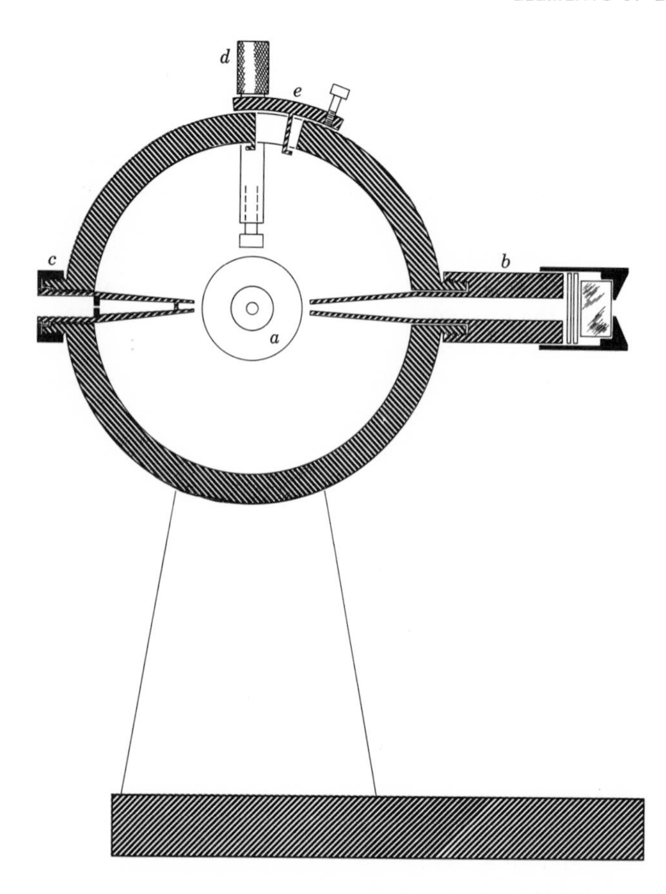

Fig. 17-8. Powder-camera design first proposed by Buerger.

in the camera. The direct-beam catcher b is then removed, and a special viewer inserted in its place. When light is passed through the collimator c, the mount appears as a horizontal line in the field of the viewer. Upon rotation of the external spindle which is connected to the specimen block a, the mount appears to move up and down. The centering of the mount is accomplished by rotating the spindle until the mount is in its highest position, and then pushing the block downward by means of the adjusting screw d, until the specimen bisects the field of the viewer. This routine is repeated until the specimen appears motionless during a complete rotation of the spindle. As soon as the specimen is centered, the adjusting screw d should be raised so that it will not interfere with subsequent

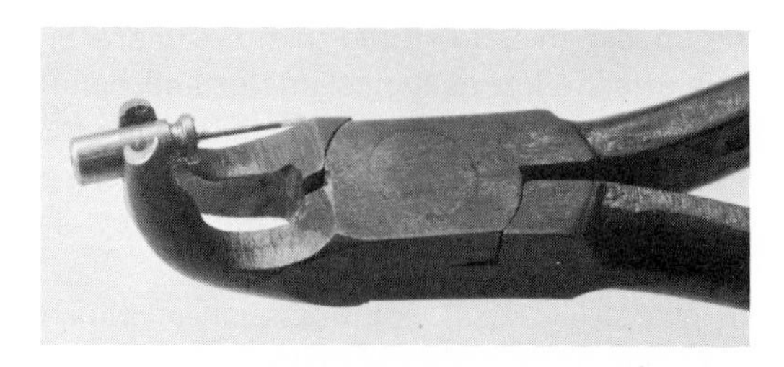

Fig. 17-9

rotation of the specimen block. The viewer is then removed, and the direct-beam catcher and camera lid are replaced.

Before placing the film in the camera, the camera cover, pinhole assembly, and direct-beam catcher must be removed. Before inserting the film, it can be marked with pencil for subsequent identification. The strip of film, cut to size and prepunched to permit subsequent insertion of the collimator and beam-catcher assemblies, is placed flush against the inside wall of the camera. It is important that, during the exposure, the film be held tightly against the inside wall of the camera. This is assured by pushing the slider e at the top of the camera as far as possible against the film. The collimator, direct-beam-catcher assemblies, and the cover are then replaced. Finally, the camera is placed on its track on the x-ray unit. The external spindle to which the powder mount is affixed is then connected, usually by means of a rubber band, to the small motor.

When the film is fully exposed, it is removed from the camera in a manner reversing the procedures described above. Before development, a clip, preferably numbered, is placed at one end of the film, and another clip with a lead weight attached to it is placed at the opposite end. (So-called dental film clips are most suitable.) The function of the lead weight is to prevent the film from curling during the development procedure. The film is developed and fixed according to the manufacturer's instructions. It should be washed for at least $\frac{1}{2}$ hour after fixing and before hanging up the film to dry.

Cylindrical cameras employing somewhat different film arrangements have been produced in Europe. The most popular of these arrangements is one suggested by A. J. Bradley and A. H. Jay and modified by Bradley and Bragg. In this modification two strips of film are placed symmetrically about the direct-beam path. The ends of these half films are secured by means of knife-edge film holders. The collimator and direct-beam catcher are inserted in the spaces between the films. The shadows cast by the knife-edge portions of the film holders overlapping the film are used to indicate predetermined values of the diffraction angles. These cameras have a certain advantage in that shorter individual film strips are utilized, particularly if oversize cameras are employed, but they have not gained the popularity of cameras using the Straumanis-Ieviņš arrangement. The principles of operation are quite similar to those of all cylindrical cameras, and for this reason they are not further discussed in this book.

As discussed in a later section, it is desirable for precise lattice-constant measurements to utilize the back-reflection region. Although the conventional cylindrical camera (Fig. 17-8) can be used for this purpose, much sharper lines can be obtained by placing the specimen on the cylindrical surface of the camera, as indicated in Fig. 17-10. In this camera the x-ray beam enters at a pinhole or slit on the surface of the cylinder, from which it diverges until it reaches the specimen on the opposite surface. The specimen is curved so as to conform with the cylindrical curvature of the surface. This geometry causes the radiation reflected from a particular plane to converge from all points in the sample to the same pair of lines on the film. In this way the specimen is said to focus the reflected radiation to the lines. The arrangement shown in Fig. 17-10 is symmetrical, so that each reflection is recorded twice on the same film, the two reflections being symmetrically

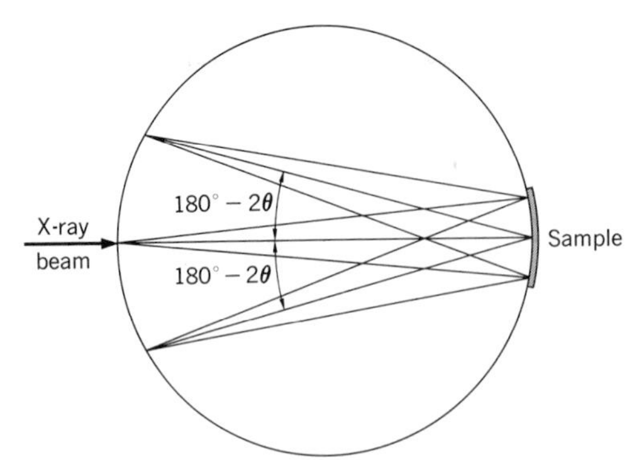

Fig. 17-10. Symmetrical back-reflection focusing arrangement.

disposed about the direct-beam entrance. Halfway between two such lines is the exact location of $2\theta = 180°$. The possibility of locating this fiducial point accurately is the main advantage of the symmetrical arrangement over unsymmetrical ones.

Use of crystal monochromators. In order to obtain a truly monochromatic beam, it is most convenient to use a single crystal having a high reflecting power and low absorption as a monochromator. Although plane-crystal monochromators are easier to construct, they cause the intensity to decline by about one order of magnitude. This loss can be alleviated somewhat by cutting the crystal at an angle to the reflecting planes, as shown in Fig. 17-11. In this arrangement, first suggested by I. Fankuchen, the asymmetric cut of the crystal serves to concentrate the diffracted beam by a preferential absorption process, as indicated.

As already described in Chaps. 6 and 13, it is possible to focus the diffracted rays by bending the reflecting planes, in which case one obtains a bent-crystal monochromator. Whether one merely bends (Johann crystal) or grinds and bends the crystal (Johansson crystal), the focusing action serves to increase the intensity to such a degree that the intensity decline, when compared with that in the incident beam, is largely offset by the virtual elimination of background scattering. A

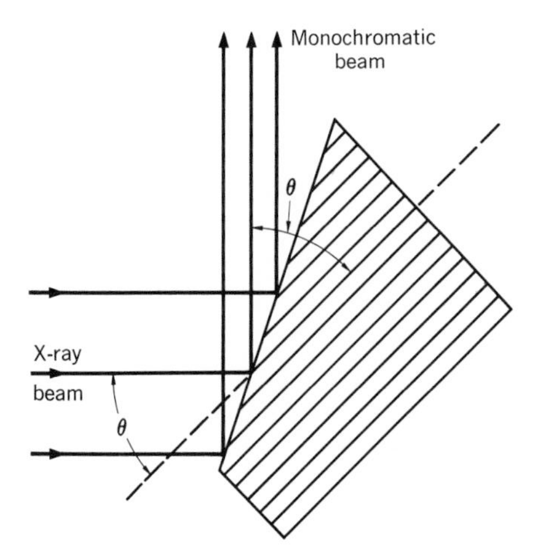

Fig. 17-11. Fankuchen condensing monochromator.

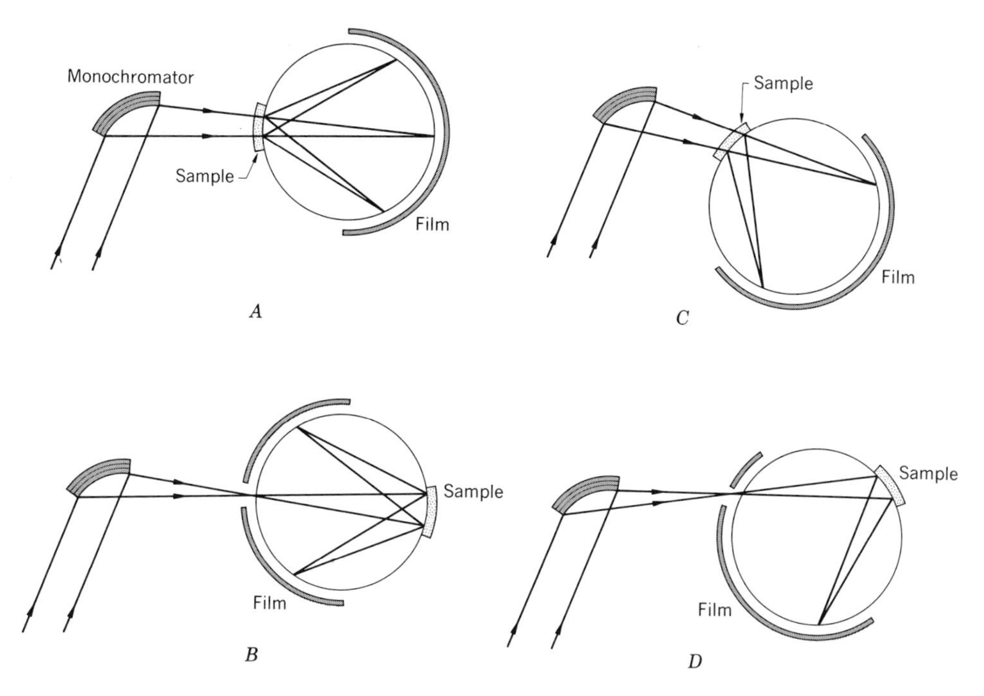

Fig. 17-12. Bent-crystal monochromators. (*A*) Symmetric transmission arrangement; (*B*) symmetric back-reflection arrangement; (*C*) Guinier asymmetric arrangement; (*D*) Seeman-Bohlin arrangement.

further benefit of using bent-crystal monochromators is that the diffraction profiles are sharpened by the focusing action, enhancing thereby the resolution of closely spaced lines. Although the virtues of bent-crystal monochromators were preached by A. Guinier for several decades, it is only fairly recently that their use in diffraction has really gained popularity.

Some of the camera arrangements employing bent-crystal monochromators are indicated in Fig. 17-12. The symmetric arrangements have the virtue that the diffracted beams are focused uniformly along the focusing circle to which the film conforms. By their very nature, however, they are limited to either the front- or the back-reflection regions. When a larger portion of the reflection spectrum is desired, the specimen must be placed asymmetrically. In this case it is not possible to produce focusing along a circle, so that a slight defocusing takes place for most reflections. Nevertheless, such focusing arrangements are very convenient, because relatively short exposures are sufficient to produce the clearly resolved reflections. In a particularly ingenious proposal made by de Wolff, metal septi are placed at right angles to the film (parallel to the plane of the drawings in Fig. 17-12), so that the camera actually can record two or more photographs on the same film. A different sample can be placed in the space of each sector, so that several diffraction spectra can be recorded for side-by-side comparison (Fig. 17-13). This is of real value when polymorphous compounds or otherwise related crystals are being studied.

In concluding this section it should be noted that it is possible to use a double curvature to increase the intensity concentration still further. In principle, it is possible to devise a monochromator that would cause x rays diverging from a point source to refocus at a point. In practice, such doubly bent crystals have been constructed, but their focusing action fails to attain the theoretical limit. A fairly popular mode of doubly bending a crystal, first proposed by B. E. Warren, refocuses x rays emanating from a line source to a fairly perfect line focus, as indicated in

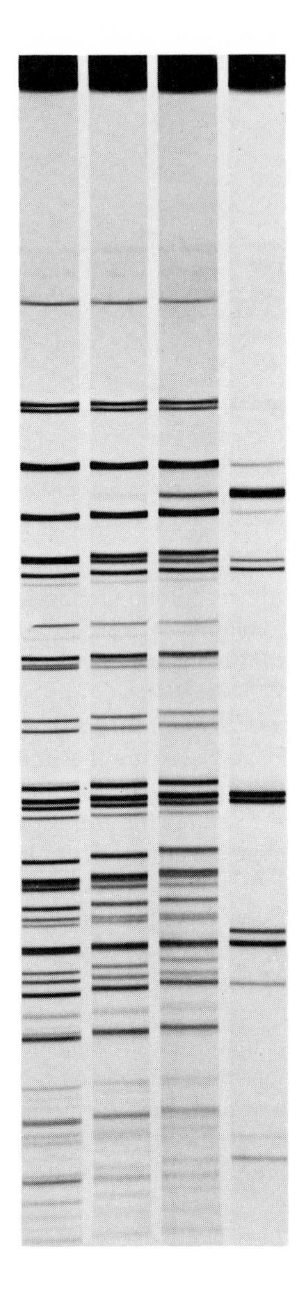

Fig. 17-13. Effect of addition of increasing amounts of CeO_2 in solid solution to pure ZrO_2 (extreme right). Note change of line positions and relative intensities. *(Original x-ray photograph kindly supplied by S. M. Lang and E. W. Franklin.)*

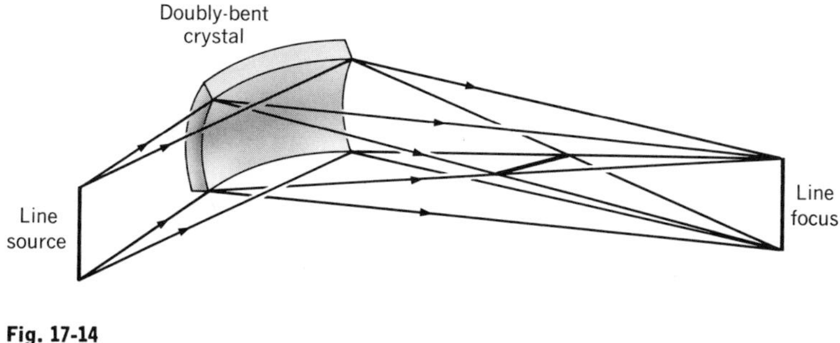

Fig. 17-14

Fig. 17-14. At first, Warren cut out thin slivers of LiF and spaced them on slightly tilted curved sectors to produce the vertical curvature. Since then, several of his former graduate students succeeded in developing techniques for doubly bending thin plates of LiF and similar crystals to the desired curvatures directly. By carrying out the bending operations in a solvent for the crystal, formation of excessive crystal imperfections is suppressed, while the plastic flow required to attain uniformity throughout the monochromator is facilitated.

DETERMINATION OF PRECISE LATTICE CONSTANTS

General considerations. Many crystallographic investigations require that the reflections from individual planes be fully identified. The procedures for indexing powder photographs can become quite involved, so that they are treated in a separate chapter. As pointed out in Chap. 18, it is possible to carry out the indexing of a powder photograph regardless of its symmetry, but the process becomes relatively more cumbersome as the symmetry decreases and requires increasingly more accurate d values. As discussed below, the errors inherent in each d value measured can be averaged out in the process of determining the most precise lattice-constant value for the crystal. It is not possible to increase the accuracy of individual d values, however, except by using careful laboratory procedures to minimize the sources of errors discussed briefly in an earlier section in this chapter.

In the study of various properties of solids, it is often desirable to establish whether the dimensions of the unit cell in a crystal have been altered as the result of some specialized treatment of the sample. This requires the measurement of the lattice constants in the most precise manner possible. There are essentially three different ways of doing this:

1. Practice of careful experimental technique
2. Averaging of measurements by graphical extrapolation
3. Averaging of measurements by analytical extrapolation

In addition, there are certain variants of these procedures involving the use of internal standards. A further discussion of them and related details can be found in some of the literature cited at the end of this chapter.

The practice of careful experimental technique has the important advantage that the desired results are obtained directly and under controlled conditions that are also known precisely. Usually, this requires, however, rather elaborate equipment, such as the apparatus shown in Fig. 17-15. The instruments shown in Fig. 17-15 have been built by Straumanis, who is one of the strongest exponents of this approach to precision. In addition to a carefully constructed camera housing a special sample holder, the apparatus shown includes thermostatically controlled enclosures to assure temperature stability during the entire exposure.

It should be noted that it is possible to achieve an adequate precision for many purposes by using only one or two d values. This is particularly true when comparisons of the same or identical samples are desired, for example, at different temperatures. For such measurements, the greater speed and versatility of the diffractometer are particularly useful. It is, of course, essential to maintain the same sample geometry in all such measurements, and one's ability to do this places the most severe limitation on this approach. It is also important, for this method, to select one or two reflections that are favorably situated. The essential requirements are that the reflection be easily resolvable and that it occur as close to $\theta = 90°$ as possible. The reason for the second requirement becomes clear when the relation between θ and d is examined.

By writing the Bragg relation

$$\frac{1}{d_{hkl}} = \frac{2}{\lambda} \sin \theta \tag{17-4}$$

it is clear that $1/d_{hkl}$ has a sinusoidal dependence on θ. This is illustrated in Fig. 17-16, which also compares the effect that the same error in measuring the angle ($\Delta\theta$) has on the corresponding error (Δd) in the spacing. Clearly, the larger the reflection angle, the smaller the error in d becomes, vanishing completely when $\theta = 90°$. This also can be seen by differentiating the Bragg equation.

$$0 = 2\Delta d \sin \theta + 2d(\cos \theta)\,\Delta\theta \tag{17-5}$$

Fig. 17-15. Equipment for precision determination of lattice constants at constant temperatures ($\pm 0.05°$) of the sample. The camera is inside a thermostated jacket on the x-ray generator shown on the right, and heating or cooling is provided by the two constant-temperature baths seen on the left and center of the photograph. *(Courtesy of M. E. Straumanis.)*

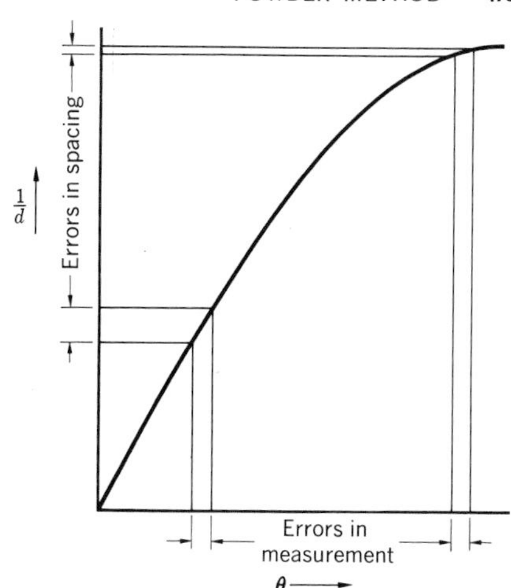

Fig. 17-16

After rearranging the terms in (17-5),

$$\frac{\Delta d}{d} = -\Delta\theta \frac{\cos \theta}{\sin \theta} = -\Delta\theta \cot \theta \tag{17-6}$$

which goes to zero as θ approaches $90°$. Consequently, other things being equal, spacing values measured in the back-reflection region are comparatively more precise, the precision increasing as θ approaches $90°$.

Graphical extrapolation. The first to suggest that a nearly error-free lattice constant can be measured at $\theta = 90°$, G. Kettman, calculated the a value of a cubic crystal using several measured d values and then plotted the a values directly against θ. By fitting the best sine curve to his points, he extrapolated the curve to $\theta = 90°$, where he then obtained an "error-free" value. The difficulty with this procedure lies in fitting a nonlinear curve to the data points. Also, as shown below, the systematic errors in d-value measurements do not depend directly on θ. In the absence of such systematic errors, however, when random (reading) errors are predominant, the Kettman extrapolation against θ directly is the correct one to use.

The first systematic analysis of the errors discussed near the beginning of this chapter was performed by Bradley and Jay. They derived explicit analytical functions for the various kinds of errors, and concluded that the lattice constant a of a cubic crystal varied linearly with $\cos^2 \theta$ at least for $\theta > 60°$. Thus they suggested that the following procedure be adopted in refining the lattice constant of a cubic crystal:

1. The reflections having $\theta > 60°$ are indexed, and the a value corresponding to each measured d value is calculated.
2. The a values are plotted against $\cos^2 \theta$ on a linear graph paper, with $\cos^2 \theta$ marked off along the abscissa.
3. The best straight line is drawn through the plotted points, and where it intersects the ordinate at $\cos^2 \theta = 0$, the error-free value of a is read off.

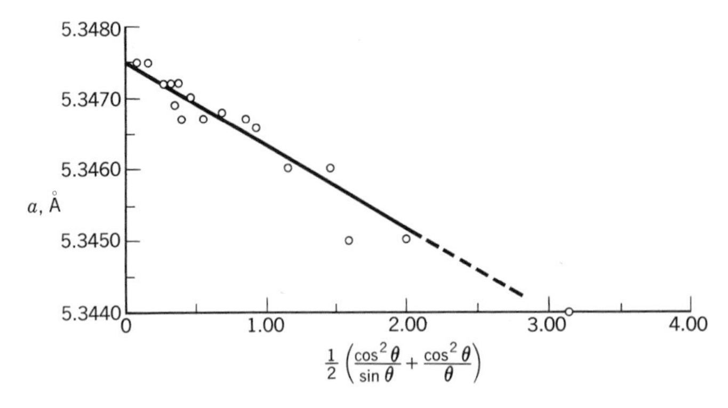

Fig. 17-17. Precision lattice-constant determination for CaF_2. *(After Azároff and Buerger.)*

The number of data points used in this extrapolation can be increased by combining experimental values determined with different-wavelength x radiation on a single graph.

Of the different sources of systematic errors considered above, only the error introduced by absorption cannot be removed by careful experimental technique. It can be minimized by preparing suitable samples, but except for very light elements, it is never truly negligible. The effect of absorption errors is very difficult to evaluate theoretically when the nonparallel x-ray beam used in practice is considered. After a reexamination of several such analyses, A. Taylor and H. Sinclair concluded that the best linear dependence was obtained for the functions $\frac{1}{2}[(\cos^2\theta)/\sin\theta + (\cos^2\theta)/\theta]$. Independently and at the same time, J. B. Nelson and D. P. Riley carried out a very careful experimental analysis and plotted their results against the various functions that had been proposed to correct for the effects of absorption. They also concluded that $\frac{1}{2}[(\cos^2\theta)/\sin\theta + (\cos^2\theta)/\theta]$ provided a linear relation. This is a most significant result, because it greatly facilitates the extrapolation procedure outlined above.

Figure 17-17 shows the kind of scatter in experimental a values that one normally obtains when an ordinary cylindrical powder camera and careful technique are employed. The foreknowledge that all the recorded points ideally should lie on a straight line greatly facilitates selecting the correct line to draw. As can be seen in Fig. 17-17, the scatter of points when $\theta > 50°$ (extrapolation function < 0.5) is so large that it is not easy to select the correct straight line. When the other points at smaller θ values are considered, however, it becomes relatively easier to point the line to the correct extrapolated value, which, as before, is read off the ordinate at the zero point corresponding to $\theta = 90°$.

Least-squares method. An appealing procedure for avoiding the need to make a subjective judgment in an extrapolation scheme is to use the well-known mathematical method of least squares. Based on the analytical expressions derived by Bradley and Jay, M. V. Cohen derived the necessary expressions for calculating the so-called *normal equations* for cubic crystals. Later, Cohen extended this procedure to include hexagonal and tetragonal crystals also. (The simultaneous refinement of more than one lattice constant from the same d values is possible only by iterative means because the relative magnitude of the errors in each parameter is not known in advance.) Because of the need for rather extensive mathematical manipulations, the least-squares method did not become really popular until the recently increased availability of high-speed digital computers.

By the same token, least-squares procedures are so commonly used nowadays to refine various kinds of experimental data (cf. Chap. 11) that a detailed discussion of its application to lattice-constant refinement is not included here. The interested reader can find a complete exposition of Cohen's method and references to later work in the literature cited at the end of this chapter.

A word of caution about the use of graphical or analytical procedures. Each of the procedures described above was devised for a very special situation. Thus the Taylor-Sinclair-Nelson-Riley function is intended for an absorbing cylindrical specimen placed in a cylindrical powder camera (Fig. 17-8). Its use for other kinds of experimental arrangements is not really warranted, therefore, unless there is additional evidence to indicate its applicability. Similarly, when a virtually nonabsorbing sample is examined, this function again is inappropriate. Similarly, when using diffractometer data, Wilson has shown that peak positions are best suited for extrapolation methods, whereas centroids give the best reproducibility when a values calculated from different lines on the same recording are compared. Since the precision sought is often about 1 part per 100,000, it is essential to use the correct procedure to assure significant results.

SUGGESTIONS FOR SUPPLEMENTARY READING

Leonid V. Azároff and Martin J. Buerger, *The powder method in x-ray crystallography* (McGraw-Hill Book Company, New York, 1958).

André Guinier, *X-ray crystallographic technology*, English translations by T. L. Tippell, edited by K. Lonsdale (Hilger and Watts Ltd., London, 1952).

Harold P. Klug and Leroy E. Alexander, *X-ray diffraction procedures for polycrystalline and amorphous materials* (John Wiley & Sons, Inc., New York, 1954).

H. S. Peiser, H. P. Rooksby, and A. J. C. Wilson, *X-ray diffraction by polycrystalline materials* (Chapman & Hall, Ltd., London, 1955).

EXERCISES

17-1. Consider the powder photograph of tungsten shown in Fig. 17-3. How can one distinguish the front- from the back-reflection regions? Qualify your answer.

17-2. Mentally position the tungsten photograph in a film reader as indicated in Fig. 17-4. Referring to the notation in that figure, the following experimental values are read off the film:

x_2	21.270	22.170	22.915	23.605	32.195	31.480	30.680	29.565
x_1	17.225	16.325	15.580	14.890	24.285	25.000	25.800	26.915

(The last two sets are for the stronger of the two lines in the doublets.) Assuming a camera diameter of 114.6 mm and that filtered copper radiation was used in the experiment, calculate the d

values for each reflection. **Hint:** The above experimental values are first read in pairs about the hole at $\theta = 0°$ and then about the hole at $\theta = 90°$.

17-3. Using the d values determined in Exercise 17-2, calculate the correct a value using the relation $d = (h^2 + k^2 + l^2)/a^2$ for each measured d value. To do this, it is necessary to postulate the appropriate hkl values for each reflection. Since tungsten has a body-centered cubic structure, these reflections are, in order, 110, 200, 211, 220, 310, 222, 321, and 400. How do the calculated a values compare with each other? Is there a better agreement between back-reflection values than there is with other values? Why should this be?

17-4. Using the a values obtained in Exercise 17-3, prepare a linear plot of a vs $\cos^2 \theta$ and a vs the Taylor-Sinclair-Nelson-Riley function. Which do you find more convenient to use?

17-5. Cylindrical powder cameras constructed in England traditionally have diameters that are round numbers, for example, 60 or 190 mm. How does this complicate the interpretation of the resulting powder photographs?

17-6. As has been discussed in earlier chapters, the combination of the Lorentz factor and the polarization effect of x-ray scattering leads to an increase in the intensity close to $\theta = 0°$ and close to $\theta = 90°$. This has the effect of increasing the background intensity on which the reflections are superimposed. Illustrate this effect graphically, and deduce from it which way this shifts a reflection in the back-reflection region. Is this classifiable with the so-called systematic errors?

17-7. Consider the powder photograph of tungsten in Fig. 17-3. How can one increase the number of reflections recorded? Using the results of Exercise 17-2, calculate the Bragg angle for the reflection having the largest recorded d values when Mo $K\alpha$ radiation is employed.

17-8. Making use of Exercise 17-3, what are the indices of the last reflection of tungsten that can be recorded using Cr $K\alpha$ radiation?

17-9. Derive an expression relating the distance from the point $\theta = 90°$ to any other point on the film to the Bragg angle for the symmetrical back-reflection focusing camera shown in Fig. 17-10. It is common practice to build such cameras with notches that provide fiducial markers at regular angular intervals along the film. How can one check on the accuracy of these markers? Would you prefer to use the relation just derived or such markers when reading an actual film?

17-10. Assume for a cylindrical specimen that the effect of absorption on the diffracted-beam intensity varies as $e^{-\mu D}$, where D is the sample diameter. At the same time, assume that the intensity increase, due to more material placed in the path of the beam, varies at D^2. Derive an expression for the optimum sample diameter for which these opposing trends just balance. For the case of tungsten, what are the optimum sample diameters when Cu $K\alpha$ and Mo $K\alpha$ radiations are used?

EIGHTEEN
Indexing procedures

WHEN THE CRYSTAL SYSTEM IS KNOWN

Interplanar spacing relations. Following the determination of the d values from a powder photograph according to the procedures described in Chap. 17, it is frequently desired to identify the corresponding crystallographic planes (hkl). The reasons for this are many. When a material is not readily obtainable in the form of single crystals, the powder method provides the only means for complete crystallographic characterization. Not only is this important for newly synthesized or discovered materials, but also in many processes involving known materials. For example, changes in the composition during isomorphic replacement can be followed by noting intensity changes in the reflections from certain planes. Even more important, when quantitative analyses of mixtures are carried out, frequently a few lines remain unidentified, for reasons discussed in the next chapter. When this occurs, it is essential to establish whether the remaining reflections originated from a phase not already identified or whether they are previously unidentified reflections from one of the recognized phases present. An easy way to answer this question is to attempt to index the remaining lines on the basis of the known unit cells. For this reason, some of the procedures that can be followed to index lines on a powder photograph when the unit cell is known are considered first.

Before one can assign indices to the observed reflections, it is necessary to establish the exact relations between the indices of a plane and the unit-cell parameters. This is a fairly easy problem in geometry when the cell angles are all $90°$, but becomes progressively more difficult as the number of arbitrary parameters increases from four in the monoclinic system to six in the triclinic system. This situation is eased considerably when one uses reciprocal-lattice coordinates, because the individual planes are then described by their respective normals σ_{hkl}, and it is much easier to analyze the angular interrelations of lines than of planes. Recalling from Chap. 7 that the dot product of a reciprocal-lattice vector with itself is equal in magnitude to $1/d_{hkl}^2$, it is conventional to designate this quantity Q_{hkl}. According to (7-22), then,

$$Q_{hkl} = \frac{1}{d_{hkl}^2} = h^2a^{*2} + k^2b^{*2} + l^2c^{*2} + 2hka^*b^* \cos \gamma^* + 2klb^*c^* \cos \alpha^* + 2lhc^*a^* \cos \beta^* \quad (18\text{-}1)$$

The other forms of (18-1), when crystal symmetry causes certain of the parameters to become equivalent, are tabulated in Table 18-1.

Making use of the relations between the coordinate systems in direct space and in reciprocal space given in Table 7-1, it is possible to transform the relations in Table 18-1 to direct space. The resulting relations are presented in Table 18-2. This procedure is purely analytical and obviates the need for a very elaborate geometrical analysis in the triclinic system. A comparison of the resulting expression in Table 18-2 with the relation in (18-1) demonstrates the obvious advantage in using reciprocal-lattice coordinates.

Graphical indexing. The advantages inherent in utilizing the reciprocal-lattice concept were not recognized by the early crystallographers, and so they concentrated their efforts in the development of procedures based on cell edges in direct space. This limited the procedures to the more symmetric crystal systems, and it was not until the middle of this century that a method for indexing powder photographs regardless of the symmetry was developed. This procedure naturally utilized the reciprocal-lattice relation (18-1), and is described in the second half of this chapter, because, when the crystal system is not known beforehand, the analysis is considerably eased by employing reciprocal-lattice coordinates from the outset.

The relations in Table 18-2 are useful for indexing a powder photograph when-

Table 18-1 Spacing as a function of reciprocal-cell edges

Crystal system	$\frac{1}{d_{hkl}^2} = \sigma_{hkl}^2 = Q_{hkl}$
Cubic	$(h^2 + k^2 + l^2)a^{*2}$
Tetragonal	$(h^2 + k^2)a^{*2} + l^2c^{*2}$
Orthorhombic	$h^2a^{*2} + k^2b^{*2} + l^2c^{*2}$
Hexagonal:	
Hexagonal indices	$(h^2 + hk + k^2)a^{*2} + l^2c^{*2}$
Rhombohedral indices	$(h^2 + k^2 + l^2)a^{*2} + 2(hk + kl + lh)a^{*2} \cos \alpha^*$
Monoclinic	$h^2a^{*2} + k^2b^{*2} + l^2c^{*2} + 2hka^*b^* \cos \gamma^*$ (first setting) $h^2a^{*2} + k^2b^{*2} + l^2c^{*2} + 2lhc^*a^* \cos \beta^*$ (second setting)
Triclinic	$h^2a^{*2} + k^2b^{*2} + l^2c^{*2} + 2hka^*b^* \cos \gamma^* + 2klb^*c^* \cos \alpha^* + 2lhc^*a^* \cos \beta^*$

Table 18-2 Spacing as a function of cell edges

Crystal system	$\frac{1}{d^2_{hkl}}$
Cubic	$\frac{1}{a^2}(h^2 + k^2 + l^2)$
Tetragonal	$\frac{h^2 + k^2}{a^2} + \frac{l^2}{c^2}$
Orthorhombic	$\frac{h^2}{a^2} + \frac{k^2}{b^2} + \frac{l^2}{c^2}$
Hexagonal:	
Hexagonal indices	$\frac{4}{3a^2}(h^2 + hk + k^2) + \frac{l^2}{c^2}$
Rhombohedral indices	$\frac{1}{a^2}\frac{(h^2 + k^2 + l^2)\sin^2\alpha + 2(hk + kl + lh)(\cos^2\alpha - \cos\alpha)}{1 + 2\cos^3\alpha - 3\cos^2\alpha}$
Monoclinic	$\frac{\frac{h^2}{a^2} + \frac{k^2}{b^2} - \frac{2hk\cos\gamma}{ab}}{\sin^2\gamma} + \frac{l^2}{c^2}$ (first setting)
	$\frac{\frac{h^2}{a^2} + \frac{l^2}{c^2} - \frac{2hl\cos\beta}{ac}}{\sin^2\beta} + \frac{k^2}{b^2}$ (second setting)
Triclinic	$\frac{\frac{h^2}{a^2}\sin^2\alpha + \frac{k^2}{b^2}\sin^2\beta + \frac{l^2}{c^2}\sin^2\gamma + \frac{2hk}{ab}(\cos\alpha\cos\beta - \cos\gamma) + \frac{2kl}{bc}(\cos\beta\cos\gamma - \cos\alpha) + \frac{2lh}{ca}(\cos\gamma\cos\alpha - \cos\beta)}{1 - \cos^2\alpha - \cos^2\beta - \cos^2\gamma + 2\cos\alpha\cos\beta\cos\gamma}$

ever the crystal system is known. For the higher-symmetry systems, graphical procedures can be devised to save considerable computational labor. Consider the interplanar spacing equation in the cubic system:

$$\frac{1}{d^2_{hkl}} = \frac{1}{a^2}(h^2 + k^2 + l^2) \tag{18-2}$$

or $$d_{hkl} = (h^2 + k^2 + l^2)^{-\frac{1}{2}}a \tag{18-3}$$

Now the sum of three integers squared can only have the values $1, 2, 3, 4, 5, 6, 8, 9, \ldots$, or in other words, all integer values except $4^p(8n + 7)$, where p and n can have all integer values, including zero. This means that (18-3) predicts the same set of d values for all cubic crystals, except for a scale factor, namely, the cell edge a. It is simple and most convenient, therefore, to construct a graph on which the relation (18-3) is plotted for each possible combination of h, k, and l.

Such a graph is shown in Fig. 18-1. To use it, simply mark off the observed d values on a strip of paper, using the scale at the bottom of the graph (Fig. 18-1) and, keeping $d = 0$ on the ordinate, place it horizontally across the appropriate a value. The hkl values of each reflection are then read off directly, as can be seen in Fig. 18-1. Note that this procedure can be used with equal ease even when the magnitude of the lattice constant a is not known beforehand. Since $d = 0$ is fixed to lie along the ordinate, and since the scale along which d is measured must remain parallel to the abscissa, a paper strip on which the d values have been plotted is moved upward in Fig. 18-1 until a match between the lines on the graph and the marks on the paper strip is obtained. The unknown value of a then can be read on the ordinate, and the indices can be assigned to each reflection, as

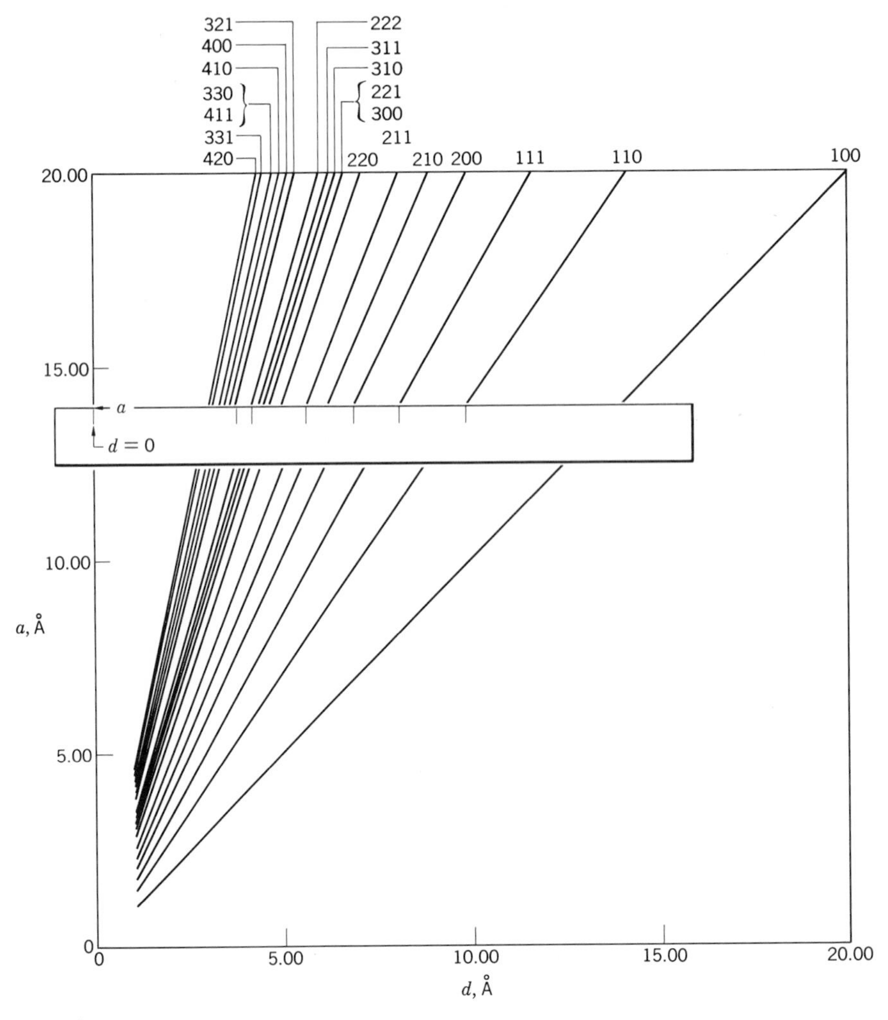

Fig. 18-1

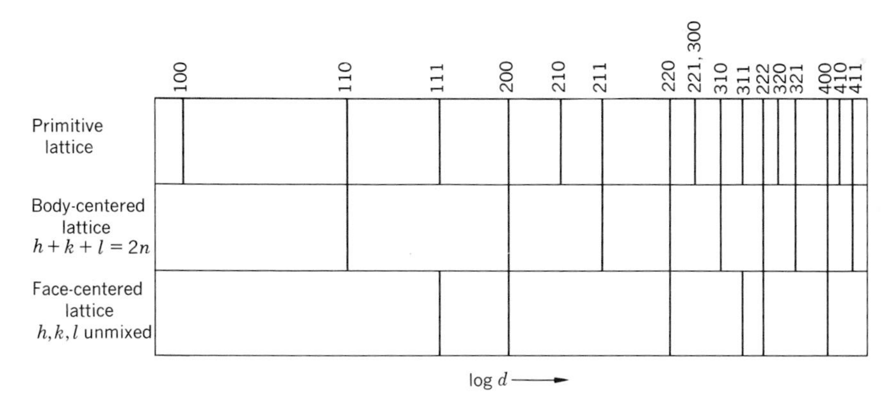

Fig. 18-2

before. When carrying out such analyses in practice, it is convenient to prepare a chart like Fig. 18-1 drawn to a scale of about 20 by 20 in.

The effect of lattice extinctions usually can be utilized to hasten the indexing of cubic crystals because the reflections appear on the powder photographs in characteristic groupings. When the cubic lattice is primitive, one observes seven regularly spaced lines, followed by a space and seven more lines, etc. The sequences that are observed when the cubic lattice is face-centered or body-centered are indicated in Fig. 18-2.

The foregoing procedure is extremely straightforward because only one cell constant is involved in the analysis. When there are two, graphical methods become more complicated. From Table 18-2, the interplanar-spacing relation for the tetragonal system is

$$\frac{1}{d_{hkl}^2} = \frac{h^2 + k^2}{a^2} + \frac{l^2}{c^2} \tag{18-4}$$

Recasting the above expression by dividing out a^2 and letting $C = c/a$,

$$d_{hkl}^2 = a^2 \left(h^2 + k^2 + \frac{l^2}{C^2}\right)^{-1} \tag{18-5}$$

It was first suggested by A. W. Hull and W. P. Davey that relation (18-5) could be used to prepare charts on which curves relating d_{hkl} to C could be plotted provided that the second parameter in (18-5), a^2, was somehow eliminated. The method adopted by Hull and Davey was to take logarithms of both sides of (18-5) so that

$$2 \log d_{hkl} = 2 \log a - \log \left(h^2 + k^2 + \frac{l^2}{C^2}\right) \tag{18-6}$$

Since the second term on the right of (18-6) already includes the value of a in $C = c/a$, it is possible to let $a = 1$ or $\log a = 0$ and to consider only the relation

$$2 \log d_{hkl} = - \log \left(h^2 + k^2 + \frac{l^2}{C^2}\right) \tag{18-7}$$

Plotting $\log d_{hkl}$ along the abscissa and C along the ordinate, Hull and Davey prepared charts for the tetragonal and hexagonal systems, including a separate chart using rhombohedral indices.

A Hull-Davey chart for the tetragonal system is shown in Fig. 18-3. Although it contains a much larger clustering of curves than the simpler cubic chart in Fig. 18-1, its use is straightforward when the crystal system and the lattice constants c and a are known in advance. The observed d values are plotted on a strip of paper, using the scale on the graph, and then the strip is moved upward to the appropriate value of the ratio C. Since the direct dependence on a has been removed in this chart, it is necessary to move the strip horizontally back and forth until a match between the plotted curves and the marks on the paper strip is obtained, all the while keeping the edge of the strip horizontally aligned at the correct C value. Once a match is found, the appropriate hkl values can be read off directly, as before. Note that the density of curves increases markedly to the left in Fig. 18-3. For this reason it is essential that a match be sought in which the largest d values line up with the lower-order indices.

Somewhat simpler charts have been developed by subsequent investigators,† in which the individual curves are more widely dispersed. Nevertheless, utilization of this procedure, when the crystal system but not the cell constants is known beforehand, becomes more tedious. The suitably prepared paper strip must now be moved sideways and also vertically in seeking an appropriate match. There are only two guidelines that one has in such a search: The strip must be kept parallel to the abscissa, and the larger d values must be assigned low-order indices. Once a match is obtained, the indexed interplanar-spacing values are used in conjunction with relation (18-4), or its equivalent, if the crystal is hexagonal, to establish the values of a and c, corresponding to the ratio C, determined at the time when the match was obtained on the chart. Clearly, this procedure involves a certain amount of trial-and-error hunting. It is equally clear that, when the crystal system is not known beforehand, it is necessary to seek a match on more than one chart without any assurance of success, since the material may not be uniaxial. Graphs similar to the one in Fig. 18-2 can be prepared for the orthorhombic system also, by plotting $\log d$ against a/c on a series of charts for which the value of b/c has been fixed. Their usefulness in practice is limited to the case when it is known that the measured d values are those of an orthorhombic crystal, and even then the analytical procedures described in the next section are far simpler to use.

Analytical procedures. When the crystal system is known in advance, it is easier to index powder reflections analytically. That is particularly true when reciprocal-lattice coordinates are used. Consider, for example, the orthorhombic system for which

$$\begin{aligned} Q_{hkl} &= h^2a^{*2} + k^2b^{*2} + l^2c^{*2} \\ &= Q_{h00} + Q_{0k0} + 0_{00l} \end{aligned} \tag{18-8}$$

† A complete discussion of all indexing procedures devised before 1958 can be found in the book by Azároff and Buerger listed at the end of this chapter.

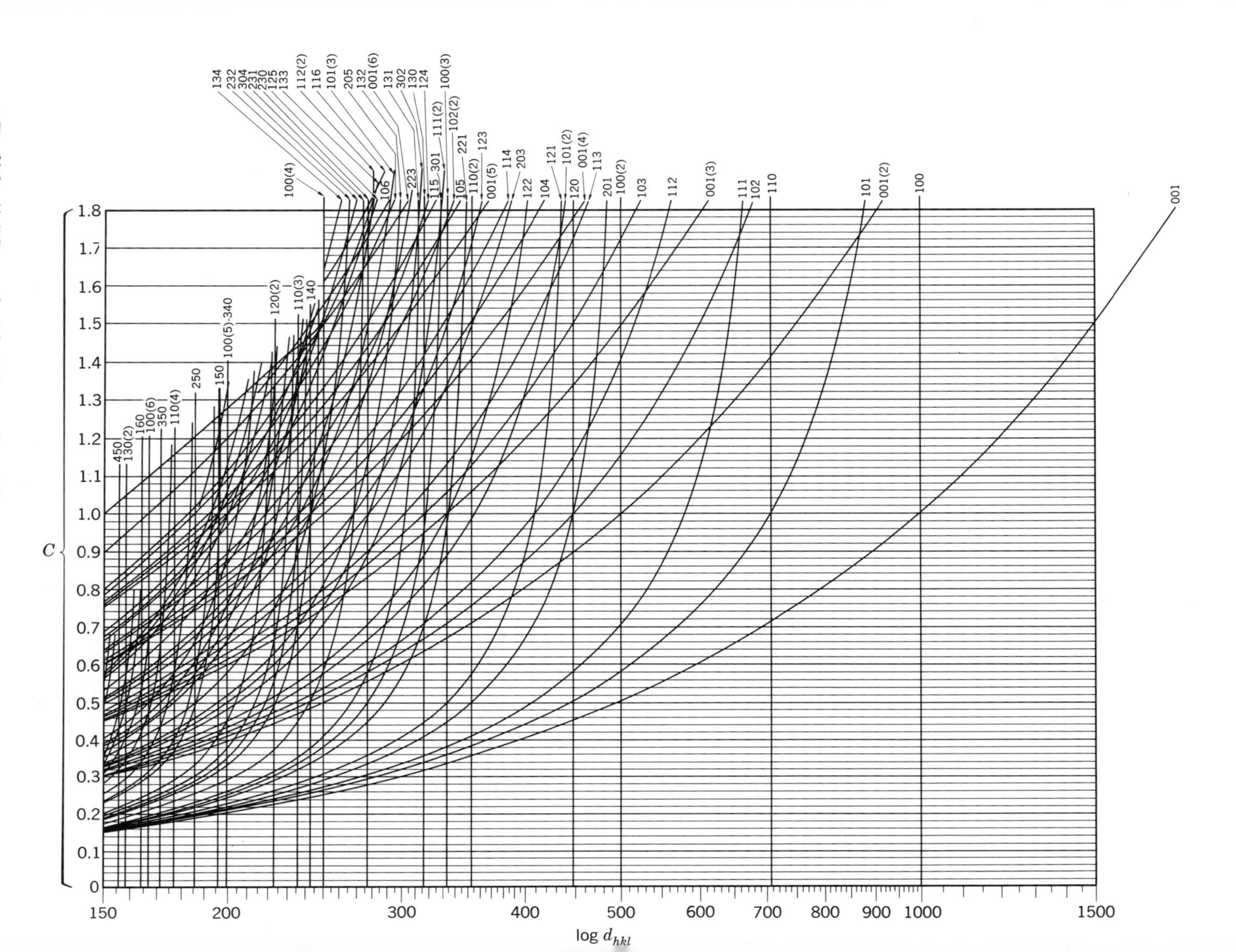

Fig. 18-3. Hull-Davey chart for tetragonal crystals.

By preparing these columns of permissible pinacoidal Q values, one for each of the three terms in (18-8), all possible Q_{hkl} can be obtained directly by simple addition of various pairs or triplets of the values listed. The only caution is not to use two numbers from the same column. (See Exercise 18-2.)

Not only is the above procedure very straightforward for orthogonal crystal systems, but also for the oblique systems. This is so because relation (18-1) is linear in terms of the Q's. For example,

$$Q_{hk0} = Q_{h00} + Q_{0k0} + 2hka^*b^* \cos \gamma^* \tag{18-9}$$

so that, in the monoclinic system,

$$Q_{hkl} = Q_{hk0} + Q_{00l} \tag{18-10}$$

where $Q_{00l} = l^2c^{*2}$. Thus four columns of Q values must be prepared, one for each pinacoidal Q and one for the cross term in (18-9). Since c^* is orthogonal to both a^* and b^*, Q_{h0l} and Q_{0kl} values are calculated by summing pairs of numbers from the appropriate columns. Then Q_{hk0} values are calculated by combining appropriate sets of three numbers, as indicated in (18-9). Finally, the remaining Q_{hkl} values are summed according to (18-10). A similar strategy is followed when the crystallites are triclinic. The above procedure involves the formation of simple products (squares) and subsequent addition. This means that the entire indexing procedure can be carried out easily and rapidly, using an ordinary desk calculator.

WHEN THE CRYSTAL SYSTEM IS UNKNOWN

Tests for cubic, tetragonal, and hexagonal systems. The graphical and analytical procedures described above can be adapted to the indexing of unknown powder photographs provided it is known that the list of d values belongs to a single phase. To ascertain this fact, it is helpful to examine the original powder microscopically. If a polarized-light microscope is available and the powder is transparent, such an examination also will establish whether the material is isotropic (cubic), uniaxial (hexagonal or tetragonal), or biaxial. When such information is not available in advance of the x-ray study, it is advisable to make a few simple tests of the x-ray data before proceeding with the more complex general procedure discussed in a later section.

If the material is cubic, this fact often can be recognized directly from an examination of the powder photograph because the reflections must occur in one of the three sequences illustrated in Fig. 18-2. This regularity is a characteristic of all cubic powder photographs. When the total number of lines is still fairly small but the reflections are clustered in somewhat less regular groups, it is quite probable that the material belongs to one of the two uniaxial systems. As the density of lines increases, the likelihood is that the symmetry decreases. Unfortunately, this is not an infallible guide since the density of lines is also proportional to cell size. Thus a large cubic cell ($a \approx 10$ Å) gives rise to as many powder lines as a somewhat smaller noncubic cell.

In the absence of any clues, therefore, it is desirable to proceed with the indexing in a systematic manner. For reasons that should be obvious by now, this can be done most conveniently by using reciprocal-lattice coordinates. Consider the form of (18-1) when the crystallites are

$$
\begin{array}{llll}
\text{Cubic:} & Q_{hkl} = N_C a^{*2} & N_C = 1, 2, 3, 4, 5, 6, 8, \text{etc.} & \\
\text{Tetragonal:} & Q_{hkl} = N_T a^{*2} + l^2 c^{*2} & N_T = 1, 2, 4, 5, 8, 9, 10, \text{etc.} & (18\text{-}11) \\
\text{Hexagonal:} & Q_{hkl} = N_H a^{*2} + l^2 c^{*2} & N_H = 1, 3, 4, 7, 9, 12, 13, \text{etc.} &
\end{array}
$$

It is clear from (18-11) that, except for the permissible values of N, the three relations are identical when $l = 0$ in the uniaxial systems. Use can be made of this fact by constructing a linear chart like the one shown in Fig. 18-4. Note that because the lattice constant is squared, a more condensed scale is used along the abscissa, so that the first line does not have a slope of $45°$. This graph is similar to the one in Fig. 18-1, except that all N values are included, and its use is quite analogous. The Q values measured on a photograph (see Fig. 17-6) are marked off along a strip of paper, and it is slid vertically to seek a match. (Note that $Q = 0$ remains on the ordinate, and the strip is kept horizontal.) If a match can be found for all experimental values, the material is cubic and the N_C values of the reflections are read off directly. If a match is not obtained, the following steps are taken.

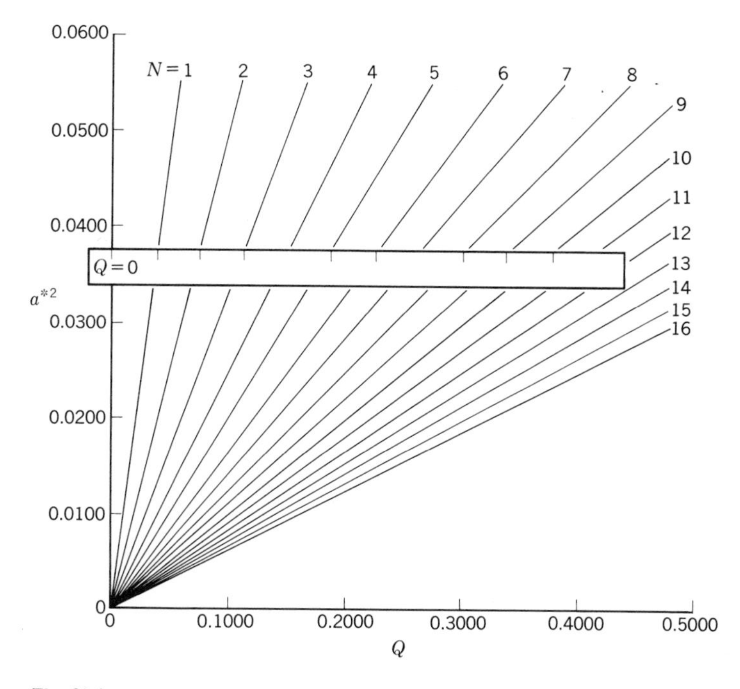

Fig. 18-4

The bottom two relations in (18-11) suggest that relations of the type

$$Q_{h_1k_1l_1} - Q_{h_2k_2l_1} = (N_1a^{*2} + l_1c^{*2}) - (N_2a^{*2} + l_1c^{*2})$$
$$\Delta Q = \Delta N\, a^{*2} \tag{18-12}$$

should exist, where ΔN is the difference between the two N values and can assume all integer values. When two Q values are subtracted from each other, it is, of course, possible that relations of the type

$$Q_{h_1k_1l_1} - Q_{h_1k_1l_2} = (N_1a^{*2} + l_1c^{*2}) - (N_1a^{*2} + l_2c^{*2})$$
$$\Delta Q = \Delta l^2\, c^{*2} \tag{18-13}$$

also may occur. Finally, it is most likely that neither pair of coefficients is the same, and

$$\Delta Q = Q_1 - Q_2 = \Delta N\, a^{*2} + \Delta l^2\, c^{*2} \tag{18-14}$$

Suppose, however, one were to form all possible differences between the experimentally determined Q values. Which differences are likely to recur more than twice? Surely, it is evident that the recurrence of differences like (18-14) would require a nearly integral ratio between the magnitudes of a^{*2} and c^{*2}. Barring this unlikely situation, the only relations that are likely to yield recurring differences are (18-12) and (18-13). Of these, (18-12) is the most probable to recur since it requires only that pairs of reflections have the same l index.

Consider the list of observed Q values in Table 18-3, which were obtained from a sample of GeO_2. A table of all possible differences is prepared by first subtracting Q_1 from Q_2, Q_3, etc.; next, by subtracting Q_2 from Q_3, Q_4, etc.; and so on, until all possible ΔQ values are obtained. (In practice, it is usually sufficient to utilize only the first 12 succeeding Q values.) A convenient way to examine these differences is to plot them as short vertical bars, as shown in Fig. 18-5. Each ΔQ value, including the original list of Q's, since they equal $\Delta Q = Q - 0$, is marked off along the horizontal scale. When the same value recurs, the height of the bar is increased correspondingly. By selecting a horizontal scale commensurate with the experimental accuracy in measuring Q values, the scatter of values is decreased,

Table 18-3 Observed Q values for germanium dioxide

Q_1	0.0536	Q_{11}	0.3396	Q_{21}	0.6567
Q_2	0.0850	Q_{12}	0.3750	Q_{22}	0.6599
Q_3	0.1606	Q_{13}	0.4067	Q_{23}	0.6729
Q_4	0.1786	Q_{14}	0.4440	Q_{24}	0.6968
Q_5	0.1919	Q_{15}	0.4959	Q_{25}	0.7281
Q_6	0.2145	Q_{16}	0.5002	Q_{26}	0.7668
Q_7	0.2456	Q_{17}	0.5139	Q_{27}	0.7817
Q_8	0.2817	Q_{18}	0.5544	Q_{28}	0.8220
Q_9	0.2860	Q_{19}	0.6075	Q_{29}	0.8356
Q_{10}	0.3357	Q_{20}	0.6431	Q_{30}	0.8767

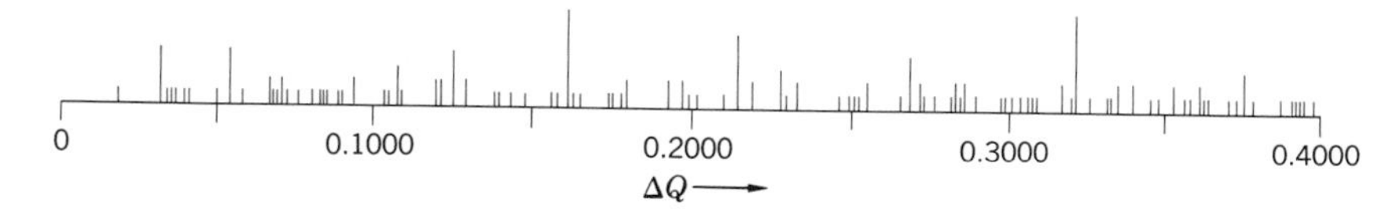

Fig. 18-5. Hesse plot for GeO_2 powder data.

and the recurring coincidences "grow" like tall trees in a brush forest. An examination of the differences plotted in Fig. 18-5 shows that

0.0314 recurs 4 times.

0.0536 recurs 4 times.

0.1070 recurs 3 times.

0.1252 recurs 4 times.

0.1608 recurs 8 times. (18-15)

0.2145 recurs 6 times.

0.2218 recurs 3 times.

0.2682 recurs 4 times.

0.3217 recurs 7 times.

0.3750 recurs 3 times.

These differences are most likely to represent pairs of reflections like those in (18-12). It is evident from the similarity of this expression to (18-11), for the cubic system, that a plot of these values on a strip of paper and a search for a match on the graph in Fig. 18-4 should identify the correct a^{*2} value for GeO_2. This turns out to be the case with $a^{*2} = 0.0536$. Note that a match is obtained at this point for all values in (18-15) except for $\Delta Q = 0.0314, 0.1252$, and 0.2281. The N values corresponding to the "indexed" Q values are $1, 2, 3, 4, 5, 6$, and 7, as can be seen by examining the list in (18-15) directly.

Knowing the value of a^{*2}, it is now possible to calculate Q_{hk0} values, except that one does not know at this stage whether to use the allowed N_T or N_H values. This uncertainty can be resolved by trying both possibilities and comparing the resulting values with the experimentally determined list. In the present case, a comparison of the values in Table 18-3 with those in (18-5) shows that

$$
\begin{aligned}
Q_1 &= 1 \times a^{*2} \\
Q_3 &= 3 \times a^{*2} \\
Q_6 &= 4 \times a^{*2} \\
Q_{12} &= 7 \times a^{*2}
\end{aligned} \qquad (18\text{-}16)
$$

from which it follows that the system must be hexagonal, since the coefficients in (18-16) correspond to the possible N_H values only.

The remaining job of determining c^{*2} can be carried out either by iteration, using the remaining Q values in Table 18-3, or as in the present case, by realizing that the second most likely recurrences are those described by (18-13). If the "unindexed" ΔQ values in (18-15) represent $\Delta l^2 c^{*2}$ values, they should form a regular sequence, since

$$\Delta l^2 = 1, 3, 4, 5, 7, 8, 9, 11, 12, \text{etc.} \tag{18-17}$$

Assuming $c^{*2} = 0.0314$, the other two ΔQ values, 0.1252 and 0.2218, clearly fit the above sequence, confirming this choice. A final check is obtained by using (18-11) to calculate all Q_{hkl} values, following the procedure described in the preceding section, and by assigning hkl indices to all the experimentally determined Q values in Table 18-3 (Exercise 18-3).

Test for orthorhombic system. The procedure described above for establishing the a^* and c^* values for uniaxial crystals is based on one devised independently by R. Hesse and H. Lipson for indexing orthorhombic crystals. Because of the linearity of relation (18-8), it is possible to form difference equations like (18-12) in the orthorhombic system also. Specifically,

$$\begin{aligned} Q_{h_1k_1l_1} - Q_{h_2k_1l_1} &= \Delta h^2 a^{*2} \\ Q_{h_1k_1l_1} - Q_{h_1k_2l_1} &= \Delta k^2 b^{*2} \\ Q_{h_1k_1l_1} - Q_{h_1k_1l_2} &= \Delta l^2 c^{*2} \end{aligned} \tag{18-18}$$

which include, of course, such special cases as

$$Q_{h_10l_1} - Q_{h_20l_1} = \Delta h^2 a^{*2} \tag{18-19}$$

and so forth. Clearly, when a list of ΔQ values is examined for frequency of recur-

Table 18-4 Observed Q values of an "unknown" orthorhombic crystal

Line	Q	*Line*	Q
1	0.0865	15	0.4956
2	0.1396	16	0.5208
3	0.1406	17	0.5472
4	0.1755	18	0.5624
5	0.1912	19	0.5973
6	0.2413	20	0.6930
7	0.2804	21	0.7020
8	0.3308	22	0.7748
9	0.3460	23	0.8538
10	0.3564	24	0.8675
11	0.3658	25	0.8875
12	0.4031	26	0.9241
13	0.4350	27	0.9616
14	0.4547	28	1.0131

Table 18-5 Partial list of differences between Q values in Table 18-4 (decimal point omitted)

No.	Q	1	2	3	4	5	6	7	8	9	10	11	12	13	14	15	16
1	0.0865																
2	0.1396	0531															
3	0.1406	0541	0010														
4	0.1755	0890	0359	0349													
5	0.1912	1047	0516	0506	0157												
6	0.2413	1548	1017	1007	0658	0501											
7	0.2804	1939	1408	1398	1049	0892	0391										
8	0.3308	2443	1912	1902	1553	1396	0895	0504									
9	0.3460	2595	2064	2054	1705	1548	1047	0656	0152								
10	0.3564	2699	2168	2158	1809	1652	1151	0760	0256	0104							
11	0.3658	2793	2262	2252	1903	1746	1245	0854	0350	0198	0094						
12	0.4031	3166	2635	2625	2276	2119	1618	1227	0723	0571	0567	0473					
13	0.4350	3485	2954	2944	2595	2438	1937	1546	1044	0892	0788	0694	0319				
14	0.4547	3682	3151	3141	2792	2635	2134	1743	1239	1087	0983	0889	0516	0197			
15	0.4956		3560	3550	3201	3044	2543	2152	1648	1496	1392	1298	0925	0606	0409		
16	0.5208			3802	3453	3296	2795	2404	1900	1748	1644	1550	1177	0858	0661	0252	
17	0.5472				3717	3560	3056	2668	2164	2012	1908	1814	1441	1122	0925	0616	0264

rences, the only ones that should recur often are those listed in (18-18). Because they must obey sequences like that in (18-17), it is possible to group the recurring ΔQ values into sets and to distinguish them from the recurrences that occur in either the tetragonal or hexagonal systems.

In practice, one first tries the tests suggested in the preceding section. If they fail to provide a match but the plot in Fig. 18-5 clearly shows a number of trees, the procedure outlined below is followed, because the material probably is orthorhombic. A fortunate coincidence comes into play here. Most orthorhombic crystals have nonprimitive cells or symmetry, giving rise to space-group extinctions. This limits the various combinations of hkl values possible and increases the probability of the recurrence of relations like those in (18-18) among the Q values in a powder photograph.

As an illustration of how the Hesse-Lipson procedure works in practice, consider the Q values of an "unknown" powder photograph in Table 18-4. Their differences

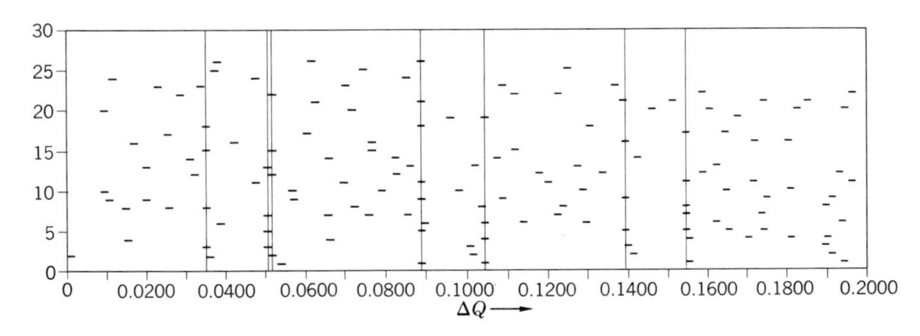

Fig. 18-6. Lipson plot for an "unknown" orthorhombic crystal.

are systematically arrayed in Table 18-5. Instead of plotting them on a bar graph like the one first suggested by Hesse and shown in Fig. 18-5, a scheme proposed by Lipson is used. The ΔQ values in each column in Table 18-5 are plotted as horizontal bars in equally spaced rows. The recurrences are then established by sliding a vertical straightedge over the assembly of bars and noting the ΔQ values at which there are many intersections. (A pair of right triangles can be used in practice.) Whereas the experimental scatter of values in the Hesse plot shows up a "bush" formation instead of a "tree," in the Lipson plot it shows up by a dispersion of the bars on both sides of the vertical line. An examination of the ΔQ values plotted in Fig. 18-6 yields the following distribution of differences:

$$\begin{aligned}
&0.0890 \text{ occurs } 7 \text{ times.}\\
&0.1548 \text{ occurs } 6 \text{ times.}\\
&0.1045 \text{ occurs } 5 \text{ times.}\\
&0.1396 \text{ occurs } 5 \text{ times.}\\
&0.0350 \text{ occurs } 4 \text{ times.}\\
&0.0505 \text{ occurs } 4 \text{ times.}\\
&0.0515 \text{ occurs } 4 \text{ times, etc.}
\end{aligned} \tag{18-20}$$

If it is assumed that the most frequently recurring difference in (18-20) almost certainly is of the type predicted by (18-18), it follows that

$$\begin{aligned}
Q_{100} &= 1^2a^{*2} = 0.0890\\
Q_{200} &= 2^2a^{*2} = 0.3560\\
Q_{300} &= 3^2a^{*2} = 0.8010\\
&\cdots\cdots\cdots\cdots
\end{aligned} \tag{18-21}$$

A check of the observed values in Table 18-4 shows that $Q_{10} = 0.3564$, so that the probability of the correctness of the above choice is increased. (Regardless of lattice type, the 200 reflection is most likely to be present.)

A careful examination of the Q values in Table 18-4 discloses, however, that the smallest Q value in that list, namely, $Q_1 = 0.0865$, is smaller than Q_{100}. This means that the next pinacoidal Q value chosen should be smaller than Q_1. Checking the frequently recurring differences in (18-20) shows a likely candidate.

$$\begin{aligned}
Q_{010} &= 1^2b^{*2} = 0.0350\\
Q_{020} &= 2^2b^{*2} = 0.1400\\
Q_{030} &= 3^2b^{*2} = 0.3150\\
Q_{040} &= 4^2b^{*2} = 0.5600\\
&\cdots\cdots\cdots\cdots
\end{aligned} \tag{18-22}$$

Referring to Table 18-4, it is found that $Q_3 = 0.1406$ and $Q_{18} = 0.5624$. Since higher-order reflections are more accurate, let $Q_{010} = 0.0351$, in which case the values in (18-22) agree more closely with the experimental ones. To verify further the correctness of the two choices made so far, consider some of the possible Q_{hk0} values that now can be calculated:

$$\begin{aligned}
Q_{110} &= Q_{100} + Q_{010} = 0.0890 + 0.0351 = 0.1241 \\
Q_{120} &= Q_{100} + Q_{020} = 0.0890 + 0.1404 = 0.2294 \\
Q_{210} &= Q_{200} + Q_{010} = 0.3560 + 0.0351 = 0.3911 \\
Q_{220} &= Q_{200} + Q_{020} = 0.3560 + 0.1404 = 0.4964 \\
&\cdots\cdots\cdots\cdots\cdots\cdots\cdots\cdots
\end{aligned} \tag{18-23}$$

The last value in (18-23) corresponds to Q_{15} in Table 18-4, supporting the selections made in (18-21) and (18-22), and so forth.

Returning to the Q values in Table 18-4, it is seen that Q_1 still has not been accounted for. It is clear that it must be either Q_{001} (unlikely since the corresponding Q_{002} is not observed) or else it must be Q_{011}, since it is larger than Q_{010}. On this assumption

$$Q_{001} = Q_{011} - Q_{010} = 0.0865 - 0.0351 = 0.0514 \tag{18-24}$$

This value occurs four times in the list of frequently recurring differences in (18-20), so that it is a likely choice for c^{*2}. A calculation of various Q_{hkl} values using all three of the above choices quickly confirms the validity of this assumption.

The above examples have illustrated some of the aspects involved in indexing procedures for noncubic crystals. The process is iterative in nature, but it can be systematized by realizing the limitations that are imposed on the Q values and, particularly on ΔQ values, by the basic interplanar-spacing equation (18-1). When systematic searches of ΔQ tabulations do not disclose several really frequent recurrences, despite a large number of observed Q values (more than 30 with Cu $K\alpha$ radiation), one may assume that the crystallites probably are either monoclinic or triclinic. In this case, the general procedure developed by Tei-Ichi Ito and described in the next section should be tried.

General procedure. In principle, Ito's method can be used for any crystal system, but because it requires a large number of Q values, it works best when the symmetry is low, so that systematic absences are infrequent. Like the above methods, it is iterative, so that some of the previously described practical steps should be followed here too. Since the method is supposed to be independent of symmetry, no attempt is made to establish the crystal system at the outset. Instead, it is decided to find a reciprocal lattice on the basis of which all the observed Q values can be indexed. Once such a lattice is known, it is, clearly, the correct lattice for this crystal, although a better choice of three cell edges may be available. This, however, is a simple matter for transformation theory (Chap. 3), and a systematic procedure to follow is described in the next section.

Suppose that the experimental values are those for $MgWO_4$ listed in Table 18-6 and that the previous tests failed to provide a match. One begins by arbitrarily assuming that the first three observed Q values are, respectively, Q_{100}, Q_{010}, and Q_{001}. In the absence of evidence to the contrary, they can be assumed to be the

Table 18-6 Observed Q values for $MgWO_4$

Line	Q	Line	Q
1	0.0310	21	0.3250
2	0.0457	22	0.3322
3	0.0730	23	0.3364
4	0.0769	24	0.3428
5	0.1165	25	0.3451
6	0.1187	26	0.3505
7	0.1239	27	0.3664
8	0.1649	28	0.3723
9	0.1699	29	0.3824
10	0.1816	30	0.4016
11	0.1957	31	0.4083
12	0.2077	32	0.4432
13	0.2123	33	0.4450
14	0.2386	34	0.4498
15	0.2436	35	0.4608
16	0.2517	36	0.4659
17	0.2563	37	0.4768
18	0.2793	38	0.4863
19	0.2884	39	0.4918
20	0.3065	40	0.4938
		41	0.5376

squares of three mutually noncoplanar reciprocal-lattice vectors, thereby defining some cell in the reciprocal lattice. Following a previously made suggestion, high-order Q values are calculated as shown in Table 18-7. The purpose of this calculation, however, is not to verify the aptness of the initial selections so much as to correct the values chosen for a^{*2}, b^{*2}, and c^{*2}, since much of what follows depends on how accurately it is possible to match calculated and observed Q values. From Table 18-7 it follows that better agreement will result if the following adjustments are made:

$$
\begin{aligned}
Q_{100} &= 0.0310 \pm 0.0000 = 0.0310 = a^{*2} \qquad & a^* &= 0.1761 \\
Q_{010} &= 0.0457 - 0.0003 = 0.0454 = b^{*2} \qquad & b^* &= 0.2131 \\
Q_{001} &= 0.0730 - 0.0009 = 0.0721 = c^{*2} \qquad & c^* &= 0.2685
\end{aligned}
\tag{18-25}
$$

Consider now a relation like (18-9), but for the general case when it is necessary to distinguish not only the three indices h, k, and l, but also their positive and

Table 18-7 Selection of Q_{100}, Q_{010}, and Q_{001} from Table 18-6

Q_{hkl}	*Computed*	*Observed*	*Error in* Q_{100}
Q_{100}	......	0.0310	
Q_{200}	0.1240	0.1239	$-\frac{0.0001}{4} = 0$
Q_{300}	0.2790	0.2793	$+\frac{0.0003}{9} = 0$
Q_{010}	......	0.0457	
Q_{020}	0.1828	0.1816	$-\frac{0.0012}{4} = -0.0003$
Q_{030}	0.4113	0.4083	$-\frac{0.0030}{9} = -0.0003$
Q_{001}	......	0.0730	
Q_{002}	0.2920	0.2884	$-\frac{0.0036}{4} = -0.0009$
Q_{003}	0.6570		

negative values. Specifically, consider

$$Q_{h0l} = Q_{h00} + Q_{00l} + 2lhc^*a^* \cos \beta^* \tag{18-26}$$

and $$Q_{h0\bar{l}} = Q_{h00} + Q_{00l} - 2lhc^*a^* \cos \beta^* \tag{18-27}$$

If (18-26) and (18-27) are subtracted from each other,

$$Q_{h0l} - Q_{h0\bar{l}} = 4lhc^*a^* \cos \beta^* \tag{18-28}$$

from which the unknown angle β^* can be determined.

If (18-26) and (18-27) are added,

$$Q_{h0l} + Q_{h0\bar{l}} = 2(Q_{h00} + Q_{00l}) = 2Q'_{h0l} \tag{18-29}$$

which suggests that (18-26) and (18-27) can be simplified to

$$Q_{h0l} = Q'_{h0l} + 2hla^*c^* \cos \beta^* \tag{18-30}$$

$$Q_{h0\bar{l}} = Q'_{h0l} - 2hla^*c^* \cos \beta^* \tag{18-31}$$

An examination of the above two relations demonstrates that for each possible Q' value there ought to exist two actual Q values which differ from it by the same fixed magnitude. If this magnitude can be established, it can be used to determine the interaxial angle. This is the entire basis for the Ito method, and it is diagramed in Fig. 18-7. Using the corrected values in (18-25), it is possible to calculate

$$Q'_{101} = Q_{100} + Q_{001} = 0.0310 + 0.0721 = 0.1031 \tag{18-32}$$

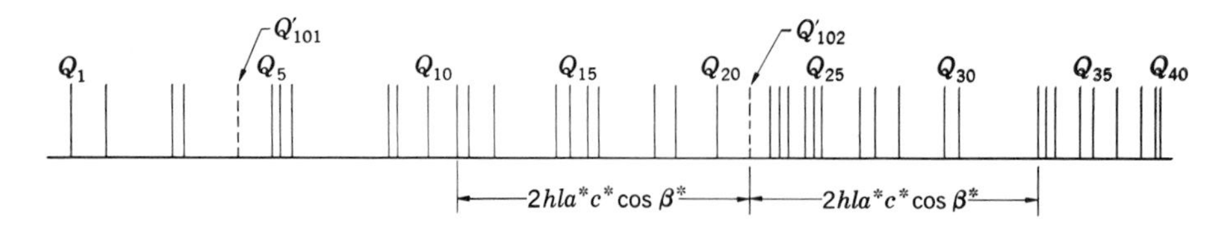

Fig. 18-7

An examination of Table 18-6 soon shows that there are not two Q values present that are, respectively, smaller and larger than (18-32) by the same magnitude. This means that either the corresponding Q_{101} or $Q_{10\bar{1}}$ value is not observed. Consequently, another value is chosen for Q_{00l}, say, Q_{002}, in which case

$$Q'_{102} = Q_{100} + Q_{002} = 0.0310 + 0.2884 = 0.3194 \tag{18-33}$$

An examination of Table 18-6 shows that Q_{11} and Q_{32} appear to satisfy the conditions in (18-30) and (18-31). This is also diagramed in Fig. 18-7. From this examination it follows that

$$\begin{aligned} Q_{102} - Q'_{102} &= Q_{32} - Q'_{102} \\ &= 0.4432 - 0.3194 = 0.1238 \end{aligned} \tag{18-34}$$

and

$$\begin{aligned} Q'_{102} - Q_{10\bar{2}} &= Q'_{102} - Q_{11} \\ &= 0.3194 - 0.1957 = 0.1237 \end{aligned} \tag{18-35}$$

Making use of (18-28) and solving for β^*,

$$\beta^* = \cos^{-1} \frac{Q_{102} - Q_{10\bar{2}}}{4 \cdot 1 \cdot 2 \cdot a^*c^*} = 48° \tag{18-36}$$

Analogous procedures can be used to determine the angles α^* and γ^*. (See Exercise 18-5.) In doing this, several points should be kept in mind. Each angle determined should be verified by calculating several Q_{h0l} values and checking them against the observed Q values. Only after several matches have been found can one be sure that the chosen angle is really correct. Similarly, the values of the lattice constants determined should be refined carefully at all stages, since the successful application of Ito's method rests on how closely it is possible to match up observed and calculated values. This also suggests that the beginning Q values should be determined as accurately as possible and at least to three significant numbers. Finally, it is necessary to keep in mind several possible coincidences, which, when present, can greatly ease the analysis. Suppose one begins by calculating Q'_{110} and then finds that Q_3, originally selected to be Q_{001}, is really of the type Q_{hk0}. It is then necessary to select a new value for Q_{001} since Q_3 is obviously coplanar with Q_1 and Q_2. Conversely, suppose the value of Q'_{hk0} actually coincides with one of the observed Q values. According to (18-9), this can occur when $\gamma^* = 90°$, so that this interaxial angle is determined directly. Before accepting this as real and not as a coincidence, other values of Q_{hk0} should be calculated and verified. All this suggests that the Ito method is best applied by exploring one zone of Q values at a time. This was realized some time ago by P. M. de Wolff, who developed several tests for systematizing this procedure even more. Further discussions and several actual examples of such analyses are given in the references at the end of this chapter.

Choice of correct cell. Once the correct reciprocal lattice for a crystal has been established, it is a fairly simple matter to select the unit cell that correctly describes the symmetry and suggests the optimum lattice choice. The necessary transformation relations, however, are less likely to cause difficulties when carried out in direct space. Accordingly, it is customary to transpose the reciprocal cell at this stage, using the relations tabulated in Table 7-1. Next it is necessary to find the unit cell in the crystal lattice that has the three shortest translations for its cell edges, called the *reduced cell*. Once the reduced cell is known, a suitable table can be consulted† to find the optimal or conventional choice for a unit cell.

The whole trick, then, is to find, first, the reduced cell in the lattice determined by Ito's method. To do this, one makes use of the geometry of lattices discussed in Chap. 3. It is possible to define any translation in a lattice as a vector.

$$\mathbf{t} = u\mathbf{a} + v\mathbf{b} + w\mathbf{c} \tag{18-37}$$

where u, v, and w are three integers representing the direction of the translation. Multiplying the vector by itself,

$$\begin{aligned}\mathbf{t}\cdot\mathbf{t} &= u^2(\mathbf{a}\cdot\mathbf{a}) + v^2(\mathbf{b}\cdot\mathbf{b}) + w^2(\mathbf{c}\cdot\mathbf{c}) + 2uv(\mathbf{a}\cdot\mathbf{b}) + 2vw(\mathbf{b}\cdot\mathbf{c}) + 2wu(\mathbf{c}\cdot\mathbf{a})\\ t^2 &= u^2s_{11} + v^2s_{22} + w^2s_{33} + 2uvs_{12} + 2vws_{23} + 2uws_{31}\end{aligned} \tag{18-38}$$

The important quantities in (18-38) are the six scalars cutomarily used to represent a cell:

$$\begin{pmatrix} s_{11} & s_{22} & s_{33} \\ s_{23} & s_{31} & s_{12} \end{pmatrix} \tag{18-39}$$

and can be diagramed in a triangular fashion as shown in Fig. 18-8.

Given two mutually noncollinear translations in a plane lattice, it is possible to establish quickly whether they are the two shortest translations in that lattice by following a procedure first developed by Buerger. Consider the plane lattice shown in Fig. 18-9, and suppose that it had been defined originally by the transla-

† A complete discussion of reduced cells is given in Chaps. 11 and 12 of the book by Azároff and Buerger listed at the end of this chapter.

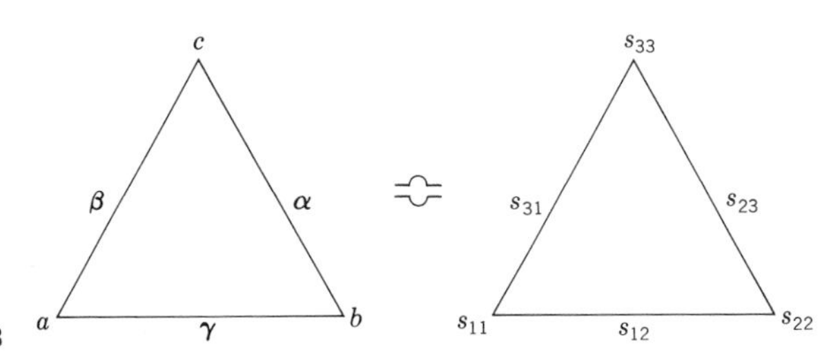

Fig. 18-8

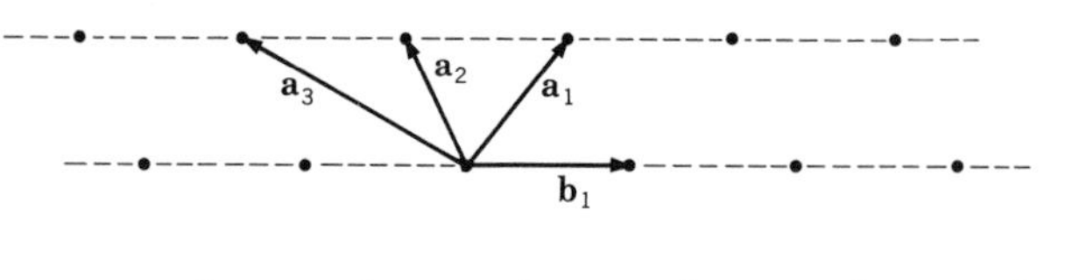

Fig. 18-9

tions $\mathbf{a}_1$ and $\mathbf{b}_1$. Then consider the scalar s_{12} formed by multiplying these two vectors.

$$\mathbf{a}_1 \cdot \mathbf{b}_1 = |\mathbf{a}_1|\,|\mathbf{b}_1| \cos \gamma \tag{18-40}$$

Next, consider some of the other choices for $\mathbf{a}$ that are delimited by the band parallel to $\mathbf{b}$ (shown dashed in Fig. 18-9). For example,

$$\begin{aligned} \mathbf{a}_2 \cdot \mathbf{b}_1 &= (\mathbf{a}_1 - \mathbf{b}_1) \cdot \mathbf{b}_1 = \mathbf{a}_1 \cdot \mathbf{b}_1 - \mathbf{b}_1 \cdot \mathbf{b}_1 \\ &= |\mathbf{a}_1|\,|\mathbf{b}_1| \cos \gamma - |\mathbf{b}_1|^2 \end{aligned} \tag{18-41}$$

and

$$\begin{aligned} \mathbf{a}_3 \cdot \mathbf{b}_1 &= (\mathbf{a}_1 - 2\mathbf{b}_1) \cdot \mathbf{b}_1 = \mathbf{a}_1 \cdot \mathbf{b}_1 - 2\mathbf{b}_1 \cdot \mathbf{b}_1 \\ &= |\mathbf{a}_1|\,|\mathbf{b}_1| \cos \gamma - 2|\mathbf{b}_1|^2 \end{aligned} \tag{18-42}$$

An examination of Fig. 18-9 and the above relations shows that *the shortest value of* $\mathbf{a}$ *relative to* $\mathbf{b}$ *results when*

$$|\mathbf{a} \cdot \mathbf{b}| \leq \tfrac{1}{2} b^2 \tag{18-43}$$

that is, the shortest value of $\mathbf{a}$ relative to $\mathbf{b}$ has a projection on $\mathbf{b}$ that is less than one-half the length of $\mathbf{b}$. Clearly, similar relations can be established for the other axial pairs.

Examination of relations (18-41) and (18-42) also shows that it is possible to satisfy condition (18-43) by subtracting a suitable number of multiples of b_1^2 from the value in (18-40). The reduction procedure can be epitomized by triangular representations like those shown in Fig. 18-8. Each pair of cell edges is tested against condition (18-43) by noting whether the magnitude of the dot product s_{12} is less than $b^2/2$ and also less than $a^2/2$. If not, a suitable number of vectors are subtracted, with the result indicated in Fig. 18-10. Note that changing the length of one cell edge alters the magnitude of two dot products, and therefore three of the scalar quantities in (18-39). It is necessary, therefore, to check the dot

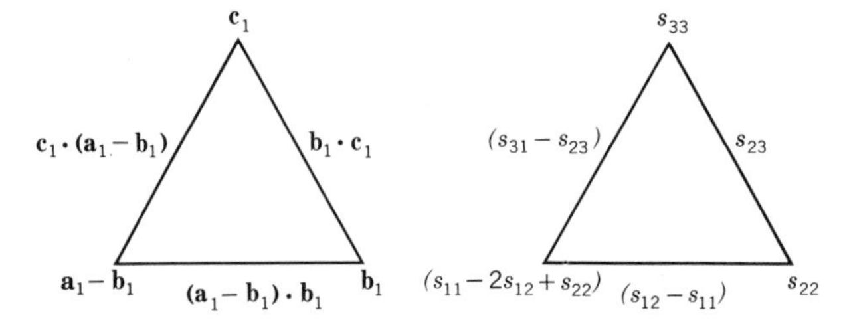

Fig. 18-10. Reduction procedure indicated by Fig. 18-9.

products on two sides of the triangle to make sure that no further reductions are possible (Exercise 18-6). When this is the case, the scalar representation (18-39) of the reduced cell can be used to look up the optimum lattice representation in a suitable table. When such a table is not handy, the optimum choice can be made from an analysis of the lattice defined by the reduced cell. Since the lattice is unaffected by which cell is chosen to represent it, this procedure is required only when the reduced cell is one of the primitive cells in a centered lattice.

SUGGESTIONS FOR SUPPLEMENTARY READING

Leonid V. Azároff and Martin J. Buerger, *The powder method in x-ray crystallography* (McGraw-Hill Book Company, New York, 1958).

P. M. de Wolff, On the determination of unit-cell dimensions from powder diffraction patterns, *Acta Crystallographica*, vol. 10 (1957), pp. 590–595.

——— Detection of simultaneous zone relations among powder diffraction lines, *Acta Crystallographica*, vol. 11 (1958), pp. 664–665.

EXERCISES

18-1. Beginning with relation (18-1) and utilizing the relations in Table 7-1, derive the correct relation between d_{hkl} and the four lattice constants in the monoclinic system. [Because the form of (18-1) is so easy to remember, this procedure provides a handy means for verifying the correctness of published interplanar-spacing equations.] Show that this relation reduces to the correct one for the hexagonal system when $\gamma = 120°$ and $a = b$.

18-2. Suppose the reciprocal cell of an orthorhombic crystal has $a^* = 0.10$, $b^* = 0.08$, and $c^* = 0.12$. Calculate the Q_{hkl} values for all permissible combinations of h, k, l, out to 6, if the crystal has an all-face-centered lattice.

18-3. Following the analysis in the text, index all the Q values listed in Table 18-3, using the procedure described earlier in this chapter.

18-4. After a list of ΔQ values had been examined, the following values were found to recur three or more times: 0.0475, 0.0945, 0.1422, 0.1615, 0.1899, and 0.3792. Deduce the correct a^{*2} value. What do you think is a likely value for c^{*2}? (These are the actual values obtained for rutile crystallites.)

18-5. Using the Q values in Tables 18-6 and 18-7, determine the reciprocal lattice of $MgWO_4$ by finding α^* and γ^*. Check your analysis by indexing the first 15 values in Table 18-6.

18-6. The reciprocal-cell constants determined in Exercise 18-5 should be those given in (18-25), with $\alpha^* = 48°$, $\beta^* = 89.5°$, and $\gamma^* = 90°$. By using the reduction procedure described in the text, deduce the reduced cell. (In this case it happens to be also the optimum cell.) Derive the necessary transformation matrix for transforming the hkl indices assigned in Exercise 18-5.

NINETEEN
Identification of unknowns

QUALITATIVE ANALYSIS

Identification procedures. The first example of the application of x-ray diffraction to the identification of unknown substances was made by Hull as early as 1919. What Hull showed was that the chemically difficult to distinguish NaF and $NaHF_2$ could be easily identified by comparing their powder photographs. Because an x-ray diffraction diagram is determined by the exact atomic arrangement in a material, it is like a fingerprint, in that no two materials give rise to identical diagrams. Conversely, if two materials do give rise to identical diagrams, they must be the same material. This immediately suggests that it would be desirable to have a large repository for powder photographs (fingerprints) of all known materials, which could be consulted when the identification of any one material was sought. In essence, what is desired is a catalog of reference films with which a particular photograph can be compared. Whenever such standard films are available, they provide a convenient way of identifying the constituents of an unknown. To become a practical method, however, the direct comparison of films would require the existence of a vast library of films of known substances. Not only is such a library difficult to obtain, but its utilization would entail a tedious comparison between the test photograph and successive standard films. Although such a procedure is obviously not suitable for routine identifications, it can be used to great advantage when a group of structurally related compounds is being studied. For example, if one is studying complex sulfosalts, it is possible to prepare a small library of films which can be used to identify an unknown sulfosalt. The common structural features in such a group often give rise to lines which appear on the photographs of the entire group. The lines not common to the group then are the ones most helpful in distinguishing between the various members of the groups. Such comparisons can be aided considerably by having an illuminated viewer available on which the films can be compared side by side. It should be recalled also from Chap. 17 that special cameras can be constructed to record the powder photographs of several substances on the same film (Fig. 17-13), thereby aiding in such analyses.

The value of a catalog, however, is determined by how easy it is to use it, and not by how elegant its format may be. Experience has shown that the most useful scheme in practice is one originally proposed by J. D. Hanawalt, H. W. Rinn, and L. K. Frevel, in 1938. In its current and only somewhat modified form, it is known as the *Powder diffraction file* and consists of an *Index* and a collection of individual cards for each of over fifteen thousand substances, as described in the next section. Before proceeding with this more detailed discussion, it should be noted that the powder method is only one of several analytical procedures that can be used. Until 1918, the two principal methods available for identifying an unknown were chemical analysis and optical crystallographic procedures, using the polarizing microscope. Chemical methods are capable of disclosing the atomic composition, but not the nature of the atomic aggregation. The polarizing microscope can be used in a manner similar to the powder method, to disclose both the composition and the state of aggregation of the atoms. This is so because the indices of refraction for visible or infrared light are as characteristic of a crystalline substance, as is its powder photograph. Optical methods cannot be used, however, if the crystallites are too small or if the substance is opaque to visible or infrared light. In addition, there are other analytical methods, such as optical spectroscopy, x-ray fluorescence analysis, and x-ray absorption analysis, that can be used also. The latter two are described briefly in the closing sections of this chapter, and are limited to the identification of the elemental composition. Although only x-ray diffraction can be used to identify uniquely all crystalline materials, these other methods constitute useful adjuncts in any real analysis, particularly when complex mixtures are examined.

Powder diffraction file. The information contained in a typical card comprising the *Powder diffraction file* is shown in Fig. 19-1. On the first line, the interplanar spacings of the three most intense lines are arranged in order of decreasing intensity; the fourth d value is the largest observed spacing. The relative intensities of these reflections, based on 100 for the strongest intensity observed, are listed on the next line, directly below the corresponding d values. The space to the right of these values is reserved for the chemical formula and the name of the material, as well as other descriptive information, such as the mineral name or structural formula of an organic compound. The next lines, which are self-explanatory, list the experimental data, crystallographic data, and optical data. The last line provides miscellaneous information pertaining to any unusual treatment of the sample, its source, or its chemical composition. The complete list of observed interplanar spacings, the relative reflection intensities, and the reflection indices, when known, are tabulated on the right-hand section of the card.

The *Powder diffraction file* is published in sets, from time to time, and currently consists of 17 sets of individual cards which can be obtained in the form of ordinary 3- by 5-in. cards (Fig. 19-1), Keysort cards, and IBM cards (also on magnetic tape), as well as in book form, in which the first 10 sets have been printed on consecutive pages of two volumes to date, with additional volumes in preparation. Within each set, the cards are numbered consecutively. They can be located quite easily by the number appearing in the upper-left-hand corner (Fig. 19-1), which first gives

7-210

d	3.10	4.76	3.07	4.76	$CaWO_4$	★
I/I_1	100	53	31	53	CALCIUM TUNGSTATE	SCHEELITE

Rad. $CuK\alpha_1$ λ 1.5405 Filter Ni Dia.
Cut off I/I_1 SPECTROMETER
Ref. NBS CIRCULAR 539, VOL 6 (1956)

Sys. TETRAGONAL S.G. $C_{4H}^{6} - I4_1/A$
a_0 5.242 b_0 c_0 11.372 A C
α β γ Z 4 Dx 6.12
Ref. IBID.

εα 1.918 nωβ 1.935 εγ Sign +
2V D mp Color COLORLESS
Ref. IBID.

SAMPLE FROM KERNVILLE, CALIFORNIA; SPECT. ANALYSIS SHOWS < 1.0% MO; < 0.1% NA; < 0.01% AL, SI, SR; < 0.001% AG, CR, CU, MG, MN.
PATTERN MADE AT 25°C.

REPLACES 1-0806

d Å	I/I_1	hkl	d Å	I/I_1	hkl
4.76	53	101	1.3358	3	217
3.10	100	112	1.3106	3	400
3.072	31	103	1.2638	2	411
2.844	14	004	1.2488	13	316
2.622	23	200	1.2284	2	109
2.296	19	211	1.2074	5	332
2.256	3	114	1.2054	5	413
2.0864	5	105	1.1901	4	404,307
1.9951	13	213	1.1728	1	420
1.9278	28	204	1.1280	5	228
1.8538	12	220	1.1096	2	415
1.7278	5	301	1.0870	5	1.1.10,424
1.6882	16	116	1.0838	8	327,501
1.6332	10	215	1.0439	3	431,2.0.10
1.5921	30	312	1.0351	2	336
1.5532	14	224	1.0140	6	1.0.11
1.4427	6	321	1.0116	4	512
1.4219	2	008	0.9699	1	521,2.2.10
1.3859	3	305	.9636	4	408
1.3577	4	323	PLUS 26 LINES TO 0.7937		

Fig. 19-1

the set number and then the card number. This numbering system also simplifies the reconstruction of the proper storage sequence when cards are returned to the file drawers. To find the number of the card bearing the desired information, it is necessary to consult first the *Index* to the *Powder diffraction file.*

Index* to *Powder diffraction file. The *Index* currently is published in two volumes, one for inorganic, the other for organic, materials. In the first part of each index volume, the cards are arranged according to the d values of their three most intense reflections. This is also the information excerpted in the upper left corner of each card (Fig. 19-1). Following a scheme suggested by Hanawalt, the cards are arranged among the groups shown in Table 19-1 according to the d value of their most intense reflection; within each group, the cards are arranged according to the declining magnitude of the second d value. Card number 7-210, shown in Fig. 19-1, is classified in the group containing cards whose most intense reflections have d values between 3.14 and 3.10 Å. Within that group it lies between a card whose second d value is 5.02 Å and one whose second $d = 4.68$ Å, despite the fact that the first d value of both of these cards is larger than 3.10 Å. The d value of the third most intense reflection is used to establish the listing sequence in case more than one card in a particular Hanawalt group has the same second d value.

For the original *File*, Hanawalt suggested that three cards be made up for each substance, listing the three most intense reflections, in the order d_1, d_2, d_3, on a master card, and in the orders d_2, d_1, d_3 and d_3, d_1, d_2, on two supplementary cards. The purpose of this was to keep the different judgments regarding the proper sequence of I_1, I_2, and I_3 from impairing the ease of locating the cards in the *Index*. This procedure has been perpetuated to the present time, despite considerable improvements in the accuracy of intensity measurement. It has the important

Table 19-1 The Hanawalt grouping of file cards according to d values

Groups arranged according to d value of most intense reflection, Å	*Number of groups*
20.00 and larger	One
19.99–18.00	One
17.99–16.00	One
15.99–14.00	One
13.99–12.00	One
11.99–11.00	One
10.99–10.00	One
9.99– 6.00	Eight (in steps of 0.50 Å)
5.99– 5.00	Four (in steps of 0.25 Å)
4.99– 3.50	Fifteen (in steps of 0.10 Å)
3.49– 1.00	Fifty (in steps of 0.05 Å)
0.99– 0.90	One
0.89– 0.80	One
0.79 and smaller	One

advantage that not only differences in judgment on the part of the investigator using the *Index* are circumvented, but also those of the supplier of the original information reproduced on the cards. Since 1949, the card information has been carefully scrutinized by special groups of editors who have succeeded in eliminating most of the dubious entries. In addition, the National Bureau of Standards has been publishing, over the years, diagrams of high quality for the more commonly recurring compounds. The cards containing such information are marked by stars in the upper right corner and also opposite the entry in the *Index*.

The second half of the *Index* contains a systematic classification of the information presented in the first half, arranged so that the card of a substance can be found when its name, but not its x-ray characteristics, is known. In the volume for inorganic compounds this part consists of two sections. The first lists the compounds alphabetically by chemical name; the second is an alphabetical listing of mineral names. In the volume for organic compounds, the second part similarly contains an alphabetical listing of compounds, followed by an arrangement by formula, in order of increasing carbon content. In all entries in the *Index*, the d values and relative intensities of the three most intense reflections are given, as well as the card number, with whose aid the complete crystallographic information available can be located.

Fink index. One of the difficulties in using the *Powder diffraction file* already mentioned is the possibility of incorrectly assigning the proper relative intensities to a triplet of reflections chosen to seek a match in the *Index*. This is a particularly serious problem when the *File* is used for identifications based on electron diffraction rather than x-ray diffraction measurements, since the relative intensity values are often somewhat different. An alternative indexing scheme was proposed for the cards, therefore, by a joint committee of several technical societies, under the chairmanship of W. L. Fink. In the *Fink index*, the d values of the eight most intense reflections are selected and arranged, not in order of decreasing relative intensity, but in order of decreasing interplanar-spacing magnitude. For each substance, eight entries are made, in which the d values are permuted according to the scheme

$$\begin{array}{l} d_1,\ d_2,\ d_3,\ d_4,\ d_5,\ d_6,\ d_7,\ d_8 \\ d_2,\ d_3,\ d_4,\ d_5,\ d_6,\ d_7,\ d_8,\ d_1 \\ d_3,\ d_4,\ d_5,\ d_6,\ d_7,\ d_8,\ d_1,\ d_2 \\ \cdot\ \ \cdot\ \ \cdot\ \ \cdot\ \ \cdot\ \ \cdot\ \ \cdot\ \ \cdot \\ d_8,\ d_1,\ d_2,\ d_3,\ d_4,\ d_5,\ d_6,\ d_7 \end{array} \tag{19-1}$$

without regard to their relative intensities. These eight entries are then grouped according to the scheme presented in Table 19-2. Within each group, the listing sequence follows decreasing magnitudes of the second d value in the entry, and when two or more have the same second d value, according to declining third d values, and so forth. Since each card is entered eight times in the *Fink index* and only three times in the original *Index* based on the Hanawalt scheme, it is often easier to make an identification using the *Fink index*. Some of the practical difficulties and generally recommended procedures are described in the next section.

Identification of mixtures. When a powder sample contains only one compound, the identification procedure is quite simple. The three most intense reflections are used to find a suitable match in the *Hanawalt index*, or the eight most intense reflections are used to seek a match in the *Fink index*. When one or more likely matches are found, the card numbers are used to locate the appropriate cards in the *Powder diffraction file*. The complete list of observed d values is compared

Table 19-2 The grouping of file cards in the *Fink index*

Groups arranged according to d value of largest spacing, Å	*Number of groups*	
10.00 and larger	One	
9.99–8.00	One	
7.99–6.00	Four	(in steps of 0.05 Å)
5.99–4.40	Eight	(in steps of 0.02 Å)
4.39–3.70	Seven	(in steps of 0.10 Å)
3.69–3.00	Fourteen	(in steps of 0.05 Å)
2.99–2.00	Fifty	(in steps of 0.02 Å)
1.99–1.50	Ten	(in steps of 0.05 Å)
1.49–1.00	Five	(in steps of 0.10 Å)
0.99 and smaller	One	

with the tabulated values, and after all observed values are thus matched up, the identification is completed. In doing this it is necessary to keep in mind that the precision with which the d values are measured in the laboratory depends on various instrumental factors, as well as on the diligence of the investigators, always remembering that the cards and the laboratory data are subject to mechanical and human errors. Some of the consequences that this has on the tactics to be used in an actual analysis are illustrated by several examples below.

The powder diagram of a mixture obviously consists of reflections from two or more substances, with the further possibility that fortuitous superpositions of reflections may occur. Since the *Fink index* was especially devised to overcome such difficulties, its utilization is considered first. The experimental data obtained for a mechanical mixture of two laboratory reagents are presented in Table 19-3. The eight most intense reflections have the following d values, listed in declining order of their magnitudes:

$$3.43,\ 2.97,\ 2.45,\ 2.12,\ 2.10,\ 1.80,\ 1.51,\ 1.284\ \text{Å} \tag{19-2}$$

At the outset it is known only that $d = 3.43$ Å is the longest spacing for one of the constituents present. Which of the succeeding values in (19-2) should be chosen as d_2, d_3, etc., cannot be predicted, so that it is necessary to try out each one, successively. Thus one begins by looking in the index volume in the group containing the largest d value. In the 3.44–3.40 Å group, one looks for possible matches assuming the second, third, or fourth value in (19-2) as d_2. If necessary, the neighboring 3.49–3.45 and 3.39–3.35 Å groups can be similarly explored. In establishing a match, the other values in (19-2) are also considered, as well as any supplementary information available regarding the possible composition of the unknown. The last point cannot be overemphasized, because a great deal of needless effort can be saved by limiting the number of choices at the outset.

Only part of this procedure is illustrated here: By looking in the 3.44–3.40 Å group in the *Fink index* for $d_2 = 2.97$ Å, four choices appear: PbS, Be_5Pt, Be_5Pd, and PbF_2. Actually, one should not stop at this point, but go on to consider all possibilities in this group for d_2 values in the range 2.97 ± 0.05, 2.45 ± 0.05, 2.12 ± 0.05 Å, as well as the two adjacent groups for d_1. For the sake of brevity, however, this discussion will be limited to considering the first four candidates only. The eight d values listed in the *Fink index* for each of the four possible matches discovered so far are reproduced in Table 19-3. Since the index volume gives d values to only two decimal places, it is difficult to judge how closely the listed values agree with observed values, particularly when the more accurate, smaller d values are compared. It is for this reason that it is most important to locate the maximum number of potential matches in the index volume. Since the necessary data are conveniently compiled in a single volume, this part of the analysis does not require much time and should be done quite thoroughly.

An examination of Table 19-3 shows that all four candidates discovered so far show equally good agreement with the experimentally determined d values. This, of course, merely means that all these compounds have dimensionally similar lattices. This is not at all unusual, because most inorganic compounds are based on closest packings of the larger atoms, whose interatomic distances dominate

Table 19-3 Experimental d values for an unknown mixture and for possible constituents

Unknown, Å	PbS, Å	Be_5Pt, Å	Be_5Pd, Å	PbF_2, Å
3.43	3.43	3.41	3.42	3.43
3.01				
2.97	2.97	2.97	2.97	2.97
2.45				
2.12				
2.10	2.10	2.10	2.10	2.10
1.80	1.79	1.80	1.80	1.79
1.70	1.71			
1.510				
1.485	1.48			
1.360	1.36		1.37	1.36
1.327	1.33		1.33	1.33
1.284				
1.229			1.22	
1.210				1.21
1.142		1.15	1.15	
1.065				
1.050				
1.004		1.01		
0.989				
0.979				
0.956				
0.936				
0.905				
0.871				
0.831				
0.823				
0.821				
.....		0.80		0.79

the lattice constants, and therefore the interplanar spacings. It further illustrates the consequence of using only half the available information. By ignoring the relative intensities, the *Fink index* provides a very rapid means for finding a multitude of possible matches. It is then necessary to consult the individual cards to establish which one of them actually agrees with the experimental data. Since a relatively large number of such cards need to be compared in this process, the details are not discussed further here. Instead, the procedure originally proposed by Hanawalt, Rinn, and Frevel, which uses the d values of the three most intense reflections in its initial searches, is discussed in full below.

Turning to the experimental data reproduced in Table 19-4, including the relative intensities, it is seen that the most intense reflections have the following values:

$\frac{I_{max}}{I} \times 100$	d, Å
100	2.97
90	2.45
85	3.43
54	2.10
50	1.510
39	1.284

(19-3)

Table 19-4 Powder data for an unknown containing a mixture of components, and comparison with the tabulated data of several possibilities

Unknown		PbS		TeO_2		NaBr	
d, Å	$\frac{I}{I_{max}} \times 100$	d, Å	$\frac{I}{I_{max}} \times 100$	d, Å	$\frac{I}{I_{max}} \times 100$	d, Å	$\frac{I}{I_{max}} \times 100$
3.43	85	3.429	84	3.40	80	3.449	64
3.01	2						
2.97	100	2.969	100	2.99	100	2.988	100
2.45	90	.....	...	2.41	16		
2.12	28						
2.10	54	2.099	57	.....	...	2.113	63
.....	...	.....	...	1.869	56		
1.80	29	1.790	35	.....	...	1.802	21
1.70	15	1.714	16	1.699	8	1.725	19
.....	...	.....	...	1.658	32		
1.510	50	.....	...	1.520	8		
1.485	10	1.484	10	1.485	25	1.495	8
1.360	9	1.362	10	.....	...	1.371	7
1.327	15	1.327	17	.....	...	1.337	15
1.284	39	.....	...	1.260	14		
1.229	4	.....	...	1.224	10	1.221	9
1.210	12	1.212	10	1.184	14		
1.142	5	1.142	6	1.114	4	1.150	4
1.065	3	.....	...	0.091	6		
1.050	2	1.049	3	.....	...	1.057	2
1.004	4	1.003	5			1.010	2
0.989	6	0.989	6			.....	...
.....	...	.....	...				

The d value of the most intense reflections lies in the Hanawalt group 2.99 to 2.95. Several choices can be made in selecting the second most intense reflection of the same set. Proceeding in a systematic manner, the next three most intense reflections are successively chosen as being the second most intense reflection of the same component, and for each choice the appropriate Hanawalt group is examined. It should be realized that the third reflection completing the set may be one of the six listed in (19-3), or it may be one of the other reflections listed in Table 19-4. Likely matches obtained by consulting the *Index* volume are listed in Table 19-5. As can be seen in this table, if $d = 2.45$ Å is chosen as the second most intense reflection of one component, it is not possible to find a set of three d's in the *Index* consistent with the experimentally observed data. This indicates that $d = 2.45$ Å is probably the most intense reflection of the second component. Selecting next $d = 3.43$ Å as the second d value of a possible set, four possibilities occur: $NaBr$, PbS, EuS, and TeO_2. Of these, europium sulfide can be eliminated from further consideration since there is good reason to believe that the rare-earth sulfide is not a likely major constituent in the mixture being investigated. Another possibility for the second d value is $d = 2.10$ Å. As can be seen in Table 19-5, only $SrCeO_3$ appears to be a likely choice, because the relative intensities of the other candidates do not agree with the observed values.† As above, the presence of a rare-earth oxide as a major constituent is deemed unlikely, so that it can be dropped from further consideration. Finally, if $d = 1.51$ Å is chosen as the second reflection, it is not possible to obtain a match. From this it can be concluded that one component is probably either $NaBr$, PbS, or TeO_2. A comparison of the complete set of observed spacings, with the d's listed on the *File* cards of the three compounds (Table 19-4), shows a satisfactory agreement for PbS only, so that it can be concluded from this that it is one of the components of the unknown mixture.

Having identified one of the components, the reflections remaining in Table 19-4 which were not accounted for are used to identify the second component. It is advisable at this point to rescale the relative intensities of the remaining reflections by setting the strongest remaining intensity equal to 100. In the example in Table 19-4, this can be done by multiplying each intensity by $\frac{100}{90}$. A list of the remaining reflections, thus scaled, is given in Table 19-6. The four most intense reflections remaining have the following d values:

$\frac{I}{I_{max}} \times 100$	d, Å
100	2.45
55	1.510
43	1.284
31	2.12

(19-4)

† This discussion ignores the possibility of overlapping reflections for $d = 2.97$ Å, so that the relative intensities listed in Table 19-5 are assumed to be genuine. This is always the assumption one should make at the outset, postponing the consideration of complications until the very last.

Table 19-5 Possible combinations of strong lines of an unknown, compared with several sets of data from the *Index*, furnishing the best match

	d, Å			$\frac{I}{I_{max}} \times 100$			Substance
Choice 1	2.97	2.45	?	100	90	?	Unknown
	A "set" having suitable relative intensities for the first three d's could not be found.						
Choice 2	2.97	3.43	?	100	85	?	Unknown
	2.99	3.45	2.11	100	64	63	NaBr
	2.97	3.43	2.10	100	80	60	PbS
	2.95	3.44	2.10	100	80	80	EuS
	2.99	3.40	1.87	100	80	56	TeO_2
Choice 3	2.97	2.10	?	100	54	?	Unknown
	2.98	2.11	1.80	80	100	60	SmS
	2.99	2.11	1.72	100	32	22	$SrCeO_3$
	2.98	2.11	1.70	100	80	80	$LaYO_3$
	2.97	2.10	1.33	90	100	90	PrTe
	2.96	2.09	3.42	100	80	60	CaC_2
Choice 4	2.97	1.51	?	100	50	?	Unknown
	A "set" having suitable relative intensities for the first three d's could not be found.						

A match for these lines is sought in the Hanawalt groups 2.49 to 2.45 and 2.44 to 2.40. The possible identifications are determined with the aid of the second, third, and fourth d values in (19-4), as above, and are listed in Table 19-7. Since only four really intense reflections are present in Table 19-6, it is most likely that only one more component remains unidentified in the mixture. Because this cannot be

Table 19-6 Remainder of powder data from unknown in Table 19-4

d, Å	$\frac{I}{I_{max}} \times 100$	d, Å	$\frac{I}{I_{max}} \times 100$
3.01	2	0.956	2
2.45	100	0.936	4
2.12	31	0.905	1
1.510	55	0.895	4
1.284	43	0.871	2
1.229	4	0.831	3
1.065	3	0.823	2
0.989	7	0.821	1
0.979	7		

taken for granted, however, it is necessary to consider the several alternatives indicated in Table 19-7. As it turns out, only the first one produces any likely candidates. Of these, the relative intensities of Mn_2N_3, Cu_2O, and U_3Si appear most likely. A comparison of all the reflections listed in the *File* cards of these three candidates shows that Cu_2O provides the best match. In fact, it agrees completely with the observed data in Table 19-6, so that it can be concluded that the two constituents of the unknown mixture have been identified.

It is important to stress at this point that a satisfactory analysis requires that all the observed reflections be identified. The relative intensities observed and reported on *File* cards may differ somewhat for a number of reasons. Solid-solution effects may cause relative-intensity changes (Chap. 11). Preferred orientation (Chap. 20), particularly when diffractometers are used, similarly can modify relative intensities. Geometric and instrumental factors, particularly when different x radiations were employed, also affect the relative intensities differently. Thus, when there are a few unidentified reflections remaining, it is necessary to find the reason for this before dismissing them out of hand. With the extremely large number of compounds included in the current *Powder diffraction file*, it is very unlikely that any unidentified reflections belong to a truly "unknown" constituent.

There are two procedures one can follow in such a case. The first is to try to index the remaining reflections, on the assumption that they were produced by

Table 19-7 Possible combinations for remainder of strong lines in Table 19-6, compared with several sets of data from the *Index*, furnishing the best match

	d, Å			$\frac{I}{I_{max}} \times 100$			*Substance*
Choice 1	2.45	2.12	?	100	31	?	Unknown
	2.44	2.12	1.50	77	100	56	TiN
	2.44	2.11	1.50	40	100	100	$CoNiO_2$
	2.40	2.10	1.45	100	32	25	Mn_2N_3
	2.41	2.09	1.48	91	100	57	NiO
	2.41	2.09	1.48	45	100	50	TiO
	2.47	2.14	1.52	100	100	72	VN
	2.47	2.14	1.51	100	37	27	Cu_2O
	2.46	2.13	1.51	75	100	50	CoO
	2.46	2.11	1.52	100	24	24	U_3Si
Choice 2	2.45	1.51	?	100	55	?	Unknown
	A "set" having suitable relative intensities for the first three d's could not be found.						
Choice 3	2.45	1.284	?	100	43	?	Unknown
	A "set" having suitable relative intensities for the first three d's could not be found.						

one of the components already identified. This is equivalent to assuming that the remaining reflection is one whose relative intensity has been increased by one of the above-cited factors, and the indexing can be carried out by following one of the procedures described in Chap. 18. If this still does not provide an identification, it is necessary to assume that the remaining reflection(s) belongs to an as yet unidentified component whose other lines overlap previously "identified" reflections. This is generally a difficult problem to deal with, since it is not known which of the other reflections to try, unless one or two of them had been noted previously to have much larger intensities than comparable values reported on *File* cards.

An illustration of some of these problems is provided by an actual example taken from a study of a relatively unstable compound, potassium ozonide, KO_3. Because KO_3 is known to decompose slowly, even when all reasonable measures have been taken to stabilize it during the x-ray diffractometer measurements, it can be expected that some side products will be observed. A complete list of the first 24 reflections observed is given in Table 19-8. At the time when this analysis was being carried out, the only previously published information available was that

Table 19-8 X-ray powder data for a KO_3 sample

I/I_{max}	d_{obs}	d_{calc}	hkl
0.7	4.30	4.30	200
0.9	3.38	3.38	211
0.7	3.03	3.04	220
0.2	2.92	....	...
0.2	2.89	....	...
1.0	2.72	2.72	202
0.3	2.67	....	...
0.9	2.31	2.31	222
0.5	2.26	2.26	321
0.8	2.15	2.15	400
0.5	2.01	2.01	213
0.1	1.97	....	...
0.7	1.920	1.921	420
0.3	1.835	1.837	402
0.6	1.770	1.770	004
0.7	1.690	1.689	422
0.4	1.636	1.637	204
0.2	1.554	1.557	521
0.4	1.528	1.529	224
0.4	1.517	1.520	440
0.4	1.396	1.396	442
0.3	1.385	1.386	611
0.5	1.365	1.366	404
0.6	1.362	1.362	532

by two Russian investigators.† They had been able to identify 17 out of 22 observed reflections by comparing their powder photograph with that of KN_3, and attributed the remaining five reflections to unidentified impurities. A similar attempt to identify the 58 reflections recorded in the subsequent investigation failed to match up over half of the peaks observed in the diffractometer tracing prepared with Cu $K\alpha$ radiation. In view of this, it was assumed that KO_3 does not have the same structure as KN_3, although it still seemed likely that they are not too dissimilar. On the assumption that both are tetragonal, but that $a_{KO_3} = \sqrt{2}a_{KN_3}$, while the c axes are closely similar, it is possible to assign indices to 54 of the observed reflections. Only a slight adjustment of the a and c values is needed to provide the extremely good agreement between the observed and calculated d values shown in Table 19-8 for the first 24 reflections.

In seeking to identify the remaining four reflections, the cards of all possible potassium compounds should be consulted. This is actually done most easily by using the alphabetic listing of compounds in the second half of the *Index*. On this basis, the identifications listed in Tables 19-9 and 19-10 are obtained. Note that the relative intensities have been rescaled from the values originally given in Table 19-8. Also note that several of the strongest reported reflections for KOH and KO_2 actually overlap KO_3 reflections. Nevertheless, the identification of both

† G. S. Zhdanova and Z. V. Zvonkova, The crystal structure of potassium ozonide (in Russian), *Zhurnal Fizichisky Khimii*, vol. 25 (1951), pp. 100–109.

Table 19-9 Observed and reported KOH reflections

Observed		*Reported*	
I/I_{max}	d, Å	I/I_{max}	d, Å
0.75	2.92	0.67	2.93
1.00	2.67	1.00	2.69
(KO_3)	2.31	0.23	2.30
0.50	1.97	0.83	1.98

Table 19-10 Observed and reported KO_2 reflections

Observed		*Reported*	
I/I_{max}	d, Å	I/I_{max}	d, Å
(KO_3)	3.38	0.55	3.35
1.00	2.89	1.00	2.87
(KO_3)	2.15	0.37	2.17
(KO_3)	2.01	0.21	2.02

compounds can be assumed to be substantiated, even though KO_2 appears to be identified by the presence of only one reflection ($d = 2.89$ Å). In this instance, the importance of the investigation justified the preparation of diffractometer tracings of the pure components in order to confirm the analysis.

QUANTITATIVE ANALYSIS

General considerations. In the discussions so far in this chapter, the identification of compounds giving rise to observed reflections has been considered without regard for the relative amounts of each component present. An obvious question is, what is the smallest relative amount that can be detected by x-ray diffraction? The answer, unfortunately, is neither simple nor unqualified. The relationships that determine the relative intensities of various materials have been considered in earlier chapters. Clearly, some compounds give rise to more intense reflections, so that they can be detected when present in smaller amounts. Another factor that must be considered is the relative absorption of the reflected intensities by the total sample. Thus an element having a large atomic number is "detectable" in considerably smaller concentrations than one having a small atomic number. This is so because its scattering power is greater and its relative absorption for the other component's reflections is larger (Exercise 19-1). The same criteria apply to compounds containing such elements.

The important factors in determining the intensity of a reflection observed in the powder method can be epitomized by the relation

$$I_{hkl} = I_0 \frac{Cj}{\mu} F_{hkl}^2 V Lp \tag{19-5}$$

where I_0 is the direct-beam intensity; C is an experimental constant having the same value for all reflections recorded on one photograph; j is the multiplicity of the reflecting planes; μ is the linear absorption coefficient; F_{hkl} is the crystal-structure factor; V is the total volume of the diffracting crystals; and Lp is the Lorentz-polarization factor.

From this a number of conclusions can be drawn. The product

$$\frac{I_0 C Lp j}{\mu} = C'_{hkl} \tag{19-6}$$

is a constant quantity for each reflection hkl recorded in the same way. If the ratio between the intensities of two reflections is considered, therefore, it depends only on the magnitudes of F_{hkl}^2 and the relative amounts of each substance present. This means that such ratios could be used, in principle, to determine the relative amounts of each substance present. In practice, the errors inherent in actual calculations (Chap. 11) and their complexity severely limit the utility of this approach. When two components in a mixture have virtually the same absorption coefficients, the only real difference from one mixture to another is the amount of each phase present. This is the case when polymorphic modifications having the same compositions are compared, for example, SiO_2 dust containing cristobalite, which is believed to cause silicosis when inhaled in excessive amounts, or retained austenite in incompletely transformed martensitic steels. Thus it is possible to form a ratio

$$\frac{{}^{A}(I/C')_{hkl}}{{}^{B}(I/C')_{hkl}} = \frac{{}^{A}(F^2_{hkl}V)}{{}^{B}(F^2_{hkl}V)} \tag{19-7}$$

which can be used to establish quantitatively the relative amounts of A and B present from a single diffractometer tracing. (See also Exercise 19-2.) Even in such cases, however, more satisfactory results are obtained by preparing, first, a graph relating the intensity ratios to composition. Note from (19-7) that such a relation is linear because the absorption of the sample is unaffected by its composition. (See also Exercise 19-3.)

Internal-standard technique. In most cases encountered in practice, however, the absorption of the sample depends rather markedly on its composition. In fact, the most usual case is one in which the amount of one constituent present needs to be determined quite precisely in the presence of several other constituents of the mixture. The internal-standard technique previously developed for optical spectroscopy provides a means for doing just this. This technique was first adapted to x-ray diffraction by L. E. Alexander and H. P. Klug. It has the important advantage that it is independent of the linear absorption coefficient; therefore the relationship between the relative intensities and relative percentage volumes is linear.

If it is desired to determine the fractional volume v_A of component A, a known amount of the standard S is added to the mixture. If the fractional volumes of A and S in the new mixture are designated v'_A and v'_S, their respective reflection intensities can be determined by

$$I_A = K_A \frac{v'_A}{\mu} \tag{19-8}$$

and

$$I_S = K_S \frac{v'_S}{\mu} \tag{19-9}$$

where

$$K_A = I_0 C V_A (jF^2_{hkl} Lp)_A \tag{19-10}$$

Dividing (19-8) by (19-9),

$$\frac{I_A}{I_S} = \frac{K_A v'_A}{K_S v_S} \tag{19-11}$$

which can be solved for v'_A:

$$v'_A = \frac{K_S v_S I_A}{K_A I_S} \tag{19-12}$$

The relation between v_A, the fractional volume sought, and v'_A, given by (19-12), can be determined as follows: Denoting by V_T the total volume of the new mixture, that is, the volume of the original mixture plus the volume V_S of the added substance,

$$v'_A = \frac{V_A}{V_T} \tag{19-13}$$

and $$v_A = \frac{V_A}{V_T - V_S} \tag{19-14}$$

Dividing (19-13) by (19-14),

$$\frac{v'_A}{v_A} = \frac{V_A/V_T}{v_A/(V_T - V_S)} = 1 - \frac{V_S}{V_T} \tag{19-15}$$

But $V_S/V_T = v_S$ by definition, and

$$\frac{v'_A}{v_A} = 1 - v_S$$

or $$\frac{v_A}{v'_A} = \frac{1}{1 - v_S} \tag{19-16}$$

Finally, $$v_A = \frac{v'_A}{1 - v_S} \tag{19-17}$$

Substituting (19-12) for v'_A in (19-17),

$$v_A = \frac{K_S v_S}{K_A(1 - v_S)} \frac{I_A}{I_S}$$

$$= K \frac{I_A}{I_S} \tag{19-18}$$

where $$K = \frac{K_S v_S}{K_A(1 - v_S)} \tag{19-19}$$

Since, according to (19-10), K does not depend on μ, Eq. (19-18) is independent of μ also. It is possible, therefore, to prepare a graph of I_A/I_S as a function of v_A, for a fixed v_S, from known mixtures. Since (19-18) is a linear equation, the graph of v_A vs I_A/I_S is a straight line whose slope is K. In preparing such a graph, it is most important to keep v_S constant in each mixture prepared. After the standard has been added, however, it is possible to add any amount of another substance, to decrease the absolute x-ray absorption of the sample, without affecting the analysis. This means that suitable linear graphs can be prepared in advance, to determine quantitatively the amount of substance present in any matrix whatever. Mechanical mixtures of the desired compound and a suitable standard are first prepared, and their intensity ratios used to construct the linear graph. As illustrated by several examples given in the books listed at the end of this chapter, the quantitative analysis then can be carried out, regardless of what the other components present may be. As in all cases where accurate intensity measurements are involved, sources of error, such as preferred orientation effects, for example, must be minimized.

SPECTROCHEMICAL ANALYSIS

Fluorescent analysis. The utilization of x-ray emission spectrography in chemical analysis begins with the pioneering discovery by Moseley that the wavelengths of atomic spectra follow a regular sequence (Chap. 6). He went on to predict in 1913 that this regularity may enable the discovery of "missing" elements whose relative positions now could be foretold. This prediction was borne out within ten years when D. Coster and G. Hevesy used x-ray spectrography to prove that they had indeed discovered a new element, hafnium. Electron excitation was used in these studies, so that the materials analyzed had to be made part of the anodes of demountable x-ray tubes, with all the experimental difficulties attendant thereto. By 1928 it was established that, for purposes of chemical analysis, x rays emitted by the target could be used equally well to excite the characteristic emission spectra in materials. In succeeding years, spectrographic analysis based on x-ray fluorescence overtook methods employing direct electron bombardment of the sample material, and the subsequent development of x-ray spectrometers having increased sensitivity has made x-ray fluorescence analysis probably the most powerful tool in analytical chemistry. As often happens in science, however, the pendulum of scientific interest is beginning to swing back to direct electron excitation. This is because it is possible to focus electrons so that very small portions of larger samples can be examined. Some of the aspects underlying the developments in microanalysis are discussed in the concluding section of this chapter.

The basic principles important in x-ray fluorescent analysis have already been discussed in this book. A schematic view of a flat-crystal spectrometer is presented in Fig. 19-2. X rays emitted from a high-intensity sealed-off tube impinge on the sample placed in a suitable container that absorbs all x radiation except that emanating from the sample in the direction of the Soller slits. In addition to some scattering of the incident radiation by the sample, this radiation is comprised primarily of fluorescent radiation from the various atoms composing the sample being analyzed. The single-crystal analyzer placed at the center of the spectrome-

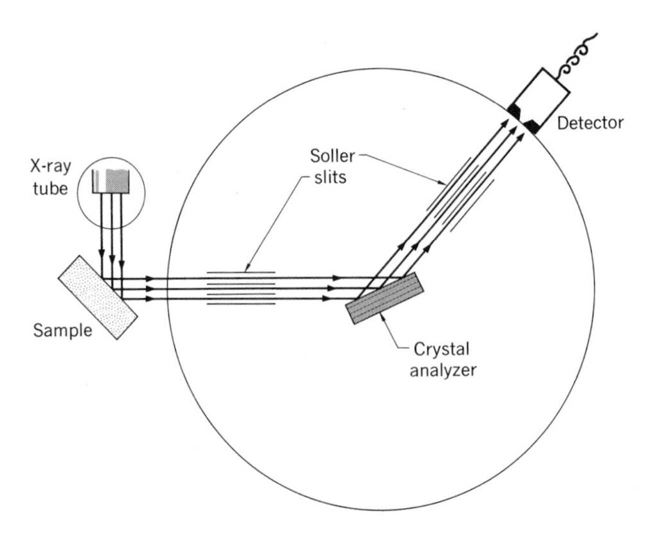

Fig. 19-2

ter then is used to examine the spectral distribution of the x rays incident on it in the usual way; that is, the crystal rotates through an angle θ, while the detector forms the angle 2θ. Since the d value of the crystal's reflecting planes is known, the λ values of x rays reflected at various 2θ angles can be established directly. Comparison with the known spectra of all the atoms in the periodic table then enables the determination of the atoms whose fluorescent spectra have been detected. Clearly, such a qualitative analysis can be carried out quite easily on an instrument that is not unlike the x-ray diffractometers discussed in Chap. 13. Instruments employing bent-crystal analyzers, either in transmission or reflection arrangements, also have been devised, but introduce no new principles requiring detailed discussion here.

In order to excite an atom in fluorescence, it is necessary that the x-ray photons incident on it have an energy sufficient to knock out an inner electron. The intensity of the fluorescent radiation produced then depends on the number of such photons absorbed. Recalling from earlier discussions that the absorption coefficient of an atom increases proportionately to λ^3, up to the wavelength of its absorption edge, it is clear that the intensity emitted by the atoms in a sample depends not only on the number of such atoms present, but also on the spectral composition of the radiation used to induce fluorescence. The fluorescent radiation produced must then pass through a spectrometer such as the one indicated in Fig. 19-2, and undergoes absorption in the sample itself, the air path, the analyzer crystal, and in the detector window. Each wavelength component present in the fluorescence spectrum is absorbed differently, and its relative degree of absorption depends on the exact components employed in the spectrometer. All this points up the fact that many more experimental variables need to be considered in x-ray fluorescent analysis than in x-ray diffraction analysis, even though the basic principles involved are identical in both methods.

It is not possible, in the brief space allotted to this discussion, to examine more than the highlights of x-ray fluorescent analysis. It is first necessary to consider the kinds of spectra that can be easily excited in fluorescence. The variation of the wavelength of the $K\alpha_1$ and $L\alpha_1$ lines with the atomic number of the elements is depicted in Fig. 19-3. When a spectrometer is operated in air, the spectral region conveniently studied has a wavelength range of 0.4 to 3.0 Å. This means that elements having $21 < Z < 55$ can be detected by their K spectra, while those having $Z > 55$ can be detected most conveniently by their L spectra. Elements having $Z < 21$ require special spectrometers, in which air absorption has been minimized by evacuation, and organic crystal analyzers having large d values and relatively low absorption for the softer radiations are employed. By such means, commercial vacuum spectrographs like the one shown in Fig. 19-4 can detect elements down to $Z = 9$ (fluorine).

The choice of x radiation to excite fluorescence in the sample becomes more critical when low-atomic-number elements are studied. Ideally, the intense $K\alpha$ line emitted by the x-ray tube target should have a wavelength just slightly less than that of the K edge of the fluorescing element. Such a choice guarantees maximum fluorescent intensity, but clearly can be satisfied for only one element at a time. It is usual, therefore, to utilize the continuous radiation emanating from

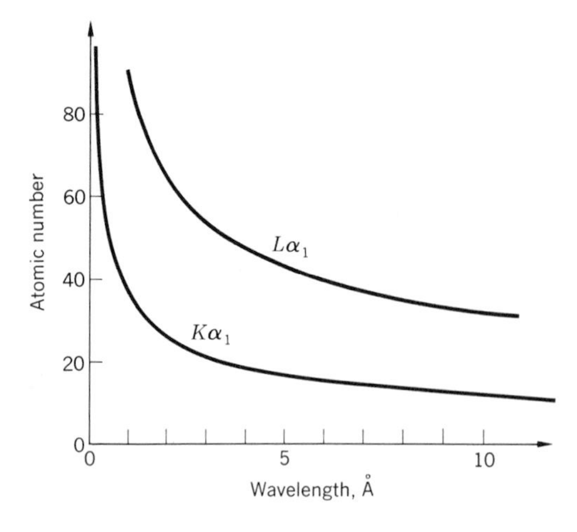

Fig. 19-3. Variation of wavelength of $K\alpha_1$ and $L\alpha_1$ with atomic number of the elements.

an x-ray tube rather than rely on its more intense characteristic spectrum. (Generally, heavy metals such as tungsten or palladium are used as anodes, because the intensity of the continuous spectrum increases with Z, as discussed in Chap. 6.) When dealing with light elements, however, it must be recalled from Chap. 8 that Compton modified scattering becomes relatively more intense as Z declines, so that the background intensity also rises. To minimize the background due to scattering of penetrating shorter-wavelength components present in the x radiation used to excite light elements in fluorescence, therefore, it is more desirable to use a lower-Z anode metal. One such metal is chromium, whose $K\alpha$ radiation can excite fluorescence in the lighter elements fairly efficiently without introducing an excessive amount of continuous radiation. Because the quantum yield for the

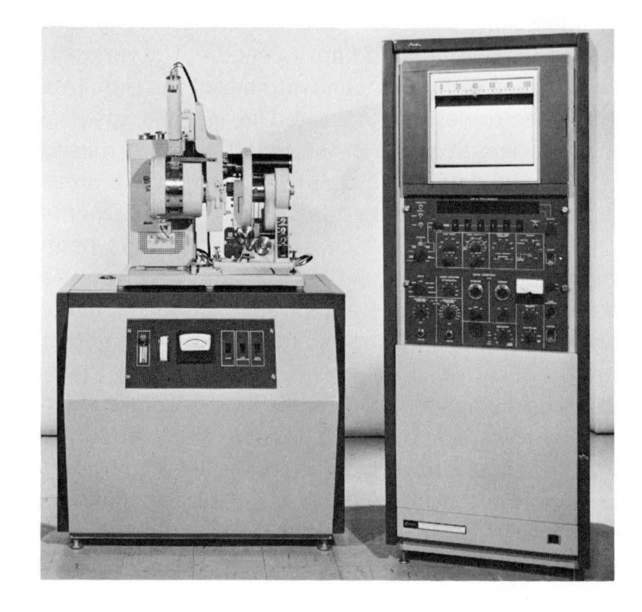

Fig. 19-4. Norelco vacuum spectrograph. *(Courtesy of Philips Electronic Instruments.)*

Fig. 19-5. Henke soft x-ray spectrograph. A continuously evacuated x-ray tube is mounted on a regular Norelco vacuum spectrograph (Fig. 19-4) that contains a special analyzer crystal and a flow-proportional detector. *(Courtesy of Philips Electronic Instruments.)*

very light elements declines rather rapidly, special high-intensity excitation radiation is required. A commercial spectrograph using demountable x-ray tubes that can withstand loads up to 200 mA is shown in Fig. 19-5. By replacing the crystal analyzer in a conventional vacuum spectrograph with a suitably assembled stack of fatty-acid (lead stearate) monolayers, it is possible to analyze for radiation having wavelengths up to 150 Å. An instrument like the one shown in Fig. 19-5 has been used successfully to analyze for beryllium ($\lambda_K = 114$ Å), and can be used in analyses for carbon ($\lambda_K = 44.6$ Å) almost routinely.

From simple considerations it can be shown that the intensity of the fluorescent radiation from a constituent in a sample is proportional to the fraction of that element present. As already noted in the case of quantitative analysis by x-ray diffraction, this proportionality is not linear, because of the nonlinear way in which absorption in a sample changes with composition. In x-ray fluorescent analysis this problem is further complicated by the fact that the absorption of the incident radiation, as well as the fluorescent radiation, must be considered. The actual analysis is additionally aggravated when the continuous spectrum emanating from an x-ray tube is used to excite fluorescence in the sample. From the preceding discussion it is possible to distinguish four groups of experimental factors that must be considered in a quantitative analysis:

1. Absorption by the sample
2. Enhancement by the sample
3. Inhomogeneities in the sample
4. Instrumental instabilities

(19-20)

Absorption and enhancement usually are considered together because both cause comparable deviations from linearity in the intensity dependence on composition. Enhancement takes place when the fluorescent radiation of some element present in the sample can excite fluorescence in the element being analyzed. The presence of such heavier elements increases the fluorescent intensity over what it would be in a reference standard not containing such heavier elements. Thus the composition of the matrix in which the "unknown" element is found plays a very significant role in fluorescent analysis. Similarly, inhomogeneities in the surface of the sample or irregularities in an element's distribution because of segregation effects cause deviations from ideal conditions which are usually not easy to recognize or correct. Similarly, instrumental instabilities are considerably more important in x-ray fluorescent analyses than in x-ray diffraction studies. Because diffraction experiments involving quantitative intensity measurements almost always require the measurement of monochromatic radiation, short-term fluctuations can be easily detected and, if need be, corrected. In the case of x-ray fluorescent analysis, transient electric currents or voltage fluctuations in the x-ray tube or detection system affect different spectral components differently, and are most difficult to identify when they occur. For this reason, the utmost stability in all components is required for maximum accuracy, including temperature and pressure stability in the mechanical components of the spectrometer.

One way to minimize the effect of instrumental variations is to compare the intensity obtained from the sample with that from a pure element. By plotting the ratio of these intensities against the fraction of element present, only the first three factors in (19-20) need be considered. For carefully homogenized alloys, the third factor also can be ignored, and only the effects of absorption and enhancement by the other elements present in the matrix must be evaluated. To see what effect absorption has, the ratio of the intensity of Ni $K\alpha$ fluorescent radiation obtained from a binary alloy, I_x, to that from pure Ni, I_{Ni}, is shown plotted in Fig. 19-6 for two alloys of nickel. Note that the fluorescent radiation displays a positive deviation from linearity in a less absorbent matrix (Ni-Al) and a negative deviation in a more absorbent matrix (Ni-Ag). The nickel emission curve in nickel-silver alloys is the resultant of absorption variations caused by composition changes and of increasing enhancement with increasing silver content, since, under usual tube operating conditions (50 kV), the continuous spectrum emanating from the x-ray tube also excites K fluorescence in the silver atoms.

The use of internal standards is made more difficult in x-ray fluorescent analysis because of the factors listed in (19-20). Ideally, the amount of an internal standard added should be commensurate with that of the concentration being analyzed. Usually, this means that concentrations up to 10 to 15 at.% are best studied with internal standards. Also, the atomic number of the standard element added should differ by no more than ± 1 from that of the element sought. (See

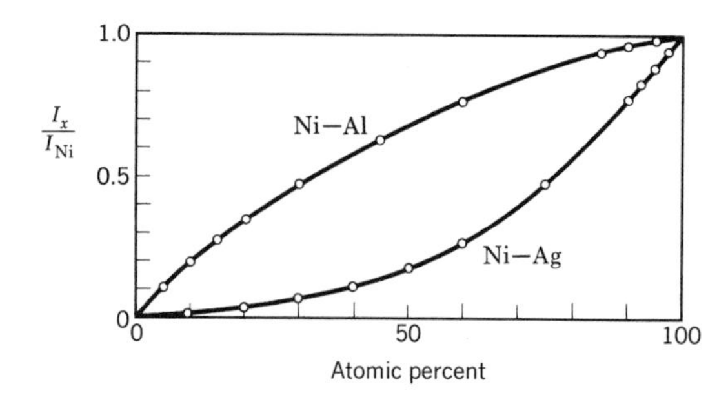

Fig. 19-6. Intensity variation of Ni $K\alpha$ fluorescent radiation in several mixtures of nickel with aluminum and with silver. The curves are normalized by dividing the mixture's intensity I_x by that of pure nickel. Note that the density of calibration points varies with the precision desired in the analysis.

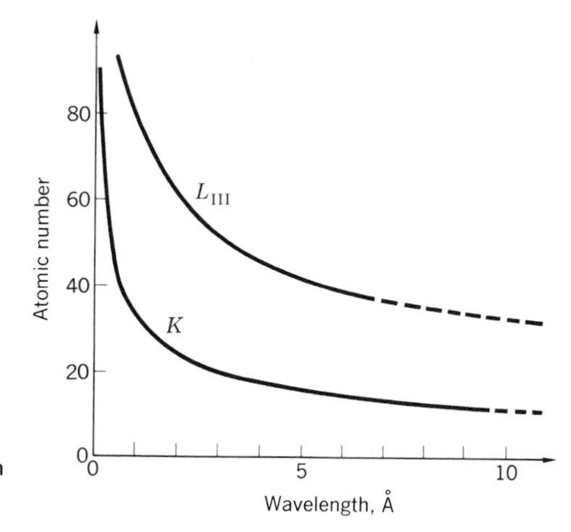

Fig. 19-7. Variation of the K and L_{III} absorption edges with atomic number of the elements.

Exercise 19-4.) This not only minimizes significant absorption variations, but also assures comparable fluorescent intensity, hence comparable accuracies. To eliminate such causes of error as incomplete mixing of the standard and the sample to be analyzed, some strong line in the incident radiation scattered by the sample can be used as a reference standard. Provided that the same x-ray tube and instrumental settings are employed, the intensity of the scattered line does not depend on the concentration of the element being analyzed. Because the wavelength of the scattered radiation must be different from that of the fluorescent line, it is differently absorbed by samples having different concentrations (μ values), so that the effect of the matrix will not be the same for the two intensities. This, in turn, will cause a deviation from linearity, so that, unlike the use of internal standards in x-ray diffraction analysis, their use in x-ray fluorescent analyses does not eliminate the dependence on absorption variations in the sample. As discussed in some of the books listed at the end of this chapter, various other schemes can be devised for analyzing samples quantitatively. Each system to be studied, however, introduces its own peculiarities, which must be clearly recognized before commencing an analysis.

Absorption analysis. In addition to detecting the intensity of an x-ray line excited in fluorescence, it is possible to analyze for an element by noting the absorption of the incident x-ray beam causing the fluorescence. As discussed in detail in Chap. 6, the mass absorption coefficient of an element increases in proportion to λ^3 until the wavelength reaches the value corresponding to an absorption edge, at which point μ drops discontinuously to a much lower value. Since the energies of absorption edges in atoms are determined by the binding energies of the electrons concerned, a plot of λ_K and λ_L values against atomic number (Fig. 19-7) shows a similar variation to that displayed by the characteristic emission lines in Fig. 19-3. By placing a suitably thin sample in the path of the x-ray beam in an ordinary spectrometer, it is possible to use the analyzer crystal as a means of establishing the relation between the intensity transmitted through the sample and the wavelength of the transmitted x-ray beam. The transmitted intensity will decline as the absorption coefficient increases up to the absorption edge of one of the constituent elements. At that point a discontinuity in the transmitted intensity occurs, and its wavelength value can be used to identify the element causing the discontinuity.

To make the procedure quantitative, it is only necessary to consider the simple relation between the mass absorption coefficients of element A and the rest of the sample, denoted by the subscript R. The mass absorption coefficient of the sample

$$\mu_S = w_A\mu_A + w_R\mu_R \tag{19-21}$$

where w_A is the weight fraction of element A sought in the analysis, and $w_R = 1 - w_A$. Denoting the value of the mass absorption coefficients at two different wavelengths (usually chosen on opposite sides of an absorption edge) by the subscripts 1 and 2, the transmitted intensity ratios at these two wavelengths can be written

$$\begin{aligned} I_1 &= \left(\frac{I}{I_0}\right)_1 = e^{-(w_A\mu_{A_1} + w_R\mu_{R_1})\rho x} \\ I_2 &= \left(\frac{I}{I_0}\right)_2 = e^{-(w_A\mu_{A_2} + w_R\mu_{R_2})\rho x} \end{aligned} \tag{19-22}$$

where ρ is the density of the entire sample. Assuming that $\mu_{R_1} \cong \mu_{R_2}$ at the two wavelengths, a condition easily satisfied by selecting λ_1 close to λ_2, it is possible to form a ratio of the two relations in (19-22) to give

$$\frac{I_1}{I_2} = e^{-w_A(\mu_{A_1} - \mu_{A_2})\rho x} \tag{19-23}$$

which is independent of the composition of the rest of the sample. The quantity $\rho x = M_S$, the mass per unit area of the sample, while $w_A M_S = M_A$, the mass of element A per unit area of the sample. Provided that all samples have a constant thickness x, (19-23) yields a simple linear relation

$$\ln\frac{I_1}{I_2} = -(\mu_{A_1} - \mu_{A_2})M_A \tag{19-24}$$

so that a "universal" graph can be prepared for each element sought, in advance. In principle, relation (19-24) can be used to measure the amount of element A present, M_A, quantitatively. In practice, the necessity to keep the thickness constant in the preparation of suitable calibration standards, and of the sample to be analyzed, imposes serious limitations on the accuracies attainable in view of the very small thicknesses required for adequate intensity transmission. The above procedure can be used more effectively in the analysis of liquids, which can be poured into standardized sample holders having uniform dimensions.

When the absorption coefficient of the matrix can be considered to remain approximately invariant, and only the concentration of one element changes from sample to sample, it is possible to use the continuous (polychromatic) radiation from an x-ray tube, with a resultant gain in intensity. This is a useful procedure,

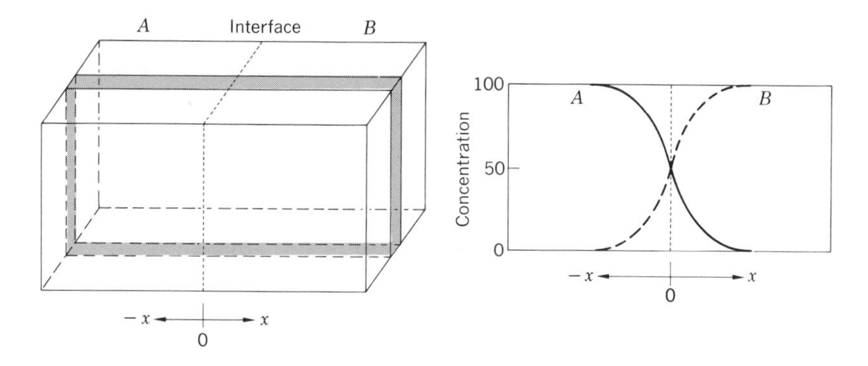

Fig. 19-8. Diffusion couple between elements *A* and *B*. The concentration of the two elements in the central slice is represented on the right by the solid and the dashed curves, respectively.

for example, when the concentration of a relatively heavy element is sought in an organic liquid. Although the presence of a large range of wavelengths makes it impossible to utilize relations like (19-23), it is possible to prepare calibration curves with samples having known concentrations. Over narrow concentration ranges, the relation between the transmitted intensity and the concentration is usually fairly linear, particularly when $\mu_x > \mu_{\text{matrix}}$. To optimize this condition, the tube voltage is usually adjusted so that the maximum in the continuous spectrum coincides with μ_K of the element sought. Absorptiometry with a polychromatic beam is a very rapid analytical method that can be adapted quite easily to many manufacturing processes, and has been used quite widely for that reason.

A very special case of absorptiometry arises when one monochromatic beam is employed to study concentration gradients in a single sample. A typical example is the analysis of a diffusion couple, produced by suitably joining two metals at an interface as shown in Fig. 19-8. The atoms from metal A diffuse across the interface into metal B at the same time that B atoms diffuse into A. The diffusion rates and penetration depths can be studied as a function of temperature and time by analyzing the concentrations on both sides of the interface. Provided that the variation of the sample's density ρ with concentration is known to a sufficient accuracy, relation (19-21) and one of the equations in (19-22) can be used to determine the weight fraction w_A directly from a measurement of the ratio between the intensity of the beam incident on, and the beam transmitted through, a transverse slice, such as the one indicated to the left in Fig. 19-8. Usually, a characteristic line in the x-ray tube spectrum lying between the absorption edges of A and B is chosen so as to provide maximum contrast. Several samples having identical thicknesses are cut from the diffusion couple and analyzed at several points along a line passing through the interface. The resulting concentration gradients are then indicated, as shown on the right in Fig. 19-8. The two curves frequently are symmetric, although they need not be.

Microanalysis. The concentration gradient across an interface could be studied equally well by emission spectrography if a suitably sharp probe could be employed.

Because it can be focused more easily, an electron beam constitutes a better probe with which the composition of adjacent regions in a sample can be examined. The first electron-probe microanalyzer was constructed by R. Castaing in 1951 by modifying an electron microscope. The basic elements of its design are indicated in Fig. 19-9. Electrons emanating from an electron gun pass through a condenser lens, followed by a deflection coil which controls the positioning of the transmitted beam. A second lens then refocuses the electron beam at the sample surface plane. With appropriate lenses it is possible to obtain electron beams having cross sections approximately one micron in diameter. The one-micron area provides an ideal point source for bent-crystal focusing (Chaps. 13 and 17) of x rays emitted by that area, so that it is usual to employ bent-crystal spectrometers in conjunction with the electron microprobe. An optical microscope is also made an integral part of the instrument, so that the particular point on the sample being bombarded can be established. An arrangement like the one shown in Fig. 19-9 can be used to analyze the composition of small grains, inclusions in such grains, grain boundaries, and so forth. This can be done by displacing the sample and analyzing it quantitatively at a series of adjacent points. Alternatively, it is possible to employ pairs of deflection coils to scan over the sample area. In fact, it is possible to synchronize the scanning motions with those of a television tube, so that

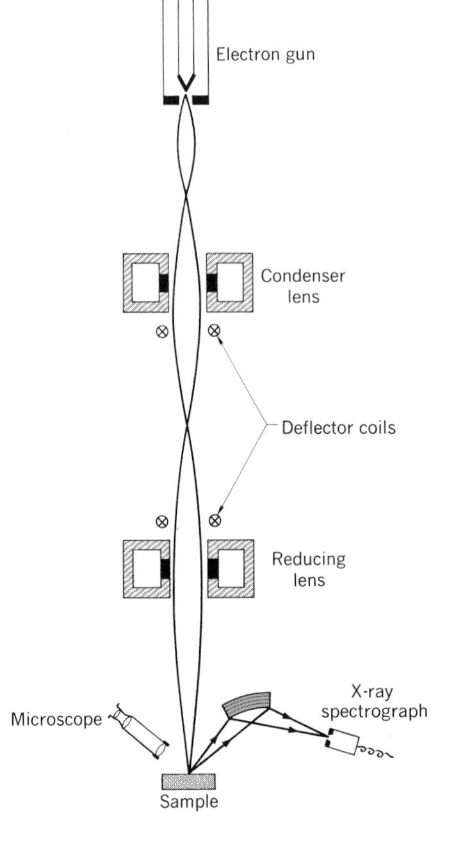

Fig. 19-9. Schematic representation of a scanning microanalyzer. The two sets of deflector coils enable the operator to scan the sample, with the spectrograph set to detect a characteristic line, so that the intensity measured indicates the concentration of the element throughout the surface area scanned.

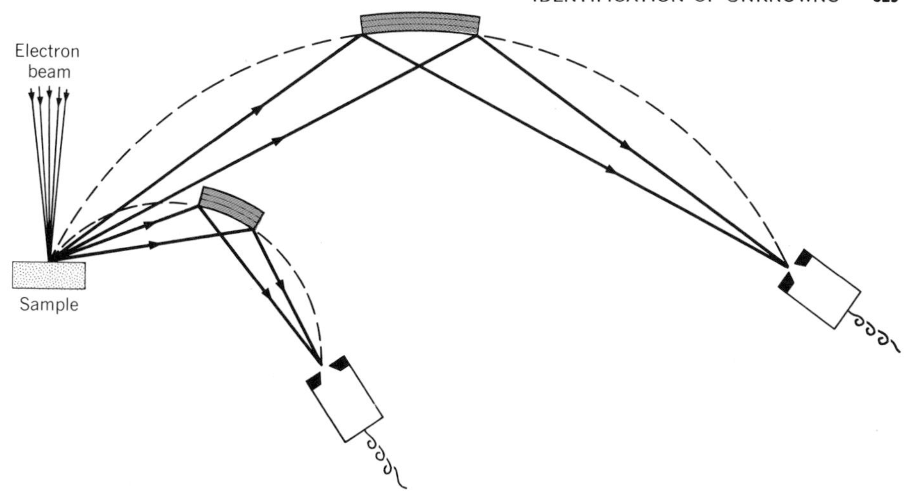

Fig. 19-10

the output of the x-ray detector can be displayed visually by shades of gray throughout the area scanned on the sample. Analogous displays can be prepared for several elements by repeating the scans while recording the outputs of different spectrographs incorporated in the instrument. It is quite usual to have two or more bent-crystal spectrographs positioned around the sample at the same time (Fig. 19-10). Each has its own curvature along a preselected focusing circle, so that one specific element can be analyzed with each spectrograph. Commercial instruments have means for interchanging crystals and detectors, so that a variety of elements can be analyzed for, quite routinely.

The surface of the sample must be quite flat, so that x rays are emitted in all directions uniformly. In a nonmetal, the sample surface must be rendered conductive by evaporating a thin metallic film onto it. This is necessary so that the electrons impinging on it are conducted away before they can build up a space charge that would deflect the incident beam. Similarly, the takeoff angle at which the x rays are viewed by the bent crystal must be selected by judicious consideration of enhancement and absorption effects during the passage of the x rays through the sample. Such factors and examples of actual applications are discussed further in some of the books listed at the end of this chapter. In concluding the discussion here, it should be noted that microbeams of x rays can be used similarly to the electron probe discussed above. Such x-ray beams have much lower intensities (one or more orders of magnitude smaller), and usually cannot be collimated or focused down to comparably small areas. When liquids or other kinds of volatile samples have to be examined, however, electron excitation cannot be used because of the need to expose the sample's surface to a vacuum. Thus electron and x-ray excitation have their appropriate roles to play in x-ray spectrochemical analysis.

SUGGESTIONS FOR SUPPLEMENTARY READING

Leonid V. Azároff and Martin J. Buerger, *The powder method in x-ray crystallography* (McGraw-Hill Book Company, New York, 1958).

L. S. Birks, *X-ray spectrochemical analysis* (Interscience Publishers, Inc., New York, 1959).
Electron probe microanalysis (Interscience Publishers, Inc., New York, 1963).
V. E. Cosslet and W. C. Nixon, *X-ray microscopy* (Cambridge University Press, New York, 1960).
Burton L. Henke, Application of multilayer analyzers to 15–150 Å fluorescence spectroscopy for chemical and valence band analysis, *Advances in X-ray Analysis*, vol. 9 (1966), pp. 431–440.
Harold P. Klug and Leroy E. Alexander, *X-ray diffraction procedures for polycrystalline and amorphous materials* (John Wiley & Sons, Inc., New York, 1954).
H. A. Liebhafsky, H. G. Pfeiffer, E. H. Winslow, and P. D. Zemany, *X-ray absorption and emission in analytical chemistry* (John Wiley & Sons, Inc., New York, 1960).

EXERCISES

19-1. Assume that the intensity of a reflection in the powder method is directly proportional to F^2 and the amount of the material present and inversely proportional to μ. Consider a mechanical mixture of two elements, such as Pb ($Z = 82$, $\mu = 230\ \text{cm}^{-1}$) and Cu ($Z = 29$ and $\mu = 51\ \text{cm}^{-1}$), and calculate the smallest weight percent of copper that can be detected in such a mixture, assuming that $I/I_{\max}$ must exceed 5% to be detectable above background. Repeat this calculation for lead. **Hint:** Prepare a suitable graph of I vs composition.

19-2. Consider the relation (19-7) in the text. Suppose one measures ${}^{A}I$ and ${}^{B}I$ for two reflections belonging to two different constituents of a binary mixture. If they both have virtually identical absorption factors, what quantities in (19-5) must be evaluated for each reflection to enable (19-7) to predict the correct ratio of ${}^{A}V/{}^{B}V$? Is there anything omitted in (19-5) that might affect the outcome?

19-3. Consider the intensity ratio of two reflections of components A and B in a binary mixture. Prepare a schematic plot of this ratio against the amount of component A present when $\mu_A > \mu_B$; when $\mu_A = \mu_B$; and when $\mu_A < \mu_B$.

19-4. Consider the two curves shown in Fig. 19-6. For the two alloy systems considered, what effect would the addition of zinc have on the shapes of the curves? Can you suggest an experimental procedure whereby the deviation caused by zinc can be minimized?

19-5. One of the on-line applications of x-ray fluorescent analysis in industry is in the determination of plating thicknesses. Supposing you have a tungsten-target x-ray tube and a single-crystal spectrometer available, describe two possible schemes by means of which you could measure plating thicknesses of the order of 0.001 in. when one metal is deposited on another. Describe the characteristic way each system depends on plating thickness.

19-6. In Exercise 19-5, which arrangement is best suited for measuring the thickness of tin plated on steel?

19-7. Consider the problem described in Exercise 19-5. Can it be handled by analogous processes using x-ray diffraction in place of fluorescent radiation? What is the chief limitation of diffraction in thickness measurement?

19-8. The mass absorption coefficients of pure copper on opposite sides of its K absorption edge are 37 and 307 cm²/g, respectively. Supposing that the smallest reliably measurable change in intensity, when measured on two sides of the K edge in a copper-nickel alloy, is 5%, what is the minimum amount of copper that can be detected in an alloy sheet 0.5 mm thick if the density can be assumed to remain constant and $\rho = 2.70$ g/cc?

TWENTY
Special methods

TEXTURE STUDIES

Types of texture. The preceding three chapters have considered the general kinds of information that can be obtained from an examination of powder diagrams. On several occasions it was suggested that the crystallites comprising the polycrystalline aggregates are not always truly random. The nature of the deviation from complete randomness is called the *texture* of the polycrystalline aggregate. Two kinds of texture should be distinguished, *shape texture*, which results from the nonrandom packing of asymmetric crystallites in a mechanically formed aggregate, and *orientation texture*, which is an intrinsic property of the polycrystalline materials and reflects its mode of creation. The way the presence of texture manifests itself in powder photographs is illustrated in Fig. 20-1, which compares the effects of the shape texture of needle-shaped and plate-shaped crystallites with the orientation texture of a cold-drawn metal wire. In x-ray diffraction experiments, shape texture is usually of importance in connection with sample-preparation procedures. As can be seen in Fig. 20-1, the uneven intensity distribution along a diffraction ring complicates many measurements. When a powder diffractometer is used, any texture present in the sample causes very pronounced changes in the

Fig. 20-1. Texture effects in powder photographs. Top photograph shows effect of plate shape (talc); center photograph shows effect of needle shape (PbO); and bottom photograph shows typical wire texture (W). *(After Azároff and Buerger.)*

Fig. 20-2. Norelco sample spinner. *(Courtesy of Philips Electronic Instruments.)*

relative intensities of reflections. (The detector in a diffractometer "sees" only the equatorial trace of the photographs in Fig. 20-1.) To minimize this effect, a mechanical device for rotating the sample (Fig. 20-2) may be used. Such a sample spinner, however, only randomizes the orientation within the sample plane, but cannot overcome the texture that arises from the packing parallel to the sample holder's surface. Some means for minimizing these effects have been suggested in Chap. 17.

There are many different causes of orientation texture in polycrystalline aggregates; in fact, truly random aggregates occur in nature very rarely. When rocks crystallize, for example, quite commonly, individual minerals orient themselves in preferential ways in response to thermal and compositional convection currents set up in the cooling melts. Subsequent recrystallization or metamorphosis causes a new texture to appear that is generally quite different from the one preceding it. Similarly, metals and alloys develop very pronounced textures when they are cooled from a melt. Castings tend to develop columnar grains perpendicularly oriented to the walls of the mold. Electrodeposited, evaporated, and sputtered films similarly develop a variety of textures which depend directly on the conditions prevailing during their formation. When metals are heated subsequent to being formed into suitable shapes, an *annealing* or *recrystallization texture* is produced. At relatively low annealing temperatures, a primary recrystallization texture appears which is caused mainly by the growth of stress-free grains at the expense of their stressed (higher-energy) neighbors. A secondary texture may be developed at considerably higher temperatures, and usually it bears no relation to the primary texture. The full understanding of why and how this takes place is still the subject of current investigations, but, as clearly demonstrated by x-ray diffraction studies, the orientation of individual crystallites in metals, as well as the relative size, changes as a result of heat-treatment, aging, and other causes. When the individual grains attain sufficient size, some of the methods described in Chap. 10 can be employed. For crystallites smaller than 10^{-3} cm, the procedures described below have been developed.

It should be realized, of course, that the physical and mechanical properties of materials depend rather markedly on their texture. This is a direct consequence of the anisotropy of most properties of single crystals, including elasticity, conductivity, magnetizability, etc. The anisotropies of individual crystallites tend to cancel out in random aggregates, but reinforce each other when they assume preferred orientations. This is particularly true when materials are manufactured into various shapes. For example, when a wire is drawn through a die, the fact that each crystallite has certain crystallographic directions along which the deformation proceeds more easily causes the crystallites to align in such a way that they all have this direction(s) parallel to the wire axis. For this reason, this is often called *deformation texture*. Organic fibers, whether occurring in nature or produced in the laboratory, similarly tend to have their constituent crystallites (large molecules) aligned parallel to the fiber axis. This kind of alignment is popularly called *wire* or *fiber texture*. Analogously, when a sheet of metal or polyethylene is formed by rolling, the individual crystallites are aligned still further into what is called *sheet texture*. Because the subsequent handling of these materials is directly affected by their texture, it is important to be able to characterize it precisely. X-ray diffraction is the principal method for determining the textures of all kinds of materials.

Ideal fiber texture. Plastic or permanent deformation of metals normally proceeds by a process called slip, in which dislocations already present, or new ones generated by the deformation forces, move along certain crystallographic planes, called slip planes. Depending on the crystal structure, dislocations can move most easily along certain crystallographic directions. Usually, but not always, these are the directions along which the atoms are most closely spaced, so that the least amount of work is required to move a dislocation one translation distance, denoted by the Burgers vector **b**. (See also beginning of Chap. 10.) When metal is forced through a circular die, the crystallites in the extended wire consequently have their slip directions parallel to each other and to the wire. Idealizing the shapes of the crystallites to perfect cubes and assuming that their slip directions are parallel to $[[100]]$, the polycrystalline aggregate in the wire looks like that shown in Fig. 20-3. Note that there is no preferential alignment about the wire axis, as can be seen in a cross-sectional view to the left in Fig. 20-3. Thus wire texture can be specified completely by indicating the crystallographic direction $[uvw]$ parallel to the wire axis.

To picture what an x-ray diffraction diagram of a wire should look like, it is only necessary to consider how the preferred orientation affects the reciprocal-lattice construction of a random aggregate. Recalling that the random orientation of the reciprocal-lattice vectors formed concentric spheres, the constraint imposed by wire texture is that they now form fixed angles with the wire axis. The normals to the reflecting planes are still randomly distributed about the wire axis, so that,

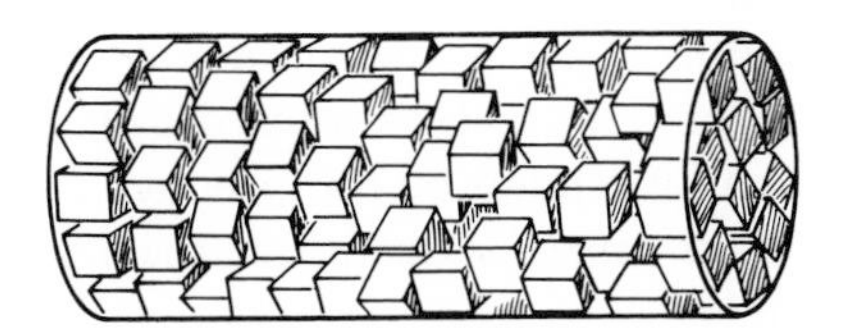

Fig. 20-3

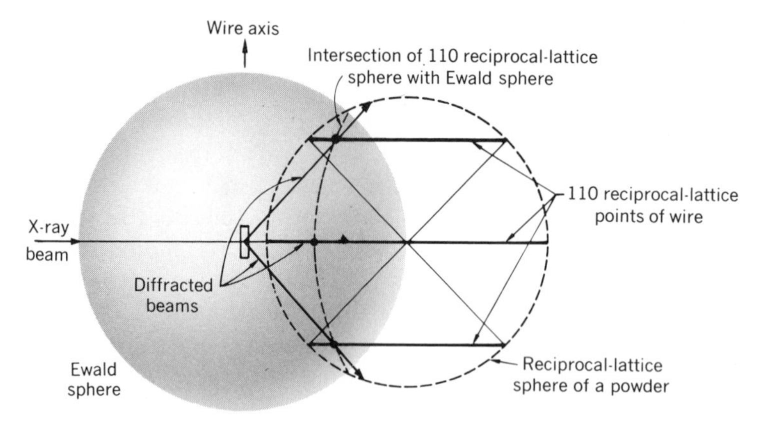

Fig. 20-4

instead of spheres, the reciprocal lattice of a wire consists of latitude circles. As an example, consider the reciprocal-lattice vectors of the $((110))$ planes in a cubic metal having $[100]$ wire texture. In a random aggregate they form a reciprocal-lattice sphere of radius σ_{110}. In a single cube, they form the angle 45 or $90°$ with $[100]$. For the wire, therefore, the reciprocal-lattice construction looks like Fig. 20-4 The reciprocal-lattice points are constrained to lie along three circles, at the equator and at $\pm 45°$ latitudes. Where these circles intersect the Ewald sphere, the diffraction conditions are satisfied, and diffracted beams are produced. As should be evident from an examination of Fig. 20-4, the reciprocal-lattice construction of a wire also can be produced by starting with that of a suitably oriented single crystal and rotating it about the wire axis.

It follows directly from the construction in Fig. 20-4 that an x-ray diffraction photograph of a wire consists of spots located at certain points along what would have been powder rings in the absence of texture. In fact, a photograph of an ideal wire is identical with that of a rotating crystal, except that the large number of crystallites already oriented (rotated) randomly about the wire axis eliminates the need to rotate the wire. The appearance of an idealized photograph of a wire having $[100]$ fiber texture is considered in Exercise 20-1. The appearance of an idealized photograph of a face-centered cubic material having $[111]$ fiber texture normal to the incident x-ray beam is depicted in Fig. 20-5. The powder rings and the layer lines are shown dashed, with the reflections indicated by dots. A comparable construction for a body-centered cubic material having $[110]$ fiber texture is shown in Fig. 20-6. Face-centered cubic metals having $[111]$ texture when drawn into wires include aluminum, copper, nickel, silver, α brass, α bronze, and other alloys. When they are extruded, however, they tend to adopt $[110]$ textures. Body-centered cubic metals like α iron and γ brass, on the other hand, adopt $[110]$ texture when drawn into wires by pulling and $[111]$ and $[100]$ textures when extruded by compressive forces. This latter case is an example of what is called *double texture*, and results when two slip directions are equally likely in a metal.

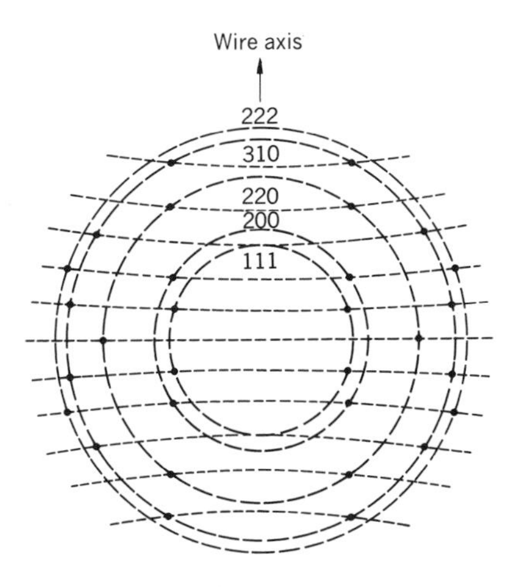

Fig. 20-5. Ideal [111] texture. Dotted curves indicate how constant ζ values would appear in a Bernal chart for a flat film.

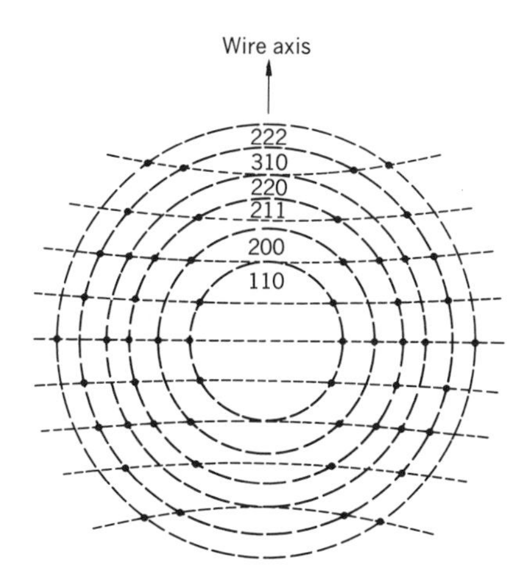

Fig. 20-6. Ideal [110] texture. (Compare with Fig. 20-5.)

The idealized representations in Figs. 20-5 and 20-6 show the appearance of a flat film placed in the transmission arrangement (Fig. 20-7) at right angles to the direct beam. Knowing the sample-to-film distance D, the diffraction angle of each powder ring can be determined by measuring its radius R on the film since

$$2\theta = \tan^{-1}\frac{R}{D} \tag{20-1}$$

Using the procedures discussed in Chap. 18, the reflections then can be indexed.

The diffracted beam forms the angle δ with the vertical plane containing the direct beam and the wire axis. This angle is indicated on the reciprocal-lattice sphere in Fig. 20-8. The NS axis passes through the reciprocal-lattice origin O and is parallel to the fiber axis (Fig. 20-7). The trace of the intersection with the Ewald sphere is indicated about the WE axis, denoting the direct-beam direction, while the location of the reciprocal-lattice points about the NS axis is shown by the horizontal circle having the colatitude ρ. Since this is the angle formed by the wire axis and the normals to the reflecting planes, a knowledge of the angle ρ for at least two sets of planes would allow one to establish the crystallographic direction parallel to the wire axis by triangulation procedures similar to those discussed in Chap. 14. The value of ρ can be determined by considering the spherical triangle WNP in Fig. 20-8. From spherical geometry

$$\begin{aligned}\cos\rho &= \sin(90° - \theta + \theta)\sin(90° - \theta)\cos\delta - \cos(90° - \theta)\cos(90° - \theta + \theta)\\ &= \sin(90° - \theta)\cos\delta + 0\\ &= \cos\theta\cos\delta\end{aligned} \tag{20-2}$$

Thus a measurement of δ on the film (Fig. 20-7), combined with the previously determined value of θ, is sufficient to determine the ρ value corresponding to each reflection. By consulting Table 2-1, listing the interplanar angles for cubic crystals, it is then possible to establish what the fiber texture is.

It is possible to display these angles also in a stereographic projection, as shown in Fig. 20-9. Consider the 110 reflections of a wire having $[100]$ texture. The latitude circles along which the 110 reciprocal-lattice points lie (Fig. 20-8) are, of course, the same as those containing the poles of the (110) reflecting planes. Projecting along the direct-beam direction WE, the trace of the Ewald sphere (powder ring) appears as a small circle about the center of the projection. Note that this circle lies $\theta°$, measured along the vertical meridian, from the north pole. Its intersections with the latitude circles containing the 110 reciprocal-lattice points [or (110) poles] mark the reciprocal-lattice points of those planes that are in reflecting position. The reflection passing through the point [pole] P (Fig. 20-8) lies in the diametral plane containing the projection direction, so that it appears as a central meridian in Fig. 20-9. The angle δ is formed by this plane and the NS axis, as shown, while the inclination angle of the (110) pole ρ must be measured along a meridian (great circle) passing through P.

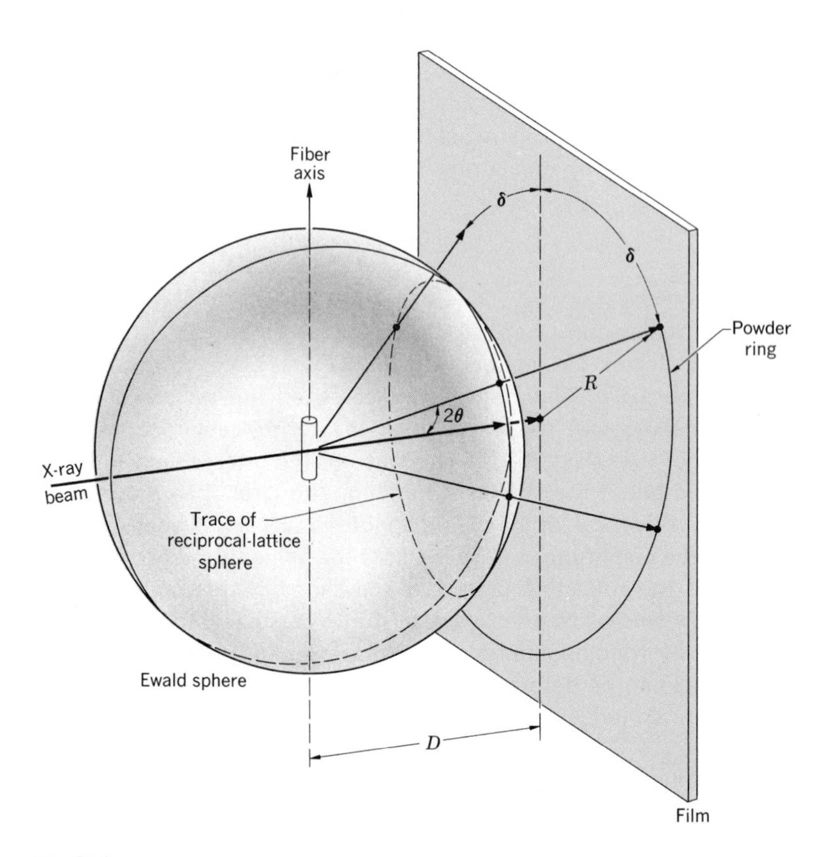

Fig. 20-7

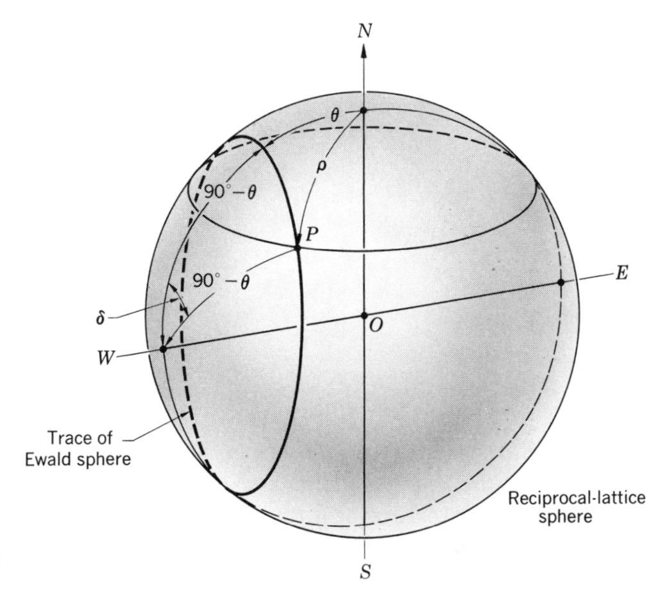

Fig. 20-8

Real fiber texture. The fiber texture encountered in real wires or fibers deviates from the idealized cases discussed above primarily in that the inclination of the planes deviates slightly from the ideal value ρ. Normally, the amount of deviation ϕ is symmetric about the ideal value, so that the reciprocal-lattice construction appears as shown to the left in Fig. 20-10, where it is seen that the idealized latitude circles have widened into bands. These bands intersect the Ewald sphere in arc segments, so that one records arcs instead of spots along the powder rings. Generally, the intensity of the arcs is largest at their center, indicating that more planes tend to have the ideal inclination ρ, while those deviating from this value

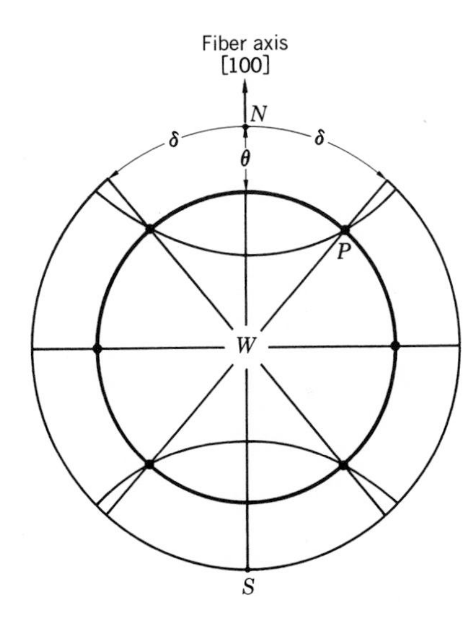

Fig. 20-9

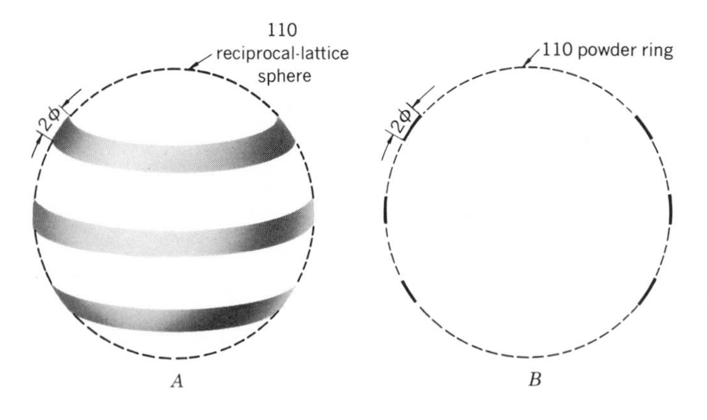

Fig. 20-10. Fiber texture: (*A*) reciprocal-lattice construction; (*B*) appearance of front-reflection photograph.

decrease in number as ϕ increases (Fig. 20-11). Strictly speaking, the arc width 2ϕ measures the change in δ, not in ρ. When using such indications to characterize further the fiber texture of a material, however, it is usual to quote 2ϕ values as measured on the film. Since $\cos\theta \approx 1$ for small diffraction angles, it is seen from (20-2) that this does not introduce a very large error, in any case.

Fig. 20-11. X-ray diffraction photograph of an aluminum wire. Mo radiation, 5-cm wire-to-film distance, 5-min exposure.

Fig. 20-12. X-ray diffraction photograph of a polytetrafluoroethylene fiber.

When a metal wire is drawn through a die, the crystallites at the center of the wire are more accurately aligned than those near the surface. Thus, if a sequence of photographs is compared, beginning with the wire as drawn and then after thinning of the wire by etching, a progressive narrowing of the arc (2ϕ) is observed. Not surprisingly, the tensile strength of the wire is found to be larger in the central portion. This zonal texture is a characteristic of most metal wires, but has not been observed in fibers of organic materials. Organic fibers or filaments display fiber texture (Fig. 20-12), even though they are not composed of actual crystallites. This is so because the large molecules comprising them tend to align parallel to each other in the fibers, and the interatomic spacings in such bundles tend to be quite regular. Sometimes, however, the molecules form helices about the fiber axis. If the molecules deviate from perfect parallelism to such helicoids, then arcs, instead of spots, appear, except that, because of the continuous spiraling about the fiber axis, the intensity maxima in the arcs do not lie at their centers, but split into two or four maxima, corresponding to the possible inclinations of the spirals.

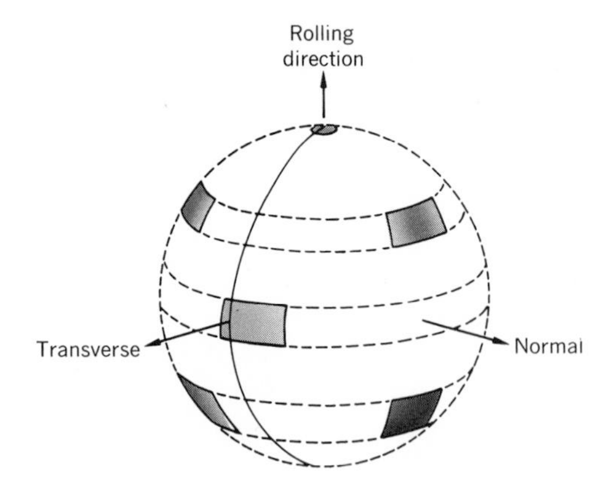

Fig. 20-13

Sheet texture. When a metal is forced to pass through a circular aperture in a die, it deforms by slipping along the drawing, or extrusion, direction, but the orientation of the slip planes (crystallites) about this direction remains random. When a metal is forced to pass between two rolls in drawing a sheet, the slip directions again tend to align parallel to the rolling direction. Because of the lateral pressure of the rolls, however, the slip planes tend to lie in the plane of the sheet. Thus the resulting sheet texture requires two symbols to identify it, the indices of the slip plane, followed by the indices of the slip direction: $(hkl)[uvw]$. In most materials, one slip plane predominates, usually the closest-packed plane. Occasionally, more than one may be active, in which case they are listed in the order of their importance. Since the slip mechanisms are an intrinsic property of a material, it is difficult to control the texture produced by mechanical means alone. Normally, one tries to minimize it, because the mechanical strength of a sheet is most anisotropic, sometimes varying by a factor of 2 to 3 between the rolling direction and the transverse direction. Mechanically, one can minimize it by cross-rolling, that is, by alternately rolling the sheet in mutually orthogonal directions. This is not only a costly manufacturing process, but it also results in the sheet having an "average" strength that is closest to the lower of the two original values. Alternatively, attempts to change the sheet texture can be made by a suitable annealing, during which a new recrystallization texture develops. This process can be particularly important when one desires to impart a very specific sheet texture. For example, a good material for transformer plates should have an "easy" direction of magnetization parallel to the expected lines of the magnetic field flux. In the case of iron, $[100]$ is an easy direction; so the ideal sheet texture for this application would be $(100)[001]$, called cube texture. In the past it was believed that the closest one could come to this was to have $(110)[001]$ texture, but in the last decade it was shown that, by suitably thinning the sheet and then annealing at over $1000°$C, recrystallization produced the desired cube texture.

If a plane in each crystallite and a direction in that plane are made parallel in all the adjacent crystallites, a single crystal results. In actual sheets, of course, this idealized state is never attained. Instead, what one finds can be thought of as a fiber texture (Fig. 20-10) parallel to the rolling direction, with a very limited range of orientations normal to it. A simplified version of the reciprocal-lattice construction for one set of planes is shown in Fig. 20-13. Generally, the amount of deviation from the ideal alignment of the slip planes is different for the rolling and the transverse directions, which is a contributing factor to the anisotropy of a sheet. For example, in hexagonal closest-packed magnesium, the (0001) slip planes deviate by as much as $25°$ in the rolling direction and by up to $15°$ at right angles to it. It follows from this that more information than that given by the symbol $(hkl)[uvw]$ is needed to completely describe sheet texture. In 1924, F. Wever suggested that a stereographic projection of the density of poles of one set of planes would accurately characterize the distribution of a set of reflecting planies (usually, but not always, chosen to be the slip planes) about the rolling direction. The construction of such a *pole figure* requires that a series of photographs be prepared of the sheet inclined at several angles to the direct beam. In turn, this ntroduces an absorption correction, which becomes increasingly serious as the inclination angle increases. When feasible, the sheet may be formed into a rod parallel to the rolling direction by cutting or etching so as not to disturb the sheet texture. More commonly, however, a sheet is used as is, and the intensity values are subsequently corrected.

Pole figures. The disposition of the incident and diffracted beams is indicated in Fig. 20-14. In the initial position, the sheet is placed at right angles to the beam, with its rolling direction vertical. The resulting photograph looks not unlike a photograph of a wire, because the sheet and its reciprocal lattice (Fig. 20-12) have mm symmetry about the sheet's normal. In subsequent photographs, the sheet is rotated about its rolling direction, usually in steps of $\beta = 5$ or $10°$. This has the effect of examining the intensity distribution in the reciprocal-lattice sphere along similarly inclined diametral planes. As can be seen in Fig. 20-13, the intensity distribution changes at the successive cross-sectional slices and ceases to be symmetrical. In fact, both the intensity and the locations of the arcs usually change as a function of the inclination β (Fig. 20-15). To display this distribution of reflecting-plane orientations, the location and extent of the arcs can be plotted on a stereographic net, similarly to the procedure used in constructing Fig. 20-9.

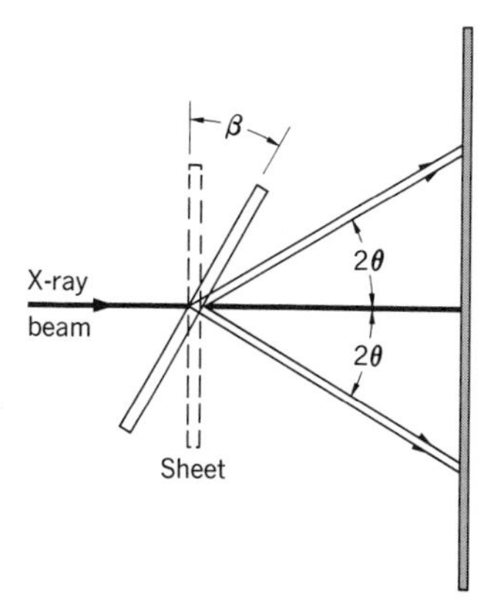

Fig. 20-14

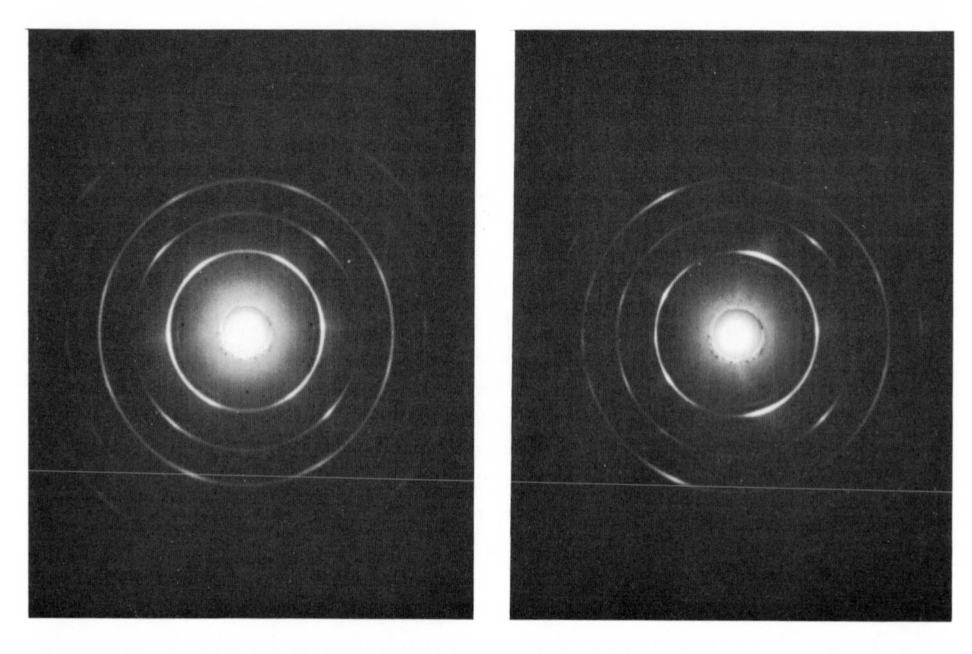

Fig. 20-15. Transmission photographs of a vanadium sheet. The photograph on the left was taken with the sheet normal to the beam, and for the one on the right $\beta = 30°$.

An idealized version of one of the powder rings recorded at a β angle of $30°$ is illustrated in Fig. 20-16, and can be compared with the photograph on the right of Fig. 20-15. Suppose that the diffraction angle for this reflection, $2\theta = 20°$, and the azimuthal limits of the two arcs recorded in Fig. 20-16,

$$\delta_1 = 45° \qquad \delta_2 = 25° \qquad \delta_3 = 40° \qquad \delta_4 = 65° \tag{20-3}$$

To transfer this information to a stereographic net, it is first necessary to locate the powder ring on the net (as in Fig. 20-9). Since $\theta = 10°$, the powder ring for the normal setting is a circle about the projection direction (sheet normal is parallel to incident beam) $10°$ inside the outer meridian. The angles can be laid off along

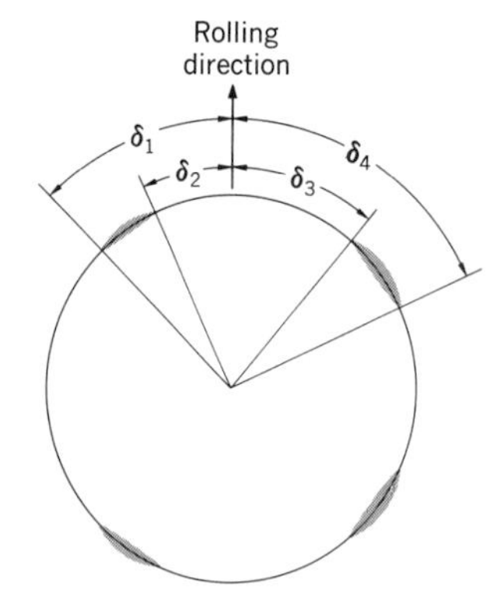

Fig. 20-16

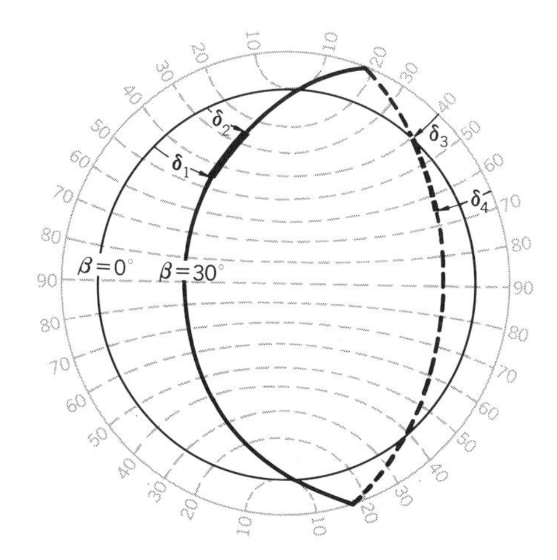

Fig. 20-17

this circle by drawing in the diametral planes, as shown in Fig. 20-17. The plane of the sheet in preparing Fig. 20-16 was inclined by $30°$ to the incident beam, however, so that the powder ring lies along a plane that is inclined by $\beta = 30°$ to the plane of the projection in Fig. 20-17. The trace of this powder ring is found by moving several points of the original circle by $30°$ along corresponding latitudes, and then passing a circle through the new points, recalling from Chap. 2 that circles project as true circles in the stereographic projection. The arc subtended by δ_1 and δ_2 now can be transposed to the correct circle by rotating along the respective latitudes, as shown in Fig. 20-17. Note that the back part of the powder ring has been rotated $30°$ also, in this case $10°$ in front of the net and then $20°$ behind it. The trace of this part of the powder ring is shown dashed in Fig. 20-17, along with the transposed arc $\delta_3\delta_4$. Because of the mm symmetry of sheet texture, it is possible to "reflect" both arcs about the orthogonal mirror planes, as shown in Fig. 20-18. Recalling that a pole figure is nothing but a plot of the density of reciprocal-

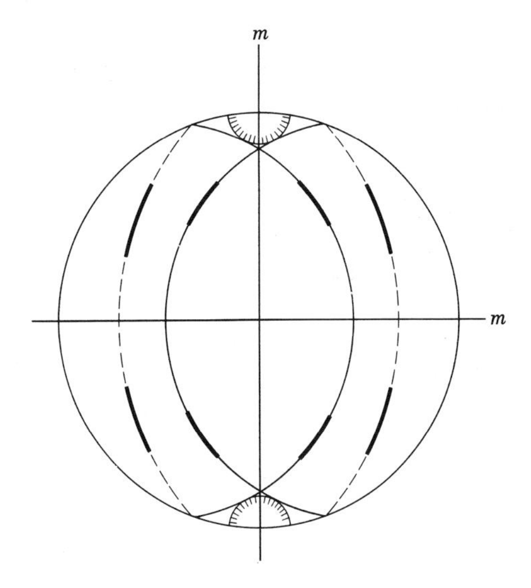

Fig. 20-18

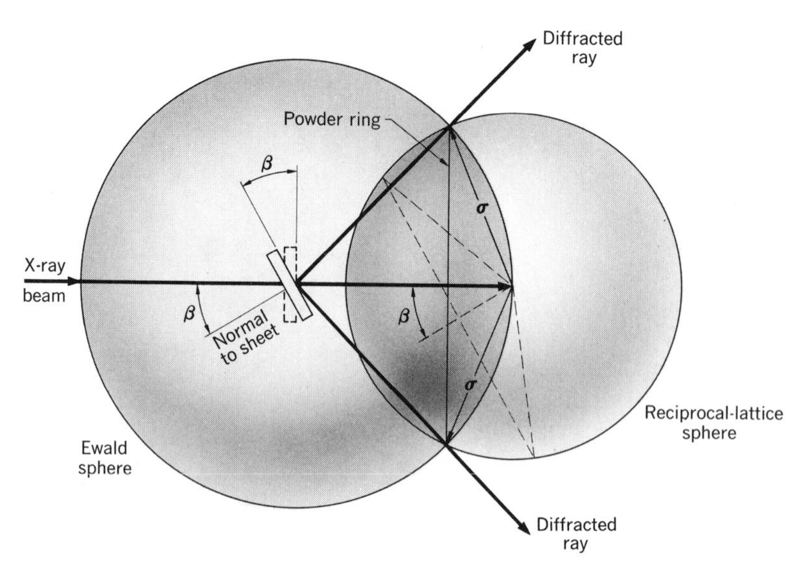

Fig. 20-19

lattice points (Fig. 20-13), the ever-present symmetry center makes the correct symmetry of the reciprocal-lattice of a sheet mmm. It is permissible, therefore, to reflect into the forward projection area the arc $\delta_3\delta_4$ lying in the back of the net in Fig. 20-18.

The reciprocal-lattice construction for the above process is illustrated in Fig. 20-19. The sheet at the center of the Ewald sphere is shown after rotation by an amount β about its rolling direction, which is normal to the drawing. The diffracted rays pass through the powder ring lying at the mutual intersection of the two spheres. The powder ring's position on the reciprocal-lattice sphere, when the sheet was normal to the incident x-ray beam, is indicated by the dashed line in Fig. 20-19. It is clear from this that the strategy used in an arrangement such as the one shown in Figs. 20-14 and 20-19 is to sample the reciprocal-lattice sphere along equivalent circles displaced about its equator in steps of β.

Diffractometer methods. Once it is recognized that a pole figure actually is a stereographic projection of one reciprocal-lattice sphere of the sheet, it becomes apparent that a more precise distribution of orientations can be specified by noting the relative intensity distributions (pole densities) along the arcs. Since diffractometers are able to measure intensities more precisely than films, the natural evolution has been to devise mechanical contrivances for duplicating the information contained in films. The basic problem is that a diffractometer can measure the intensity only for a very short portion of the powder ring since it is constrained to move along the equator of the Ewald sphere. An examination of Fig. 20-19 immediately suggests that the entire intensity distribution along the powder ring can be measured by the detector, provided that the sample is rotated about the incident x-ray beam. This is equivalent to point-by-point scanning of the intensity distribution along rings like those shown in Fig. 20-15, so that the procedure for constructing a pole figure remains unchanged from the one already described. The rotation about the x-ray beam is a measure of the angle δ along the ring, and such an instrument can be easily devised on the basis of the geometry of three-circle goniometers, discussed in Chap. 13. If the transmission arrangement is used, the absorption correction for the diffracted beams remains exactly the same as that in the film arrangement. Whereas, in the film arrangement, a blind spot develops near the N and S poles of the figure (Exercise 20-3), requiring a remounting of the sheet, the goniometer motions would allow rotation about the rolling or transverse direction directly.

The designers of the first commercially constructed diffractometer adaptor chose to rotate the sheet within its own plane. The reciprocal-lattice construction for this arrangement is shown in Fig. 20-20. When the sample sheet is normal to the beam ($\beta = 0$), rotation about the sheet's normal is equivalent to rotation about the x-ray beam, and the intensity distribution along the powder ring is measured as before. When the sample is inclined by an amount β, however, the sampling circle lies along a slice through the reciprocal-lattice sphere that is parallel to the sheet, since rotation proceeds about the normal to the sheet. Note that the diameter of the sampling circle changes as the inclination angle β changes. It is not convenient, therefore, to use the same projection procedure as employed in Fig. 20-17, and one uses a polar projection instead. By convention, the inclination angle of the sheet's normal is called α (in place of β), and it is chosen to be zero when the sheet bisects the angle formed by the incident and reflected beams ($\beta = \theta$). As shown in Fig. 20-21, the sampling circle is perpendicular to the sheet normal in this position, and contains the reciprocal-lattice vectors σ. The reciprocal-lattice points being sampled at this setting lie along the outermost circle in the polar net shown in Fig. 20-22. To measure their density (intensity of beam diffracted at 2θ),

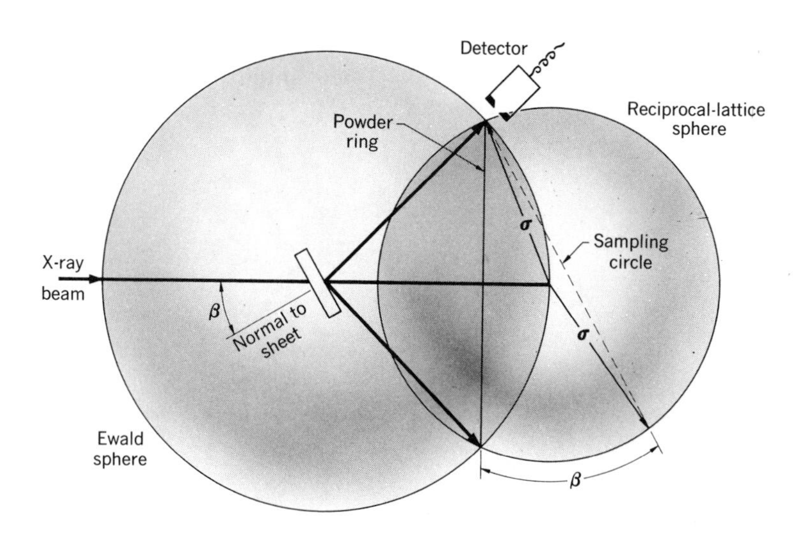

Fig. 20-20

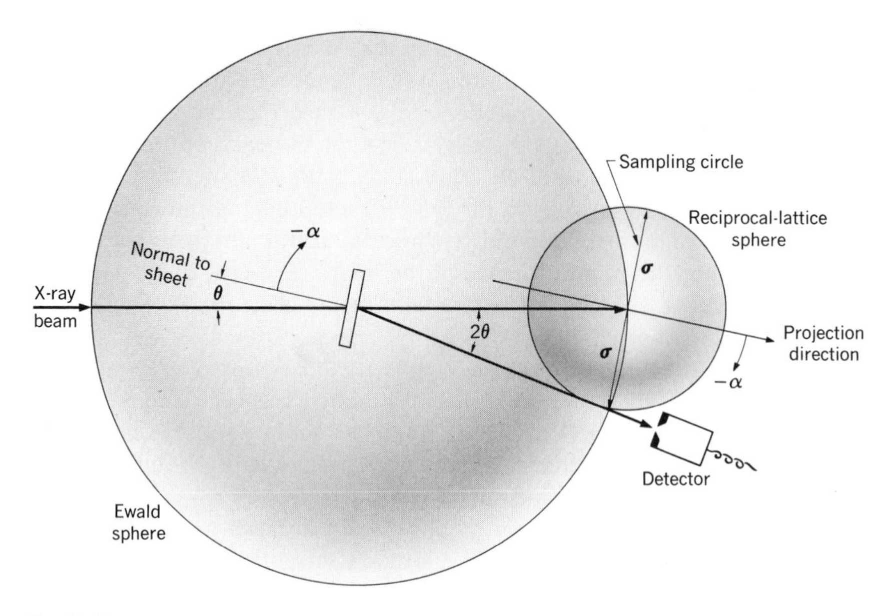

Fig. 20-21

the sample sheet is rotated about its normal by an amount δ. (This rotation is not indicated in Fig. 20-21 because it occurs in the vertical plane.) The angle δ is generally measured using the transverse direction as the starting point. Designating rotation in a clockwise sense about the vertical rolling direction (perpendicular to the plane of Fig. 20-21) by $-\alpha$, it is seen in Fig. 20-21 that such rotation decreases the size of the sampling circle (Exercise 20-4). One can easily see from this that the inclination angle $-\alpha$ is measured inward along the radial meridians in the polar net (Fig. 20-22). In fact, in the operation of the commercial instrument, the angle δ within the sheet is fixed, and successive intensity readings are made at regular intervals $-\alpha$, usually in steps of 5 or $10°$. The intensity readings, corrected for

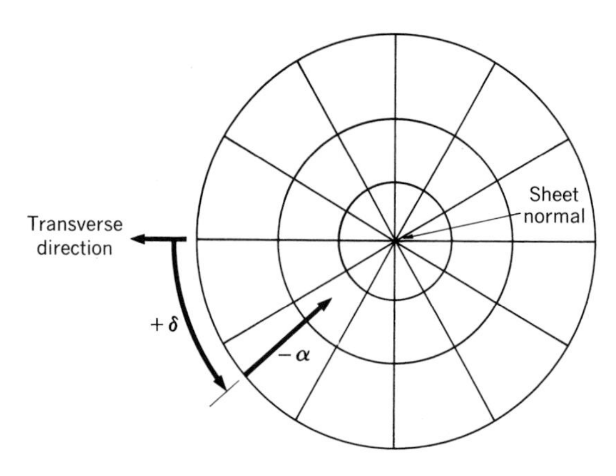

Fig. 20-22

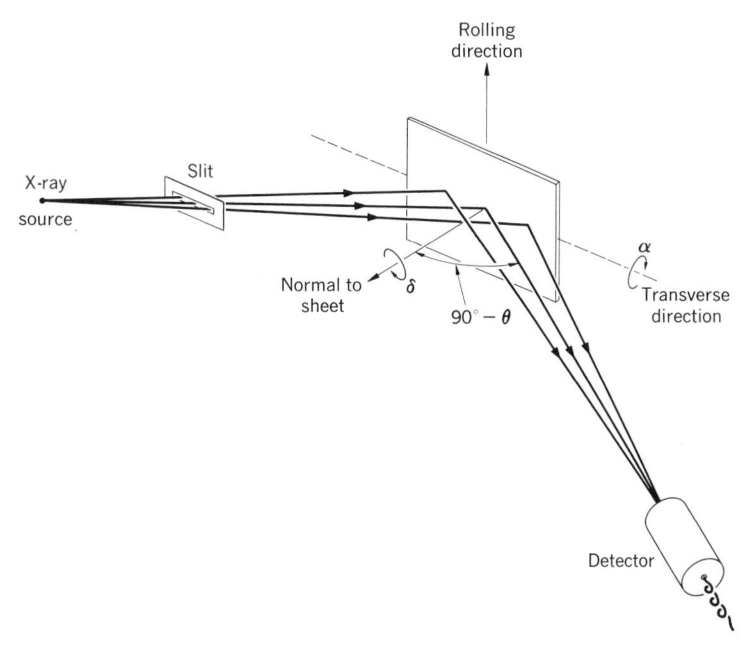

Fig. 20-23

absorption (Exercise 20-5), are then entered along the radial meridians, as indicated in Fig. 20-22.

The transmission arrangement described so far has two important limitations: There is an angular range of $-\alpha$ (Exercise 20-4) in which the sample holder interferes with the x-ray beams, and the absorption correction (Exercise 20-5) becomes inaccurate. This occurs approximately in the range $50 < -\alpha < 90°$. The other limitation is occasioned by the divergence in the x-ray beam. It is clear from earlier discussions in this book that the parafocusing geometry of a diffractometer is not satisfied in the transmission arrangement. Thus the diffracted beam diverges, and may cause overlapping of neighboring reflections, particularly when more penetrating, shorter-wavelength radiation is employed. This difficulty can be overcome by placing the sample sheet in the reflecting position, as indicated in Fig. 20-23. A diverging beam passes through a horizontal-line slit placed parallel to the transverse direction of the sheet (initially). The reflected beam then refocuses at the detector as shown. The central ray forms the angle $90° - \theta$ with the sheet's normal when the rolling direction is initially vertical and perpendicular to the focusing plane. The inclination of the sheet α, relative to the incident beam, takes place by rotation about the horizontal axis, as indicated. Since the polar projection is along the sheet's normal, in Fig. 20-22, this means that the inclination is upward along the vertical meridian. When the plane is rotated about its normal by an amount δ, as indicated, this is equivalent to moving around a latitude circle in Fig. 20-22. Thus it is clear that the mechanism of data gathering

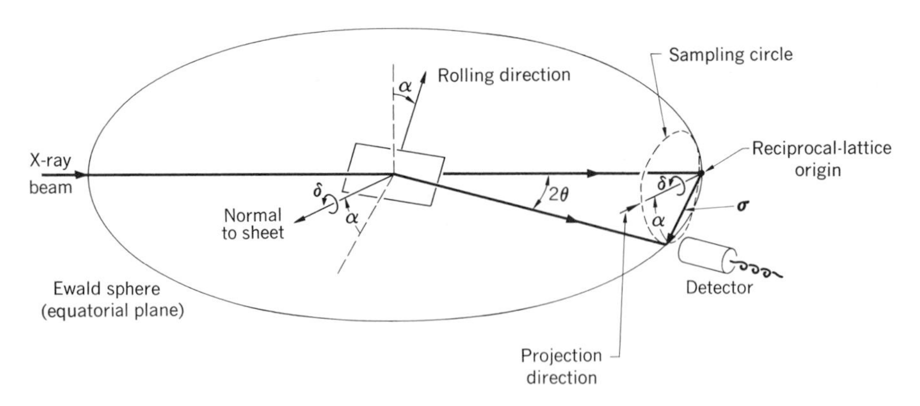

Fig. 20-24

and plotting is quite similar to that already discussed for the transmission arrangement.

The reciprocal-lattice construction for the reflection arrangement is given in Fig. 20-24. The sheet is shown inclined by an amount α from the equatorial plane of the Ewald sphere (focusing plane of diffractometer). The projection direction in the reciprocal-lattice sphere is similarly inclined by α deg to the horizontal plane containing the reciprocal-lattice vector σ in reflecting position. Rotation of the plane about its normal by δ is then equivalent to rotating the reciprocal-lattice sphere about the projection direction. Note that the sampling circle (powder ring) has a diameter equal to zero when $\alpha = 0°$ and increases as α increases. By comparison with the transmission arrangement (Fig. 20-21), the reflection arrangement is particularly well suited to examining the inner portion of the polar projection (Fig. 20-22). It has the advantage that the parafocusing arrangement eliminates the need for an absorption correction for α angles up to $50°$, provided that the horizontal slit shown in Fig. 20-23 is sufficiently narrow to limit the vertical divergence. It is usual practice, therefore, to employ the reflection arrangement to measure the pole density (intensity) in the inner $50°$ and the transmission arrangement for the outer $50°$. The overlap of $10°$ serves to check the relative accuracy of the measurements and helps establish a common intensity scale.

The use of diffractometers has increased considerably the accuracy with which pole figures can be determined. More generalized diffraction geometries, based on the Eulerian angles discussed in Chap. 13, can be employed, and the automation of intensity measurement developed for single-crystal diffractometry can be used equally well for pole-figure determinations. A number of automatic intensity-measuring devices have, in fact, been constructed, including automatic pole-figure plotters that transcribe the angular settings and intensities (expressed as number of counts) onto a polar net. The only thing remaining for the operator is the job of connecting up like numbers, representing like pole densities, with contours in the final pole figure. A large number of actual pole figures, along with a detailed discussion of their utilization in metallurgy, can be found in several of the books listed in the literature section of this chapter.

CRYSTALLITE–SIZE ANALYSIS

Effect on powder diffraction. The effect that crystallite size has on powder rings is illustrated in Fig. 20-25. The "ideal" size for powder diffraction depends on the relative perfection of the polycrystalline material, but usually lies in the range 10^{-3} to 5×10^{-5} cm. If the crystallites are smaller than this, the number of parallel planes available is too small for a sharp diffraction maximum to build up, and the lines in a powder photograph become broadened, as shown in the left photograph of Fig. 20-25. Below about 10^{-6} cm, the crystallites are too small for diffraction at the Bragg angles, so that their only constructive interference effects are limited to small angles. This is similar to the effects discussed in Chap. 8 for scattering by gas molecules, and procedures for studying the sizes and shapes of very small crystallites using specialized small-angle scattering instruments have been devised.†

When the crystallites exceed the "ideal" size, the powder rings become spotty (part C of Fig. 20-25), because the number of crystallites in the irradiated portion of the sample is insufficient to reflect to each portion of the ring. The larger the crystallite size, the more spotty the ring becomes, until the crystallites are so large ($>10^{-2}$ cm) that only a few are irradiated at a time, and the powder photograph is replaced by one comprised of a superposition of just a few single-crystal photographs. Procedures have been devised for counting the number of spots in a ring and deducing the crystallite size from this. Alternatively, comparisons with previously prepared photographs of similar crystallites having known sizes can be used. None of these methods are very accurate, and require extreme care in sample preparation and treatment during the exposure. Since the crystallite size in this range is large enough to be visible under an optical microscope, such examination is normally preferred to x-ray studies. It should be noted, however, that the x-ray method can be used to evaluate crystallite size in the "ideal" size range, even though, normally, sharp rings are observed. To produce a spotty ring in this case, a very fine x-ray beam is required to decrease the irradiated sample volume. Such beams can be produced from conventional x-ray sources by using tapered lead-glass capillaries as collimators, or special microfocus tubes can be employed.

The effects of crystallite size are even more important in the case of powder diffractometry. In the photographic method, any deviation from ideal conditions is immediately apparent, whereas in the case of diffractometer measurements, one literally operates in the dark. Because the parafocusing geometry requires a fairly large number of randomly oriented crystallites to provide accurate statistics for the intensity measurements, crystallite sizes less than 5×10^{-4} cm are usually needed to achieve 1% reproducibility. This is a matter that can be controlled when powdered materials are examined, but becomes quite serious when polycrystalline aggregates, notably metals, are being studied. Note that this is an important consideration for texture studies, discussed in the preceding sections. In routine diffractometry, one way to "increase" the number of reflecting grains

† André Guinier and Gérard Fournet, *Small-angle scattering of x-rays,* translated by Christopher B. Walker (John Wiley & Sons, Inc., New York, 1955).

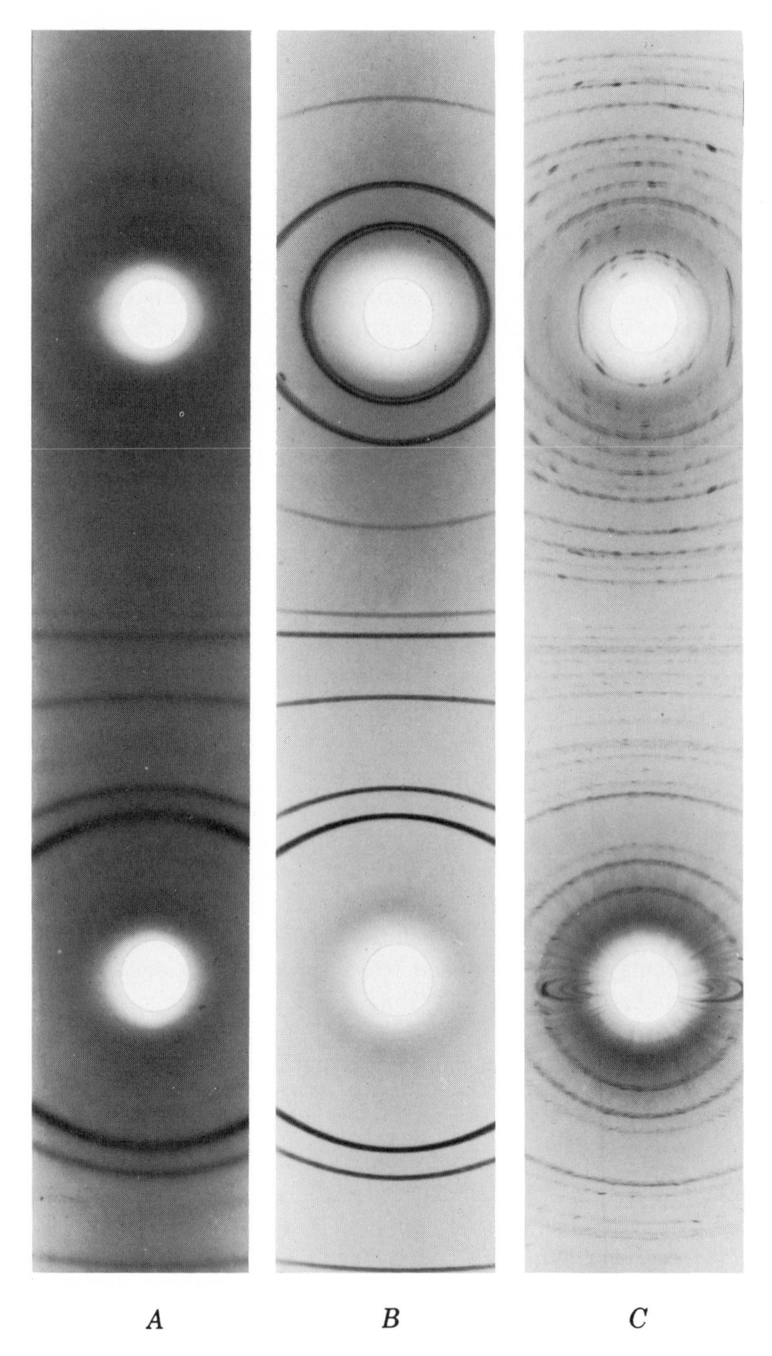

Fig. 20-25. Crystallite-size effects. *(After Azároff and Buerger.)* (*A*) Raney nickel, about 10^{-6} cm average diameter; (*B*) nickel powder of "ideal" size, about 10^{-3} cm; (*C*) rutile, TiO_2 powder insufficiently ground, about 10^{-2} cm.

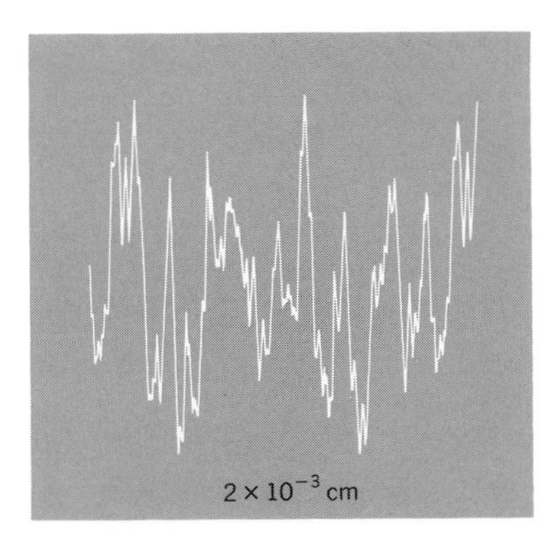

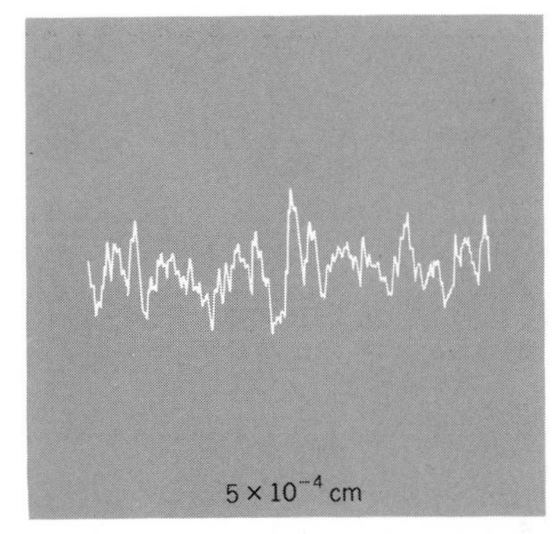

Fig. 20-26. Crystallite-size effect on diffractometer intensity displayed by slowly rotating a silicon powder within its own plane. [*After P. M. de Wolff, J. M. Taylor, and W. Parrish, Journal of Applied Physics, vol.* 30 (1959), *pp.* 63–69.]

is to use a sample spinner and rotate the sample during the measurement. In the case of texture studies, however, this is not possible, so that lateral motions have to be provided to oscillate the sample back and forth without disturbing its orientation relative to the focusing plane of the diffractometer.

An example of the kind of effect that crystallite size can have on diffractometer measurements is given in Fig. 20-26. The left tracing shows the statistical variation of a slowly rotating silicon sample when the crystallite size is too large. The right tracing illustrates the maximum tolerable fluctuation for precisions of 1%. A comparison of complete scans through the reflection position of the same two samples showed that the peak intensity of the coarser one had dropped by 22%. The difference between the two samples can be reduced by increasing the speed of the spinner or by using more penetrating radiation. In the absence of a careful and lengthy examination of each sample, however, it is not possible to tell a priori what one may expect, since some of the instrumental factors discussed in Chap. 13 similarly can affect the appearance of a diffractometer tracing and the precision of intensity measurements. Whenever one suspects that the intrinsic properties of the sample may be responsible for unusual effects, the most satisfactory procedure is to remove the sample from the diffractometer and examine it photographically.

Theoretical analysis. The role that the number of parallel planes N spaced a apart have on the intensity of x-ray reflection by such planes was examined in an approximate way in Chap. 9. According to that treatment, the breadth of the reflection peak at half-maximum intensity,

$$\epsilon_{\frac{1}{2}} = \left(\frac{\ln}{\pi}\right)^{\frac{1}{2}} \frac{\lambda}{D \cos \theta} \tag{9-28}$$

By convention, one defines the halfwidth $\beta_{\frac{1}{2}}$ in terms of the diffraction angle 2θ

(measured in radians), so that

$$\beta_{\frac{1}{2}} = \epsilon_{\frac{1}{2}} = \frac{0.94\lambda}{D \cos \theta} \tag{20-4}$$

where D is the "average" dimension of the crystallites normal to the reflecting planes.

Relation (20-4) actually was first derived by Scherrer in 1918. Since then a number of investigators have confirmed this relation. In 1926, Laue considered the *integral width* β_i, defined in the usual way,

$$\beta_i = \frac{\int_{-\infty}^{+\infty} y(x)\, dx}{y_{\max}} \tag{20-5}$$

Assuming that the shape of the reflection can be represented by

$$y(\epsilon_{\frac{1}{2}}) = I_{\max}\, e^{-(\pi/\lambda^2)(D\epsilon_{\frac{1}{2}} \cos \theta)^2}$$

as was done in Chap. 9, then

$$\beta_i = \frac{\int_{-\infty}^{+\infty} I_{\max}\, e^{-(\pi/\lambda^2)(D\epsilon_{\frac{1}{2}} \cos \theta)^2}\, d\epsilon_{\frac{1}{2}}}{I_{\max}}$$

$$= \frac{\lambda}{D \cos \theta} \tag{20-6}$$

since $\int_{-\infty}^{+\infty} e^{-ax^2}\, dx = (\pi/a)^{\frac{1}{2}}$.

Others have considered what effect the shape of the crystallites has on the halfwidth. (Scherrer had assumed cubic crystallites.) In general, one can write

$$\beta = \frac{k\lambda}{D \cos \theta} \tag{20-7}$$

where k varies from 0.89 to 1.39, but, for most cases, is close to 1.00. Since, as discussed in the next section, the precision of crystallite-size analysis is, at best, of the order of $\pm 10\%$, the assumption of $k = 1.0$ in (20-7) is generally justifiable.

Method mixtures. In addition to the crystallite-size broadening considered above, various instrumental factors also affect the width of the reflection. In order to estimate their magnitude, it is usual to mix the sample with some well-crystallized standard powder whose crystallite size causes no broadening. In this case, the width of a reflection from the standard, β_{instr}, can be assumed to be caused exclusively by instrumental factors. Let the shape of this curve be given by some function,

$$y = f(x) \tag{20-8}$$

where x is a measure of the "distance" from the $y_{\max}$ position at $x = 0$ (Fig. 20-27).

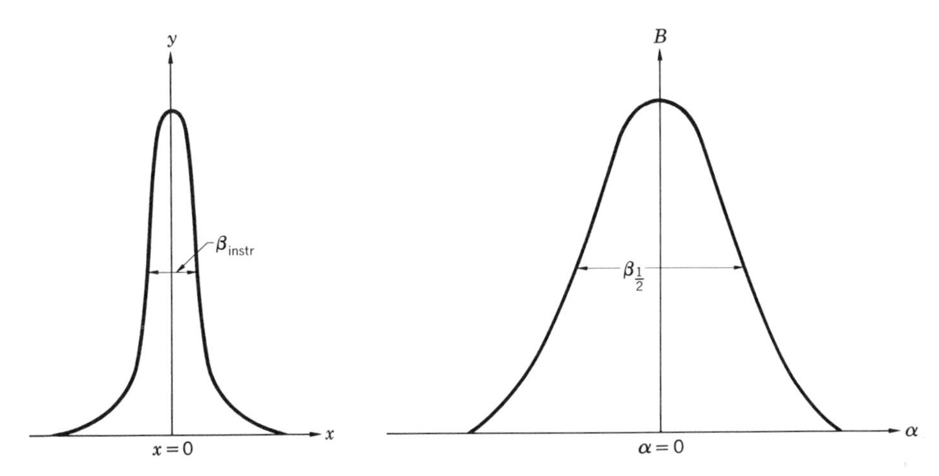

Fig. 20-27

The effect that small crystallites have on the diffraction profile is to broaden each area element of the curve on the left in Fig. 20-27 by some amount determined by a broadening function $g(\alpha)$. The total crystallite-size broadening $B(\alpha)$ then can be represented by a curve, such as the one shown to the right in Fig. 20-27, whose halfwidth $\beta_{\frac{1}{2}}$ is the quantity sought in the analysis. The functional description of this curve is

$$B = Ag(\alpha) \tag{20-9}$$

where A is the area of the instrumental curve.

The reflection one actually observes is a composite of the instrumental effects described by $y(x)$ and the crystallite-size broadening $B(\alpha)$. One way to look at this is to think that each portion of the pure instrumental curve is being broadened by $g(\alpha)$. This produces the observed reflection, which can be described by some function

$$Y = h(X) \tag{20-10}$$

where $X = x + \alpha$, and the function is portrayed in Fig. 20-28.

The relationship between the functions in (20-8) to (20-10) can be understood by considering how the curve in Fig. 20-28 comes about. At some value x, the small portion of the instrumental curve of height y and width dx is broadened by $g(\alpha)$, as shown by the small dashed curve in Fig. 20-28. At some distance α from the origin of the curve (which is located at the point x), this makes a contribution to the composite (actually observed) curve of amount

$$dY = (y\,dx)g(\alpha) \tag{20-11}$$

Substituting $y = f(x)$ and $\alpha = X - x$ in (20-11), the total contribution at X can be obtained directly by integration.

$$Y(X) = \int_{-\infty}^{+\infty} f(x)g(X - x)\,dx \tag{20-12}$$

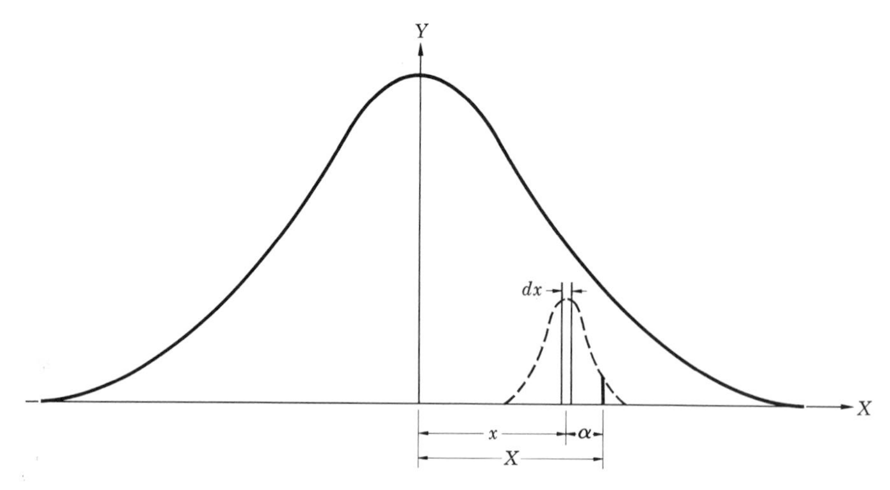

Fig. 20-28

This is called the *convolution function* of $f(x)$ and $g(X - x)$, alternatively described as the folding of $g(X - x)$ into $f(x)$.

The evaluation of (20-12) requires that the exact functional forms of the instrumental function (20-8) and the crystallite-size broadening $g(\alpha)$ be known. There are several possible functions that closely match the actual profiles of x-ray reflections. Three of them are compared in Fig. 20-29. Suppose one lets

$$y = f(x) = \frac{1}{a^2 + x^2} \qquad \text{and} \qquad g(\alpha) = \frac{1}{b^2 + \alpha^2} \tag{20-13}$$

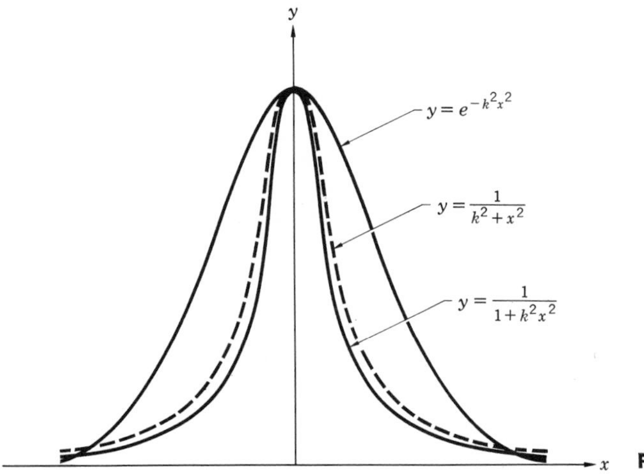

Fig. 20-29

Then the constants a and b can be determined as follows:

$$\text{When } x = a \qquad y = f(a) = \frac{1}{2a^2} = \frac{1}{2}\left(\frac{1}{a^2}\right) = \tfrac{1}{2}y_{\max}$$
$$\text{and } \alpha = b \qquad B = g(b) = \frac{1}{2b^2} = \frac{1}{2}\left(\frac{1}{b^2}\right) = \tfrac{1}{2}B_{\max} \tag{20-14}$$

by direct substitution in (20-13). The meaning of the relations in (20-14) is that when the two curves are at one-half of their maximum heights, their widths are $2x = 2a$ and $2\alpha = 2b$, respectively, so that

$$\beta_{\text{instr}} = 2a$$
$$\text{and} \qquad \beta_{\text{cs}} = 2b \tag{20-15}$$

Substituting the functions assumed in (20-13) into the convolution function (20-12),

$$Y(X) = \int_{-\infty}^{+\infty} (a^2 + x^2)^{-1}[b^2 + (X - x)^2]^{-1}\, dx$$
$$= \frac{\pi(b + a)}{ab[(a + b)^2 + X^2]} \tag{20-16}$$

as can be checked by consulting a table of definite integrals. As before, when $X = a + b$, $Y(a + b) = \frac{1}{2}Y_{\max}$, so that the halfwidth of the experimentally measured curve,

$$\beta_{\text{meas}} = 2(a + b)$$
$$= \beta_{\text{instr}} + \beta_{\text{cs}} \tag{20-17}$$

according to (20-15). Insofar as the assumptions made in (20-13) are valid, (20-17) says that the measured breadth is simply the sum of the instrumental and the crystallite-size breadth. Interestingly, this somewhat naive-sounding proposal was originally made by Scherrer in his original analysis of crystallite-size effects. Recasting (20-17) to solve for the crystallite-size breadth,

$$\beta_{\text{cs}} = \beta_{\text{meas}} - \beta_{\text{instr}} \tag{20-18}$$

Suppose one assumes that the error curve $e^{-a^2x^2}$ in Fig. 20-29 more correctly represents the actual reflection and instrumental profiles. This assumption was made by Warren, and leads to the relation (Exercise 20-7)

$$\beta_{\text{cs}}^2 = \beta_{\text{meas}}^2 - \beta_{\text{instr}}^2 \tag{20-19}$$

In concluding this section it should be noted that the function in (20-12) can be deconvoluted directly, without any assumptions regarding $f(x)$ or $g(\alpha)$, by making use of Fourier theory. This approach was first utilized in diffraction-profile analysis by A. R. Stokes in 1948, and detailed descriptions of its development are given in several of the books listed at the end of this chapter. The ready availability of high-speed computers has taken much of the drudgery out of such procedures, although the practical results obtained seldom justify such an extensive analysis.

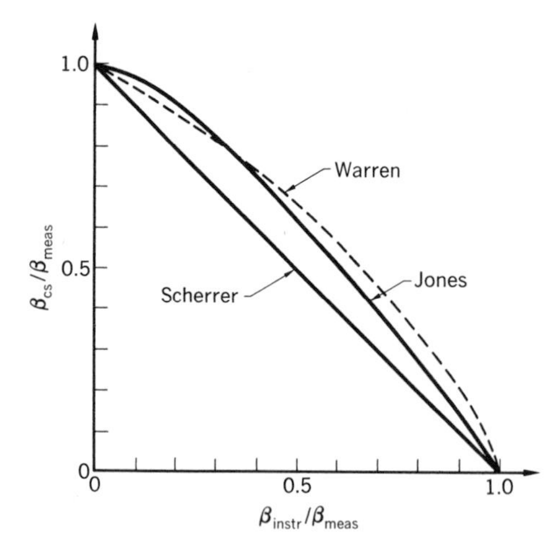

Fig. 20-30. Comparison of conversion curves proposed for crystallite-size analysis.

Practical analysis. In the absence of clear-cut evidence whether the Scherrer relation (20-18) or his relation (20-19) should be used, Warren suggested that the geometric mean,

$$\beta_{cs} = [(\beta_{meas} - \beta_{instr}) \sqrt{\beta_{meas}^2 - \beta_{instr}^2}]^{\frac{1}{2}} \tag{20-20}$$

might prove to be more nearly correct. In an independent analysis of the suitability of the three functions shown in Fig. 20-29, F. W. Jones concluded that the function $1/(1 + k^2x^2)$ was probably the most reliable one. In his analysis, he utilized an experimentally measured reflection curve for the instrumental curve $f(x)$ and assumed $g(\alpha) = (1 + k^2\alpha^2)^{-1}$ as probably the best functional relation. (See also Exercise 20-8.) On this basis he obtained the curve shown in Fig. 20-30 for the relation between β_{cs}/β_{meas} and $\beta_{instr}/\beta_{meas}$. For comparison purposes, similarly recast relations (20-18) and (20-20) are also shown plotted in Fig. 20-30.

To determine the crystallite size in a powder sample, it should be intimately mixed with a suitable standard. The latter's suitability is determined by its crystallite size and the fact that it gives rise to one or more reflections lying close to reflections from the test sample. Proximity of the two reflections chosen (without overlap) is desired, so that the instrumental broadening can be reliably assumed to be the same for both. The halfwidths of both reflections are then measured, and the ratio $\beta_{instr}/\beta_{meas}$ used to find β_{cs}/β_{meas} with the aid of Fig. 20-30. These values can be read directly on a diffractometer tracing. (In the case of powder photographs, a photometer tracing of the reflection must be prepared.) Once β_{cs} is known, relation (20-7) can be used to calculate D, the average dimension of the crystallites. If the crystallites have pronouncedly anisotropic shapes but are randomly oriented in the sample, reflections from suitable planes can be used to establish what their exact dimensions are. It is important to realize, however, that the x-ray beam irradiates a large number of crystallites, so that the value of D obtained represents the mean value of the actual size distribution present.

The accuracy of crystallite-size analysis is considered in Exercise 20-9. It is easy to show that the experimental accuracy is not very high unless the crystallite-size effect is quite pronounced. Moreover, it should be noted that other characteristics of a powder can cause reflections to broaden, some of which have been considered in Chap. 10. The effect of residual stresses that may be present in the crystallites is considered in the next section. A variable strain in the crystallite causes the average d value of the reflecting planes to vary by some small increment Δd. Its effect can be assessed by differentiating the Bragg equation

$$\Delta\lambda = 2\,\Delta d \sin\theta + 2d\,\Delta(\sin\theta) \tag{20-21}$$

and by dividing (20-21) by the Bragg equation, after noting that the differential of the constant wavelength is zero.

$$0 = \frac{2\Delta d \sin\theta + 2d\,\Delta(\sin\theta)}{2d\sin\theta}$$

$$= \frac{\Delta d}{d} + \frac{\Delta(\sin\theta)}{\sin\theta}$$

from which
$$\frac{\Delta d}{d} = -\frac{\cos\theta\,\Delta\theta}{\sin\theta} = \frac{-2\,\Delta\theta}{2\tan\theta} \tag{20-22}$$

Assuming that for small values $2\,\Delta\theta = \Delta(2\theta)$ and rearranging the terms in (20-22) gives

$$\beta_{\mathrm{strain}} = \Delta(2\theta) = -2\frac{\Delta d}{d}\tan\theta \tag{20-23}$$

The angular dependence of (20-23) is compared with that of the Scherrer equation (20-4) in Fig. 20-31. An interesting feature of the two curves is that both increase rapidly with diffraction angle. Thus, although it is tempting to carry out crystallite-size analyses using reflections at large angles, it should be realized that the presence

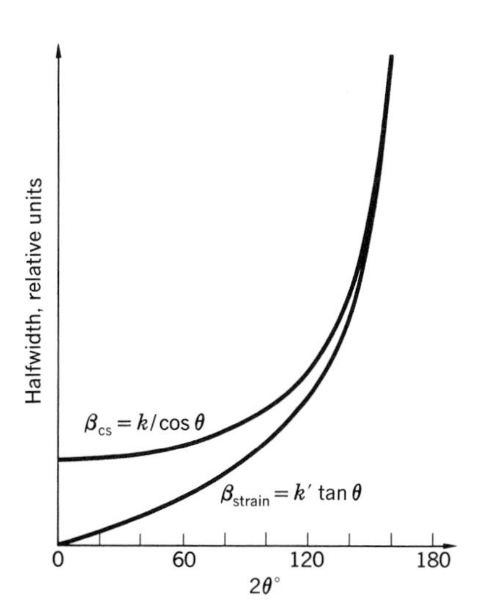

Fig. 20-31. Comparison of angular dependence of reflection halfwidths, caused by crystallite-size effects and by strains in the crystallites.

of strains or other causes of variable variations in the interplanar spacings renders such analyses less reliable. Keeping in mind that crystallite-size effects and strain broadening frequently go together, the utilization of smaller diffraction angles is indicated. For more precise analyses, both effects can be evaluated by realizing that the composite profile can be obtained by a convolution of the individual curves. For example, F. R. L. Schoening assumed that the crystallite-size effect has the shape of a Cauchy distribution, $(1 + k^2x^2)^{-1}$, while the strain broadening has a Gaussian shape, $(1 + k^2x^2)^{-2}$. Substituting these functions in (20-12) and using Fourier transform theory then gives the following relation between the two kinds of breadths:

$$\beta_{\text{meas}} = \frac{(2\beta_{\text{strain}} + \beta_{\text{cs}})^2}{4\beta_{\text{strain}} - \beta_{\text{cs}}} \tag{20-24}$$

This expression was derived for integral breadths by Schoening, who also has shown that, by utilizing the Scherrer equation (20-4) and relation (20-23) and doing some algebra, it is possible to employ (20-24) and a pair of observed reflections to determine the individual breadths and the average crystallite size, even in the presence of strain broadening or some other causes of reflection broadening.

RESIDUAL-STRESS ANALYSIS

Elastic deformation. The application of an external stress to a single crystal can produce a temporary deformation which disappears when the stress is removed. Whether a stressed crystal deforms elastically or plastically depends on the magnitude of the stress and the nature of the interatomic bonds in the crystal. Under normal loads, metal crystals rarely deform elastically without some plastic deformation also taking place. On the other hand, many nonmetallic crystals can be deformed elastically quite easily.

When a crystal is stressed, it exerts equal and opposite forces that resist further deformation, so that, at equilibrium, the two sets of forces are balanced and equal. If a particular plane in the crystal is considered, it is possible to resolve the external stress into three mutually orthogonal components. The one normal to the plane is called the *normal stress* σ, and the two lying in the plane are called *shear stresses* τ. It is always possible to select three mutually perpendicular planes such that the shear stress for these planes is zero. These three planes are called the three principal planes, and the normal stresses acting on each plane are the three *principal stresses* of the crystal. For example, consider a cylindrical crystal that is being compressed along the cylinder axis. The plane normal to the compressive stress and any two mutually orthogonal planes that are parallel to the compression direction can be selected as the principal planes. In this simple example, the principal stresses on the last two planes are zero, while the principal stress on the plane normal to the compression direction is equal to the applied force divided by the area of the plane. If any other plane in such a crystal is selected, both the normal and the shear stresses have nonzero values.

The *normal strain* in the crystal, ε, is defined as the ratio of the change in its length Δl to the original length l,

$$\varepsilon = \frac{\Delta l}{l} \tag{20-25}$$

According to Hooke's law, the normal strain produced by an applied stress, $\sigma = F/A$, is

$$\varepsilon = \frac{\sigma}{E} \tag{20-26}$$

where E is Young's modulus.

The principal stresses σ_1, σ_2, σ_3 in an isotropic body are related to the principal strains by

$$\begin{aligned}\varepsilon_1 &= \frac{1}{E}[\sigma_1 - \nu(\sigma_2 + \sigma_3)] \\ \varepsilon_2 &= \frac{1}{E}[\sigma_2 - \nu(\sigma_3 + \sigma_1)] \\ \varepsilon_3 &= \frac{1}{E}[\sigma_3 - \nu(\sigma_1 + \sigma_2)]\end{aligned} \tag{20-27}$$

where ν is Poisson's ratio,

$$\nu = \frac{E}{2G} - 1 \tag{20-28}$$

and G is the shear modulus of elasticity. Values of the different elastic constants for various materials and a further discussion of elasticity theory can be found in some of the books listed at the end of this chapter.

When an external stress is applied to a polycrystalline material, the strains set up in the individual crystallites depend on a number of factors, not all of which are fully understood. During usual deformations, a considerable amount of plastic deformation takes place, and after the external forces are removed, internal stresses set up in the crystallites tend to remain. There is no nondestructive way to measure the strains produced by these *residual stresses*, except by x-ray diffraction procedures. Mechanical methods employing strain gauges require successive sectioning of the sample to establish the stress distributions. X-ray diffraction methods, however, allow one to measure the elongations (20-25), caused by the strains in individual grains, by noting changes in the interplanar spacings of appropriately oriented planes. As an example, consider a bar in tension, with the tensile stress applied in the z direction, as shown in Fig. 20-32. To measure ε_z according to (20-25) one should measure the change in the interplanar spacing of some planes oriented at right angles to the z axis.

$$\varepsilon_z = \frac{\Delta d}{d_0} = \frac{d - d_0}{d_0} \tag{20-29}$$

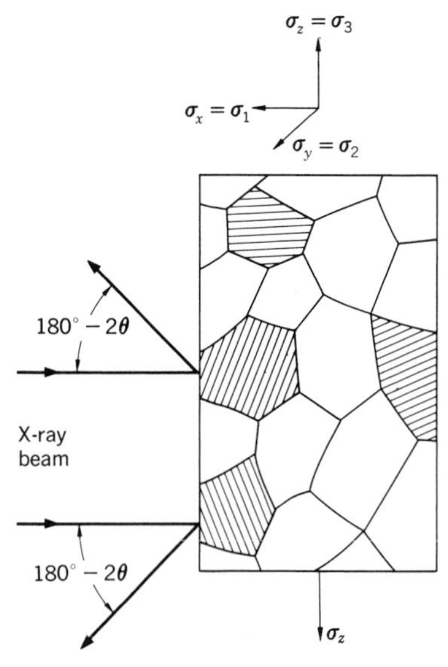

Fig. 20-32

where d is the spacing in the stressed crystallite, and d_0 in a stress-free one. Unfortunately, it is usually not possible to obtain reflections from such planes, so that one makes use of the relations (20-27) and measures one of the other principal strains. In the present case of a uniaxial stress parallel to z (Fig. 20-32), it follows that $\sigma_x = \sigma_y = 0$ and $\varepsilon_x = \varepsilon_y = -(\nu/E)\sigma_z$, while $\varepsilon_z = (1/E)\sigma_z$, so that

$$\varepsilon_z = -\frac{1}{\nu}\varepsilon_x \tag{20-30}$$

To measure ε_x one merely needs to measure the interplanar spacing d of a crystallite whose reflecting planes are normal to the x direction. This can be done most conveniently using a back-reflection camera and a reflection having 2θ as close to $180°$ as possible. The experimental parameters usually are chosen to optimize this condition, as discussed in the last section of this chapter. It should be noted here that the magnitude of Δd in (20-29), produced even by the largest stresses, is nevertheless quite small in most hard materials. Therefore the use of back-reflection methods is dictated also by the need to attain maximum precision in the measurements. (See Chap. 17.) The value of d_0 in the stress-free crystallites can be determined on the same material before the external stress is applied, or it can be calculated, whenever the precise lattice constant of the sample is known beforehand.

In a more general case than the one considered in the above example, the relation between the principal stress and the stress along a particular direction σ is determined by the stress ellipsoid shown in Fig. 20-33. Suppose that σ_3 is directed normal to the surface of a polycrystalline specimen. Since there is no stress possible at a free surface (in the absence of an external load), $\sigma_3 = 0$. According to (20-27), the strain is not zero, and

$$\varepsilon_3 = -\frac{\nu}{E}(\sigma_1 + \sigma_2) \tag{20-31}$$

From this it can be seen that a measurement of the interplanar-spacing change in planes parallel to the free surface (20-30) measures the sum of the other two principal stresses.

Normally, one wishes to measure the strain ε_ϕ parallel to the surface of the sample or, by measuring the strains along two mutually orthogonal directions, determine the principal strains. (Note that one can construct a strain ellipsoid similar to the stress ellipsoid in Fig. 20-33.) The relationship between the strain ε_ψ in a direction that forms the angle ψ with ε_3 and lies in the plane containing ε_ϕ (Exercise 20-10) is given by

$$\varepsilon_\psi - \varepsilon_3 = \frac{1+\nu}{E}(\sigma_1 \cos^2\phi + \sigma_2 \sin^2\phi)\sin^2\psi \tag{20-32}$$

Examination of the stress ellipsoid in Fig. 20-33 shows that

$$\sigma_\phi = \sigma_1 \cos^2\phi + \sigma_2 \sin^2\phi \tag{20-33}$$

so that combining (20-33) with (20-32),

$$\sigma_\phi = (\varepsilon_\psi - \varepsilon_3)\left(\frac{E}{1+\nu}\sin^2\psi\right) \tag{20-34}$$

The above stress-strain relationships are strictly applicable only to elastically isotropic crystallites and polycrystalline bodies. Although these conditions are rarely met in practice, the relations prove to be a most satisfactory approximation, and the elastic constants thus determined provide useful design criteria.

The relationship between the interplanar spacing $d_\perp$ and in the direction parallel to ε_3 (normal to the sample surface) and d_0 in the stress-free case is $(d - d_0)/d_0$, according to (20-29). Similarly, $\varepsilon_\psi = (d_\psi - d_0)/d_0$, according to (20-29). Substituting these quantities in (20-34) gives

$$\begin{aligned}\sigma_\phi &= K_\psi\left(\frac{d_\psi - d_0}{d_0} - \frac{d_\perp - d_0}{d_0}\right)\\ &= K_\psi\left(\frac{d_\psi - d_\perp}{d_0}\right)\end{aligned} \tag{20-35}$$

where $$K_\psi = \frac{E}{1+\nu}\sin^2\psi \tag{20-36}$$

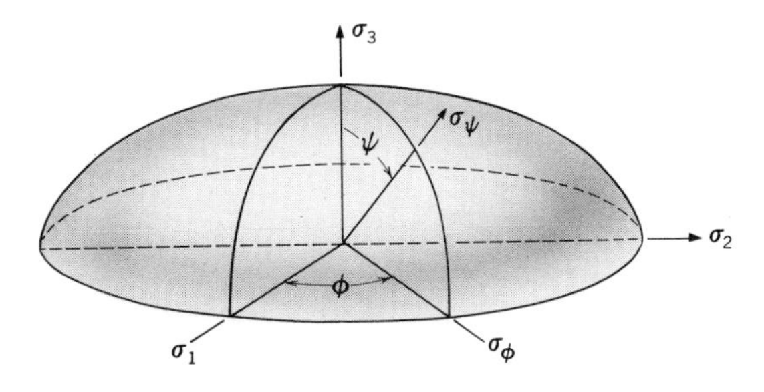

Fig. 20-33

Two-exposure method. The difference between d_ψ and $d_\perp$ in (20-35) compared with the magnitude of d_0 is so small that an accurate knowledge of d_0 is not as essential as that of the two d values in the numerator. For this reason it is possible to replace d_0 in the denominator of (20-35) by $d_\perp$ to give

$$\sigma_\phi = K_\psi \frac{d_\psi - d_\perp}{d_\perp} \tag{20-37}$$

An examination of this relation shows that only two measurements are necessary. One establishes $d_\perp$ from a back-reflection photograph at normal incidence, and then determines d_ψ from a back-reflection photograph with the incident beam inclined at an angle ψ to the normal direction. To the extent that $d_\perp \cong d_0$, a measurement using an unstressed material is not necessary, so that the entire analysis can be made on the sample directly. This means that such studies can be carried out equally easily on test samples subjected to known loads in the laboratory, to provide an accurate calibration of K_ψ, or on metal girders in existing structural units, provided that suitably portable x-ray equipment is available.

The two-exposure method requires a very simple camera, as shown in Fig. 20-34. By suitably curving the back of the camera, it is possible to obtain a parafocusing condition (Exercise 20-11) which helps the precise determination of the diffraction angle by sharpening the reflection. This is an important consideration because the presence of "macro" stresses, in addition to the residual "micro" stresses, usually causes line broadening in the back-reflection region. Note in

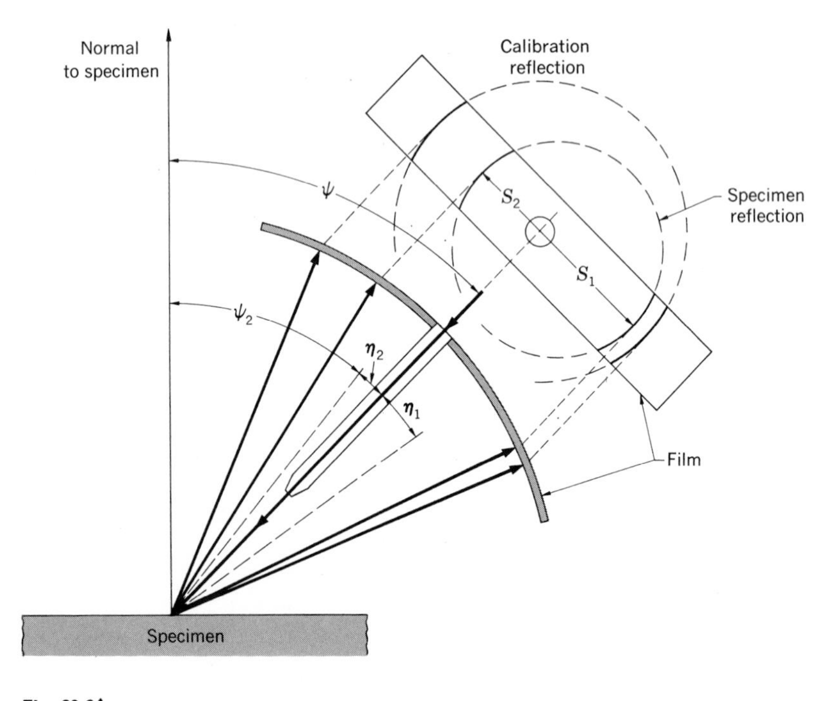

Fig. 20-34

the ray diagram and the plan view in Fig. 20-34 that the powder ring diffracted by the sample is not circular. This is so because the strains in the crystallites contributing to the two extreme portions of the ring are slightly different. Usually, the more sensitive (larger-strain) value of S_1 is used. To simplify the analyses in practice, it is convenient to define a new constant in (20-37),

$$K = \frac{E \cot \theta}{2R(1 + \nu) \sin^2 \psi} \tag{20-38}$$

in terms of which (Exercise 20-12) the stress is given by

$$\sigma_\phi = K(S_\psi - S_\perp) \tag{20-39}$$

where S_ψ and $S_\perp$ are the arc lengths of the reflection, for which K was derived, in the inclined and in the normal orientations, respectively. [Note that, for a flat-film cassette, $K_{ff} = K \cos^2 2\theta$, as can be readily verified by differentiating $S = R \tan 2\theta$ in the derivation of (20-38).]

The two-exposure method also can be adopted to a diffractometer arrangement. As illustrated in Fig. 20-35, at the appropriate diffraction angle the sample is inclined so that its normal bisects the supplement of 2θ. This means that the reflection comes from planes parallel to the sample surface, and $d_\perp = d_0$ exactly. With the sample inclined by ψ, it can be seen to the right in Fig. 20-35 that the diffracted beams come to a focus at a point considerably removed from the diffractometer circle. This means that the detector must be moved in closer so that it can accurately locate the peak position, and this procedure is strongly recommended for precise measurements. A calibrated bracket for doing this and a special sample holder for residual-stress analysis are shown mounted on a diffractometer in Fig. 20-36.

When a diffractometer is used, the experimental constant in (20-37) remains unchanged, but as shown in Exercise 20-14, the stress is given by

$$\sigma_\phi = K(2\theta_\perp - 2\theta_\psi) \tag{20-40}$$

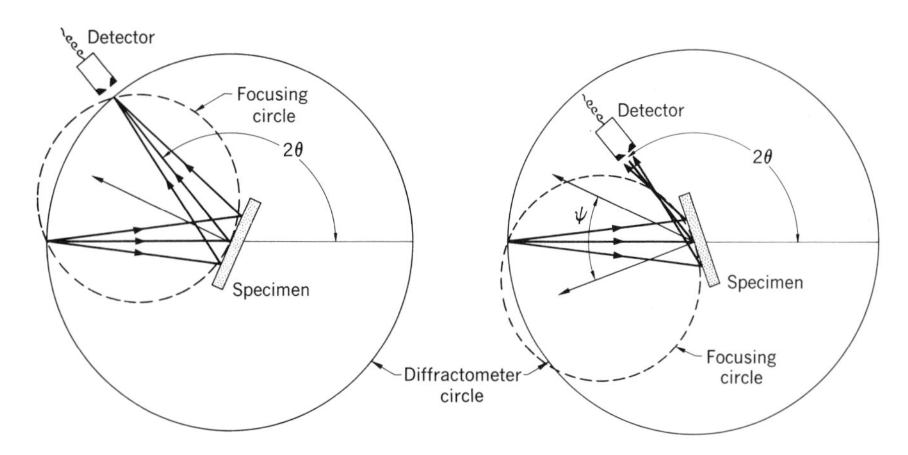

Fig. 20-35

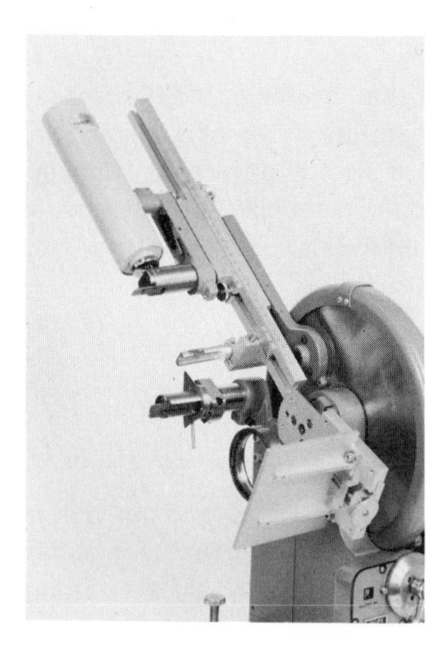

Fig. 20-36. AMR stress attachment for Norelco diffractometer. *(Courtesy of Advanced Metals Research Corporation.)*

where $2\theta_\perp$ and $2\theta_\psi$ are the diffractometer angles in the normal and inclined orientations. Note that the convention regarding the sign of the stress is preserved in (20-40) since the arc lengths in (20-39) are measures of the supplement of 2θ. Because the accuracy of the stress analysis depends rather markedly on the precision with which diffraction angles can be measured (Exercises 20-13 and 20-14), it is advisable to calibrate the diffractometer using a stress-free sample reflecting in the same angular range. This is particularly necessary when the detector is inclined by ψ from its usual orientation. Although, in general, photographic methods are easier to use to achieve maximum precision in lattice-constant measurements (Chap. 17) when changes in the diffraction angle of a single reflection are sought, the better resolution of the diffractometer makes it more precise. Unfortunately, it is not possible to retain this precision in a portable diffractometer, so that back-reflection cameras and portable x-ray tubes (Fig. 20-37) are commonly used for residual-stress analyses in practice.

Single-exposure method. An examination of the angular relations in the back-reflection region (Fig. 20-34) and in relation (20-35) shows that it is possible to determine the stress also from a single photograph. Consider the two angles

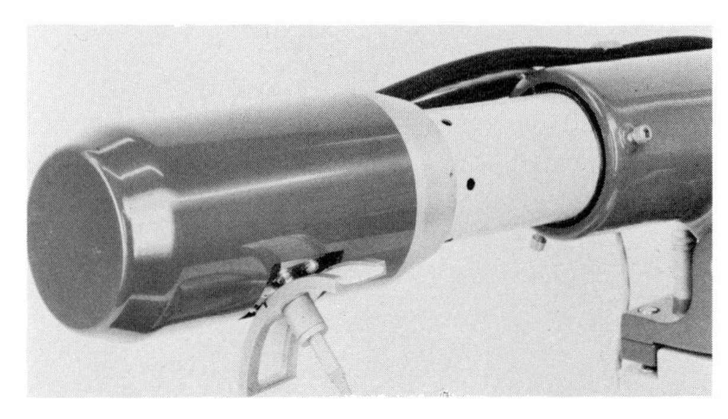

Fig. 20-37. AMR portable x-ray residual-stress analyzer. *(Courtesy of Advanced Metals Research Corporation.)*

$\psi_1 = \psi + \eta_1$ and $\psi_2 = \psi - \eta_2$ shown in Fig. 20-34. Substituting them in (20-35) gives

$$\sigma_\phi = \frac{K_{\psi_1}(d_1 - d_\perp)}{d_0}$$

$$\text{and} \qquad \sigma_\phi = \frac{K_{\psi_2}(d_2 - d_\perp)}{d_0} \tag{20-41}$$

where the appropriate value of ψ must be substituted in (20-36) to determine the coefficients. By making the approximation $\eta = \eta_1 = \eta_2 = 90° - \theta$, it is possible to combine the two simultaneous equations in (20-41) to give

$$\sigma_\phi = \frac{K'_\psi(d_1 - d_2)}{d_0} \tag{20-42}$$

where $$K'_\psi = \frac{E}{1+\nu} \sin 2\psi \sin 2\theta \tag{20-43}$$

as shown in Exercise 20-15. Note that when $\psi = 45°$, $\sin 2\psi = 1.0$ and (20-43) is further simplified. To use (20-42), a knowledge of d_0 in a stress-free sample is required. In most cases this can be calculated from the known lattice constants of the material.

A comparison of (20-43) and (20-36) shows that they have somewhat different angular dependencies on ψ and on θ. Selecting $\psi = 45°$ for experimental and analytical convenience,

$$K'_{45°} = \frac{K_{45°}}{2 \sin 2\theta} \tag{20-44}$$

Relationship (20-44) shows that, when $2\theta = 150°$, the two coefficients are equal. As the diffraction angle increases, however, so does K'_ψ, whereas K remains constant. An examination of the more convenient form of this coefficient in (20-38) shows that, in fact, it decreases as the diffraction angle increases. This means that the two-exposure method is more accurate, at least in principle. Its chief drawback lies in the necessity to know the correct value of R in (20-38) which must be established for each exposure separately. Since the same value of R is used in calculating the d values for (20-42), the one-exposure method may give comparable accuracy when a portable instrument is used in a difficult location.

Practical considerations. In order to eliminate the dependence on knowing R precisely, it is possible to place a material whose lattice constant is known in contact with the sample's surface and to record reflections from both on the same film. Suitable materials are aluminum and the noble metals, because their lattice constants are accurately known as a function of temperature. Either foils, suitably cross-rolled to minimize preferred orientation and annealed for stress relief, or powders can be used for this purpose. (When the surface is vertical, a paste of the metal powder in collodion or some other binder can be used.) One then measures the distance along the film from the standard powder ring (Fig. 20-34) to determine S_1 and S_2.

The condition of the sample's surface is extremely important. It should be as flat as possible because excrescences from the surface contribute more to the reflection intensity than does the underlying bulk, but they are less strained and so give rise to inaccuracies. In preparing the surface by successive polishing and etching, to remove the disturbed layers, care must be exercised not to adulterate the stresses being sought. Electropolishing procedures should be used whenever possible. It can be shown that the depth of x-ray penetration in the sample's surface goes approximately as

$$x \cong \frac{-\ln(1-g)}{2\mu} \tag{20-45}$$

where g measures the fraction of the total intensity reflected by a layer of thickness x. As an order of magnitude, suppose that the location of the reflection's maximum is fixed by approximately 87% of the reflection intensity, so that $g = 0.87$ and $x = 1/\mu$. When one then considers the change in pathlength as a result of inclining the beam, it is clearly evident that a relatively thin layer near the sample's surface contributes the bulk of the reflection intensity. Note that 50% of it comes from a layer only approximately one-third as thick.

As a typical example of some of the practical factors one should consider, Table 20-1 lists some of the values that can be utilized in the residual-stress analysis of steels. Thus it is seen that the maximum penetration and highest accuracy can be attained by using Mo $K\alpha$ radiation. One of the drawbacks to this choice is that the relative intensity of the 800 reflection is lower than that of some of the alternatives listed. Also, the presence of a larger hump in the continuous spectrum, emanating from a molybdenum target, increases the background intensity due to fluorescence by the sample. Also, Mo $K\alpha$ radiation is absorbed by the photographic film less efficiently, requiring longer exposure times, a condition that is aggravated even further when an aluminum foil is interposed to absorb most of the iron-fluorescence radiation. An important consideration favoring the shorter-wavelength radiation is its better dispersion. Thus, in a camera having a 50-mm sample-to-film distance, the $K\alpha_1\alpha_2$ doublet is separated by 4.5 mm for the 800 reflection and Mo radiation, but only by 1.5 mm for 310 and Co or 1.0 mm for 211 and Cr radiation. Because of the stresses present in the sample studied, however, wider separation of somewhat broadened $\alpha_1\alpha_2$ peaks often means that the center of a relatively broader band must be located in place

Table 20-1 X-ray values for residual-stress analysis of steels

Target	$\lambda_{K\alpha}$, Å	hkl	2θ	μ, cm^{-1}
Mo	0.708	800	163.48°	38.5
Co	1.785	310	161.26	58.2
Fe	1.932	220	145.50	71.2
Cr	2.285	211	158.02	115.0

of a narrower one. Thus, as in all practical aspects for x-ray diffraction, it is necessary to consider *all* the factors affecting the experiment before deciding on what the optimum conditions should be.

SUGGESTIONS FOR SUPPLEMENTARY READING

Charles S. Barrett and T. B. Massalski, *Structure of metals*, 3d ed. (McGraw-Hill Book Company, New York, 1966).

B. D. Cullity, *Elements of x-ray diffraction* (Addison-Wesley Publishing Company, Inc., Reading, Mass., 1956).

H. R. Isenberg, *Bibliography on x-ray stress analysis* (St. John X-Ray Laboratory, Califon, N.J., 1949).

Harold P. Klug and Leroy E. Alexander, *X-ray diffraction procedures for polycrystalline and amorphous materials* (John Wiley & Sons, Inc., New York, 1954).

H. S. Peiser, H. P. Rooksby, and A. J. C. Wilson, *X-ray diffraction by polycrystalline materials* (Reinhold Publishing Corporation, New York, 1955).

F. R. L. Schoening, Strain and particle size values from x-ray line breadths, *Acta Crystallographica*, vol. 18 (1965), pp. 975–976.

A. Taylor, *X-ray metallography* (John Wiley & Sons, Inc., New York, 1961).

B. E. Warren, X-ray studies of deformed metals, *Progress in Metal Physics*, vol. 8 (1959), pp. 147–202.

EXERCISES

20-1. Assuming an x-ray beam normal to the wire, construct an idealized powder photograph of a cubic metal wire having $[100]$ fiber texture and a primitive lattice. How does this photograph differ if the metal has a face-centered cubic lattice?

20-2. Assuming an x-ray beam parallel to the wire in Exercise 20-1, construct an idealized powder photograph for the case of a face-centered cubic lattice.

20-3. A transmission photograph of a cold-drawn aluminum wire was prepared with Cu $K\alpha$ radiation ($\lambda = 1.54$ Å). From the radius of the innermost ring it was established that $2\theta = 38.6°$ and the four symmetrically placed arcs were inclined to the wire axis by $\delta = 69°$. The second ring had $2\theta = 44.6°$, and the arcs were inclined by $\delta = 51°$. Given that the lattice constant of aluminum $a = 4.05$ Å, what are the indices of the two observed reflections? What are the inclinations of their planes to the wire axis? Finally, what is the fiber texture?

20-4. By means of a suitable sketch, show that the film method normally employed in studies of sheet texture (Fig. 20-13) is equivalent to taking a stepwise-rotation photograph of a single crystal. Is it possible to explore the entire reciprocal-lattice sphere by this means? What is the latitude at which the blind region appears in the pole figure if the diffraction angle $2\theta = 20°$? If $2\theta = 36°$?

20-5. Repeat the reciprocal-lattice construction shown in Fig. 20-21 for the case when $-\alpha = 20$, 40, 60, and 80°. At about what point does the path through the sample become too large to be useful in this arrangement? Suppose the sheet is held in a sample holder four times as thick as the sheet, what is the maximum inclination angle that is practical when $2\theta = 20°$? Does this situation improve or get worse as θ increases? What can you suggest to minimize this problem?

20-6. When the transmission arrangement shown in Fig. 20-21 is employed, rotation about the sheet normal has no effect but changing the inclination by an amount α does. Show that the intensity of the beam transmitted through a sheet of thickness t when $\alpha = 0$ is given by

$$I_{\alpha=0} \propto \frac{t}{\cos\theta} I_0 e^{-\mu t/\cos\theta} \tag{20-46}$$

and at other angles of α by

$$I_{\alpha\neq 0} \propto \left(\frac{I_0}{\mu}\right) \frac{e^{-\mu t/\cos(\theta-\alpha)} - e^{-\mu t/\cos(\theta+\alpha)}}{[\cos(\theta-\alpha)/\cos(\theta+\alpha)] - 1}$$

What one normally requires is a plot of $I_{\alpha \neq 0}/I_{\alpha=0}$ for the particular values of the sample being studied, so that one can correct the measured $I_{\alpha \neq 0}$ values. For Cu $K\alpha$ and an aluminum sheet, typical values for the 111 reflection are $\mu t = 1.0$ and $\theta = 19.25°$. Plot this ratio for aluminum for $-\alpha$ ranging 0 to 90°. At what α angle does the correction exceed 50%?

20-7. Show that the assumption of an error curve for the functions in (20-13) leads to (20-19).

20-8. Derive the relation between the various breadths by assuming for the functions in (20-13) $f(x) = (1 + x^2)^{-1}$ and $g(\alpha) = (1 + k^2y^2)^{-1}$.

20-9. Suppose that $\beta_{\text{instr}}/\beta_{\text{meas}}$ turns out to be 0.8 and that the experimentally measured breadths can be measured repeatedly to within ±5%. What is the error in determining β_{cs}? Compare your answer using the three curves shown in Fig. 20-30.

20-10. Construct a strain ellipsoid similar to Fig. 20-33. The normal strain ε in any chosen direction is then given by the approximate equation

$$\varepsilon = q_1^2\varepsilon_1 + q_2^2\varepsilon_2 + q_3^2\varepsilon_3 \qquad \text{(20-47)}$$

where the direction cosines in terms of similarly chosen angles ϕ and ψ are given by

$$q_1 = \sin\psi \cos\phi$$
$$q_2 = \sin\psi \sin\phi \qquad \text{(20-48)}$$
$$q_3 = \cos\psi$$

Making use of these relations, derive (20-32) in the text.

20-11. It is possible to obtain parafocusing at one angle in the back-reflection region using a flat-plate camera. Supposing that a steel sample is to be examined with Co $K\alpha$ radiation in a camera whose film-to-sample distance is 50 mm, at what distance from the sample should one place the pinhole defining the incident beam?

20-12. To simplify the measurement of back-reflection lines for stress analysis, it is convenient to use an equation containing the diffraction-arc distances $S_\perp$ and S_ψ directly. By differentiating the Bragg equation and the arc-length relation $S = R(\pi - 2\theta)$, where R is the radius of the camera, and letting $d_\psi - d_\perp = \Delta d$, derive relations (20-38) and (20-39) in the text.

20-13. For steel, typical values of the elastic constants are $E = 30 \times 10^6$ psi, $\nu = 0.28$. Supposing that a curved back-reflection camera is used for a residual-stress analysis by the two-exposure method, what is the magnitude of the stress if the arc lengths differ by 1.2 mm between two exposures at $\psi = 90°$ and $\psi = 45°$? What is the experimental error of the measurement if $R = 57.3 \pm 0.5$ mm? Does the magnitude of the error depend on the magnitude of the stress present?

20-14. Differentiating the Bragg equation and expressing it in terms of $\Delta 2\theta = 2\Delta\theta$ and following the procedure in Exercise 20-12, derive Eq. (20-40). Using the values given in Exercise 20-13, what is the experimental error in the measured stress if 2θ can be read to an accuracy of 0.01°? How does this compare with the error using film methods? What explains the difference in values?

20-15. By substituting ψ_1 and ψ_2, respectively, in the two relations in (20-41), expanding the trigonometric functions, and combining them to eliminate $d_\perp$, derive relations (20-42) and (20-43).

APPENDIXES

APPENDIX **ONE**

Mathematical relations

VECTOR ALGEBRA

Properties of vectors. It is possible to distinguish two kinds of mathematical quantities: One is called a *scalar* quantity and *possesses only a magnitude*, for example, the volume of a unit cell. The other, called a *vector* quantity, *possesses a magnitude as well as a direction*, for example, one of the three unit-cell vectors chosen to define the cell. It is customary to use boldface type to denote a vector, for example, the cell edge $\mathbf{a}$, and italic type to denote a scalar quantity, for example, the length a of the cell edge. In general, the vector $\mathbf{a}$ extends from some point O to another, A (Fig. A-1), so that it has a length and a direction. To add to the vector $\mathbf{a}$ another vector $\mathbf{b}$ (extending from A to B) then requires placing the second vector at the tip of the first. As can be seen in Fig. A-1, one can arrive at B, the termination of the second vector, alternatively, by following the vector $\mathbf{c}$ from O to B directly. This means that

$$\mathbf{a} + \mathbf{b} = \mathbf{c} \tag{A-1}$$

As shown to the right in Fig. A-1, vector addition is commutative, so that

$$\mathbf{a} + \mathbf{b} = \mathbf{b} + \mathbf{a} \tag{A-2}$$

After inverting the direction of a vector without changing its magnitude, it becomes the negative of the initial vector. Thus the vector $-\mathbf{a}$ in Fig. A-1 is the vector A to O. Just as in ordinary algebra, one can subtract one vector from

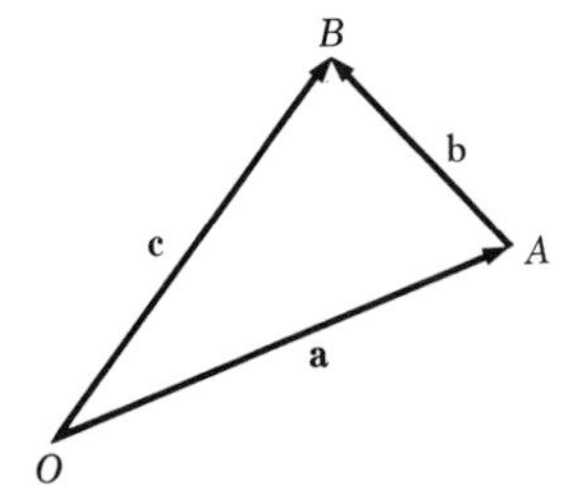

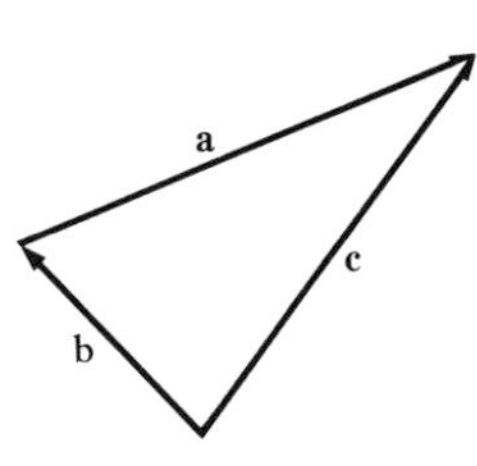

Fig. A-1

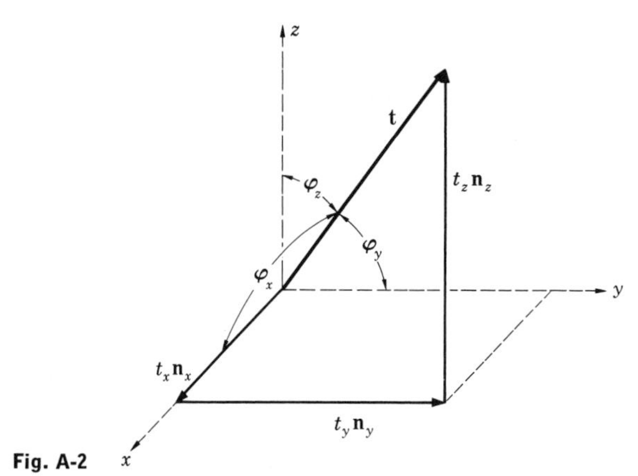

Fig. A-2

another by adding a negative vector. For example,

$$\mathbf{c} - \mathbf{a} = \mathbf{c} + (-\mathbf{a}) = \mathbf{b} \tag{A-3}$$

according to Fig. A-1.

Since vectors possess directions, it is necessary to refer them to a coordinate system so as to describe their orientations in space. Choosing a right-handed rectangular (cartesian) coordinate system (Fig. A-2), it is possible to describe a vector **t** in terms of its components (projections) t_x, t_y, and t_z, respectively, along the three axes x, y, and z.

$$\mathbf{t} = t_x\mathbf{n}_x + t_y\mathbf{n}_y + t_z\mathbf{n}_z \tag{A-4}$$

where $\mathbf{n}_x$, $\mathbf{n}_y$, and $\mathbf{n}_z$ are *unit vectors* along the three axes. (A unit vector is a vector whose length is equal to unity, so that any vector can be represented as a product of a scalar, its length, with a unit vector denoting its direction.)

If the vector **t** forms the angles φ_x, φ_y, and φ_z with the three axes shown in Fig. A-2, its three components have the lengths

$$t_x = t \cos \varphi_x \qquad t_y = t \cos \varphi_y \qquad t_z = t \cos \varphi_z \tag{A-5}$$

The length of the vector then can be determined with the aid of the geometrical relationship

$$t^2 = t_x^2 + t_y^2 + t_z^2 \tag{A-6}$$

from which it is seen that the three direction cosines in (A-5) must obey the relation

$$\cos^2 \varphi_x + \cos^2 \varphi_y + \cos^2 \varphi_z = 1 \tag{A-7}$$

Combining (A-5) and (A-6),

$$\begin{aligned}
\cos \varphi_x &= \frac{t_x}{(t_x^2 + t_y^2 + t_z^2)^{\frac{1}{2}}} \\
\cos \varphi_y &= \frac{t_y}{(t_x^2 + t_y^2 + t_z^2)^{\frac{1}{2}}} \\
\cos \varphi_z &= \frac{t_z}{(t_x^2 + t_y^2 + t_z^2)^{\frac{1}{2}}}
\end{aligned} \tag{A-8}$$

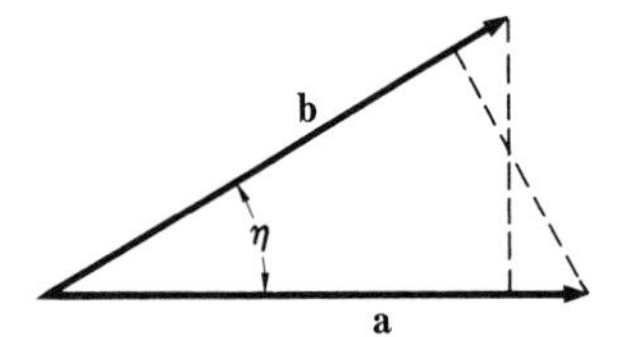

Fig. A-3

Scalar products. In vector multiplication, it is necessary to distinguish two kinds of products, one of which is a scalar, the other a vector. A *scalar product* between two vectors is denoted by placing a dot between them, and *is equal to the arithmetic product of their magnitudes multiplied by the cosine of the included angle.* Considering the two vectors in Fig. A-3, this can be expressed by

$$\mathbf{a} \cdot \mathbf{b} = ab \cos \eta \tag{A-9}$$

Note in Fig. A-3 that this means that the scalar or *dot product* in (A-9) is numerically equal to the length of $\mathbf{a}$ times the projection of $\mathbf{b}$ on $\mathbf{a}$, and alternatively, it is equal to the length of $\mathbf{b}$ times the projection of $\mathbf{a}$ on $\mathbf{b}$. It follows, therefore, that the scalar product is commutative,

$$\mathbf{a} \cdot \mathbf{b} = \mathbf{b} \cdot \mathbf{a} \tag{A-10}$$

and distributive,

$$\mathbf{a} \cdot (\mathbf{b} + \mathbf{c}) = \mathbf{a} \cdot \mathbf{b} + \mathbf{a} \cdot \mathbf{c} \tag{A-11}$$

Finally, from (A-9) one obtains the orthogonality condition that, if two vectors are mutually perpendicular, their dot product vanishes. This means, for example, that the dot products between the three unit vectors in Fig. A-2 can have the following values:

$$\begin{aligned} \mathbf{n}_x \cdot \mathbf{n}_x &= \mathbf{n}_y \cdot \mathbf{n}_y = \mathbf{n}_z \cdot \mathbf{n}_z = 1 \\ \mathbf{n}_x \cdot \mathbf{n}_y &= \mathbf{n}_x \cdot \mathbf{n}_z = \mathbf{n}_y \cdot \mathbf{n}_z = 0 \end{aligned} \tag{A-12}$$

Quite frequently, the dot is omitted, so that

$$\mathbf{a}^2 = \mathbf{a} \cdot \mathbf{a} = |\mathbf{a}|^2 = a^2 \tag{A-13}$$

where the sign $||$ means the absolute magnitude of $\mathbf{a}$.

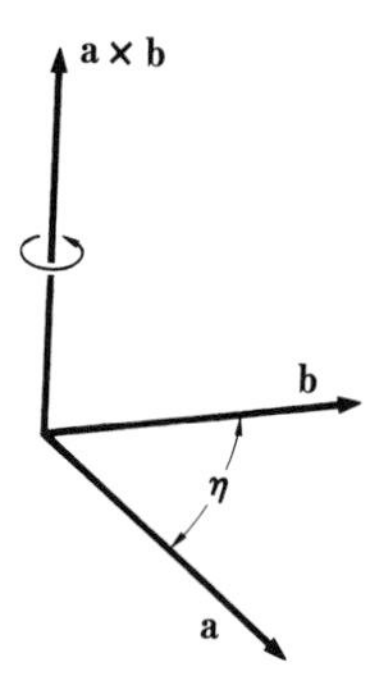

Fig. A-4

Vector products. The *vector* or *cross product* between two vectors $\mathbf{a} \times \mathbf{b}$ is defined as a vector perpendicular to the plane containing $\mathbf{a}$ and $\mathbf{b}$ whose magnitude is the arithmetic product of the lengths of the two vectors times the sine of the included angle. The positive direction of the vector $\mathbf{a} \times \mathbf{b}$ is determined by the right-hand rule, as illustrated in Fig. A-4. Accordingly,

$$\mathbf{a} \times \mathbf{b} = ab\,|\sin \eta| \tag{A-14}$$

so that the length of $\mathbf{a} \times \mathbf{b}$ is equal to the area of the parallelogram bounded by $\mathbf{a}$ and $\mathbf{b}$. Because the direction of the cross product is determined by the right-hand rule (Fig. A-4), it is not commutative, so that

$$\mathbf{b} \times \mathbf{a} = -(\mathbf{a} \times \mathbf{b}) \tag{A-15}$$

By considering the relative areas, it is easy to show that the cross product is distributive.

$$\mathbf{a} \times (\mathbf{b} + \mathbf{c}) = \mathbf{a} \times \mathbf{b} + \mathbf{a} \times \mathbf{c} \tag{A-16}$$

Finally, according to (A-14), the cross product of two parallel vectors is zero, so that, for the three unit vectors in Fig. A-2,

$$\begin{gathered} \mathbf{n}_x \times \mathbf{n}_x = \mathbf{n}_y \times \mathbf{n}_y = \mathbf{n}_z \times \mathbf{n}_z = 0 \\ \mathbf{n}_x \times \mathbf{n}_y = -\mathbf{n}_y \times \mathbf{n}_x = \mathbf{n}_z \qquad \mathbf{n}_y \times \mathbf{n}_z = \mathbf{n}_x \qquad \mathbf{n}_z \times \mathbf{n}_x = \mathbf{n}_y \end{gathered} \tag{A-17}$$

Multiple products. There are, essentially, three types of products possible when three vectors are multiplied by each other. Of these, $(\mathbf{a} \cdot \mathbf{b})\mathbf{c}$ is simply the product of the scalar $(\mathbf{a} \cdot \mathbf{b})$ with the vector $\mathbf{c}$ and serves to increase its magnitude without affecting its direction. The *triple vector product* $\mathbf{a} \times \mathbf{b} \times \mathbf{c}$ is a vector whose direction can be worked out by the reader. Since it has no importance in x-ray crystallography, its properties are not discussed further here. The *triple scalar product*

$$(\mathbf{a} \times \mathbf{b}) \cdot \mathbf{c} \tag{A-18}$$

is the scalar product between the two vectors $\mathbf{a} \times \mathbf{b}$ and $\mathbf{c}$. Recalling that the magnitude of the cross product $\mathbf{a} \times \mathbf{b}$ equals the area of the parallelogram bounded by $\mathbf{a}$ and $\mathbf{b}$, while its dot product with c equals the length of $(\mathbf{a} \times \mathbf{b})$ times the projection of $\mathbf{c}$ onto it (Fig. A-5), it follows that the triple scalar product in (A-18) is numerically equal to the volume of the parallelepipedon bounded by $\mathbf{a}$, $\mathbf{b}$, and $\mathbf{c}$.

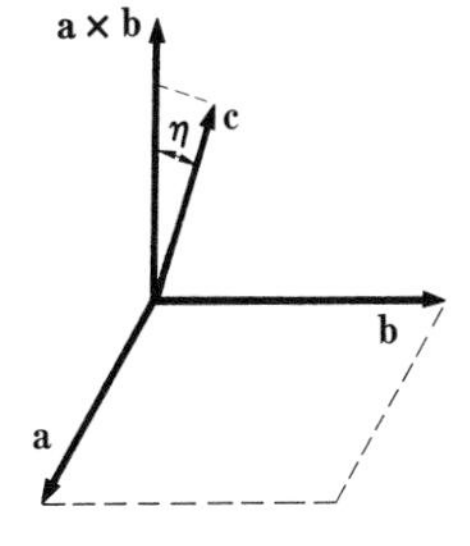

Fig. A-5

This has the consequence that the three vectors in (A-18) can be permuted cyclically without affecting the product.

$$(\mathbf{a} \times \mathbf{b}) \cdot \mathbf{c} = (\mathbf{b} \times \mathbf{c}) \cdot \mathbf{a} = (\mathbf{c} \times \mathbf{a}) \cdot \mathbf{b} \tag{A-19}$$

as can be readily verified by the reader with the aid of constructions like Fig. A-5.

COMPLEX VARIABLES

Complex plane. A complex quantity consists of a real component and an imaginary component which is usually expressed through the imaginary unit.

$$i = \sqrt{-1} \qquad \text{so that} \qquad i^2 = -1 \tag{A-20}$$

If the magnitude of the real part is x and that of the imaginary part is y, the complex quantity z can be expressed by

$$z = x + iy \tag{A-21}$$

It is most convenient to represent the complex quantity in (A-21) by a point x,y in a rectangular coordinate system called the *complex plane*. In this system (Fig. A-6), the real numbers are located along the x axis, positive numbers to the right of zero and negative numbers to the left, and the pure imaginary numbers along the y axis, with the positive sense upward. It is often convenient, therefore, to think of a complex quantity as a vector having the real part as its x component and the imaginary part as its y component. The length of this vector is called the absolute value of the complex quantity.

$$|z| = |x + iy| = (|x|^2 + |iy|^2)^{\frac{1}{2}} = (x^2 + y^2)^{\frac{1}{2}} \tag{A-22}$$

as can be seen directly in Fig. A-6.

It is often useful to consider the *conjugate* of a complex quantity, which is obtained by multiplying i by -1. Thus the complex conjugate of (A-21) is denoted

$$z^* = x - iy \tag{A-23}$$

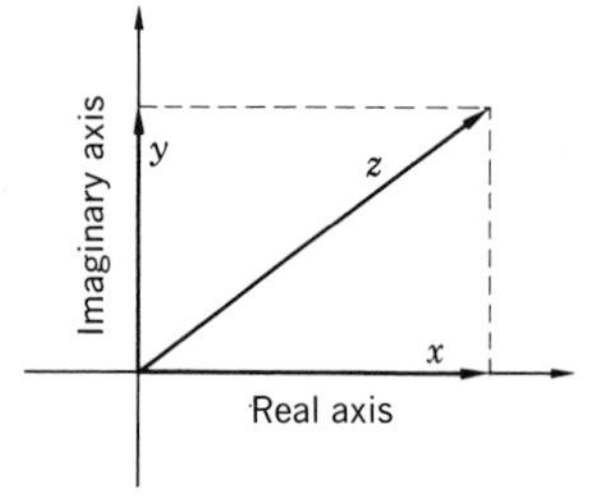

Fig. A-6

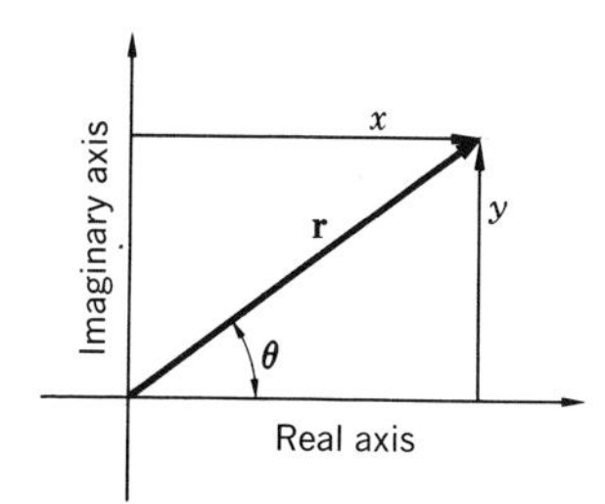

Fig. A-7

Multiplying a complex quantity (A-21) by its conjugate (A-23),

$$\begin{aligned} zz^* &= (x + iy)(x - iy) \\ &= x^2 - i^2y^2 \\ &= x^2 + y^2 \\ &= z^2 \end{aligned} \tag{A-24}$$

according to (A-22). This is an important result, because it means that *the magnitude of a complex quantity can be obtained most easily by determining the square root of the product with its conjugate.* Note also that the addition and subtraction of complex quantities obey the same rules as do the addition and subtraction of vectors, since one adds the x and y components separately before forming the resultant vector.

Circular functions. It is also possible to express a complex quantity in the complex plane using polar coordinates r and θ, as shown in Fig. A-7. With r always positive,

$$x = r\cos\theta \qquad y = r\sin\theta \tag{A-25}$$

so that

$$z = x + iy = r(\cos\theta + i\sin\theta) \tag{A-26}$$

Note that the minus sign is introduced by the sine or cosine values as r moves into the various quadrants in Fig. A-7.

According to Euler's theorem, the exponential of a complex quantity α,

$$e^{i\alpha} = \cos\alpha + i\sin\alpha \tag{A-27}$$

so that (A-26) can be written

$$z = r(\cos\theta + i\sin\theta) = re^{i\theta} \tag{A-28}$$

which is an alternative and most useful way of describing a complex quantity. By combining (A-27) with its complex conjugate $e^{-i\alpha}$, it is possible to obtain many other useful relations involving circular functions, for example,

$$\sin\alpha = \frac{e^{i\alpha} - e^{-i\alpha}}{2i} \tag{A-29}$$

and

$$\cos\alpha = \frac{e^{i\alpha} + e^{-i\alpha}}{2} \tag{A-30}$$

FOURIER THEORY

Sine series. Whenever one deals with periodically varying functions, for example, the electron density distribution in crystals, it is convenient to represent them by means of *Fourier series* which are also periodic. In effect, a Fourier series is the sum of an infinite number of harmonics, each suitably weighted by a *Fourier coefficient*, so that it is actually much more general than the introductory statement suggests, and Fourier series can be used to describe any well-behaved function by a suitable choice of coefficients. In x-ray crystallography Fourier series find many applications, so that their detailed characteristics are systematically reviewed below.

Suppose that a function of x can be represented in the interval $0 \leq x \leq 1$ by the series

$$f(x) = A_1 \sin \pi(1)x + A_2 \sin \pi(2)x + \cdots$$

$$= \sum_{n=1}^{\infty} A_n \sin \pi n x \qquad \text{(A-31)}$$

where n takes on all positive integer values.

The coefficients A_n necessary to describe the function correctly then can be determined by multiplying both sides of (A-31) by $\sin \pi m x \, dx$, where m is an integer, and integrating over the interval from $x = 0$ to $x = 1$.

$$\int_0^1 f(x) \sin \pi m x \, dx = \sum_{n=1}^{\infty} \int_0^1 A_n \sin \pi n x \sin \pi m x \, dx \qquad \text{(A-32)}$$

By direct integration it can be shown that

$$\int_0^1 \sin \pi n x \sin \pi m x \, dx = \begin{cases} 0 & \text{if } m \neq n \\ \frac{1}{2} & \text{if } m = n \end{cases} \qquad \text{(A-33)}$$

so that the coefficients in (A-31) are given by

$$A_n = 2 \int_0^1 f(x) \sin \pi n x \, dx \qquad n = 1, 2, 3, \ldots \qquad \text{(A-34)}$$

as can be seen by combining the result of (A-33) with (A-32). (Note that $A_0 = 0$ because $\sin 0 = 0$.)

The meaning of (A-34) is that the Fourier coefficients of a sine series are given by twice the average value of the product $f(x) \sin \pi n x$ in the interval $x = 0$ to $x = 1$. As is well known, the sine function has a natural period of 2π; it follows that each Fourier coefficient repeats itself at regular intervals that are twice as large as the interval of representation (integration) chosen above. Moreover, the original functional representation in (A-31) has the same period. Note also in (A-31) that, if x is replaced by $-x, f(-x) = -f(x)$. Functions having this property are called *odd functions*, to distinguish them from *even functions*, for which $f(-x) = f(x)$. From this it is clear that although (A-31) is an odd function, the Fourier coefficients in (A-34) are even functions, since the product of two odd functions is an even function.

In crystallographic applications, it is usually convenient to define the periodically repeated unit (in one dimension) by a so that the unique interval of representation becomes $a/2$. By defining the fractional variable $2x/a$ in place of x in (A-31), where the new x now takes on all values from $x = 0$ to $x = a$,

$$f(x) = \sum_{n=1}^{\infty} A_n \sin 2\pi n \frac{x}{a} \tag{A-35}$$

where $$A_n = \frac{2}{a} \int_0^a f(x) \sin 2\pi n \frac{x}{a}\, dx \qquad n = 1, 2, 3, \ldots \tag{A-36}$$

as can be verified by repeating the steps used to derive (A-34).

Cosine series. When it is desired to represent an even function of x in the interval $x = 0$ to $x = a$, it is necessary to use a Fourier cosine series.

$$f(x) = A_0 + A_1 \cos 2\pi(1) \frac{x}{a} + A_2 \cos 2\pi(2) \frac{x}{a} + \cdots$$

$$= \sum_{n=0}^{\infty} A_n \cos 2\pi n \frac{x}{a} \tag{A-37}$$

To determine the Fourier coefficients in (A-37), multiply both sides by $\cos 2\pi m(x/a)\, dx$ and integrate from $x = 0$ to $x = a$.

$$\int_0^a f(x) \cos 2\pi m \frac{x}{a}\, dx = \sum_{n=0}^{\infty} \int_0^a A_n \cos 2\pi n \frac{x}{a} \cos 2\pi m \frac{x}{a}\, dx \tag{A-38}$$

Again, it can be seen by direct integration that

$$\int_0^a \cos 2\pi n \frac{x}{a} \cos 2\pi m \frac{x}{a}\, dx = \begin{cases} 0 & \text{if } m \neq n \\ \dfrac{a}{2} & \text{if } m = n = 1, 2, 3, \ldots \\ a & \text{if } m = n = 0 \end{cases} \tag{A-39}$$

so that $$A_0 = \frac{1}{a} \int_0^a f(x)\, dx$$

$$A_n = \frac{2}{a} \int_0^a f(x) \cos 2\pi n \frac{x}{a}\, dx \qquad n = 1, 2, 3, \ldots \tag{A-40}$$

from which it follows that A_0 is the average value of $f(x)$ in the interval $x = 0$ to $x = a$, while the other coefficients A_n equal twice the average value of the product $f(x) \cos 2\pi nx$ in the same interval.

Exponential series. It should be clear that any function can be represented as a sum of an even function and an odd function. For example, let

$$f(x) = \tfrac{1}{2}[f(x) + f(-x)] + \tfrac{1}{2}[f(x) - f(-x)] \tag{A-41}$$

where the quantity in the first brackets clearly is an even function, while that in the second brackets is an odd function. Consider, then, a function whose even-function representation is given by $\sum_n A_n \cos nx$ and whose odd representation is $\sum_n B_n \sin nx$. The complete function, then, is given by

$$\rho(x) = \sum_n A_n \cos nx + \sum_n B_n \sin nx \tag{A-42}$$

Making use of (A-29) and (A-30), it can be shown that

$$\rho(x) = \sum_n [\tfrac{1}{2}(A_n + iB_n)e^{-inx} + \tfrac{1}{2}(A_n - iB_n)e^{inx}] \tag{A-43}$$

Defining $\quad F_n = A_n + iB_n \qquad$ (A-44)

it follows that

$$\rho(x) = \tfrac{1}{2}\sum_n F_n e^{-inx} + \tfrac{1}{2}\sum_n F_n^* e^{inx} \tag{A-45}$$

Finally, if $\rho(x)$ represents the electron-density distribution as a function of x, and F_n the one-dimensional complex "structure factor," the summations (A-45) should extend from minus to plus infinity. Note that identical terms will occur in each of the two sums for n and $-n$, so that the final sum can be represented equally well by

$$\rho(x) = \sum_{n=-\infty}^{\infty} F_n e^{-inx} \tag{A-46}$$

Note that the imaginary term in F has a plus sign in (A-46). Alternatively,

$$\rho(x) = \sum_{n=-\infty}^{\infty} F_n^* e^{inx} \tag{A-47}$$

according to (A-45). Thus it is seen that the imaginary parts of the structure factor and of the exponential term in a Fourier series representing an electron density must have opposite signs.

Fourier integral. Suppose that, in place of an infinite series, it is possible to use an infinite integral to represent a function

$$f(x) = \int_{-\infty}^{\infty} A(u)e^{iux}\,du \tag{A-48}$$

Assuming that such a representation exists, it remains to determine $A(u)$ for all values of u in the interval $u = -\infty$ to $u = \infty$. Consider some particular value of $f(x)$, say, when $x = s$ and multiply both sides of (A-48) by $e^{-ivs}\,ds$ and integrate over some interval ranging from $s = -S$ to $s = +S$.

$$\begin{aligned}\int_{-S}^{+S} f(s)e^{-ivs}\,ds &= \int_{-S}^{+S} e^{-ivs}\left[\int_{-\infty}^{\infty} A(u)e^{ius}\,du\right] ds \\ &= \int_{-\infty}^{\infty} A(u)\left(\int_{-S}^{+S} e^{-ivs}e^{ius}\,ds\right) du \end{aligned} \tag{A-49}$$

provided that the order of integration may be interchanged. It can be demonstrated quite rigorously that the double integration on the right of (A-49) equals

$(\frac{1}{2}\pi)A(v)$ in the limit as $S \rightarrow \infty$, so that (A-49) can be written, after interchanging the dummy variables u and v,

$$A(u) = \frac{1}{2\pi}\int_{-\infty}^{\infty} f(s)e^{-ius}\, ds \tag{A-50}$$

so that the *Fourier integral representation* of $f(x)$ in (A-48) becomes

$$f(x) = \frac{1}{2\pi}\iint_{-\infty}^{\infty} f(s)e^{-ius}e^{iux}\, ds\, du \tag{A-51}$$

Defining the *Fourier transform* of $f(s)$,

$$\varphi(u) = \int_{-\infty}^{\infty} f(s)e^{-ius}\, ds \tag{A-52}$$

it follows directly from (A-51), upon substituting $x = s$, that

$$f(s) = \frac{1}{2\pi}\int_{-\infty}^{\infty} \varphi(u)e^{ius}\, du \tag{A-53}$$

Relations (A-52) and (A-53) form an extremely useful pair for many kinds of crystallographic problems since they demonstrate that, when the Fourier transform of a function (A-52) is known, relation (A-53) can be used to determine the original function.

APPENDIX **TWO**

Physical constants

Table A-1 General constants

Mass of an electron, m	9.108×10^{-28} g
Charge of an electron, e	4.803×10^{-10} esu
e/m	5.269×10^{17} esu/g
Velocity of light, c	2.998×10^{10} cm/sec
Electron volt, eV	1.606×10^{-10} esu
Atomic mass unit, amu	1.600×10^{-24} g
One gram equivalent	8.987×10^{20} ergs
Planck's constant, h	6.628×10^{-27} erg-sec
Avogadro's number, N_0	6.025×10^{23} amu/g
Boltzmann constant, k	1.380×10^{-16} erg/°K
Rydberg (gas) constant, R	8.137×10^{7} erg/(mole)(°K)
Angstrom unit, Å	10^{8} cm
	1.00202 kxu

Table A-2 Periodic chart of the elements

Group / Period	I	II	III	IV	V	VI	VII	VIII			Zero
1	1 H Hydrogen 1.008										2 He Helium 4.003
2	3 Li Lithium 6.940	4 Be Beryllium 9.013	5 B Boron 10.82	6 C Carbon 12.01	7 N Nitrogen 14.01	8 O Oxygen 16.000	9 F Fluorine 19.00				10 Ne Neon 20.18
3	11 Na Sodium 22.99	12 Mg Magnesium 24.32	13 Al Aluminum 26.98	14 Si Silicon 28.09	15 P Phosphorus 30.97	16 S Sulfur 32.06	17 Cl Chlorine 35.45				18 Ar Argon 39.94
4	19 K Potassium 39.10	20 Ca Calcium 40.08	21 Sc Scandium 44.96	22 Ti Titanium 47.90	23 V Vanadium 50.95	24 Cr Chromium 52.01	25 Mn Manganese 54.94	26 Fe Iron 55.85	27 Co Cobalt 58.94	28 Ni Nickel 58.71	36 Kr Krypton 83.80
	29 Cu Copper 63.54	30 Zn Zinc 65.38	31 Ga Gallium 69.72	32 Ge Germanium 72.60	33 As Arsenic 74.91	34 Se Selenium 78.96	35 Br Bromine 79.91				
5	37 Rb Rubidium 85.48	38 Sr Strontium 87.63	39 Y Yttrium 88.92	40 Zr Zirconium 91.22	41 Nb Niobium 92.91	42 Mo Molybdenum 95.95	43 Tc Technetium 99	44 Ru Ruthenium 101.1	45 Rd Rhodium 102.9	46 Pd Palladium 106.4	54 Xe Xenon 131.3
	47 Ag Silver 107.9	48 Cd Cadmium 112.4	49 In Indium 114.8	50 Sn Tin 118.7	51 Sb Antimony 121.8	52 Te Tellurium 127.6	53 I Iodine 126.9				
6	55 Cs Cesium 132.9	56 Ba Barium 137.4	57 La Lanthanum **A** 138.9	72 Hf Hafnium 178.5	73 Ta Tantalum 180.9	74 W Tungsten 183.9	75 Re Rhenium 186.2	76 Os Osmium 190.2	77 Ir Iridium 192.2	78 Pt Platinum 195.1	86 Rn Radon 222
	79 Au Gold 197.0	80 Hg Mercury 200.6	81 Tl Thallium 204.4	82 Pb Lead 207.2	83 Bi Bismuth 209.0	84 Po Polonium 210	85 At Astatine 210				
7	87 Fr Francium 223	88 Ra Radium 226	89 Ac Actinium **B** 227								

Lanthanide series	*Actinide series*
A	**B**
58 Ce Cerium	90 Th Thorium
59 Pr Praseodymium	91 Pa Protactinium
60 Nd Neodymium	92 U Uranium
61 Pm Promethium	93 Np Neptunium
62 Sm Samarium	94 Pu Plutonium
63 Eu Europium	95 Am Americium
64 Gd Gadolinium	96 Cm Curium
65 Tb Terbium	97 Bk Berkelium
66 Dy Dysprosium	98 C Californium
67 Ho Holmium	99 Es Einsteinium
68 Er Erbium	100 Fm Fermium
69 Tm Thulium	101 Md Mendelevium
70 Yb Ytterbium	102 No Nobelium
71 Lu Lutetium	103 Lw Lawrencium

Table A-3 X-ray K spectra

Element		Emission lines, Å				Absorption edge, Å
No.	Symbol	α_2	α_1	β	γ	
1	H					
2	He					
3	Li	240				226.5
4	Be	113				
5	B	67				
6	C	44				$43._{68}$
7	N	31.60				$30._{99}$
8	O	23.71				$23._{32}$
9	F	18.31				
10	Ne	14.616		14.464		
11	Na	11.909		11.617		
12	Mg	9.8889		9.558		9.5117
13	Al	8.33916	8.33669	7.981		7.9511
14	Si	7.12773	7.12528	6.7681		6.7446
15	P	6.1549		5.8038		5.7866
16	S	5.37471	5.37196	5.0316		5.0182
17	Cl	4.73050	4.72760	4.4031		4.3969
18	A	4.19456	4.19162	(3.8848)		3.8707
19	K	3.74462	3.74122	3.4538		3.43645
20	Ca	3.36159	3.35825	3.0896		3.07016
21	Sc	3.03452	3.03114	2.7795		2.757_{3}
22	Ti	2.75207	2.74841	2.51381		2.497_{30}
23	V	2.50729	2.50348	2.28434		2.269_{02}
24	Cr	2.29351	2.28962	2.08480		2.070_{12}
25	Mn	2.10568	2.10175	1.91015		1.896_{36}
26	Fe	1.93991	1.93597	1.75653		1.743_{34}
27	Co	1.79278	1.78892	1.62075		1.608_{11}
28	Ni	1.66169	1.65784	1.50010	1.48861	1.488_{02}
29	Cu	1.54433	1.54051	1.39217†	1.38102	1.380_{43}
30	Zn	1.43894	1.43511	1.29522	1.28366	1.283_{3}
31	Ga	1.34394	1.34003	1.20784	1.19595	1.195_{67}
32	Ge	1.25797	1.25401	1.12890	1.11682	1.116_{52}
33	As	1.17981	1.17581	1.05726	1.04498	1.044_{97}
34	Se	1.10875	1.10471	0.99212	0.97986	0.979_{78}
35	Br	1.04376	1.03969	0.93273	0.92064	0.91995

Table A-3 X-ray K spectra (continued)

Element		Emission lines, Å				Absorption edge, Å
No.	Symbol	α_2	α_1	β	γ	
36	Kr	0.9841	0.9801	0.87845	0.86609	0.86547
37	Rb	0.92963	0.92551	0.82863	0.81641	0.81549
38	Sr	0.87938	0.87521	0.78288	0.77076	0.76969
39	Y	0.83300	0.82879	0.74068	0.72874	0.72762
40	Zr	0.79010	0.78588	0.70170	0.68989	0.68877
41	Nb	0.75040	0.74615	0.66572	0.65412	0.65291
42	Mo	0.71354	0.70926	0.63225	0.62099†	0.61977
43	Tc	(0.67927)	(0.67493)	(0.60141)	(0.59018)	(0.5891)
44	Ru	0.64736	0.64304	0.57246	0.56164	0.560_{47}
45	Rh	0.61761	0.61324	0.54559	0.53509†	0.533_{78}
46	Pd	0.58980	0.58541	0.52052	0.51021	0.509_{15}
47	Ag	0.56377	0.55936	0.49701	0.48701	0.4858_{2}
48	Cd	0.53941	0.53498	0.47507	0.46531	0.46409
49	In	0.51652	0.51209	0.45451	0.44496	0.44388
50	Sn	0.49502	0.49056	0.43521	0.42590	0.42468
51	Sb	0.47479	0.47032	0.41706	0.40795	0.40663
52	Te	0.45575	0.45126	0.39997	0.39108	0.38972
53	I	0.43780	0.43329	0.38388	0.37547	0.373_{79}
54	X	0.42043	0.41596	0.36846	0.35989	0.35849
55	Cs	0.40481	0.40026	0.35434	0.34608	0.34474
56	Ba	0.38964	0.38508	0.34078	0.33274	0.33137
57	La	0.37527	0.37070	0.32795	0.32024†	0.31842
58	Ce	0.36166	0.35707	0.31579	0.30826†	0.30647
59	Pr	0.34872	0.34412	0.30423	0.29690†	0.29516
60	Nd	0.35648	0.33182	0.29327	0.28631	0.28451
61	Pm	0.3249	0.3207	0.28209	(0.2761)	(0.2743)
62	Sm	0.31365	0.30895	0.27305	0.26629	0.26462
63	Eu	0.30326	0.29850	0.26360	0.25697	0.25552
64	Gd	0.29320	0.28840	0.25445	0.24812	0.24680
65	Tb	0.28343	0.27876	0.24601	0.23960	0.23840
66	Dy	0.27430	0.26957	0.23758	0.23175	0.23046
67	Ho	0.26552	0.26083	(0.2302)	(0.2244)	0.22290
68	Er	0.25716	0.25248	0.22260	0.21715	0.21566
69	Tu	0.24911	0.24436	0.21530	(0.2101)	0.2089
70	Yb	0.24147	0.23676	0.20876	0.20363	0.20223

Table A-3 X-ray K spectra (continued)

Element		Emission lines, Å				Absorption edge, Å
No.	Symbol	α_2	α_1	β	γ	
71	Lu	0.23405	0.22928	0.20212	0.19689	0.19584
72	Hf	0.22699	0.22218	0.19554	0.19081	0.18981
73	Ta	0.22029	0.21548	0.19007	0.18508†	0.18393
74	W	0.21381	0.20899	0.18436	0.17950†	0.17837
75	Re	0.20759	0.20277	0.17887	0.17415†	0.17311
76	Os	0.20162	0.19678	0.17360	0.16899†	0.16780
77	Ir	0.19588	0.19103	0.16853	0.16404†	0.16286
78	Pt	0.19037	0.18550	0.16366	0.15928†	0.15816
79	Au	0.18506	0.18018	0.15897	0.15471†	0.15344
80	Hg	(0.17992)	(0.17504)	(0.15439)	(0.15020†)	0.14923
81	Tl	0.17502	0.17013	0.150133	(0.1461)	0.14470
82	Pb	0.17028	0.16536	0.145980	0.14201†	0.14077
83	Bi	0.16570	0.16077	0.141941	0.13807†	0.13706
84	Po	(0.1608)	(0.1559)	0.1382†	(0.1333)	(0.1332)
85	At	(0.1570)	(0.1521)	(0.1343)	(0.1307)	(0.1295)
86	Rn	(0.1529)	(0.1479)	(0.1307)	(0.1271)	(0.1260)
87	Fr	(0.1489)	(0.1440)	(0.1272)	(0.1236)	(0.1225)
88	Ra	(0.1450)	(0.1401)	(0.1237)	(0.1203)	(0.1192)
89	Ac	(0.1414)	(0.1364)	(0.1205)	(0.1172)	(0.1161)
90	Th	0.13782	0.13280	0.11738	0.11416†	0.11293
91	Pa	(0.1344)	(0.1294)	(0.1143)	(0.1112)	(0.1101)
92	U	0.13096	0.12594	0.11138	0.10864	0.10680
93	Np	(0.1278)	(0.1226)	(0.1085)	(0.1055)	(0.1045)
94	Pu	(0.1246)	(0.1195)	(0.1058)	(0.1029)	(0.1018)
95	Am	(0.1215)	(0.1165)	(0.1031)	(0.1003)	(0.0992)
96	Cm	(0.1186)	(0.1135)	(0.1005)	(0.0978)	(0.0967)
97	Bk	(0.1157)	(0.1107)	(0.0980)	(0.0953)	(0.0943)
98	Cf	(0.1130)	(0.1079)	(0.0956)	(0.0930)	(0.0920)
99	Es	(0.1103)	(0.1052)	(0.0933)	(0.0907)	(0.0897)
100	Fm	(0.1077)	(0.1026)	(0.0910)	(0.0885)	(0.0875)

† Mean value of doublet.

Values shown are those appearing in *International tables for x-ray crystallography*, vol. 3 (The Kynoch Press, Birmingham, England, 1962). Doubtful values are enclosed in parentheses. More complete listings are given by J. A. Bearden, *X-ray wavelengths* (U.S. Department of Commerce, National Bureau of Standards, Clearinghouse for Federal Scientific and Technical Information, Springfield, Va., 1964).

Table A-4 X-ray L spectra

Element		Emission lines, Å				Absorption edge, Å
No.	Symbol	α_2	α_1	β_1	β_2	
1	H					
2	He					
3	Li					
4	Be					
5	B					
6	C					
7	N					
8	O					
9	F					
10	Ne					
11	Na		407.6			(400) ($L_{II,III}$)
12	Mg		251.0			251 (L_{III})
13	Al		169.8			170 ($L_{II,III}$)
14	Si		123			126.5 (L_{III})
15	P					
16	S					(76.05)
17	Cl					(62.93)
18	A					(50.60)
19	K		(42.7)			(42.17)
20	Ca		36.32	35.95		35.$_{49}$
21	Sc		31.33	31.01		(30.53)
22	Ti		27.39	27.02		(27.37)
23	V		24.26	23.85		(24.26)
24	Cr		21.67	21.28		20.$_7$
25	Mn		19.45	19.12		(19.40)
26	Fe		17.567	17.255		(17.53)
27	Co		15.968	15.667		(15.93)
28	Ni		14.566	14.279		14.579
29	Cu		13.330	13.053		13.288$_8$
30	Zn		12.257	11.985		12.130$_6$
31	Ga		11.290	11.023		11.149
32	Ge		10.435	10.174		10.229
33	As		9.671	9.414		9.3671
34	Se		8.990	8.736		8.645$_6$
35	Br		8.375	8.125		(7.989)

Table A-4 X-ray L spectra (continued)

Element		Emission lines, Å				Absorption edge, Å
No.	Symbol	α_2	α_1	β_1	β_2	
36	Kr	(7.822)		(7.574)		(7.395)
37	Rb	7.3249	7.3181	7.0757		6.863_3
38	Sr	6.8694	6.8625	6.6237		6.386_8
39	Y	6.4555	6.4485	6.2117		5.961_8
40	Zr	6.0776	6.0702	5.8358	5.5861	5.582_9
41	Nb	5.7317	5.7240	5.4921	5.2377	5.222_6
42	Mo	5.41406	5.40625	5.17679	4.9230	4.912_5
43	Tc	(5.1212)	(5.1126)	(4.8782)	(4.635)	(4.629)
44	Ru	4.85343	4.84552	4.62041	4.3715	4.368_9
45	Rh	4.60528	4.59727	4.37392	4.1305	4.129_6
46	Pd	4.37572	4.36760	4.14596	3.9087	3.908_1
47	Ag	4.16269	4.15412	3.93443	3.70307	3.698_3
48	Cd	3.96482	3.95628	3.73808	3.51407	3.503_8
49	In	3.78060	3.77191	3.55520	3.33832	3.324_4
50	Sn	3.60873	3.59987	3.38478	3.17519	3.155_9
51	Sb	3.44828	3.43915	3.22559	3.02333	2.999_9
52	Te	3.29835	3.28909	3.07666	2.88207	2.855_4
53	I	3.15780	3.14849	2.93733	2.75043	2.719_4
54	X	(3.026)	(3.016)	(2.807)	(2.627)	2.592_4
55	Cs	2.9016	2.8920	2.6834	2.5115	2.473_9
56	Ba	2.7849	2.7752	2.5674	2.4042	2.362_8
57	La	2.6743	2.6651	2.4583	2.3026	2.258_3
58	Ce	2.5703	2.5612	2.3558	2.2086	2.163_9
59	Pr	2.4726	2.4627	2.2584	2.1191	2.077_0
60	Nd	2.3804	2.3701	2.1666	2.0355	1.994_7
61	Pm	2.2925	2.282	2.0796	1.9557	(1.918)
62	Sm	2.2102	2.1994	1.9976	1.8819	1.844_5
63	Eu	2.1316	2.1206	1.9202	1.8118	1.775_3
64	Gd	2.05677	2.0460	1.8462	1.7454	1.709_4
65	Tb	1.9863	1.9755	1.7763	1.6824	1.648_6
66	Dy	1.91983	1.90875	1.7100	1.6231	1.579
67	Ho	1.8558	1.8447	1.6468	1.5669	1.535_3
68	Er	1.79564	1.78428	1.58729	1.51399	1.482_{19}
69	Tu	1.7374	1.7263	1.5299	1.4631	1.432_8
70	Yb	1.68268	1.6719	1.47556	1.41530	1.386_{09}

Table A-4 X-ray L spectra (continued)

Element		Emission lines, Å				Absorption edge, Å
No.	Symbol	α_2	α_1	β_1	β_2	
71	Lu	1.63019	1.61943	1.42350	1.37002	1.34130
72	Hf	1.58038	1.56955	1.37402	1.32631	1.29711
73	Ta	1.53283	1.52187	1.32697	1.28447	1.25511
74	W	1.48742	1.47635	1.28176	1.24458	1.21545
75	Re	1.44387	1.43286	1.23853	1.20658	1.1769_7
76	Os	1.40226	1.39113	1.19723	1.16978	1.14043
77	Ir	1.36250	1.35130	1.15783	1.13534	1.10565
78	Pt	1.32438	1.31298	1.11984	1.10196	1.07239
79	Au	1.28777	1.27639	1.08356	1.07021	1.03994
80	Hg	1.25258	1.24114	1.04861	1.03966	1.0089_8
81	Tl	1.21879	1.20735	1.01519	1.01033	0.97930
82	Pb	1.18644	1.17504	0.98222	0.98297	0.95029
83	Bi	1.15531	1.14385	0.95197	0.95514	0.92336
84	Po	1.12556	1.11377	0.92219	0.92930	(0.8970)
85	At	(1.0966)	(1.0850)	(0.8936)	(0.9043)	(0.8720)
86	Rn	(1.0689)	(1.0572)	(0.8659)	(0.8805)	(0.8479)
87	Fr	(1.0421)	1.030	0.840	0.858	(0.8248)
88	Ra	1.01650	1.00468	0.81370	0.83532	0.8027
89	Ac	(0.9917)	(0.9799)	(0.7890)	(0.8140)	(0.7813)
90	Th	0.96771	0.95598	0.76517	0.79352	0.7606
91	Pa	0.9446	0.9328	0.7422	0.7737	(0.7411)
92	U	0.92242	0.91053	0.71999	0.75466	0.7222
93	Np	0.9010	0.8893	0.6984	0.7362	0.7042
94	Pu	0.8802	0.8682	0.6777	0.7185	0.6867
95	Am	0.8602	0.8481	0.6576	0.7014	0.6700
96	Cm	(0.8406)	(0.8287)	(0.6388)	(0.6849)	(0.6532)
97	Bk	(0.8219)	(0.8098)	(0.6203)	(0.6688)	(0.6375)
98	Cf	(0.8036)	(0.7917)	(0.6023)	(0.6534)	(0.6223)
99	Es	(0.7861)	(0.7740)	(0.5850)	(0.6385)	(0.6076)
100	Fm	(0.7691)	(0.7570)	(0.5682)	(0.6236)	(0.5935)

Values shown are those appearing in *International tables for x-ray crystallography*, vol. 3 (The Kynoch Press, Birmingham, England, 1962). Doubtful values are enclosed in parentheses. More complete listings are given by J. A. Bearden, *X-ray wavelengths* (U.S. Department of Commerce, National Bureau of Standards, Clearinghouse for Federal Scientific and Technical Information, Springfield, Va., 1964).

Table A-5 Mass absorption coefficients of the elements for $K\alpha$ radiation of selected targets

Absorber		μ/ρ, cm^2/g				
No.	Symbol	Ag	Mo	Cu	Fe	Cr
1	H	*0.371*	*0.38*	*0.43*	*0.48*	*0.54*
2	He	*0.195*	*0.20*	*0.38*	*0.56*	*0.81*
3	Li	*0.187*	*0.21*	*0.71*	*1.25*	*1.96*
4	Be	*0.229*	*0.29*	*1.50*	*2.80*	*4.50*
5	B	*0.279*	*0.39*	*2.39*	*4.55*	7.38
6	C	*0.400*	*0.62*	*4.60*	8.90	14.5
7	N	*0.544*	*0.91*	7.52	14.6	23.9
8	O	*0.740*	*1.31*	11.5	22.4	36.6
9	F	*0.976*	*1.80*	16.4	32.1	52.4
10	Ne	*1.31*	*2.47*	22.9	44.6	72.8
11	Na	*1.67*	*3.21*	30.1	58.6	95.3
12	Mg	*2.12*	*4.11*	38.6	74.8	121
13	Al	*2.65*	5.16	48.6	93.9	152
14	Si	*3.28*	6.44	60.6	117	189
15	P	*4.01*	7.89	74.1	142	229
16	S	*4.84*	9.55	89.1	170	272
17	Cl	5.77	11.4	106	200	318
18	A	6.81	13.5	123	232	366
19	K	8.00	15.8	143	266	417
20	Ca	9.28	18.3	162	299	463
21	Sc	10.7	21.1	184	336	*513*
22	Ti	12.3	24.2	208	377	*571*
23	V	14.0	27.5	233	419	68.4
24	Cr	15.8	31.1	260	463	79.8
25	Mn	17.7	34.7	285	57.2	93.0
26	Fe	19.7	38.5	308	66.4	108
27	Co	21.8	42.5	313	76.8	125
28	Ni	24.1	46.6	45.7	88.6	144
29	Cu	26.4	50.9	52.9	103	166
30	Zn	28.8	55.4	60.3	117	189
31	Ga	31.4	60.1	67.9	131	212
32	Ge	34.1	64.8	75.6	146	235
33	As	36.9	69.7	83.4	160	258
34	Se	39.8	74.7	91.4	175	281
35	Br	42.7	79.8	99.6	190	305

Table A-5 Mass absorption coefficients of the elements for $K\alpha$ radiation of selected targets (continued)

Absorber		μ/ρ, cm^2/g				
No.	*Symbol*	Ag	Mo	Cu	Fe	Cr
36	Kr	45.8	84.9	108	206	327
37	Rb	48.9	90.0	117	221	351
38	Sr	52.1	95.0	125	236	373
39	Y	55.3	100	134	252	396
40	Zr	58.5	15.9	143	268	419
41	Nb	61.7	17.1	153	284	441
42	Mo	64.8	18.4	162	300	463
43	Tc	67.9	19.7	172	316	485
44	Ru	10.7	21.1	183	334	*509*
45	Rh	11.5	22.6	194	352	*534*
46	Pd	12.3	24.1	206	371	*559*
47	Ag	13.1	25.8	218	391	*586*
48	Cd	14.0	27.5	231	412	*613*
49	In	14.9	29.3	243	432	*638*
50	Sn	15.9	31.1	256	451	*662*
51	Sb	16.9	33.1	270	472	*688*
52	Te	17.9	35.0	282	490	*707*
53	I	19.0	37.1	294	*506*	*722*
54	Xe	20.1	39.2	306	*521*	*763*
55	Cs	21.3	41.3	318	*534*	*793*
56	Bd	22.5	43.5	330	*546*	*461*
57	La	23.7	45.8	341	*557*	202
58	Ce	25.0	48.2	352	*601*	219
59	Pr	26.3	50.7	363	*359*	236
60	Nd	27.7	53.2	374	*379*	252
61	Pm	29.1	55.9	386	172	268
62	Sm	30.6	58.6	397	182	284
63	Eu	32.2	61.5	*425*	193	299
64	Gd	33.8	64.4	*439*	203	314
65	Tb	35.5	67.5	*273*	214	329
66	Dy	37.2	70.6	*286*	224	344
67	Mo	39.0	73.9	128	234	359
68	Er	40.8	77.3	134	245	373
69	Tm	42.8	80.8	140	255	387
70	Yb	44.8	84.5	146	265	401

Table A-5 Mass absorption coefficients of the elements for $K\alpha$ radiation of selected targets (continued)

Absorber		μ/ρ, cm^2/g				
No.	*Symbol*	Ag	Mo	Cu	Fe	Cr
71	Lu	46.8	88.2	153	276	416
72	Hf	48.8	91.7	159	286	430
73	Ta	50.9	95.4	166	297	444
74	W	53.0	99.1	172	308	458
75	Re	55.2	103	179	319	473
76	Os	57.3	106	186	330	487
77	Ir	59.4	110	193	341	*502*
78	Pt	61.4	113	200	353	*517*
79	Au	63.1	115	208	365	*532*
80	Hg	64.7	117	216	377	*547*
81	Tl	66.2	119	224	389	*563*
82	Pb	67.7	120	232	402	*579*
83	Bi	69.1	120	240	415	*596*

Reproduced from *International tables for x-ray crystallography*, vol. 3 (The Kynoch Press, Birmingham, England, 1962). Doubtful values are shown in italics.

APPENDIX **THREE**

Suggestions for a crystallographic library

A compilation of all books devoted to crystallographic topics is published periodically by the Commission on Crystallographic Teaching of the International Union of Crystallography. The latest version is edited by Helen D. Megaw, *Crystallographic book list* (N.V.A. Oosthoek's Vitgevers Mij, Utrecht, 1965). In addition to an alphabetical listing by authors, it gives cross references by topical content and includes a listing of conference and serial publications. Out of this wealth of printed matter, the budding crystallographer has to decide which books to acquire for a personal library. It is difficult to advise on this matter because the amount of use that a book receives varies with its owner. For this reason, the book list compiled below is arranged by topical content, additional references having been cited at the ends of each chapter. It is no more and no less than one man's opinion of which English-language books in each category will find most general utility.

GEOMETRICAL CRYSTALLOGRAPHY

M. J. Buerger, *Elementary crystallography* (John Wiley & Sons, Inc., New York, 1956).

X-RAY PHYSICS

A. H. Compton and S. K. Allison, *X-rays in theory and experiment*, 2d ed. (D. Van Nostrand Company, Inc., Princeton, N.J., 1935).

X-RAY DIFFRACTION THEORY

R. W. James, *The optical principles of the diffraction of x-rays* (G. Bell & Sons, Ltd., London, 1950).

W. H. Zachariasen, *Theory of x-ray diffraction in crystals* (John Wiley & Sons, Inc., New York, 1945).

POWDER METHOD

L. V. Azároff and M. J. Buerger, *The powder method in x-ray crystallography* (McGraw-Hill Book Company, New York, 1958).

H. P. Klug and L. E. Alexander, *X-ray diffraction procedures* (John Wiley & Sons, Inc., New York, 1954).

SINGLE-CRYSTAL METHODS

M. J. Buerger, *X-ray crystallography* (John Wiley & Sons, Inc., New York, 1942).

STRUCTURE ANALYSIS

M. J. Buerger, *Crystal-structure analysis* (John Wiley & Sons, Inc., New York, 1960).

H. Lipson and W. Cochran, *The determination of crystal structures*, 2d ed. (G. Bell & Sons, Ltd., London, 1966).

SMALL-ANGLE SCATTERING

A. Guinier and G. Fournet, *Small-angle scattering of x-rays*, translated by Christopher B. Walker (John Wiley & Sons, Inc., New York, 1955).

BROAD COVERAGE

C. S. Barrett and T. B. Massalski, *Structure of metals*, 3d ed. (McGraw-Hill Book Company, New York, 1966).

H. S. Peiser, H. P. Rooksby, and A. J. C. Wilson, *X-ray diffraction by polycrystalline materials* (Chapman & Hall, Ltd., London, 1955).

APPENDIX FOUR
Answers to selected problems

CHAPTER 1

1-4: $2m$. **1-7:** Four 3's, three 2's, three m's, and one $\bar{1}$; class is $\frac{2}{m}\bar{3}$ in the cubic system. **1-9:** (101) (011) $(10\bar{1})$, $(01\bar{1})$; since all faces belong to a single form, choice of a_1 and a_2 pair is arbitrary. **1-11:** $((100))$ and $((110))$; $[[100]]$; $[[110]]$.

CHAPTER 2

2-6: $R = CA = \csc \Phi$; $OC = \cot \Phi$. **2-7:** Any circles passing through the plane of vision and appearing as straight lines may be regarded as circles of infinite radius.

CHAPTER 3

3-3: Yes, halfway between translation-equivalent mirror reflections new mirrors arise; $p2mm$. **3-5:** 4 is $[[100]]$; 3 is $[[111]]$; 2 is $[[110]]$. **3-11:** Modulus $= 3$.

CHAPTER 4

4-3: Unique space groups are Pm, Pa, Im $(= In)$, Ia. **4-6:** Subgroups of $6mm$ are 622, 6, $3m$, 3, $2mm$, 222, m, 2, 1; subgroups of mmm are $2mm$, 222, $\frac{2}{m}$, m, 2, $\bar{1}$, 1. **4-10:** 2.66×10^{-8} cm; 2.91×10^{-8} cm. **4-11:** Diamond cell contains 8 atoms; graphite cell contains 4 atoms.

CHAPTER 5

5-4: 13.70, 15.90, 22.73, 26.90, 27.96, and 33.11°. **5-6:** $n = 6$. **5-7:** Because of errors in density measurements (imperfect wetting, etc.) and possible defects present in crystal.

CHAPTER 6

6-2: 3×10^{23} electrons/g; 0.5×10^{23} atoms/g. **6-5:** $\mu_m = 23.21$ cm^2/g; $\mu_l = 114$ cm^{-1}. **6-7:** $W_K = 1.1 \times 10^{-7}$ erg; $W_L = 1.6 \times 10^{-8}$ erg; $V_K = 6.3 \times 10^4$ V. **6-15:** $\lambda \cong 1.53$ Å. **6-22:** $W_K = 9{,}000$ eV; $\Delta\lambda = 7.6 \times 10^{-4}$ Å; $\Delta\theta = 0.0072°$.

CHAPTER 7

7-7: 130, 420, 530; all of them lie in the front-reflection region. **7-9:** $a = 7.52$ Å, $b = 4.69$ Å, $c = 4.94$ Å, $\alpha = 90°40'$, $\beta = 131°0'$, $\gamma = 89°34'$. **7-11:** $\theta = L/4R$.

CHAPTER 8

8-4: Cu $K\beta$. **8-5:** No, it is not, because the alternating planes are equally spaced and $\phi = \pi$. **8-7:** The scattering factors of spherical atoms fall off with $(\sin \theta)/\lambda$, while those of point atoms remain constant.

CHAPTER 9

9-3: Nonzero F_{hkl} when $h + k + l = 2n$. **9-11:** $I_Z/I_{2Z} = 4$ for ideally imperfect, and 8 for ideally perfect, crystal. **9-12:** $t_0 = 5 \times 10^{-4}$ and 2.5×10^{-3} cm for Cu and 3.7×10^{-3} and 1.8×10^{-2} cm for W. **9-14:** 0.05, 1.95, 0.33, and 1.67μ, respectively, for waves a to d.

CHAPTER 10

10-3: $2B = 1.48$; at 500°C attenuation is 22%; at 20°C it is 57%. **10-7:** 144.3°K. **10-10:** For $h + k + l = 2n$, $F = f_{Cu} + f_{Zn}$, and for $h + k + l = 2n + 1$, $F = (f_{Cu} - f_{Zn})S$; no, because $f_{Cu} \approx f_{Zn}$ for Mo $K\alpha$; anomalous dispersion must be utilized. **10-17:** $P\frac{6_3}{m}mc$; 1.633; 2 atoms per attice point. **10-18:** (100); the (100) planes are shifted relatively by $b/2$.

CHAPTER 11

11-1: $k + l = 2n$; applies to all reflections. **11-9:** Yes; $mmmPbc$- and $mmmP$-cb or $mmmPba$-. **11-16:** Yes; yes. **11-21:** Absorption, temperature, and extinction corrections.

CHAPTER 12

12-1: 400 W; 26.7 kV. **12-3:** $3\frac{1}{3}$ mA at 2100°K and 10 mA at 2200°K. **12-4:** 10×10 mm. **12-8:** 4,435; 30,906 counts.

CHAPTER 13

13-3: $2\theta = 8.50, 17.0, 72.8°$; 4.8, 2.4, and 0.6 Å for Mo $K\alpha$ and 10.4, 5.2, and 1.3 Å for Cu $K\alpha$. **13-7:** Mount to rotate about c to give densest reciprocal-lattice nets normal to rotation axis. Inclination angle is the same since it depends on $\zeta = \lambda/d_{001}$. (Note that corner points of face-centered reciprocal cell are $2a^*$, etc., away from origin.) **13-10:** See paper by U. Arndt in literature list of this chapter for a complete discussion.

CHAPTER 14

14-2: Parabola; straight line; no, because one end of the zone axis will form a smaller angle. **14-10:** $OP = R \tan(45° - \theta/2)$, but $OX = D \tan(180° - 2\theta)$. **14-11:** 160; 150; 140; 130; 120 (240); 110 (220); 250; 230; 210 (420); 350; 340; 320; 310; 430; 410; 520; 510; and symmetry-equivalent reflections in other four quadrants.

CHAPTER 15

15-4: $x_{opt} = 1/\mu$; $x_{opt} = 3/\mu$. **15-6:** The shorter ζ value corresponds to rotation about [110]; no, for a body-centered crystal, rotation about a has shortest ζ value because a^* has one-half the length of the face-centered reciprocal cell. **15-9:** Parallel arc has no effect; changing other arc will skew the layer lines.

CHAPTER 16

16-4: For $I4$ the reciprocal lattice is all-face-centered. Thus $\sigma_{110} = a^*/\sqrt{2}$ and lies along the densest central rows spaced $\omega = 90°$ apart on the photograph. (The a^* rows are $\omega = 45°$ away.) **16-5:** 6 cm. **16-8:** $\Delta\xi - |\Delta\xi'| = (2 \sin^2 2\epsilon \sin \bar{\mu})/(\cos^2 2\epsilon - \sin^2 \mu)$.

CHAPTER 17

17-1: The back-reflection region contains the resolved $\alpha_1\alpha_2$ doublets. **17-3:** $a = 3.166$ Å. **17-7:** $\theta = 9.16°$. **17-10:** $D_{opt} = 2/\mu$.

CHAPTER 18

18-5: $\alpha^* = 89.5°$, $\gamma^* = 90°$; the first sixteen Q values have the indices 100, 010, 001, 110, $01\bar{1}$, 011, 200, $20\bar{2}$, 101, 210, 020, $10\bar{2}$, $21\bar{2}$, 120, $11\bar{2}$, and $1\bar{1}\bar{2}$. **18-6:** $a = 4.693$ Å, $b = 4.933$ Å, $c = 5.680$ Å, $\alpha = \beta = 90°$, $\gamma = 90.5°$.

CHAPTER 19

19-5: The $K\alpha$ emission by the plating metal increases with thickness until a plateau is reached at "infinite" thickness. Absorption of the $K\alpha$ line emitted by the substrate metal increases with plating thickness till insufficient intensity is transmitted. **19-6:** Transmission of Fe $K\alpha$ for thickness below $1/\mu$. **19-8:** 8.2 wt%.

CHAPTER 20

20-3: 111 ($\rho = 70°$); 200 ($\rho = 55°$); [111] texture. **20-5:** At $-\alpha = 80°$ the sample is parallel to the incident beam; this condition becomes aggravated by increasing θ; a practical solution is to use short-wavelength radiation. **20-9:** Error in ratio is 7%, so that the range is 0.74 to 0.86 for the measured ratio. **20-13:** $\sigma_\phi = 56{,}400 \pm 5{,}100$ psi; error is constant for experiment.

INDEX

INDEX